Useful Books

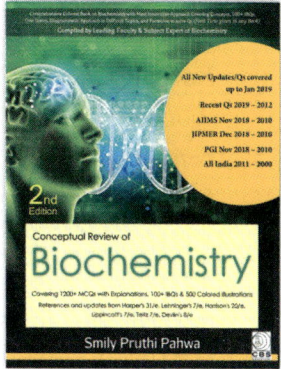

About the Author

Karthikeyan Pethusamy, *MD, DNB, MNAMS*, is a PhD Scholar (Department of Biochemistry) at All India Institute of Medical Sciences, New Delhi. He **had** completed his MBBS with three distinctions and a gold medal in Medicine from Kanyakumari Government Medical College, Tamil Nadu.

After MBBS, he opted to study MD Biochemistry because of his passion towards the subject. Pertinently speaking, he was the first one to take MD (Biochemistry) in All India Counseling 2012. He joined AIIMS as an SR and PhD schol after completing his MD from Maulana Azad Medical College, New Delhi. He did his MD thes under the mentorship of Dr Sarita Agarwal on *the effect of inflammatory gene polymorphis on stroke.* Currently, he is studying the epigenetic changes in Acute Myeloid Leukemia und the mentorship of Dr Subhradip Karmakar. While doing both senior residency and Ph simultaneously, he also passed DNB theory and practical examinations at the first attempt.

He is popular among his students in MAMC as well as AIIMS for his clarity and studen friendly method of teaching. His grip on all subjects is firm, and he brings it into the teaching the subject of his choice—Biochemistry. Making use of the modern technology, he is active engaged in teaching students through his Android app, YouTube channel and his Faceboo **page.** Recently, he has been acknowledged in the 31st edition of the Harper's illustrate Biochemistry textbook for his suggestions and corrections.

Besides Biochemistry, he loves nature. He has been riding bicycle to commute around for th past ten years and motivates others to do the same. During his tenure, as Green Care Secretar of Kanyakumari Government Medical College, he was instrumental in planting aroun thousand trees on his vast college campus.

He is an active contributor to Tamil Wikipedia and Quora. Writing scientific books in Tam language and starting an NGO that will work for environmental conservation projects ar some of his dream projects.

Passing MBBS Series

Biochemistry

for Undergraduates

Based on the Competency-Based Medical Education (CBME) Curriculum (2019)

Karthikeyan Pethusamy MD, DNB, MNAMS

PhD Scholar, Department of Biochemistry
All India Institute of Medical Sciences, New Delhi

CBS
Dedicated to Education

CBS Publishers & Distributors Pvt Ltd

• New Delhi • Bengaluru • Chennai • Kochi • Kolkata • Mumbai
• Hyderabad • Nagpur • Patna • Pune • Vijayawada

Passing MBBS Series

Biochemistry
for Undergraduates

ISBN: 978-93-88725-97-2

First Edition: 2020

Published by **Satish Kumar Jain** and produced by **Varun Jain** for

CBS Publishers & Distributors Pvt Ltd

4819/XI Prahlad Street, 24 Ansari Road, Daryaganj, New Delhi 110 002, India.

Ph: +91-11-23289259, 23266861, 23266867   Website: www.cbspd.com

Fax: 011-23243014

e-mail: delhi@cbspd.com; cbspubs@airtelmail.in.

Corporate Office: 204 FIE, Industrial Area, Patparganj, Delhi 110 092

Ph: +91-11-4934 4934   Fax: 4934 4935

e-mail: feedback@cbspd.com; bhupesharora@cbspd.com

Branches

- **Bengaluru:** Seema House 2975, 17th Cross, K.R. Road
  Banasankari 2nd Stage, Bengaluru 560 070, Karnataka
  Ph: +91-80-26771678/79   Fax: +91-80-26771680   e-mail: bangalore@cbspd.con

- **Chennai:** 7, Subbaraya Street, Shenoy Nagar, Chennai 600 030, Tamil Nadu
  Ph: +91-44-26680620, 26681266   Fax: +91-44-42032115   e-mail: chennai@cbspd.com

- **Kochi:** 68/1534, 35, 36-Power House Road, Opp. KSEB, Cochin-682018, Kochi, Kerala
  Ph: +91-484-4059061-65   Fax: +91-484-4059065   e-mail: kochi@cbspd.com

- **Kolkata:** 6/B, Ground Floor, Rameswar Shaw Road, Kolkata-700 014, West Bengal
  Ph: +91-33-22891126, 22891127, 22891128   e-mail: kolkata@cbspd.com

- **Mumbai:** 83-C, Dr E Moses Road, Worli, Mumbai-400018, Maharashtra
  Ph: +91-22-24902340/41   Fax: +91-22-24902342   e-mail: mumbai@cbspd.com

Representatives

- **Hyderabad**   +91-9885175004
- **Pune**   +91-9623451994
- **Patna**   +91-9334159340
- **Vijayawada**   +91-9000660880

Printed At : Goyal Offset Works (P) Limited

REVIEWERS' LIST

EXPERT REVIEWERS

FROM ALL INDIA INSTITUTE OF MEDICAL SCIENCES (AIIMS), NEW DELHI

- **Dr Archana Singh**, Associate Professor
- **Dr Ashikh Seethy**, Senior Resident and PhD Scholar
- **Dr Rakhee Yadav**, Assistant Professor
- **Dr Sajib Kumar Sarkar**, Senior Resident and PhD Scholar

FROM OTHER INSTITUTIONS

- **Dr Amarpreet Kaur**, Assistant Professor, AIIMS, Gorakhpur, Uttar Pradesh
- **Dr Ankita Raj**, Tutor, Government Medical College, Nizamabad, Telangna
- **Dr Chitra Dhume**, Professor and Head, Goa Medical College, Goa
- **Dr Dholariya Sagar Jayantilal,** Associate Professor, RD Gardi Medical College, Ujjain, Madhya Pradesh
- **Dr Diravya Seelan**, Senior Resident, Vardhman Mahavir Medical College, New Delhi
- **Dr Elza Boby**, Assistant Professor, KMCT Medical College, Calicut, Kerala
- **Dr Maheswari K**, Assistant Professor, Maulana Azad Medical College, New Delhi
- **Dr Montosh Chakroborty**, Associate Professor, AIIMS, Mangalagiri, Andhra Pradesh
- **Dr Pinky Garg**, Associate Professor, North Delhi Medical Corporation Medical College, New Delhi
- **Dr Prajwal Athreya,** Senior Resident, NIMHANS, Bengaluru, Karnataka
- **Dr Sarita Agarwal**, Professor and Head, Hamdard Institute of Medical Sciences and Research, New Delhi

STUDENT REVIEWERS

FROM ALL INDIA INSTITUTE OF MEDICAL SCIENCES (AIIMS), NEW DELHI

- **Alapan Das**, II year MBBS
- **Alen Joe Joseph**, III year MBBS
- **Amandeep Sahu**, III year MBBS
- **Amit Kumar Garg**, III year MBBS
- **Ananthakrishna J**, II year MBBS
- **Ansh Gupta**, Intern
- **Anshul Gupta**, III year MBBS
- **Aranya Dutta**, III year MBBS
- **Aruna Arumugam**, PhD Scholar
- **Ashutosh Behra**, III year MBBS
- **Astha Sachan**, Junior Resident
- **Ayushi Jain**, PhD Scholar

- **David Raja**, PhD Scholar
- **Faisal Saif**, PhD Scholar
- **Gaurav Kumar**, Graduate Student
- **Jasmeen Gupta**, Junior Resident
- **Kushagra Pandey**, III year MBBS
- **Madhav Mohata**, Intern
- **Mohamed Shuaib**, II year MBBS
- **Nidhi Thakur**, Junior Resident
- **Priya Narwal**, Intern
- **Rajkumar Kulandaisamy**, PhD Scholar
- **Sagar Mishra,** II year MBBS
- **Sakthivel**, Junior Resident
- **Sankeerth S**, III year MBBS
- **Sanket Katpara**, Junior Resident
- **Theeepan,** Junior Resident
- **Vadanya Shrivastava**, Junior Resident
- **Vasishta Polisetty**, Intern

FROM OTHER INSTITUTIONS

- **Akshay Munjal**, PhD Scholar, Jawaharlal Nehru University, New Delhi
- **Ashwin AJ**, Intern, Government Villupuram Medical College, Tamil Nadu
- **Asutosh Tiwary**, Graduate Student, Delhi University, New Delhi
- **Hari Kishan Yadav**, Intern, Maulana Azad Medical College, New Delhi
- **Manas Kamakshi Sharma**, I year MBBS, SGT Medical College, Hospital and Research Institute, Gurugram, Haryana
- **Pradeep Singh**, Graduate Student, Hamdard Institute of Medical Sciences and Research, New Delhi
- **Prasanth Kumar**, Final MBBS, Vardhman Mahavir Medical College, New Delhi
- **Rahul Yadav**, PhD Scholar, IISc, Bengaluru, Karnataka
- **Shubham Goyal**, Intern, Maulana Azad Medical College, New Delhi

PREFACE

Dear Friends,

I am delighted to bring you this competency-based, concise, colorful, conceptual yet exam-oriented book. This book is the culmination of my seven years of faithful study of the Medical Biochemistry. I can say for sure that I have put my best efforts in bringing out this book. So, you can put your trust in this book.

I am glad that you continue to read the preface. Let me tell you why I have written this book. Compared to the Biochemistry I studied during my first year of MBBS, a present first-year student is required to learn a myriad of details. So, it is humanly impossible to revise the subject before any exam, be it qualifying or competitive. Therefore, I have decided to come up with a concise book that will be of great help as it is based on the new curriculum. I assure you that the content of this book is on par with the Indian as well as the International curriculum.

I am glad to inform you that the purchase of this book gives you advertisement-free access to my android app "Biochemistry with Dr Karthi". In the app, you will get the explanations for multiple-choice questions. You will also be able to watch topic-wise lectures. This app is a portal for you to contact me and get your doubts clarified.

I have referred to the list of MCI competencies and ensured that this book covers almost all of them. To keep the book concise, I have deliberately skipped physiology topics, like immunity, muscle contraction, digestion and absorption. If you want me to cover any other topic, let me know through the android app or YouTube channel. I will make a video and give a handout.

I made it a point to write this book only when I was in a peaceful and pleasant state of mind. I have spent the best hours of my life writing this book. I hope learning Biochemistry will be one of the joyful times of your life.

I welcome comments, suggestions, corrections and constructive criticism.

Karthikeyan Pethusamy

BIOCHEMISTRY

Numbers	Competency The student should be able to:	Domain K/S/A/C	Level K/KH/ SH/P	Core (Y/N)	Suggested teaching learning method	Suggested assessment method	Number required to certify P	Vertical integration	Horizontal integration
	Topic: Basic Biochemistry	**Number of competencies: (01)**			**Number of procedures that require certification: (NIL)**				
BI1.1	Describe the molecular and functional organization of a cell and its sub-cellular components.	K	KH	Y	Lecture, Small group discussion	Written assessment/ Viva voce			Physiology
	Topic: Enzyme	**Number of competencies: (07)**			**Number of procedures that require certification: (NIL)**				
BI2.1	Explain fundamental concepts of enzyme, isoenzyme, alloenzyme, coenzyme & co-factors. Enumerate the main classes of IUBMB nomenclature.	K	KH	Y	Lecture, case discussion	Written assessment/ Viva voce			
BI2.2	Observe the estimation of SGOT & SGPT.	K	K	Y	Demonstration	Viva voce			
BI2.3	Describe and explain the basic principles of enzyme activity.	K	KH	Y	Lecture, case discussion	Written/ Viva voce			
BI2.4	Describe and discuss enzyme inhibitors as poisons and drugs and as therapeutic enzymes.	K	KH	Y	Lecture, small group discussion	Written/ Viva voce		Pathology, General Medicine	
BI2.5	Describe and discuss the clinical utility of various serum enzymes as markers of pathological conditions.	K	KH	Y	Lecture, small group discussion	Written/ Viva voce		Pathology, General Medicine	
BI2.6	Discuss use of enzymes in laboratory investigations (Enzyme-based assays).	K	KH	Y	Lecture, small group discussion	Written/ Viva voce		Pathology, General Medicine	
BI2.7	Interpret laboratory results of enzyme activities & describe the clinical utility of various enzymes as markers of pathological conditions.	K	KH	Y	Lecture, small group discussion, doap sessions	Written/ Viva voce		Pathology, General Medicine	

Contd…

Numbers	Competency The student should be able to:	Domain K/S/A/C	Level K/KH/ SH/P	Core (Y/N)	Suggested teaching learning method	Suggested assessment method	Number required to certify P	Vertical integration	Horizontal integration
	Topic: Chemistry and Metabolism of Carbohydrates	**Number of competencies: (10)**			**Number of procedures that require certification: (NIL)**				
BI3.1	Discuss and differentiate monosaccharides, di-saccharides and polysaccharides giving examples of main carbohydrates as energy fuel, structural element and storage in the human body.	K	KH	Y	Lecture, small group discussion	Written/ Viva voce			
BI3.2	Describe the processes involved in digestion and assimilation of carbohydrates and storage.	K	KH	Y	Lecture, small group discussion	Written/ Viva voce			
BI3.3	Describe and discuss the digestion and assimilation of carbohydrates from food.	K	KH	Y	Lecture, small group discussion	Written/ Viva voce			
BI3.4	Define and differentiate the pathways of carbohydrate metabolism, (glycolysis, gluconeogenesis, glycogen metabolism, HMP shunt).	K	KH	Y	Lecture, small group discussion	Written/ Viva voce		General Medicine	
BI3.5	Describe and discuss the regulation, functions and integration of carbohydrate along with associated diseases/disorders.	K	KH	Y	Lecture, small group discussion	Written/ Viva voce		General Medicine	
BI3.6	Describe and discuss the concept of TCA cycle as a amphibolic pathway and its regulation.	K	KH	Y	Lecture, small group discussion	Written/ Viva voce			
BI3.7	Describe the common poisons that inhibit crucial enzymes of carbohydrate metabolism (e.g. fluoride, arsenate).	K	KH	Y	Lecture, small group discussion	Written/ Viva voce			Physiology
BI3.8	Discuss and interpret laboratory results of analytes associated with metabolism of carbohydrates.	K	KH	Y	Lecture, small group discussion	Written/ Viva voce		Pathology, General Medicine	
BI3.9	Discuss the mechanism and significance of blood glucose regulation in health and disease.	K	KH	Y	Lecture, small group discussion	Written/ Viva voce		General Medicine	

Contd…

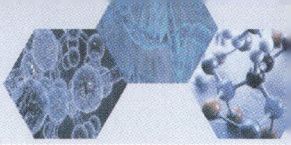

Numbers	Competency The student should be able to:	Domain K/S/A/C	Level K/KH/ SH/P	Core (Y/N)	Suggested teaching learning method	Suggested assessment method	Number required to certify P	Vertical integration	Horizontal integration
BI3.10	Interpret the results of blood glucose levels and other laboratory investigations related to disorders of carbohydrate metabolism.	K	KH	Y	Lecture, small group discussion	Written/ Viva voce		General Medicine	
	Topic: Chemistry and Metabolism of Lipids	**Number of competencies: (07)**			**Number of procedures that require certification: (NIL)**				
BI4.1	Describe and discuss main classes of lipids (Essential/ non-essential fatty acids, cholesterol and hormonal steroids, triglycerides, major phospholipids and sphingolipids) relevant to human system and their major functions.	K	KH	Y	Lecture, small group discussion	Written/ Viva voce		General Medicine	
BI4.2	Describe the processes involved in digestion and absorption of dietary lipids and also the key features of their metabolism.	K	KH	Y	Lecture, small group discussion	Written/ Viva voce		General Medicine	
BI4.3	Explain the regulation of lipoprotein metabolism & associated disorders.	K	KH	Y	Lecture, small group discussion	Written/ Viva voce		General Medicine	
BI4.4	Describe the structure and functions of lipoproteins, their functions, interrelations & relations with atherosclerosis.	K	KH	Y	Lecture, small group discussion	Written/ Viva voce		General Medicine	
BI4.5	Interpret laboratory results of analytes associated with metabolism of lipids.	K	KH	Y	Lecture, small group discussion	Written/ Viva voce		General Medicine	
BI4.6	Describe the therapeutic uses of prostaglandins and inhibitors of eicosanoid synthesis.	K	KH	Y	Lecture, small group discussion	Written/ Viva voce		General Medicine	
BI4.7	Interpret laboratory results of analytes associated with metabolism of lipids.	K	KH	Y	Lecture, small group discussion	Written/ Viva voce		General Medicine	
	Topic: Chemistry and Metabolism of Proteins	**Number of competencies: (05)**			**Number of procedures that require certification: (NIL)**				
BI5.1	Describe and discuss structural organization of proteins.	K	KH	Y	Lecture, small group discussion	Written/ Viva voce			

Contd…

Syllabus

Numbers	Competency The student should be able to:	Domain K/S/A/C	Level K/KH/ SH/P	Core (Y/N)	Suggested teaching learning method	Suggested assessment method	Number required to certify P	Vertical integration	Horizontal integration
BI5.2	Describe and discuss functions of proteins and structure-function relationships in relevant areas e.g. hemoglobin and selected hemoglobinopathies.	K	KH	Y	Lecture, small group discussion	Written/ Viva voce		Pathology, General Medicine	Physiology
BI5.3	Describe the digestion and absorption of dietary proteins.	K	KH	Y	Lecture, small group discussion	Written/ Viva voce		Pediatrics	
BI5.4	Describe common disorders associated with protein metabolism.	K	KH	Y	Lecture, small group discussion	Written/ Viva voce		Pediatrics	
BI5.5	Interpret laboratory results of analytes associated with metabolism of proteins.	K	KH	Y	Lecture, small group discussion	Written/ Viva voce		General Medicine	
Topic: Metabolism and Homeostasis		**Number of competencies: (15)**			**Number of procedures that require certification: (NIL)**				
BI6.1	Discuss the metabolic processes that take place in specific organs in the body in the fed and fasting states.	K	KH	Y	Lecture, small group discussion	Written/ Viva voce		General Medicine	
BI6.2	Describe and discuss the metabolic processes in which nucleotides are involved.	K	KH	Y	Lecture, small group discussion	Written/ Viva voce			
BI6.3	Describe the common disorders associated with nucleotide metabolism.	K	KH	Y	Lecture, small group discussion	Written/ Viva voce			Physiology
BI6.4	Discuss the laboratory results of analytes associated with gout & Lesch Nyhan syndrome.	K	KH	Y	Lecture, small group discussion	Written/ Viva voce		General Medicine	
BI6.5	Describe the biochemical role of vitamins in the body and explain the manifestations of their deficiency.	K	KH	Y	Lecture, small group discussion	Written/ Viva voce		General Medicine	
BI6.6	Describe the biochemical processes involved in generation of energy in cells.	K	KH	Y	Lecture, small group discussion	Written/ Viva voce			
BI6.7	Describe the processes involved in maintenance of normal pH, water & electrolyte balance of body fluids and the derangements associated with these.	K	KH	Y	Lecture, small group discussion	Written/ Viva voce		General Medicine	Physiology
BI6.8	Discuss and interpret results of Arterial Blood Gas (ABG) analysis in various disorders.	K	KH	Y	Lecture, small group discussion	Written/ Viva voce		General Medicine	

Contd…

Numbers	Competency The student should be able to:	Domain K/S/A/C	Level K/KH/ SH/P	Core (Y/N)	Suggested teaching learning method	Suggested assessment method	Number required to certify P	Vertical integration	Horizontal integration
BI6.9	Describe the functions of various minerals in the body, their metabolism and homeostasis.	K	KH	Y	Lecture, small group discussion	Written/ Viva voce		General Medicine	Physiology
BI6.10	Enumerate and describe the disorders associated with mineral metabolism.	K	KH	Y	Lecture, small group discussion	Written/ Viva voce		General Medicine	
BI6.11	Describe the functions of haem in the body and describe the processes involved in its metabolism and describe porphyrin metabolism.	K	KH	Y	Lecture, small group discussion	Written/ Viva voce		Pathology, General Medicine	Physiology
BI6.12	Describe the major types of haemoglobin and its derivatives found in the body and their physiological/pathological relevance.	K	KH	Y	Lecture, small group discussion	Written/ Viva voce		Pathology, General Medicine	Physiology
BI6.13	Describe the functions of the kidney, liver, thyroid and adrenal glands.	K	KH	Y	Lecture, small group discussion	Written/ Viva voce		Pathology, General Medicine	Physiology, Human Anatomy
BI6.14	Describe the tests that are commonly done in clinical practice to assess the functions of these organs (kidney, liver, thyroid and adrenal glands).	K	KH	Y	Lecture, small group discussion	Written/ Viva voce		Pathology, General Medicine	Physiology, Human Anatomy
BI6.15	Describe the abnormalities of kidney, liver, thyroid and adrenal glands.	K	KH	Y	Lecture, small group discussion	Written/ Viva voce		Pathology, General Medicine	Physiology, Human Anatomy
	Topic: Molecular Biology	**Number of competencies: (07)**			**Number of procedures that require certification: (NIL)**				
BI7.1	Describe the structure and functions of DNA and RNA and outline the cell cycle.	K	KH	Y	Lecture, small group discussion	Written/ Viva voce			
BI7.2	Describe the processes involved in replication & repair of DNA and the transcription & translation mechanisms.	K	KH	Y	Lecture, small group discussion	Written/ Viva voce			
BI7.3	Describe gene mutations and basic mechanism of regulation of gene expression.	K	KH	Y	Lecture, small group discussion	Written/ Viva voce		Pediatrics	
BI7.4	Describe applications of molecular technologies like recombinant DNA technology, PCR in the diagnosis and treatment of diseases with genetic basis.	K	KH	Y	Lecture, small group discussion	Written/ Viva voce		Pediatrics, General Medicine	

Contd…

Syllabus

Numbers	Competency The student should be able to:	Domain K/S/A/C	Level K/KH/ SH/P	Core (Y/N)	Suggested teaching learning method	Suggested assessment method	Number required to certify P	Vertical integration	Horizontal integration
BI7.5	Describe the role of xenobiotics in disease.	K	KH	Y	Lecture, small group discussion	Written/ Viva voce			
BI7.6	Describe the anti-oxidant defence systems in the body.	K	KH	Y	Lecture, small group discussion	Written/ Viva voce			
BI7.7	Describe the role of oxidative stress in the pathogenesis of conditions such as cancer, complications of diabetes mellitus and atherosclerosis.	K	KH	Y	Lecture, small group discussion	Written/ Viva voce		General Medicine, Pathology	
	Topic: Nutrition	**Number of competencies: (05)**			**Number of procedures that require certification: (NIL)**				
BI8.1	Discuss the importance of various dietary components and explain importance of dietary fibre.	K	KH	Y	Lecture, small group discussion	Written/ Viva voce		General Medicine, Pediatrics, Pathology	
BI8.2	Describe the types and causes of protein energy malnutrition and its effects.	K	KH	Y	Lecture, small group discussion	Written/ Viva voce		General Medicine, Pediatrics, Pathology	
BI8.3	Provide dietary advice for optimal health in childhood and adult, in disease conditions like diabetes mellitus, coronary artery disease and in pregnancy.	K	KH	Y	Lecture, small group discussion	Written/ Viva voce		General Medicine	
BI8.4	Describe the causes (including dietary habits), effects and health risks associated with being overweight/obesity.	K	KH	Y	Lecture, small group discussion	Written/ Viva voce		General Medicine, Pathology	
BI8.5	Summarize the nutritional importance of commonly used items of food including fruits and vegetables (macro-molecules & its importance).	K	KH	Y	Lecture, small group discussion	Written/ Viva voce		Community Medicine, General Medicine, Pediatrics	
	Topic: Extracellular Matrix	**Number of competencies: (03)**			**Number of procedures that require certification: (NIL)**				
BI9.1	List the functions and components of the extracellular matrix (ECM).	K	KH	Y	Lecture, small group discussion	Written/ Viva voce			
BI9.2	Discuss the involvement of ECM components in health and disease.	K	KH	Y	Lecture, small group discussion	Written/ Viva voce		General Medicine	
BI9.3	Describe protein targeting and sorting along with its associated disorders.	K	KH	N	Lecture, small group discussion	Written/ Viva voce			

Syllabus

Contd...

Numbers	Competency The student should be able to:	Domain K/S/A/C	Level K/KH/ SH/P	Core (Y/N)	Suggested teaching learning method	Suggested assessment method	Number required to certify P	Vertical integration	Horizontal integration
	Topic: Oncogenesis and Immunity	**Number of competencies: (05)**			**Number of procedures that require certification: (NIL)**				
BI10.1	Describe the cancer initiation, promotion oncogenes & oncogene activation. Also focus on p53 & apoptosis.	K	KH	Y	Lecture, small group discussion	Written/ Viva voce		Obstetrics & Gynae-cology, General Surgery, Pathology	
BI10.2	Describe various biochemical tumor markers and the biochemical basis of cancer therapy.	K	KH	Y	Lecture, small group discussion	Written/ Viva voce		Obstetrics & Gynae-cology, General Surgery, Pathology	
BI10.3	Describe the cellular and humoral components of the immune system & describe the types and structure of antibody.	K	KH	Y	Lecture, small group discussion	Written/ Viva voce		Obstetrics & Gynae-cology, General Surgery, Pathology	
BI10.4	Describe & discuss innate and adaptive immune responses, self/non-self recognition and the central role of T-helper cells in immune responses.	K	KH	Y	Lecture, small group discussion	Written/ Viva voce		General Medicine, Pathology	Physiology
BI10.5	Describe antigens and concepts involved in vaccine development.	K	KH	Y	Lecture, small group discussion	Written/ Viva voce		Pathology, Pediatrics, Microbio-logy	
	Topic: Biochemical Laboratory Tests	**Number of competencies: (24)**			**Number of procedures that require certification: (05)**				
BI11.1	Describe commonly used laboratory apparatus and equipments, good safe laboratory practice and waste disposal.	K	KH	Y	Lecture, small group discussion	Written/ Viva voce			
BI11.2	Describe the preparation of buffers and estimation of pH.	K	KH	Y	Lecture, small group discussion	Written/ Viva voce			
BI11.3	Describe the chemical components of normal urine.	K	KH	Y	Lecture, small group discussion	Written/ Viva voce			
BI11.4	Perform urine analysis to estimate and determine normal and abnormal constituents.	S	P	Y	Doap session	Skill assessment	1	General Medicine	Physiology
BI11.5	Describe screening of urine for inborn errors & describe the use of paper chromatography.	K	KH	Y	Lecture, small group discussion	Written/ Viva voce		General Medicine	

Contd…

Syllabus

Numbers	Competency The student should be able to:	Domain K/S/A/C	Level K/KH/ SH/P	Core (Y/N)	Suggested teaching learning method	Suggested assessment method	Number required to certify P	Vertical integration	Horizontal integration
BI11.6	Describe the principles of colorimetry.	K	KH	Y	Lecture, small group discussion	Written/ Viva voce			
BI11.7	Demonstrate the estimation of serum creatinine and creatinine clearance.	S	P	Y	Practical	Skills assessment	1		
BI11.8	Demonstrate estimation of serum proteins, albumin and A:G ratio.	S	P	Y	Practical	Skills assessment	1		
BI11.9	Demonstrate the estimation of serum total cholesterol and HDL-cholesterol.	S	P	Y	Practical	Skills assessment			
BI11.10	Demonstrate the estimation of triglycerides.	S	P	Y	Practical	Skills assessment			
BI11.11	Demonstrate estimation of calcium and phosphorous.	S	P	Y	Practical	Skills assessment			
BI11.12	Demonstrate the estimation of serum bilirubin.	S	P	Y	Practical	Skills assessment			
BI11.13	Demonstrate the estimation of SGOT/SGPT.	S	P	Y	Practical	Skills assessment			
BI11.14	Demonstrate the estimation of alkaline phosphatase.	S	P	Y	Practical	Skills assessment			
BI11.15	Describe & discuss the composition of CSF.	K	KH	Y	Lecture, small group discussion	Written/ Viva voce			
BI11.16	Observe use of commonly used equipments/ techniques in biochemistry laboratory including: • pH meter • Paper chromatography of amino acid • Protein electrophoresis • TLC, PAGE • Electrolyte analysis by ISE • ABG analyzer • ELISA • Immunodiffusion • Autoanalyser • Quality control • DNA isolation from blood/tissue	S	KH	Y	Demonstration	Skill assessment			

Contd…

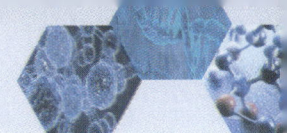

Numbers	Competency The student should be able to:	Domain K/S/A/C	Level K/KH/ SH/P	Core (Y/N)	Suggested teaching learning method	Suggested assessment method	Number required to certify P	Vertical integration	Horizontal integration
BI11.17	Explain the basis and rationale of biochemical tests done in the following conditions: • Diabetes mellitus • Dyslipidemia • Myocardial infarction • Renal failure, gout • Proteinuria • Nephrotic syndrome • Edema • Jaundice • Liver diseases, pancreatitis, disorders of acid-base balance • Thyroid disorders.	K	KH	Y	Lecture, small group discussion	Written/ Viva voce		General Medicine, Pathology	
BI11.18	Discuss the principles of spectrophotometry.	K	KH	Y	Lecture, small group discussion	Written/ Viva voce			
BI11.19	Outline the basic principles involved in the functioning of instruments commonly used in a biochemistry laboratory and their applications.	K	KH	Y	Lecture, small group discussion	Written/ Viva voce			
BI11.20	Identify abnormal constituents in urine, interpret the findings and correlate these with pathological states.	S	SH	Y	Doap sessions	Skill assessment	1		
BI11.21	Demonstrate estimation of glucose, creatinine, urea and total protein in serum.	S	SH	Y	Doap sessions	Skill assessment	1		
BI11.22	Calculate albumin: globulin (AG) ratio and creatinine clearance.	K	KH	Y	Lecture, small group discussion	Written/ Viva voce		General Medicine	
BI11.23	Calculate energy content of different food items, identify food items with high and low glycemic index and explain the importance of these in the diet.	K	KH	Y	Lecture, small group discussion	Written/ Viva voce		General Medicine	
BI11.24	Enumerate advantages and/or disadvantages of use of unsaturated, saturated and trans fats in food.	K	KH	Y	Lecture, small group discussion	Written/ Viva voce		General Medicine	

Column C: K – Knowledge, S – Skill, A – Attitude/professionalism, C – Communication.
Column D: K – Knows, KH – Knows how, SH – Shows how, P – Performs independently.
Column F: DOAP session – Demonstrate, Observe, Assess, Perform.
Column H: If entry is P: indicate how many procedures must be done independently for certification/graduation.

Syllabus

Almost all of the competencies are covered inside the book. The list of competencies achieved are mentioned at the end of each chapter. In addition to Biochemistry, competencies of other subjects are also achieved in this book as a part of horizontal and vertical integration. The following abbreviations will be used:

Legend	Subject
BI	Biochemistry
PY	Physiology
PA	Pathology
PE	Pediatrics
IM	General Medicine
PH	Pharmacology
AN	Human Anatomy
DR	Dermatology

COMPLIANCE TO THE NEW UG CURRICULUM

Medical Council of India (MCI) has laid down a set of competency goals for the Indian medical graduates[1]. I have tried my best to incorporate all these goals in this book.

MCI recommendations	Actions taken
• Obtaining competencies recommended	• Competencies achieved are given at the end of every chapter.
• Demonstrate the ability to perform an objective self-assessment of knowledge	• Self-assessment questions are given at the end of every chapter. • The explanations for the MCQs are given in the companion Android app.
• Demonstrate ability to apply newly-gained knowledge or skills to the care of the patient	• Biochemical basis of 170 diseases has been given. • Mechanism of action of 70 drugs given.
• Horizontal integration	• Topics, like karyotyping, plasma proteins are discussed with Biochemistry point of view for the horizontal integration. • I recommend the students to read topics, like digestion and muscle contraction from physiology books.
• Vertical integration	• Harrison's Corner and Clinical Correlation boxes are given to help in vertical integration.
• Case-based discussion in an appropriate format ensuring that elements in the same phase (horizontal) and from other phases are addressed	• Clinical Case-based Questions are given at the end of appropriate chapters. • Case discussion video lectures are given in the Android app
• To be familiar with biomedical waste disposal	• Biomedical Waste Management has been explained in detail in Chapter 46.

Definition of competency: "An observable ability of a health professional, integrating multiple components such as knowledge, skills, values and attitudes"[2].

REFERENCES

1. Medical Council of India, Regulations on Graduate Medical Education, 1997 (Amended up to May, 2018), 2018, pg 6-8.
2. Medical Council of India, Competency-based Undergraduate Curriculum for the Indian Medical Graduate, 2018. Vol. 1; pg 38.

ACKNOWLEDGEMENTS

My teachers of biochemistry at Kanyakumari Government Medical College, Tamil Nadu:

- Dr Lalita
- Dr Vijayalakshmi

My teachers of biochemistry at Maulana Azad Medical College, New Delhi:

- Late Professor Dr PC Ray
- Professor Dr Alka Khaneja
- Professor Dr Alpana Saxena
- Professor Dr BC Koner
- Professor Dr Lal Chandra
- Professor Dr P Lali
- Professor Dr SK Gupta
- Professor Dr Sarita Agarwal (my thesis guide)
- Professor Dr Smita Kaushik
- Professor Dr TK Mishra

My PhD mentor who always encourage my academic activities in addition to research:

- Dr Subhradip Karmakar, Associate Professor, Department of Biochemistry, AIIMS, New Delhi

My supportive PhD doctoral committee members:

- Dr Anita Chopra, Associate Professor, Department of Laboratory Oncology, AIIMS, New Delhi
- Dr Jayanth Kumar, Associate Professor, Department of Biochemistry, AIIMS, New Delhi
- Dr Sameer Bakhshi, Professor, Department of Medical Oncology, AIIMS, New Delhi
- Dr Shyam S Chauhan, Professor and Head, Department of Biotechnology, AIIMS, New Delhi
- Dr Subrata Sinha, Professor and Head, Department of Biochemistry, AIIMS, New Delhi
- Dr V Sreenivas, Professor, Department of Biostatistics, AIIMS, New Delhi

My lab members who have always been my constant support:

- Dr Ruby Dhar, DBT-BioCARe fellow
- Dr Rashmi, Research Associate
- Miss Bharoti Sengupta, Research Fellow
- Miss Indrani Mukherjee, PhD Scholar
- Mr Dheeraj Bhist, Laboratory Attendant
- Mr Gaurav Kumar, Laboratory Technician
- Mr Sunil Singh, PhD Scholar
- Mr Tryambak Srivatsava, PhD Scholar
- Mr Vivek Raj, Laboratory Attendant

My seniors and friends who are my source of inspiration:

- Dr Ayush Jain, Senior Resident, Radiology, King Edward Memorial Hospital, Mumbai, Maharashtra
- Dr Dnyanesh Amle, Assistant Professor, AIIMS, Nagpur, Maharashtra
- Mr Dinesh, Nursing Officer, AIIMS, New Delhi
- Dr Janvie Manhas, AIIMS, New Delhi
- Dr Jhuma Das, Senior Resident, Maulana Azad Medical College, New Delhi

- Dr Hariprasad, Associate Professor, Biophysics, AIIMS, New Delhi
- Dr Pankhuri Dudani, Dermatology Resident, AIIMS, New Delhi
- Dr Parthiban N, Genesis Diabetes Management Centre, Mavelikara, Kerala
- Dr Radhey Natung, OSD to Minister, Sports and Youth Affairs and Department of Water Resources, Arunachal Pradesh
- Dr Vineet Sehgal, Glaucoma Specialist, Sharp Eye Centre, Delhi

I express my sincere gratitude to my parents and family members for their constant support. I thank all my students at MAMC and AIIMS for their love and support. I thank all my batchmates and juniors of Kanyakumari Government Medical College, Tamil Nadu.

I thank Team Viewer, a free remote access software which enabled me to work with the CBS production team from the comfort of my room. I acknowledge that I have used many images from the Wikimedia Commons, the non-profit, multilingual, free-content repository and have given the necessary attributions inside the book.

I appreciate the support of **Mr Satish Kumar Jain** (Chairman) and **Mr Varun Jain** (Managing Director), M/s CBS Publishers and Distributors Pvt Ltd for their wholehearted support in the publication of this book. I have no words to describe the role, efforts, inputs and initiatives undertaken by **Mr Bhupesh Arora,** (Vice President – Publishing and Marketing, PGMEE and Nursing Division), for his endeavor toward the development of the book.

I sincerely thank the entire CBS team for bringing out the book with utmost care and attractive presentation. I would like to thank Dr Mrinalini Bakshi (Editorial Head & Content Strategist) for her editorial support and Ms Nitasha Arora (Production Head & Content Strategist), Dr Anju Dhir (Project Manager & Senior Scientific Coordinator), Mr Shivendu Bhushan Pandey (Senior Editor), Mr Ashutosh Pathak (Senior Proof Reader) and all the production team members Mr Bunty Kashyap, Mr Phool Kumar, Mr Chaman Lal, Mr Prakash Gaur, Mr Chander, Ms Tahira Parveen, Ms Babita Verma, Ms Manorama, Mr Raju Sharma, Mr Manoj Chaudhary, Mr Vikram Chaudhary, Mr Manoj Malakar, Mr Arun Kumar and Mr Rahul Negi for devoting laborious hours in designing and typesetting of the book.

Acknowledgements

CONTENTS

SECTION IV: NUTRITION

SECTION V: SPECIAL TOPICS

Contents

SECTION VI: REVIEW

Contents

KEY CONTENTS

BIOCHEMICAL BASIS OF DISEASES

As per the recommendation of Medical Council of India, "The knowledge acquired in biochemistry should help the students to integrate molecular events with structure and function of the human body in health and disease[1]." So, I have given the biochemical basis of many common as well as uncommon diseases inside the book. Here is an alphabetical list. The page numbers on which you can find them in the book are as follows:

MECHANISM OF ACTION OF DRUGS, POISONS AND TOXIN

As per the MCI recommendation "The broad goal of the teaching of undergraduate students in biochemistry is to make them understand the scientific basis of the life processes at the molecular level and to orient them toward the application of the knowledge acquired in solving clinical problems[2]."

Key Contents

TECHNIQUES AND TESTS

Key Contents

DID YOU KNOW?

In this book, I have explained many important and interesting concepts. Here is a selected list of few.

REFERENCES

1. Medical Council of India, Regulations on Graduate Medical Education, 1997 (Amended up to May, 2018), 2018, pg 38.
2. Medical Council of India, Regulations on Graduate Medical Education, 1997 (Amended up to May, 2018), 2018, pg 36.

Key Contents

Fundamental Concepts of Biochemistry

Section Outline

Chemistry of Amino Acids

Amino acids are the monomeric units of proteins. In addition, they give rise to important molecules like heme, purine and pyrimidine nucleotides, polyamines, nitric oxide and creatine.

STANDARD AMINO ACIDS

Standard amino acids have their own codon(s) and they are found in proteins. Although 300 amino acids are found in nature, only 21 are coded by codons. The 21st amino acid selenocysteine utilizes a stop codon.

ALL THE AMINO ACIDS IN HUMAN PROTEINS ARE L-AMINO ACIDS

An asymmetric/chiral carbon is a carbon atom that is attached to 4 different types of atoms or groups of atoms. Except for glycine, all amino acids possess chiral centers and they can exist as either L or D enantiomer.

All the amino acids in human proteins are in L-form (biological homochirality). Free D-amino acids (D-serine, D-glutamate) are found in brain as neurotransmitters. (Compare: All the monosaccharides in human body are in D-form, except L-fucose).

L-amino acid   D-amino acid

- ❑ Amino acid without a chiral carbon—Glycine
- ❑ Amino acids possessing two chiral carbons—Isoleucine and threonine.

STRUCTURE OF AMINO ACIDS

Learning the structure of amino acids is quite easy and fun. You will find it valuable when you go downstream in this book and you will clearly understand why "Structure is related to function".

All the protein-forming amino acids in our body are α-amino acids. In α-amino acids, the amino group is attached to the same carbon to which carboxyl group is attached.

I am going to teach you how to draw the structure of amino acids. Take a pen and paper and follow my instructions. You should remember that on adding CH₂ group to R group of glycine, you will get alanine as a result and this is an example to understand the structure of amino acids. This does not talk about the ways in which they are synthesized in the body. The synthesis of amino acids will be discussed in Chapter 11.2. (Refer to page no. 188)

Look at the general structure of amino acids:

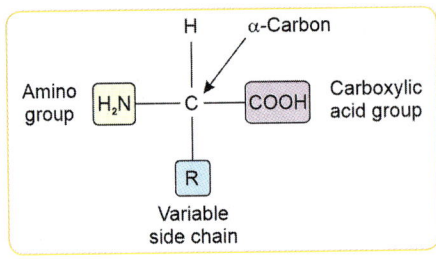

Glycine

- ❑ If the R-group is nothing but a hydrogen, the amino acid is glycine.
- ❑ It is the **only amino acid without a chiral carbon**. So, it is optically inactive. There is no D or L-form of glycine
- ❑ As glycine is the smallest amino acid, it fits into the narrow gaps in between the three alpha chains of collagen.
- ❑ As this is the smallest amino acid, it gives flexibility to protein chains.
- ❑ As the R-group is just a hydrogen, glycine is the least polar among the polar amino acids.

$$COO^-$$
$$H_3N^+-C-H$$
$$H$$
Glycine (G)

If you notice the structure of glycine COOH is written as COO⁻ and NH₂ is written as NH₃⁺. Why so?
At physiological pH, COOH group tends to lose its proton and the NH₂ group tends to accept a proton. Thus, amino acids are amphoteric molecules (carrying both acidic and basic groups).

Alanine

- ❑ Just add a CH₂ to the R-group of glycine, you will get alanine.
- ❑ Transamination of alanine produces CH₃—CO—COOH, i.e. pyruvate.
- ❑ As the R-group is the methyl group, alanine is nonpolar.

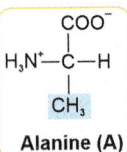

Serine

Add a hydroxyl group to the R-group of alanine, you will get serine with hydroxymethyl R-group.

The OH-group can make hydrogen bond with water and thus serine is polar but uncharged.

In proteins, OH-group of serine can be phosphorylated by protein kinases and glycosylated (enzymatic addition of carbohydrate) by glycosyl transferase.

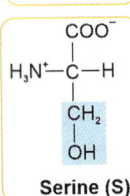

Threonine

- Threonine is a homolog of serine with a —CH$_2$ group extra.
- OH group of threonine can also be phosphorylated and glycosylated.
- Like serine, R-group of threonine is polar but uncharged.
- Threonine contains two chiral carbons (marked with asterisks in figure).

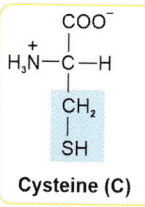

Threonine (T)

Cysteine

Put an 'S' in place of 'O' in the R-group of serine. You will get cysteine with methyl mercaptan R-group.

SH group is known as thiol, thio alcohol, sulfhydryl or mercaptan. SH group is polar but uncharged at physiological pH.

Thiol group of cysteine can form thioester bonds with acyl groups, e.g. Fatty acyl-CoA, Acetyl-CoA.

Cysteine (C)

 Clinical Correlation

Sulfur of cysteine has the property of binding to heavy metals—like arsenic and lead. This is why arsenite poisoning inactivates SH group containing enzymes and SH group containing British Anti-Lewisite (dimercaprol) is an antidote for heavy metal poisoning.

Glutathione

Glutathione (Chapter 2) *(Refer to page no. 18)* is a tripetpide containing cysteine. That is why it is written as G-SII. When glutathione reduces some substance, it gets oxidized. Oxidized glutathione is written as GSSG.

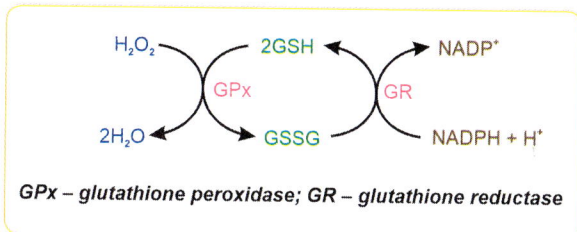

GPx – glutathione peroxidase; GR – glutathione reductase

Homocysteine

Homocysteine is nothing but cysteine with one CH$_2$ group extra. So, it is a higher structural homolog of cysteine. It is neither coded by codons nor found in proteins. It is a non-protein amino acid produced in the metabolism of methionine.

Cystine

Oxidation of 2-molecules of cysteine produces a disulfide linkage between them and the resultant molecule is known as cystine (Remember: there is only one 'e' in cystine).

Homocysteine

Chapter 1 Chemistry of Amino Acids

Cysteine

Cysteine ⇌ (oxidation / reduction) Cystine + 2H⁺ + 2e⁻

Cysteine

Cystine

Similar sounding terms:
- Cysteine - Amino acid
- Cystine - Dipeptide
- Cytosine - Pyrimidine base
- Cytidine - Nucleoside

Clinical Correlation

Bond, Disulphide Bond

Cystinuria (Chapter 11.6) *(Refer to page no. 216)* is an inherited condition with excessive urinary excretion of cysteine and few other amino acids. In the acidic pH of urine, cysteine is converted into insoluble cystine which forms stone in the urinary tract. Alkalinization of urine is done to prevent the conversion of cysteine to cystine.

Methionine

- Methionine is a homolog of homocysteine with an extra CH_2 group.
- R-group of methionine is $CH_3\text{-}S\text{-}CH_2\text{-}CH_2$.
- Because of this methyl group, methionine is nonpolar.
- Methionine contains sulfur, does not contain SH group.

Methionine (M)   S-Adenosyl methionine (SAM)

- When adenosyl group is added to the sulfur of methionine, S-Adenosyl methionine (SAM) is produced.
- SAM is also known as **activated methionine** and is the donor of methyl group in methylation reactions.

Selenocysteine

In the structure of cysteine, replace the 'sulfur' with 'selenium', you get selenocysteine. But cysteine is not involved in selenocysteine biosynthesis.

Selenocysteine is synthesized from serine, OH group of serine is replaced by SeH. The stop codon UGA codes for selenocysteine.

Selenocysteine is present in **thioredoxin reductase, glutathione peroxidase, 5' deiodinase and selenoprotein P.**

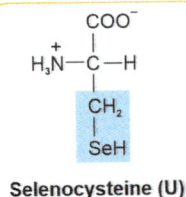

Selenocysteine (U)

We will discuss the biosynthesis of selenocysteine in Chapter 25 (Refer to page no. 437).
Note: Pyrrolysine is the 22nd amino acid coded by UAG. It is **not** found in humans.

Aromatic Amino Acids—Phenylalanine, Tyrosine and Tryptophan

- ❑ Add a phenyl (C_6H_5) group to alanine, you will get phenyl alanine.
- ❑ As there is an aromatic ring, this is an aromatic amino acid and it is nonpolar.
- ❑ Adding an OH group in the para position of phenylalanine will give you tyrosine.
- ❑ In our body, tyrosine is synthesized from phenylalanine.
- ❑ Similar to other OH group containing amino acids, tyrosine residues of proteins undergo phosphorylation and glycosylation.

| Phenylalanine (F) | Tyrosine (Y) | Tryptophan (W) |

By adding an indole group to alanine, you will get tryptophan.
Indole is nothing but 2,3 benzopyrrole.
You will come across an enzyme tryptophan 2,3 dioxygenase/tryptophan pyrrolase that will degrade tryptophan.
You will also study about a metabolite 5-hydroxyindoleacetic acid which is a degradation product of serotonin, a tryptophan derivative.

Aromatic Amino Acids Absorb UV Light at 280 nm

Compounds with conjugated double bonds absorb UV light that is why aromatic amino acids, porphyrins, purine and pyrimidine bases absorb UV light. Among the aromatic amino acids, tryptophan absorbs the maximum.

Light-absorbing capacity of aromatic amino acids is used to estimate the protein concentration with spectrophotometer (instrument that can measure the light absorbance from infrared to ultraviolet range). With the absorbance value of a known concentration of protein solutions, a standard curve is made. Concentration of the protein is calculated using the standard curve.

Compound	λ_{max} (nm) (Wavelength at which maximal light absorption occurs)
Peptide bond	190–230
Purine and pyrimidine bases	260
Aromatic amino acids	280
NADH and NADPH	340
Porphyrin (Soret Band)	400

Amino Acids with Basic Side Chain—Histidine, Lysine and Arginine

Histidine

- ❑ When you add 1, 3-diazole/imidazole group to alanine, you get histidine.
- ❑ Histidine with pKa value of 6 is a basic amino acid with the capacity to protonate or deprotonate depending on the surrounding amino acids.
- ❑ All the enzymes that act through acid-base catalysis have one or more histidine residues in their active site.

Chapter 1 Chemistry of Amino Acids

7

❏ Histidine contributes to the *buffering capacity of proteins*.
  ○ The pKa of histidine is 6.0, so histidine is best at buffering at pH 6.0. The acidic amino acids have lower pKa values if compared to histidine, and the other basic amino acids have greater pKa values. Hence, the pKa of histidine, amongst all amino acids is the closest to the physiological pH of 7.4.
❏ Histidine binds to nickel. This property is used in affinity chromatography to separate histidine tagged proteins *(This will be explained in Chapter 43)* *(Refer to page no. 630)*.

Histidine (H)

Lysine (K)

Lysine

You know that the carbon to which functional group is attached is the alpha carbon.

In general structure of amino acids, extend the side chain up to epsilon carbon and adds an extra amino group. You will get lysine.

Because of the extra amino group, lysine side chain is positively charged.

This side-chain protrudes out from the peptide backbone of proteins. In proteins, epsilon amino group of lysine can undergo various enzymatic modifications, like acetylation, ubiquitination and nonenzymatic modification, like glycation (nonenzymatic addition of sugars).

In certain enzymes, cofactors like pyridoxal phosphate (derivative of vitamin B_6) and biotin are covalently attached to epsilon amino group of lysine.

Arginine (R)   **Guanidine**   **Urea**

Arginine

Arginine contains a guanidino group at the δ position

Guanidine is NH_2—(C=NH_2)—NH_2. Can you notice the similarity between guanidine and urea?

Arginine is the amino acid that produces urea in our body on hydrolysis.

δ guanidino group has the pKa value of 12.48. Thus, **arginine is the most basic**, polar and positively charged R-group containing amino acid.

Proteins like histone are rich in arginine and lysine. They will be positively charged even at the pH of 8.

His Basics are Loose - Histidine, Lysine and Arginine are basic amino acids

Section I Fundamental Concepts of Biochemistry

Acidic Amino Acids

Now, we will look at the 2 amino acids with polar, negatively charged side chain.

Aspartate and glutamate are the dicarboxylic amino acids with extra COOH group. Thus, they are polar and negatively charged at physiological pH.

Aspartate and glutamate are not essential in diet since they can be synthesized from transamination of oxaloacetate and α-ketoglutarate respectively in the body.

Aspartate (D)   **Glutamate (E)**

Asparagine and Glutamine

Asparagine and glutamine are the amides of aspartate and glutamate.

Because of the amide group, they are polar but uncharged.

Amide group of asparagine is glycosylated by enzymes to produce N-Glycoproteins.

Glutamine is the source of NH_3 in the kidney for buffering action and it is the transport form of NH_3 from the brain to liver. Glutamine is also involved in many biosynthetic reactions as a source of nitrogen, e.g. Aspargaine synthesis.

Asparagine (N)   **Glutamine (Q)**

Branched Chain Amino Acids

❑   Valine, isoleucine and leucine are the 3 amino acids with branched, aliphatic and non-polar side chains.
❑   Valine contains isopropyl R-group. Leucine contains isobutyl R-group. Isoleucine is nothing but a chain isomer of leucine.
❑   Unlike other amino acids, *branched chain amino acids are not degraded in liver.* They are utilized by the muscles.

Valine (V)   **Leucine (L)**   **Isoleucine (I)**

Proline is an Imino Acid

❑   To be precise, proline is not an amino acid. It is an imino acid.
❑   Cyclic side chain of proline is rigid.
❑   Because of absence of amino group, it cannot form hydrogen bond when placed in an alpha helix and it will disrupt the chain. Thus, proline is an α-helix breaker.
❑   Imino group of proline does not undergo transamination.

Proline (P)

CLASSIFICATION OF AMINO ACIDS

Amino acids can be classified based on the following:
❑   Chemical nature of the side chain
❑   Polarity of the side chain
❑   Dietary requirement
❑   Metabolic fate

Chapter 1 Chemistry of Amino Acids

Classification Based on the Chemical Nature of the Side Chain

Type	Example
Simple, aliphatic	Glycine
CH_3 group, aliphatic	Alanine
Branched chain, aliphatic	Valine Leucine Isoleucine
OH group containing	Serine Threonine
Sulfur containing	Cysteine Methionine
SH group containing, polar	Cysteine
Thioether containing, nonpolar	Methionine
Aromatic amino acids	Phenylalanine Tyrosine Tryptophan
Acidic/Dicarboxylic, polar negatively charged	Aspartate Glutamate
Amides (polar but uncharged)	Asparagine Glutamine
Basic, polar, positively charged	Histidine Lysine Arginine
Imino acid, nonpolar	Proline

Classification Based on the Polarity of the Side Chain

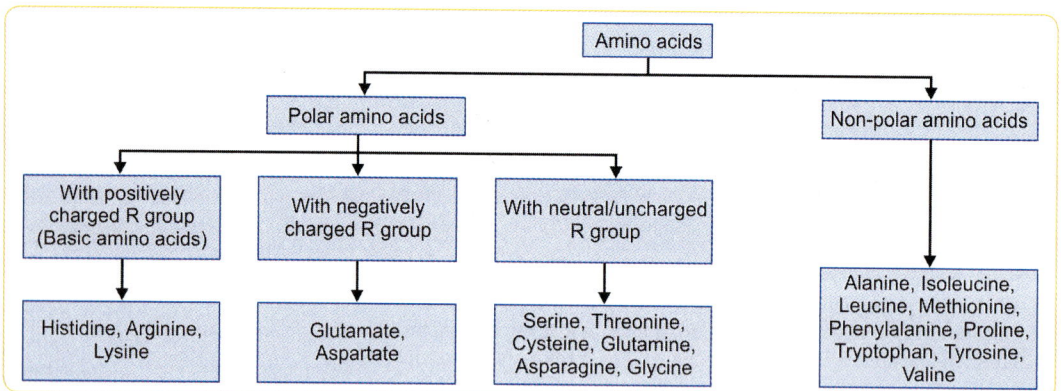

CONCEPT CORNER

Which amino acids are polar?
- ❑ All amino acids carrying a charged R group (positive/negative).
- ❑ All amino acids with SH, OH or amide R group which enable them to make hydrogen bond with water.

Section I Fundamental Concepts of Biochemistry

Classification Based on the Dietary Requirement

Those amino acids which cannot be synthesized in our body are known as dietarily essential amino acids. **Essential and nonessential amino acids** are discussed in Chapter 11.2 *(Refer to page no. 188)*.

Classification Based on the Metabolic Fate

Based on the fate of carbon skeleton during metabolism, amino acids are classified into glucogenic and ketogenic amino acids. This is discussed in Chapter 11.3 *(Refer to page no. 191)*.

DERIVED AMINO ACIDS—THREE TYPES

Amino Acids due to Post-translational Modifications

❑ Hydroxylysine, hydroxyproline, gamma carboxyglutamate are some derived amino acids found in proteins. These amino acids are not coded by the genetic code. They arise out of the modification of existing amino acids in the proteins.
❑ Hydroxylysine and hydroxyproline arise from the hydroxylation of lysine and proline residues by vitamin C dependent hydroxylases.
❑ γ carboxyglutamate arises out of vitamin K dependent γ-carboxylation of glutamic acid. (For more details, refer to the translation Chapter 25) *(Refer to page no. 428)*

 High Yield

Gamma Carboxyglutamic Acid (Gla) is a Result of Post-translational Modification

Gamma carboxyglutamic acid (GLA) contains one more carboxyl group at the γ carbon which enables the *dicarboxylate* group to bind *divalent* calcium. γ-carboxylation is aided by vitamin K. GLA containing proteins are:
❑ Factor II, VII, IX, X
❑ Protein C, S, Z
❑ Osteocalcin, matrix GLA protein
❑ Nephrocalcin, transthyretin, periostin

Non-protein Amino Acids

Non-protein amino acids are found in the body but not in proteins. Ornithine, citrulline and homocysteine are some nonprotein amino acids formed during the metabolism of amino acids but are not incorporated in proteins.

CONCEPT CORNER

Although many textbooks consider citrulline as a nonprotein amino acid, it is actually found in proteins normally. Post-translational deimination of arginine produces citrulline in proteins. In certain autoimmune diseases, antibodies against these citrullinated proteins are found. For example, anticitrullinated peptide antibodies are more specific in the diagnosis of rheumatoid arthritis.

Non-alpha Amino Acids

Non-alpha amino acids are also found in our body but not in proteins. Alpha carbon is the one to which carboxylic group is attached. Non-alpha amino acids are those in which amino group is attached to the non-alpha carbon.

Chapter 1 **Chemistry of Amino Acids**

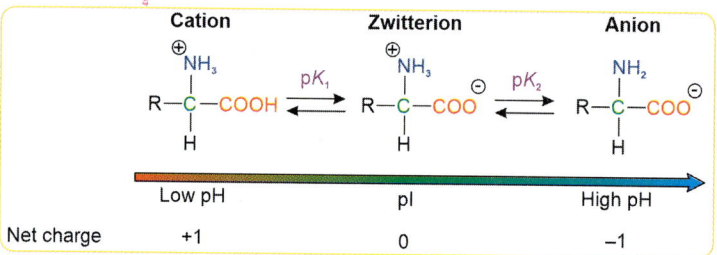

β-alanine   γ-amino butyric acid (GABA)   δ-amino levulinic acid

- ❑ β-alanine—derived from catabolism of cytosine/uracil
- ❑ β-amino isobutyric acid—derived from catabolism of thymine
- ❑ γ-amino butyric acid—derived from decarboxylation of glutamate; inhibitory neurotransmitter
- ❑ δ-amino levulinic acid—derived from condensation of glycine with succinyl-CoA; involved in heme synthesis.

ISOELECTRIC pH OF AMINO ACIDS

We have seen that at physiological pH and alpha carboxyl group of amino acids lose its proton and amino group tends to accept protons.

Zwitterion is a molecule carrying equal number of positive and negative charges; net charge is zero.

Isoelectric pH (pI) is the pH at which a molecule exists in Zwitterion form. In German, Zwitter means hermaphrodite.

- ❑ When the pH is above the isoelectric pH, molecule will carry net negative charge.
- ❑ When the pH is below the isoelectric pH, molecule will carry net positive charge.
- ❑ At pI, the molecule will carry no net charge.

	Cation	Zwitterion	Anion
	$\overset{\oplus}{NH_3}$	$\overset{\oplus}{NH_3}$	NH_2
	R—C—COOH	R—C—COO$^\ominus$	R—C—COO$^\ominus$
	H	H	H
	Low pH	pI	High pH
Net charge	+1	0	−1

pK_1 pK_2

Calculation of isoelectric pH of amino acids, pH and buffer systems are discussed in Chapter 37 (Refer to page no. 572).

THREE-LETTER AND ONE-LETTER CODE OF AMINO ACIDS

Three-letter code is the first three letters of the amino acid. Isn't it simple? Yes, it is? The one-letter code is the first letter of a few but not all amino acids. The name of many amino acids begins with the same letter. For example, alanine, aspartate, asparagine and arginine have the same first letter. So, the following strategies are used in the allocation of one-letter code:

- ❑ Using the second letter of the 3-letter code – R for arginine
- ❑ Using the third letter of the 3-letter code – N for Asparagine
- ❑ Using the similar-sounding letter – F for Phenylalanine
- ❑ Using the previous alphabet – K for Lysine
- ❑ Using an unassigned alphabet – Q for Glutamine

Amino acid	Three-letter code	One-letter code
Alanine	Ala	A
Arginine	Arg	R
Asparagine	Asn	N
Aspartic acid	Asp	D

Contd...

Section I   Fundamental Concepts of Biochemistry

Amino acid	Three-letter code	One-letter code
Cysteine	Cys	C
Glutamic acid	Glu	E
Glutamine	Gln	Q
Glycine	Gly	G
Histidine	His	H
Isoleucine	Ile	I
Leucine	Leu	L
Lysine	Lys	K
Methionine	Met	M
Phenylalanine	Phe	F
Proline	Pro	P
Serine	Ser	S
Threonine	Thr	T
Tryptophan	Trp	W
Tyrosine	Tyr	Y
Valine	Val	V

 Clinical Correlation

Most common mutation in cystic fibrosis is ΔF508 which indicates the deletion of phenylalanine (F) at the 508th position of the CFTR protein. *This is why you need to be familiar with one-letter code of amino acids.*

SELF-ASSESSMENT

Short Answer Questions

1. Using your knowledge of one-letter code of amino acids, what can you infer from the following terms in relation to mutations found in proteins?

   a. ΔF508
   b. K-RasG12D

2. What is cystine? What is its biomedical importance?

3. Why are amino acids called 'amphoteric molecules'? Why do some amino acids like histidine have better buffering capacity at body pH, compared to glycine?

4. What are non-protein amino acids? Name any two.

5. Name the 21st amino acid. Which is the codon for this? Name any two enzymes containing the 21st aminoacid.

6. Classify the amino-acids based on the polarity and the charge on their R-group (side chain).

7. Define isoelectric pH (pI). What will be the charge on the amino acid if the surrounding pH is above, below and same as that of the pI of the amino acid?

8. What is γ-carboxy glutamic acid (Gla)? Which vitamin is required for the formation of Gla? Name three Gla containing proteins.

9. Give three examples of amino acids with a non-alpha amino group, that have important functions in mammalian metabolism.

10. Give two examples each of acidic, basic and aromatic amino acids respectively.

Multiple Choice Questions

1. Which one of the following amino acids is most likely to be found in the transmembrane region of a protein?
   a. Lysine
   b. Arginine
   c. Leucine
   d. Aspartate

2. When the following amino acids are separated by running them on agarose gel at pH 7, which one of them will migrate slowest towards the anode?
   a. Glycine
   b. Valine
   c. Aspartic acid
   d. Lysine

3. Which of the following amino acids contain polar but uncharged side-chain?
   a. Cysteine
   b. Leucine
   c. Methionine
   d. Glutamate

4. Which of the following amino acids can be phosphorylated?
   a. Cysteine
   b. Leucine
   c. Methionine
   d. Serine

5. Which of the following clotting factor does not contain gamma carboxyglutamate?
   a. Factor II
   b. Factor IV
   c. Factor IX
   d. Factor X

6. Which one of following amino acids is polar?
   a. Isoleucine
   b. Methionine
   c. Glutamic acid
   d. Tryptophan

7. **Which one of the following amino acids is nonpolar?**
   a. Glutamate
   b. Glutamine
   c. Histidine
   d. Methionine

8. **Amino acid with aliphatic side chain is:**
   a. Serine
   b. Leucine
   c. Threonine
   d. Aspartate

9. **Nonprotein amino acid is:**
   a. Aspartate
   b. Histidine
   c. Ornithine
   d. Tyrosine

10. **Which of the following amino acid contains two chiral carbons?**
    a. Leucine
    b. Valine
    c. Threonine
    d. Methionine

11. **Cystine is formed by:**
    a. Hydroxylation of cysteine molecule
    b. Carboxylation of cysteine molecule
    c. Peptide bond between two cysteine molecules
    d. Disulfide bond between cysteine molecule

12. **Free SH group is present in:**
    a. Cysteine
    b. Methionine
    c. Taurine
    d. Homoserine

13. **Substitution of which one of the following amino acids in place of alanine would increase the absorbance of protein at 280 nm?**
    a. Leucine
    b. Arginine
    c. Tryptophan
    d. Proline

ANSWERS

1. c.	2. d.	3. a.	4. d.	5. b.	6. c.
7. d.	8. b.	9. c.	10. c.	11. d.	12. a.
13. c.					

Direct One-liner Type Questions

1. Optically inactive amino acid is _____
2. Amino acids with indole group is called _____
3. Amino acids with guanidino group is called ____
4. Amino acids with imidazole group is known as _
5. Most basic amino acid is _____
6. Aromatic amino acids absorb UV light at the wave length of ___ nm
7. Amino acid from which selenocysteine is derived: _____
8. 22nd amino acid is _____
9. Amino acid involved in N-glycosylation is ____
10. Amino acid acting as neurotransmitter is called _____
11. Imino acid is _____
12. One-letter code for phenylalanine is _____
13. The codon for selenocysteine is ____
14. Activated methionine is ____

ANSWERS

1. Glycine	2. Tryptophan	3. Arginine	4. Histidine
5. Arginine	6. 280	7. Serine	8. Pyrrolysine
9. Asparagine	10. Glycine, D-serine, D-Glutamate	11. Proline	12. F
13. UGA	14. S-Adenosyl methionine		

Protein Structure

Proteins are polypeptides, i.e. series of amino acids which are covalently linked via peptide bonds. A peptide bond is a special type of amide bond between the α-carboxyl group of an amino acid and the α-amino group of the adjacent amino acid (Fig. 1).

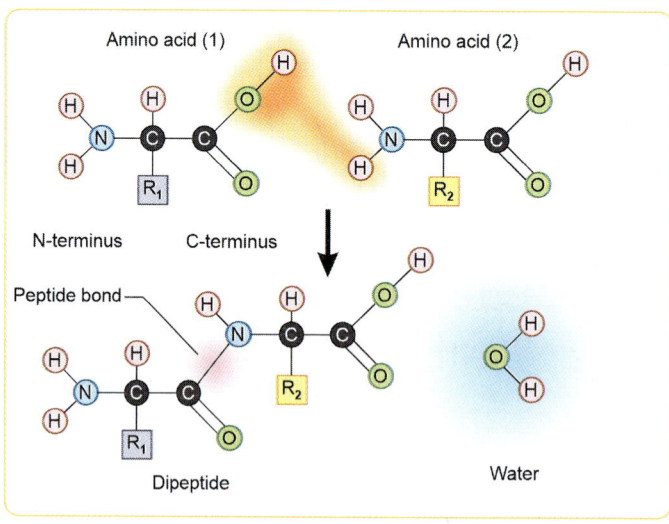

Fig. 1: Illustration of peptide bond formation

Source: GYassineMrabet - Own work, Public Domain, https://commons.wikimedia.org/w/index.php?curid=2552310

CHARACTERISTICS OF PEPTIDE BOND

❏ Covalent amide (–CONH–) bond.
❏ -C=O and -NH groups of the peptide bond are uncharged, but polar because of the ability to form hydrogen bonds with water; Moreover, -C=O and -NH groups can form hydrogen bonds with each other. This *ability of peptide bond gives rise to the secondary structure of proteins.*
❏ Mostly *Trans* in configuration to prevent steric clash.
❏ Rigid and planar because of partial double bond character (because of delocalization of electrons). The C-N bond has the property in between single and double bonds.

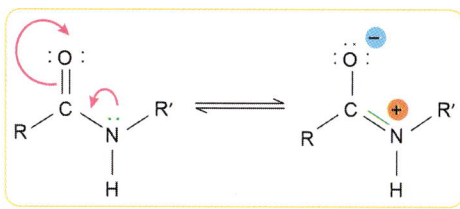

ISOPEPTIDE BOND

A peptide bond involving either non-α amino group or non-α carboxyl group is called an isopeptide bond.
❏ Lysine contains an additional amino group at epsilon carbon. When this ∈-amino group forms a peptide bond with α-carboxyl group of another amino acid, we call that bond as isopeptide bond. (e.g., Ubiquitination of proteins; discussed in Chapter 11.1.) *(Refer to page no. 183)*
❏ Glutamate contains an additional carboxyl group attached to the gamma carbon. When this γ-carboxyl group forms a peptide bond with α-amino of another amino acid, we call that bond as isopeptide bond. (e.g., Glutathione; discussed below).
▶ youtu.be/GOYLvRcvj7Q or scan QR code

THE FOUR LEVELS OF PROTEIN STRUCTURE

PRIMARY STRUCTURE

The primary structure of a protein is the linear sequence of amino acids (from the amino terminal to carboxy terminal) linked by covalent peptide bonds. According to certain authors, the position of disulfide bonds (if any) between cysteine residues of the same chain also comes under the primary structure (Fig. 2).
❏ 1° structure is always written from amino-terminal (left) to carboxy-terminal (right)
❏ Three-letter or one-letter code is used to depict the constituent amino acids
❏ Position of disulfide bonds are indicated
Bonds stabilizing: Peptide bond and disulfide bond
Sanger was the first person to deduce the primary structure of (bovine) insulin. He was awarded the Nobel prize for this.
Human insulin is made up of two polypeptide chains, A and B.

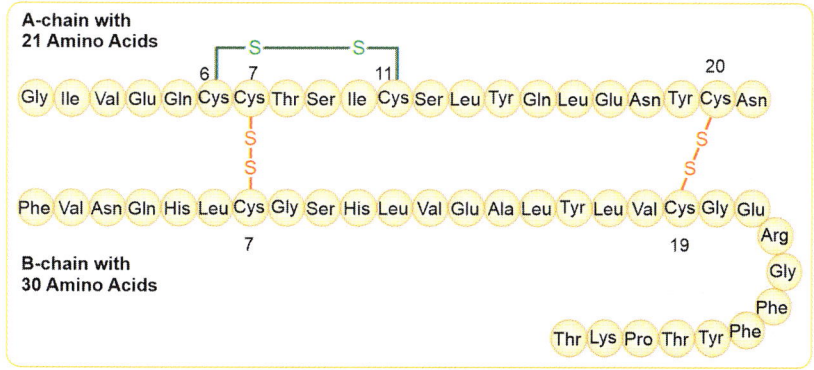

Fig. 2: Interchain disulfide bonds are shown in red; intrachain disulfide bonds are shown in green

Chapter 2 **Protein Structure**

Chain	Number of amino acids	Position of intra-chain disulfide bond
A	21	Between 6th and 11th cysteine
B	30	Nil

There are three disulfide bonds in insulin. Two are interchain, and one is intrachain.

Methods to Deduce 1° Structure

The primary structure is the linear sequence of amino acids. Analysis of primary structure is nothing but sequencing of amino acids. The following techniques are used for sequencing of a polypeptide. Some techniques are analyzed from the amino-terminal while others are analyzed from carboxy terminal of the polypeptide. Sanger got the first of his two Nobel prizes for the sequencing of bovine insulin.
- ❏ Edman degradation, Sanger sequencing and dansyl chloride methods are for N-terminal analysis
  - ○ Sanger's reagent is -fluoro-2,4-dinitrobenzene (FDNB)
  - ○ Edman's reagent is phenylisothiocyanate (C_6H_5-NCS) (PITC)
- ❏ Carboxypeptidase assay and Akabori reaction for carboxy-terminal analysis
- ❏ Tandem mass spectrometry (MS/MS)—In addition to the primary structure, it can detect post-translational modifications, like glycosylation, phosphorylation.

Dipeptide Contains a Single Peptide Bond

- ❏ The word "di" in dipeptide denotes the number of amino acid residues present in a peptide chain.
- ❏ Glycyl Glycine is a dipeptide made up of two glycine residues linked by a single peptide bond.

Glutathione is a Tripeptide

- ❏ Glutathione is γ-glutamyl cysteinyl glycine, i.e. glutathione is a tripeptide made up of 3 aminoacyl residues – Glutamate, Cysteine, Glycine (Fig. 3).
- ❏ Glutathione is the major intracellular reducing substance/anti-oxidant.
- ❏ SH group of glutathione is responsible for its antioxidant function which we will study when we explore biochemistry further.

The biological role of glutathione will be discussed in Chapter 11.4. The role of glutathione in Paracetamol poisoning will be discussed in Chapter 17 (Refer to page no. 348).

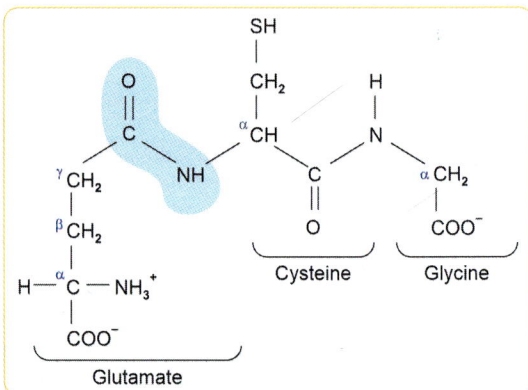

Fig. 3: The bond between glutamic acid and cysteine is an isopeptide bond (marked in blue)

Primary Structure Determines the Higher-order of Structures

Anfinsen's protein unfolding and folding experiments with Ribonuclease proved that 1° structure of a protein determines its higher levels of structural organization. He was awarded the Nobel prize for this discovery.

Replacement of a single amino acid in a polypeptide may drastically affect the structure and function. This can be understood with the example of sickle cell anemia. Replacement of glutamate with valine in the 6th position of β-globin chain drastically alters the structure and function of hemoglobin during hypoxic exposure *(detailed in Chapter 3 (Refer to page no. 27).*

SECONDARY STRUCTURE

Secondary structure refers to the regular, recurring spatial arrangement of amino acids that are close to one another in the linear sequence. Hydrogen bonding between peptide bonds maintains the secondary structure.

φ and ψ Angles (Dihedral Angles)

The peptide bond (C-N) is rigid and does not rotate. Rotation is possible only at the following angles (Fig. 4):
☐ φ (phi) angle is the angle between α-carbon and amide nitrogen.
☐ ψ (psi) angle is the angle between α-carbon and carbonyl carbon.

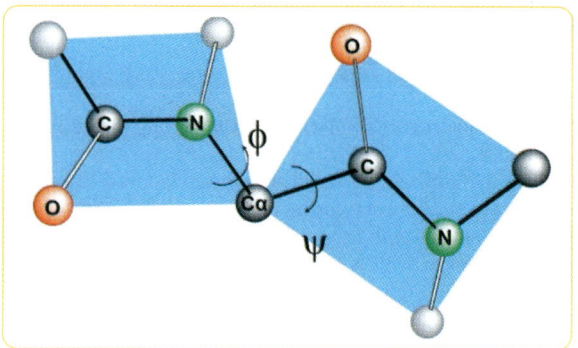

Fig. 4: Illustration of φ and ψ dihedral angles

Ramachandran Plot

G. N. Ramachandran (Gopalasamudram Narayana Iyer Ramachandran) developed a graphical plot to visualize the energetically allowed regions of protein secondary structures. He plotted φ angle against ψ angle. From this plot, sterically permitted secondary structures, right-handed α-helix, and β-sheets (marked in blue and pink) can be found. No secondary structure is possible in the white regions (Fig. 5).

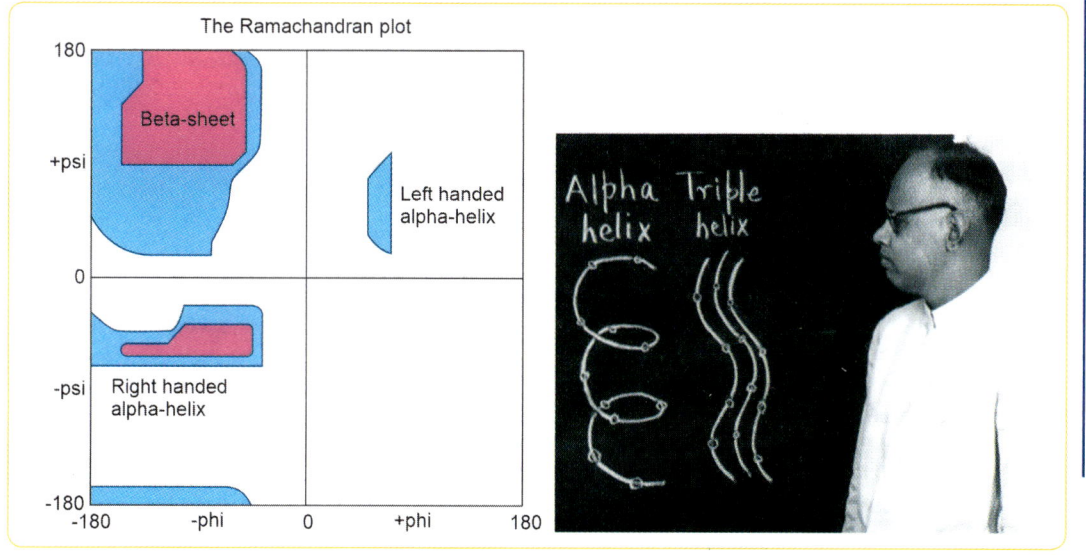

Fig. 5: G.N.Ramachandran explaining the protein secondary structure

Classes of Secondary Structures

- α-helix
- β-pleated sheet
- Turns and loops

Alpha-helix is the Most Common Secondary Structure

- α-helix is a rod-like helical structure where the peptide backbone winds around a long axis with the side chains (R groups) radiating outward.
- N–H group of one amino acid is hydrogen bonded to the C=O group of the amino acid located four residues away along the protein sequence.
- This hydrogen bonding creates a right-handed spiral or helix (coiled, rod-like structure).
- Most of the naturally occurring α-helices are right-handed.
- One turn of α-helix contains 3.6 amino acids (amino acid residues, to be precise).
- Distance between two adjacent amino acids is 1.5 A°
- Many globular proteins like hemoglobin and myoglobin are mainly made up α helices, and certain proteins contain both α-helix and β sheets. Chymotrypsin is devoid of the α-helix.

D-amino Acids and Glycine Favor the Formation of Left-handed α-helix

- In the human body, proteins are exclusively made up of L-amino acids.
- L-amino acids favor right-handed α-helix.
- D-amino acids and Glycine favor left-handed α-helix (Glycine does not have an asymmetric carbon).

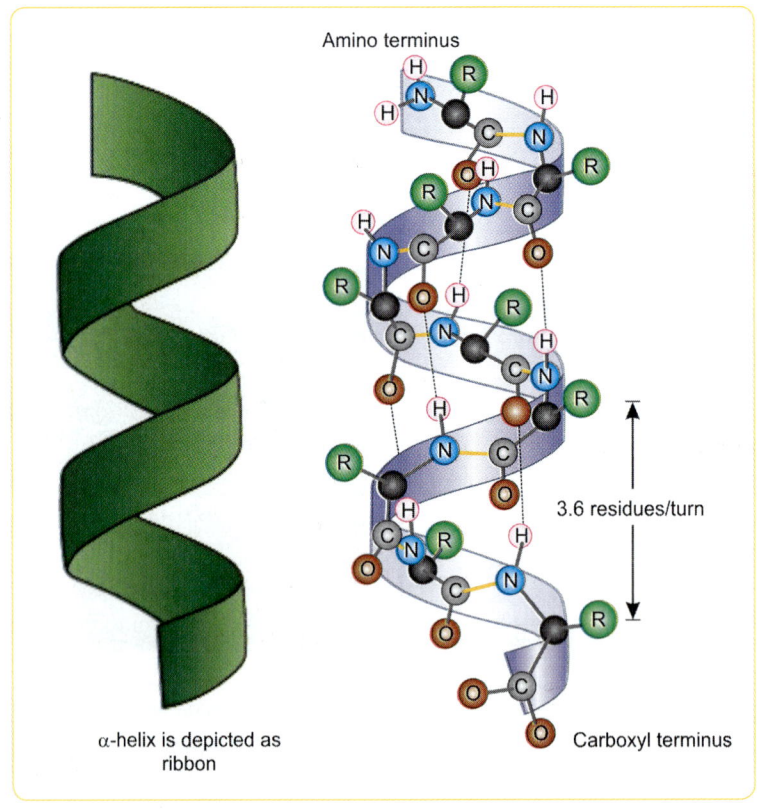

Amino terminus

3.6 residues/turn

α-helix is depicted as ribbon

Carboxyl terminus

Proline is α-Helix Breaker

Proline can't be accommodated in the α-helix, because,
- ❏ Imino (NH) group of proline can't form hydrogen bonds.
- ❏ Cyclic nature of the R-group of proline prevents the attainment of Ramachandran angles necessary for the secondary structure. So, proline causes a bend in α-helix.

This does not mean that proline is absent in α-helix containing proteins. Proline is usually seen at the beginning and end of the α-helix. Proline is important in turns and bends which will be discussed shortly.

Glycine is also disfavoured in α-helix because of its small-sized side chain (Hydrogen). Presence of multiple glycine in a protein makes a coiled structure quite different from α-helix (e.g. collagen).

Amino acids with bulky R-groups can't be placed nearby each other in α-helix because of the steric hindrance of side-chains (Fig. 6).

β-Pleated Sheet

- ❏ β-sheet is almost fully extended structure.
- ❏ Hydrogen-bonded β strands produce β-sheets.
- ❏ The distance between 2 adjacent amino acids is 3.5 Å.
- ❏ β-sheet can be either parallel, anti-parallel or mixed.
- ❏ If both the strands involved in hydrogen bonding run in the same direction (amino to carboxyl), the β sheet is said to be parallel.

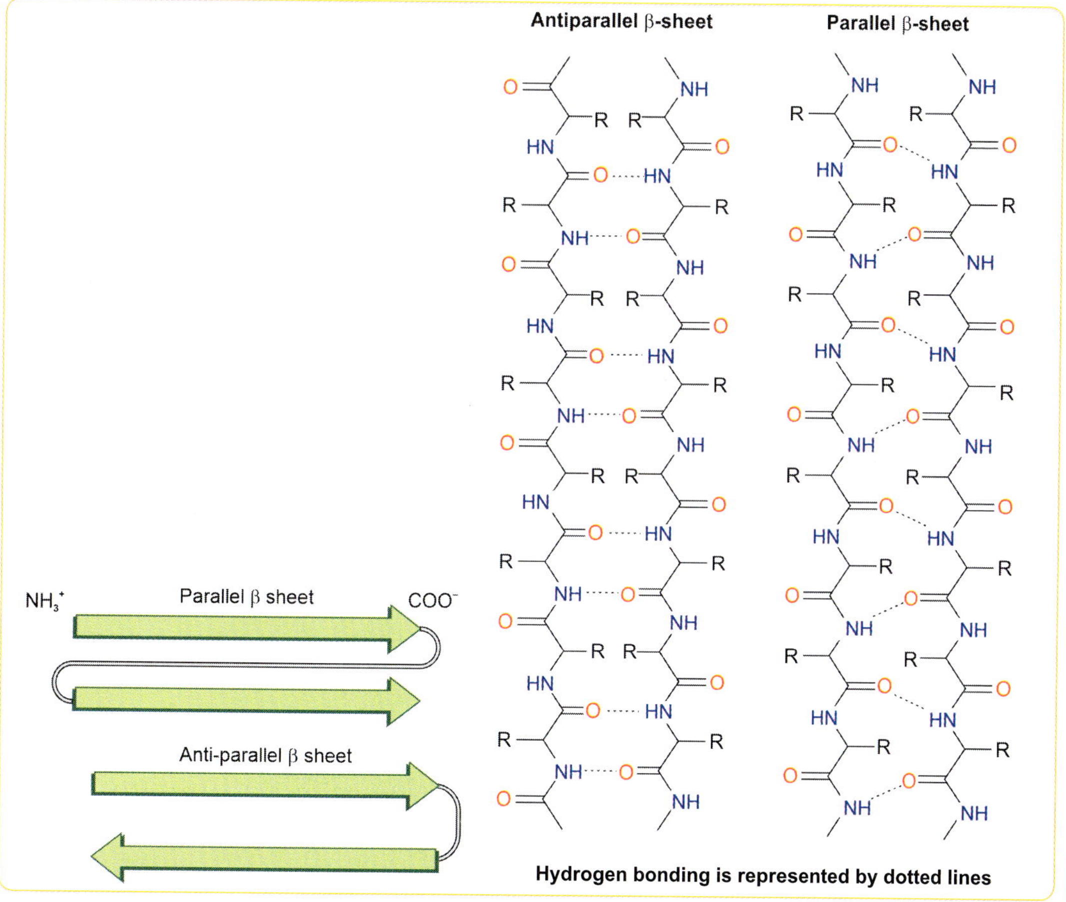

Antiparallel β-sheet   **Parallel β-sheet**

NH₃⁺   Parallel β sheet   COO⁻

Anti-parallel β sheet

Hydrogen bonding is represented by dotted lines

Fig. 6: Schematic representation of β-pleated sheet

α-helix Versus β-pleated Sheet

α-helix	β-pleated sheet
Rigid, rod-like, usually right-handed	Extended sheet, either parallel or antiparallel
Hydrogen bonding (i+4) in the same strand (intrastrand)	Hydrogen bonding to the neighboring strand (interstrand)
Is a single chain	Minimum of 2 **strands** are needed to make a sheet
Distance between two adjacent amino acids is 1.5 Å	Distance between two adjacent amino acids is 3.5 Å

Super Secondary Structure

Turns and bends are short segments of aminoacyl residues that join two secondary structures.

β-turn is usually made up four amino acid residues and contains glycine and proline. The first amino acid is hydrogen bonded to the fourth amino acid. Glycine and proline provide the extended conformation for the turn.

Loop is also a structure connecting secondary structures, but the number of the amino acids is more than that required for connecting the secondary structures. Loop is more extended than turn.

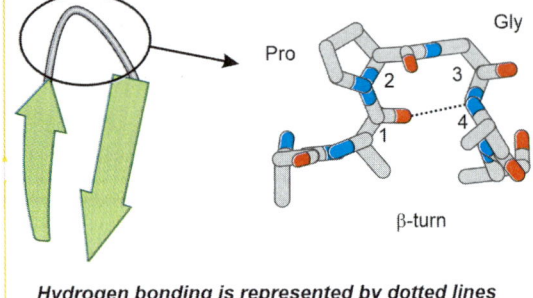

Hydrogen bonding is represented by dotted lines

Loops are found in **super-secondary** motifs like
- ❏ Helix-turn-helix (DNA binding motif) *(Chapter 27)* *(Refer to page no. 448).*
- ❏ Helix-loop-helix

A protein domain is the part of the protein that is independent of the rest of the protein structurally and functionally. Motifs can be a part of the domain.

Example:
- ❏ Rossman fold—NAD(P)⁺ binding domain of certain dehydrogenase enzyme
- ❏ Zinc finger motif - DNA binding motif

TERTIARY STRUCTURE

Tertiary structure refers to the three-dimensional shape of the protein. It explains how the secondary structures are assembled to form functional regions (domains) of the protein. Amino acids which are far away in the primary structure can come close in the tertiary structure. Proteins fold in such a way that hydrophobic residues avoid contact with aqueous environment. So, hydrophobic residues are found in the interior or core of a globular protein. *Protein folding will be discussed in Chapter 25 (Refer to page no. 438).*

> **Biochemical Basis of Hair-Straightening**
>
> α-helices of keratin are extensively cross-linked with disulfide bonds. Number of disulfide bonds varies between hair and nail. Curly hair is due to more number of disulfide bonds. Application of heat breaks this and makes the hair straight.

QUATERNARY STRUCTURE

Quaternary structure refers to the non-covalent association of different polypeptides. The subunits can be either same or different, e.g. Hemoglobin is a heterotetramer made up of two identical subunits.
Monomeric proteins like albumin and myoglobin cannot have quaternary structure.

Protein	Quaternary structure
Adult Hb	$\alpha_2\beta_2$
Insulin receptor	$\alpha_2\beta_2$
Protein kinase A	R_2C_2 (two regulatory and two catalytic subunits)

Contd...

Protein	Quaternary structure
Collagen	Quarter staggered structure
Myoglobin, Albumin	Nil

SUPRAMOLECULAR ASSEMBLY

A supramolecular assembly is a multimeric protein complex held together by non-covalent bonds. The process by which a supramolecular assembly forms is called **molecular self-assembly,** e.g. components of electron transport chain, Ribosome, Spliceosome, Proteasome

Review of Levels of Protein Structure

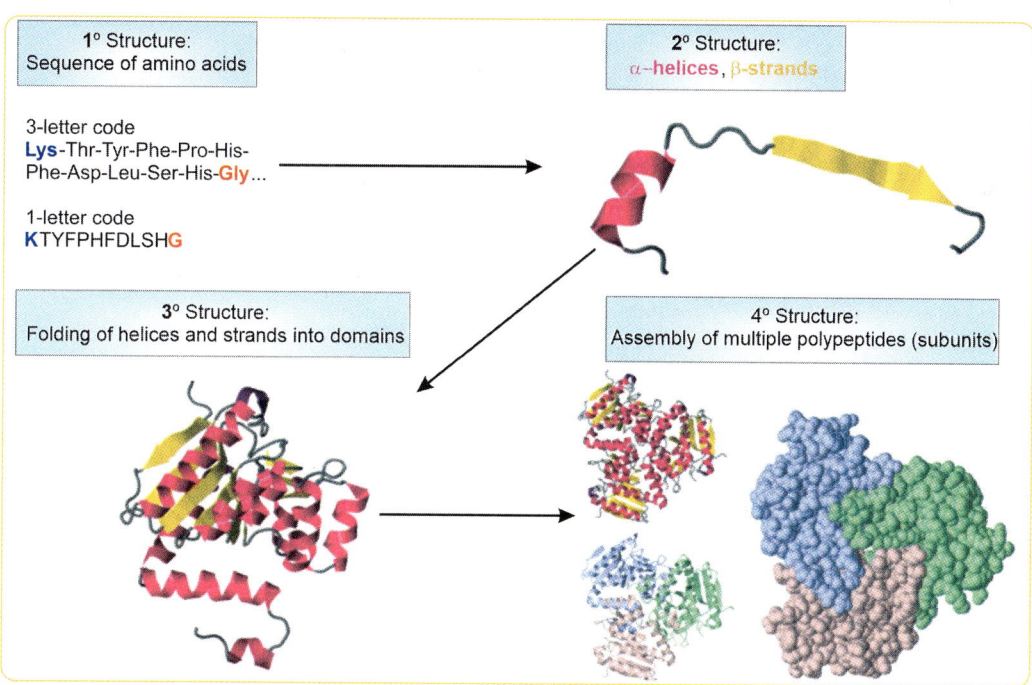

1° Structure:
Sequence of amino acids

3-letter code
Lys-Thr-Tyr-Phe-Pro-His-
Phe-Asp-Leu-Ser-His-**Gly**...

1-letter code
KTYFPHFDLSH**G**

2° Structure:
α–helices, β-strands

3° Structure:
Folding of helices and strands into domains

4° Structure:
Assembly of multiple polypeptides (subunits)

Level	Description	Force stabilizing	Methods to detect
1°	Linear sequence of amino acids of a polypeptide chain that are linked by covalent peptide bonds and disulfide bonds between cysteine residues of the same chain.	Covalent bonds—Peptide bond and *disulfide bond if it is present in the same polypeptide*	Protein Sequencing: ❑ Sanger's method ❑ Edman's degradation method ❑ Akabori method (C-terminal) ❑ Mass spectrometry
2°	The arrangement of **amino acids that are near one another** in the linear sequence linked by hydrogen bonds.	Hydrogen bond (non-covalent)	Circular dichroism (CD) spectroscopy
3°	Relation of **amino acids which are far away from each other** in the linear sequence. Hydrophobic amino acids are buried in the core of the protein.	Covalent disulfide linkages, non-covalent electrostatic interactions and Van der Waals forces.	❑ X-ray crystallography ❑ NMR spectroscopy
4°	Spatial arrangement of subunits and the nature of their interactions	Non-covalent interactions	X-ray crystallography NMR spectroscopy Cross-linking studies

Chapter 2 **Protein Structure**

23

PROTEIN DENATURATION

Denaturation is the loss of 2°, 3° and 4° structures of the protein resulting in the loss of biological activity. The primary structure remains intact after denaturation.

Denaturation is not degradation. Protein degradation refers to the loss of primary structure, i.e. breakage of peptide bonds.

Protein denaturing agents:

❑ **Physical agents:** Heat, Ultraviolet Light, Ultrasound, High pressure, etc.
❑ **Chemical agents:** Acids, Alkalis, Heavy Metals, Urea, Ethanol, Guanidine, Detergents, etc.

DENATURATION VERSUS COAGULATION

Coagulation is similar to denaturation. Coagulum, the product of coagulation is more visible than denaturation. Denaturation is said to be sometimes reversible. Coagulation is always irreversible, e.g. Boiled egg.

Precipitation reactions of proteins are discussed in Chapter 46 (Refer to page no. 658).

CLASSIFICATION OF PROTEINS

Proteins are classified into many ways based on their shapes, conjugated group, function and biological value.

Based on	Types
Shape	Globular and Fibrous (see the table below)
Function	Structural proteins—Collagen Transport proteins—Hemoglobin, Transferrin Catalytic proteins—Enzymes
Conjugated group	Hemoproteins—Hemoglobin, Catalase Flavoproteins—Succinate dehydrogenase (contains FAD) Glycoproteins—All plasma proteins except human albumin, LDL receptor Lipoproteins—Chylomicron, HDL, LDL, VLDL Nucleoproteins—Histone Phosphoproteins—Casein Metalloproteins—Ceruloplasmin, Ferritin
Biological value	Complete and Incomplete proteins: The egg protein is a complete protein which can provide all the 20 essential amino acids that support growth and convalescence. Most of the plant proteins are incomplete proteins.

Simple proteins are those which yield only amino acids and occasional small carbohydrate compounds on hydrolysis, e.g. albumin, globulins, glutelin, albuminoids. I hope the classification is self-explanatory, and we will encounter these types throughout the book. Let us differentiate the globular and fibrous proteins.

	Fibrous proteins	Globular proteins
Axial ratio (the longer axis divided by, the shorter)	≥10	<3
Solubility	Insoluble in water, acids, etc.	Soluble
2° structure	Either α-helix or β pleated sheet	Mixed
Example	Collagen	Albumin

Proteomics and the techniques of proteomics like protein electrophoresis, chromatography, western blotting will be discussed in Chapter 43 (Refer to page no. 624). Protein folding and prion diseases will be discussed in Chapter 25 (Refer to page no. 438).

Competency

| ❑ BI5.1 | Describe and discuss structural organization of proteins |

Key Points

❏ A peptide bond is polar, rigid with partial double bond nature, planar, and trans in configuration.
❏ Sanger's reagent is 1-fluoro-2,4-dinitrobenzene (FDNB), Edman's reagent is phenylisothiocyanate (PITC).
❏ Mass spectrometry can detect post-translational modifications, like glycosylation and phosphorylation.
❏ Primary structure determines the higher-order of structures.
❏ A dipeptide contains a single-peptide bond.
❏ Glutathione is γ-glutamyl cysteinyl glycine.
❏ Secondary structure is stabilized by hydrogen bonds.
❏ α-helix is a rod-like helical structure whereas β-sheet is an almost fully extended structure.
❏ Proline is an α-helix breaker.
❏ The primary structure is preserved during protein denaturation.
❏ Helix-Loop-Helix and Helix-Turn-Helix are examples of super-secondary structures.

SELF-ASSESSMENT

Short Answer Questions

1. Enumerate the characteristics of the peptide bond.

2. Differentiate a dipeptide from tripeptide. Give an example for each.

3. What is the biochemical basis of hair straightening?

4. What is an isopeptide bond? Name a molecule containing it.

5. Write the primary structure of glutathione. Mention the nature of bonds between the amino acid residues of glutathione.

6. Justify the statement—Primary structure of a protein determines its higher levels of structural organization of a protein. Which Nobel-prize winning experiment proved this?

7. Name the amino acids involved in the formation of glutathione. What do you understand by G-SH and G-S-S-G? Mention any two biological functions of glutathione.

8. Differentiate between α-helix and β-sheets.

9. Mention the salient features of the alpha helix. Explain why glycine and proline residues are not commonly found in alpha-helix.

10. What is a super secondary structure? Give an example.

11. What is a supramolecular assembly? Give an example.

12. What types of bonds (or forces) are the most important for primary, secondary and tertiary structure respectively?

13. In a tabular format, write the following regarding the four levels of protein structures:
   a. Description
   b. Force stabilising
   c. Methods of detection

14. Classify proteins based on their functions with an example for each class.

15. Define protein denaturation. Differentiate denaturation from coagulation.

16. What is protein precipitation? Name any two precipitating agents (Chapter 46).

Multiple Choice Questions

1. **The peptide bond has all the following characteristics, EXCEPT:**
   a. Covalent in nature
   b. Planar in nature
   c. Partial double-bond character
   d. Free to rotate

2. **Glutathione is made up of all the following, EXCEPT:**
   a. Glycine
   b. Glutamate
   c. Cysteine
   d. Arginine

3. **Which of the following bond maintains the primary structure?**
   a. Peptide bond
   b. Hydrogen bond
   c. Van der Walls interaction
   d. Hydrophobic interaction

4. **How many interchain disulfide bonds are present in insulin molecule?**
   a. 1
   b. 2
   c. 3
   d. 4

5. **Sanger's reagent is:**
   a. Flouro dinitro benzene
   b. Phenyl isothiocyanate
   c. Diaminobenzidine
   d. Tetramethylbenzidine

6. **How many amino acids are present in a single turn of α-helix?**
   a. 3.3
   b. 3.6
   c. 10.5
   d. 11.5

7. **Which of the following amino acid cannot be accommodated in α-helix?**
   a. Alanine
   b. Serine
   c. Proline
   d. Cysteine

8. **Which of the following amino acids are abundant in β-turn?**
   a. Glycine and valine b. Proline and valine
   c. Valine and alanine d. Glycine and proline

9. **Which of the following protein cannot have quaternary structure?**
   a. Albumin
   b. Immunoglobulin
   c. Hemoglobulin
   d. Collagen

10. **Which of the following is an example of supramolecular assembly?**
   a. Albumin
   b. Myoglobin
   c. Ribosome
   d. Ubiquitin

11. **Which of the following method is used for the analysis of C-terminal end of a polypeptide?**
   a. Sanger's method
   b. Edman's degradation method
   c. Akabori method
   d. None of the above

12. **Which of the following amino acids shows a tendency to form a left-handed helix?**
   a. Cysteine
   b. Glycine
   c. Alanine
   d. Histidine

ANSWERS

1. d.	2. d.	3. a.	4. b.	5. a.	6. b.
7. c.	8. d.	9. a.	10. c.	11. c.	12. b.

Hemoglobin and Myoglobin

The atmosphere of the primordial earth was completely devoid of oxygen and the metabolism of all the primitive organisms was anaerobic. When the oxygen appeared in the environment, it was toxic to the primitive organisms. So, they produced proteins that can bind and limit oxygen toxicity. Later, aerobic organisms got evolved. With the endosymbiotic incorporation of mitochondria, oxygen could be used as a final acceptor of electrons in the electron transport chain. But, the solubility of oxygen in aqueous medium is poor. Thus, organisms required an oxygen transporter and hemoglobin functions as that transporter. Myoglobin is the reservoir of oxygen.

Hemoglobin and myoglobin are conjugated proteins containing the heme prosthetic group. Let us discuss the structure of heme briefly:

Heme is the iron containing protoporphyrin IX/III, the word IX and III indicates the series in which Fischer described the structure of porphyrins (Fig. 1). *Protoporphyrin IX and III are nothing but the same.* Different protoporphyrins contain different side-chains. Protoporphyrin IX/III contains methyl, vinyl, methyl, vinyl, methyl, propionate, propionate and methyl (M, V, M, V, M, Pr, Pr, M) groups.

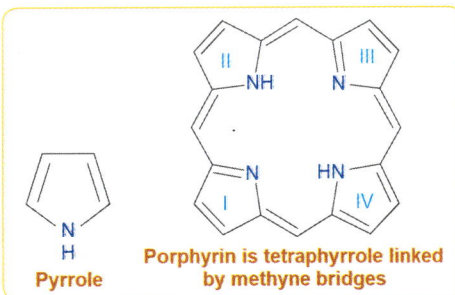

Pyrrole

Porphyrin is tetraphyrrole linked by methyne bridges

Fig. 1: Porphyrins: Four pyrrole groups linked by methyne/methine bridges

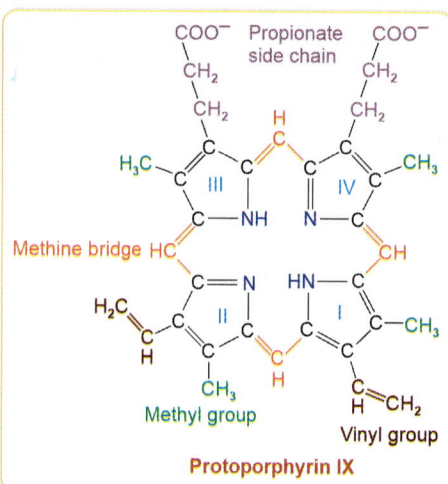

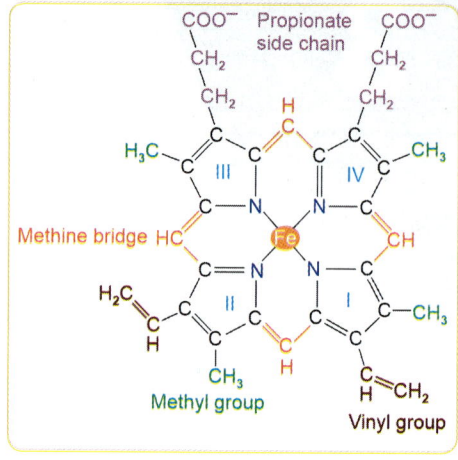

Protoporphyrin IX

Fig. 2: Heme = Ferrous iron + Protoporphyrin IX

STRUCTURE OF HEMOGLOBIN

Let us understand the structure of hemoglobin using the knowledge obtained after reading Chapter 2 *(Refer to page no. 16).*

Hemoglobin is a heterotetramer with 4 subunits of two different types (Fig. 3):

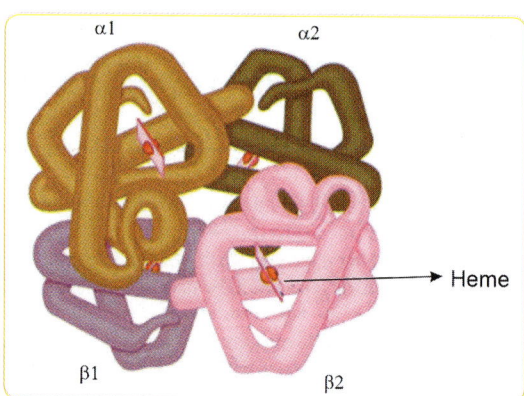

Fig. 3: One hemoglobin molecule contains four heme molecules. Each heme can bind to one molecule of oxygen

1° structure	α chain is made up of 141 amino acids; coded by chromosome 16.
	β, γ, δ → 146 amino acids; coded by chromosome 11.
2° structure	α chain → 7 α helices
	β chain → 8 α helices; *similar* to myoglobin in both secondary and tertiary structure.
3° structure	Stabilized by interaction with heme prosthetic group. No disulfide bonds are involved in hemoglobin.
4° structure	**Tetramer**
	HbA → $\alpha_2\beta_2$
	HbA$_2$ → $\alpha_2\delta_2$

Quaternary Structure of Hemoglobin Differs from Embryo to Adult

Different types of globin chains are produced during different times of human life (Fig. 4). In adults, majority of the hemoglobin is $\alpha_2\beta_2$.

A locus on chromosome 16 also known as α globin cluster codes for ζ and α chains.

A locus on chromosome 11 also known as β globin cluster codes for β, γ, δ and ϵ chains.

In early embryonic life, ζ chains is produced from alpha globin cluster and ϵ chain is produced from beta globin cluster to produce Gower type I embryonic hemoglobin.

In later period, the expression pattern differs to produce different types of hemoglobin as shown below:

Embryonic hemoglobins (produced from Yolk sac)	Gower I $\rightarrow \zeta_2\epsilon_2$ Gower II $\rightarrow \alpha_2\epsilon_2$ HbE Portland I$\rightarrow \zeta_2\gamma_2$ HbE Portland II $\rightarrow \zeta_2\beta_2$
Fetal (produced from fetal liver)	HbF $\rightarrow \alpha_2\gamma_2$
Adult (produced from bone marrow)	HbA $\rightarrow \alpha_2\beta_2$ (major) HbA$_2$ $\rightarrow \alpha_2\delta_2$ (minor)

Normal adult blood contains 97% HbA, 2% HbA$_2$ and 1% HbF.

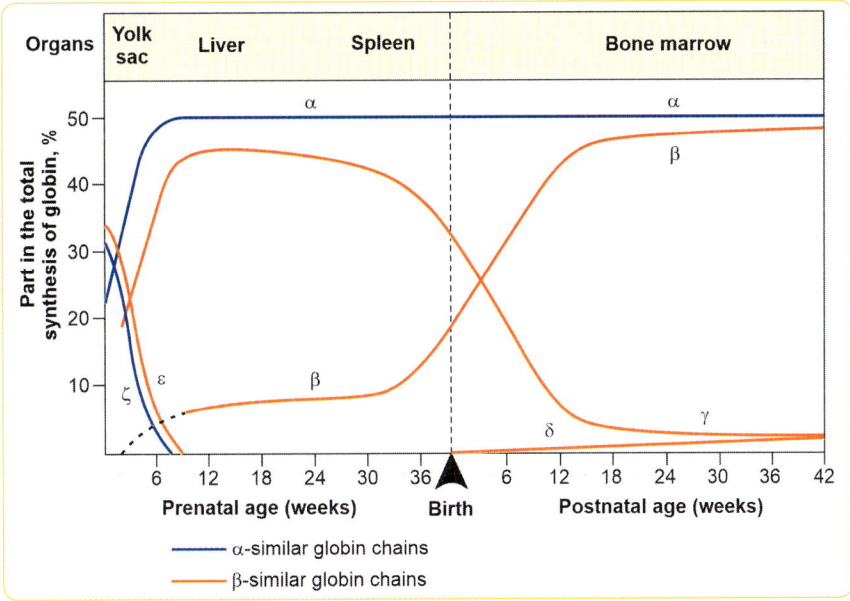

Fig. 4: Expression of different types of globin chains during different time points of prenatal and postnatal life

Globin is Made up of Many Alpha Helices

Alpha globin is made up of 7 α-helices and beta globin is made up of 8 α-helices. These helices are named alphabetically. (A to G/F). Heme prosthetic group is positioned in a hydrophobic pocket between helix E and helix F.

Hemoglobin is a tetramer. There is one heme prosthetic group per globin. Each heme can bind to one molecule of oxygen. Thus, one molecule of hemoglobin can bind to 4 molecules of oxygen.

Hemoglobin Contains Polar Histidine Residues in the Inner Core

Globular proteins usually contain polar amino acids at their surface and non-polar amino acids in the inner core. In hemoglobin, heme is located in a hydrophobic pocket of globin. But, two histidine residues are also found in the hydrophobic pocket. One histidine is located at the 7th position of the helix E and is known as HisE7. The other histidine is located at the 8th position of the helix F and is known as HisF8.

Distal Histidine E7

Fe

F8

Proximal Histidine

   HisF8 is closer to the heme and is known as proximal histidine.

   HisE7 is away from the heme and is known as distal histidine.

   Unless they have some unique role, nature would not have positioned these 2 polar amino acids in the hydrophobic pocket.

Proximal Histidine (HisF8) is Bound to Heme

- ❑ Iron in heme is held by coordinate bonds.
- ❑ Ferrous iron of the heme possesses 6 valances.
- ❑ 4 valences are occupied by nitrogen of porphyrin ring.
- ❑ **5th valance is occupied by proximal histidine (HisF8).**
- ❑ 6th valance is free to bind O_2/CO/NO.
- ❑ In methemoglobin, the ferrous iron is oxidized to ferric iron and the 6th valance is lost. This is why methemoglobin cannot bind to oxygen.
- ❑ The proximal histidine also pulls the iron in heme out of the plane of the heme molecule. This iron will be pulled back into the plane when it is oxygenated (see the image below).

Distal Histidine Modulates the Binding of Carbon Monoxide with Heme

In the iron bound oxygen molecule, the second oxygen is oriented 120° to the first oxygen. Binding of oxygen to ferrous iron is favored by distal histidine by hydrogen bonding. By making hydrogen bond, the distal histidine prevents oxidizing molecules from oxidizing the heme iron.

Protein

'Distal' histidine

Protein

(Hydrogen bond)

O_2

Protein

Protein

Binding of oxygen to ferrous iron is favored by distal histidine (HisE7)

But, in the iron bound carbon monoxide, C, O and N are in the same plane. Distal histidine (HisE7) provides steric hindrance to the binding of CO to heme.

Affinity of CO to free heme is 25,000 higher compared to oxygen. This affinity is reduced to 200 times in hemoglobin because of the steric hindrance explained above.

This is also the basis of hyperbaric (high-pressure) oxygen therapy in carbon monoxide poisoning. Oxygen at high pressure, can replace CO from heme because binding of O_2 is favored compared to CO.

Hemoglobin binds to nitric oxide also.

Oxygen Binding to One Heme Alters the Affinity of Other Heme Because of Changes in the Globin (Allosteric Effect)

Hemoglobin is an allosteric protein.
It exists in 2 states:

> **Taut or T state (Low-affinity for oxygen):** Because of salt-bridges between globin chains, hemoglobin is taut. **Oxygen unloading** is favored, i.e. deoxy Hb.
>
> **Relaxed or R state (High-affinity for oxygen):** Because of disruption of salt-bridges, hemoglobin is relaxed. **Oxygenation** of hemoglobin is favored, i.e. oxy Hb.

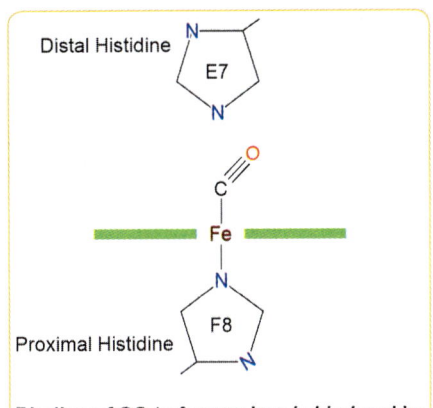

Binding of CO to ferrous iron is hindered by the distal histidine (HisE7)

Binding of first oxygen to iron of heme causes 0.04 nm movement of heme beyond the plane. This movement causes 15° rotation of α/β subunits. This results in disruption of salt-bridges of the other globin molecules and favors the conversion of T state to R state.

Binding of oxygen to one heme increasing the oxygen affinity of other heme molecules is known as 'cooperative binding'. This cooperative binding is possible only because of the quaternary structure, i.e. presence of 4 heme containing globin chains.

Oxygen Dissociation Curve (ODC) of Hemoglobin Demonstrates the Cooperativity

In oxygen dissociation curve of Hb, partial pressure of oxygen is plotted on the X-axis and percentage of hemoglobin saturated with oxygen is plotted on the Y-axis. We get a **sigmoidal curve** for hemoglobin.
- p50 refers to the partial pressure of oxygen at which 50% of hemoglobin is saturated with oxygen. Adult hemoglobin gets 50% saturated with oxygen at **26 mm Hg** or Torr (1 mm Hg = 1 Torr).
- In the lungs, the maximum oxygen saturation of systemic arterial blood averages 97%.
- Shift of oxygen dissociation curve to right refers to the increased unloading of oxygen.
- *Imagine a person doing exercise in a warm room.* **Raised CO_2, Raised temperature, Raised 2, 3-BPG and Decreased pH causes shift of ODC to right.**

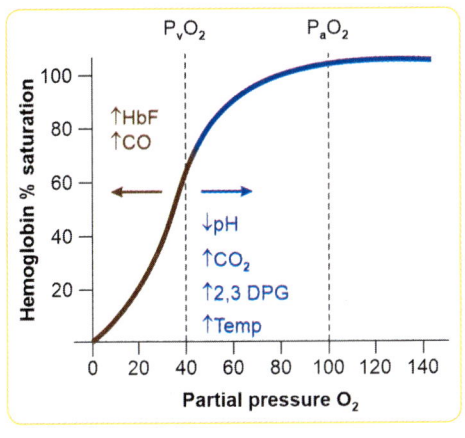

Chapter 3 Hemoglobin and Myoglobin

BOHR EFFECT: BINDING OF CO_2 AND PROTONS TO Hb CAUSES UNLOADING OF OXYGEN TO TISSUES

CO_2 produced in the tissues enters the RBC and is rapidly converted to carbonic acid. Spontaneous dissociation of carbonic acid produces proton.

Binding of carbon dioxide to hemoglobin produces carbaminohemoglobin which favors T-state of hemoglobin.

Binding of protons favors the T-state of hemoglobin. So, oxygen is released from hemoglobin to tissues.

In the lungs, oxygen binding to hemoglobin, releases the CO_2—this is known as *Haldane effect*.

Most of the CO_2 produced in the metabolism is transported in the form of HCO_3^- in the blood. Some amount of CO_2 binds to the amino group of the N-terminal of the α and β globins to produce carbaminohaemoglobin which facilitates the formation of salt-bridges and T-state stabilisation.

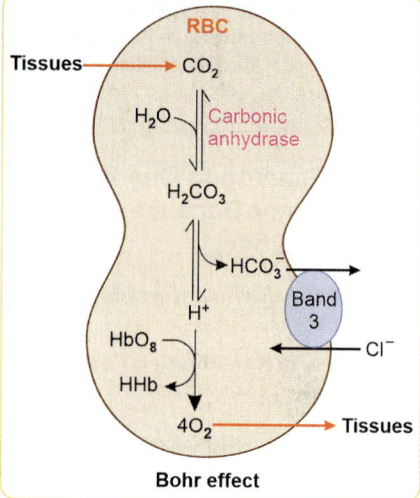

Bohr effect

2,3 BISPHOSPHOGLYCERATE HELPS IN UNLOADING OF O_2

2,3 bisphosphoglycerate (2,3 BPG) is produced in erythrocytes by a glycolytic shunt known as **Rapoport–Luebering shunt**. Concentration of 2,3 BPG in RBCs is approximately equal to the concentration of hemoglobin. One molecule of 2,3 BPG binds with one hemoglobin tetramer.

Negatively charged 2,3 BPG binds to positively charged amino acid residues (Histidine H21, Lysine EF6) of β chain in the central cavity of **deoxyHb**. This binding stabilizes the deoxyHb in 'T' state and causes **unloading of oxygen** from hemoglobin.

In lung, $\uparrow pO_2$ results in T → R transition and release of 2,3 BPG from central cavity.

$\uparrow$2,3 BPG levels	$\downarrow$2,3 BPG levels
Hypoxic conditions:	❑ Stored blood in
❑ High altitude	blood banks
❑ Severe anemia	❑ Diabetes mellitus
❑ Congenital heart	❑ Hypophosphatemia
diseases	
❑ Smokers	
❑ Chronic lung diseases	

RAPOPORT–LUEBERING SHUNT

❑ Rapoport–Luebering shunt is a glycolytic shunt pathway unique to RBCs (Fig. 5).

❑ Glycolytic intermediate 1,3 BPG is converted to 2,3 BPG by a bifunctional enzyme—2, 3-bisphosphogylcerate synthase (mutase)/2-phosphatase

❑ 2,3 BPG produced is an **allosteric regulator of hemoglobin**. It stabilizes the deoxyHb in 'T' state and helps in **unloading of oxygen to tissues**.

❑ Another enzyme **multiple inositol polyphosphate phosphatase (MIPP)** hydrolyse the 2,3 BPG to 3-phosphoglycerate and glycolysis proceeds.

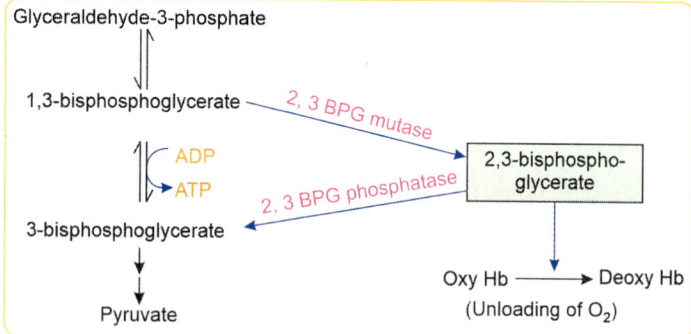

Fig. 5: Rapoport–Luebering shunt

❑ This shunt **compromises the 1 ATP** production by phosphoglycerate kinase reaction of glycolysis.

Blood Stored for Prolonged Period of time cannot Deliver Oxygen
2,3 BPG levels get reduced in blood stored for long time in the blood banks. If this blood is transfused to an anemic person, it may not serve the purpose of increasing the oxygen-delivering capacity. That is why adenine, dextrose, mannitol, sodium citrate and inorganic phosphate are used in the preservative solution to maintain the level of ATP and 2,3 BPG.

High Affinity of Fetal Hemoglobin can be Explained by the Inability of 2,3 BPG to Bind to γ Chain

We have seen that 2,3 BPG binds to some positively charged residues of β chain (H21). 21st amino acid in the H helix of γ chain is serine, not histidine. Unlike histidine, the R-group of serine is uncharged. As a result, negatively charged 2,3 BPG can make only very less ionic interaction with γ chain of HbF. So, affinity of HbF to oxygen is high and P_{50} value is low. **This is why oxygen is transported to placenta from maternal circulation to fetal circulation not the vice versa.**

Fetal Hb versus Adult Hb

	HbF	HbA
4° structure	$\alpha_2\gamma_2$	$\alpha_2\beta_2$
P_{50} value	20	26
Oxygen affinity	30% higher than HbA	Lesser than HbF
2,3 BPG effect	Minimal	More
In fetoplacental circulation	Extracts O_2 from HbA	Gives oxygen to HbF
Reaction with NaOH (Apt-Downey test)	Resistant to alkali denaturation	Susceptible to alkali denaturation

Apt Test or Alkali Denaturation Test
Fetal hemoglobin is resistant to alkali (basic) denaturation, whereas adult hemoglobin is susceptible to such denaturation. This is the basis of the apt test which is used to differentiate fetal or neonatal blood from maternal blood found in a newborn's stool or vomiting, or from maternal vaginal blood.

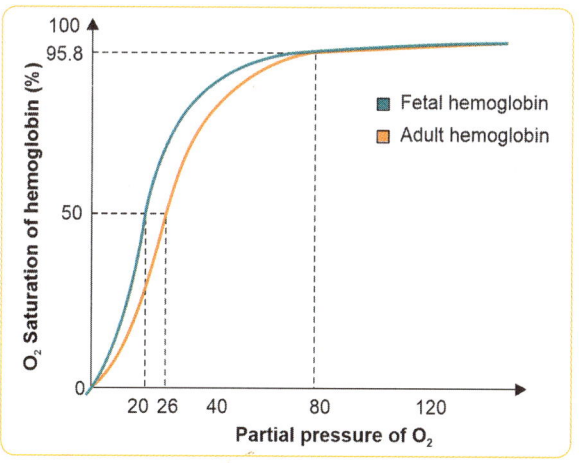

Fig. 6: Oxygen dissociation curve of fetal and adult Hb.

Look at the difference in the P50 value (Fig. 6)

Chapter 3 Hemoglobin and Myoglobin

Double Bohr Effect

When the Bohr Shift operates in one direction in the maternal blood and in other direction in the fetal blood, this phenomenon is called double Bohr effect, e.g. In placenta, oxygen dissociation curve for maternal Hb shifts to right whereas that of fetal Hb shifts to left.

Chloride Shift a.k.a Hamburger Phenomenon can be Explained by Donnan Membrane Equilibrium

In the venous side of capillaries, CO_2 enters the RBCs from tissues → Production of H_2CO_3 inside RBCs → Dissociation into H^+ and HCO_3^- → H^+ is buffered by deoxyHb → **HCO_3^- comes out** of the RBCs to blood via band 3 protein → To maintain electrical neutrality most abundant anion of ECF goes inside RBCs via the same **band 3 anion exchanger** of RBC membrane (Fig. 7). *Buffering action of hemoglobin will be discussed in Chapter 37 (Refer to page no. 575).*

Fig. 7: Diagrammatic representation of chloride shift

MYOGLOBIN

Myoglobin is a Reservoir of Oxygen

- Myoglobin is a single polypeptide (monomer) with a heme prosthetic group.
- It is made up of 153 amino acids arranged in 8 alpha helices (A to H).
- As there is only one chain, there is **no quaternary structure** in myoglobin and in turn there is **no cooperativity**. P_{50} value of myoglobin for oxygen is 2 mm Hg which means myoglobin readily gets saturated with oxygen and it would not release oxygen easily.
- Shape of the myoglobin oxygen dissociation curve is **hyperbolic**.
- As myoglobin has very high affinity for oxygen, it releases oxygen only under critical hypoxic conditions. Thus, myoglobin is a **reservoir of oxygen**, not a carrier.

Hemoglobin versus Myoglobin

Myoglobin	Hemoglobin
Single polypeptide chain with one heme prosthetic group	Multisubunit protein made up of 4 polypeptides with 4 heme
Oxygen **reservoir**	Oxygen **transporter**
As there is a single chain → no 4° structure → no co-operative binding	As 4° structure is present, cooperative binding of oxygen occurs
Oxygen dissociation curve (ODC)	
ODC curve is **hyperbolic** in shape	*Sigmoid* due to cooperative binding of O_2
P_{50} value: 2 mm Hg or 2 Torr	26 mm Hg for adult Hb

PATHOLOGY ASSOCIATED WITH HEMOGLOBIN

METHEMOGLOBINEMIA

Methemoglobin (MetHb) is Hemoglobin with Ferric Iron

Normally iron of hemoglobin is oxygenated but not oxidized. Oxidation of ferrous iron of heme to ferric form results in MetHb. Coordination number of ferric iron of hemoglobin is 5. As there is loss of 6th valance of iron, metHb cannot bind to O_2.

MetHb is present in a very small amount (<1%) in normal RBCs as a result of spontaneous oxidation. MetHb is reduced back to normal by 3 systems:
1. **NADH** dependent MetHb reductase also known as cytochrome b5 reductase—the major one responsible for the 75% of reduction
2. **NADPH** dependent reduction—20% of reduction
3. **Glutathione** dependent reduction—5% of reduction

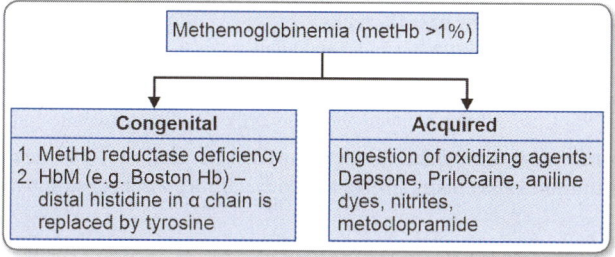

Clinical features: Headache, cyanosis, dizziness

Diagnosis: MetHb produced gives blood a characteristic **chocolate brown** color. Examination of blood color on white filter paper after exposure to room air or after aerating a tube of blood with 100% oxygen; if the blood remains dark even after this methemoglobinemia is suspected.

Treatment: Hyperbaric oxygen therapy (Oxygen at high pressure), methylene blue dye (reducing agent that reduces the metHb back to normal with the help of metHb reductase) (Fig. 8).

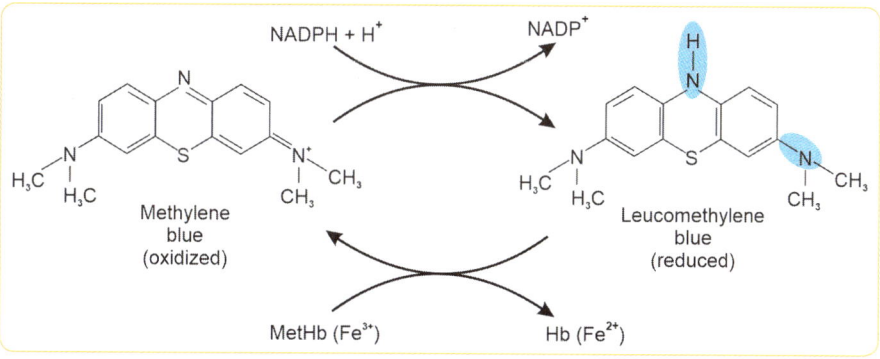

Fig. 8: Methylene blue helps in the conversion of MetHb to Hb

GLYCATED HEMOGLOBIN AND HBA$_{1C}$

Glycated hemoglobin is a product of **nonenzymatic** addition of carbohydrate to amino acid side chains of hemoglobin (Glycosylation is enzymatic). Addition of sugar molecules alter the charge and isoelectric pH of the hemoglobin. Therefore, these molecules can easily be separated out by electrophoresis or chromatography.

There are many types of glycated hemoglobins named on the order of their chromatographic separation.

Type	Molecule attached
HbA$_1$a$_1$	Fructose-1,6-bisphosphate
HbA$_1$a$_2$	Glucose-6-phosphate
HbA$_1$b	Pyruvate
HbA$_1$c	Glucose

HbA1 denotes the adult Hb which is glycated. a, b and c denote their order of elution in chromatography.

- HbA1c is a type of glycated hemoglobin in which glucose gets non-enzymatically attached to **N-terminal valine and epsilon amino group of lysine of β chains** of adult hemoglobin (HbA).
- HbA1c is the most abundant glycated hemoglobin.
- As glycation is a non-enzymatic reaction, HbA1c formation is directly proportional to the concentration of blood glucose during the life span of RBC.
- Percentage of HbA1c out of the total adult hemoglobin thus provides an estimate of average blood glucose concentration of a person during the past 12 weeks.
  Estimated Average Glucose (eAG) = $(28.7 \times A_{1c}) - 46.7$ mg%.

Detection of HbA1c:

Routinely used: High-performance liquid chromatography (HPLC) using cation-exchange columns.
Others: Electrophoresis, immunoassay using HbA1c antibody, affinity chromatography.

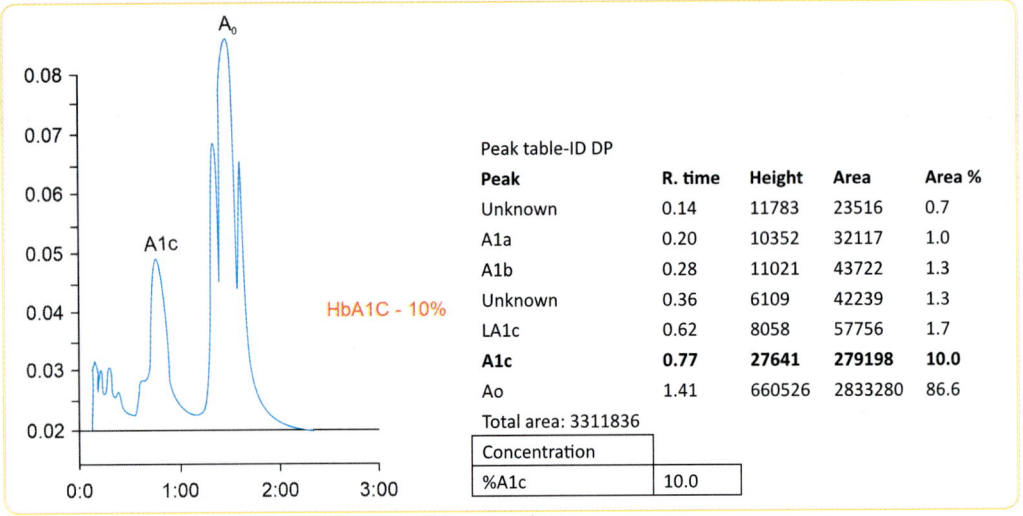

Peak table-ID DP

Peak	R. time	Height	Area	Area %
Unknown	0.14	11783	23516	0.7
A1a	0.20	10352	32117	1.0
A1b	0.28	11021	43722	1.3
Unknown	0.36	6109	42239	1.3
LA1c	0.62	8058	57756	1.7
A1c	**0.77**	**27641**	**279198**	**10.0**
Ao	1.41	660526	2833280	86.6

Total area: 3311836

Concentration	
%A1c	10.0

Fig. 9: Chromatogram

Chromatogram is shown in Fig. 9. Retention time is in the X-axis and signal intensity is in the Y-axis. Various hemoglobins elute at different times. Retention time is used to identify the type of Hb. Area under the curve is used to quantify them. Here you can notice that HbA1c elutes earlier than HbA. The person is uncontrolled diabetic since his/her HbA1c is 10%.

Uses of HbA1c Measurement

- **Diagnosis:** HbA1c ≥6.5% is one of the diagnostic criteria for diabetes mellitus.
- **Follow up (Glycaemic control):** Diabetic patients are advised to maintain their HbA1c <7%.

HbA1c measurement is of no use in hemoglobinopathies and severe anemia. In these conditions, fructosamine (glycated albumin) can be used.

Note: Fructosamine refers to glycated labile proteins. As albumin is the most abundant labile protein, fructosamine can be considered as glycated albumin.

HEMOGLOBINOPATHIES

Hemoglobinopathies are genetic diseases with abnormal structure of Hb.
- HbS
- HbC
- HbM

Sickle Cell Hemoglobin (HbS) (β6 Glu → Val)

HbS is due to a **point mutation** in the beta globin gene—the 6th codon GAG is converted to GUG.

GAG codes for glutamate, but GUG codes for valine—thus, this is a **partially acceptable, missense** mutation.

We know that primary structure is the important determinant of the higher order structures. Negatively charged polar glutamate on the surface of the hemoglobin is replaced by non-polar valine. This results in formation of a **sticky patch at the surface of Hb (Fig. 11).**

There is a **complementary hydrophobic pocket (sticky patch receptor) on the deoxyHb which binds to the sticky patch**.

Under hypoxic conditions, level of deoxy Hb is high. So, sticky patch of HbS binds to its receptor on deoxyHb. This leads to polymerization and precipitation of Hb. Polymerised Hb leads to activation of ion channels and loss of water from RBCs. So, RBCs become sickle-shaped. As the polymerized hemoglobin fibrils make the RBC membrane rigid, sickle cells are hemolyzed in large numbers.

Oxy HbA contains neither sticky patch nor the receptor. Deoxy HbA contains only the receptor. Oxy HbS contains only the sticky patch. Deoxy HbS contains both the sticky patch and its receptor (Fig. 10)

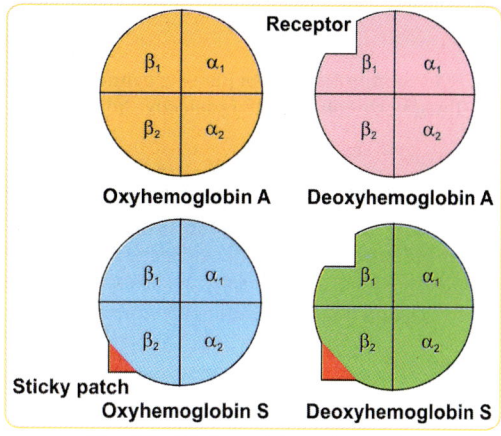

Fig. 10: Sticky patch and receptor

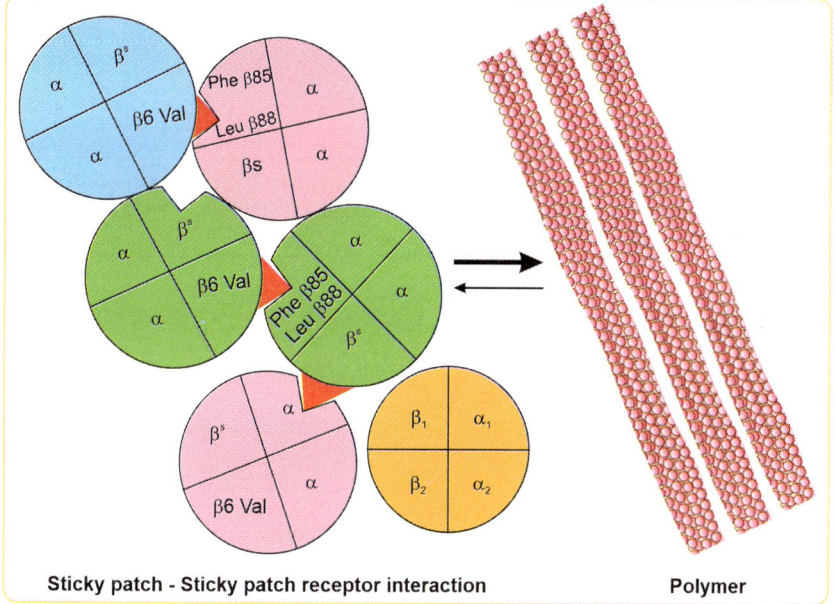

Sticky patch - Sticky patch receptor interaction   Polymer

Fig. 11: Sticky patch and sticky patch receptor interaction leading to polymerization. Can you observe that oxy HbA1 is not involved in the process of sickling?

Sickle Cell Crisis

- ❑ Massive hemolysis leads to vasospasm.
- ❑ This causes **bone pain** due to bone infarcts.
- ❑ **Autosplenectomy** happens due to splenic infarcts.

Sickle Cell Trait Versus Sickle Cell Disease

- ❑ Heterozygote condition with HbS and HbA is known as sickle cell trait.
- ❑ These patients are asymptomatic until they are exposed to hypoxia.
- ❑ Sickle cell disease is a severe condition with homozygous HbS.
- ❑ Patients may require multiple blood transfusions.

Sickle Cell Trait gives Protection Against Severe Malaria

- ❑ *Plasmodium falciparum* parasite cannot complete its life cycle inside RBCs containing HbS.
- ❑ Thus, sickle cell trait offers selective advantage in malaria endemic coastal regions.

Diagnosis

- ❑ Findings of hemolytic anemia
- ❑ **Sickling test—sodium metabisulfite** induces sickling of RBCs
- ❑ **Hb electrophoresis:** Reduced negative charge also alters the electrophoretic mobility. Order of electrophoretic mobility toward anode (positive end): **HbA > HbS > HbC**
- ❑ Restriction fragment length polymorphism for prenatal diagnosis.

Treatment

- ❑ Frequent blood transfusions are needed in sickle cell disease.
- ❑ **Hypoxic conditions should be avoided**. Any infection should be properly treated.
- ❑ Sickle cell crisis is treated with supportive management. As **hydroxyurea induces HbF**, it is used to prevent sickling.
- ❑ Bone marrow transplantation for sickle cell disease.
- ❑ **Vaccination against capsulated bacteria** in children with autosplenectomy.

 Clinical Correlation

Hydroxyurea is used in sickle cell disease
Hydroxyurea increases the level of fetal hemoglobin. HbF contains γ chain in place of β chain. So, there is no sticky patch and sticky patch receptor in HbF. Thus, it disrupts the polymerization of HbS. This is why hydroxyurea is used in patients of sickle cell anemia to reduce sickling.

HbC (β6 Glu → Lys)

- ❑ HbC is due to a point mutation in the gene coding for β chain of Hb in which negatively charged polar glutamate at 6th position is replaced by positively charged lysine.
- ❑ HbC reduces the sickling of HbS.

HbM

- ❑ HbM is methemoglobin. This can be caused by multiple mutations.
- ❑ In HbM Boston—the distal histidine (58th residue in the α-chain) is replaced by tyrosine.
- ❑ In HbM iwata—the proximal histidine (87th residue in the alpha chain) is replaced by tyrosine.
- ❑ This leads to oxidation of ferrous heme iron to ferric.

Hemoglobinopathies can be detected by hemoglobin electrophoresis and high-performance liquid chromatography (HPLC)

Alkaline Hemoglobin Electrophoresis

At alkaline pH, hemoglobin is negatively charged. Hemoglobinopathies alter the net mass/charge ratio of hemoglobin and the electrophoretic mobility of Hb.

HbA has the fastest electrophoretic mobility toward the anode (Fig. 12).

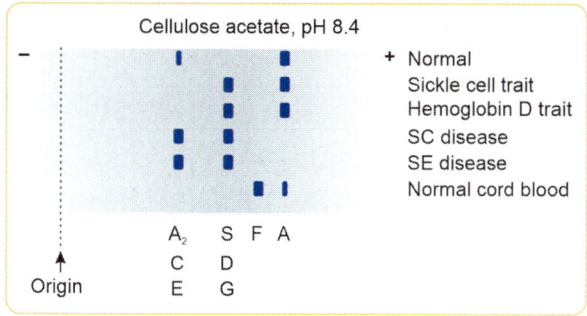

Cellulose acetate, pH 8.4

–	+ Normal
	Sickle cell trait
	Hemoglobin D trait
	SC disease
	SE disease
	Normal cord blood

A₂ S F A
C D
Origin E G

Fig. 12: Electrophoretic mobility of various hemoglobins during alkaline electrophoresis. The sample is placed at the cathode (negative end)

THALASSEMIA

The thalassemias are inherited disorders due to **quantitative defect** in globin chains. Defect in production of α-globin from its gene on chromosome 16 leads to α-thalassemia. Defect in β-globin from its gene on chromosome 11 leads to β-thalassemia.

There are two functional alpha genes in each of the chromosome 16. So, all normal individuals have 4 functional alpha globin genes.

Fetal hemoglobin does not contain β-chain. **So β-thalassemia is compatible with fetal life.** As α-chain is important in fetal Hb, **complete absence of α-chains leads to intrauterine death.** As there are 4 alleles for α-chain, quite a large size deletion is involved in α-thalassemia.

Nomenclature

Genotype		Phenotype	Clinical	Hb electrophoresis
$\alpha\alpha/\alpha-$	One gene is deleted	Silent carrier	Normal	Normal
$\alpha\alpha/--$ (or) $\alpha-/\alpha-$	Deletion of two genes on the same loci or deletion of two genes of different loci	α-thalassemia minor (trait)	Mild anemia	Normal
$\alpha-/--$	Presence of only one gene	HbH disease	Severe anemia	HbH (β_4 tetramers)
$--/--$	Complete absence of globin chain	α-thalassemia major	Hydrops fetalis	Hb Barts (γ_4 tetramers)

Beta Thalassemia

β° indicates complete absence of globin chain production. β^+ indicates some but reduced production. β indicates the normal production.

Genotype	Phenotype	Anemia	Hb electrophoresis finding
β/β°, β/β^+	β-thalassemia minor (trait)	Nil to mild	Mild elevation of HbA_2 and HbF
β^+/β°, β^+/β^+	β-thalassemia intermedia	Mild to moderate	$\uparrow$ HbF; HbA_2 may be normal or $\uparrow$
β°/β°	β-thalassemia major	Severe	$\uparrow$ HbA_2 and HbF

Diagnosis of Beta Thalassemia

❑ Microcytic hypochromic anemia in hemogram
❑ $\uparrow$ HbF and HbA_2 levels in High-Performance liquid chromatography (HPLC)
❑ RFLP analysis

Cation exchange HPLC is routinely used in the detection of thalassemia.

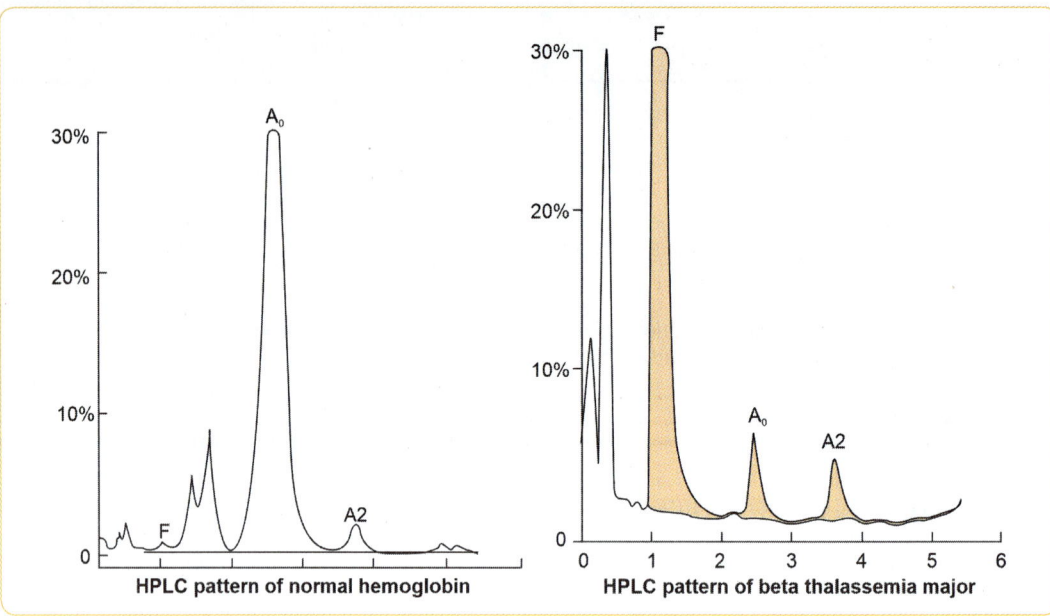

HPLC pattern of normal hemoglobin

HPLC pattern of beta thalassemia major

Competency

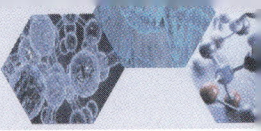

Key Points

- ❏ Myoglobin is made up of single polypeptide, so it does not have quaternary structure. It acts as a storage form of oxygen in muscles.
- ❏ HisF8 is proximal histidine and HisE7 is distal histidine.
- ❏ In hemoglobin, the innate affinity of heme for carbon monoxide is diminished by the steric hindrance provided by HisE7 (distal histidine).
- ❏ Imidazole nitrogen of proximal Histidine (HisF8) is covalently attached to the iron of heme.
- ❏ In deoxy hemoglobin, iron of heme is away from the plane of heme. Binding of oxygen shifts the heme iron toward the plane (0.04 nm). This movement causes 15° rotation of α/β subunits.
- ❏ Oxygen dissociation curve of myoglobin is hyperbolic.
- ❏ T (Taut/tense) state has less affinity to oxygen; favours unloading of oxygen. 2,3 BPG stabilizes the T state and releases oxygen. R (state) has more affinity to oxygen.
- ❏ Bohr effect is decreased oxygen affinity of Hb in response to raised pCO_2 or decreased pH.
- ❏ Oxidation of ferrous iron to ferric iron in hemoglobin leads to methemoglobin formation.
- ❏ Amyl nitrite promotes formation of methemoglobin, which combines with cyanide to form non-toxic cyanmethemoglobin – basis of use in cyanide poisoning.
- ❏ Since fetal hemoglobin does not have β chain, 2,3 BPG can't act on HbF. This is the reason for high oxygen affinity of HbF.
- ❏ In sickle cell anemia, the sticky patch is generated as a result of replacement of a polar (β 6 glutamate) residue with a non-polar (valine) residue.
- ❏ HbA1c >6.5% is diagnostic of Diabetes mellitus.
- ❏ 1 gm of Hb carries 1.34 ml of oxygen.
- ❏ Order of electrophoretic mobility of hemoglobin toward anode: HbA > HbS > HbC

SELF-ASSESSMENT

Short Answer Questions

1. The core of globular proteins is usually made up of hydrophobic residues. But hemoglobin contains two polar histidine residues in the core. What is the role of these two residues?

2. What is the role of proximal and distal histidine in hemoglobin?

3. The affinity of carbon monoxide to free heme is 25,000 times higher compared to oxygen. But the heme in hemoglobin has 100-fold reduced affinity for CO compared to free heme. Explain how.

4. Tabulate any four differences between hemoglobin and myoglobin.

5. Why the P50 value of myoglobin is ten-fold lower compared to hemoglobin?

6. How does the structure of HbF differ from HbA? Explain why oxygen flows from the mother's blood (HbA) to the fetal blood (HbF), and not in the reverse direction?

7. Explain the statement "Structural difference of fetal hemoglobin from maternal hemoglobin is useful to the fetus but harmful after birth".

8. The P50 value of HbF for oxygen (20 mmHg) is lower than that of HbA (26 mmHg). Explain its molecular basis. Name a clinical condition in which HbF concentration is high in postnatal life. Why the lower P50 value of HbF is useful for the fetus, but has deleterious consequences after birth?

9. Specify the molecular defect in Sickle cell hemoglobin. Why exposure to hypoxic conditions may precipitate an acute hemolytic crisis in these patients?

10. Explain why low oxygen tension may precipitate an acute hemolytic crisis in patients with sickle cell disease? Why the harmful HbS mutation has not been eliminated by nature's selection process?

11. What is glycated hemoglobin? Explain its clinical importance.

12. Explain the role of the Bohr effect in the efficient transport of oxygen by hemoglobin.

13. Explain Chloride shift with a schematic illustration.

14. What is methemoglobin? Name the protective mechanisms present in our body against methemoglobin formation. Mention one inherited and one acquired cause of methemoglobinemia. What is the biochemical basis of treatment of this condition?

15. What is the rationale behind using hydroxyurea in sickle cell disease?

16. What is the biochemical basis of Apt test used for the differentiation of fetal blood from maternal blood?

17. HbS mutation has not been eliminated by nature. What is the selective advantage provided by this mutation so that it is prevalent?

18. Explain why people with sickle cell disease are advised to avoid hilly regions.

Clinical Case-based Question

1. A 16-year-old boy is presented in the emergency department after he had collapsed as a result of a dental extraction in which prilocaine hydrochloride 3% was used as anaesthesia. His mother tells the emergency physician that he felt unwell after the procedure and gradually became more and more drowsy. The patient is deeply cyanosed despite a good respiratory effort, a clear airway, and bilateral breath sounds. Although he initially responds to vocal commands, his level of consciousness soon deteriorates. Oxygen saturation is 60% with the patient breathing room air. An endotracheal tube is introduced, and assisted ventilation started, but this does not improve his cyanosis or level of consciousness. A blood specimen is drawn. The blood is brownish-red in colour.

    a. What is your diagnosis and explain the biochemical basis of the boy's condition?
    b. Why oxygen saturation is only 60%?
    c. How will you treat this boy?

Multiple Choice Questions

1. The highest level of protein structure seen in myoglobin is:
    a. Primary
    b. Secondary
    c. Tertiary
    d. Quaternary

2. The major hemoglobin present in RBCs in normal adult is which one of the following?
    a. HbA2
    b. HbA
    c. HbA1c
    d. HbA1b

3. Quaternary structure of Portland hemoglobin is:
    a. $\zeta2\epsilon2$
    b. $\zeta2\gamma2$
    c. $\alpha2\epsilon2$
    d. $\alpha2\gamma2$

4. The shape of oxygen dissociation curve of myoglobin is:
    a. Parabola
    b. Hyperbola
    c. Sigmoid
    d. Circular

5. Binding of carbon monoxide to iron of hemoglobin is unfavored because of the steric hindrance provided by:
    a. Histidine E7
    b. Histidine E8
    c. Histidine F7
    d. Histidine F8

6. How much oxygen can be carried by one deciliter of blood with 15 g of hemoglobin?
    a. 7.5 mL of oxygen
    b. 15 mL of oxygen
    c. 20 mL of oxygen
    d. 30 mL of oxygen

7. Increased oxygen affinity of fetal hemoglobin compared to HbA can be explained by:
    a. α chain of fetal hemoglobin cannot bind to 2,3 BPG efficiently
    b. α chain of fetal hemoglobin binds to 2,3 BPG more efficiently
    c. γ chain of fetal hemoglobin binds to 2,3 BPG more efficiently
    d. γ chain of fetal hemoglobin cannot bind to 2,3 BPG efficiently

8. Cyanosis appears if deoxy hemoglobin level is more than:
    a. 5 g/dL
    b. 7 g/dL
    c. 9 g/dL
    d. 10 g/dL

9. Carboxy hemoglobin is formed in the blood due to exposure to:
    a. CO
    b. CO_2
    c. O_2
    d. O_3

10. Alpha globin is made up of _____ amino acids and coded by chromosome_____
    a. 146 and 11
    b. 146 and 16
    c. 141 and 16
    d. 141 and 11

11. To which of the following amino acid, heme iron is linked?
    a. Leucine
    b. Histidine
    c. Isoleucine
    d. Valine

12. **Porphyrin of normal erythrocyte hemoglobin is:**

    a. Coproporphyrin I b. Uroporphyrin III
    c. Protoporphyrin III d. Uroporphyrin I

13. **TRUE statement about adult hemoglobin is:**

    a. Homotetramer
    b. Two alpha and two beta subunits
    c. Each hemoglobin molecule binds to only one O_2 molecule
    d. Each hemoglobin has one heme molecule

14. **Rapoport-Luebering shunt takes place in:**

    a. Liver   b. RBCs
    c. Kidney   d. Brain

15. **Which of the following statements about hemoglobin is correct?**

    a. 2,3-Bisphosphoglycerate (BPG) increases the affinity of hemoglobin for oxygen.
    b. Deoxygenated hemoglobin has a higher binding affinity for protons than has oxyhemoglobin
    c. Hemoglobin has a higher affinity for oxygen than does myoglobin
    d. One molecule of hemoglobin binds sixteen molecules of oxygen -4 per subunit

16. **As hemoglobin binds oxygen molecules, its affinity for oxygen increases, driving the binding of further oxygen molecules. Which term best describes this phenomenon?**

    a. Catalysis
    b. Saturation
    c. Allostery
    d. Isomerism

17. **All of the following will cause a shift to the right of the oxygen-hemoglobin dissociation curve, EXCEPT:**

    a. Hypoxia
    b. 2,3 BPG
    c. Increase in pH
    d. Increase in temperature

18. **All of the following are true about methemoglobinemia, EXCEPT:**

    a. Ferrous iron is oxidized to ferric iron
    b. Either congenital or acquired
    c. Leads to hemolysis
    d. Oxygen binding capacity is lost

ANSWERS

1. c.	**2.** b.	**3.** b.	**4.** b.	**5.** a.	**6.** c.
7. d.	**8.** a.	**9.** a.	**10.** c.	**11.** b.	**12.** c.
13. b.	**14.** b.	**15.** b.	**16.** c.	**17.** c.	**18.** c

CHAPTER 4

Bioenergetics

The Sun is the source of all the energy in the world. Directly or indirectly, all living organisms are harvesting the solar energy. The way we convert this energy into various other forms of energy to maintain our cellular processes is really a marvellous feat.

BASIC THERMODYNAMIC TERMINOLOGIES

Let us recall the laws of thermodynamics from our school days.

- **First law (law of conservation):** Energy can neither be created nor destroyed; one form of energy can be converted to other form; total amount of energy in the universe remains constant.
- **Second law:** The universe always favors increasing disorderness (entropy).
- **Entropy (S)** is the degree of disorderness or randomness. According to the big-bang theory, the size of this entire universe was a pinhead at the beginning. From the day of big-bang, entropy of the universe is constantly increasing as the universe keeps expanding. Moreover, we are all contributing to the randomness of the universe by many ways.
- **Free energy (G)** is the energy available for doing work. This is named after the scientist Gibbs and is known as G. Change in free energy is denoted by ΔG (Δ denotes difference).
- **Enthalpy (H)** is the heat content of the reacting system. In other words, enthalpy reflects the number and nature of chemical bonds. **Compared to glucose, fatty acids have more bonds (electrons) and release more energy**.

Free Energy Change (ΔG)

$$\Delta G = \Delta H - T\Delta S$$

ΔG = Difference in the free energy of products and substrates
ΔH = Difference in the enthalpy of products and substrates

T = Absolute temperature ($-273°C$)

ΔS = Change in the entropy of the reacting system, i.e. product and substrate.

Exergonic reaction is the one in which energy is released into universe; free energy of the product is lower than that of the substrate. Thus, ΔG is negative. Exergonic reactions can occur spontaneously when other thermodynamic influences are favorable.

Endergonic reaction is the one in which energy is consumed (i.e. ΔG is positive). These reactions cannot occur spontaneously. Endergonic reactions are made possible by coupling with exergonic reactions, which we will discuss in brief.

CONCEPT CORNER

To understand the entropy better, let me tell you my reply when people advise me to keep things in my room arranged.

"When you arrange things in your room, you are decreasing the entropy. Nature favors disorderness. That is why I keep my room messy."

In the famous sitcom, The Big Bang Theory, Sheldon refers Penny's disorganized room as "Swirling vortex of entropy".

Standard Free Energy Change ($\Delta G°$)

$\Delta G°$ is the free energy change under standard conditions (temperature: $25°C$, 1 atmospheric pressure, all reactants are at 1 mole concentration)

The relationship between ΔG and $\Delta G°$ for the reaction $A + B \rightleftharpoons C + D$

$$\Delta G = \Delta G° + RT \ln \frac{[C][D]}{[A][B]}$$

$$\Delta G = \Delta G° + RT \ln Keq \qquad \text{(ln - natural log)}$$

Keep this relationship in mind. We will encounter the application of this many times.

$\Delta G°'$ is standard free energy change ($\Delta G°$) at pH 7 which is close to the bodily pH. So, hereafter, we will be using only $\Delta G°'$.

THERMODYNAMIC COUPLING

Standard free energy changes in a metabolic pathway are additive. In case of two sequential chemical reactions,

$$A \rightleftharpoons B \qquad \Delta G°'_1$$
$$B \rightleftharpoons C \qquad \Delta G°'_2$$

Since the two reactions are sequential, we can write the overall reaction as

$$A \rightleftharpoons C \qquad \Delta G°'_{total}$$

The $\Delta G°'$ values of sequential reactions are the sum of $\Delta G°'$ of individual reactions.

$$\Delta G°'_{total} = \Delta G°'_1 + \Delta G°'_2$$

This explains how a thermodynamically unfavorable (endergonic) reaction can be made favorable by coupling it with a highly exergonic reaction.

Remember: All endergonic reactions are coupled with some exergonic reaction. There is an energy cycle in cells that link anabolic and catabolic reactions.

Hydrolysis of ATP is the most common coupling reaction.

Let us take the well-known example—first step of glycolysis.

$$\text{Glucose} + P_i \rightarrow \text{Glucose-6-phosphate} + H_2O$$

$\Delta G°'$ of this reaction is $+3.3$ kilo calories/mol.

$\Delta G°' > 0$, so, this is an endergonic reaction. Then, how is this reaction made possible?

The reaction is made possible by coupling to the hydrolysis of ATP.

$$ATP + H_2O \rightarrow ADP + P_i$$

$\Delta G°'$ of this reaction is -7.3 kilo calories/mol

$\Delta G°' < 0$, so, this is an exergonic reaction.

$$\text{Glucose} + P_i \rightarrow \text{Glucose 6-phosphate} + H_2O$$
$$ATP + H_2O \rightarrow ADP + P_i$$

Sum: Glucose + ATP → Glucose 6-phosphate + ADP + H⁺

Apply the $\Delta G^{\circ\prime}_{total}$ $= \Delta G^{\circ\prime}_1 + \Delta G^{\circ\prime}_2$

$= 3.3 - 7.3$

$= -4 \text{ kcal/mol}$

Thus, overall reaction is made exergonic. Energy stored in ATP is used to drive the synthesis of glucose 6-phosphate.

Structure of ATP

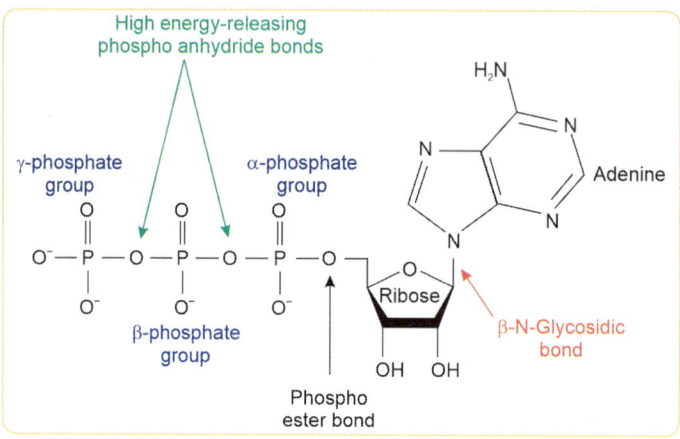

Fig. 1: Structure of adenosine triphosphate (ATP)

ATP (Fig. 1) is adenosine triphosphate (adenine + ribose + 3 phosphate groups)

❑ The bond between N_9 of adenine and C_1 ribose is β-N-Glycosidic bond

❑ The bond between α phosphate group and OH of group of 5th carbon of ribose is phosphoester bond.

❑ The bond between alpha phosphate and beta phosphate and beta phosphate and gamma phosphate are phosphoanhydride bonds. These are the high-energy bonds.

Hydrolysis of ATP

ATP to ADP conversion is exergonic, because:

❑ Hydrolysis of ATP^{4-} to ADP^{3-} relieves the charge repulsion on ATP.

❑ The phosphate released is stabilized by resonance stabilization.

This does not mean that ATP can spontaneously undergo hydrolysis in our body. Only enzymes can mediate this.

The free energy change for hydrolysis of ATP to ADP and P_i is -7.3 kcal/mol.

The free energy change for hydrolysis of ATP to AMP and pyrophosphate (PP_i) is -7.7 kcal/mol. Moreover, hydrolysis of inorganic pyrophosphate also releases huge amount of energy. Thus, those reactions in which ATP to AMP hydrolysis takes place are essentially irreversible.

Reactions in which 2 High-energy Phosphate Bonds of ATP to AMP Conversion is Utilized

❑ Amino acyl-tRNA synthetase

❑ Asparagine synthetase

❑ Benzoyl-CoA synthetase

❑ Fatty acyl-CoA synthetase

❑ Ubiquitin activating enzyme (E1) and Ubiquitin ligase (E3)

❑ Argininosuccinate synthetase

The Only Two Reactions where all the 3 Phosphate Bonds of ATP are Utilized in our Body

Here, the adenosyl group is transferred to the substrates.
1. Synthesis of S-Adenosyl methionine
2. Synthesis of S-Adenosyl cobalamin

High-energy Phosphate Compounds

High-energy phosphate compounds are those compounds having free energy of hydrolysis/phosphoryl transfer potential higher than that of ATP.

Some of the high-energy phosphate compounds can generate ATP through substrate-level phosphorylation *(discussed below)*.

Four High-energy Phosphate Compounds

1. Phosphoenolpyruvate ($\Delta G^{0\prime}$ of –14.8 kcal/mol)—contains the highest-energy
2. Carbamoyl phosphate
3. 1, 3-bisphosphoglycerate
4. Phosphocreatine/Creatine phosphate ($\Delta G^{0\prime}$ of –10.3 kcal/mol)

ATP is the Energy Currency of the Cell

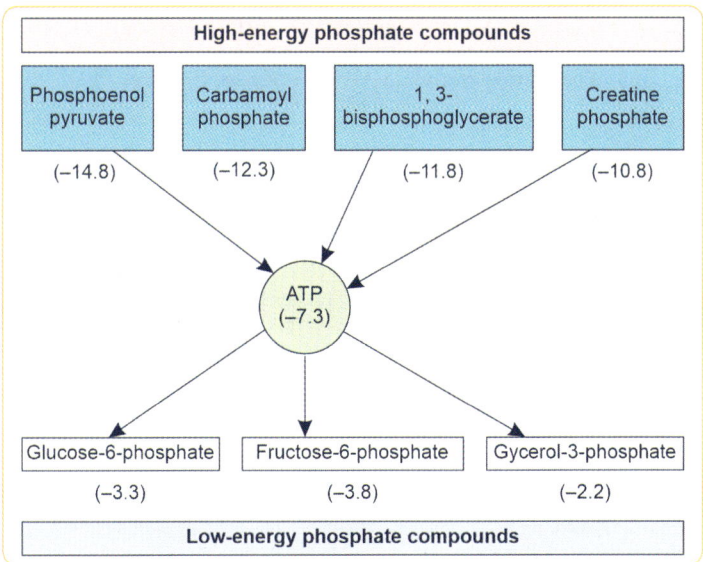

Note: Free energy of hydrolysis value is given within brackets. Look at the strategic position of ATP. ATP can be generated from high-energy phosphate compounds and ATP can generate low-energy phosphate compounds. This is the reason why ATP is utilized as energy currency. The currency should not be too weak and too strong.

Creatine Phosphate is a Phosphagen in Vertebrates

Phosphagens are high-energy phosphate compounds which supply immediate but limited energy during sudden demand for skeletal muscle, heart, sperm and brain. Phosphocreatine is the phosphagen in vertebrates and phosphoarginine is the phosphagen in invertebrates.

$$\text{Creatine phosphate + ADP} \underset{}{\overset{\text{Creatine kinase}}{\rightleftharpoons}} \text{Creatine + ATP}$$

This reaction is known as **Lohmann reaction.**

 **Chapter 4 Bioenergetics**

 47

Some Thioesters are also High-energy Compounds

Thioesters like succinyl-CoA are also high-energy compounds and can produce ATP at substrate-level phosphorylation reaction in TCA cycle.

Substrate-level Phosphorylation

❑ Main site of production of ATP is the electron transport chain by oxidative phosphorylation.
❑ Substrate-level phosphorylation is phosphorylation of ADP without the involvement of electron transport chain.
❑ Only high-energy compounds with free energy of hydrolysis more than that of ATP can produce ATP by substrate level phosphorylation.

Substrate-level phosphorylation reactions are displayed in the following manner:

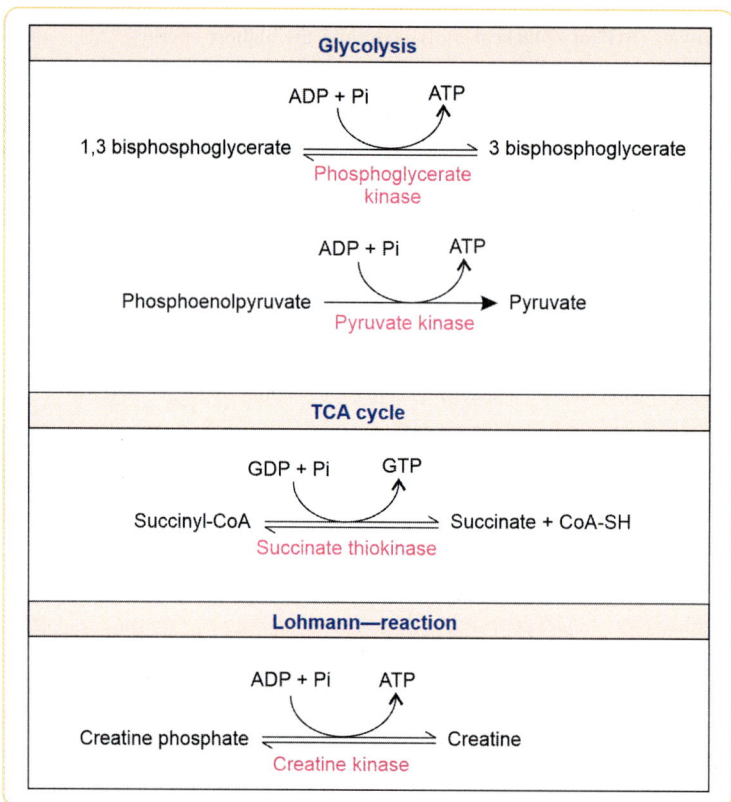

ATP and GTP are Interconvertible

ATP and GTP are interconvertible with the help of the enzyme nucleoside-diphosphate kinase. That is why GTP to GDP hydrolysis is considered as one ATP equivalent.

$$\text{GTP} + \text{ADP} \xrightleftharpoons{\text{Nucleoside-diphosphate kinase}} \text{ATP} + \text{GDP}$$

Note: ATP to AMP hydrolysis is considered as two ATP equivalents. To be precise, ATP to AMP hydrolysis should be considered as use of 2 high-energy phosphate bonds.

How can ATP Produce Creatine Phosphate?

Phosphoryl transfer potential of ATP is lesser than that of creatine phosphate.
❑ Free energy of hydrolysis of ($\Delta G°$) ATP is –7.3 kcal/mol
❑ Free energy of hydrolysis of creatine phosphate ($\Delta G°$) is –10.3 kcal/mol
This implies that creatine phosphate can transfer its phosphoryl group to ADP but ATP can't transfer its phosphoryl group to creatine.
Then, how is creatine phosphate produced in our body?
It is the time to apply what we learnt already.

$$\Delta G = \Delta G° + RT \ln \frac{[C][D]}{[A][B]}$$

In our body, ΔG determines the direction of reaction. Look at the equation. Based on the law of mass action, concentration of substrate and products affect the value of ΔG.

When ATP is in excess and creatine phosphate is getting utilized continuously, ATP can transfer its phosphoryl group to creatine.

Competency

❑ BI6.6   Describe the biochemical processes involved in generation of energy in cells

Key Points

❑ Entropy is the degree of disorderness or randomness.
❑ ΔG is negative in exergonic reactions and they can occur spontaneously.
❑ All endergonic reactions are coupled with some exergonic reactions.
❑ Creatine phosphate and arginine phosphate are the phosphagens in vertebrates and invertebrates respectively.
❑ ATP is the energy currency of the cell.
❑ Free energy of hydrolysis of ATP to ADP is –7.3 kcal/mol.
❑ Synthesis of S-Adenosyl methionine and synthesis of S-Adenosyl cobalamin are the only 2 reactions where all the phosphate bonds of ATP are utilized.

SELF-ASSESSMENT

Short Answer Questions

1. Define high-energy phosphate compounds. Give three examples.

2. What are phosphagens? Which molecule acts as a phosphagen in the vertebrate animals? Write the enzymatic reaction.

3. What is substrate-level phosphorylation? Give two examples with complete reaction.

4. What is the advantage of having ATP as the energy currency of the cell? Why can't carbamoyl phosphate or glucose-6-phosphate be used as energy currency?

5. Name any four biochemical reactions in which 2 High-energy Phosphate Bonds of ATP to AMP conversion is used.

6. Phosphoryl transfer potential of ATP is lesser than that of creatine phosphate. Then, how is creatine phosphorylated by ATP in the human body?

7. Explain why nature selected ATP as the energy currency" even though creatine phosphate (CP), Phosphoenolpyruvate (PEP), GTP and ATP are all high energy phosphate compounds.

Multiple Choice Questions

1. **All of the following are high-energy phosphate compounds, EXCEPT:**
   a. Creatine phosphate
   b. Carbamoyl phosphate
   c. Glucose-6-phosphate
   d. Phosphoenolpyruvate

2. **Which of the following TCA cycle enzyme catalyzes the substrate level phosphorylation?**
   a. Isocitrate dehydrogenase
   b. Succinate thiokinase
   c. Fumarase
   d. Malate dehydrogenase

3. **The bond between adenine and ribose in the ATP molecule is:**
   a. Glycosidic bond
   b. Thioester bond
   c. Phosphoester bond
   d. Phosphoanhydride bond

4. **In glucose to glucose-6-phosphate conversion, which of the following phosphate group of ATP is transferred to glucose?**
   a. Alpha
   b. Beta
   c. Gamma
   d. None

5. **All of the following endergonic reactions are made possible with coupling of ATP to AMP hydrolysis, EXCEPT:**
   a. Glutamine synthetase
   b. Asparagine synthetase
   c. Fatty acyl-CoA synthetase
   d. Aminoacyl tRNA synthetase

6. **Which of the following is the phosphagen in invertebrates?**
   a. Creatine phosphate
   b. Arginine phosphate
   c. Carbamoyl phosphate
   d. Glucose-6-phosphate

7. **Which of the following enzyme is involved in the interconversion of ATP and GTP?**
   a. Adenylyl cyclase
   b. Myokinase
   c. Nucleoside-diphosphate kinase
   d. Purine nucleoside phosphorylase

8. **How many phosphoanhydride bonds are present in ATP?**
   a. 1   b. 2
   c. 3   d. 4

9. **Entropy is the:**
   a. Degree of orderness
   b. Heat content of the system
   c. Degree of randomness
   d. Double bond loving nature

ANSWERS

1. c.	2. b.	3. a.	4. c.	5. a.	6. b.
7. c.	8. b.	9. c.			

5
CHAPTER

Enzymology

Enzymes are biological catalysts. They increase the rate of otherwise impossible and slower biochemical reactions up to trillion-fold without being themselves consumed, i.e. amino acid residues of enzymes may undergo some changes during catalysis. But, enzymes revert to their original state at the end of the catalysis.

DIFFERENCE BETWEEN CHEMICAL CATALYSTS AND ENZYMES

Unlike chemical catalysts,
- Enzymes are highly specific
- They work in the milder conditions of the body (within a narrow pH and temperature limit)
- Enzyme activity can be fine-tuned or regulated.

RIBOZYMES: CATALYTIC RNAs

All enzymes were initially thought to be proteins. The breakthrough came when Thomas Cech and Sidney Altman described the catalytic activity of RNA molecules independently.

Ribozymes are RNAs with catalytic activity, i.e. catalytic RNAs.

Examples

- 23S rRNA in prokaryotes and 28S rRNA in eukaryotes are ribozymes with peptidyl transferase activity. Thus, ribosome is a ribozyme.
- Small nuclear RNAs (SnRNA) are ribozymes with splicing activity (Chapter 23) *(Refer to page no. 415).*
- RNase P is a ribozyme involved in post-transcriptional modification of tRNA.
- Self-splicing introns are ribozymes.

ABZYMES: ANTIBODIES WITH CATALYTIC ACTIVITY

- Abzymes are antibodies which can specifically bind to transition state of the reaction.
- As abzymes specifically bind and stabilize transition state, they have catalytic property, e.g. some anti-ds DNA antibodies found in patients of systemic lupus erythmatosus are abzymes.

CLASSIFICATION OF ENZYMES

INTERNATIONAL UNION OF BIOCHEMISTRY AND MOLECULAR BIOLOGY (IUBMB) SYSTEMATIZED THE NOMENCLATURE OF ENZYMES

We can guess the action of the enzymes with names like urease, sucrase, carbonic anhydrase, carboxy peptidase, ribonuclease, etc. But, what can we infer from names like pepsin, trypsin, subtilisin? That is why systematic classification of enzymes was developed by International Union of Biochemistry and Molecular Biology. **There are seven enzyme classes (EC).** *They can be remembered as Over The Hill Lies the Isle of Ligases and Translocases.*

Class	Reaction catalyzed	Examples
EC1. Oxidoreductase	Oxidation/reduction reactions	❑ Hydroperoxidase ❑ Oxidase ❑ Oxygenase ❑ Dehydrogenases (e.g. Alcohol dehydrogenase) Mnemonic: HOOD
EC2. Transferase	**Group transfer** reactions	❑ Kinases (phosphotransferases) ❑ Aminotransferases (Alanine aminotransferase and aspartate aminotransferase)
EC3. Hydrolase	Cleavage of bonds with the use of water (hydrolysis)	❑ All proteases ❑ Urease ❑ Glucose-6-phosphatase
EC4. Lyase	Cleavage **without using water** → double bond formation	❑ Aconitase ❑ Aldolase ❑ Decarboxylase ❑ Enolase ❑ Fumarase ❑ Hydratase
EC5. Isomerase	Rearrangement of atoms within the molecule	❑ Isomerase ❑ Epimerase ❑ Mutase
EC6. Ligase	ATP mediated joining of molecules	❑ Biotin-dependent carboxylases ❑ Glutamine synthetase ❑ tRNA synthetase ❑ E3 Ubiquitin ligase ❑ DNA ligase

Contd…

EC7. Translocase	Transport of ions or molecules	Ornithine translocase, Carnitine-acylcarnitine translocase, Na⁺/K⁺-ATPase

Lyase Versus Hydrolase

Both are involved in cleaving of bonds. Hydrolases use water while lyases do not require water.

Synthases Versus Synthetases

Synthase	Synthetase
Catalyzes a synthetic process without ATP	Requires ATP
Comes under lyase or transferase classification	Comes under ligase classification
For example, HMG-CoA synthase, ATP synthase	For example, Succinyl-CoA synthetase, Glutamine synthetase, Carbamoyl phosphate synthetase I

ACTIVE SITE

ACTIVE SITE IS THE SITE OF ENZYMATIC CATALYSIS

Most enzymes are proteins and are made up of many amino acid residues. But, the catalysis takes place only in a cleft or crevice of the enzyme. Rest of the protein portion acts as a scaffold and provides stability to the enzyme. In case of multisubunit enzymes, non-active site amino acids are involved in oligomerization and binding to allosteric regulators.

ACTIVE SITE CONTAINS SUBSTRATE BINDING SITES AND CATALYTIC SITE

❑ Catalytic site is the site involved in the actual catalytic mechanism. Substrate binding site binds to the substrates and cofactor.
❑ Amino acid residues in the substrate binding sites determine the specificity of the enzyme (*explained in the following table*).
❑ Substrate binding sites of enzymes are specialized motifs.
  ○ **Example:**
    ➢ NAD binding site of many dehydrogenases (e.g. alcohol dehydrogenase) is made up of Rao-Rossmann fold

Properties of Active Site

❑ The active site is a three-dimensional cleft or pocket formed by amino acid residues that come from different parts of the primary structure of proteins.
❑ The active site of multimeric enzymes may be located at the interface between subunits and residues from more than one monomer may be involved.
❑ Substrates are bound to enzymes by multiple, weak, usually noncovalent interactions like electrostatic interactions, hydrogen bonds, van der Waals forces, and hydrophobic interactions. This binding is mediated by nonactive site amino acids also.
❑ Initially it was thought that the active site is complementary to the substrate. Now, it is known that the active site is actually complementary to the transition state *which we will discuss in the coming pages*.
❑ Allosteric site is different from active site.
❑ Water is excluded from the active site until and unless it is a reactant (Desolvation of active site).

Identification of Active Site Residues

- Diisopropyl fluorophosphate inactivates trypsin, chymotrypsin and elastase. This is due to the inactivation of serine residues of these enzymes. As serine plays a crucial role in the catalysis, these three enzymes are known as **serine proteases.**
- Replacement or removal of an amino acid by **site directed mutagenesis** (codon for an amino acid is changed or removed at the gene level) is used to establish the importance of a particular amino acid residue in enzyme catalysis.

Levels of Specificity of Enzymes

Level of specificity	Acts on	Examples
Stereospecificity	Substrates with specific optical rotation	D-amino acid oxidases L-amino acid oxidases
Bond/linkage specificity	Substrates with same bond	Amylase acts on all the molecules with α (1,4) glycosidic bond, i.e. starch, glycogen, dextrin
Functional group specificity	Substrates with specific groups around the bond	Even though trypsin, chymotrypsin and elastase are endopeptidases and act on peptide bond, they differ in their specificity based on the amino acids in the substrate-binding pocket (*explained below*).
Absolute specificity	Single substrate	Urease is specific for only urea.

Note: Glycine can be acted upon by both L and D amino acid oxidases since it does not contain a chiral carbon.

Certain Enzymes can have more than One Active Site

Most of the enzymes have only one active site per enzyme molecule. Certain enzymes can have more than one active site with different catalytic activities. These are known as bifunctional or multifunctional enzymes.

- Bifunctional enzymes make the regulation of metabolic pathway easier. For example, phosphorylation of PFK-2 reciprocally regulates glycolysis and gluconeogenesis.
- Bifunctional enzymes have the different enzyme commission numbers as they catalyze two different chemical reactions.
- They might have formed through the fusion of two genes during evolution.

Some bifunctional enzymes are as follows:

Bifunctional enzyme	Activities
PFK-2	Phosphofructokinase-2 Fructose 2,6-bisphosphatase
2,3-bisphosphoglycerate mutase (BPGM)	2,3-bisphosphogylcerate synthase/2-phosphatase
Glycogen debranching enzyme	Glucanotransferase Amylo-1,6-glucosidase
Prostaglandin endoperoxide H synthase (PGHS)	Cyclooxygenase Endoperoxidase
UMP synthase	Orotate phosphoribosyltransferase OMP decarboxylase

MULTIFUNCTIONAL ENZYME VERSUS MULTIENZYME COMPLEX

Multifunctional enzymes should be differentiated from multienzyme complex.

Multifunctional enzyme	Multienzyme complex
Enzymes that contain two or more different catalytic activities within a single polypeptide chain, e.g., fatty acid synthase	Stable, noncovalent assembly of different enzymes involved in consecutive reactions of a metabolic pathway, e.g., PDH complex, α-keto glutarate dehydrogenase complex

Note: Fatty acid synthase is considered as an example of both multienzyme complex and multifunctional enzyme.

FISHER'S LOCK AND KEY HYPOTHESIS AND KOSHLAND'S INDUCED FIT HYPOTHESIS

In 1894, Emil Fischer assumed enzyme active sites as lock and the substrate to be key. Thus, it requires exact fit.

In 1959, Daniel Koshland postulated the induced fit model. Binding of substrate induces conformational change in the active site of the enzyme which becomes complementary to the substrate. This can be imagined to the process of wearing gloves. *Induced fit mechanism of hexokinase is explained in the Chapter 10.2.* *(Refer to page no. 121)*

As per the current understanding, **active site is actually complementary to the transition state** rather than substrate.

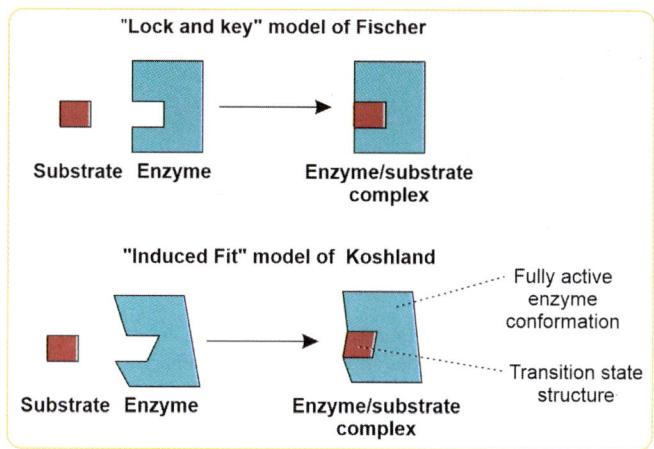

Amino Acid Residues in the Substrate Binding Site Determine the Specificity of the Enzyme

To understand this, let us consider the example of serine proteases—Chymotrypsin, Trypsin and Elastase (Fig. 1).

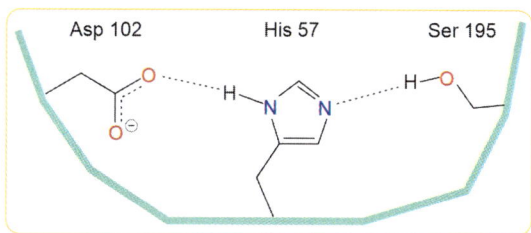

Fig. 1: Components of catalytic triad of chymotrypsin— The number indicates the position of amino acid in the primary structure. Can you appreciate that the amino acids which are far away in the primary structure come closer in tertiary structure?

Catalytic site of all these three enzymes contain a specific set of 3 amino acids (Serine, histidine and aspartate) involved in electron relay and covalent catalysis. These 3 amino acids form the catalytic triad of these serine proteases.

❏ Substrate binding site of trypsin contains negatively charged amino acid—aspartate. So, trypsin binds to and cleaves after positively charged amino acids like lysine and arginine.

❏ Look at the substrate binding site of elastase. It is made up of valine and threonine. R-group of these amino acids makes the substrate binding site shallow. So, small amino acids like alanine can be bound.

❏ Substrate binding pocket of chymotrypsin is wide enough to bind to bulky aminoacids like phenylalanine and tyrosine (Fig. 2).

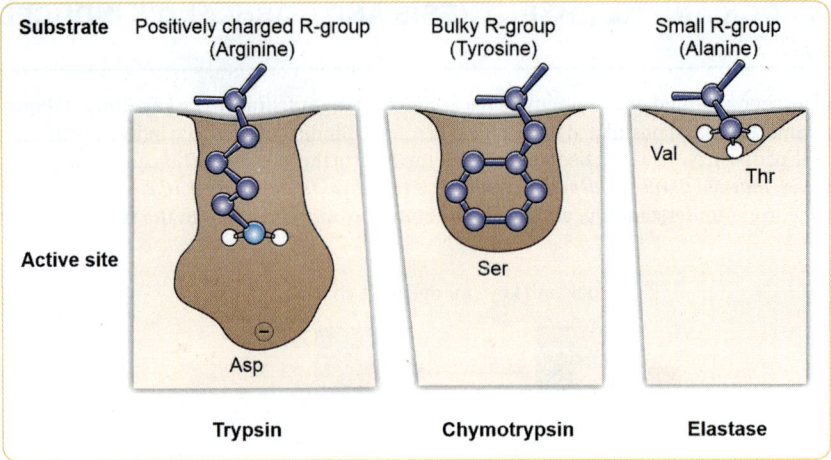

Fig. 2: Explanation for substrate specificity of serine proteases

Most Enzymes Require Additional Organic or Inorganic Molecules for their Action

- ❑ Enzymes require additional molecules for their activities.
- ❑ These additional molecules can be subdivided into metals and small organic molecules called coenzymes.
- ❑ Coenzymes are usually vitamin derivatives.
- ❑ Prosthetic groups are tightly bound. Loosely associated coenzymes are more like cosubstrates because, like substrates and products, they bind to the enzyme and are released from it.
- ❑ A single coenzyme can be employed by many enzymes.

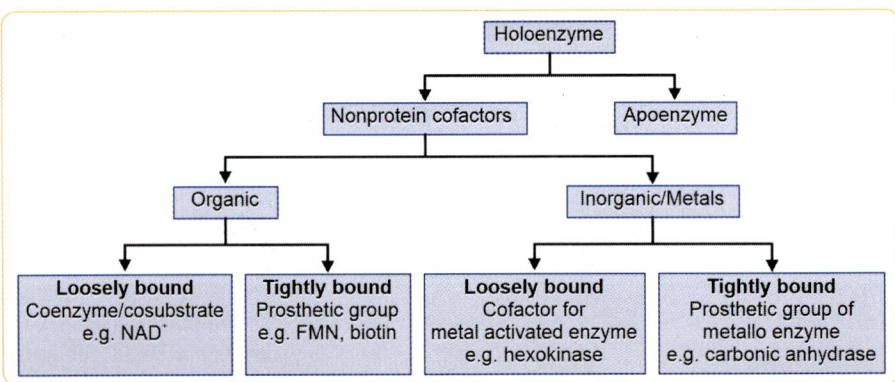

Cofactor or prosthetic group	Vitamin source	Group transferred	Enzyme
Thiamine pyrophosphate	Vitamin B₁	Aldehyde	Pyruvate dehydrogenase
Flavin adenine dinucleotide prosthetic group	Vitamin B₂ (Riboflavin)	Electrons	Succinate dehydrogenase
Flavin adenine mononucleotide (FMN) prosthetic group	Vitamin B₂ (Riboflavin)	Electrons	Complex I of electron transport chain
Nicotinamide adenine dinucleotide	Niacin	Hydride ion	Lactate dehydrogenase
Pyridoxal phosphate	Vitamin B₆	❑ Amino group ❑ Phosphate	❑ Transaminases ❑ Glycogen phosphorylase

Contd...

Cofactor or prosthetic group	Vitamin source	Group transferred	Enzyme
Coenzyme A (CoA)	Vitamin B_5 (Pantothenic acid)	Acyl group	Pyruvate dehydrogenase
5'-Deoxyadenosylcobalamin	Vitamin B_{12}	Alkyl groups and hydrogen atoms	Methylmalonyl-CoA mutase
Tetrahydrofolate	Folic acid (B_9)	One-carbon groups	Thymidylate synthase
Biocytin prosthetic group	Biotin	Carbon dioxide/ Bicarbonate	Pyruvate carboxylase

Metal cofactor	Enzyme
Zinc	❑ Carbonic anhydrase ❑ Carboxy peptidase ❑ Alcohol dehydrogenase ❑ Alkaline phosphatase ❑ Lactate dehydrogenase ❑ ALA dehydratase ❑ Glutamate dehydrogenase ❑ Superoxide dismutase ❑ Phospholipase C ❑ RNA polymerase
Copper	❑ Monoamine oxidase ❑ Ferroxidase (ceruloplasmin) ❑ **Cytochrome c oxidase** ❑ **Tyrosinase** ❑ Superoxide dismutase ❑ Dopamine hydroxylase ❑ **Lysyl oxidase**
Molybdenum	❑ Xanthine oxidase ❑ Sulfite oxidase
Magnesium	❑ All kinases ❑ Alkaline phosphatase ❑ DNA and RNA polymerase
Selenium	❑ Thioredoxin reductase ❑ Glutathione peroxidase ❑ 5' deiodinase
Nickel	Urease (absent in humans)
Potassium	Thiolase, Pyruvate kinase

MECHANISM OF ENZYME ACTION

In a nutshell, mechanism of action of enzyme can be described as, **"Enzymes increase the rate of the reaction by reducing the activation energy of a reaction and selectively stabilizing the transition state. Reduction in activation energy is contributed by binding energy."**

To understand this better, let us consider the bioenergetics of a reaction where the substrate (S) is converted to the product (P). For the substrate to become the product, it should first gain some energy and reach a transition state.

Transition state is a fleeting, momentary, unstable chemical species which can be converted to product or substrate depending upon the equilibrium constant of the reaction.

The free energy difference between the substrate and the transition state is known as activation energy.

Activation energy ($\Delta G^{\ddagger}$) = Free energy of substrate (G_S) – free energy of transition state ($G_{S^{\ddagger}}$)

Even if the substrate to product formation is exergonic, substrate to transition state formation is always endergonic. This creates an energy barrier for the substrate to become product.

ENZYMES HELP IN THE FORMATION OF TRANSITION STATE BY REDUCING ΔG‡

Enzymes bind to substrate to produce enzyme-substrate (ES) complex. In the ES complex, binding of substrates to enzyme is by multiple noncovalent bonds. The energy released upon binding of substrates to enzymes is known as binding energy. This binding energy accounts for the reduction of activation energy and stabilization of transition state by enzymes. If the equilibrium constant of the reaction is favorable for the product formation, the substrate in the ES complex becomes product, i.e. EP complex is formed. EP complex dissociates into enzyme and product.

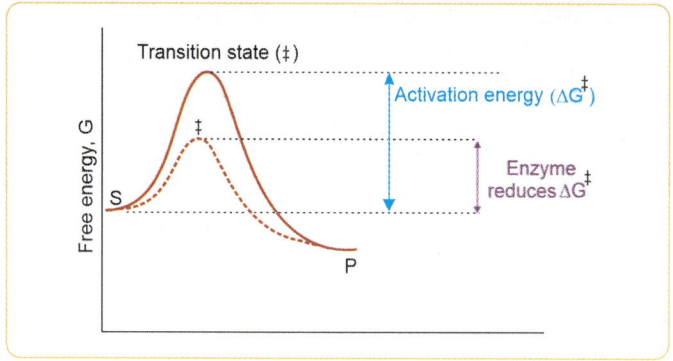

Note: Double dagger (‡) is used to denote the transition state.

Having said these, it is important to remember that enzymes cannot alter the equilibrium of a reaction. It increases the rate of both forward and backward reaction in the equal amount. The direction of the reaction is determined by the thermodynamics.

Let us look at the mechanisms taking place at the active site of the enzymes.

Mode of Catalysis

❑ Acid base catalysis
❑ Covalent catalysis
❑ Metal ion catalysis
❑ Catalysis by proximity and orientation/entropy reduction
❑ Catalysis by strain

Acid Base Catalysis

Enzymes utilising acid-base catlaysis have histidine at their active site. Remember the peculiarity of histidine we studied in the first chapter. Histidine is a basic amino acid with the capacity to protonate or deprotonate depending on the surrounding environment (pKa = 6), e.g. Ribonucleases, HIV protease.

Covalent Catalysis

Enzymes employing covalent catalysis involve the formation of an intermediate which is covalently bound to the active site residues of the amino acids. The covalent bond is transient and the enzyme is regenerated back at the end of catalysis, e.g. Serine proteases.

Metal Ion Catalysis

One-third of all the known enzymes utilize metals for the catalysis. Metals help in binding of substrates in the:
❑ Proper orientation of substrate and enzyme through formation of enzyme-metal-substrate bridge.
❑ Metals can serve as electron donors or acceptors and thus help in oxidation-reduction reactions.
❑ Electrostatically stabilizing or shielding negative charges.

For example, Mg^{2+} stabilizes the negative charges of β and γ phosphates of ATP. **All kinases require magnesium or manganese (Fig. 3).**

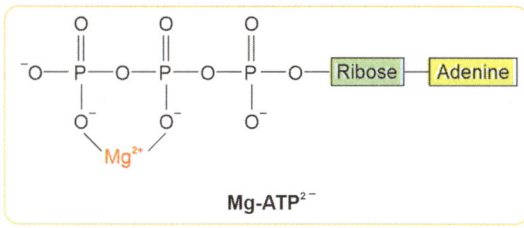

Mg-ATP^{2-}

Fig. 3: Magnesium stabilizing ATP

Carbonic Anhydrase Reaction is a Typical Example for Metal Ion Catalysis

Zinc prosthetic group of carbonic anhydrases lowers the pKa of water from 14 to 7, generating a hydroxide ion (OH⁻). This hydroxyl ion attacks CO_2 forming HCO_3^-.

Catalysis by Proximity and Orientation/Entropy Reduction

Enzyme brings the reacting substrate molecules together and aligns them in an optimal geometry, which increases the rate of the reaction. This reduces the entropy of the reactants.

Catalysis by Strain

Conformational change in the enzyme brings about a strain in the bonds present in substrate. This leads to breaking of bonds in the substrate and release of products.

This mode of catalysis is used by enzymes like hydrolases. The example of sucrase is illustrated in Figure 4.

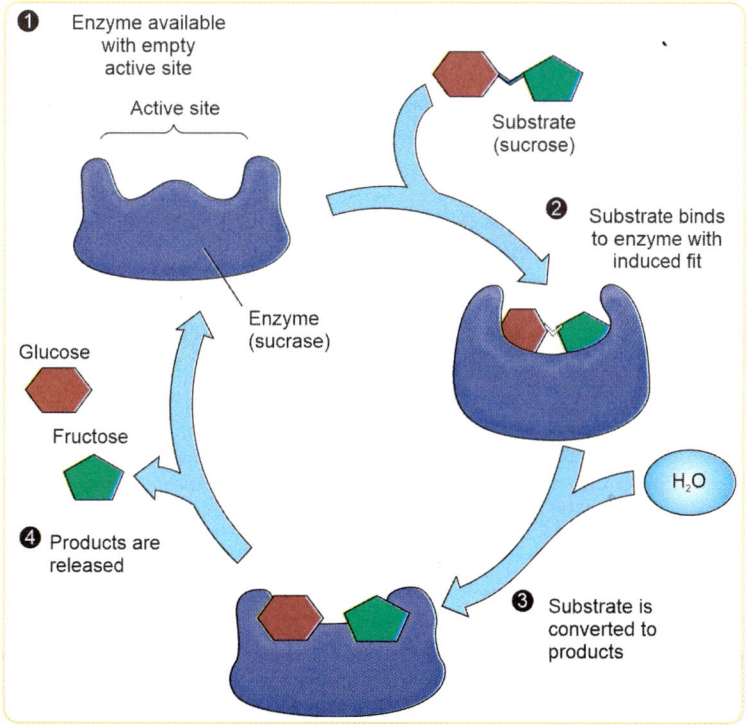

Fig. 4: Sucrase is a disaccharidase present in the intestinal brush border. Action of sucrase an example for catalysis by substrate-strain

ENZYME KINETICS

Enzyme kinetics is the study of the chemical reactions that are catalyzed by enzymes. In enzyme kinetics, the reaction rate is measured and the effects of varying conditions on the reaction rate are investigated.

Michaelis-Menten kinetics is one of the simplest models to explain the kinetics of non-allosteric enzymes.

Let us assume V_0 as the initial rate of the reaction, [S] is the concentration of substrate and Vmax is the maximum velocity or maximum rate at which the enzyme catalyzes a reaction.

Michaelis and Menten described their relationship in the following equation.

$$V_0 = V_{max} \frac{[s]}{K_m + [s]}$$

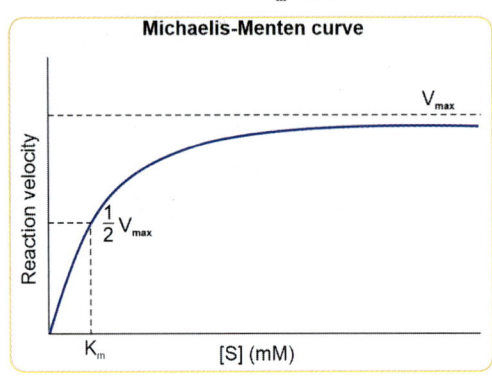

K_m Value

K_m is the concentration of the substrate at half maximal velocity

Unit of K_m: As K_m is the concentration of substrate, it is usually expressed in mM (millimoles/L) or μM (micromoles/L)

Significance

- K_m is a measure of affinity of enzyme for its substrates. **K_m is inversely proportional to substrate affinity of the enzyme**.
- If an enzyme acts on multiple substrates, it can have different K_m values for different substrates. For example, K_m of hexokinase for glucose is 0.1 mM and for fructose is 1.5 mM. Thus, only when fructose is at very high concentration, hexokinase can act on it.
- Isoenzymes can also have different K_m value for same substrate.
- K_m value of an enzyme for a substrate can be altered in the presence of inhibitor (*discussed under the section enzyme inhibition*).

Specific Activity of Enzyme

Specific activity is a **measure of enzyme purity**. Enzymes are either isolated from tissues or synthesized by recombinant DNA technology. Tissue proteins will have other enzymes and proteins. Enzymes synthesized by recombinant DNA technology will be relatively pure. Specific activity of enzyme tells us how much of the protein is actually the enzyme of our interest.

$$\text{Specific activity of enzyme} = \frac{\text{Enzyme activity}}{\text{Protein concentration}}$$

Unit: Units/mg ($\mu mol\ min^{-1}mg^{-1}$)

Turnover Number (K_{cat})

$$K_{cat} = \frac{V_{max}}{\text{Concentration of enzyme (number of active sites)}}$$

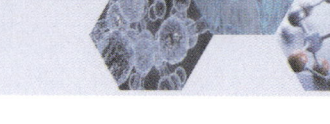

Turnover number (K_{cat}) is the number of substrate molecules converted into products by one enzyme active site per unit time (usual enzymes have one active site per enzyme molecule).

Unit of K_{cat}: s^{-1} (per second)

Enzyme	Turnover number (s^{-1})
Catalase	40 million
Carbonic anhydrase	4 lakhs
Fumarase	800

Catalytic Efficiency

For an enzyme to be efficient, it should have low K_m and high turnover number.

$$\text{Catalytic efficiency} = \frac{K_{cat}}{K_m}$$

International Unit of Enzyme Activity

International unit (IU) is the amount of enzyme required to convert 1 μ mole substrate to product in 1 minute under standard conditions.

Katal is the SI Unit for Enzyme Activity

Katal is the amount of enzyme required to convert one mole of substrate into product per second.

$$1\ U = 1/60\ \mu\ \text{katal}$$

REACTION ORDER

Zero order kinetics: The reaction is independent of the concentration of the reactant(s). This is observed when an enzyme is saturated by reactants.

First order kinetics: The rate of a first-order reaction varies linearly on the concentration of one reactant.

Second order kinetics: The rate of a second-order reaction varies linearly with the square of concentrations of one reactant (or with the product of the concentrations of two reactants).

Pseudo-first order kinetics: When a reaction involves two substrates and one of the substrates is in a huge quantity, the reaction effectively becomes a first order reaction. This is known as pseudo-first order kinetics.

BI-BI REACTION

Bi-Bi reaction is a reaction catalyzed by a single enzyme and involving two substrates and two products. 60% of the known biochemical reactions belong to this type. Bi-Bi reaction is of 2-types which is shown in the following schematic diagram.

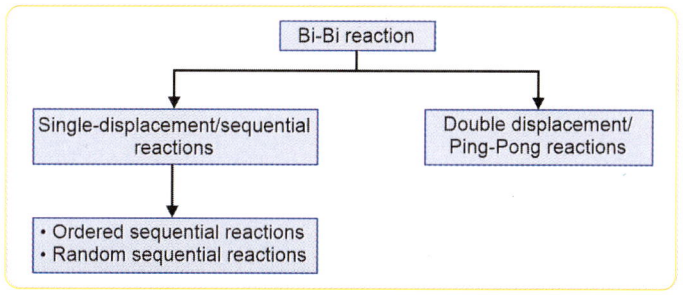

Chapter 5 Enzymology

Ordered Sequential Reactions

Binding of substrate and release of products from the enzyme follow a predetermined compulsory order.
Example: NAD dependant oxidoreductases like lactate dehydrogenases. Look at the following illustration. NADH binds to the enzyme first and then pyruvate. There is formation of a ternary complex. During release, the last to bind (pyruvate) is the first to be released as product (lactate). NAD^+ is released and the active enzyme is regenerated.

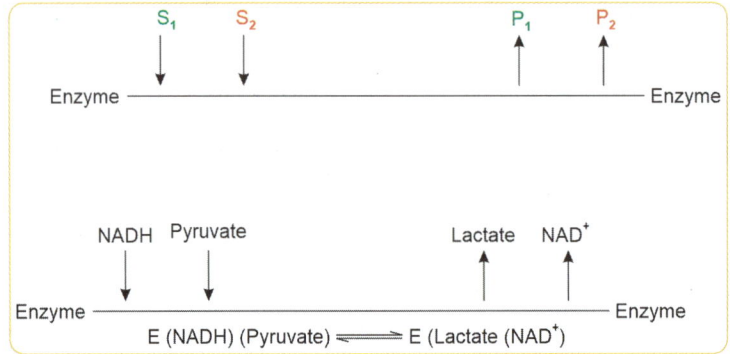

Random Sequential Reaction

The order of binding of substrates and release of products is random. Example: creatine kinase reaction.

Ping-pong/Double Displacement Reaction

Transaminase (amino transferase) reaction is the classical example for this.

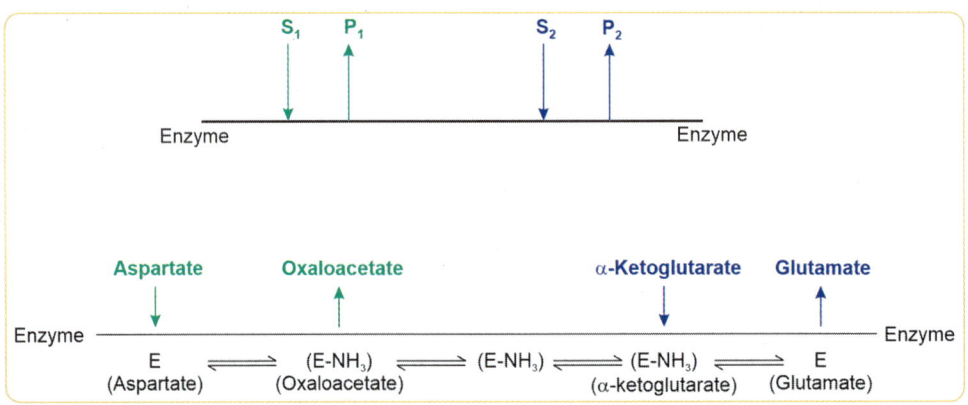

- ❑ Aspartate enters and is released as oxaloacetate giving its amino group to the enzyme. (Enzyme is modified and there is no ternary complex formation).
- ❑ Then, alpha keto glutarate comes and receives the amino group from enzyme and is released as glutamate.
- ❑ Active enzyme is regenerated.
Note: Ping-Pong is another name of table tennis.

FACTORS INFLUENCING ENZYME ACTIVITY

EFFECT OF TEMPERATURE

↑temperature → ↑energy of substrate molecules → ↑molecular collisions → substrates can overcome the energy barrier and become product.

Section I **Fundamental Concepts of Biochemistry**

Q_{10} **(temperature coefficient)** is the measure of the temperature sensitivity of an enzymatic reaction. It is the factor by which the rate of a reaction increases for every 10°C rise in the temperature. For most enzymes, Q10 value is 2.

But very high rise in temperature leads to *denaturation* of the enzyme and loss of activity. For human enzymes, optimal temperature for activity is 37°C. That is why hyperthermia is dangerous to humans.

Bacteria living in hot spring are equipped with thermostable enzymes, e.g. Taq polymerase.

EFFECT OF pH

Change in pH → protonation/deprotonation of residues involved in catalysis → altered enzyme activity.

Large change in pH will lead to loss of tertiary structure and denaturation of enzyme. Maximal rate of enzyme catalysis occurs at optimum temperature and optimum pH. Optimal pH for activity of human enzymes is around 6 to 8. There are enzymes which are active in extremes of pH.

For example, Pepsin is active at pH of 2 and alkaline phosphatase is active at pH of 10.

EFFECT OF SUBSTRATE CONCENTRATION

Look at the Michaelis-Menten equation:

$$V_0 = V_{max} \frac{[s]}{K_m + [s]}$$

So, when we increase the substrate concentration, the reaction rate will also increase initially. But, after a certain point of time, all the enzyme molecules will be saturated with the enzyme and the enzyme concentration becomes limiting. Thus, increasing the substrate concentration does not affect the reaction rate after the point of saturation.

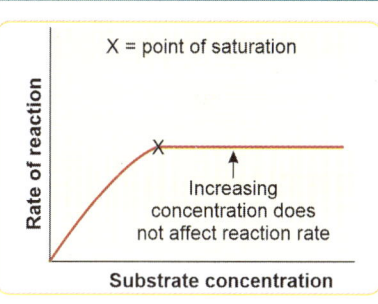

EFFECT OF ENZYME CONCENTRATION

When the substrate concentration is not limiting, increasing the enzyme concentration will increase the rate of reaction, as more enzymes will be colliding with substrate molecules.

This is the reason why excess substrate is added during the measurement of enzyme activity in clinical laboratories to avoid the substrate concentration being the limiting factor. *Enzyme assays will be discussed in Chapter 45* *(Refer to page no. 647).*

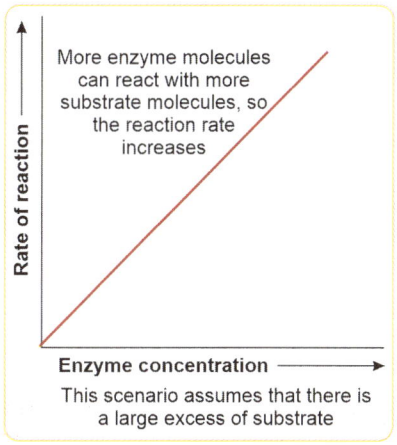

Chapter 5 Enzymology

❑ **Enzyme inhibition is discussed in detail in the following section.**

ENZYME INHIBITION

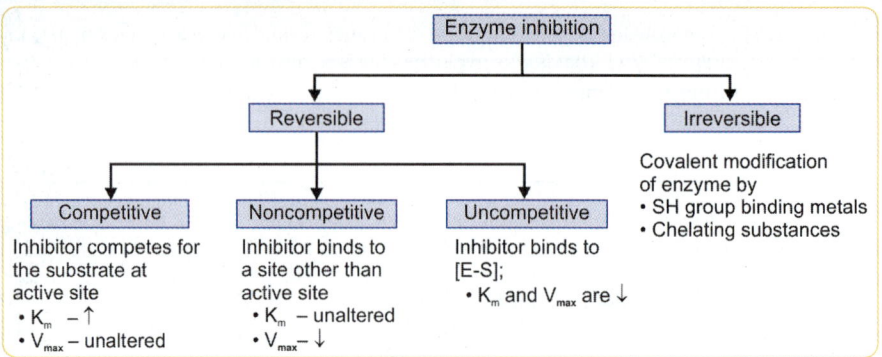

COMPETITIVE INHIBITION VERSUS NONCOMPETITIVE INHIBITION

Competitive inhibition	Noncompetitive inhibition
Inhibitor resembles the substrate, i.e. they are structural analogues	May not resemble
Binds to the active site and competes for the substrate	Binds to a site other than active site; forms Enzyme-inhibitor complex
Increasing the concentration of substrate can reverse the inhibition since the competition is for the active site	Cannot reverse
K_m is increased; V_{max} is unaltered	V_{max} is decreased; K_m is unaltered
Examples: ❑ Malonate inhibits succinate dehydrogenase ❑ Ibuprofen inhibits COX1 and 2 ❑ Ethanol competes with methanol for alcohol dehydrogenase ❑ Statins inhibits HMG CoA reductase ❑ Methotrexate inhibits DHFR	**Example:** ❑ Lithium inhibits inositol monophosphatase in brain

Uncompetitive Inhibition

❑ Inhibitor binds only to the enzyme-substrate complex (Fig. 5)
❑ Both K_m and V_{max} are decreased
❑ For example, L-phenyl alanine inhibits placental and intestinal alkaline phosphatase

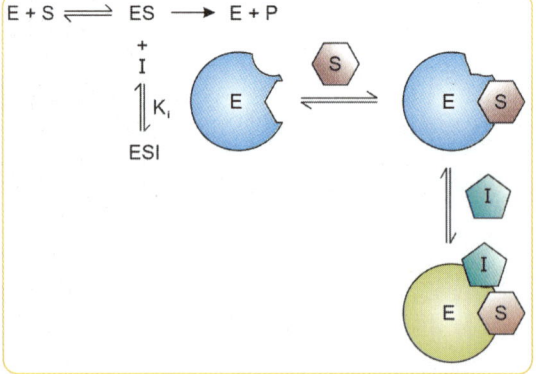

Fig. 5: Uncompetitive inhibitor binds to ES complex

Comparison of 3-types of Enzyme Inhibition

	Competitive inhibition	Noncompetitive	Uncompetitive
Inhibitor binds to	Active site of the free enzyme	Free enzyme but other than the active site	ES complex
Competition for the active site	Yes	No	No
K_m	Increased	Unaltered	Decreased
V_{max}	Same	Decreased	Decreased

Lineweaver-Burk Plot is a Double-Reciprocal Plot

Because of practical problems with Michaelis-Menten plot, Lineweaver-Burk plot was developed. Both [S] and velocity values are taken in reciprocal to make a double-reciprocal plot.

X axis $\rightarrow$ 1/[S]

Y axis $\rightarrow$ 1/V_i

Using the y = mx + c equation, we get the following inference:

$$\frac{1}{V_0} = \frac{K_m}{V_{max}} \cdot \frac{1}{S} + \frac{1}{V_{max}}$$

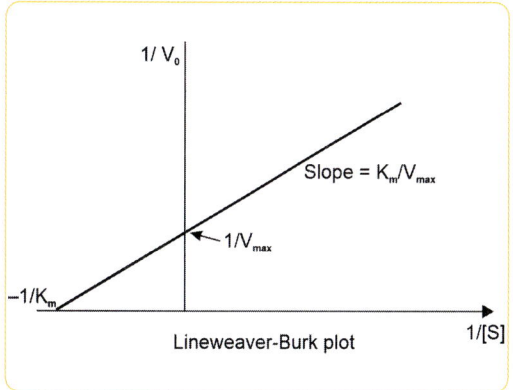

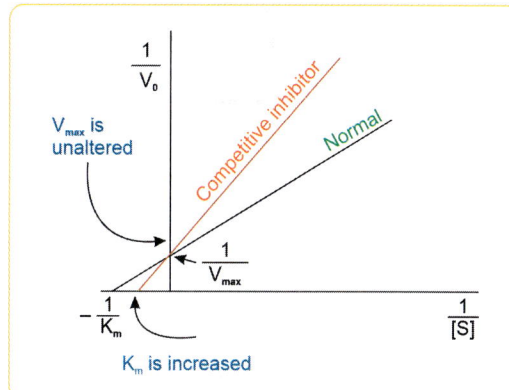

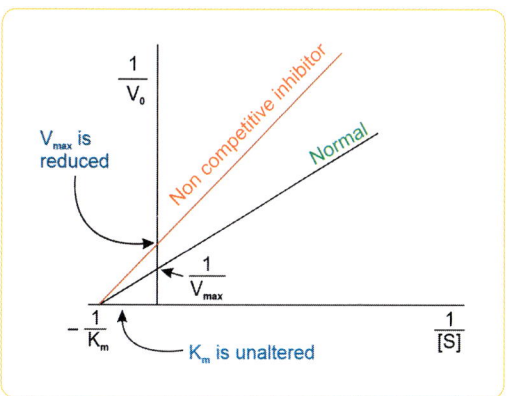

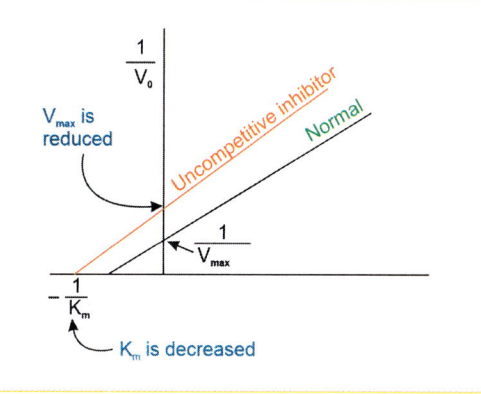

Dixon Plot is an Alternative to the Lineweaver-Burk Plot

Dixon plot is an alternative to the Lineweaver-Burk plot for determining inhibition constants; mainly used by pharmaceutical industry.

Suicidal Inhibition or Mechanism-based Inhibition

- ❑ Suicidal inhibitors resemble the actual substrate of the enzyme.
- ❑ Enzymes bind to suicidal inhibitor and catalytic process starts.
- ❑ In the process of catalysis, enzyme undergoes irreversible covalent modification by the inhibitor and active enzyme cannot be regenerated (enzyme catalyzes its own inactivation).
- ❑ Neither the Lineweaver-Burk plot nor the Dixon plot can be used to express the action of suicidal inhibitors.

Suicidal inhibitor	Enzyme
Allopurinol	Xanthine oxidase
Aspirin	Cyclooxygenase
5-FU	Thymidylate synthase
Penicillin	Transpeptidase

Note: Suicidal enzymes need to be differentiated from suicidal inhibition. Cyclooxygenase is a suicidal enzyme. The enzyme is regulated by self-destruction.

Irreversible Enzyme Inhibitors or Enzyme Poisons

1. **Iodoacetate**—Inhibits glyceraldehyde-3-phosphate dehydrogenase
2. **Diisopropyl fluorophosphate**—Phosphorylates active site serine of serine proteases and esterases.
3. **Heavy metals**—Heavy metal ions such as mercury (Hg) and lead (Pb) block sulfhydryl group of enzymes.
   Lead poisoning:
   - ○ Lead (Pb) inhibits SH group of two of the enzymes of the heme synthesis, namely ALA dehydratase and ferrochelatase.
   - ○ Characterized by anemia with basophilic stippling and ↑ RBC free protoporphyrinogen levels and excretion of δ-ALA in urine.

   Treatment of heavy metal poisons:
   - ○ Oral egg white and milk are given in acute poisoning. Heavy metals denature proteins forming insoluble complexes.
   - ○ Chelating agents like dimercaprol/BAL (British anti-Lewisite) are used in chronic poisoning.
4. Suicidal inhibition is also an irreversible inhibition.

Fig. 6: SH group of dimercaprol chelates heavy metals

REGULATION OF ENZYME ACTIVITY

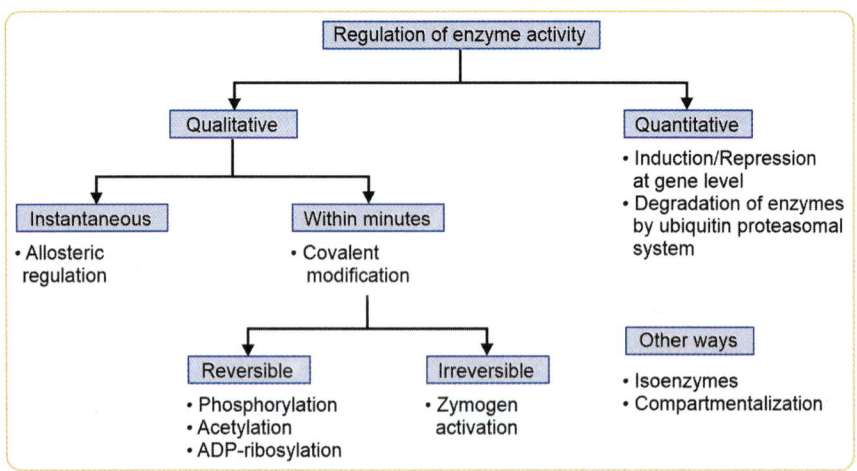

Section I Fundamental Concepts of Biochemistry

The above is the overview of regulation of enzymatic activity. Regulation can be either qualitative or quantitative. Qualitative regulation involves change in conformation of the enzyme induced by covalent modification or allosteric molecules. Quantitative regulation involves change in the amount of enzyme. This is achieved by inducing or repressing the transcription of enzyme genes and by degradation of enzyme. Let us discuss each regulation in detail.

ALLOSTERIC REGULATION

❑ Allo means other.
❑ Allosteric enzymes also contain **a site other than active site** which binds to allosteric regulator.
❑ Allosteric enzymes are usually **multisubunit** enzymes.
❑ Binding of allosteric regulator to the allosteric site alters the affinity of active site. The affinity can be increased or decreased.
❑ When the substrate of the enzyme itself acts as an allosteric regulator, it is known as homotropic allosteric regulator, e.g. ATP is the negative homotropic allosteric regulator of phosphofructokinase-1.
❑ Positive allosteric regulators increase the affinity and negative allosteric regulators decrease the affinity.
❑ **Sigmoidal curve** is obtained on plotting velocity against substrate concentration similar to that of hemoglobin-oxygen dissociation curve.
❑ Allosteric enzymes do not obey Michaelis-Menten equation. They do **obey Hill's equation**.

Fig. 7: Note the sigmoidal nature of the curve and the effect of activators and inhibitors

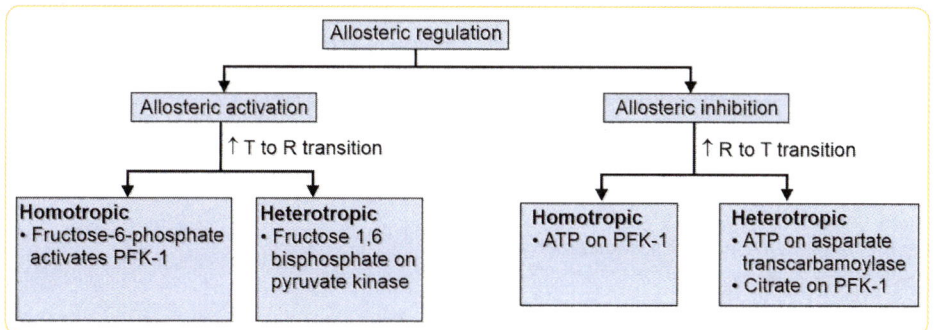

Note: Like hemoglobin, allosteric enzymes exist in 2 states – T (Taut/tensed) and R (Relaxed) states. T state is more active

Two Classes of Allosterically Regulated Enzymes

1. K-series—allosteric inhibitor raises the K_m without affecting V_{max}
2. V-series—allosteric inhibitor lowers the V_{max} without affecting K_m

COVALENT MODIFICATION

Phosphorylation is the Most Common Covalent Modification Involved in the Regulation of Enzyme Activity

❑ Serine, threonine and tyrosine residues of enzymes are phosphorylated by various **protein kinase** enzymes.
❑ ATP is the phosphate donor.

- A single protein kinase can phosphorylate multiple target enzymes and the target enzyme can catalyze thousands of reactions. This leads to the amplification of initial signal.
- Phosphoprotein phosphatases hydrolyze the phosphate groups.
- Glucagon promotes phosphorylation. Insulin mostly promotes dephosphorylation.
- Phosphorylation brings about conformational change in the enzyme.

Enzymes active in phosphorylated state	Enzymes active in dephosphorylated state
□ Glycogen phosphorylase	□ Glycogen synthase
□ Hormone sensitive lipase	□ Acetyl-CoA carboxylase
□ Phosphorylase b kinase	□ HMG-CoA reductase
□ HMG-CoA reductase kinase/AMP activated protein kinase	□ **Pyruvate dehydrogenase**
□ **ATP-Citrate lyase**	

Remember

- All the enzymes catalyzing catabolic reactions are active in the phosphorylated state except pyruvate dehydrogenase which is active in the dephosphorylated state.
- All the enzymes catalyzing anabolic reactions are active in the dephosphorylated state except ATP-citrate lyase (fatty acid synthesis) which is active in the phosphorylated state.

Zymogen Activation

- Certain enzymes are needed only in certain anatomical sites and certain physiological states.
- These enzymes are synthesized as proenzymes called zymogens and activated by proteolytic cleavage by a protease at appropriate place and time. Proteolysis serves the function of unmasking aminoacyl residues involved in catalysis.
- Zymogen activation can also be considered as a type of irreversible covalent modification of enzyme regulation.

Enzymes Regulated by Zymogen Activation

- Pepsinogen of stomach
- Pancreatic enzymes
- Proteases of coagulation and complement pathway
- Caspases involved apoptosis
- Matrix metalloproteinase of extracellular matrix

 Clinical Correlation

- Aberrant activation of pancreatic enzymes within the substance of pancreas leads to pancreatitis.
- 'Immunoreactive trypsinogen' is used in newborn screening program to screen for an increased risk of cystic fibrosis.

INDUCTION AND REPRESSION OF ENZYMES

- Constitutive enzymes are those which are needed at the same level all of the time for a cell. They are not induced or repressed, e.g. Hexokinases is constitutively active while glucokinase is induced by insulin.
- Induction refers to increased production of enzyme due to increased transcription and translation of the gene coding for the enzyme.
- Induction/repression takes time but the effects are persistent.
- Many hormones mediate induction and repression.

Let us look at the enzymes induced and repressed by insulin.

Enzymes induced by insulin	Enzymes repressed by insulin
❑ Glucokinase	❑ Glucose-6-phosphatase
❑ Pyruvate kinase	❑ Phosphoenol pyruvate carboxy kinase
❑ Phosphofructokinase	❑ Fructose-1,6-bisphophatase
❑ Acetyl-CoA carboxylase	

DEGRADATION OF ENZYMES

❑ Ubiquitin mediated proteasomal degradation, (Chapter 11.1) *(Refer to page no. 183)* plays an important role in the regulation of certain enzymes like cyclin dependent kinases.

ISOENZYMES

❑ Isoenzymes are the different enzymes catalyzing the same chemical reaction.
❑ Hexokinase and glucokinase both are involved in conversion of glucose to glucose-6-phosphate.
❑ But, glucokinase has high K_m value—it can act only when glucose level is high. Moreover, it is induced by insulin also. Thus, glucokinase helps in funneling the excess glucose from the blood and also liver that store it in the form of glycogen.

COMPARTMENTALIZATION

❑ All the cellular reactions take place in certain well-defined compartments which are separated from each other by semipermeable membranes.
❑ Compartmentalization helps in the optimization of reaction. For example, lysosomal enzymes are active only in the acidic pH of the lysosomes.
❑ Compartmentalization of fatty acid oxidation and TCA cycle in the mitochondrial matrix is well suited because electron transport chain is also located in the inner mitochondrial membrane.

At this point the discussion on regulation of enzyme activities comes to its end. But wait, Enzyme chapter is not over. These are is just the basics. Enzymes as the target of drugs, therapeutic enzymes, isoenzymes and their clinical applications will be discussed in the Chapter 44 *(Refer to page no. 638)*.

Competency

❑	BI2.1	Explain fundamental concepts of enzyme, isoenzyme, alloenzyme, coenzyme & co-factors. Enumerate the main classes of IUBMB nomenclature
❑	BI2.3	Describe and explain the basic principles of enzyme activity
❑	BI2.4	Describe and discuss enzyme inhibitors as poisons and drugs and as therapeutic enzymes

Chapter 5 Enzymology

Key Points

- ❏ Except for ribozymes, all enzymes are proteins.
- ❏ Ribozymes are RNA molecules with catalytic activity (non-protein enzymes), e.g., peptidyl transferase is a ribozyme present in ribosome, small nuclear RNAs are ribozymes with splicing activity.
- ❏ RNase P is a ribozyme; RNase H is not a ribozyme.
- ❏ Abzymes are antibodies with catalytic activity.
- ❏ All the digestive enzymes are hydrolases (EC3).
- ❏ Aldolase, enolase, aconitase and fumarase are lyases.
- ❏ All kinases require magnesium/manganese. Pyruvate kinase requires potassium also.
- ❏ Enzymes bind to substrates with multiple, weak, non-covalent interactions. But enzymes bind more tightly to the transition state.
- ❏ Site directed mutagenesis is used to establish the importance of a specific amino acid residue in enzyme catalysis.
- ❏ Rossmann fold is a $NAD(P)^+$ binding domain present in certain dehydrogenases, e.g. Lactate dehydrogenase.
- ❏ Allosteric enzymes are usually multi-subunit enzymes and show the phenomenon of cooperativity. They do not obey Michaelis-Menten equation, they follow Hill's equation.
- ❏ An allosteric modulator acts instantaneously; participates in feed-forward or feedback regulation.
- ❏ Lowering of activation energy by the enzymes is due to binding energy contributed by enzyme-substrate interactions.
- ❏ Enzymes do not change the equilibrium constant of a reaction. They just increase the rate of forward and reverse reaction equally.
- ❏ Specific activity of enzyme = Enzyme activity/protein concentration
- ❏ K_m is the concentration of substrate at ½ V_{max}
- ❏ $K_{cat} = V_{max}$/enzyme concentration (number of active sites)
- ❏ Catalytic efficiency = K_{cat}/K_m
- ❏ Transaminase reaction is an example of ping-pong Bi-Bi or double displacement reaction
- ❏ Phosphorylation leads to conformational changes in the enzyme
- ❏ Phosphorylation of enzyme is carried out by protein kinases and reversed by protein phosphatases
- ❏ Recently, acetylation is found to be a ubiquitous post-translational modification of many enzymes. Acetyl group is added to the ε amino group of lysine of the enzyme.
- ❏ Isoenzymes are the different enzymes catalyzing the same chemical reaction. They have different primary structure, kinetic properties, electrophoretic mobility and tissue of origin.
- ❏ Isoenzymes have the same Enzyme Commission number as they catalyze the same chemical reaction.
- ❏ Aspirin is the suicidal inhibitor of cyclooxygenase. Allopurinol is the suicidal inhibitor of xanthine oxidase.
- ❏ Suicidal inhibition cannot be represented using neither Lineweaver-Burk nor the Dixon plot.
- ❏ Dixon plot is an alternative to the Lineweaver-Burk plot for determining inhibition constants; mainly used by pharmaceutical industry.
- ❏ Multifunctional enzyme is the one having more than one enzymatic activity in a single polypeptide chain, e.g. phosphofructokinase-2

Short Answer Questions

1. What are ribozymes? Give an example.

2. What do you understand by the term 'active site' of an enzyme? Mention its salient features.

3. What are zymogens? Mention two examples to illustrate their physiological advantages.

4. What is the advantage of multienzyme complexes? Name one example of a multienzyme complex, its components and the coenzymes required by it.

5. Give one example for each – Enzymes as the target of drugs, Therapeutic enzymes, Enzymes in the diagnosis of disease.

6. Differentiate between cofactors, coenzymes and prosthetic group with one example each.

7. Define Km of the enzyme. What is its significance?

8. What is suicide inhibition or mechanism-based inhibition of enzymes? Give two examples of mechanism-based inhibitors.

9. Describe the different levels of enzyme specificity with suitable examples.

10. What are bifunctional enzymes? Give two examples.

11. Give one example each for the enzymes containing copper, zinc, molybdenum, iron and selenium.

12. Define transition state with reference to enzyme catalysis. What is the pharmacological importance of transition state?

13. What is the specific activity of the enzyme? What is its unit?

14. What is 'turn over number' of enzymes? What is its unit?

15. What is catalytic efficiency of an enzyme?

16. Describe the properties of allosteric enzymes. What do you understand by the terms T state and R state? Which state of the enzyme is more active?

17. What are isoenzymes? Give two examples.

18. What is the diagnostic significance of the isoenzymes? Explain with the help of examples (Chapter 44).

19. What is the advantage of compartmentalization of enzymes inside the cell? Provide an appropriate example.

20. Draw a Lineweaver Burke plot to illustrate (a) competitive and (b) uncompetitive inhibition. Explain the change in Km which occurs in both cases.

21. Explain how covalent modification – both reversible and irreversible help to regulate enzyme activity. Give their physiological significance.

22. Differentiate

 a. Multifunctional enzyme and multienzyme complex
 b. Ribozymes and Abzymes
 c. Active site and substrate-binding site
 d. Lock and key hypothesis and induced fit hypothesis
 e. Coenzyme and Prosthetic group
 f. Homotropic allosteric inhibitor and heterotropic allosteric inhibitor
 g. Competitive and non-competitive inhibition
 h. Lyase and Ligase
 i. Synthase and synthetase
 j. Isoenzymes and alleloenzymes (Chapter 44).
 k. Michaelis Menten plot and Lineweaver Burke plot

Long Essay

1. What are enzymes? Give the IUBMB classification of enzymes with one example for each. Explain any four factors that affect enzyme activity.

2. Describe the different types of enzyme inhibitions and its role in clinical medicine.

3. Explain the mechanisms by which the enzyme activity is regulated in the short term and long term. Give suitable examples.

Multiple Choice Questions

1. **All of the following are ribozymes, EXCEPT:**
 a. Ribosome
 b. RNase P
 c. RNase H
 d. SnRNA

2. **Hexokinase belongs to:**
 a. Oxidoreductase   b. Transferase
 c. Hydrolase   d. Lyase

3. **To which of the family does the enzyme Akt belong to?**
   a. Oxidoreductase
   b. Transferase
   c. Hydrolase
   d. Lyase

4. **All of the following are lyases, EXCEPT:**
   a. Aldolase
   b. Enolase
   c. Fumarase
   d. Acetyl-CoA carboxylase

5. **Activation energy is the free energy difference between the ($\Delta G\ddagger$):**
   a. Substrate and product
   b. Substrate and transition state
   c. Transition state and product
   d. Sum of all the above

6. **Which of the following statement about the active site of an enzyme is correct?**
   a. The active site of an enzyme binds the substrate of the reaction it catalyzes more tightly than it does in the transition state
   b. The active site of an enzyme binds the substrate of the reaction it catalyzes less tightly than it does in the transition state
   c. The active site of an enzyme binds the product of the reaction it catalyzes more tightly than it does in the transition state
   d. The active site of an enzyme is complementary to the substrate of the reaction it catalyzes.

7. **Which of the following techniques is used to establish the importance of a particular amino acid residue in enzyme catalysis?**
   a. ELISA
   b. Spectrophotometry
   c. Affinity chromatography
   d. Site directed mutagenesis

8. **All of the following mechanisms are utilized by enzymes to reduce the activation energy, EXCEPT:**
   a. Entropy reduction
   b. Desolvation of active site
   c. Decrease in molecular collisions
   d. Conformational changes in the enzyme and substrate after interaction

9. **An allosteric inhibitor of an enzyme usually:**
   a. Denatures the enzyme
   b. Causes enzyme to work faster
   c. Binds to active sites
   d. Participates in feedback regulation

10. **Which of the following statements about the control of enzyme activity by phosphorylation is correct?**
    a. Phosphorylation of an enzyme is not reversible process since it is a covalent modification
    b. Phosphorylation of an enzyme occurs by protein phosphatases
    c. Phosphorylation of an enzyme is an intracellular process and cannot occur in response to external signals
    d. Phosphorylation of an enzyme results in a conformational change

11. **5 micrograms of carbonic anhydrase (MW = 30000) in the presence of excess substrate gives a reaction rate of 6.82 × 10³ μmol/min. At [S] = 0.012 M, the rate is 3.41 × 10³ μmol/min. Calculate the K_m:**
    a. 3.14 M
    b. 0.012 M
    c. 6.82 M
    d. 0.024 M

12. **Which of the following is not a covalent modification?**
    a. Activation by divalent cation
    b. Dephosphorylation
    c. Phosphorylation
    d. Proteolytic cleavage

13. **The best measure of enzyme efficiency is:**
    a. K_m
    b. K_{cat}
    c. K_m/V_{max}
    d. K_{cat}/K_m

14. **All of the following statements about enzymes are true EXCEPT:**
    a. Multienzyme complexes greatly reduce the rates of reactions
    b. Isoenzymes differ in their amino acid sequence
    c. Allosteric regulators usually do not bind to the active site of the enzyme
    d. Transcriptional regulation is method of long-term regulation of enzymatic activity

15. **Which of the following is the least preferred method to separate isoenzymes?**
    a. Agarose gel electrophoresis
    b. Gel filtration chromatography
    c. Ion-exchange chromatography
    d. Affinity chromatography using the substrate as ligand

16. All of the following statements about competitive inhibitors are true EXCEPT:
    a. Their effect can be nullified by increasing substrate concentration
    b. They bind to the active site of the enzyme
    c. They increase the apparent K_m of a reaction
    d. They reduce the V_{max} of the reaction

17. Which of the following is the effect of an uncompetitive inhibitor?
    a. Increase V_{max} but no increase in K_m
    b. Decrease K_m but no effect on V_{max}
    c. Decrease V_{max} but no change on K_m
    d. Decrease both V_{max} and K_m

18. Rossman fold is present in:
    a. Lactate dehydrogenase
    b. Pyruvate kinase
    c. Pyruvate carboxylase
    d. Alanine transaminase

19. The unit of specific activity of enzyme is:
    a. units/mg   b. mmol/L
    c. mg/dL   d. units/L

20. All of the following are mechanism-based inhibitors, EXCEPT:
    a. Aspirin
    b. Allopurinol
    c. Malonate
    d. 5-Fluorouracil

21. Alanine transaminase catalyzed reaction is an example of:
    a. Single displacement reactions
    b. Random sequential reaction
    c. Ordered sequential reaction
    d. Double displacement or Ping-pong bi-bi reaction

22. Which of the following enzyme has the highest degree of substrate specificity?
    a. Alcohol dehydrogenase
    b. Aldose reductase
    c. Hexokinase
    d. Urease

ANSWERS

1. c.	**2.** b.	**3.** b.	**4.** d.	**5.** b.	**6.** b.
7. d.	**8.** c.	**9.** d.	**10.** d.	**11.** b.	**12.** a.
13. d.	**14.** a.	**15.** d.	**16.** d.	**17.** d.	**18.** a.
19. a.	**20.** c.	**21.** d.	**22.** d.		

6

CHAPTER

Biological Oxidation and Electron Transport Chain

Chapter Outline

- Classes of Oxidoreductases
  - Hydroperoxidases
  - Oxidases
  - Oxygenases
  - Dehydrogenases
- Monooxygenase versus Dioxygenase
- NADPH Generating Reactions
- $FADH_2$ Producing Reactions
- Oxidative Phosphorylation
- Electron Transport Chain and Components
  - Sites of Proton Pumping
  - Mobile Electron Carriers
  - Mitchell's Chemiosmotic Hypothesis
  - Mechanism of ATP Synthase
- Uncouplers and Inhibitors of ETC
  - Dinitrophenol
  - Thermogenin
- P:O Ratio
- Glycerol-3-phosphate Shuttle & Malate-aspartate Shuttle
- Disorders of the Electron Transport Chain

Macromolecules present in the food are oxidized by the body in intermediary metabolism. Oxidation is loss of electrons. Electrons extracted from the macromolecules are transferred to NAD^+ or FAD and these electrons are presented to electron transport chain to generate an electrochemical gradient which in turn produces ATP. Electrons transferred to $NADP^+$ are later used in reductive biosynthetic reactions.

All the enzymes involved in biological oxidation belong to the oxidoreductase (EC 1) family.

Four classes of oxidoreductases - HOOD

1. Hydroperoxidases
2. Oxidases
3. Oxygenases
4. Dehydrogenases

It is very important to differentiate them.

Oxidases	Oxygenases	Dehydrogenases
Catalyse the oxidation of a substrate **without incorporating oxygen** into the substrate	**Incorporate one or two oxygen** atom(s) into the substrate.	Removes hydrogen with the help of coenzymes (NAD^+/FAD/$NADP^+$)

Contd...

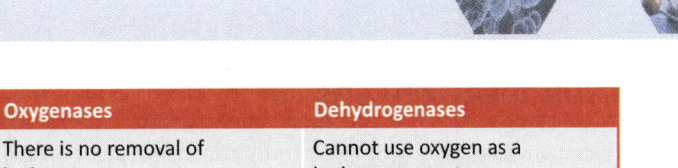

Oxidases	Oxygenases	Dehydrogenases
Removes hydrogen from substrates and uses oxygen as acceptor to produce H_2O or H_2O_2	There is no removal of hydrogen.	Cannot use oxygen as a hydrogen acceptor
For example, Cytochrome c oxidase, Xanthine oxidase	For example, Phenylalanine hydroxylase	For example, Succinate dehydrogenase

Dehydrogenases transferring electrons to NAD^+ are numerous. So, I advise you to study the dehydrogenases transferring electrons to FAD and $NADP^+$ first.

NADPH PRODUCING DEHYDROGENASES

(Government Medical College)
- ❑ **G**lucose-6-phosphate dehydrogenase and 6-phosphogluconate dehydrogenase of HMP shunt
- ❑ **M**alic enzyme (malate → pyruvate)
- ❑ **C**ytosolic Isocitrate dehydrogenase

FADH$_2$ PRODUCING DEHYDROGENASES

(Foreign Graduate Students)
- ❑ Fatty acyl-CoA dehydrogenase
- ❑ Glycerol-3-phosphate dehydrogenase (Mitochondrial)
- ❑ Succinate dehydrogenase

Note: Glutamate dehydrogenase can use either NAD^+ or $NADP^+$

CONCEPT CORNER

Not all FAD Containing Enzymes Produce FADH$_2$

Enzymes with FAD/FMN prosthetic group are known as flavoenzymes. Amino acid oxidases, Glutathione reductase, Pyruvate dehydrogenase, α ketoglutarate dehydrogenase, Branched chain ketoacid dehydrogenase and Xanthine oxidase—All these enzymes contain FAD prosthetic group. FAD is an intermediate in the electron transfer. These enzymes do not release FADH$_2$.

TYPES OF OXYGENASES – MONOOXYGENASE AND DIOXYGENASE

Monooxygenases incorporate only a single atom of O_2 into a substrate with the other being reduced to a water molecule. That is why they are also know as mixed function oxidases.

For example, All hydroxylases, Cytochrome P_{450} mixed function oxidase.

Dioxygenases incorporate both the oxygen atoms into substrate. They are mainly involved in the **breaking of aromatic ring of aromatic amino acids** which we will study later.

Example: Tryptophan pyrrolase, Homogentisate oxidase, Cyclooxygenase, Cysteine dioxygenase

HYDROPEROXIDASES

Hydroperoxidase uses hydrogen peroxide or organic peroxide as a substrate.
For example, Catalase and Glutathione peroxidase —oxidizes H_2O_2 to water and oxygen

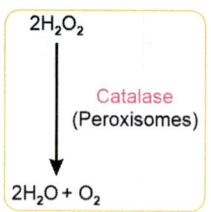

MITOCHONDRIA - THE POWERHOUSE OF THE CELL

Our earth is 4.5 billion years old. Life begun 3.7 billion years ago. The primitive earth was devoid of oxygen and early organisms were anaerobic. When oxygen appeared in the atmosphere 2 billion years ago, the organisms slowly adapted to it. Certain unicellular organisms were most efficient in utilizing oxygen. One such organism (probably ancestor of *Rickettsia*) which was engulfed by another organism finally got merged with the host. The engulfed organism made an endosymbiotic relationship with the host and evolved into mitochondria. Mitochondria enabled eukaryotic organisms to harvest the full potential of oxygen and to diversify into various life forms.

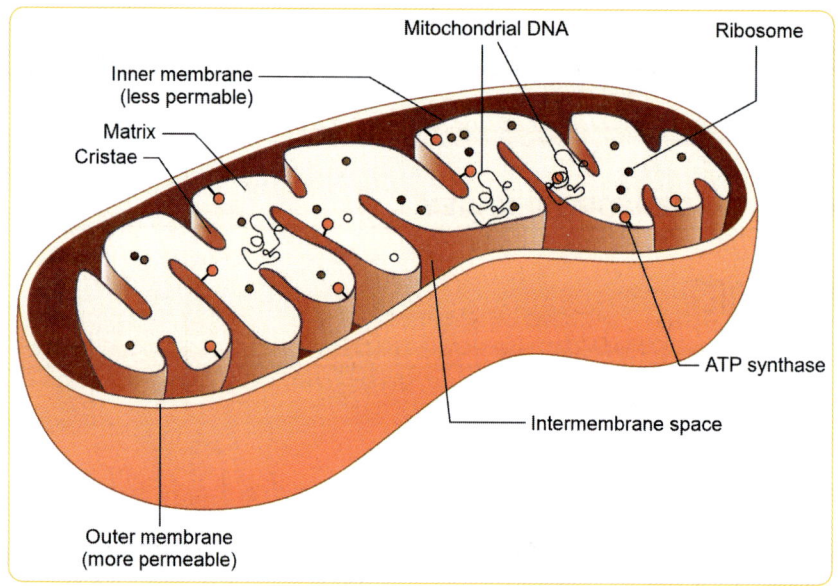

Fig. 1: Components of mitochondria

Note: Mitochondria contains its own DNA and protein synthesising machinery

Source: Kelvinsong: Own work, CC0, https://commons.wikimedia.org/w/index.php?curid=27715320

Oxidative Phosphorylation Takes Place in the Inner Mitochondrial Membrane

Oxidation: Loss of electrons from NADH and $FADH_2$. These electrons flow in the electron transport chain (Complex I to IV).

Phosphorylation: Synthesis of ATP by complex V.

Phosphorylation is made possible by the oxidation. Oxidation can happen without phosphorylation. But, phosphorylation can't happen without oxidation.

When we extract electron from a donor, we need an acceptor and if there are multiple acceptors, they need to be arranged in a proper way.

Movement of Electrons Depends upon the Redox Potential

A redox couple is a reducing species and its corresponding oxidizing form, e.g. Fe^{2+}/Fe^{3+} and $NADH/NAD^+$

Redox potential is the relative tendency of reductant to donate electrons as compared to standard hydrogen electrode. The redox potential of standard hydrogen electrode (pH = 0) is considered as zero. Redox potential

determines the flow of electrons. The more positive the redox potential is, greater the tendency to accept electrons. Electrons flow from a substance with lower potential to substance with higher potential. Thus, substances with negative reduction potential tend to donate electrons.

Redox couple (pH =7)	Number of electrons transferred	Standard redox potential (V)
2 H$^+$/H$_2$	2	−0.42
NAD$^+$/ NADH	2	−0.32
NADP$^+$/ NADPH	2	−0.29
FAD/FADH$_2$	2	−0.22
Cytochrome a (+3)/Cytochrome a (+2)	1	+0.29
Fe^{3+}/Fe^{2+}	1	+0.77
½ O$_2$/ H$_2$O	2	+0.82

If you see the table, you can notice that oxygen has the more positive reduction potential and this is the reason it is the **final acceptor of electrons in the electron transport chain**.

Electron Transport Chain (ETC) is Made up of 5 Complexes Embedded in the Inner Mitochondrial Membrane

Transfer of electrons from NADH to oxygen in a single reaction will release tremendous energy and will be difficult to capture. There are chances of dissipation also. To prevent this, electrons are transferred in a sequential manner and each step, the released energy is used to pump protons into the intermembranous space.

There are 5 complexes and 2 mobile electron carriers in the electron transport chain. Out of these 5 complexes proton pumping takes place only in 3 complexes (I, III and IV). Complex II is nothing but the TCA Cycle enzyme succinate dehydrogenase. ATP synthesis takes place at the 5th complex using the proton motive force created by the above said 3 complexes.

Complex	Components	No of H$^+$ pumped
I (NADH-CoQ oxidoreductase)	FMN-FeS	4
II (succinate dehydrogenase or succinate-CoQ oxido reductase)	FAD-FeS	0
Ubiquinone (CoQ or Q$_{10}$) is made up of 10 isoprene rings; lipid-soluble; transfers electrons from **complex I and II to III within the membrane**		
III (CoQ-Cytochrome c oxidoreductase)	Cyt b-Cyt C$_1$ - contains **Reiske** Protein (special Fe-S protein)	4
Cytochrome c is a small soluble protein; Carries electrons from complex III to IV.		
IV (Cytochrome c Oxidase)	Cyt a and a$_3$ Cu$_{A/B}$	2
V (ATP synthase)	F$_o$-F$_1$	1 ATP per 3 protons.

COMPARISON OF THE ELECTRON CARRIERS OF ELECTRON TRANSPORT CHAIN

	Ubiquinone (Coenzyme Q)	Cytochrome c
Chemical nature	Lipid – made up of 10 isoprene units, hence the name Q_{10}	Soluble protein
Accepts electrons from	Complex I and II	Complex III
Delivers electrons to	Complex III	Complex IV
Number of electrons that can be accepted at a time	Two	One
Other points:	Statins inhibit the synthesis of coenzyme Q10 and result in muscle cramps	Cytochrome c in cytosol mediates **apoptosis**.

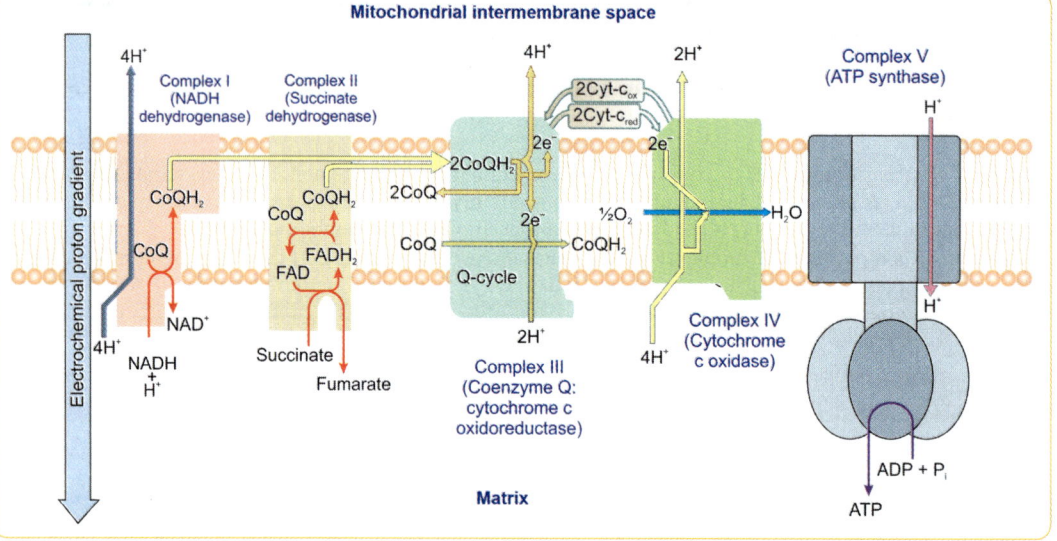

Fig. 2: Components of electron transport chain—Movement of two electrons lead to pumping of ten protons and production of a molecule of water. Note that there is no pumping of protons in complex II. Ubiquinone carries two electrons. But, cytochrome c can accept only one electron. Thus, Q cycle operates to efficiently transfer electrons from ubiquinone to cytochrome c.

Q CYCLE

Ubiquinone (Coenzyme Q) carries two electrons. But cytochrome c can accept only one electron at a time. So, Coenzyme Q gets reduced to ubiquinol and oxidised to ubiquinone sequentially (as shown i n the image). In this reduction-oxidation cycle, electrons are transferred from ubiquinone to cytochrome c.

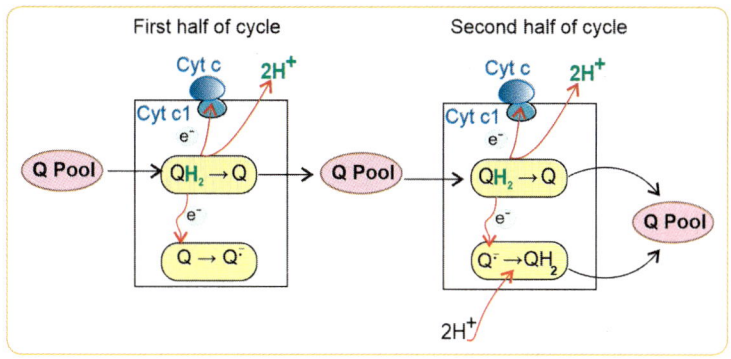

CYTOCHROMES PLAY CRUCIAL ROLE IN ELECTRON TRANSPORT CHAIN

Cytochromes are heme containing proteins. Cytochromes are different from hemoglobin and myoglobin. Iron of cytochromes must interchange between ferrous and ferric oxidation states for the electron transport whereas the iron of hemoglobin needs to remain only in ferrous state for the oxygen transport.

Cytochromes are involved in the transfer of electrons down the reduction potential.

Cytochromes of electron transport chain	
Cytochrome	Function
b and c 1	Component of complex III
a and a 3	Component of complex IV
Cyt c	Soluble mobile carrier of electrons

Cytochrome P450 involved in steroid synthesis and detoxification is discussed in the Chapter 17 (Refer to page no. 345).

Cytochrome c structure is highly conserved
Cytochrome c is a small protein. Its structure has been highly conserved throughout the evolution. Wheat-germ cytochrome c can react with human cytochrome c oxidase. Such is the level of conservation. What is conserved most is the tertiary structure. In primary structure, only certain key amino acids are conserved.

Cytochrome c Oxidase Reduces Molecular Oxygen to Water

Complex IV (cytochrome c oxidase) incorporates four electrons to molecular oxygen to produce water. **The requirement of oxygen for this complex makes us aerobic organisms. This process is also known as internal respiration.**

Mitchell's Chemiosmotic Theory Explains the Coupling of Oxidation and Phosphorylation

Peter D. Mitchell's chemiosmotic theory explains how oxidation and phosphorylation are coupled by proton motive force.

Energy derived from electron transport is used to pump protons into the inter-mitochondrial space by the complexes I, III and IV. Thus, each of this complex acts as a proton pump. Inner mitochondrial membrane is impermeable to protons. This creates both electrical and chemical (pH) gradient between the matrix and inner mitochondrial membrane.

Proton motive force = Electrical gradient + Chemical (pH) gradient

Accumulated protons enter through the F_0 subunit (which is similar to osmosis; chemiosmosis – entry of proton) of ATP synthase and the energy is utilized for chemical synthesis of ATP by F_1 subunit.

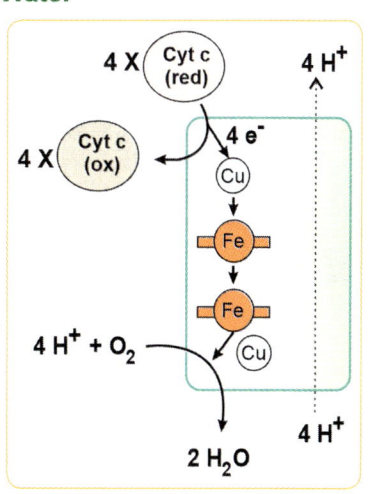

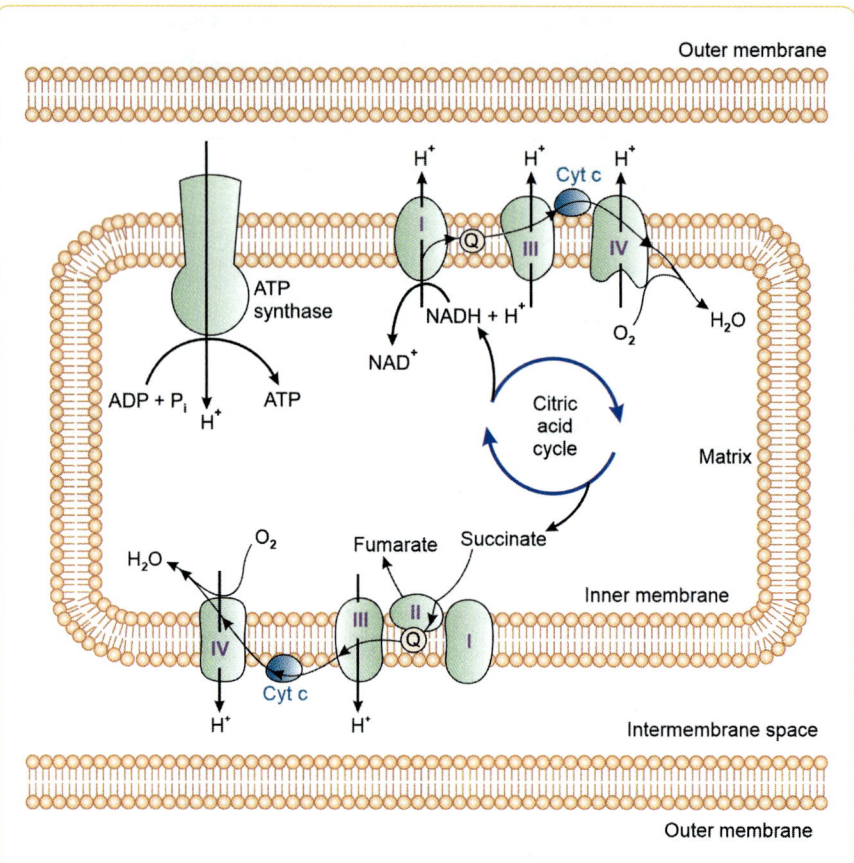

Fig. 3: Chemiosmotic hypothesis theory

COMPLEX V IS A MOLECULAR MOTOR

Complex V or ATP synthase is a molecular motor. It uses the electrochemical gradient to pump protons inside and by doing so it phosphorylates ADP to ATP by **binding change mechanism**.

ATP synthase is a supra-molecular assembly. This can be visualized as a ball on a stick. Fo (o in Fo is not zero; This indicates that F_0 is inhibited by oligomycin) subunit forms the stick and F_1 forms the balls.

The F1 subunit consists of five types of polypeptide chains (α_3, β_3, γ, δ and $\in$).

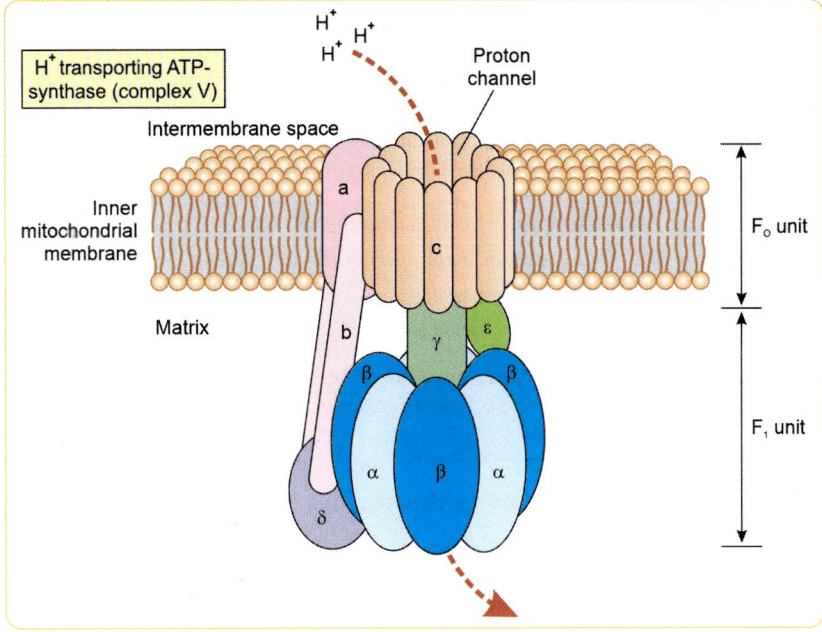

Fig. 4: ATP synthase is made up of α, β, γ, δ and ε subunits. F_1 has a water-soluble part that can hydrolyze ATP. F_O on the other hand has mainly hydrophobic regions. $F_O F_1$ creates a pathway for protons across the membrane.

INHIBITORS OF ELECTRON TRANSPORT CHAIN

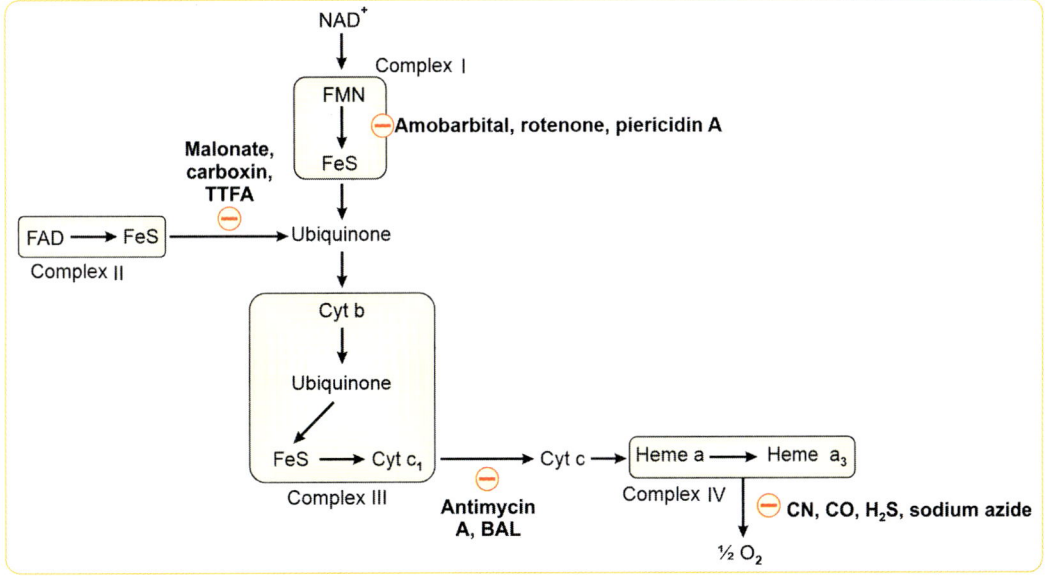

- ❑ Complex I inhibitors block the oxidation of the Fe-S clusters of complex I.
- ❑ Malonate is a competitive inhibitor of succinate dehydrogenase (complex II).

- ❑ Complex III inhibitor antimycin A blocks the transfer of electrons between Cyt b_H and coenzyme Q.
- ❑ Complex IV inhibitors inhibit electron transfer by binding tightly with the iron coordinated in Cyt a3.
  - ○ Azide and cyanide bind to ferric iron.
  - ○ CO binds to the ferrous iron.
- ❑ Complex V inhibitor oligomycin binds to the Fo (NB: *pronounced as F O, not F zero*) component of the ATP synthase complex and blocks the entry of proton through the Fo channel.
- ❑ Atractyloside is a plant glycoside that blocks the ATP-ADP translocase and stops the oxidative phosphorylation.

Complex I	Amobarbital, rotenone (fish poison), piericidin A
Complex II	Malonate, carboxin, thenoyltrifluoroacetone (TTFA)
Complex III	Antimycin A, dimercaprol (British anti lewisite)
Complex IV	CO, Cyanide, H_2S, Sodium azide
Complex V (FoF1 complex)	**Oligomycin** inhibits **Fo** component therefore inhibits the proton flow Venturicidin inhibits Fo and aurovertin inhibits F_1
ATP-ADP translocase	Atractyloside

Difference between Uncouplers and Inhibitors

- ❑ Uncouplers should be differentiated from inhibitors.
  - ○ Inhibitors inhibit the electron transport in one or more complexes
  - ○ Uncouplers doesn't inhibit the electron transport. Ionophores act as uncouplers.
  Ionophores are lipid soluble molecules which can insert themselves into lipid bilayer and transport a specific molecule across the membrane. Thus, they abolish the electrochemical gradient necessary for the proton entry.

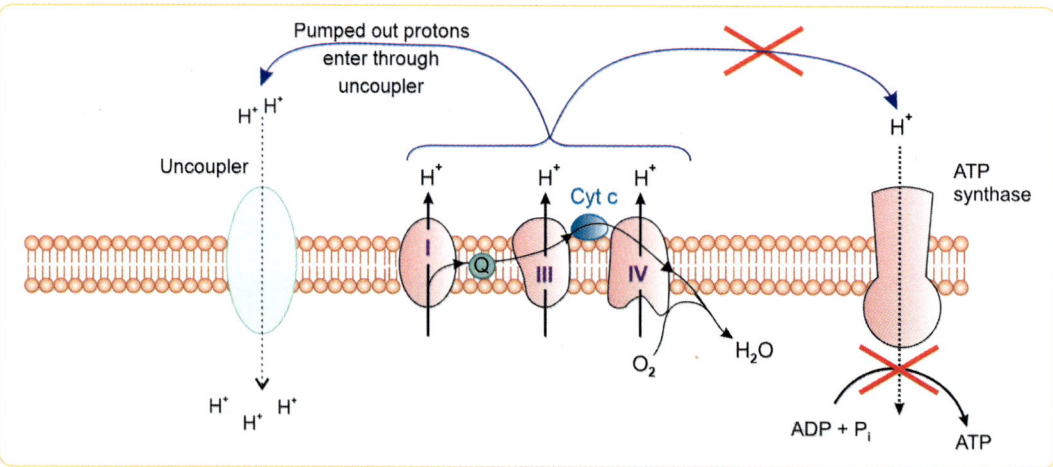

2,4 dinitrophenol is an ionophore which allows proton to pass through and dissipates the electrochemical gradient generated by proton pumps. This uncouples the oxidation from phosphorylation. Oxidation continues without phosphorylation. Therefore, all ionophores are uncouplers of oxidative phosphorylation.

2,4 dinitrophenol was once used for weight-reduction but discontinued later because of adverse effects.

H^+ Ionophores:
- ❑ **Physiological:** Thermogenin (UCP-1), Thyroxin, Long chain fatty acids
- ❑ **Chemical:** 2,4,-dinitrophenol, Dinitro cresol, High dose aspirin
K^+ Ionophore: Valinomycin
Both H^+ and K^+: Nigericin, Gramicidin A

THERMOGENIN MEDIATES NON-SHIVERING THERMOGENESIS

- ❑ Thermogenin also known as uncoupling protein 1 (UCP1) is found in the mitochondria of brown adipose tissue (BAT).
- ❑ Brown adipose tissue is abundant in the back mainly interscapular region.
- ❑ This is a physiological uncoupler of protons and dissipates the energy as heat.
- ❑ Thermogenin mediates the generation of heat by **non-shivering thermogenesis** in infants.
- ❑ Infants have high surface area to volume ratio. So, there is more chance of heat loss.
- ❑ Uncoupling protein-2, a homologue of thermogenin is present in many cell types.

P:O RATIO

Phosphate/Oxygen ratio is the amount of ATP produced when one atom of oxygen is reduced.

It is estimated that 4 protons are required for the generation of 1 ATP (3 protons for complex V and 1 proton for the entry of inorganic phosphate).

When NADH is oxidised 10 protons are pumped. But, $FADH_2$ oxidation leads to pumping of only 6 protons. Thus, 10 protons can produce around 2.5 ATPs.

Reducing equivalent	P : O ratio
NADH	2.5
$FADH_2$	1.5

CYTOSOLIC NADH IS TRANSPORTED INTO MITOCHONDRIA BY TWO SHUTTLE MECHANISMS

Processes like TCA cycle and fatty acid oxidation take place inside the mitochondrial matrix. NADH produced in these reactions can directly go to complex I of electron transport chain.

Inner mitochondrial membrane is impermeable to NADH. So, NADH produced in the cytosol (e.g. by the glycolytic enzyme glyceraldehyde 3 phosphate dehydrogenase) needs to be transported inside mitochondria somehow.

This is achieved by two shuttle mechanisms
1. Glycerol-3-phosphate shuttle or
2. Malate-aspartate shuttle

Glycerol-3-phosphate Shuttle

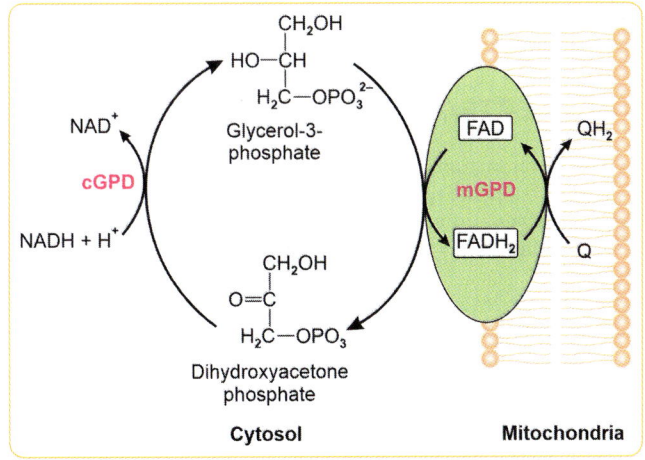

Fig. 5: Glycerol-3-phosphate shuttle. cGPD – cytosolic Glycerol-3-phosphate dehydrogenase. mGPD – mitochondrial Glycerol-3-phosphate dehydrogenase. Q- Ubiquinone. QH2 – Ubiquinol (reduced form of ubiquinone). Can you appreciate that mGPD produces FADH2?

Chapter 6 Biological Oxidation and Electron Transport Chain

The illustration of glycerol-3-phosphate shuttle is simple and self-explanatory. One important point which should be noted in the illustration is the transfer of electrons of cytoplasmic NADH to mitochondrial FADH2. So, instead of 2.5 ATP, only 1.5 ATPs will be produced. Complete oxidation of glucose yields 30 ATP instead of 32 when the two cytoplasmic NADH are transported by glycerol-3-phosphate shuttle.

Malate-Aspartate Shuttle

1. The NADH produced in the glycolysis reduces oxaloacetate to malate.
2. Malate moves inside the mitochondria through an antiporter transport system in exchange for alpha-ketoglutarate.
3. In the mitochondrial matrix, malate is oxidized back into oxaloacetate. In this process, NADH is regenerated. This NADH can now transfer two electrons to complex I.
4. The inner mitochondrial membrane is impermeable to oxaloacetate. So, it is transaminated to aspartate using the amino group of glutamate.
5. Aspartate goes to the cytosol via an antiporter system in exchange for the entry of glutamate.
6. Aspartate transported into the cytosol undergoes transamination to produce oxaloacetate.

Note: We started with oxaloacetate in step 1 and in step 6, you can clearly see that oxaloacetate is regenerated.

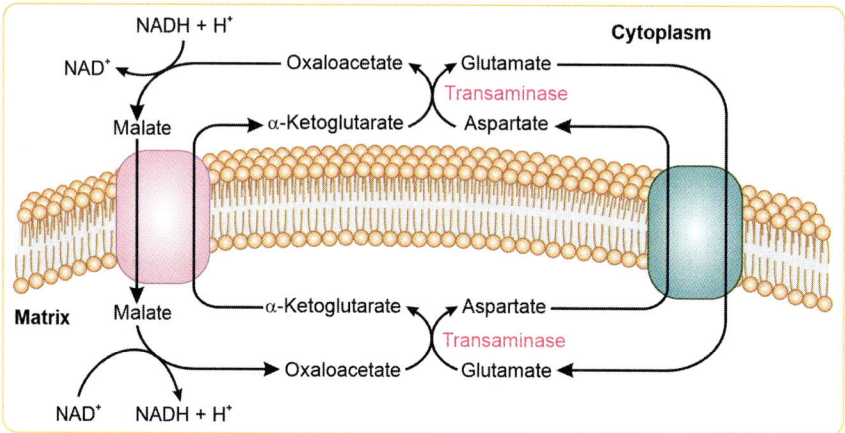

Note: Cardiac myocytes and hepatocytes utilize the Malate-Aspartate Shuttle. So, complete oxidation of glucose in these tissues generate 32 ATPs whereas skeletal muscle utilizes the glycerol 3-phosphate shuttle. So, complete oxidation of glucose in skeletal msucle generates 30 ATP molecules.

DISORDERS OF THE ELECTRON TRANSPORT CHAIN

There are 89 proteins in the electron transport chain. 13 out of 89 are coded by the mitochondrial DNA (Chapter 20) *(Refer to page no. 378)* and the rest are coded by the nuclear DNA. In addition to this, mitochondria has its own tRNA and rRNAs. Mutation in any of these mitochondrial proteins leads to various disorders, e.g. MELAS (Mitochondrial Encephalomyopathy, Lactic acidosis and Stroke-like episodes) is due to defect in leucyl-tRNA (tRNA[Leu]).

Impaired electron transport chain leads to energy failure in the cell affecting brain and muscles. As NADH can not be oxidized in the electron transport chain, accumulated NADH converts pyruvate to lactate leading to lactic acidosis.

Competency

□ BI6.6 Describe the biochemical processes involved in generation of energy in cells

Key Points

- All hydroxylases are monooxygenases.
- Glutamate dehydrogenase can use either NAD^+ or $NADP^+$.
- NADPH is used in reductive biosynthesis reactions. It can't produce ATPs.
- Oxygen is the final acceptor of electrons in the electrons transport chain.
- Ubiquinone (CoQ or Q10) is made up of 10 isoprene rings; lipid-soluble; transfers electrons from complex I and II to III.
- Cytochrome c is a small soluble protein; Carries electrons from complex III to IV.
- There is no pumping of protons in complex II.
- P: O ratio of NADH and FADH2 is 2.5 and 1.5 respectively.
- Thermogenin is a physiological uncoupler mediating the non-shivering thermogenesis in brown adipose tissue.
- Cyanide inhibits cytochrome oxidase: Atractyloside inhibits ATP-ADP translocase; Oligomycin inhibits Fo component of complex V.
- Malate-aspartate shuttle and glycerol-3-phosphate shuttle are the two mechanisms for the movement of NADH from the cytoplasm into the mitochondria. Glycerol-3-phosphate shuttle compromises one ATP.

SELF-ASSESSMENT

Short Answer Questions

1. What is the importance of brown adipose tissue?

2. Cyanide causes histotoxic hypoxia – Why?

3. Explain the biochemical basis of profuse sweating and increased body temperature on ingestion of uncouplers. What happens to the P:O ratio in the presence of uncouplers?

4. What is the P:O ratio for NAD+ and FAD-linked dehydrogenases? Explain why it is different in the two cases.

5. Mention an inhibitor of ATP synthase and its mechanism of action. Will it also inhibit electron transport? Explain with reasons.

6. Name two mobile carriers in the respiratory chain and indicate their positions in the electron transport pathway.

7. Describe the role of metals in the enzymatic catalysis.

8. Name three dehydrogenases that produce NADPH.

9. Mention the name of the complexes of electron transport chain which directly contributes to ATP formation. How many protons are pumped in each of these complexes?

10. Name the two shuttle mechanisms which are involved in the transport of cytosolic NADH into mitochondria. Illustrate with a schematic diagram. Is the final ATP yield the same for both shuttle mechanisms? Justify your answer.

11. With a schematic diagram illustrate the order of transfer of electrons in the transport chain.

12. Explain why impairment of electron transport chain leads to lactic acidosis.

13. Write the mechanism of action of the following compounds in the inhibition of ATP synthesis:
    a. Oligomycin
    b. 2, 4 Dinitrophenol (2, 4 DNP)
    c. H_2S
    d. Nigericin
    e. Atractyloside

14. Differentiate
    a. Oxidase and Oxygenase
    b. Monooxygenase and Dioxygenase
    c. Ubiquinone and Cytochrome c
    d. Malate-Aspartate shuttle and Glycerol-3-phosphate shuttle

Multiple Choice Questions

1. All of the following are true statements, except:
   a. Oxidases catalyze reactions involving hydrogen peroxide
   b. Oxidases catalyze reaction using oxygen as a hydrogen acceptor
   c. Oxidases catalyze reactions using FMN as prosthetic group
   d. Oxidases catalyze reactions of direct incorporation of oxygen in to the substrate

2. All of the followings are NAD+ requiring enzymes EXCEPT one:

   a. Fatty acyl-CoA dehydrogenase
   b. Glyceraldehyde-3-P dehydrogenase
   c. Pyruvate dehydrogenase complex
   d. Malate dehydrogenase

3. Which of the following is the (water) soluble mobile carrier of electrons in the electron transport chain?

   a. Ubiquinone   b. Cytochrome c
   c. NADH-Q oxidoreductase
   d. Succinate dehydrogenase

4. All of the following are flavoproteins EXCEPT:

   a. Xanthine oxidase
   b. NADH dehydrogenase –Q reductase
   c. Succinate dehydrogenase
   d. Cytochrome c

5. Which of the following enzyme can utilize both NAD+ or NADP+ as a coenzyme?

   a. Aldehyde dehydrogenase
   b. Alcohol dehydrogenase
   c. Glutamate dehydrogenase
   d. Glycerol-3-P dehydrogenase

6. All of the following enzymes are located in the mitochondrial matrix EXCEPT:

   a. Enzymes of fatty acid oxidation
   b. Creatine kinase
   c. Enzymes of TCA cycle
   d. Pyruvate dehydrogenase complex

7. Which of the following is NOT a haemoprotein?

   a. Catalase   b. Peroxidase
   c. Ubiquinone   d. Cytochrome c

8. Which of the components of electron transport chain does not contain Iron sulfur center?

   a. NADH-CoQ oxidoreductase
   b. Cytochrome c oxidase
   c. Succinate dehydrogenase
   d. Coenzyme Q – cytochrome c reductase

9. Which of the following best describes the biochemical basis of hyperthermia associated with Aspirin toxicity?

   a. Increased fatty acid oxidation
   b. Increased muscular activity
   c. Elevated consumption of ATP to support muscle contraction
   d. Uncoupling of oxidative phosphorylation

10. The electron flow from complex I to complex III is through:

    a. Cytochrome c   b. Ubiquinone
    c. Complex II   d. Complex IV

11. Which of the following is a dioxygenase?

    a. Tryptophan pyrrolase
    b. Lactate dehydrogenase
    c. Cytochrome oxidase
    d. L-amino acid oxidase

12. In electron transport chain, for each water molecule formed, how many protons are pumped into the inter membranous space?

    a. 4   b. 2
    c. 10   d. 5

13. How many ATP molecules are produced when one molecule of FADH2 is oxidized in the electron transport chain?

    a. 1
    b. 2.5
    c. 1.5
    d. 2

14. Which one of the following enzymes can use molecular oxygen as a hydrogen acceptor?

    a. Cytochrome c oxidase
    b. Isocitrate dehydrogenase
    c. Homogentisate dioxygenase
    d. Superoxide dismutase

15. All of the following are true about cytochromes, EXCEPT:

    a. They are hemoproteins
    b. Iron oscillates between Fe3+ and Fe2+ during the redox reactions
    c. They also have an important role in the hydroxylation of steroids in the endoplasmic reticulum.
    d. They are all dehydrogenase enzymes

16. An obese young woman ingested a banned drug which was once used for weight reduction and developed high fever. The drug is known to affect the ATP formation in electron transport chain. What could be that drug?

    a. Barbiturate
    b. Malonate
    c. 2,4 dinitrophenol
    d. Rotenone

17. Which of the following is an inhibitor of complex I of ETC?

    a. Rotenone
    b. H2S
    c. British anti lewisite (BAL)
    d. Cyanide

18. Which of the following is an uncoupler of oxidative phosphorylation?

    a. Cyanide   b. Oligomycin
    c. Thermogenin   d. Carbon monoxide

19. **All are true about oxygenase EXCEPT:**
    a. Incorporate one atom of O_2
    b. Incorporate both atoms of O_2
    c. Involved in hydroxylation of steroids
    d. Involved in carboxylation of drugs

20. **The correct order of transfer of electrons in the electron transport chain is:**
    a. NADH → Q → cytochrome b → cytochrome c1 → cytochrome c → cytochrome a → cytochrome a3 → O_2
    b. NADH → Q → $FADH_2$ → cytochrome b → cytochrome c1 → cytochrome c → cytochrome a → cytochrome a3 → O_2

    c. NADH → Q → cytochrome c → cytochrome b1 → cytochrome a → cytochrome c → cytochrome a3 → O_2
    d. FAD → Q → cytochrome c → cytochrome c1 → cytochrome b → cytochrome a → cytochrome a3 → O_2

21. **Cytochrome c of the bacteria has 50% identity of amino acid sequence with that of human. Which of the following is the most conserved parameter in these two proteins?**
    a. Primary structure b. Secondary structure
    c. Tertiary structure d. Quaternary structure

ANSWERS

1. d.	**2.** a	**3.** b.	**4.** d.	**5.** c.	**6.** b.
7. c.	**8.** b.	**9.** d.	**10.** b.	**11.** a.	**12.** c.
13. c.	**14.** a.	**15.** d.	**16.** c.	**17.** a.	**18.** c.
19. d.	**20.** a.	**21.** c.			

CHAPTER 7

Chapter Outline

TCA CYCLE: THE FINAL COMMON OXIDATIVE PATHWAY

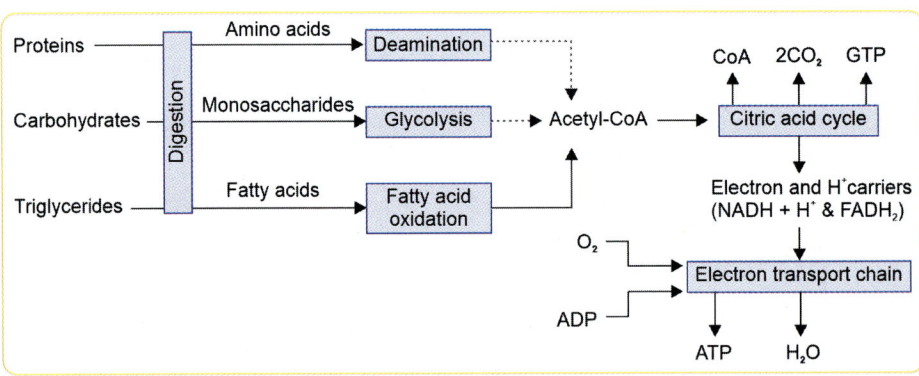

Let us analyze the term "Final common oxidative" pathway word by word.

- **Final:** This means TCA cycle is the final pathway for extraction of electrons. Consider the carbohydrates as example. Carbon atoms of glucose/fructose undergo initial pathways like glycolysis to produce pyruvate. Pyruvate is converted to acetyl-CoA which enters the TCA cycle. There is complete oxidation of acetyl-CoA in the TCA cycle.
- **Common:** Preliminary oxidation of carbohydrate, protein and fat produces acetyl-CoA which enters TCA cycle. Thus, this is a common pathway for all macromolecules.
- **Oxidative:** TCA cycle extracts electrons from the molecules (Loss of electrons is oxidation). Electrons are transferred to NAD^+ and FAD. These electrons are presented to the electron transport chain to generate ATP.

Subcellular Location of TCA Cycle Enzymes

- All the TCA cycle enzymes are present in the mitochondrial matrix except succinate dehydrogenase which is in fact the complex II of electron transport chain.
- Succinate dehydrogenase is embedded in the inner mitochondrial membrane.

Reactions of TCA Cycle

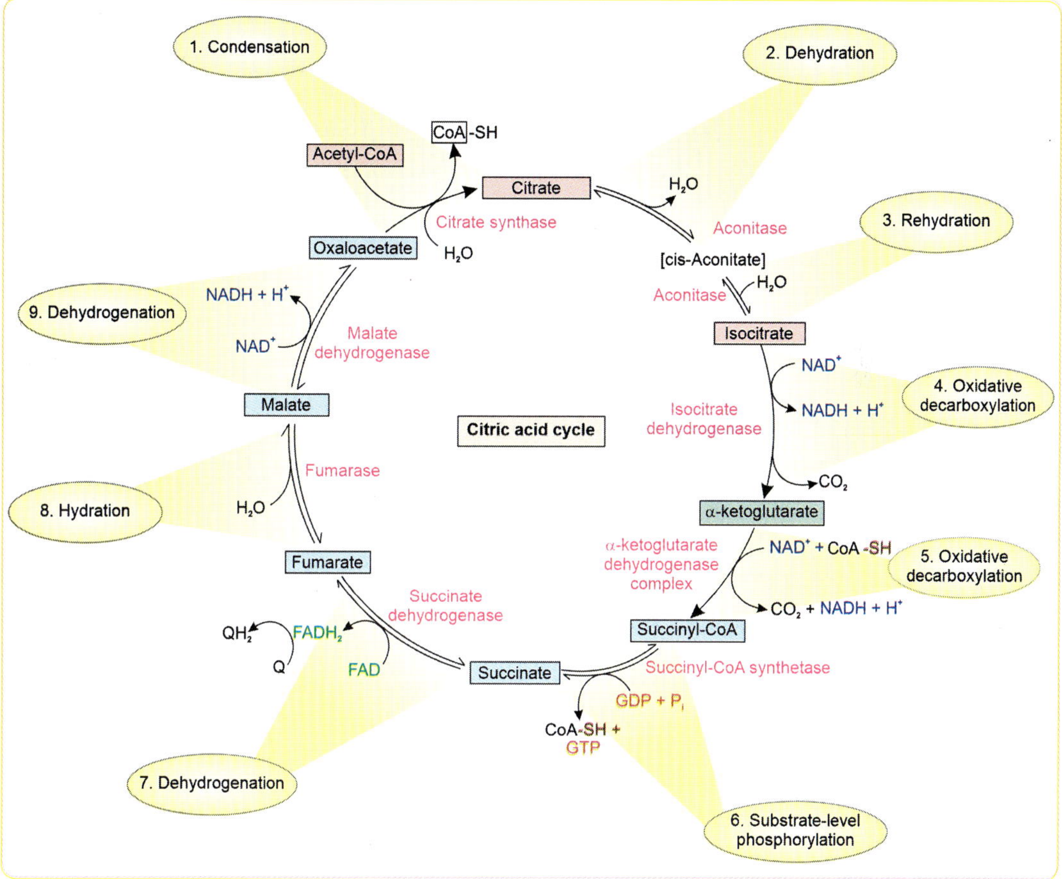

1. Condensation

2. Dehydration

3. Rehydration

CoA-SH

Acetyl-CoA

Citrate

H_2O

Citrate synthase

Aconitase

Oxaloacetate

H_2O

[cis-Aconitate]

Aconitase

H_2O

9. Dehydrogenation

NADH + H⁺

NAD⁺

Malate dehydrogenase

Isocitrate

NAD⁺

4. Oxidative decarboxylation

Malate

Citric acid cycle

Isocitrate dehydrogenase

NADH + H⁺

Fumarase

CO_2

8. Hydration

H_2O

α-ketoglutarate

α-ketoglutarate dehydrogenase complex

NAD⁺ + CoA-SH

5. Oxidative decarboxylation

Fumarate

Succinate dehydrogenase

CO_2 + NADH + H⁺

QH₂

FADH₂

Succinyl-CoA

FAD

Succinate

Succinyl-CoA synthetase

Q

GDP + Pᵢ

7. Dehydrogenation

CoA-SH + GTP

6. Substrate-level phosphorylation

Fig. 1: Succinate dehydrogenase is a part of electron transport chain. It contains FAD prosthetic group and transfers electrons to the mobile electron carrier coenzyme Q through iron-sulfur complex.

 Mnemonic

TCA Cycle Intermediates
Can I Ask Some Sweets For MOm? (Citrate → Isocitrate → Alpha keto glutarate → Succinyl-CoA → Succinate → Fumarate → Malate → Oxaloacetate)
Alternate mnemonic: Chitra Is A Sweet Smart Female Medical Officer

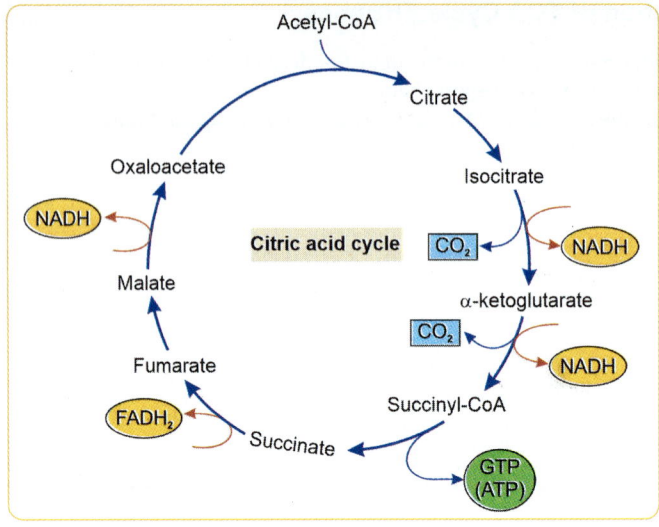

Fig. 2: Simplified view of TCA cycle

CONCEPT CORNER

Biochemical pathways may be different but the common theme in biochemistry is same and recurring. Look at the final steps of TCA cycle starting from succinate dehydrogenase. This same theme, **FAD dependent oxidation, hydration, NAD dependent oxidation** is also utilized in β-oxidation of fatty acids.

TCA Cycle: An Amphibolic Pathway

An amphibolic pathway is the one which is both catabolic and anabolic depending on the energy status of the cell. In addition to being the final common oxidative pathway for catabolism, TCA cycle intermediates are utilized in anabolic reactions like heme synthesis, fatty acid synthesis, gluconeogenesis and amino acid synthesis as illustrated in Figure 3.

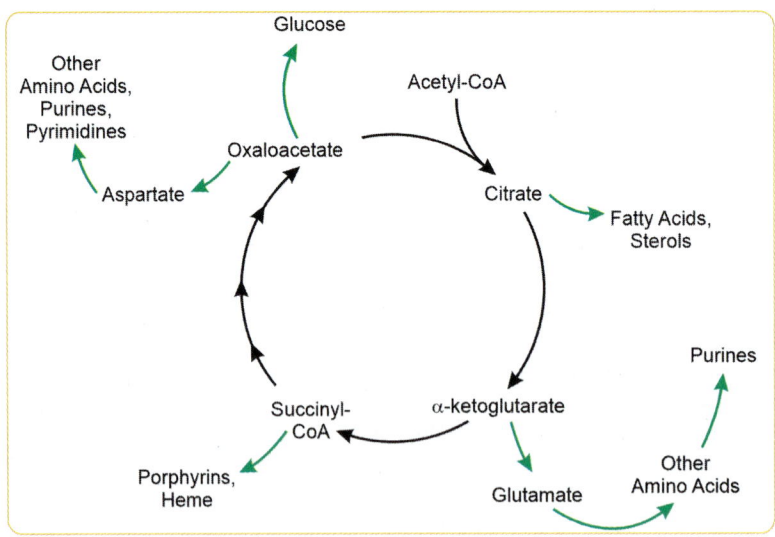

Fig. 3: Involvement of TCA cycle in biosynthetic reactions

Anaplerotic Reactions Replenishes the Intermediates of TCA Cycle

Anaplerotic (gap-filling) reactions are the reactions that replenishes the depleted TCA cycle intermediates that were used up in the biosynthetic reactions.

Intermediate	Depleted (used) by	Replenished by
Oxaloacetate	Amino acid synthesis	**Pyruvate carboxylase**—Major anaplerotic enzyme
Succinyl-CoA	Heme synthesis	From propionyl-CoA by carboxylase and methylmalonyl-CoA mutase
Oxaloacetate, fumarate	Gluconeogenesis	From amino acids

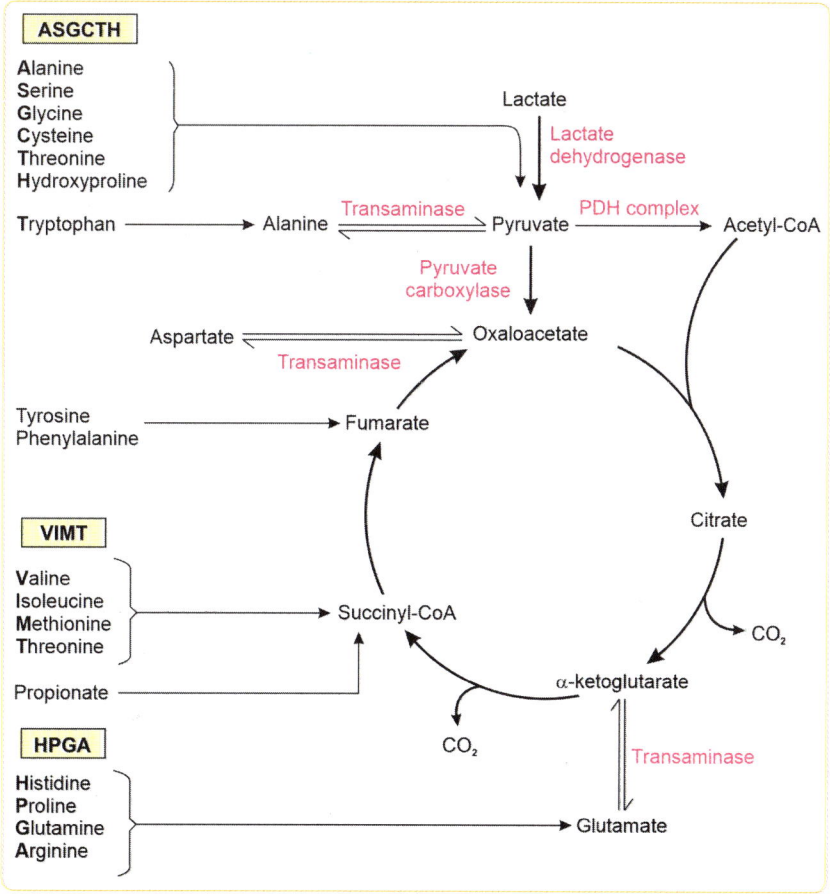

Fig. 4: Anaplerotic reactions

Succinate Thiokinase Reaction is the Substrate-level Phosphorylation

- ❑ The reaction catalyzed by succinate thiokinase (Other names: Succinyl-CoA synthetase and succinate-CoA ligase) is a substrate level phosphorylation reaction producing either ATP or GTP.
- ❑ Only gluconeogenic tissues contain both ATP and GTP producing isoforms of succinate thiokinase. The GTP produced by this reaction is used by the enzyme phosphoenolpyruvate carboxykinase (PEPCK).

Role of B-complex Vitamins as TCA Cycle Cofactors

Coenzyme	Vitamin	TCA cycle enzyme
Thiamine pyrophosphate (TPP)	B_1 (Thiamine)	α-ketoglutarate dehydrogenase (also requires other 4 cofactors similar to pyruvate dehydrogenase)
Nicotinamide adenine dinucleotide (NAD⁺)	B_3 (Niacin)	☐ Isocitrate dehydrogenase ☐ α-ketoglutarate dehydrogenase ☐ Malate dehydrogenase
Flavin adenine dinucleotide (FAD)	B_2 (Riboflavin)	Succinate dehydrogenase
Coenzyme A (CoA)	B_5 (Pantothenic acid)	—

Regulation of TCA Cycle

Overall high (ATP)/(ADP) ratio and high (NADH)/(NAD) ratio will inhibit the TCA cycle. When ATP supply is adequate, availability of ADP for phosphorylation will be less. Oxidation of reducing equivalents and phosphorylation of ADP are tightly coupled. High ATP will thus lead to less oxidation and NADH will accumulate. Dehydrogenases require NAD⁺ for their activity. NADH excess will inhibit the dehydrogenases.

Enzyme	Activators	Inhibitors
Citrate synthase	Acetyl-CoA, Oxaloacetate, Insulin	ATP, Long chain fatty acyl-CoA, Citrate, NADH
Isocitrate dehydrogenase	Calcium, ADP, ↓ NADH/NAD ratio	↑ NADH/NAD ratio
α-ketoglutarate dehydrogenase	Calcium, ADP, ↓ NADH/NAD ratio	↑ NADH/NAD ratio, Succinyl-CoA

Role of Citrate as the Indicator of High-Energy Status Inside the Cell

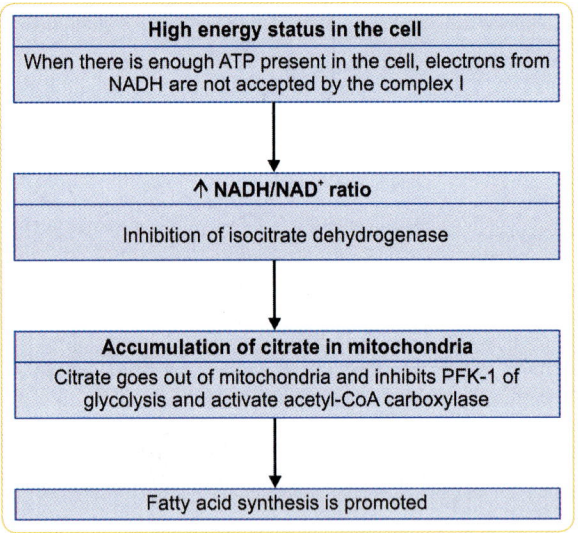

Role of Citrate in Fatty Acid Synthesis

In addition to its role as signal for high-energy status, citrate plays a crucial role in fatty acid synthesis. Citrate activates the rate-limiting enzyme of fatty acid synthesis, i.e. acetyl-CoA carboxylase and produces acetyl-CoA and NADPH in the cytoplasm (see the following figure).

Remember

Citrate comes out to cytoplasm only when there is adequate ATP. Biosynthetic reactions happen only when there is adequate energy.

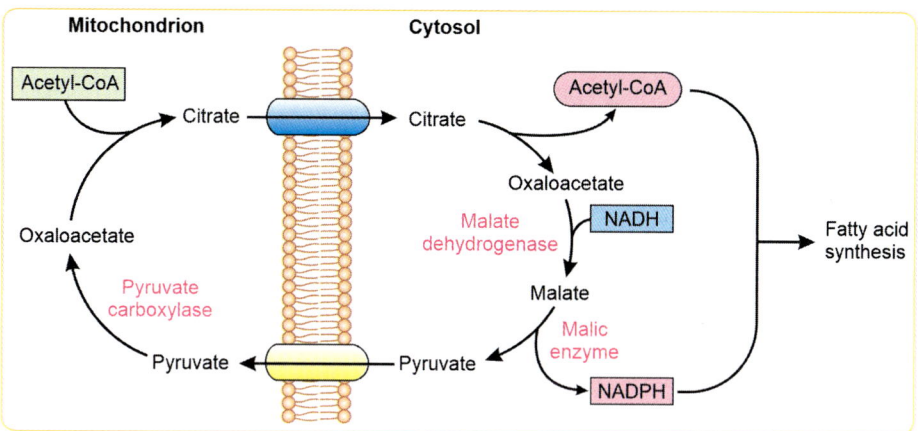

Inhibitors of TCA Cycle

Inhibitor	Enzyme inhibited	Mechanism of inhibition
Fluoroacetate (rat poison)	Aconitase	Fluoroacetate combines with oxaloacetate to produce fluorocitrate which inhibits aconitase
Arsenite	α-ketoglutarate dehydrogenase	Arsenite binds to SH group of lipoic acid cofactor
High concentrations of NH_3 (Hepatic encephalophathy)	α-ketoglutarate dehydrogenase	Substrate depletion - α-ketoglutarate is converted to glutamic acid by glutamate dehydrogenase
Malonate	Succinate dehydrogenase	Competitive inhibition

Energy Yield from TCA Cycle-10 ATP per Acetyl-CoA

Reaction	Reducing equivalent generated	Number of ATPs produced
Isocitrate dehydrogenase	NADH	2.5
α-ketoglutarate dehydrogenase	NADH	2.5
Succinate thiokinase	Substrate-level	1 ATP/GTP
Succinate dehydrogenase	$FADH_2$	1.5
Malate dehydrogenase	NADH	2.5
Total		10

Chapter 7 Tricarboxylic Acid (TCA) Cycle

Complete Oxidation of 1 Mole of Glucose to CO_2 and H_2O Under Aerobic Conditions Produces a Total of 32 ATPs

Pathway	Reactions	ATP equivalents
Glycolysis	Substrate level phosphorylation ❑ Phosphoglycerate kinase ❑ Pyruvate kinase	2 2
	Glyceraldehyde-3-phosphate dehydrogenase 2 NADH	5*
	Energy invested in hexokinase and PFK	− 2
Net		7
Pyruvate dehydrogenase reaction	2 × 1 NADH	5
TCA cycle	2 acetyl-CoA (*See above for explanation*)	20
Net Total		**32 ATP**

*Assumed that NADH formed in glycolysis is transported into mitochondria by the **Malate-Aspartate shuttle**. Reduce 1 ATP per NADH, if glycerophosphate shuttle is involved.*

Oxaloacetate Plays a Catalytic Role in TCA Cycle

After each turn of TCA cycle, oxaloacetate is regenerated. For every acetyl-CoA entered in the cycle, there is net release of 2 CO_2. If we imagine the TCA cycle as a combustion engine, oxaloacetate is the spark. One molecule of oxaloacetate can catalyze the oxidation of thousands of acetyl-CoA molecules.

Acetyl-CoA (2 Carbon)   CoA

Oxaloacetate (4 Carbon)

Citrate (6 Carbon)

CO_2 CO_2

Fats are Burnt in the Flame of Carbohydrates

❑ Fatty acid oxidation produces acetyl-CoA. Burning/oxidation of acetyl-CoA takes place in TCA cycle.
❑ Oxaloacetate is derived from carboxylation of the glycolytic end product pyruvate.
❑ Oxidation/burning of acetyl-CoA (fat) requires oxaloacetate (Carbohydrate).

Ketogenic Diet Contains Very Less Carbohydrate

❑ If there is no oxaloacetate (when all the oxaloacetate is used for gluconeogenesis), acetyl-CoA cannot be burnt and is converted to ketones.
❑ Thus, carbohydrate intake prevents ketogenesis.
❑ This is the reason why ketogenic diets contain high-fat, high-protein but low carbohydrates.
❑ Ketogenic diet is used for weight loss and in patients with refractory epilepsy.
I highly recommend you to read this Chapter again when you complete the metabolism of lipids and amino acids. The next reading will really make a difference in your understanding.

Competency

❑ BI3.6   Describe and discuss the concept of TCA cycle as a amphibolic pathway and its regulation

Key Points

- Tricarboxylic acid (TCA) cycle is the final common oxidative pathway for carbohydrate, protein and fat metabolism.
- All the TCA cycle enzymes are located in the mitochondrial matrix except succinate dehydrogenase (complex II of ETC) which is embedded in the inner mitochondrial membrane.
- Citrate is the first tricarboxylic acid formed in the TCA cycle.
- Citrate produced in the TCA cycle is involved in fatty acid synthesis.
- TCA cycle is an amphibolic pathway involved in both anabolism and catabolism.
- TCA cycle intermediates utilized in other synthetic pathways are replenished by anaplerotic reactions.
- Pyruvate carboxylase catalyzed reaction is the major anaplerotic reaction.
- Hyperammonemia inhibits the TCA cycle by depleting α-ketoglutarate.
- Fumarate is the link between urea cycle and TCA cycle.
- There are 4 dehydrogenase enzymes in the TCA cycle producing 3 NADH and 1 FADH2.
- Complete oxidation of one molecule of acetyl-CoA in TCA cycle produces 10 ATPs in electron transport chain.
- Complete oxidation of one molecule of glucose under aerobic conditions can produce a maximum of 32 ATPs.
- Fluoroacetate inhibits aconitase - used as rat poison.
- Malonate is a competitive inhibitor of succinate dehydrogenase.
- Arsenite inhibits α-ketoglutarate dehydrogenase.
- Succinate thiokinase reaction of the TCA cycle is a substrate-level phosphorylation reaction.
- Oxaloacetate acts as a catalyst in the TCA cycle.

SELF-ASSESSMENT

Short Answer Questions

1. TCA cycle is the final common oxidative pathway – Justify this statement.

2. What are all the intermediates of the TCA cycle that can contribute to metabolic pathways in the cytoplasm? Illustrate with two examples. Discuss two mechanisms by which the intermediates of the TCA cycle can be replenished.

3. What are the anaplerotic reactions of the TCA cycle? Why these reactions are needed? Mention any two.

4. Explain the role of B-complex Vitamins as TCA Cycle Cofactors

5. Name any two inhibitors of TCA Cycle along with the enzyme inhibited and the mechanism of inhibition.

6. How many ATPs are produced from the complete oxidation of one molecule of acetyl-CoA in the TCA cycle? Explain your answer.

7. Explain the adage – "Fats are Burnt in the Flame of Carbohydrates"

8. A minimal supply of glucose is necessary for extrahepatic tissues to maintain the integrity of the TCA cycle - Explain.

9. In which way does succinate dehydrogenase differ from other enzymes in the citric acid cycle?

10. "Citric acid cycle provides the substrates for (a) glucose (b) fatty acid (c)amino acid synthesis". Support the statement by giving one example each.

Multiple Choice Questions

1. Tricarboxylic acid cycle does not occur in:
   - a. Myocyte
   - b. Red blood cell
   - c. Neuron
   - d. Hepatocyte

2. All of the following are intermediates of TCA cycle, EXCEPT:
   - a. Malonate
   - b. α-ketoglutarate
   - c. Succinate
   - d. Fumarate

3. **Which of the following is the major anaplerotic enzyme?**
   a. Pyruvate carboxylase
   b. Acetyl-CoA carboxylase
   c. Pyruvate dehydrogenase
   d. Succinate dehydrogenase

4. **How many dehydrogenase enzymes are operative in the TCA cycle?**
   a. 2   b. 3
   c. 4   d. 5

5. **Which of the following TCA cycle intermediate is the link between urea cycle and TCA cycle?**
   a. Argininosuccinate   b. Fumarate
   c. Oxaloacetate   d. Succinate

6. **Hyperammonemia inhibits TCA cycle by depleting:**
   a. Alpha ketoglutarate
   b. Fumarate
   c. Oxaloacetate
   d. Succinate

7. **Which of the following is the FAD linked dehydrogenase of TCA cycle?**
   a. Isocitrate dehydrogenase
   b. Malate dehydrogenase
   c. Succinate dehydrogenase
   d. α ketoglutarate dehydrogenase

8. **Which of the following substance binds to CoA and condenses with oxaloacetate to inhibit the TCA cycle?**
   a. Malonate   b. Arsenite
   c. Fluoride   d. Fluoroacetate

9. **Vitamin required in TCA cycle include all, EXCEPT:**
   a. Biotin   b. Niacin
   c. Riboflavin   d. Thiamine

10. **All of the following TCA cycle enzymes are located in the mitochondrial matrix, EXCEPT:**
    a. Alpha ketoglutarate dehydrogenase
    b. Isocitrate dehydrogenase
    c. Succinate dehydrogenase
    d. Malate dehydrogenase

11. **In gluconeogenic tissues, which of the following enzyme catalyzed reaction provides the GTP required for the phosphoenolpyruvate carboxy kinase at substrate level?**
    a. Pyruvate kinase
    b. Phosphoglycerate kinase
    c. Creatine kinase
    d. Succinate thiokinase

12. **Decreased energy production in thiamine deficiency is due to:**
    a. It is a cofactor for oxidative reduction
    b. It is required for transamination reactions
    c. It is a coenzyme for alpha ketoglutarate dehydrogenase and pyruvate dehydrogenase in citric acid cycle
    d. It is a coenzyme for transketolase in hexose monophosphate shunt

13. **How many ATPs are produced from the complete oxidation of 1 molecule of acetyl-CoA in TCA cycle?**
    a. 10   b. 12
    c. 22   d. 32

ANSWERS

1. b.	2. a.	3. a.	4. c.	5. b.	6. a.
7. c.	8. d.	9. a.	10. c.	11. d.	12. c.
13. a.					

8
CHAPTER

Introduction to Metabolism

Metabolism is the process through which living systems acquire and use free energy that sustains life. A metabolic pathway is a series of connected enzymatic reactions.

Metabolic reactions are either anabolic or catabolic.

ANABOLIC REACTIONS

- ❏ Assembly of macromolecules from small molecules.
- ❏ Always require energy, e.g. fatty acid synthesis—synthesize fatty acids from 2 carbon acetyl-CoA.
- ❏ Enzymes catalyzing anabolic reactions are active in the dephosphorylated state except **ATP-citrate lyase** which is active in the phosphorylated state.

CATABOLIC REACTIONS

- ❏ Degradation of macromolecules into small molecules.
- ❏ Energy is often released. The released energy is conserved in NADH, $FADH_2$, NADPH or substrate level phosphorylation, e.g. fatty acid oxidation—breaks the fatty acids to 2 carbon acetyl groups linked to CoA.
- ❏ Enzymes catalyzing catabolic reactions are active in the phosphorylated state except **pyruvate dehydrogenase** which is active in the dephosphorylated state.

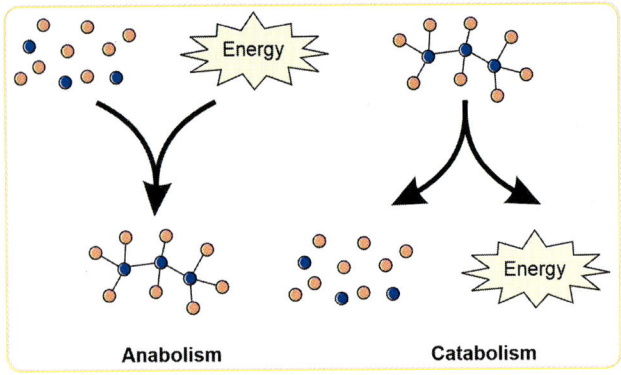

| Anabolism | Catabolism |

Amphibolic pathways are involved in both anabolism and catabolism, e.g. TCA cycle.

FLUX-GENERATING REACTION

- ❑ The flux-generating reaction is the first reaction in a pathway that is saturated with the substrate.
- ❑ This reaction is an irreversible (nonequilibrium) reaction.
- ❑ K_m of the flux generating enzyme is lower than the usual concentration of substrate.
- ❑ For example, hexokinase is the flux-generating enzyme of glycolysis. K_m of hexokinase for glucose is just 0.05 mmol/L. This means all the available glucose will be channeled to glycolysis by hexokinase.
- Video: https://youtu.be/gV8Eqa6bFEY or scan QR code

RATE-LIMITING REACTION

- ❑ Rate-limiting reaction is the rate and direction determining reaction of an entire metabolic pathway.
- ❑ Reaction is *always irreversible*/unidirectional/nonequilibrium.
- ❑ Rate-limiting step is always one of the initial steps of the pathway (not necessarily the first step).
- ❑ These reactions are the *targets for regulation* by various modes.
- ❑ Rate-limiting enzymes have the *highest K_m* value—lowest substrate affinity. So, the catalyzed reaction is *the slowest* among the metabolic pathway.
- ❑ For example, phosphofructokinase 1 is the rate-limiting enzyme of glycolysis.

AVOIDING FUTILE CYCLE IN METABOLISM

- ❑ When two metabolic pathways run simultaneously in opposite directions, there will be dissipation of energy in the form of heat. This is known as futile cycle.
- ❑ That is why, when glycogenolysis takes place, glycogen synthesis is inhibited; when fatty acid oxidation takes place, fatty acid synthesis is inhibited.

DEFECT IN METABOLIC PATHWAYS LEADS TO VARIOUS DISORDERS

A defect in an enzyme catalyzing a metabolic pathway leads to accumulation of substrate and defective formation of the product. Metabolic pathways are not always linear. So, the accumulated substrate can enter some alternate pathways often to produce undesirable compounds. We will discuss these disorders throughout our study of metabolism.

Simplified view of metabolic process we are going to explore under this section is shown in the following illustration:

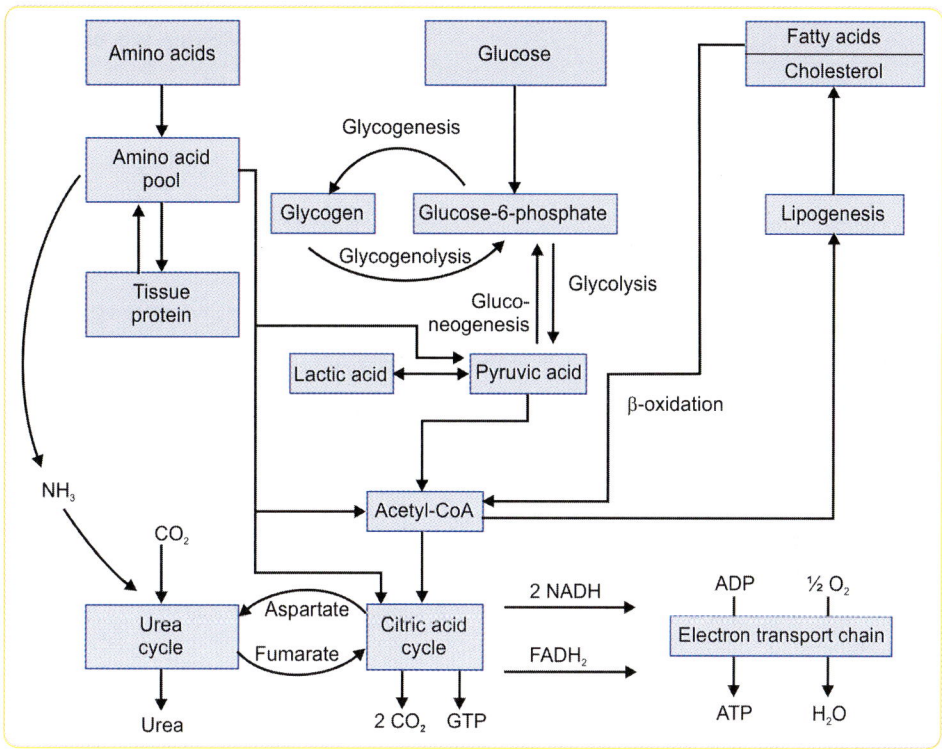

Key Points

- ❏ Enzymes catalyzing anabolic reactions are active in the dephosphorylated state except ATP-citrate lyase which is active in the phosphorylated state.
- ❏ Enzymes catalyzing catabolic reactions are active in the phosphorylated state except pyruvate dehydrogenase which is active in the dephosphorylated state.
- ❏ Amphibolic pathways are involved in both anabolism and catabolism, e.g. TCA cycle.
- ❏ The flux-generating reaction is the first reaction in a pathway that is saturated with the substrate, e.g Hexokinase of glycolysis.
- ❏ Rate-limiting reaction is the rate and direction determining reaction of an entire metabolic pathway, e.g. PFK-1 of glycolysis.
- ❏ Simultaneous operation of two metabolic pathways in opposite directions leads to dissipation of energy in the form of heat. This is known as futile cycle.

Short Answer Questions

1. Differentiate between anabolic and catabolic reactions.

2. Define flux-generation reaction of a metabolic pathway. What are the properties of the enzyme catalyzing this reaction? Provide an example.

3. Define the rate-limiting reaction of a metabolic pathway. What are the properties of the enzyme catalyzing this reaction? Provide an example.

4. What are futile cycles in metabolism and why should they be avoided?

Multiple Choice Questions

1. **Flux-generating enzyme of glycolysis is:**

   a. Hexokinase   b. PFK-1

   c. PFK-2   d. Pyruvate kinase

2. **The rate-limiting enzyme of glycolysis is:**

   a. Hexokinase   b. PFK-1

   c. PFK-2   d. Pyruvate kinase

3. **Which of the following is a purely catabolic reaction?**

   a. Fatty acid oxidation

   b. Fatty acid synthesis

   c. Glycolysis

   d. TCA cycle

4. **All of the following are true about the enzyme catalyzing the rate-limiting step of a metabolic pathway, EXCEPT:**

   a. K_m of the enzyme is lower than the usual concentration of the substrate

   b. Reaction is irreversible

   c. Reaction is regulated

   d. The reaction is the slowest among the metabolic pathway

5. **All of the following are true about catabolic reactions, EXCEPT:**

   a. Degradation of macromolecules into small molecules

   b. Energy is often released

   c. Fatty acid oxidation is an example

   d. NADPH is often used

ANSWERS

1. a.	2. b.	3. a.	4. a.	5. d.

Chemistry of Carbohydrates

BIOMEDICAL IMPORTANCE OF CARBOHYDRATES

- Carbohydrates are the most abundant dietary source of energy.
- Neurons and RBCs are exclusively dependent on glucose as the energy source.
- Glucose is the only molecule that can be utilized anaerobically.
- Glycogen serves as the source of energy during fasting.
- Carbohydrates are precursors for other biomolecules, like fats, amino acids.
- Glycoconjugates are important for the structure of cell membrane and cellular functions (cell growth, adhesion and fertilization).
- Ribose and deoxyribose are the components of nucleic acids (DNA and RNA) and nucleotides.

DEFINITION OF CARBOHYDRATES

Carbohydrates are polyhydroxy aldehydes or ketones or compounds which produce them on hydrolysis.
- Polyhydroxy aldehydes → Glucose, Mannose

- ❏ Polyhydroxy ketones → Fructose, Ribulose
- ❏ Compounds which produce these on hydrolysis → Lactose, Starch, Glycogen, etc.

CLASSIFICATION OF CARBOHYDRATES

- ❏ Monosaccharides: Polyhydroxy aldehydes or ketones
- ❏ Disaccharides: Two monosaccharides linked by glycosidic linkage
- ❏ Oligosaccharides: 2 to 10 monosaccharides linked by glycosidic linkage (*oligosaccharides include disaccharides also!*)
- ❏ Polysaccharides: >10 monosaccharides linked by glycosidic linkage

MONOSACCHARIDES

All monosaccharides can be considered as the hydrates of carbon. Their molecular formula is $C_n(H_2O)_n$.

For a molecule to be a polyhydroxy aldehyde or ketone, it should have at least 3 carbons (one carbon for the carbonyl group and the other two for hydroxyl groups). Glyceraldehyde and dihydroxy acetone are the simple sugars.

Glyceraldehyde is a 3-carbon carbohydrate with an aldehyde functional group while dihydroxy acetone is a ketone.

Glyceraldehyde, an aldotriose

Dihydroxyacetone, a ketotriose

Classification of Monosaccharides Based on the Functional Group and Number of Carbons

Number of carbons	Functional group	Name	Example
3	Aldehyde	Aldotriose	Glyceraldehyde
	Ketone	Ketotriose	Dihydroxyacetone
4	Aldehyde	Aldotetrose	Erythrose
5	Aldehyde	Aldopentose	Ribose
	Ketone	Ketopentose	Ribulose
6	Aldehyde	Aldohexose	Glucose
	Ketone	Ketohexose	Fructose
7	Ketone	Ketoheptose	Sedoheptulose
9	Deoxy sugar	Neuraminic acid	

ISOMERISM IN CARBOHYDRATES

To understand the concept of isomerism, we should first understand the asymmetric carbon.

An asymmetric carbon is a carbon which is attached to four different groups. Look at the structures of Glyceraldehyde and Dihydroxyacetone in Figure 1. The second carbon of glyceraldehyde is attached to 4 different groups while dihydroxyacetone does not contain any asymmetric carbon.

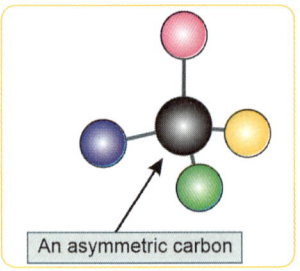

An asymmetric carbon

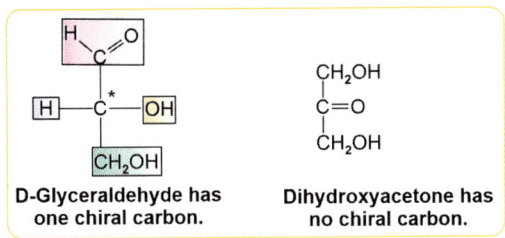

| D-Glyceraldehyde has one chiral carbon. | Dihydroxyacetone has no chiral carbon. |

Fig. 1: In our body, glyceraldehyde-3-phosphate and dihydroxyacetone phosphate are produced from the break down of glucose in the glycolytic pathway

Functional Isomers or Constitutional Isomers

Look at the structures of glyceraldehyde and dihydroxyacetone again. They have the same molecular formula but the functional groups are different, i.e. the order of attachment of atoms is different. So, these two are functional isomers of each other. Isomerases are a class of enzymes that catalyse the reversible interconversion of isomers. e.g. Phosphoglucose isomerase catalyses the interconversion of glucose-6-phosphate to fructose-6-phosphate.

Other Functional Isomers

- Glucose and Fructose
- Ribose and Ribulose
- Xylose and Xylulose

Stereoisomers

Stereoisomers have the same functional group but the atoms differ in their spatial arrangement around one or multiple carbons.

The number of possible stereoisomers of a molecule is known by the formula 2^n. n is the number of chiral carbons.

Glyceraldehyde has one asymmetric carbon. So, it can have two stereoisomers.

Stereoisomers are of 2 types.
1. Enantiomers
2. Diastereoisomers

Enantiomers

Enantiomers are pair of stereoisomers that are non-superimposable mirror images of each other, e.g. D and L forms are enantiomers.

D and L isomers are usually represented by Fischer projection *(see the adjacent illustration)*.

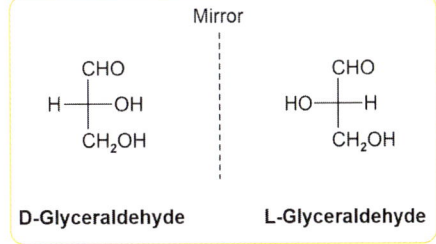

Diastereoisomers

Diastereoisomers (diastereomers) are stereoisomers which are NOT mirror images of each other. There are two special types of diastereoisomers:
1. Epimers
2. Anomers

Epimers

Epimers are diastereoisomers that differ around the configuration of a single asymmetric carbon other than the anomeric carbon.

Mnemonic

Epimers of glucose can be remembered as MAGI. Mannose is the C2 epimer, Allose is the C3 epimer, Galactose is the C4 epimer, and L-Idose is the C5 epimer.

Note:

❑ Glucose and galactose are epimers of each other. Glucose and mannose are epimers of each other.

❑ But, mannose and galactose are not epimers of each other. They are diastereomers because they differ in the orientation of more than one carbon atoms.

❑ Epimerases are a type of isomerase enzymes which convert one epimer to the other. e.g. UDP-glucose 4-epimerase catalyses the reversible conversion of UDP-glucose to UDP-galactose.

Anomers

Anomers are special types of diastereomers that differ only around the anomeric carbon—first carbon in glucose, second carbon in fructose. **Anomerism is seen only in ring structure.**

Anomerism is due to Cyclization of Sugars

To understand the anomerism, we need to first understand the cyclization of sugars. To understand the cyclization, we need to understand the hemiacetal and hemiketal formation.

An alcohol reacts with an aldehyde to produce hemiacetal. An alcohol reacts with a ketone to produce hemiketal.

Glucose is a polyhydroxy aldehyde. So, it has both alcohol and aldehyde group required for the formation of hemiacetal. Hydroxyl group of 5th carbon of glucose reacts with aldehyde group (1st carbon) to produce a intramolecular hemiacetal. This results in the formation of a 6- membered cyclic structure known as pyranose ring (pyran is a 6-membered cyclic hydrocarbon).

Cyclic structures are represented by Haworth projection (Fig. 3).

Fig. 2: Note that the previously achiral carbon of aldehyde or ketone group has become chiral after hemiacetal or hemiketal formation.

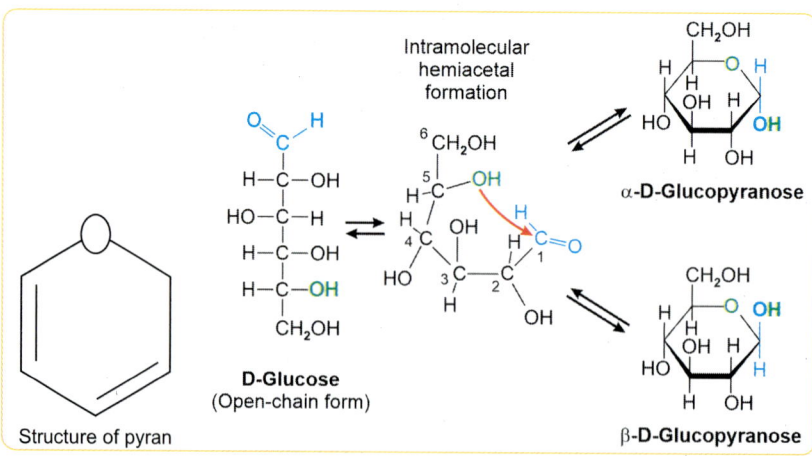

Fig. 3: Appreciate the intramolecular hemiacetal formation and the anomers represented by Haworth projection

Section II Intermediary Metabolism

This cyclization converted the previously symmetric carbon of the linear chain to an asymmetric carbon. This new asymmetric carbon is the anomeric carbon.

❏ If the OH group of the anomeric carbon is above the plane of the pyranose ring or at the same plane of the C6, it is known as β anomer.

❏ If the OH group of the anomeric carbon is below the plane of the pyranose ring or to the opposite plane of C6, it is known as α anomer.

When the C5 OH group of fructose reacts with the keto group at the 2nd carbon, a 5-membered furanose ring is formed. Fructose can form both pyranose and furanose rings. Phosphorylated fructose exists exclusively as furanose ring.

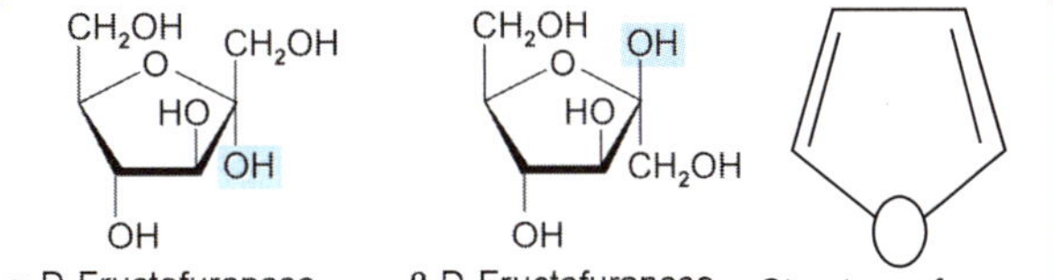

α-D-Fructofuranose   β-D-Fructofuranose   Structure of pyran

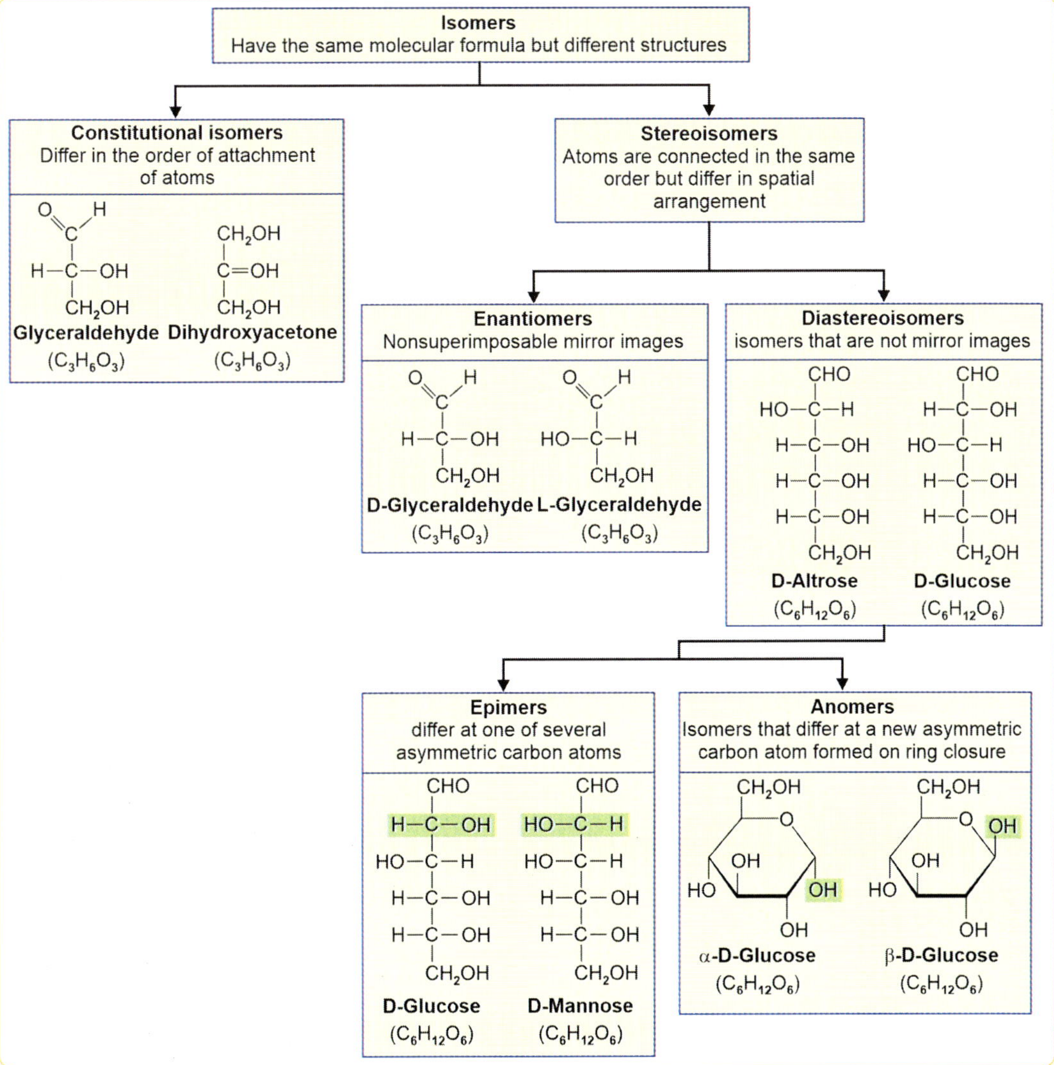

Isomers
Have the same molecular formula but different structures

Constitutional isomers
Differ in the order of attachment of atoms

Glyceraldehyde Dihydroxyacetone
$(C_3H_6O_3)$ $(C_3H_6O_3)$

Stereoisomers
Atoms are connected in the same order but differ in spatial arrangement

Enantiomers
Nonsuperimposable mirror images

D-Glyceraldehyde L-Glyceraldehyde
$(C_3H_6O_3)$ $(C_3H_6O_3)$

Diastereoisomers
isomers that are not mirror images

D-Altrose   D-Glucose
$(C_6H_{12}O_6)$ $(C_6H_{12}O_6)$

Epimers
differ at one of several asymmetric carbon atoms

D-Glucose   D-Mannose
$(C_6H_{12}O_6)$ $(C_6H_{12}O_6)$

Anomers
Isomers that differ at a new asymmetric carbon atom formed on ring closure

α-D-Glucose   β-D-Glucose
$(C_6H_{12}O_6)$ $(C_6H_{12}O_6)$

Mutarotation: The Interconversion of Anomers

Crystalline glucose is in the α-D-glucopyranose form. When dissolved in water, the ring structure opens up to form linear chain and closes again to from β-D-glucopyranose. After the final equilibrium, 36% of α anomer and 64% of β anomer exist in the solution.

This spontaneous anomeric interconversion in solution is known as mutarotation.

Mutarotation is accelerated by acids/bases. Mutarotation is prevented when the anomeric carbon is involved in the formation of glycosidic bond (Fig. 4).

```
┌──────────────────────────────────────────────────────────┐
│        Crystalline glucose (α-D-glucopyranose) (+112°)     │
└──────────────────────────────────────────────────────────┘
                    │  Dissolved in water
┌──────────────────────────────────────────────────────────┐
│ Spontaneous conversion of α anomer to β → change in        │
│ optical rotation (+18.7°)                                   │
└──────────────────────────────────────────────────────────┘
                    │
┌──────────────────────────────────────────────────────────┐
│ Final Equilibrium reached → optical rotation stable (+52.7°).│
└──────────────────────────────────────────────────────────┘
```

This is not the average of the above 2 values as equilibrium mixture is NOT equimolar of α (36%) and β (64%).

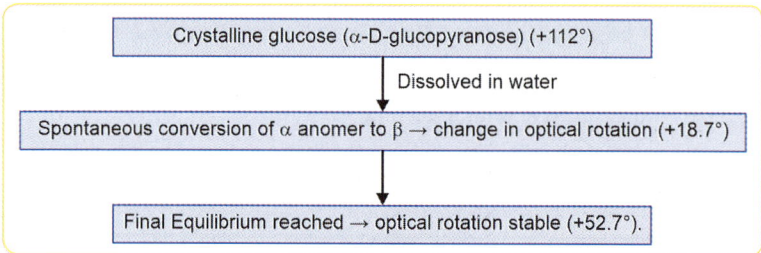

Fig. 4: Optical rotation of α-D-glucose is +112° and that of β-D-glucose is +18.7°. As the final equilibrium contains 36% of α anomer and 64% of β anomer, optical rotation will be (112 X 0.36) + (18.7 X 0.64). If you do the maths, you will get the value around +52°.

Cyclic forms can Exist Either as Chair or Boat Conformation

Cyclic forms of sugars do not exist as shown in Haworth projection. They actually exist either in chair or boat conformation. **Chair form is the most stable form**. More details are not required for undergraduate level.

DERIVATIVES OF MONOSACCHARIDES

Monosaccharides are not just energy sources. They also undergo various chemical modifications in the body to produce derivatives.

- ❏ Aldonic acid: Gluconic acid
- ❏ Uronic acid: Glucuronic acid
- ❏ Alditols: Sorbitol
- ❏ Deoxy sugars: Fucose
- ❏ Amino sugar: N-acetyl glucosamine
- ❏ N-acetyl Neuraminic acid (Sialic acid)

For ease of understanding, let me differentiate the first three derivates listed above in a tabular form.

Gluconic acid	Glucuronic acid	Sorbitol
Oxidation of C_1 aldehyde group to carboxylic acid	**Oxidation** of C_6 hydroxyl group to carboxylic acid	**Reduction** of C_1 aldehyde group will produce the polyol, sorbitol with 6 alcohol groups.

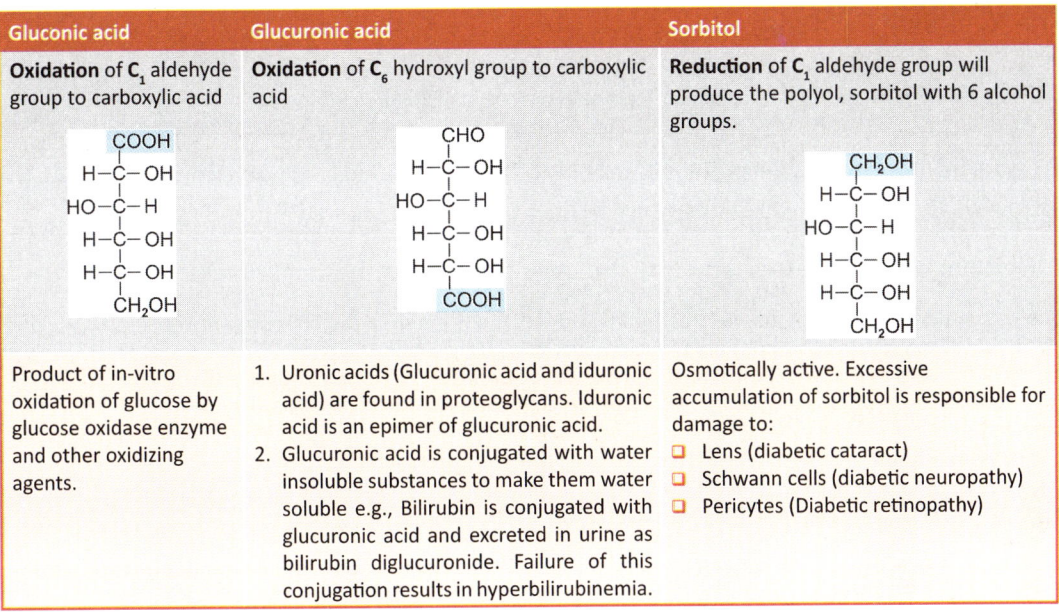

Product of in-vitro oxidation of glucose by glucose oxidase enzyme and other oxidizing agents.	1. Uronic acids (Glucuronic acid and iduronic acid) are found in proteoglycans. Iduronic acid is an epimer of glucuronic acid. 2. Glucuronic acid is conjugated with water insoluble substances to make them water soluble e.g., Bilirubin is conjugated with glucuronic acid and excreted in urine as bilirubin diglucuronide. Failure of this conjugation results in hyperbilirubinemia.	Osmotically active. Excessive accumulation of sorbitol is responsible for damage to: ❑ Lens (diabetic cataract) ❑ Schwann cells (diabetic neuropathy) ❑ Pericytes (Diabetic retinopathy)

Inositol is a Polyhydroxy Alcohol

❑ Inositol is nothing but cyclohexane-1, 2, 3, 4, 5, 6-hexol *(Draw cyclohexane. Add OH group to every carbon) (Fig. 5)*.
❑ Inositol is synthesized in the body from glucose.
❑ Phosphatidylinositol is a component of the cell membrane.
❑ Phytic acid is nothing but inositol hexaphosphate. It interferes with the absorption of many minerals and is considered as an antinutrient.

Fig. 5: Compare the structures of inositol and glucose. Glucose is a polyhydroxy aldehyde. Inositol is a polyhydroxy alcohol

Amino Sugars are Components of Glycoconjugates

❑ In amino sugars, one hydroxyl group is replaced with an amino group, e.g. Glucosamine (GlcN).
❑ Amino sugars are usually acetylated, e.g. N-Acetylglucosamine (GlcNAc).
❑ **Fructose-6-phosphate is the immediate precursor of amino sugars.**
❑ Amino sugars are found in glycoconjugates like glycoproteins, proteoglycans and gangliosides

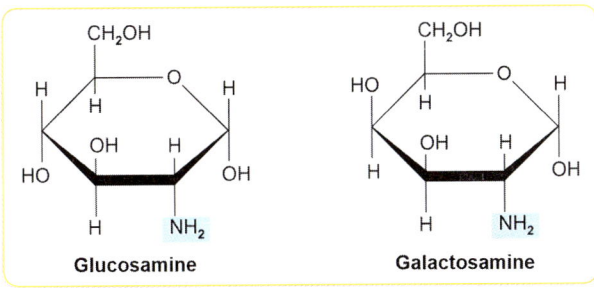

Deoxy Sugars

Deoxy sugars are types of sugars that have a hydroxyl group replaced with a hydrogen atom. For example, deoxy ribose is ribose in which OH group at the 2nd carbon is replaced with hydrogen.

Deoxy sugar	Role
Deoxyribose	Constituent of DNA
Fucose	Constituent of glycoproteins e.g. blood group substance

Fucose is a Deoxy Sugar and it is the only L-sugar found in Humans

Fucose is a deoxy sugar with a methyl group instead of CH_2OH group in the 6th position. We learnt that all the sugars in the body are D-sugars. Fucose is an exception.

GDP-fucose is synthesized from GDP-mannose. Fucosyl transferase enzymes transfer fucose to various glycoconjugates like glycoproteins and proteoglycans, e.g. ABO blood group antigens contain L-fucose.

People with Bombay blood group lack L-fucosyl transferase enzyme.

L-Fucose

N-acetyl Neuraminic acid (Sialic Acid)

- ❏ Neuraminic acids are 9-carbon compounds derived from the condensation of **pyruvate and D-mannosamine**.
- ❏ N or O-substituted derivatives of neuraminic acid are known as sialic acids.
- ❏ N-acetyl Neuraminic acid (NANA) is the predominant sialic acid in our body.
- ❏ NANA is a component of glycoproteins and gangliosides.
- ❏ Loss of NANA leads to uptake of plasma glycoprotein via asialoglycoprotein receptor of liver and destruction of glycoprotein.
- ❏ Ganglioside classification is based on the number of sialic acid residues – GM (Monosialo ganglioside), GD (Disialo ganglioside)

Harrison's Corner

Human Influenza virus binds to a receptor in the airway epithelium containing sialic acid (neuraminic acid) through its outer membrane hemagglutinin. During the maturation of the virus, neuraminidase enzyme of virus destroys sialic acid of the infected host cell's plasma membrane so that newly-released virus does not get stuck to the dying cell. (*Ref. Harrison 20th Ed., p. 1329*)
Oseltamivir and Zanamivir are inhibitors of viral neuraminidase.

GLYCOSIDIC LINKAGE

Remember the hemiacetal formation we discussed under the cyclization of sugars. When a hemiacetal reacts with another alcohol, it produces acetal. When a hemiketal reacts with another alcohol, it produces ketal. Formation of acetal and ketal is the basis of a glycosidic bond.

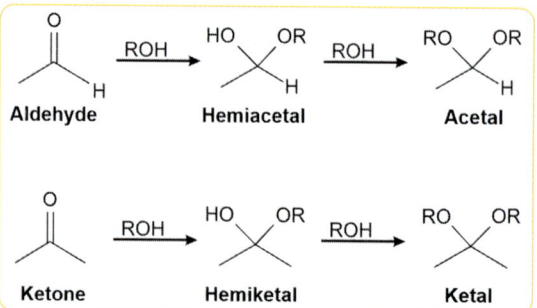

A glycosidic bond is a covalent bond formed between the hemiacetal or hemiketal group of a sugar and the hydroxyl group of another sugar or non-sugar compounds (aglycone) or amino/amide group. Glycosidic bond can be α or β depending upon the configuration of the sugar that donates the hemiacetal group.

Types of Glycosidic Bond

Based on the Linkage

- ❑ **O-glycosidic bonds:** Linkage between all the disaccharides and polysaccharides, O-glycoproteins
- ❑ **N-glycosidic bonds:** Linkage between sugar and base in nucleosides, N-glycoproteins

Based on the Anomeric Carbon of Hemiacetal Donor

- ❑ α glycosidic bond - hemiacetal contains α anomer.
- ❑ β glycosidic bond – hemiacetal contain β anomer.

> *Remember*
>
> - ❑ Glycosyltransferases are the enzymes that form glycosidic bonds.
> - ❑ Glycosidases are the enzymes that hydrolyse the glycosidic bonds.
> - ❑ Glycation is the non-enzymatic addition of sugars.

Therapeutic Glycosides

Digoxin

- ❑ Digoxin is a cardiac glycoside derived from *Digitalis purpurea* (foxglove).
- ❑ It is an inhibitor of $Na^+ K^+$ ATPase - used mainly in atrial fibrillation with congestive cardiac failure and also in other cardiac failures.
- ❑ Ouabain is a poison that also inhibits the $Na^+ K^+$ ATPase pump.
- ❑ Structure of adenosine is shown. Ribose and adenine are linked via β-N (Fig. 6) glycosidic linkage.

Fig. 6: Structure of adenosine

DISACCHARIDES

Disaccharides are two monosaccharides covalently joined together by an O-glycosidic linkage. Based on the donor of the hemiacetal or hemiketal group, the glycosidic bond of disaccharide can be classified as glucoside, galactoside, fructoside, etc.

- ❑ Glucoside is a disaccharide where hemiacetal group for glycosidic bond is given by glucose, e.g. Maltose, Trehalose
- ❑ Galactoside is disaccharide where hemiacetal group for glycosidic bond is given by galactose, e.g. Lactose.

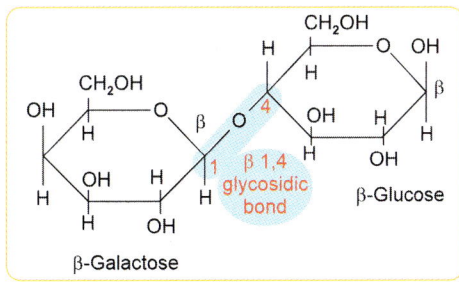

Fig. 7: Structure of lactose

- ❑ Structure of lactose is shown. Hemiacetal donor is beta galactose. That is why lactose is a beta galactoside (Fig. 7).

Chapter 9 Chemistry of Carbohydrates

Disaccharides	Components	Linkage	Role
Sucrose (cane sugar)	Glucose and Fructose	α (1→2)	Non-reducing sugar; Rare deficiency of sucrase leads to sucrose intolerance
Maltose (malt sugar)	Glucose and Glucose	α (1→4)	Amylase hydrolyse linear chain of starch to generate maltose
Isomaltose	Glucose and Glucose	α (1→6)	Amylase hydrolyse branch point of starch to generate isomaltose
Trehalose (mushroom sugar)	Glucose and Glucose	α (1→1)	Non-reducing sugar; Rare deficiency of trehalase leads to trehalose intolerance.
Lactose (milk sugar)	Galactose and Glucose	β (1→4)	Deficiency of lactase leads to lactose intolerance
Lactulose	Galactose and Fructose	β (1→4)	Synthetic sugar; can't be digested by human enzymes; fermented by intestinal bacteria. used as a mild-laxative.

Sucrose, Trehalose and Raffinose: Non-reducing Sugars

❑ Fehling's and Benedict's tests are based on the reducing property of sugars.
❑ Reducing property of sugars is due to the presence of **free aldehyde or ketone group (anomeric carbons).**
❑ All monosaccharides are reducing sugars.
❑ Disaccharides with free aldehyde or ketone group are reducing sugars, e.g. Lactose, Maltose

C$_1$ of glucose and C$_2$ of fructose condense to form sucrose. As there is no free aldehyde or keto group present in sucrose, sucrose is a non-reducing sugar.

Trehalose, in which 2 glucose molecules are linked by α (1→1) linkage, is also a non-reducing disaccharide.

Raffinose is a non-reducing trisaccharide.

Look at the structure of sucrose – glycosidic bond involves the anomeric carbons of both glucose and fructose. Anomeric carbon of glucose is C$_1$. Anomeric carbon of fructose is C$_2$ (Fig. 8).

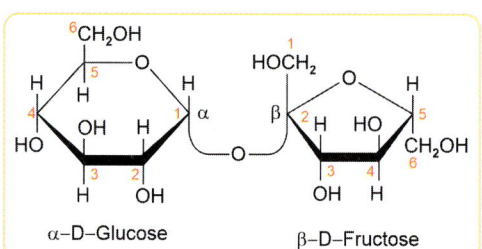

Fig. 8: Structure of sucrose

OLIGOSACCHARIDES

Oligosaccharides are short chains of carbohydrates (2 to 10) linked by glycosidic linkage. In our body, oligosaccharides are usually found in association with other macromolecules like lipid and protein. Oligosaccharides covalently linked to proteins are known as glycoproteins. Oligosaccharides covalently linked to lipids are known as glycolipids.

Fermentable Oligosaccharides

These are the oligosaccharides which can't be digested by the human digestive enzymes but can be fermented by the intestinal flora. Fructo-oligosaccharides (FOS) are short chains of fructose. Galacto-oligosaccharides are short chains of galactose.

arrison's
Corner

Fermentable oligosaccharides, disaccharides, monosaccharides, and polyols (FODMAPs) are poorly absorbed by the small intestine and fermented by bacteria in the colon to produce gas and osmotically active carbohydrates. A diet low in FODMAPs has been shown to be helpful in IBS patients. (*Ref. Harrison 20th Ed., p. 2280*)

POLYSACCHARIDES

Polysaccharides are large chains of sugar (>10) units joined together. They may be linear or branched. The sugar units may be same or different.

Cellulose is a purely linear polysaccharide while glycogen, starch and dextran are branched polysaccharides.

❑ Homoglycans (homopolysaccharides) are polymers made up of only one type of monosaccharide.
  ○ Polymers of glucose are known as glucosans, e.g. Starch and cellulose
  ○ Polymers of fructose are known as fructosans, e.g. Inulin
❑ Heteroglycans (heteropolysaccharides) are polymers made up of more than one type of monosaccharide, e.g. Glycosaminoglycans

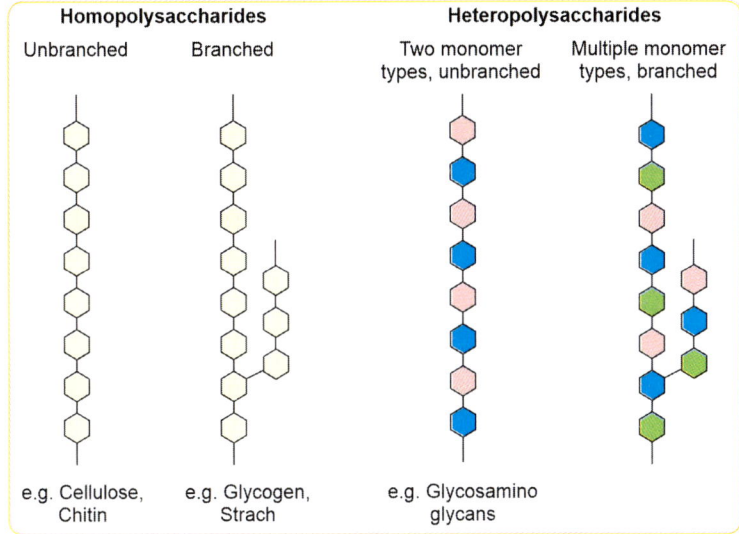

Polysaccharides can be also classified into structural and storage polysaccharides.
❑ Cellulose and Chitin are the structural polysaccharides of plants and insects respectively.
❑ Starch and glycogen are storage polysaccharides of plants and animals respectively.

Polysaccharide	Monomeric unit	Linkage	Role
Cellulose	Glucose	$\beta\,(1 \rightarrow 4)$	Structural polysaccharide of plants Dietary fiber
Starch	Glucose	$\alpha\,(1 \rightarrow 4)$ and less $\alpha\,(1 \rightarrow 6)$	Storage polysaccharide of plants Major carbohydrate in diet
Glycogen	Glucose	$\alpha\,(1 \rightarrow 4)$ and more α $(1 \rightarrow 6)$ compared to starch	Storage polysaccharide of animals
Dextran	Glucose	$\alpha\,(1 \rightarrow 6)$ and $\alpha\text{-}(1 \rightarrow 3)$	Used as plasma expander, lubricant in eye drops and vaccine preservative
Chitin	N-acetyl glucosamine	$\beta\,(1 \rightarrow 4)$	Structural polysaccharide of insects and fungi
Inulin	Fructose	$\beta\,(1 \rightarrow 2)$	Dietary fiber
Pectin	D-Galacturonic acid	$\alpha\,(1 \rightarrow 4)$	Dietary fiber

Human body doesn't contain any enzyme to break down the β-glucosidic bonds. These polysaccharides constitute the dietary fibre. *(Dietary fibers will be elaborately discussed in Chapter 32)* *(Refer to page no. 503).*

Chapter 9 Chemistry of Carbohydrates

Inulin	☐ Inulin is a homopolysaccharide of fructose (fructosan) present in garlic, onion, etc.
	☐ Breakdown of inulin by bacteria produces fermentable fructo oligosaccharides which may cause discomfort to patients with irritable bowel syndrome.
	☐ Injected inulin is neither absorbed nor secreted by renal tubules.
	☐ Inulin clearance provides a near-accurate estimate of Glomerular Filtration Rate.
Chitin	☐ Chitin is a homopolysaccharide made up of N-Acetylglucosamine linked β (1 → 4) glycosidic linkage.
	☐ It is a structural polysaccharide found in exoskeleton of insects and cell wall of fungi.

Glycogen structure will be described in Chapter 10.4 (Refer to page no. 141).
Glycoproteins and Proteoglycans will be discussed in detail in Chapter 41 (Refer to page no. 603).

Competency

☐ BI3.1   Discuss and differentiate monosaccharides, disaccharides and polysaccharides giving examples of main carbohydrates as energy fuel, structural element and storage in the human body

Key Points

- ☐ Dihydroxy acetone is the only carbohydrate without a chiral carbon.
- ☐ L-fucose is the only L-sugar present in the body.
- ☐ Enantiomers are structures which are non-superimposable mirror images of each other, e.g. D and L forms.
- ☐ Epimers differ in the orientation of substituents around one of their chiral carbons. Mannose is a C–2 epimer of glucose. Galactose is a C–4 epimer of glucose. But, Mannose and galactose are diastereoisomers of each other.
- ☐ When the C_5 OH group of glucose reacts with the C_1 aldehyde group, an intramolecular hemiacetal is formed, and the linear glucose gets cyclized.
- ☐ Anomeric carbon of glucose is C_1 and that of fructose is C_2.
- ☐ α-D-glucose and β-D-glucose are anomers.
- ☐ α-D-glucopyranose is the naturally occurring form of glucose and it is the one which can be acted upon by our body enzymes.
- ☐ Mutarotation is the spontaneous interconversion of α and β anomers in solution.
- ☐ Invert sugar is the equimolar mixture of D-glucose and D-fructose formed by hydrolysis of sucrose.
- ☐ Inulin is a polymer of fructose linked by β (1→2) linear linkage. This β linkage can't be broken in the body. So, inulin like fructans are prebiotics because of the β configuration of anomeric carbon.
- ☐ Our body can break β-galactosidic linkage of lactose but not β-glucosidic linkage of cellulose.
- ☐ Trehalose is a non-reducing disaccharide found in mushrooms. The enzyme trehalase is present in brush borders of the small intestine.
- ☐ Dextran is a plasma volume expander. Dextrin is a derivative of starch hydrolysis. Dextrose is the dextro rotatory D-Glucose.
- ☐ 1 mole of glucose = 180 gram of glucose (5 millimoles/L of glucose is equal to 90 mg/dL)
- ☐ Lyxose is a pentose sugar predominantly found in cardiac muscles.

SELF-ASSESSMENT

Short Answer Questions

1. Differentiate between epimers and anomers with the help of suitable examples.

2. What is a glycosidic bond? Name a therapeutic drug which is a glycoside.

3. Describe the structure of glycogen. How is it different from the structure of starch? How is this useful for the function of glycogen?

4. Why can't humans digest cellulose?

5. Name two non-reducing sugars. Explain why they are non-reducing.

6. Name any two disaccharides and mention the bond connecting the individual monosaccharide.

7. Classify polysaccharide with suitable examples.

8. What are the monomeric units of inulin? Mention the dietary sources and clinical utility.

Multiple Choice Questions

1. **Which of the following is NOT seen in the human body?**
   a. L-Fucose
   b. L-Fructose
   c. D-Glucose
   d. D-Fructose

2. **Which of the following pair is an example of enantiomers?**
   a. D-Galactose and L-Glucose
   b. D-Glucose and D-Galactose
   c. D-Glucose and L-Galactose
   d. D-Glucose and L-Glucose

3. **α-D-glucose and β-D-glucose are _____ of each other:**
   a. Epimers
   b. Anomers
   c. Diastereoisomers
   d. Functional isomers

4. **Mutarotation is due to:**
   a. Interconversion of glucose and fructose
   b. Hydrolysis of amylase
   c. Conversion of glucose to galactose
   d. Anomeric interconversion of α and β D-glucose

5. **Select the wrong statement:**
   a. Mannose and galactose are epimers of each other
   b. Xylulose and Ribulose are epimers of each other
   c. Glucuronic acid and Iduronic acid are epimers of each other
   d. Glucose and Galactose are epimers of each other

6. **Hydrolysis of which of these molecules will produce 4 molecules of glucose in total?**
   a. Maltose and Sucrose
   b. Maltose and Trehalose
   c. Sucrose and Lactose
   d. Lactose and Maltose

7. **Alpha amylase acts on:**
   a. $\alpha 1 \rightarrow 4$ bond
   b. $\alpha 1 \rightarrow 6$ bond
   c. $\beta 1 \rightarrow 4$ bond
   d. $\beta 1 \rightarrow 6$ bond

8. **All of the following are heteropolysaccharides, except:**
   a. Chitin
   b. Heparin
   c. Hyaluronic acid
   d. Chondroitin sulfate

9. **The structural polysaccharide chitin is a polymer of which of the following:**
   a. Galactosamine
   b. Glucosamine
   c. N-Acetyl galactosamine
   d. N-Acetyl glucosamine

10. **A deoxy-sugar characteristic of blood group antigens is:**
    a. L-Raffinose
    b. L-Fucose
    c. 2-deoxy-D-ribose
    d. 6-deoxy-L-mannose

11. **Which is the predominant monomeric unit of pectin?**
    a. Glucose
    b. Galactose
    c. Glucuronic acid
    d. Galacturonic acid

12. **Which of the following cardiac glycosides inhibits Na⁺K⁺ATPase pump?**
    a. Streptomycin
    b. Diosgenin
    c. Digoxin
    d. Auxin

13. **Aglycone in digoxin is:**
    a. Protein
    b. Steroid
    c. Alkaloid
    d. Purine

14. **Neuraminic acid is derived from:**
    a. Lysine and methionine
    b. Pyruvate and Mannosamine
    c. Pyruvate and methionine
    d. Lysine and mannosamine

ANSWERS

1. b.	2. d.	3. b.	4. d.	5. a.	6. b.
7. a.	8. a.	9. d.	10. b.	11. d.	12. c.
13. b.	14. b.				

10 CHAPTER

Carbohydrate Metabolism

10.1 INTRODUCTION

Chapter Outline

- Glucose Transporters (GLUT)
  - Glucose Transporters – Facilitated Diffusion
  - GLUT versus SGLT
- The fate of Glucose Entered Inside the Cell
- Pseudogene

ENTRY OF GLUCOSE INTO THE CELLS

Glucose transporters (GLUT) mediate the facilitated diffusion of glucose across the membrane. They traverse the membrane 12 times.

Type	Organ distribution	Function and other points
Glucose transporters—facilitated diffusion		
GLUT 1	Brain, RBC, placenta, colon	<ul><li>Basal glucose uptake</li><li>Receptor for human T-cell leukemia virus type 1 (HTLV-1)</li></ul>
GLUT 2	Liver, pancreatic β cells, kidney, intestine	<ul><li>Glucose sensor function</li><li>Defect leads to Fanconi Bickel syndrome</li></ul>
GLUT 3	Brain, kidney, placenta	Basal glucose uptake
GLUT 4	Skeletal and cardiac muscle, adipose tissue	Insulin dependent uptake
GLUT 5	Small intestine, sperm	Fructose transporter
GLUT 6	Product of pseudogene*	
GLUT 12	Heart, skeletal muscle, brown adipose tissue, and prostate	Basal and insulin dependent uptake
Sodium-dependent glucose transporter—secondary active transport		
SGLT 1	Small intestine > kidney	<ul><li>Uptake of glucose/galactose against a concentration gradient</li><li>Phlorizin inhibits both SGLT 1 and 2</li></ul>

Contd…

Type	Organ distribution	Function and other points
SGLT 2	Kidney > small intestine	❑ Uptake of glucose against a concentration gradient ❑ Inhibited by Gliflozin

Pseudogene is a mutated version of functional gene arising in due course of evolution.

▶ *Video on Pseudogene: https://youtu.be/eY0UldK_xoE or scan QR code*

Previously described endoplasmic reticulum transporter GLUT7 for glucose-6-phosphate is actually a hexose transporter in the plasma membrane. 2° Active Transport is explained in Chapter 36 (Refer to page no. 567).

Glucose entered entering into the cell can have various fates depending upon the type of cell and the energy status of the cell.

Pathway	Biochemical role
Glycolysis	Major oxidative pathway of glucose—pyruvate, NADH and ATP are produced
Hexose monophosphate shunt	❑ Alternative oxidative pathway for glucose ❑ Production of NADPH and pentose phosphates ❑ There is no production of ATP
Sorbitol pathway	Conversion of glucose to fructose
Uronic acid pathway	❑ Alternate oxidative pathway for glucose ❑ Production of glucuronic acid
Glycogen synthesis	Storage of glycogen for short-term use

When glucose is needed, it can be produced by the following pathways:

Glycogenolysis	**Liver:** Release of glucose in between meals **Muscle:** Release of glucose for its own use during muscular activity.
Gluconeogenesis	Synthesis of glucose from non-carbohydrate sources during prolonged fasting and starvation

Let us study each pathway. We will begin with glycolysis which has been described under Chapter 10.2 (Refer to page no. 119).

⊂ompetency

❑ Bl3.4   Define and differentiate the pathways of carbohydrate metabolism (glycolysis, gluconeogenesis, glycogen metabolism, HMP shunt).

Key Points

❑ Carbohydrates constitute the bulk of our routine diet.

❑ Starch is the major dietary carbohydrate.

❑ Amylase breaks the α (1,4) glycosidic bonds of starch to produce glucose and maltose.

❑ Small intestinal brush border enzyme maltase breaks maltose into two molecules of glucose.

SELF-ASSESSMENT

Short Answer Questions

1. In a tabular format mention the Glucose transporters (GLUTs), their tissue distribution and functional role.

2. Justify the statement – 'Glucose entering into the cell can have various fates depending upon the type of cell and the energy status'.

Multiple Choice Questions

1. Which of the following glucose transporter in myocytes requires insulin for translocation to the membrane?

   a. GLUT 1   b. GLUT 2

   c. GLUT 3   d. GLUT 4

2. After overnight fasting, glucose transporters of which of the following organ will be reduced?

   a. Brain   b. RBC

   c. Liver   d. Skeletal muscle

3. Which of the following transporter is specific for fructose?

   a. GLUT 3
   b. GLUT 5
   c. GLUT 7
   d. GLUT 12

4. Which of the following glucose transporters are involved in the glucose uptake in the brain?

   a. GLUT 1 and GLUT 2
   b. GLUT 1 and GLUT 3
   c. GLUT 2 and GLUT 3
   d. GLUT 3 and GLUT 4

5. Transport by the sodium-dependent glucose transporter is an example of:

   a. Facilitated diffusion
   b. Primary active transport
   c. Secondary active transport
   d. Simple diffusion

ANSWERS

1. d.	2. d.	3. b.	4. b.	5. c.

10.2 GLYCOLYSIS

Glycolysis also known as Embden-Meyerhof pathway is the major oxidative pathway for the metabolism of glucose.

SALIENT FEATURES

- Glycolysis is the breakdown of glucose into pyruvic acid
- Only pathway that can operate in anaerobic conditions—evolutionarily ancient
- Operates only in the cytoplasm
- Glycolysis is the only source of energy for RBC, cornea, lens, and major source of energy for the brain

CONCEPT CORNER

Why glycolysis, why not glucolysis?
Glycolysis is breakdown of sugars. But, as you see glucose is the sugar broken down in the pathway. Then, why are we not calling the pathway as glucolysis? The reason is that sugars like galactose can be converted into glucose. Fructose enters the glycolytic pathway bypassing the regulatory step.

STEPS OF GLYCOLYSIS

Glycolysis can be visualized in two stages:
i. Energy investment stage
ii. Energy generating stage

Energy Investment Stage

- As the name suggests, this stage involves the hydrolysis of 2 ATPs.
- Glucose is split into two 3-carbon molecules – Dihydroxyacetone phosphate (DHAP) and Glyceraldehyde-3-phosphate (G3P).
- DHAP is converted to G3P. So, two molecules of G3P enter the energy generating stage.

Energy Generating Stage

- Each molecule of G3P produces two ATPs in the consequent oxidation reactions.
- At the end, two molecules of pyruvate are produced.

Chapter 10 Carbohydrate Metabolism

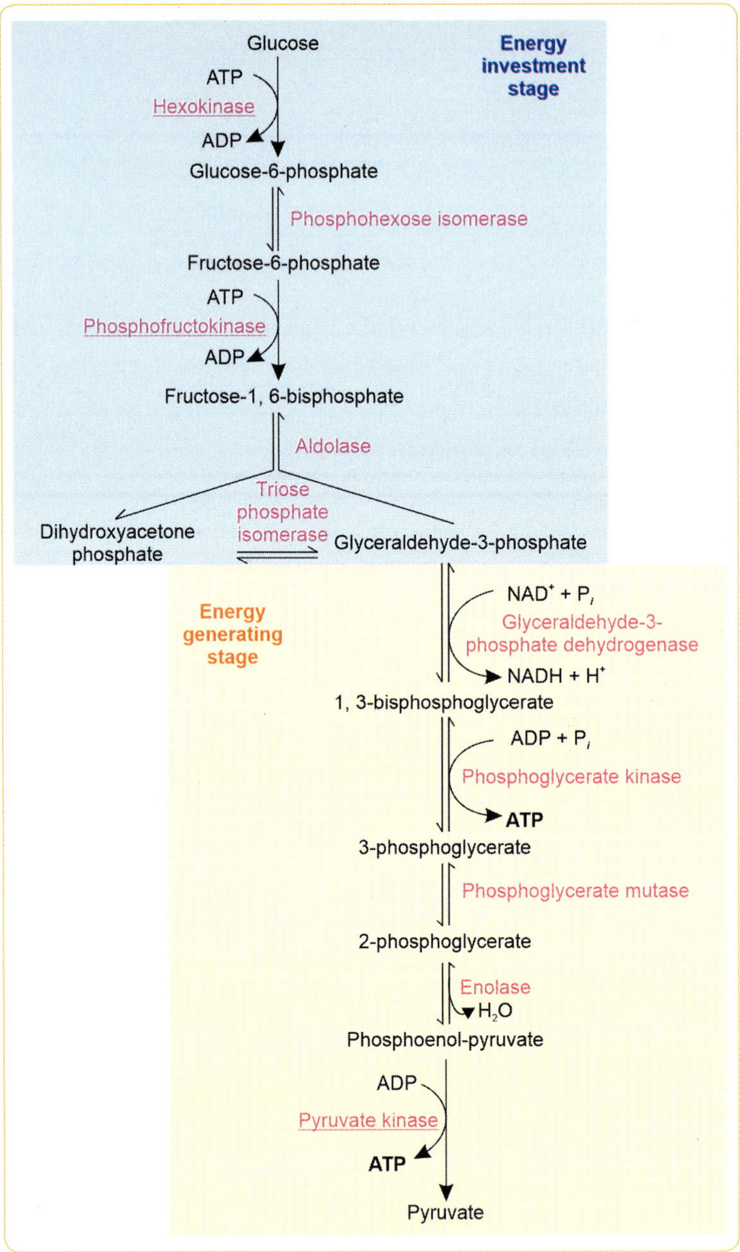

Glucose

ATP ⤵

Hexokinase

Energy investment stage

ADP ⤴

Glucose-6-phosphate

Phosphohexose isomerase

Fructose-6-phosphate

ATP ⤵

Phosphofructokinase

ADP ⤴

Fructose-1, 6-bisphosphate

Aldolase

Triose phosphate isomerase

Dihydroxyacetone phosphate ⇌ Glyceraldehyde-3-phosphate

Energy generating stage

NAD⁺ + P$_i$

Glyceraldehyde-3-phosphate dehydrogenase

NADH + H⁺

1, 3-bisphosphoglycerate

ADP + P$_i$

Phosphoglycerate kinase

ATP

3-phosphoglycerate

Phosphoglycerate mutase

2-phosphoglycerate

Enolase
H$_2$O

Phosphoenol-pyruvate

ADP

Pyruvate kinase

ATP

Pyruvate

We will discuss the reactions of glycolysis thoroughly. This will help you to understand enzymes and general aspects of metabolism.

First, apply your knowledge of enzymes to identify the class of each and every enzyme of glycolytic pathway.

Enzyme	Reaction catalyzed	Enzyme class
Hexokinase	Transfer of phosphoryl group from ATP to OH group of 6th carbon of glucose	(Phospho) transferase—EC2
Phosphohexose isomerase	Reversible conversion of Glucose-6-phosphate to Fructose-6-phosphate	Isomerase—EC5

Section II Intermediary Metabolism

Contd...

Enzyme	Reaction catalyzed	Enzyme class
Phosphofructokinase 1	Transfer of phosphoryl group from ATP to OH group of 1st carbon of fructose-6-phosphate to produce fructose-1, 6-bisphosphate	(Phospho) transferase—EC2
Aldolase A	❑ Aldol cleavage—breaking the bond between C3 and C4 of F-1, 6,-BP to produce an aldose and a ketose ❑ Aldol condensation occurs when the reaction is reversed	Lyase—EC4
Triose phosphate isomerase	Reversible conversion of aldo and keto triose phosphates	Isomerase—EC5
Glyceraldehyde-3-phosphate dehydrogenase	Oxidation of aldehyde to carboxylic acid; electrons are accepted by NAD^+	Oxidoreductase—EC1
Phosphoglycerate kinase	Reversible transfer of phosphate group from 1, 3 BPG to ADP	(Phospho) transferase—EC2
Phosphoglycerate mutase	Shift of phosphate group from hydroxyl C_3 to C_2	Isomerase—EC5
Enolase	Dehydration—removal of water	Lyase—EC4
Pyruvate kinase	Transfer of phosphate from phosphoenolpyruvate to ADP	(Phospho) transferase—EC2

DETAILED DESCRIPTION OF STEPS OF GLYCOLYSIS

Hexokinase is the Flux-Generating Reaction of Glycolysis

$$Glucose + ATP-Mg^{++} \rightarrow Glucose-6-P + ADP-Mg^{2+}$$

Hexokinase has the low K_m value (high affinity) for glucose. Thus, it converts all the available glucose to glucose-6-phosphate → generation of flux (constant flow).

Conversion of Glucose to G6P Serves Two Functions

1. Glucose is activated.
2. Addition of phosphate group gives negative charge to glucose-6-phosphate. Charged molecules cannot freely cross the cell membrane. Thus, G6P is trapped inside for further metabolism.

Have you noticed the magnesium in the reaction? Recall the role of metals in catalysis (Chapter 5) *(Refer to page no. 58-59)*.
1. Magnesium ion neutralizes some negative charges of ATP and thus prevents unnecessary, interfering ionic interaction between ATP and other enzyme residues during catalysis.
2. Magnesium ion helps to properly orient the ATP in the catalytic site.
3. Magnesium ion provides additional point of contact with the active site residues thus providing more binding energy. *(Remember: Lowering of activation energy is contributed by binding energy)*

Hexokinase Enzyme and the Induced Fit Mechanism

Hexokinase enzyme utilizes the induced fit mechanism. Hexokinase transfers phosphoryl group of ATP to OH group of glucose or other hexose sugars. You know that biochemical reactions are happening in the aqueous environment and water (HOH) also possess the OH group. If water is present in the active site, phosphoryl group of ATP may be transferred to water. This will be a total waste of ATPs.

To avoid this, hexokinase employs the induced fit mechanism (Chapter 5) *(Refer to page no. 51)*. When glucose binds to the enzyme, conformational change takes place and water is excluded from the active site. This is true for many enzymes. **Water is excluded from the active site unless it is one of the reactants (desolvation of active site).**

Glucose + ATP →(Hexokinase) Glucose-6-phosphate + ADP + H⁺

Glucose → **Glucose-6-phosphate**

Specificity of Hexokinase

❏ Hexokinase can phosphorylate not only glucose but also D-fructose, 5-keto-D-fructose, 2-deoxy-D-glucose, D-mannose and D-glucosamine.
❏ Glucokinase, an isoenzyme of hexokinase has high K_m value. Concentration of glucose is higher in the body compared to other hexoses. So, Glucokinase acts mainly on glucose.

	Hexokinase	Glucokinase (Hexokinase type IV)
Location	All the tissues	Liver and pancreatic β cell
K_m for glucose	0.1 mM (1.8 mg/dL)	10 mM (180 mg/dL)
Relationship between glucose concentration and reaction rate	Hyperbolic curve	Sigmoidal curve
Feedback inhibition by glucose-6-phosphate	Yes	No
Function	Basal metabolism of glucose	Glucose sensor—uses glucose only at high concentration
Induction by insulin	No	Yes

Need of Phosphohexose Isomerase Reaction

Glucose-6-phosphate →(Phosphoglucoisomerase)← Fructose 6-phosphate

Glucose-6-phosphate → **Fructose 6-phosphate**

Gluose-6-phosphate is in 6-membered pyranose ring with 5 carbons in the ring. Fructose-6-phosphate is in 5-membered furanose ring with 4 carbons in the ring. 1ˢᵗ carbon is also out of the ring and can easily be acted upon by the next enzyme in the pathway.

Phosphofructokinase 1 (PFK-1) Catalyzes the Committed Step of Glycolysis

❏ PFK-1 catalyzed reaction is the *committed* step of glycolysis as it is the *first irreversible reaction unique to glycolysis.*
❏ Glucose-6-phosphate (product of hexokinase) is not unique to glycolysis. It can enter other pathways like HMP shunt. But the product of PFK-1, i.e. Fructose-1,6 bisphosphate is unique to glycolysis.

Fructose 6-phosphate $\xrightarrow{\text{PFK-1}}$ Fructose-1, 6-bisphosphate

$+ATP \longrightarrow + ADP + H^+$

Note: *The term bisphosphate is used when two phosphate groups are attached to two different carbon atoms.*

❑ **PFK-1 is the rate-limiting enzyme of glycolysis.** Inhibition of this enzyme significantly reduces the operation of the pathway.

❑ This is an irreversible (nonequilibrium) reaction.

❑ One high-energy phosphate bond from ATP is utilized.

Aldolase A Cleaves F-1, 6-BP

Aldolase A breaks the 6-carbon compound to two 3-carbon compounds.

Fructose-1, 6-bisphosphate $\xrightleftharpoons{\text{Aldolase}}$ Dihydroxyacetone phosphate + Glyceraldehyde 3-phosphate

Triose Phosphate Isomerase is a Catalytically Perfect Enzyme

Enzymes like triose phosphate isomerazes (TPI) mastered the art of biological catalysis. The reaction is so efficient that the enzyme is considered as "catalytically perfect". The reaction is limited only by the rate of diffusion of substrate across the enzyme's active site and thus TPI is also known as "diffusion limited enzyme".

The high-rate of catalysis achieved by catalytically perfect enzymes cannot be explained by the mechanism of catalysis we studied in the enzymology chapter. Scientists attribute the 'quantum tunneling' mechanism for this catalytic perfection. *Details of quantum tunnelling is not necessary for exam purpose.*

Dihydroxyacetone phosphate $\xrightleftharpoons{\text{Triose phosphate isomerase}}$ Glyceraldehyde 3-phosphate

Aldolase and Triose Phosphate Isomerase Reactions have Positive ΔG°

ΔG° value of the reactions catalyzed by aldolase and triose phosphate isomerase are +5.7 kcal/mol and +1.8 kcal/mol respectively. Then how come these reactions proceed forward? Remember that it is ΔG which determines the reaction, not ΔG°.

$$\Delta G = \Delta G^\circ + RT \ln K_{eq} \text{ (In-natural log)}$$

$$K_{eq} = \frac{[Products]}{[Substrates]}$$

Chapter 10 **Carbohydrate Metabolism**

Continuous removal of the products in glycolysis reduces the equilibrium constant and drives the reaction forward by making the ΔG negative.

In gluconeogenesis, the direction of reactions is reversed since the metabolic conditions favor glucose production.

Glyceraldehyde 3-phosphate Dehydrogenase Reaction Produces NADH and Adds Inorganic Phosphate to the Substrate

Glyceraldehyde 3-phosphate dehydrogenase (GAPDH) catalyzes the conversion of glyceraldehyde 3-phosphate to 1,3-bisphosphoglycerate(1,3-BPG). This reaction serves two purposes.
1. Generation of the reducing equivalent NADH from the oxidation of the aldehyde
2. Addition of inorganic phosphate to 3-phoshphoglycerate is driven by the energy released by the oxidation reaction.

Another important point to note is the requirement of NAD^+ for the proper functioning of this enzyme and the continuation of glycolysis.

NAD^+ required for the GAPDH can be regenerated by two ways:
1. Oxidation of NADH in electron transport chain—NADH gives e^- to complex I
2. Lactate dehydrogenase reaction—NADH gives its e^- to pyruvate

1,3-bisphosphoglycerate: A High-energy Phosphate Compound

Phosphoryl transfer potential of 1,3-BPG is higher than that of ATP. Thus, 1,3-BPG can transfer its phosphate group to ADP to produce ATP in the phosphoglycerate kinase catalyzed reaction.

We have already studied this substrate-level phosphorylation reaction in the Chapter 4 *(Refer to page no. 48).*

2,3 BPG is produced by *Luebering–Rapoport shunt* of glycolysis in RBCs; helps in unloading of oxygen. It compromises the ATP produced by the 1,3 BPG kinase reaction. *This has been explained in Chapter 3 (Refer to page no. 32).*

Another ATP is Produced in the Subsequent Reactions

❑ 3-phosphghlycerate is low-energy phosphate compound. It cannot transfer its phosphate to ADP. So, it is converted into a high-energy phosphate compound in two steps.
❑ First, phosphoglycerate mutase catalyses the conversion of 3-phosphoglycerate to 2-phosphoglycerate.
❑ In the second reaction, enolase removes a water molecule from 2-phosphoglycerate, creating an enol phosphate, phosphoenolpyruvate. (en—double bond; ol—alcohol group)
❑ Phosphoenolpyruvate has the highest phosphoryl-transfer potential compared to the other high-energy phosphate compounds.
❑ Pyruvate kinase catalyze the 2nd substrate-level phosphorylation reaction of glycolysis. Phosphoryl group is transferred from phosphoenolpyruvate to ADP and pyruvate and ATP are produced.
❑ Pyruvate is the end-product of glycolysis when oxygen is available and the electron transport chain is functioning. *Energy yield of complete oxidation of glucose has been already discussed in Chapter 7 (Refer to page no. 93).*

Conversion of Pyruvate to Lactate Regenerates NAD+

Glyceraldehyde-3-phosphate dehydrogenase is NAD^+ dependent. NADH produced by this enzyme is transported via shuttle systems to mitochondria to generate free NAD^+. In ETC, NADH gives electron to the complex I. Oxygen is the final acceptor of these electrons. RBC does not contain mitochondria and vigorous muscular exercise compromises oxygen availability.

Without NAD^+, Glycolysis cannot proceed.

Under these circumstances, regeneration of NAD^+ is achieved by lactate dehydrogenase (LDH) reaction.

In this reaction, production of NAD$^+$ is more important than the production of lactate. That is why NADH is not just a coenzyme. It is known as **_cosubstrate_** of the lactate dehydrogenase reaction.

The LDH reaction is reversible. Lactate produced in muscle is converted to glucose in liver via Cori cycle (discussed in gluconeogenesis).

CONCEPT CORNER

The concept of oxygen debt: After a strenuous exercise, we need to breathe hard. This extra amount of oxygen consumed is necessary to clear lactate by reconverting it to glucose. How does it happen?

Conversion of lactate to pyruvate by lactate dehydrogenase produces NADH which goes to electron transport chain. Oxygen is the final acceptor of electrons. Thus, to convert the excess lactate produced in strenuous exercise to pyruvate, we need extra oxygen to repay the oxygen debt.

REGULATION OF GLYCOLYSIS

Regulation of glycolysis and gluconeogenesis should always be studied together. So, study gluconeogenesis also. Then, come back here.

Enzymes Catalyzing Irreversible Steps of Glycolysis

- ❏ Hexokinase
- ❏ Phosphofructokinase 1 (PFK-1)
- ❏ Pyruvate kinase

Enzyme	Inhibitors	Activators
Hexokinase	❏ Glucose-6-phosphate (Feedback allosteric inhibition)	–
Phosphofructokinase 1	❏ ATP (Homotropic negative allosteric regulator) ❏ Citrate (Heterotropic allosteric inhibitor)	❏ AMP ❏ Fructose-2,6-BP (Heterotropic positive allosteric regulator) ❏ Fructose-6-Phosphate (Homotropic positive allosteric regulator)
Pyruvate kinase	❏ ATP (Feedback inhibition) ❏ Alanine (Formed from pyruvate)	Frucotse-1,6-bisphosphate (Feed forward allosteric activation)

CONCEPT CORNER

Alanine Inhibits Glycolysis
Alanine is released from skeletal muscles in between meals. Alanine is transaminated to pyruvate and converted to glucose in the liver. Glycolysis and gluconeogenesis should not happen simultaneously. This is why alanine inhibits glycolytic enzyme.

Chapter 10 Carbohydrate Metabolism

Phosphofructokinase is the Rate-limiting Enzyme of Glycolysis

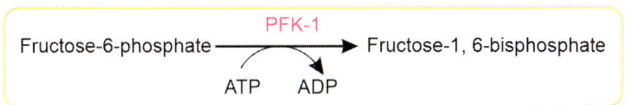

This step is regulated by
- ❏ Allosteric regulation
- ❏ Induction and repression by hormones

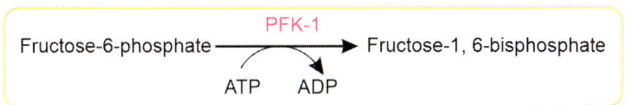

Allosteric Regulation

❑ **Activators:** Fructose-6-phosphate, 5'-AMP, fructose 2,6 bisphosphate
❑ **Inhibitors:** ATP, citrate

When ATP level is reduced, AMP is formed by adenylate kinase reaction.

$$2\,ADP \leftrightarrow ATP + AMP$$

Citrate is an indicator of high-energy status inside cell. It is also a negative allosteric modulator of PFK.

Fructose 2,6 bisphosphate is a potent activator of PFK-1 and inhibitor of gluconeogenic enzyme fructose 1,6 bisphosphatase *(discussed in detail in Chapter 10.3)* *(Refer to page no. 136)*. This helps in the reciprocal regulation of glycolysis and gluconeogenesis.

Induction and Repression

❑ Insulin, the hypoglycemic hormone increases the transcription of PFK-1 gene (induction).
❑ Glucagon, the hormone of fasting state decreases the transcription of PFK-1 gene (repression).

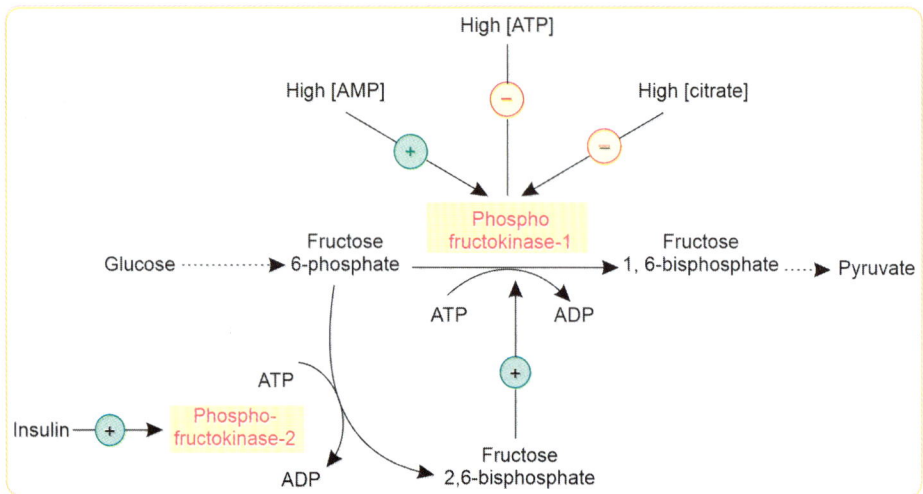

Feed Forward Activation

Product of PFK-1, i.e. Fructose 2,6 bisphosphate activates pyruvate kinase. So, the system ensures that metabolite produced in the previous step is metabolized further down in the pathway.

Inhibitors of Glycolytic Enzymes

❑ Fluoride inhibits enolase and Iodoacetate inhibits glyceraldehyde-3-phosphate dehydrogenase.

Clinical Correlation

Fluoride vials are used to collect blood samples for glucose estimation.
For estimation of glucose, blood sample is usually collected into a tube containing sodium fluoride (NaF) and potassium oxalate (1:3 mixture). In addition to the weak anti-coagulant function, *fluoride inhibits glycolytic enzyme enolase* in blood cells.

Section II Intermediary Metabolism

Arsenate Uncouples Substrate-Level Phosphorylation

Arsenate resembles phosphate and competes with phosphate incorporation in *Glyceraldehyde-3-phosphate dehydrogenase* catalyzed reaction. **Arsenate does not inhibit the progression of glycolysis**. But, no ATP can be produced from the arsenate containing molecule as it produces 3-phosphoglycerate by spontaneous hydrolysis. So, arsenate is the uncoupler of substrate-level phosphorylation.

Pyruvate Kinase Deficiency: A Common Cause of Inherited Hemolytic Anemia

❑ Pyruvate kinase reaction produces ATP.
❑ ATP is important to maintain $Na^+K^+ATPase$ in RBCs.
❑ ↓ ATP leads to failure of pumps → Osmotic imbalance → RBC swell and lyse.

PYRUVATE DEHYDROGENASE LINKS GLYCOLYSIS TO TCA CYCLE

Pyruvate dehydrogenase enzyme complex is a mitochondrial multienzyme complex that catalyzes the oxidative decarboxylation of pyruvate to acetyl-CoA.

    Pyruvate enters the mitochondria by mitochondrial pyruvate carrier (MPC) symport along with H^+
    PDH complex is made up of 3 enzymes and it requires 5 coenzymes.

3 Enzymes

1. Pyruvate dehydrogenase component with decarboxylase activity (E_1)
2. Dihydrolipoyl transacetylase component (E_2)
3. Dihydrolipoyl dehydrogenase component (E_3)

5 Coenzymes

1. Thiamine pyrophosphate prosthetic group (Vitamin B_1)
2. Lipoic acid
3. CoA (Vitamin B_5)
4. FAD prosthetic group (Vitamin B_2)
5. Free NAD^+ (Vitamin B_3)

Catalytic Mechanism

Cofactor	Location	Function
Thiamine pyrophosphate (TPP)	Bound to E_1	Decarboxylates pyruvate
Lipoic acid	Covalently linked to Lys on E_2	Accepts acetyl group from TPP
Coenzyme A	Substrate for E_2	Accepts acetyl group from lipoamide
FAD	Bound to E_3	Reduced by lipoamide
NAD^+	Substrate for E_3	Reduced by $FADH_2$

 **Mnemonic**

Tender Lovely Care For Nancy. Thiamine pyrophosphate (TPP), Lipoic acid, Coenzyme A, FAD, NAD^+ *(Memorize in the same order)*. This is the order of transfer of e^-.

Decarboxylation: E_1 activity decarboxylate pyruvate to hydroxyethyl intermediate which is bound to thiazole ring of TDP.

Oxidation (E_2): Hydroxyethyl group is oxidized by oxidized lipoamide to acetyl group. Lipoamide gets reduced to reduced lipoamide. Acetyl group is accepted by CoA. Thus, lipoamide is a carrier of acyl group.

Regeneration of active enzyme (E_3): Reduced lipoamide is oxidized back by bound FAD which gets reduced to $FADH_2$. $FADH_2$ is oxidized back by donating its electron to NAD^+ which leaves as NADH.

Electron transfer: There is a net transfer of two electrons.

$$Pyruvate + NAD^+ + CoA \rightarrow Acetyl\text{-}CoA + NADH + H^+ + CO_2$$

Regulation of PDH

End Product Inhibition

End product inhibition is inhibition of enzyme by its own products. Acetyl-CoA and NADH inhibits PDH.

Covalent Modification

Phosphorylated E_1 is inactive.
PDH kinases phosphorylate three serine residues of E_1. PDH phosphatase removes the phosphate groups.

Enzyme	Activators		
PDH kinases	[NADH]	[ATP]	[Acetyl CoA]
	[NAD$^+$]	[ADP]	[CoA]
PDH phosphatase	Ca^{2+} and Mg^{2+}		
	Insulin (in adipocyte to convert pyruvate to acetyl-CoA for lipogenesis)		

Inhibitor

Arse**nite** binds to the SH group of lipoic acid cofactor and inhibits PDH.

> **Pasteur effect:** Inhibition of glycolysis by oxygen.
> **Crabtree effect:** Inhibition of oxygen consumption by the addition of glucose to tissues.
> **Warburg effect:** Aerobic glycolysis in cancer cells.

WARBURG EFFECT

Warburg effect refers to the conversion of glucose to lactate even in the presence of oxygen in the cancer cells.
Warburg effect is also known as <u>aerobic</u> glycolysis. Do not get confused by the term 'aerobic glycolysis'. This term actually refers to the conversion of glucose to lactate even in the presence of oxygen.

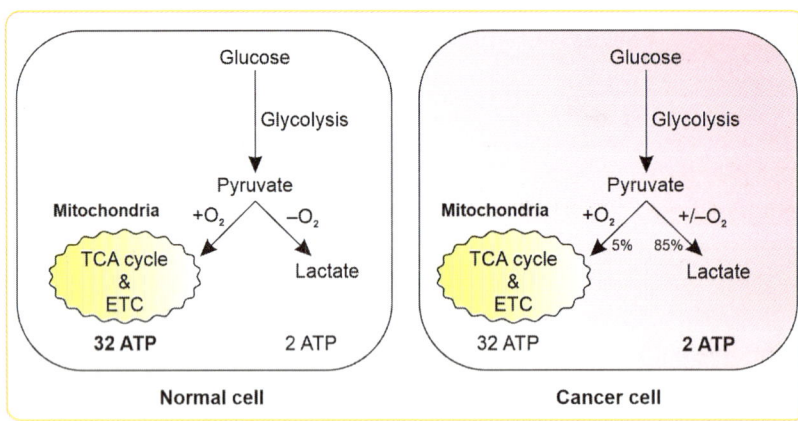

Cancer cells convert 85% of pyruvate to lactate even in the presence of O_2. Lactic acid acidifies the local environment and thus helping in metastasis. As the energy from anaerobic glycolysis is just 2 ATPs, cancer cells take up a large amount of glucose compared to normal cells. This enhanced uptake is easily visualized by PET scan using FDG.

Molecular basis:

❑ Presence of mitochondria-bound hexokinase isoenzymes which is resistant to feedback inhibition.

❑ Presence of PKM2 (isoenzyme of pyruvate kinase).

⎫ ↑Rate of Glycolysis

USE OF ¹⁸FDG IN POSITRON EMISSION TOMOGRAPHY

18-Fluoro deoxyglucose (FDG) is a glucose molecule labelled with positron-emitting ¹⁸F. On administration, FDG enters inside the cell via GLUT-1 and phosphorylated by hexokinase to ¹⁸FDG phosphate (Fig. 1). But, it cannot undergo further metabolism like glucose. Hence, it is effectively trapped inside the cell and releases positrons which can be measured by radioactivity trackers. Actively metabolizing cells take up more glucose. FDG uptake is high in areas of inflammation. Because of the Warburg effect, tumor cells need to take up more glucose since glucose to lactate conversion provides just 2 Adenosine triphosphates. This is the basis of using ¹⁸FDG to find primary cancer and secondary metastasis.

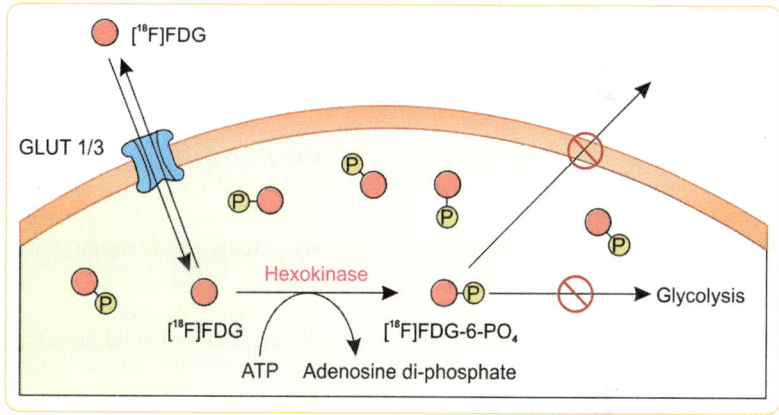

Fig. 1: FDG-6-phosphate getting trapped inside the cell

Abbreviation: HKII, Hexokinase II isoenzyme

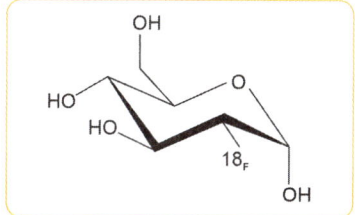

Chapter 10 Carbohydrate Metabolism

Competency

❑ BI3.4   Define and differentiate the pathways of carbohydrate metabolism (glycolysis)

❑ BI3.7   Describe the common poisons that inhibit crucial enzymes of carbohydrate metabolism (e.g. fluoride, arsenate)

Key Points

❑ Glycolysis is the only pathway that can produce ATPs anaerobically.

❑ Hexokinase is the flux-generating enzyme of glycolysis; PFK-1 is the rate-limiting.

❑ Hexokinase, PFK-1 and pyruvate kinase are the enzymes catalyzing the 3 irreversible steps of glycolysis.

❑ Phosphoglycerate kinase and pyruvate kinase are the enzymes catalyzing the substrate-level phosphorylation reactions.

❑ Arsenate uncouples the substrate level phosphorylation reaction catalyzed by phosphoglycerate kinase.

❑ Thiamine requirement increases with excessive intake of carbohydrates.

❑ NAD^+ is the final acceptor of electrons in the pyruvate dehydrogenase (PDH) complex. Oxygen is the final acceptor of electrons in the ETC.

❑ Thiamine deficiency leads to lactic acidosis since excess pyruvate cannot be converted into acetyl-CoA.

❑ Fructose-2,6-bisphosphate regulates both glycolysis and gluconeogenesis reciprocally.

❑ Fluoride inhibits enolase of glycolysis. Potassium oxalate + sodium fluoride are the anticoagulant used for blood samples for glucose estimation.

❑ Warburg effect is aerobic glycolysis. Increased uptake of glucose by the cancer cells is the principle of PET scan.

SELF-ASSESSMENT

Short Answer Questions

1. What is the biochemical basis of taking samples for blood glucose estimation in fluoride-containing blood collection tube?

2. Why does pyruvate kinase deficiency lead to hemolytic anemia?

3. Why is the phosphorylation of glucose essential before it can undergo any metabolic reaction?

4. Differentiate between glucokinase and hexokinase. Name any two tissues in which the enzyme catalyzing the reverse reaction is present and indicates its subcellular location.

5. In RBCs, glycolysis cannot continue if there is no formation of lactic acid – Explain how?

6. What is Luebering-Rapoport shunt? What is the significance of this shunt?

7. Which metabolite levels are elevated in the RBCs of people living at high altitudes, as a part of acclimatization? Mention the biochemical significance.

8. Explain the concept of oxygen debt.

9. Name the three irreversible steps of glycolysis. How these steps are regulated?

10. With the help of a glycolytic reaction, explain feed-forward activation.

11. 'Arsenate is an uncoupler of substrate-level phosphorylation' – Justify the statement.

12. Name the three enzymes and 5 cofactors of pyruvate dehydrogenase complex. Explain how the PDH complex is regulated.

13. Cancer cells convert glucose to lactate even in the presence of oxygen. What is the name of this effect? What is the advantage gained by cancer cell through this metabolic switch?

14. In the field of nuclear medicine, 18-Fluoro deoxyglucose (FDG) is used to detect the presence of primary tumor and metastasis. Explain the biochemical basis of this using your knowledge on glycolysis.

Multiple Choice Questions

1. How many ATPs are utilized in the energy investment phase of glycolysis?
   a. 1
   b. 2
   c. 3
   d. 4

2. Hexokinase belongs to:
   a. EC 1
   b. EC 2
   c. EC 3
   d. EC 4

3. **Enzyme catalyzing the flux-generating steps of glycolysis is:**
   a. Glyceraldehyde-3-phosphate dehydrogenase
   b. Hexokinase
   c. Phosphofructokinase
   d. Pyruvate kinase

4. **Which of the following kinase reaction is reversible?**
   a. Hexokinase
   b. Phosphofructokinase
   c. Phosphoglycerate kinase
   d. Pyruvate kinase

5. **Enzyme catalyzing the committed step of glycolysis is:**
   a. Hexokinase
   b. Phosphofructokinase-1
   c. Phosphoglycerate kinase
   d. Pyruvate kinase

6. **Which of the following enzyme pair is involved in feed-forward activation of glycolysis?**
   a. Hexokinase and enolase
   b. PFK-1 and pyruvate kinase
   c. PFK-1 and enolase
   d. G3PDH and pyruvate kinase

7. **Which of the following enzymes is involved in both glycolysis and gluconeogenesis?**
   a. Pyruvate kinase
   b. Phosphofructokinase
   c. Bisphosphoglycerate kinase
   d. Phosphoenolpyruvate carboxykinase

8. **Which of the following is the negative homotropic allosteric modulator of the enzyme PFK-1?**
   a. Citrate   b. ATP
   c. AMP   d. ADP

9. **Which of the following is the uncoupler of substrate-level phosphorylation?**
   a. Arsenite   b. Arsenate
   c. Arsenic   d. Arsenobetaine

10. **Which of following is used in the blood collection vials for glucose estimation?**
    a. NaF and heparin
    b. Potassium oxalate and heparin
    c. NaF and potassium oxalate
    d. NaF

11. **Warburg effect is:**
    a. Aerobic glycolysis
    b. Anaerobic glycolysis
    c. Inhibition of glycolysis by oxygen
    d. Inhibition of oxygen uptake by glycolysis

12. **Which of the following statement about glucokinase is not true?**
    a. Not inhibited by glucose 6 phosphate
    b. Acts as a glucose sensor
    c. Found in all tissues
    d. Induced by insulin

13. **Pyruvate dehydrogenase multienzyme complex carries out all the following reactions EXCEPT:**
    a. A decarboxylation reaction
    b. Producing a molecule of ATP
    c. Producing an acetyl group from pyruvate
    d. Combining the acetyl group with a cofactor

14. **PET scan uses which of the following tracer material?**
    a. FDG   b. CDF
    c. ADP   d. MIBG

15. **Which of the following enzyme catalyzes the formation of AMP from two molecules of ADP?**
    a. Adenylate kinase
    b. Adenylyl cyclase
    c. Adenosine kinase
    d. Adenosine deaminase

16. **All of the following vitamins are involved as cofactors in pyruvate dehydrogenase reaction, EXCEPT:**
    a. Vitamin B_1 b. Vitamin B_3
    c. Vitamin B_5 d. Vitamin B_{12}

17. **Which glycolytic enzyme is inhibited by fluoride?**
    a. Aldolase   b. Enolase
    c. Hexokinase   d. Pyruvate kinase

18. **All of the following tissues convert glucose to predominantly lactate, EXCEPT:**
    a. Brain   b. Cornea
    c. Lens   d. RBCs

19. **Final acceptor of electrons in the PDH complex is:**
    a. FAD   b. Molecular oxygen
    c. NAD^+
    d. Thiamine pyrophosphate

20. **Thiamine requirement increases with excessive intake of:**
    a. Carbohydrates   b. Lipids
    c. Proteins   d. Dietary fibers

21. **ATP is an allosteric regulator of:**
    a. Hexokinase and PFK 1
    b. PFK 1 and PFK 2
    c. PFK 1 and pyruvate kinase
    d. Glyceraldehyde 3 phosphate dehydrogenase and PFK 1

22. **Epinephrine does not play a role in regulation of:**
    a. Hexokinase
    b. Phosphofructokinase-2
    c. Pyruvate kinase
    d. Phosphoenolpyruvate carboxykinase

23. **The compound that regulates both glycolysis and gluconeogenesis is:**
    a. Fructose-1,6-bisphosphate
    b. Fructose-2,6-bisphosphate
    c. Citrate
    d. Glucose-6-phosphate

24. **Which of the following is a carrier of acyl group?**
    a. Thiamine pyrophosphate
    b. Lipoamide
    c. Biotin
    d. NAD^+

ANSWERS

1. b.	**2.** b.	**3.** b.	**4.** c.	**5.** b.	**6.** b.
7. c.	**8.** b.	**9.** b.	**10.** c.	**11.** a.	**12.** c.
13. b.	**14.** a.	**15.** a.	**16.** d.	**17.** b.	**18.** a.
19. c.	**20.** a.	**21.** c.	**22.** a.	**23.** b.	**24.** b.

Chapter Outline

Gluconeogenesis also known as neoglucogenesis is the synthesis of glucose from noncarbohydrate compounds.

GLUCONEOGENIC ORGANS

- ❏ Liver—major gluconeogenic organ
- ❏ Kidney
- ❏ Small intestine

Subcellular Compartment

Reactions of gluconeogenesis take place in both *cytoplasm and mitochondria*.

Substrates for Gluconeogenesis and their Sources

Gluconeogenic substrate	Source
Pyruvate	❏ Oxidation of lactate ❏ Transamination of alanine
Lactate	Anaerobic glycolysis in RBC, muscles, etc.
Glycerol	Triglyceride hydrolysis
All TCA cycle intermediates or molecules producing TCA cycle intermediates, i.e. glucogenic amino acids	❏ Propionyl CoA produces succinyl-CoA ❏ Aspartate produces oxaloacetate by transamination

Sources and Fate of Propionyl-CoA

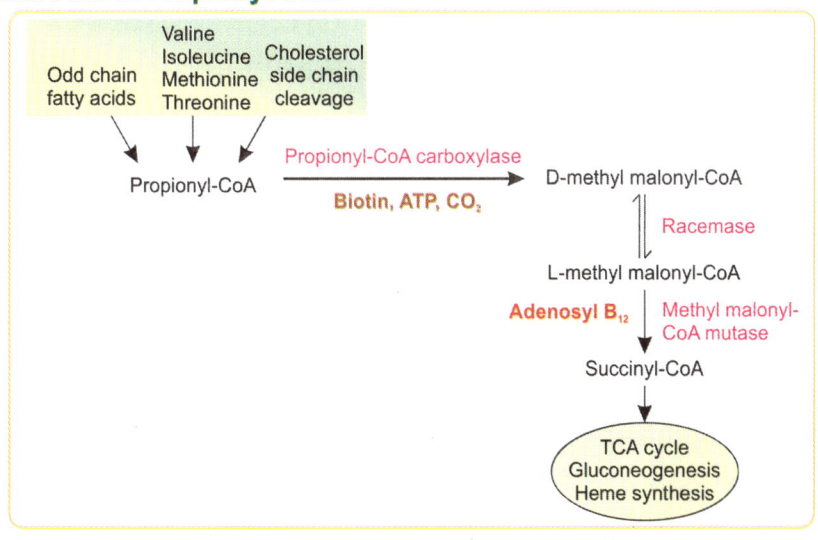

Cori Cycle Allows Reutilization of Lactate

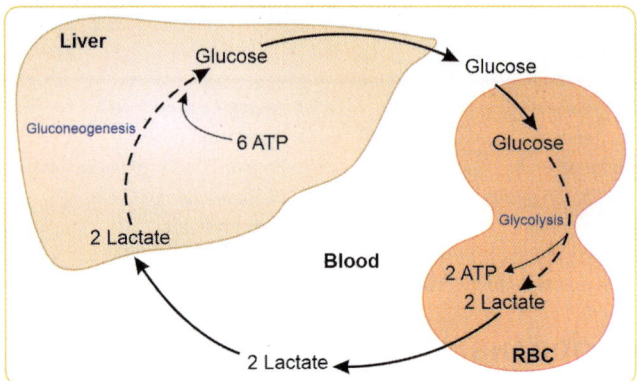

- ❏ Cori cycle is a cyclic process in which lactic acid produced by anaerobic glycolysis in exercising muscles and RBCs is converted to glucose in liver by gluconeogenesis.
- ❏ **Cori cycle allows the reutilization of lactate** which is otherwise a metabolic dead end.
- ❏ **2 lactate molecules are converted to glucose with the use of 6 ATP equivalents**.
- ❏ Liver failure leads to interruption of Cori's cycle and results in lactic acidosis and hypoglycemia.

Acetyl-CoA is NOT a Substrate for Gluconeogenesis in Animals

There are 2 reasons:
1. Pyruvate dehydrogenase reaction is irreversible. So, acetyl CoA cannot be converted back to pyruvate.

$$\text{Pyruvate} + \text{CoA-SH} + \text{NAD}^+ \rightarrow CO_2 + \text{Acetyl-CoA} + \text{NADH} + H^+$$

2. Acetyl-CoA (2-carbons) enters the TCA cycle by condensing with oxaloacetate (4 carbons). Two molecules of CO_2 are released and the oxaloacetate is regenerated. Thus, there is no net production of oxaloacetate in the TCA cycle.

So, fatty acids cannot be converted to glucose. But, glucose can be converted into fatty acids.

Animals Cannot Convert Fat (Triglycerides) into Glucose with Two Exceptions

1. Propionyl CoA derived from odd chain fatty acids is converted into succinyl CoA → Glucogenic
2. Glycerol derived from triglycerides is glucogenic.

Gluconeogenesis is NOT a Simple Reversal of Glycolysis

- ❏ There are **3 irreversible steps** in glycolysis. To synthesize glucose, these 3 steps need to be reversed by a different set of enzymes. So, gluconeogenesis is not just a simple reversal of glycolysis.
- ❏ **Four enzymes** are needed to reverse the 3 irreversible steps of glycolysis. Pyruvate kinase reaction cannot be reversed by a single enzyme. It needs 2 enzymes.
- ❏ Other enzymes catalyzing equilibrium reactions of glycolysis are used in gluconeogenesis also. Thus, the enzymes catalyzing the reversible steps of glycolysis are common to both glycolysis and gluconeogenesis.

Key Enzymes of Gluconeogenesis

Key glucogenic enzyme	Subcellular location
Pyruvate carboxylase—**requires ATP**	Mitochondria
Phosphoenolpyruvate carboxykinase—**requires GTP**	Cytosol
Fructose 1,6-bisphosphatase reverses the PFK-1 reaction	
Glucose-6-phosphatase reverses the hexokinase reaction	Luminal side of the endoplasmic reticulum

Section II Intermediary Metabolism

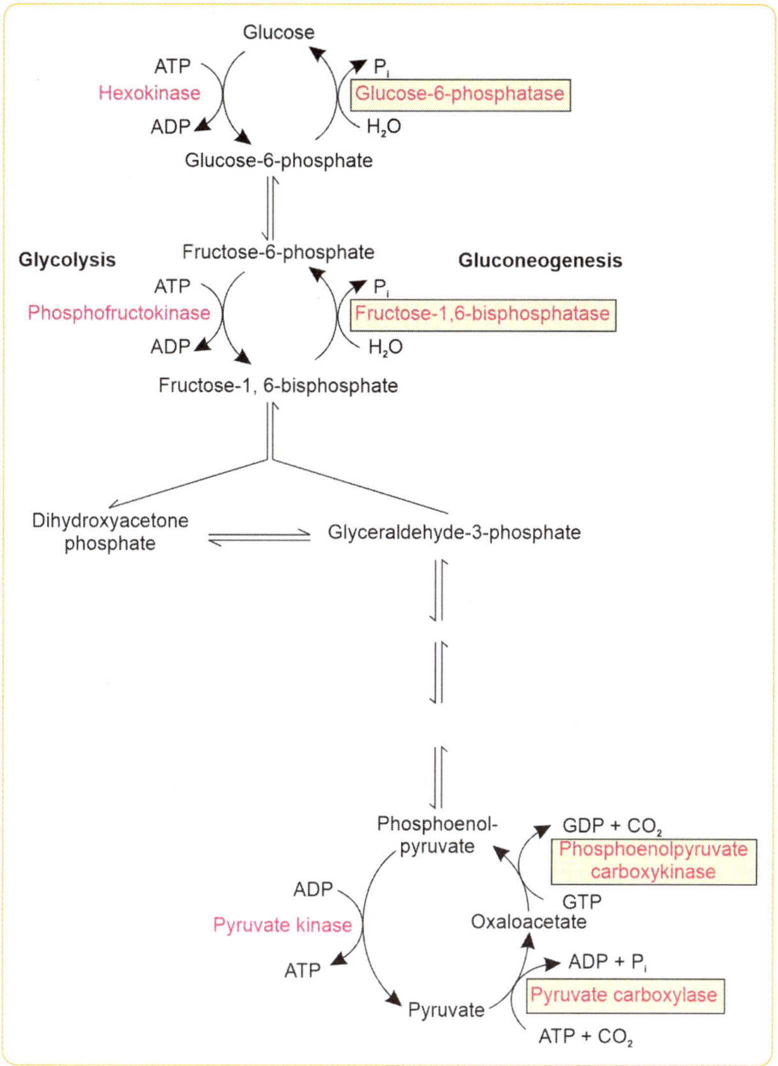

Glucose

ATP
Hexokinase
ADP

P_i
Glucose-6-phosphatase
H_2O

Glucose-6-phosphate

Glycolysis　　Fructose-6-phosphate　　**Gluconeogenesis**

ATP
Phosphofructokinase
ADP

P_i
Fructose-1,6-bisphosphatase
H_2O

Fructose-1, 6-bisphosphate

Dihydroxyacetone phosphate　　Glyceraldehyde-3-phosphate

Phosphoenol-pyruvate

GDP + CO_2
Phosphoenolpyruvate carboxykinase
GTP

Oxaloacetate

ADP
Pyruvate kinase
ATP

Pyruvate

ADP + P_i
Pyruvate carboxylase
ATP + CO_2

Inner Mitochondrial Membrane is Impermeable to Oxaloacetate

Oxaloacetate produced by the pyruvate carboxylase reaction cannot exit through the inner mitochondrial membrane since the inner mitochondrial membrane is impermeable to oxaloacetate. So, oxaloacetate comes out in the form of malate. Oxaloacetate is reduced to malate by mitochondrial malate dehydrogenase, exported into the cytoplasm, and oxidized by cytoplasmic malate dehydrogenase to regenerate oxaloacetate.

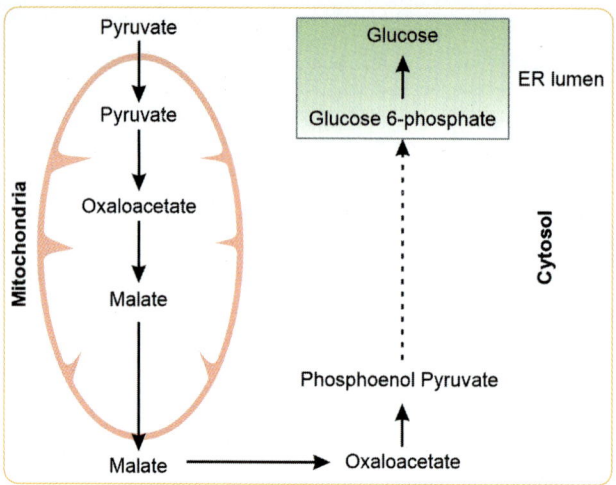

Gluconeogenesis is Energetically Expensive

Conversion of 2 molecules of lactate to one molecule of glucose requires 6 ATP equivalents.

Reaction	ATP equivalents required
Pyruvate carboxylase 2-pyruvate → 2-oxaloacetate	2
Phosphoenolpyruvate carboxykinase 2-oxaloacetate → 2-phosphoenolpyruvate	2 GTPs/ATPs
Phosphoglycerate kinase 2 (3-phosphoglycerate) → 2 (1,3 BPG)	2
Total	6 ATP equivalents

TCA Cycle Provides ATPs and GTPs Required for Gluconeogenesis

❑ GTP required for the phosphoenolpyruvate carboxykinase reaction is provided by the succinate thiokinase reaction of the TCA cycle.
❑ Even though fatty acids cannot be converted to glucose, β-oxidation of fatty acids provide the ATP equivalents for gluconeogenesis.
❑ This is why *defects in fatty acid oxidation lead to hypoglycemia.*
▶ (https://youtu.be/WFobOxfk4Ds) or scan QR code

REGULATION

❑ Gluconeogenesis is regulated by the availability of gluconeogenic substrates and the expression of gluconeogenic enzymes.
❑ *Glucagon is produced in the fasting state.* Glucagon increases the expression of key gluconeogenic enzymes while insulin decreases. *More details on insulin action is given in Chapter 10.7 (Refer to page no. 166).*

Fructose-2,6-Bisphosphate Regulates Glycolysis and Neoglucogenesis in a Reciprocal Manner

❑ Fructose-2,6-bisphosphate (F26BP) is produced from fructose-6-phosphate by phosphofructokinase 2 (PFK-2).
❑ PFK-2 is a **bifunctional or tandem enzyme** containing both kinase and phosphatase activity in the **same polypeptide chain.**

- When fructose-6-phosphate is in excess, it allosterically activates the kinase activity of PFK-2 to produce F26BP.
- **F26BP is a strong allosteric activator of PFK-1**, the rate-limiting enzyme of glycolysis. **F26BP inhibits key gluconeogenic enzyme fructose-1,6-bis phosphatase**. Thus, F26BP favors glycolysis and inhibits gluconeogenesis.

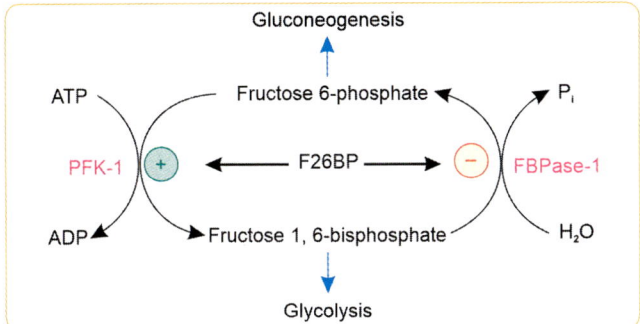

PFK-2 is Regulated by Phosphorylation and Dephosphorylation

We all know that glucagon favors gluconeogenesis and insulin (the only hypoglycemic hormone in our body) favors glycolysis.

Glucagon increases the level of cyclic AMP which activates protein kinase A. Protein kinase A phosphorylates PFK-2. Kinase activity of PFK-2 is inactive at phosphorylated state but phosphatase is active. Thus, glucagon favors gluconeogenesis.

Insulin stimulates a phosphoprotein phosphatase which removes the phosphate from PFK-2. Kinase activity of PFK-2 gets activated in unphosphorylated state and F26BP is produced → Glycolysis is favored.

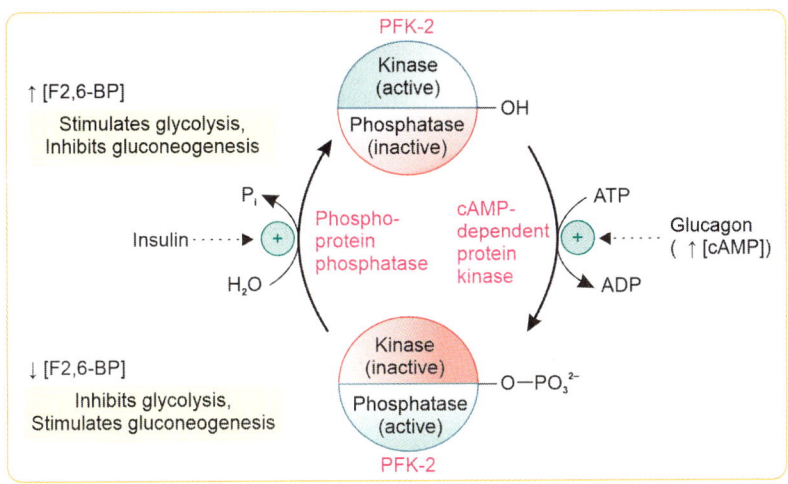

Glucose-alanine Cycle (Cahill Cycle): Indirect Way of Utilizing Muscle Glycogen for Hepatic Neoglucogenesis

During fasting, minimal breakdown of muscle protein happens. Amino acids transfer their amino group to pyruvate to form alanine. Then alanine enters liver to produce pyruvate by transamination and pyruvate is converted to glucose. Thus, pyruvate produced by muscle glycogenolysis enters the liver in the form of alanine and gets converted to glucose. Amino group of alanine is excreted as urea.

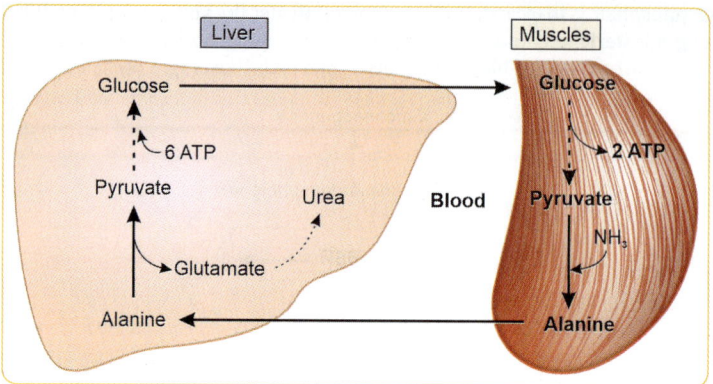

Causes of Hypoglycemia

- ❑ Diabetic patient skipping meal after insulin injection or oral hypoglycemic—commonly encountered in clinical practice
- ❑ Insulin overdose or oral hypoglycemic overdose
- ❑ Starvation and alcohol intoxication → **Ketotic** hypoglycemia
- ❑ Liver failure
- ❑ Severe sepsis
- ❑ Fatty acid oxidation defects (**Nonketotic** hypoglycemia)
  - ○ **M**edium **C**hain **A**cyl CoA **D**ehydrogenase deficiency—leads to **S**udden **I**nfant **D**eath **S**yndrome
  - ○ Carnitine deficiency *(Chapter 13.1)* *(Refer to page no. 253)*
- ❑ Defective glycogenolysis (**Fasting** hypoglycemia)
  - ○ Type I, III and IV glycogen storage diseases *(Chapter 10.5)* *(Refer to page no. 147)*
  - ○ Fanconi-Bickel syndrome (GLUT2 mutation)
- ❑ Defective gluconeogenesis (**Postprandial** hypoglycemia)
  - ○ Reactive hypoglycemia (Idiopathic postprandial syndrome)
  - ○ Hereditary fructose intolerance *(Chapter 10.6)* *(Refer to page no. 161)*
  - ○ Galactosemia *(Chapter 10.6)* *(Refer to page no. 163)*
- ❑ Endocrine diseases:
  - ○ Glucocorticoid insufficiency (Addison's disease)
  - ○ Growth hormone insufficiency
  - ○ Hypothyroidism
- ❑ Hyperinsulinemic conditions—**Nonketotic hypoglycemia** since insulin prevents ketogenesis
  - ○ Insulinoma—tumor of pancreatic β cells → ↑ level of C-peptides
  - ○ Beckwith-Wiedemann syndrome
  - ○ Dumping syndrome after gastric surgery

The reason why alcohol causes hypoglycemia is explained in Chapter 14 (Refer to page no. 309).

Competency

❑ BI3.4	Define and differentiate the pathways of carbohydrate metabolism (gluconeogenesis)

Key Points

- Acetyl-CoA cannot be converted into glucose in animals.
- Activation of pyruvate carboxylase by acetyl-CoA is a stimulus for gluconeogenesis.
- Gluconeogenesis involves both cytoplasm and mitochondria.
- Glucose-6-phosphatase is located in the luminal side of the endoplasmic reticulum membrane.
- Glucose-6-phosphatase is an enzyme of both gluconeogenesis and glycogenolysis.
- Cori cycle allows reutilization of lactate.
- Synthesis of one molecule of glucose forms 2 molecules of lactate which require 6 ATP equivalents.
- Phosphoenolpyruvate carboxykinase reaction requires GTP.
- Defective fatty acid oxidation leads to hypoketotic hypoglycemia.

SELF-ASSESSMENT

Short Answer Questions

1. Do you agree with the statement – 'Animals cannot convert fats into glucose'. Justify your answer.

2. What is the biochemical importance of gluconeogenesis? Write down the steps involved in the conversion of two molecules of lactose to one molecule of glucose. Calculate the energy required for this. Explain the reciprocal control of gluconeogenesis versus glycolysis in fed and starvation conditions.

3. Define gluconeogenesis. How is alanine converted to glucose?

4. Why acetyl-CoA cannot be converted into glucose?

5. Differentiate the Cori cycle from Cahill cycle.

6. Explain whether glycolysis and gluconeogenesis can occur in the same cell in humans at the same time.

7. What is the importance of gluconeogenesis when other sources of calories exist? Explain why gluconeogenesis cannot be a reversal of the glycolytic pathway. Outline the specific reactions with the names of enzymes catalyzing these steps that are different in the two pathways.

8. How many ATPs are required for the conversion of two molecules of pyruvate into one molecule of glucose? Justify your answer.

9. Justify the statement – 'TCA Cycle Provides the ATPs and GTPs Required for Gluconeogenesis'

10. Defect in fatty acid oxidation leads to hypoglycemia – Why?

11. Explain the reciprocal regulation of glycolysis and gluconeogenesis by fructose-2,6-bisphosphate.

12. Mention two causes of ketotic and non-ketotic hypoglycemia.

13. Explain why carbohydrates can be converted to fats while fats cannot be converted to glucose.

Multiple Choice Questions

1. Which of the following is the major gluconeogenic organ?
   a. Liver
   b. Muscle
   c. Heart
   d. Intestine

2. Which of the following is not a gluconeogenic substrate?
   a. Alanine   b. Acetyl-CoA
   c. Pyruvate   d. Glycerol

3. Which of the following statements about the process of gluconeogenesis is correct?
   a. Pyruvate is first converted to phosphoenolpyruvate by phosphoenolpyruvate carboxykinase
   b. Fructose-1,6-bisphosphatase converts fructose-1,6-bisphosphate into fructose-1-phosphate
   c. Glucose-6-phosphatase hydrolyses glucose-6-phosphate to release glucose into the blood
   d. Glucose-6-phosphatase hydrolyses glucose-6-phosphate and is found in liver and muscle

4. All of the following statements about gluconeogenesis are correct, EXCEPT:
   a. Lactate from vigorous muscle activity can be used as a carbon source in gluconeogenesis.
   b. Glycerol from the hydrolysis of triacylglycerols is converted to glucose in gluconeogenesis
   c. Lactate from red blood cells can be used as a carbon source in gluconeogenesis
   d. Fatty acids from the hydrolysis of triacylglycerols can be used as a carbon source in gluconeogenesis

5. Which of the following enzyme is common to both glycolysis and gluconeogenesis?
   a. 3-phosphoglycerate kinase
   b. Glucose 6-phosphatase
   c. Hexokinase
   d. Phosphofructokinase-1

6. How many ATPs are needed for the synthesis of 1 molecule of glucose from 2 molecules of lactate?
   a. 2   b. 3
   c. 4   d. 6

7. All of the following are true about fructose-1-6-bisphosphatase, EXCEPT:
   a. Key gluconeogenic enzyme
   b. Fructose-2, 6, bisphosphate is an allosteric activator
   c. Catalyzes the hydrolysis reaction
   d. Requires magnesium for the catalysis

8. Which of the following amino acid released from muscle helps to maintain the blood glucose levels in between meals?
   a. Glutamine
   b. Glutamate
   c. Aspartate
   d. Alanine

9. In fasting state, increased levels of alanine indicate:
   a. Increased breakdown of muscle protein
   b. Decreased utilization for gluconeogenesis
   c. Leak of alanine through plasma membrane
   d. Decreased uptake of alanine by liver for gluconeogenesis

10. All of the following are associated with non-ketotic hypoglycemia, EXCEPT:
    a. Von-Gierke disease
    b. Insulinoma
    c. Carnitine deficiency
    d. Medium chain acyl-CoA dehydrogenase deficiency

11. Which of the following metabolic disorders cause postprandial hypoglycemia?
    a. Glycogen storage disease type I
    b. Glycogen storage disease type III
    c. Fanconi-Bickel syndrome
    d. Hereditary fructose intolerance

ANSWERS

| **1.** a. | **2.** b. | **3.** c. | **4.** d. | **5.** a. | **6.** d. |
| **7.** b. | **8.** d. | **9.** a. | **10.** a. | **11.** d. | |

10.4 GLYCOGEN METABOLISM

STRUCTURE OF GLYCOGEN

Glycogen is the storage carbohydrate in animals. That is why it is also known as animal starch. Glycogen is a highly-branched homopolymer of α-D-glucopyranose. Glycogenin, a protein located in the inner core of the glycogen, is the primer for glycogen synthesis.

Glycogen is more branched than starch. Branching points are located after every 8–12 linear glucose residue. Linear chain consists of glucose molecules linked by α (1 → 4) glucosidic bond. The glucose at the branch point is linked by α (1 → 6) glucosidic bond. Glycogen is stored as granules in the cytoplasm.

ORGANS INVOLVED IN GLYCOGEN STORAGE

Glycogen is stored in liver and skeletal muscles (Fig. 1). Approximately 250 g of glycogen is stored in the skeletal muscles and 100 g is stored in liver. But, liver contains more glycogen content per gram of dry tissue weight. 10% of liver is glycogen and only 2% of muscle is glycogen.

Glycogen stored in the muscle is used only by the muscle. But, the glycogen stored in liver is released for the utilization of extra hepatic tissues during fasting. After 12–18 hours of fast, the glycogen content is almost depleted in liver.

Note: You may be aware of the presence of glycogen in platelets. Researchers have found small amounts of glycogen in the kidneys, and even smaller amounts in certain glial cells in the brain and WBCs. **Periodic acid–Schiff (PAS) staining is used to detect glycogen** and glycoconjugates.

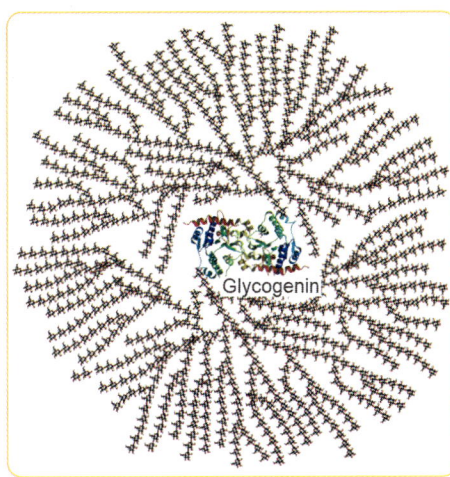

Fig. 1: Highly-branched structure of glycogen; glycogenin protein at the center

Reducing End versus Nonreducing End

❑ Reducing property of glucose is due to aldehyde group of first carbon of glucose (anomeric carbon). Formation of glycosidic linkage involves the 1st carbon. So, all the ends in the periphery of glycogen are nonreducing (Fig. 2).

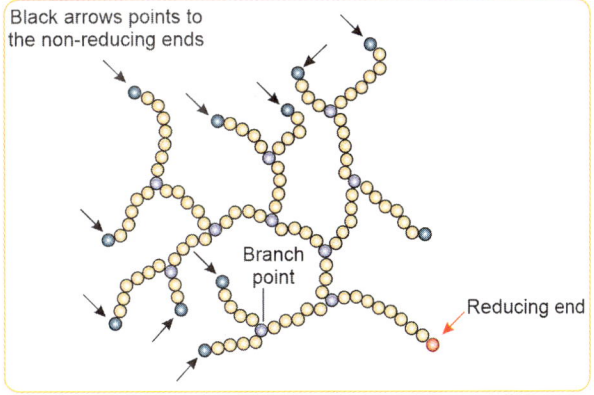

Fig. 2: Schematic illustration shows the reducing and nonreducing end. Reducing end is attached to the glycogenin protein

- There is only one reducing end at the glycogen core. That end is also linked to tyrosine of glycogenin via o-glycosidic linkage.
- Enzymes acting on the glycogen act on the nonreducing end.

Advantages of Storing Glycogen

Why cannot our body store glucose as such?
- *You know that osmotic pressure depends upon the number of particles.*
- Glycogen is osmotically inactive compared to glucose. *If we have to store glucose, we need huge amount of water to dissolve it (maintain osmolarity).*
- Branched pattern of glycogen makes it readily mobilizable when glucose is needed by the body.
- **Phosphorolysis of glycogen yields 3 ATP when it undergoes anaerobic glycolysis.** We gain 1 ATP more compared to anaerobic glycolysis of glucose (will be explained subsequently).

GLYCOGEN SYNTHESIS

Combined Action of GLUT2 and Glucokinase is Crucial for Glycogen Synthesis

Glycogen is synthesized when blood glucose level is high, as in fed state, when glucose enters the liver cell via GLUT2.

GLUT2 has low affinity but high capacity for glucose.

Glucose is converted into glucose-6-phosphate by the action of *glucokinase* (in the liver) and hexokinase (in liver and muscle).

Glucokinase also has low affinity for glucose. Both GLUT2 and glucokinase can act on glucose only when the concentration of glucose is high, as happens after a carbohydrate-rich meal. Thus, GLUT2 acts as a funnel that takes up all the excess glucose from the portal blood, and glucokinase helps in trapping all the glucose immediately into glucose-6-phosphate.

Insulin is secreted in response to high glucose after a meal. Insulin induces glucokinase enzyme in liver and it causes translocation of GLUT4 into plasma membrane in muscle. This is one-way in which insulin promotes glycogen synthesis (Fig. 3).

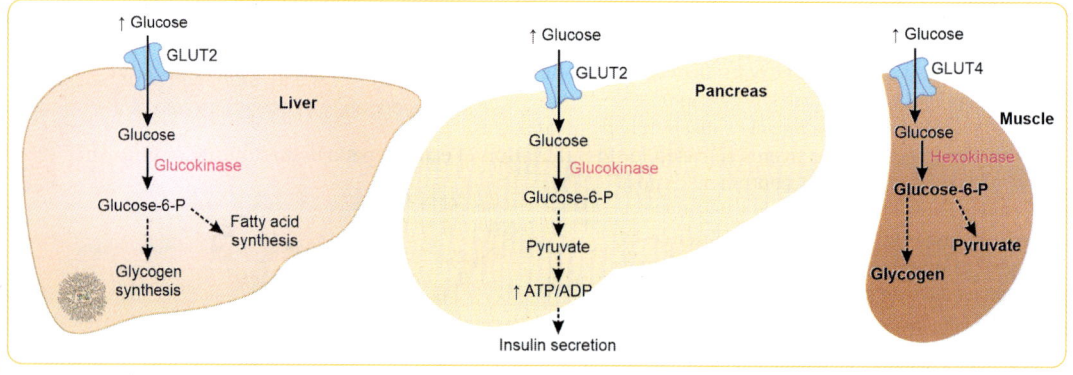

Fig. 3: Different role of GLUT2 in liver and pancreas. In liver, GLUT2 mediated glucose transport facilitate storage of glucose as glycogen. In pancreas, GLUT2 mediated glucose transport leads to secretion of insulin

UDP-glucose is the Source of Glucose for Glycogenesis

For any biosynthetic reaction, the molecules need to be activated. UDP-glucose is the activated form of glucose for glycogen synthesis.

Glucose-6-phosphate is first converted into glucose-1-phosphate by the action of phosphoglucomutase.

Glucose-1-phosphate is converted into UDP-glucose by the action of the enzyme glucose-1-phosphate uridylyl-transferase also known as UDP–glucose pyrophosphorylase. Hydrolysis of inorganic pyrophosphate (PP_i) makes this reaction irreversible.

$$\text{Glucose-1-phosphate} + \text{UTP} \rightleftharpoons \text{UDP-glucose} + PP_i$$
$$PP_i \rightarrow 2\,P_i$$

Glycogen Synthase Requires a Primer

Glycogen synthase, the rate-limiting enzyme of glycogen synthesis requires a primer of oligosaccharide with at least 8 glucose residues. Glycogenin protein acts as the primer. Glycogenin has autocatalytic activity. It glucosylates its own tyrosine residues. It can add up to 10 to 20 glucose residues via α (1→ 4) linkage.

> **Glycation:** Nonenzymatic addition of sugars
> **Glycosylation:** Enzymatic addition of sugars
> **Glucosylation:** Enzymatic addition of glucose

Once a chain of 10 to 20 glucose residues is formed by glycogenin, glycogen synthase binds to it and adds UDP-glucose to the 4th hydroxyl group of the glucose residue on the **nonreducing end** of the glycogen chain, forming α (1→4) linkages. *Glycogen synthase cannot indefinitely add glucose residues in α(1→4) linkages. If it does, we will get a linear chain-like cellulose. We need a highly branched polymer.*

Branches are introduced by the **branching enzyme** also known as amylo-α (1:4)→α (1:6) transglycosylase. When a linear chain is at least 11 glucose residues long, branching enzyme transfers six to seven glucose residues of the linear chain and links it to a nearby chain via α (1 → 6) linkage, creating a branch. Next, Glycogen synthase elongates this linearly and the cycle repeats creating the highly complex glycogen molecule (Fig. 4).

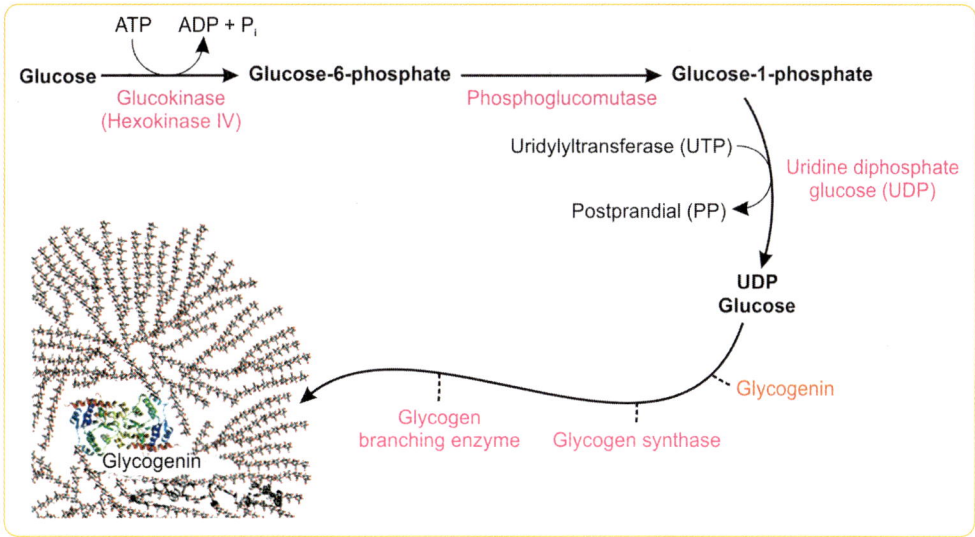

Fig. 4: Glycogen synthesis

Source: Mark Cidade – Own work, CC BY-SA 4.0, https://commons.wikimedia.org/w/index.php?curid=41182960

GLYCOGENOLYSIS

Glycogenolysis is Not the Reverse of Glycogen Synthesis

Glycogenolysis requires 2 enzymes—Glycogen phosphorylase and debranching enzyme. Glycogen phosphorylase removes the glucose from the nonreducing ends of glycogen via phosphorolysis. Hydrolysis and phosphorolysis are different.

Hydrolysis	Phosphorolysis
Breaking of covalent bonds using water R-O-R' + HOH → R-OH + R'-OH	Breaking covalent bonds using orthophosphate. R-O-R' + HO-PO$_3$$^{2-}$ → R-OH + R'-O-PO$_3$$^{2-}$
For example, amylase and all protease enzymes. All hydrolase enzymes belong to EC3.	For example, glycogen phosphorylase; belongs to EC2 (transferase)

Advantage of phosphorolysis over hydrolysis:
❑ Glucose produced is already in phosphorylated form. So, 1 ATP is gained/conserved if it enters glycolysis.
❑ Glucose-6-phosphate cannot diffuse across the cell membrane freely. So, glucose-6-phosphate in muscle can directly enter glycolysis as glucose-6-phosphatase is absent in myocyte.

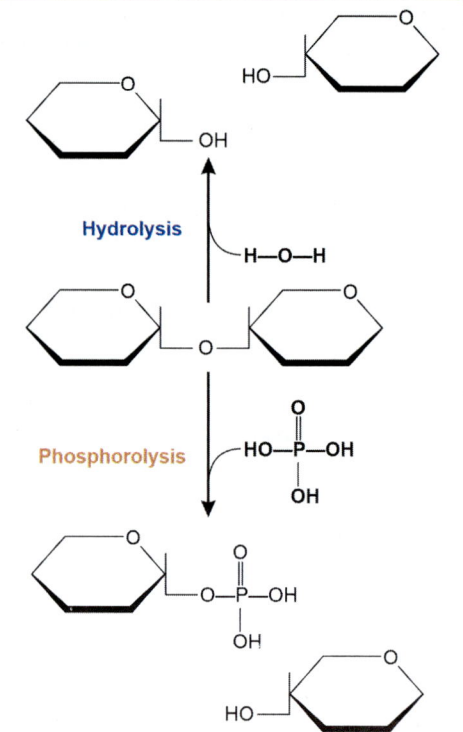

Glycogen Phosphorylase is the Rate-Limiting Enzyme of Glycogenolysis

❑ Glycogen phosphorylase acts from the nonreducing end of the glycogen molecule; breaks the α (1→4) glucosidic bond and releases glucose-1-phosphate. **Pyridoxal phosphate (Vitamin B$_6$) is required as the cofactor**.
❑ Glucose-1-phosphate is converted into glucose-6-phosphate by phosphoglucomutase.
❑ In the liver, glucose-6-phosphate is transported inside the lumen of the endoplasmic reticulum by a transporter (initially believed to be GLUT7) and is acted upon by the glucose-6-phosphatase to produce free glucose.

Muscle Glycogenolysis does NOT Release Free Glucose

❑ **Muscle tissue lacks glucose-6-phosphatase.** Gluocose-6-phosphate is used to meet the own energy need of the muscle.
❑ So, muscle glycogenolysis cannot release free glucose into blood.

Debranching Enzyme is a Bifunctional Enzyme

❑ Glycogen phosphorylase cannot act when it reaches 4 residues near the branching point.
❑ Debranching enzyme takes over from here. It is a bifunctional enzyme having 2 enzymatic activities (α 1→4 to α 1→4 glucan transferase and α 1→6 glucosidase) within the same polypeptide.
❑ As the name indicates, α 1→4 to α 1→4 glucan transferase transfers 3 glucose linear residues to nearby linear chain to form α 1→4 linkage. Transfer of this linear chain to another linear chain results in elongation of the chain which can now be acted upon by glycogen phosphorylase.
❑ The single glucose residue at the branch point that is linked by α 1→6 linkage is hydrolyzed by α 1→6 glucosidase activity of the bifunctional enzyme. This releases free glucose both in liver and muscle.
❑ The ratio of glucose-1-phosphate to free glucose during glycogenolysis is around 8:1 (because of the branching ratio).

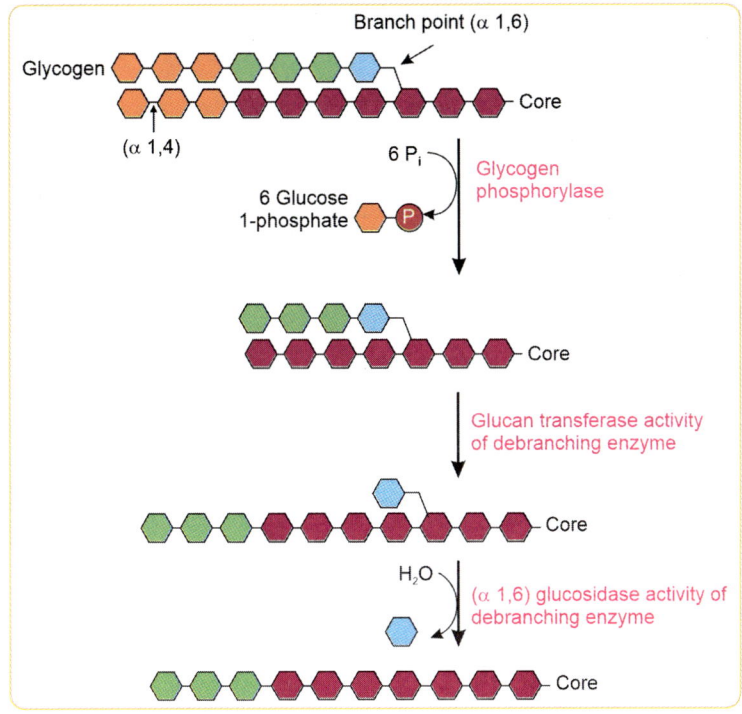

Fig. 5: Glycogen degradation by glycogen phosphorylase and glycogen debranching enyzme

REGULATION OF GLYCOGEN METABOLISM

Glycogenolysis and glycogen synthesis are **regulated reciprocally** to avoid futile cycling. Glycogenolysis takes place during fasting. Glycogen synthesis takes place during fed-state.

Regulation of Glycogen Synthase

Allosteric Activation by Glucose-6-phosphate

- ❑ Allosteric enzymes exist in T and R states. R state is active and T state is inactive.
- ❑ Glycogen synthase is allosterically activated by glucose-6-phosphate, i.e. binding of glucose-6-phosphate to a site other than active site and produces a conformational change. This converts the T state of enzyme to R state.
- ❑ Glucose-6-phosphate will accumulate only when the glycolytic pathway is saturated.

Covalent Modification

Glycogen synthase exists in two states—glycogen synthase 'a' and 'b'. 'a' form is usually active and 'b' form is usually inactive. Phosphorylation promotes the transition of 'a' to 'b' while dephosphorylation promotes the reverse.

- ❑ Glycogen synthase is active in dephosphorylated state.
- ❑ Phosphorylation is done by protein kinase A.
- ❑ Protein kinase A is activated by the second messenger cAMP.
- ❑ cAMP is produced by adenylyl cyclase.
- ❑ Adenylyl cyclase is activated by GTP bound α_s subunit which dissociates from β and γ complex.
- ❑ Dissociation is triggered by the binding of glucagon or epinephrine to their respective G-Protein coupled receptors.
- ❑ Muscle lacks glucagon receptors.

□ Dephosphorylation is mediated by protein phosphatase-1 which is regulated by cAMP-dependent protein kinase.

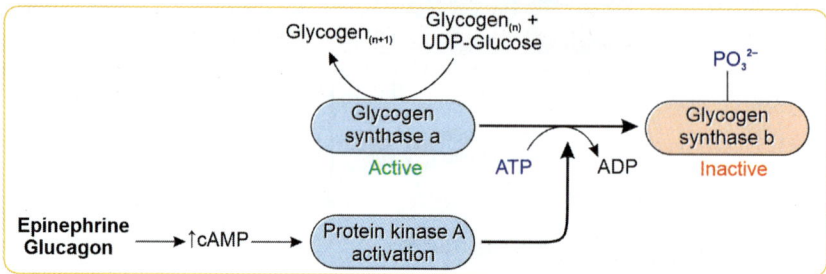

Regulation of Glycogen Phosphorylase

Glycogen phosphorylase, the rate-limiting enzyme of glycogenolysis is an allosteric enzyme. It exists in 2 forms— phosphorylase *a* and *b*. Phosphorylase *a* is usually active and *b* is usually inactive. Phosphorylation converts the 'b' form to 'a' form.

$$\text{Phosphorylase } b \xrightarrow[\text{ATP} \rightarrow \text{ADP} + \text{P}_i]{\text{Phoshporylase kinase}} \text{Phosphorylase } a$$

Allosteric Control

□ ATP and glucose-6-phosphate in muscle and glucose in liver allosterically inactivate glycogen phosphorylase.
□ In skeletal muscle (not in liver), AMP can allosterically activate glycogen phosphorylase (Fig. 6). We know that AMP is the indicator of low-energy state. Even the usually inactive b form can be activated by AMP by bringing out conformational change.

Covalent Modification and Hormonal Control

□ Glucagon and epinephrine in liver and epinephrine in muscle favor the conversion of phosphorylase b to a.
□ As we have already seen glucagon and epinephrine both can act through the second messenger cAMP.
□ cAMP activates protein kinase A which in turn phosphorylates phosphorylase kinase.
□ Insulin favors dephosphorylation through protein phosphatase I

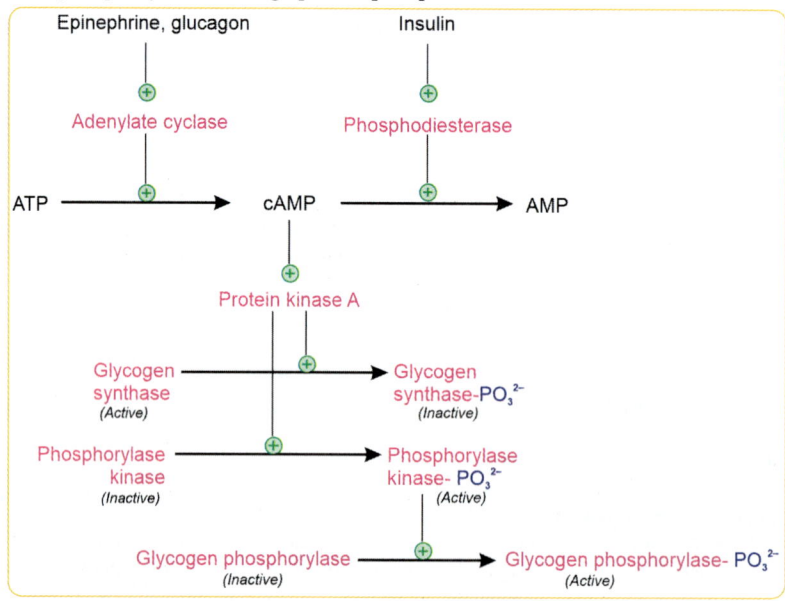

Fig. 6: Reciprocal regulation of glycogen synthase and phosphorylase

Regulation by Calcium

- ❑ In addition to its role in muscle contraction, calcium is involved in activation of glycogenolysis.
- ❑ Phosphorylase kinase enzyme is a multisubunit enzyme with four subunits – α, β, γ and δ. **The delta subunit is nothing but calmodulin which binds to 4 calcium units.**
- ❑ Binding of calcium to δ subunit activates phosphorylase kinase which in turn activates glycogen phosphorylase.
- ❑ This way muscle contraction is coupled to glycogenolysis.

Different Glycogen Metabolism in Muscle and Liver

	Muscle	Liver
Contribution to free glucose in blood	No; Muscle lacks glucose-6-phosphatase.	Yes
Entry of glucose for glycogen synthesis	Via GLUT4	Via GLUT2
Effect of glucagon	**No effect since there is no glucagon receptor**	Glycogenolysis
Allosteric activator of glycogen phosphorylase	AMP	—
Allosteric inhibitor of glycogen phosphorylase	ATP and glucose-6-phosphate	Free glucose
Source of calcium which promotes glycogenolysis	Neural stimulation mediated release of calcium from sarcoplasmic reticulum	Epinephrine mediated calcium uptake via α-adrenergic receptors

GLYCOGEN STORAGE DISORDERS

Defect of glycogen metabolism leads to various glycogen storage diseases. The important ones are listed here:

GSD	Name	Deficient enzyme	Features
0	Glycogen synthase deficiency	Glycogen synthase	Reduction of glycogen synthesis. Hepatic and muscle type have been described
Ia	Von Gierke disease	Glucose-6-phosphatase	Hypoglycemia, Lactic acidosis, Ketosis, Hyperlipidemia, glycogen accumulation in liver and kidney
Ib		Glucose 6-phosphate translocase	Neutropenia and all the features of Von-Gierke disease
II	Pompe disease	Lysosomal acid α-glucosidase	Death due to cardiac failure around the age of 2 years
	Danon's disease	Lysosome-associated membrane protein-2 (LAMP-2)	Similar to Pompe disease
III	**C**ori-Forbe disease (limit dextrinosis)	**D**ebranching enzyme	Fasting hypoglycemia, no muscle weakness
IV	**A**ndersen disease	**B**ranching enzyme	Muscle weakness, death due to cardiac or hepatic failure around the age of 5 years
V	**M**cArdle disease	**M**uscle phosphorylase	Exercise intolerance, absence of raise of blood lactate after exercise, Second wind phenomenon
VI	**H**ers disease	**H**epatic phosphorylase	Hepatomegaly with mild hypoglycemia
VII	Tarui disease	Muscle and RBC PFK-I	Exercise intolerance, hemolytic anemia
XI	Fanconi–Bickel disease	Glucose transporter-2	Hepatomegaly, failure to thrive, proximal renal tubular dysfunction and its consequences like glucosuria, phosphaturia, generalized aminoaciduria, acidosis and hypophosphatemia

Very Personal Chats Are Mostly Hidden To FB
Von-Gierke, Pompe, Cori, Anderson, McArdle, Hers, Tarui, Fanconi-Bickel

GSD with major hepatic involvement: Types 1, 3, 4, 6, 9
GSD with major muscle involvement: Types 2, 5, 7
GSD with skeletal and cardiac muscle dystrophy: Pompe and Danon's disease

VON GIERKE DISEASE (TYPE I GSD)

Inheritance: Autosomal recessive
Defective enzyme: Glucose-6-phosphatase
Glucose-6-phsophatase is an enzyme of both glycogenolysis and gluconeogenesis.

Features

Feature	Biochemical basis
Severe 'fasting' hypoglycemia	No glycogenolysis as well as gluconeogenesis. So, blood glucose level would not raise **even after the administration of glucagon and epinephrine**
High anion gap acidosis: ❑ Lactic acidosis ❑ Ketoacidosis	Lactic acid cannot be converted to glucose. Hypoglycemia reduces the insulin secretion and increases glucagon secretion→ Lipolysis in adipose tissue → ↑ Free fatty acids → Ketone bodies
Hyperlipidemia	Hypoglycemia leads to glucagon and epinephrine secretion → Lipolysis in adipose tissue → increased free fatty acids ↑ Glucose-6-phophate in liver → ↑ Acetyl-CoA → ↑ Fatty acid synthesis
Hyperuricemia due to both over production and under excretion of uric acid	**Over production:** ↑ Glucose-6-phophate in liver → diversion of glucose-6-phosphate to HMP shunt → ↑ Pentose sugars →↑ PRPP → ↑ Uric acid **Underexcretion:** Lactic acid competes with uric acid for renal excretion. Instead of uric acid, lactic acid gets excreted and uric acid accumulates in blood.
Raised bleeding time	Platelet dysfunction
Premature ovarian failure in females	—

Treatment

Primary aim is to prevent the fasting-associated hypoglycemia.
❑ In infants, feeding through nasogastric tube at night to prevent hypoglycemia.
❑ In young children, frequent feeding with low glycemic index foods like uncooked corn starch. This will release glucose slowly over a prolonged time.
❑ Avoid any kind of fasting.

Competency

❑ BI3.4   Define and differentiate the pathways of carbohydrate metabolism (Glycogen metabolism)
❑ BI3.5   Describe and discuss the regulation, functions and integration of carbohydrate along with associated diseases/disorders

Key Points

❑ Glycogen has multiple nonreducing ends thus allowing either glycogen synthase or glycogen phosphorylase to act at multiple sites.

❑ Glycogen has only one reducing end that is also attached to glycogenin. All the other ends are nonreducing ends.

❑ Glycogenin is the primer for glycogen synthesis.

❑ UTP is used to activate glucose for glycogen synthesis.

❑ Glycogenin has autocatalytic activity. Glucose molecules are added to the –OH group of tyrosine via O-glycosidic linkage.

❑ Glucose-6-phosphatase is absent in muscle. So, muscle glycogenolysis cannot increase the blood glucose.

❑ Conversion to glycogen is the major fate of glucose-6-phosphate in liver in well-fed state.

❑ In case of fasting, glycogen stored in the body can last up to 12–18 hours.

❑ Glycogen phosphorylase is active in phosphorylated state while glycogen synthase is active in dephosphorylated state.

❑ Pyridoxal phosphate (Vitamin B$_6$) is the cofactor for glycogen phosphorylase.

❑ Phosphoglucomutase is common to both glycogen synthesis and glycogenolysis.

❑ Glucagon causes phosphorylation through cAMP and protein kinase A. Insulin causes dephosphorylation by activating protein phosphatase-1.

❑ Fanconi-Bickel disease is due to defect in GLUT-2.

❑ Anaerobic glycolysis of glycogen derived glucose in muscle produces 3 ATP.

SELF-ASSESSMENT

Short Answer Questions

1. What is the advantage of glycogen over starch as a storage polysaccharide?

2. What is the advantage of converting glucose to glycogen for it to be stored in the body? What are the differences between glycogen breakdown occurring in the liver compared to glycogen breakdown occurring in muscles?

3. What are reducing end and non-reducing ends in a glycogen molecule? How many reducing ends are present in one glycogen molecule with 15000 residues and a branch at every 12 residues?

4. Explain the advantage of the phosphorolytic cleavage of glycogen over its hydrolytic cleavage.

5. Explain how allosteric regulation of liver phosphorylase and skeletal muscle phosphorylase is suited to the corresponding function of glycogen in each tissue.

6. List the different ways in which glycogen synthesis is regulated. Briefly describe the mechanism by which different kinases control the activity of glycogen synthase.

7. How calcium couples glycogenolysis with muscle contraction?

8. Glycogenolysis in muscle cannot increase the blood glucose levels – Why?

9. Differentiate hepatic glycogenolysis from muscle glycogenolysis.

10. What is the enzymatic defect in Von-Gierke's disease? Mention any four clinical features along with the biochemical basis. Add a note on treatment.

11. Differentiate hydrolysis and phosphorolysis.

12. Justify the statement – 'Phosphorolysis of glycogen yields 3 ATP when it undergoes anaerobic glycolysis.'

13. What is the biochemical basis of hyperuricemia seen in patients of type I glycogen storage disorder?

Multiple Choice Questions

1. **Glycogen phosphorylase is a:**
   a. Oxidoreductase b. Transferase
   c. Hydrolase d. Lyase

2. **In case of fasting, glycogen stored in the body can last up to:**
   a. 7 days b. 72–96 hours
   c. 6–8 hours d. 12–18 hours

3. **All of the following are reasons for converting glucose to glycogen for storage EXCEPT:**
   a. Glycogen is more compact than glucose
   b. Glycogen has multiple nonreducing ends thus allowing either glycogen synthase or glycogen phosphorylase to act at multiple sites
   c. Storing glucose without converting to glycogen creates increased intracellular osmotic pressure
   d. Glycogen binds to no water at all

4. **How many ATPs are produced from the conversion of one glucose residue in the linear chain of glycogen to lactic acid (anaerobic glycolysis)?**
   a. 1 b. 2
   c. 3 d. 4

5. **In which of the following glycogen storage disorder, severe muscle weakness is seen:**
   a. Cori disease b. Andersen disease
   c. McArdle disease d. Tarui disease

6. **Glycogen phosphorylase can be regulated by all of the following, EXCEPT:**
   a. cAMP b. Calmodulin
   c. Protein kinase A d. Glycogenin

7. **Which one of the following enzyme is involved in both glycogen synthesis and glycogenolysis?**
   a. Branching enzyme
   b. Debranching enzyme
   c. Phosphohexose isomerase
   d. Phosphoglucomutase

8. **Glycogenin is a protein with self-glucosylation capacity. To which amino acid of glycogenin, glucose molecules are attached?**
   a. Serine b. Threonine
   c. Tyrosine d. Hydroxylysine

9. **The following vitamin acts as a co-factor for glycogen phosphorylase:**
   a. Thiamine pyrophosphate (B_1)
   b. Riboflavin (B_2)
   c. Pyridoxal phosphate (B_6)
   d. Methyl cobalamin (B_{12})

10. **While breaking down glycogen, muscles cannot contribute to the blood glucose as they lack the following enzyme:**
    a. Glycogen phosphorylase
    b. Debranching enzyme
    c. Phosphoglucomutase
    d. Glucose 6-phosphatase

11. **The Glycogen storage disorder, von Gierke's disease, is characterized by glucose 6-phosphatase deficiency leading to excess accumulation of glucose-6-phosphate that may lead to:**
    a. Hypolipidemia b. Hyperglycemia
    c. Lactic acidosis d. Hemolysis

12. **Which of the following enzyme leads to release of free glucose from glycogen during glycogenolysis in muscle?**
    a. Glycogen phosphorylase
    b. Glucose-1-phosphatase
    c. Glucose-6-phosphatase
    d. Debranching enzyme

13. **The immediate product of glycogenolysis in muscle and liver is:**
    a. Free glucose
    b. Glucose-1-phosphate
    c. More glucose-1-phophate and less free glucose
    d. Equal amount of glucose and glucose-1-phosphate

14. **Fanconi-Bickel disease is due to defect in:**
    a. GLUT-1 b. GLUT-2
    c. GLUT-3 d. GLUT-4

15. **On exercise testing (exercising the forearm by squeezing a rubber ball), a patient with McArdle disease exhibits:**
    a. Exercise endurance
    b. Increased blood glucose in the blood drawn from the exercising forearm vein
    c. Increased blood lactate in the blood drawn from the exercising forearm vein
    d. Relatively decreased blood lactate in the blood drawn from the exercising forearm vein

16. **Which of the following enzyme involved in glycogen metabolism is a bifunctional enzyme?**
    a. Glycogen synthase
    b. Glycogen phosphorylase
    c. Branching enzyme
    d. Debranching enzyme

17. **Deficiency of lysosomal maltase causes:**
    a. McArdle disease b. Andersen disease
    c. Cori disease d. Pompe disease

18. A 4-month-old infant was brought by his parents with the complaints of episodes of tremors, lethargy, irritability and multiple episodes of seizure. On examination, the child was found to have a doll-like face. Abdomen was enlarged. Investigations revealed low plasma glucose and elevated serum triglycerides, low density lipoprotein (LDL), free fatty acids, lactate, and uric acid. Ultrasound scan of the abdomen revealed enlarged liver and kidneys. Liver biopsy was done. The liver tissue was found to contain large prominent lipid vacuoles, increased glycogen content. What is the most probable enzyme deficiency?

    a. Glucose-6-phosphatase
    b. Branching enzyme
    c. Debranching enzyme
    d. Muscle phosphorylase

19. Why glucose-6-phosphate produced in cytoplasm of hepatocyte is not acted upon by glucose-6-phosphatase as soon as it is formed?

    a. Thermodynamically possible only when gluconeogenesis occurs
    b. Need protein kinase for its activation
    c. Enzyme is present in endoplasmic reticulum
    d. Steric inhibition of phosphatase by albumin

20. A female infant appeared normal at birth but developed jaundice, vomiting, abdominal distension and muscular weakness at 3 months. She had episodes of tremors, lethargy, irritability, particularly on awakening. Examination revealed an enlarged liver. Laboratory analyzes following fasting revealed increased ketone bodies, blood pH 7.25 and elevated alanine transaminase (ALT), aspartate transaminase (AST) and creatine kinase (CK). No elevation of serum lactate and uric acid was seen. Administration of glucagon following a carbohydrate meal elicited a normal rise in blood glucose, but glucose levels did not rise when glucagon was administered following an overnight fast. Electrocardiogram revealed features of ventricular hypertrophy and an abnormal cardiac rhythm. Liver biopsy revealed an increase in the glycogen content in hepatocytes. Which of the following enzyme is most likely to be defective in this patient?

    a. Glucose-6-phosphatase
    b. Branching enzyme
    c. Debranching enzyme
    d. Hepatic phosphorylase

21. After a normal term delivery and unremarkable nursery stay, a 2-month-old female infant was admitted to the hospital because of decreased activity, and feeding and breathing difficulties. On examination, the child was under-weight and showed signs of respiratory distress. Liver was enlarged and muscle tone was reduced. Heart rate was found to be increased. Testing with electrocardiogram (ECG) and chest X-ray showed ventricular hypertrophy. Which of the following enzyme is most likely to be defective in this patient?

    a. Glucose-6-phosphatase
    b. Branching enzyme
    c. Debranching enzyme
    d. Lysosomal acid maltase

Chapter 10 Carbohydrate Metabolism

10.5 OTHER PATHWAYS FOR OXIDATION OF GLUCOSE

PENTOSE PHOSPHATE PATHWAY/HMP SHUNT

The HMP shunt or pentose phosphate pathway is an alternate oxidative pathway for glucose.

Why is it called as 'shunt'?
- ❑ HMP shunt begins with glucose-6-phosphate (hexose monophosphate) and ends with glyceraldehyde-3-phosphate and fructose-6-phosphate which are intermediates of glycolysis.
- ❑ This is why it is named as shunt but the products of HMP shunt do not always necessarily feed into glycolysis.

Why it is called pentose phosphate pathway?
- ❑ Since it is producing pentose phosphate, i.e. Ribose-5-phosphate

HMP shunt is active in:

❑ RBC	❑ Liver	❑ Adrenal cortex	❑ Adipose tissue
❑ Lactating mammary gland		❑ Thyroid gland	❑ Testis

Subcellular location: Cytosol

The reactions of HMP shunt can be divided into the following two phases:

1. **Irreversible, oxidative phase → produces NADPH (reactions catalyzed by G6PD and 6-phosphogluconate dehydrogenase)**
2. **Reversible, nonoxidative phase → produces ribose-5-phosphate**, hence the name pentose phosphate pathway.

Rate-limiting enzyme: Glucose-6-phosphate dehydrogenase (G6PD)

Stoichiometry of HMP Shunt

3 Glucose-6-phosphate + 6 $NADP^+$ → 3 CO_2 + 6 NADPH + 6 H^+ + 2 fructose-6-phosphate + Glyceraldehyde-3-phosphate.

SIGNIFICANCE OF HMP SHUNT

- ❑ **Generation of NADPH:** NADPH is important for maintaining glutathione in reduced state, reductive biosynthesis and xenobiotic metabolism. The following are some of the enzymes utilizing NADPH:
  - ○ HMG-CoA reductase
  - ○ Folate reductase
  - ○ Ribonucleotide reductase
  - ○ Glutathione reductase
  - ○ Cytochrome P450 mixed function oxidase
  - ○ NADPH met Hb reductase
  - ○ NO synthase
  - ○ NADPH oxidase
  - ○ Aldose reductase
- ❑ **Generation of ribose-5-phosphate**—Synthesis of the nucleotides and coenzymes like NAD, FAD.

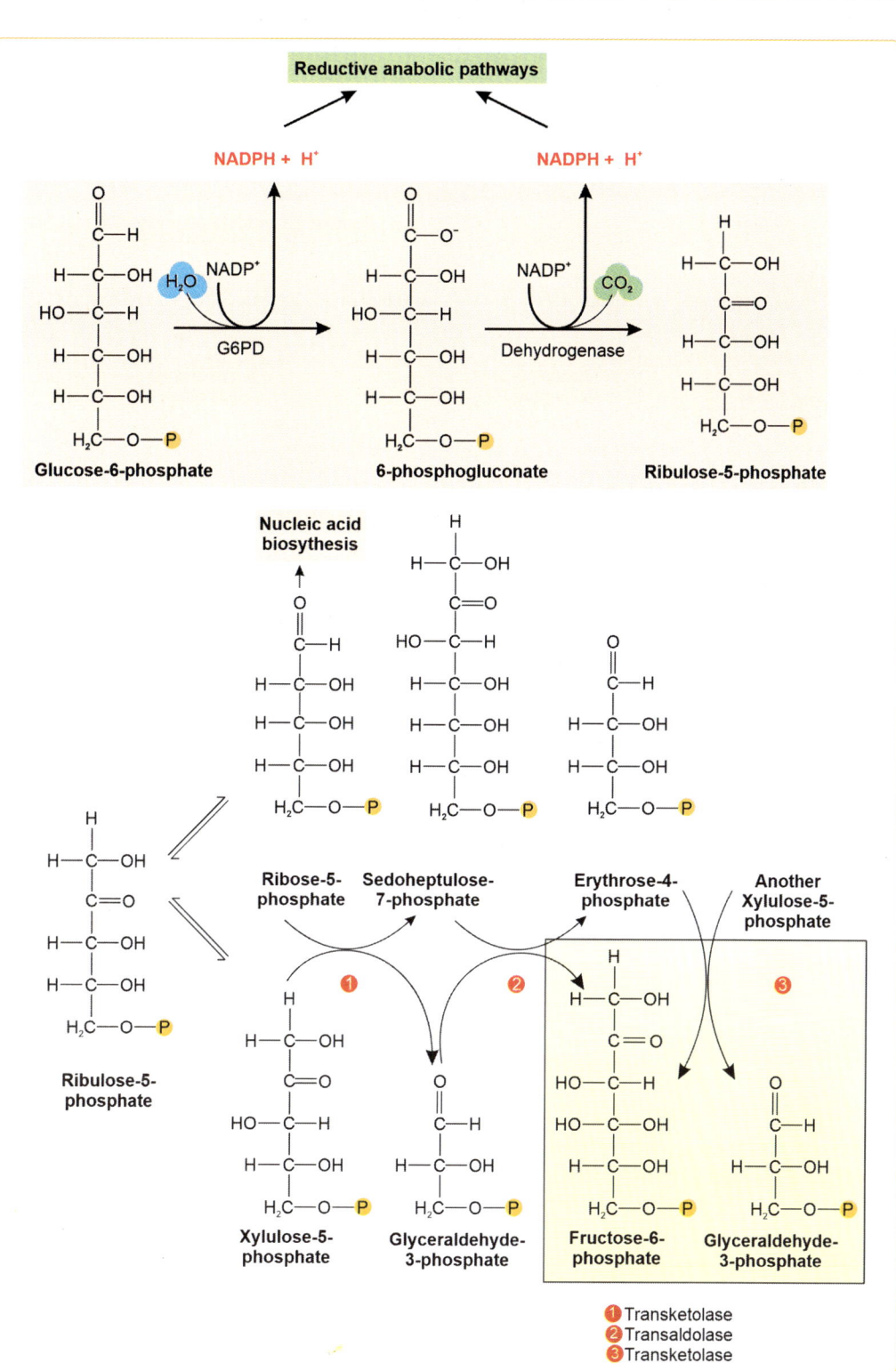

$$\dot{C}_5 + C_5 \rightleftharpoons C_3 + C_7 \qquad \text{(transketolase)}$$
$$C_3 + C_7 \rightleftharpoons C_6 + C_4 \qquad \text{(transaldolase)}$$
$$C_4 + C_5 \rightleftharpoons C_6 + C_3 \qquad \text{(transketolase)}$$

Transketolase enzyme requires thiamine pyrophosphate cofactor. Vitamin B_1 deficiency leads to decreased transketolase activity. Increased RBC transketolase activity after addition of TPP *in vitro* is diagnostic of thiamine deficiency (Fig. 1).

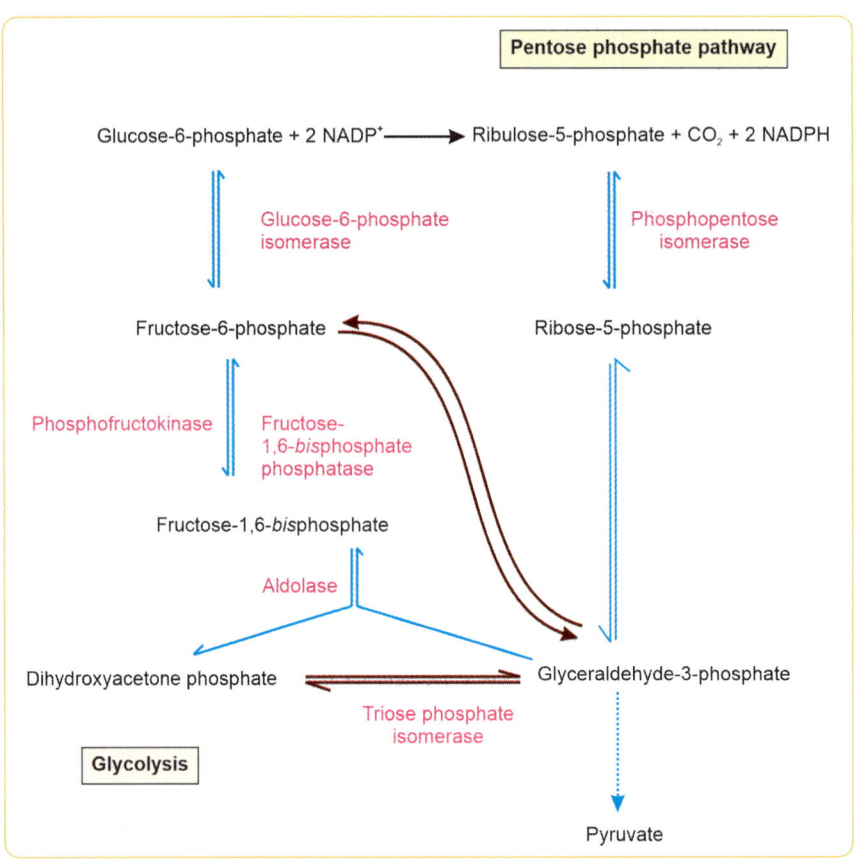

Fig. 1: Interlink between HMP shunt and glycolysis

 High Yield

Role and mode of operation of HMP shunt in various organs:

Oxidative phase is active in RBC, adipocytes, adrenal cortex, lactating mammary gland:

- In RBCs, NADPH is important for maintaining glutathione in reduced state.
- In phagocytes, NADPH is needed to produce free radicals via respiratory burst.
- NADPH is needed for lipid synthesis in adipocytes, adrenal cortex and lactating mammary gland.
- Thus, oxidative phase of the HMP shunt is active in these tissues.

Nonoxidative phase operates in skeletal muscle:

- In tissues which do not require NADPH like skeletal muscle, reversible phase operates from the glycolytic intermediates to generate pentose sugars for nucleic acid synthesis.

Link between HMP shunt and glycolysis:

❑ Xylulose 5-phosphate increases the rate of glycolysis by activating the protein phosphatase that dephosphorylates PFK-2 bifunctional enzyme. Dephosphorylation leads to activation of kinase activity and inactivation of the phosphatase activity.

Regulation of HMP shunt:

❑ The rate limiting enzyme of HMP shunt is Glucose-6-phosphate dehydrogenase (G6PD). The enzyme requires NADP+.

❑ As given above, oxidative and nonoxidative pathway can take place independent of each other.

❑ In liver, pentose phosphate pathway is active during well-fed state.

 Clinical Correlation

Glucose 6-phosphate dehydrogenase deficiency

❑ G6PD is a cytoplasmic enzyme catalyzing the rate limiting step of HMP shunt.

$$\text{Glucose 6-phosphate} + NADP^+ \rightarrow \text{6-phospho Gluconate} + NADPH + H^+$$

❑ G6PD is the **major source of NADPH**

❑ NADPH is the cofactor for glutathione reductase enzyme which **maintains glutathione in the reduced state (G-SH).**

❑ Reduced glutathione is the major defence against oxidative damage of RBC membrane and protein.

G6PD deficiency is an X-linked disorder predominantly affecting males. Affected person is usually **asymptomatic unless he is exposed to oxidative stress** like antimalarial drugs (primaquine), fava beans and severe infection. As there is deficiency of NADPH, glutathione cannot be maintained in reduced state during oxidative stress. This leads to free radical mediated damage of RBC membrane and hemolysis. Free radical damage to hemoglobin causes precipitation of Hb inside RBCs as inclusions known as **Heinz bodies.** G6PD deficiency gives protection against malaria since there is hemolysis and death of parasite. This is why this condition is more common in coastal regions and regions where malaria is epidemic because it confers survival advantage.

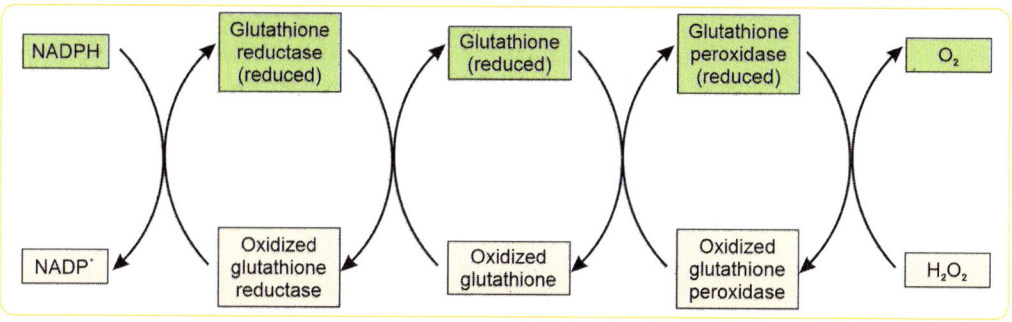

Fig. 2: NADPH is necessary to maintain glutathione in reduced state. Glutathione is involved in detoxification of reactive oxygen species

 **CONCEPT CORNER**

Why RBCs are mainly affected in G6PD deficiency, why not other cells?

❑ G6PD is the major source of NADPH. There are other sources like malic enzyme and cytosolic isocitrate dehydrogenase. Malate and isocitrate are derived in cytoplasm through the transfer of TCA cycle intermediates. RBC lacks mitochondria and thus entirely depends on G6PD for NADPH.

❑ Nucleated cells can continue to produce the mutated enzyme which may have some activity. This cannot happen in the anucleated RBCs.

Chapter 10 Carbohydrate Metabolism

URONIC ACID PATHWAY

Uronic acid pathway is alternate pathway for oxidation of glucose without ATP formation.

Recall the structure of uronic acid. Oxidation of the 6th carbon of glucose produces glucuronic acid.

Products of Uronic Acid Pathway

- ❑ Glucuronic acid
- ❑ Pentose sugars, e.g. Xylulose
- ❑ Ascorbic acid in non-primates

Role of Glucuronic Acid

UDP-glucuronic acid is the activated form of uronic acid involved in:
- ❑ Phase II conjugation reactions, e.g. Bilirubin diglucuronide, Cortisol diglucuronide
- ❑ Glycosaminoglycan biosynthesis

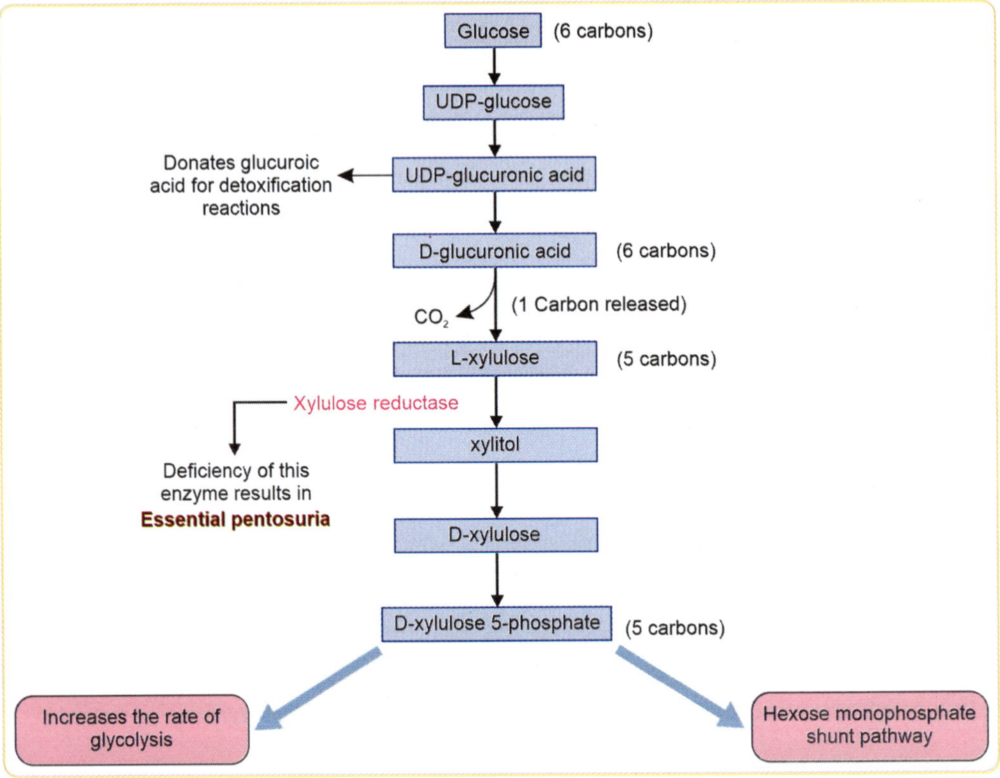

```
                        Glucose   (6 carbons)
                          │
                          ▼
                      UDP-glucose
                          │
                          ▼
Donates glucuroic ◄── UDP-glucuronic acid
acid for detoxification
reactions
                          │
                          ▼
                    D-glucuronic acid   (6 carbons)
                          │
              CO₂ ◄───────┤  (1 Carbon released)
                          ▼
                      L-xylulose   (5 carbons)

        ─── Xylulose reductase
                          │
                          ▼
     │                 xylitol
     ▼                    │
Deficiency of this        ▼
enzyme results in      D-xylulose
Essential pentosuria      │
                          ▼
                  D-xylulose 5-phosphate   (5 carbons)
```

CO_2 (1 Carbon released)

Increases the rate of glycolysis

Hexose monophosphate shunt pathway

Role of Xylulose-5-phosphate

- ❑ Xylulose-5-phoshpate produced by the uronic acid pathway can enter the HMP shunt.
- ❑ In addition, xylulose-5-phosphate *increases the rate of glycolysis* by inactivating the bifunctional enzyme fructose 2,6-bisphosphatase by activating protein 5 phosphatase. (Interlink between HMP shunt and glycolysis)

Ascorbic Acid cannot be Synthesized in Higher Primates Including Humans

- ❑ Due to *deficiency of L-gulonolactone oxidase* of uronic acid pathway.
- ❑ So, Vitamin C (ascorbic acid) cannot be synthesized in humans and must be taken in diet.

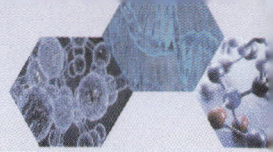

Essential Pentosuria

❑ Essential pentosuria is a benign condition due to the deficiency of the enzyme xylulose reductase/ xylitol dehydrogenase which is inherited by autosomal recessive mode.
❑ Unmetabolised xylulose (pentose) is excreted in the urine, hence the name pentosuria.
❑ The term 'essential' refers to the absence of any secondary cause of pentosuria.

Secondary Pentosuria

❑ Drug-induced pentosuria – due to intake of barbiturates and some drugs
❑ Alimentary pentosuria - due to intake of xylulose or arabinose-rich fruits

Corner

Enzyme Deficiency States in Order of Prevalence

Glucose 6-phosphate dehydrogenase (G6PD) > Pyruvate kinase > Glucose-6-phosphate isomerase. *(Ref. Harrison, 20th ed. p. 709)*

Competency

❑ BI3.5   Describe and discuss the regulation, functions and integration of carbohydrate along with associated diseases/disorders
❑ BI3.4   Define and differentiate the pathways of carbohydrate metabolism (HMP shunt)

Key Points

❑ Hexose monophosphate (HMP) shunt takes place exclusively in cytosol; produces NADPH and pentose sugar.
❑ HMP shunt is least active in skeletal muscle.
❑ There is no ATP or NADH production in the HMP shunt.
❑ NADPH is produced by Glucose-6-phosphate dehydrogenase and 6-phosphogluconate dehydrogenase of HMP shunt; oxidation of 1 mol of glucose produces 2 mol of NADPH.
❑ Transketolase is involved in transfer of 2-carbon groups and it requires thiamine pyrophosphate (vitamin B$_1$).
❑ Transaldolase is involved in the transfer of 3-carbon groups and does not require TPP.
❑ HMP shunt is the source of NADPH during respiratory burst in phagocytes that produces free radicals for killing pathogens by NADPH oxidase.
❑ Uronic acid pathway produces glucuronic acid and xylulose.

SELF-ASSESSMENT

Short Answer Questions

1. Explain the biochemical importance of HMP shunt in the (i) Skeletal muscle and (ii) adrenal cortex.

2. What are the two phases of HMP shunt? Mention the metabolic end-product of each phase.

3. Glycolysis and HMP shunt are two pathways for catabolism of glucose. Indicate the differing biochemical role of these 2 pathways in adipose tissue and erythrocytes.

4. Which is the rate-limiting reaction of the pentose phosphate pathway? What is the consequence of its deficiency? Is there a survival benefit in carriers of this deficiency?

5. List two chief sources and three uses of NADPH in the cell. How do they support the antioxidant defence in the human body?

6. Which enzyme of glucose metabolism plays a key role in the protection against reactive oxygen species in erythrocytes? Explain how?

7. Why RBCs are mainly affected in G6PD deficiency, not other cells?

8. Primaquine administration in G6PD deficient patient can precipitate hemolytic anemia – Why?

9. Which pathway is involved in the production of ascorbic acid in lower animals? Deficiency of which enzyme results in the inability to produce ascorbate in humans?

10. Mention any two metabolic roles of glucuronic acid.

11. What is the metabolic role of xylulose-5-phosphate produced in the uronic acid pathway?

12. What do you understand by the term 'essential' in essential pentosuria? Name the enzymatic defect and the mode of inheritance. What are the secondary causes of pentosuria?

13. Explain why glucose-6-phosphate dehydrogenase (G-6-PD) deficiency leads to hemolytic anemia?

14. Deficiency of Glucose-6-phosphate dehydrogenase leads to deficiency of NADPH and not of ribose-5-phosphate. Why?

Multiple Choice Questions

1. Which of the following links HMP shunt with polyol pathway?
   a. 6-Phosphogluconate
   b. Glucose
   c. NADPH
   d. Ribitol

2. Essential pentosuria is due to deficiency of:
   a. Xylitol dehydrogenase
   b. Xylitol reductase
   c. Xylulose dehydrogenase
   d. Xylulose oxidase

3. Absence of which of the following enzyme is responsible for the inability to synthesize ascorbic acid?
   a. L-Glucuronic acid oxidase
   b. L-Gulonic acid reductase
   c. L-Gulonolactone oxidase
   d. L-Gulonolactone reductase

4. All of the following enzymes utilize NADPH, EXCEPT:
   a. Glutathione reductase
   b. HMG-CoA reductase
   c. Aldose reductase
   d. Xylulose reductase

5. Oxidative phase of HMP shunt pathway is least active in:
   a. Adrenal cortex
   b. Lactating mammary gland
   c. RBC
   d. Skeletal muscle

6. Vitamin cofactor required for the transaldolase is:
   a. B_1 b. B_2
   c. B_3 d. None

7. The number of ATPs produced by HMP shunt is:
   a. Zero   b. One
   c. Two   d. Four

ANSWERS					
1. c.	**2.** a.	**3.** c.	**4.** d.	**5.** d.	**6.** d.
7. a.					

Section II Intermediary Metabolism

10.6 METABOLISM OF OTHER SUGARS

MONOSACCHARIDES ARE INTERCONVERTIBLE

We have studied how electrons are extracted from glucose and how ATP is directly produced in glycolysis at substrate-level. In addition to the metabolic role, glucose and other sugars have structural role too, as a component of glycoconjugates (glycoproteins, proteoglycans and glycolipids). In this chapter, we will learn how various monosaccharides are interconvertible in our body.

GLUCOSE CAN BE CONVERTED TO FRUCTOSE BY POLYOL PATHWAY

Glucose cannot be directly isomerized to fructose in human body by any enzyme.

Fructose is the main source of energy for sperm. Thus, we need a pathway to convert glucose to fructose. This is achieved by *polyol pathway*.

First carbon of glucose is reduced to produce sorbitol by aldose reductase and then the second carbon is oxidized to produce fructose by sorbitol dehydrogenase.

Aldose reductase uses NADPH as electron donor. Sorbitol dehydrogenase uses NAD^+ as electron acceptor.

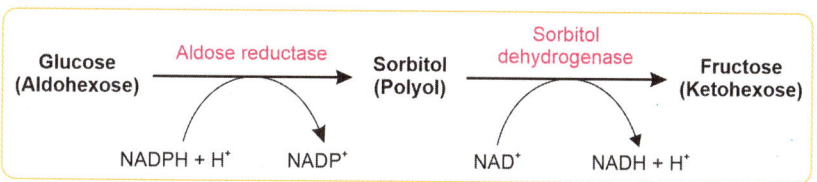

DIABETIC CATARACT IS DUE TO EXCESSIVE ACCUMULATION OF SORBITOL

In diabetes, excess glucose accumulates in the blood. Because of this, excess glucose enters lens and is converted to sorbitol by aldose reductase. **Production of sorbitol is faster than the conversion of sorbitol to fructose**. Sorbitol is osmotically more active and attracts water inside the lens. This disrupts the regular arrangement of lens fibers that is responsible for the maintenance of transparency.

Sorbitol accumulation is also responsible for diabetic neuropathy and nephropathy.

Sorbitol, Lactulose and fructose are frequently malabsorbed in the intestine. Osmotic diarrhea can happen on ingestion of medications, gum, or candies sweetened with these poorly or incompletely absorbed sugars. *(Ref. Harrison 20th Ed., p. 264)*

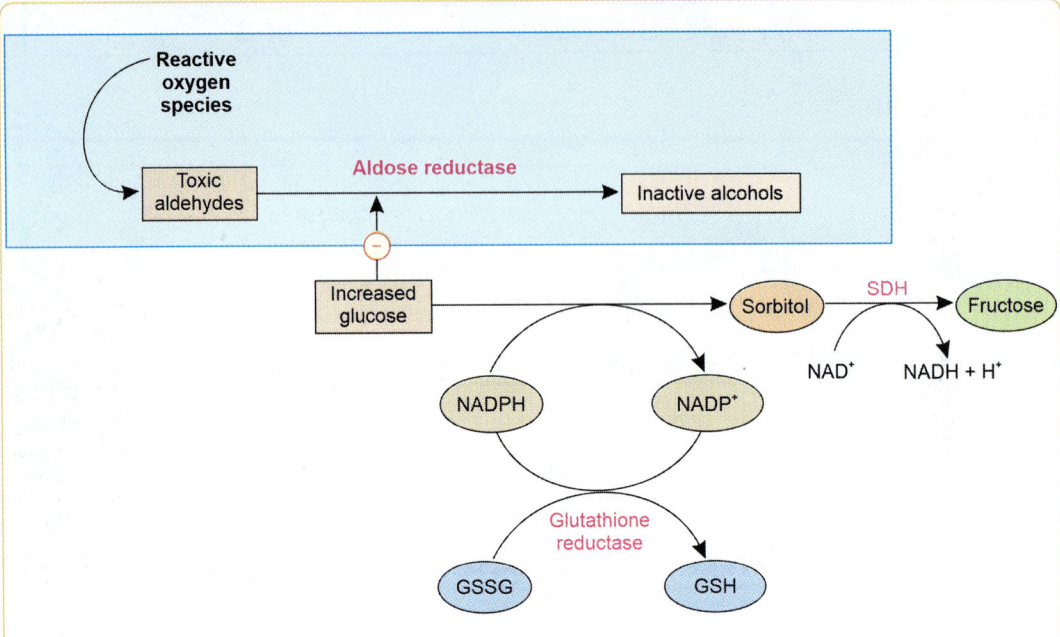

Fig. 1: In our body, toxic aldehydes are produced by reactive oxygen species (ROS). These toxic aldehydes are detoxified by aldose reductase. Excess sorbitol competes for this enzyme and leads to accumulation of toxic aldehydes and depletion of NADPH

FRUCTOSE ENTERS THE GLYCOLYSIS BYPASSING THE REGULATORY STEP OF GLYCOLYSIS

GLUT5 is the specific transporter of fructose. In liver, fructose is phosphorylated to fructose-1-phosphate by the fructokinase enzyme.

Fructose-1-phosphate is cleaved by aldolase B into dihydroxyacetone phosphate and glyceraldehyde. DHAP can directly enter glycolysis. Glyceraldehyde is converted to glyceraldehyde-3-phosphate by triose kinase and then enters glycolysis.

Will hexokinase act on fructose?

Yes. Hexokinase can act on all hexoses. But the K_m for each hexose is different. K_m of hexokinase for fructose is high.

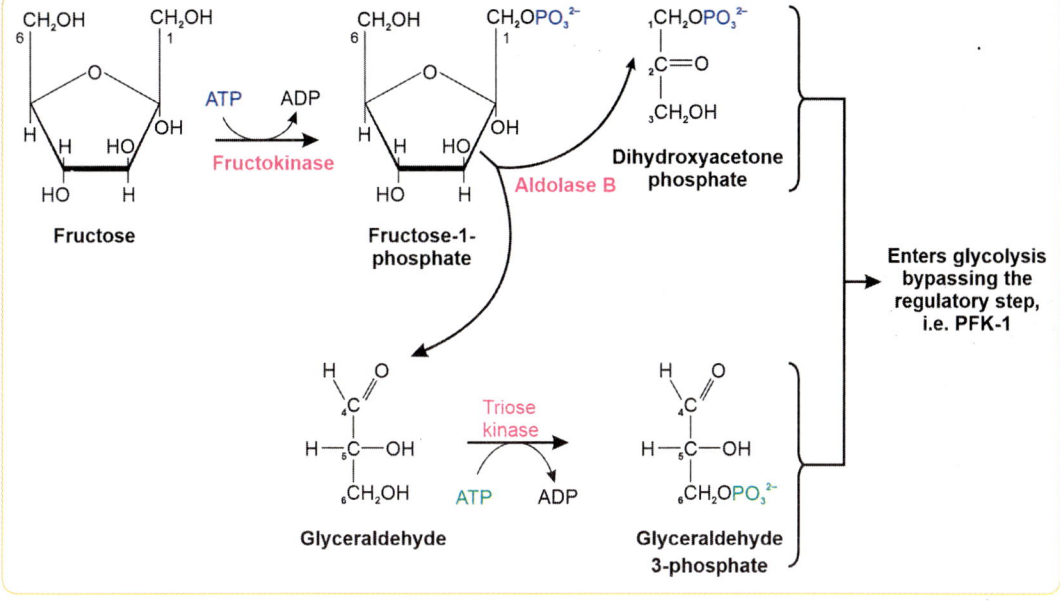

Fig. 2: Fructolysis

The above pathway is known as fructolysis. **ATP yield of fructose is same as that of glucose**. One may argue that fructose enters glycolysis bypassing the energy investment step and thus 1 ATP should be obtained extra. Look at the ATP used in the triose kinase enzyme (Fig. 2).

Essential Fructosuria due to Fructokinase Deficiency is a Benign Condition

Deficiency of fructokinase → inability to trap fructose in the form of fructose-1-phosphate inside the hepatocytes→ accumulation of fructose in blood → excess fructose appears in urine. This is a completely benign condition and is usually an incidental finding during urine examination.

Ingestion of fructose (sugar, fruits, fructose polymers, etc.) → Liver cannot use fructose because of absence of fructokinase → Hexokinase uses fructose very slowly → Accumulation of fructose in blood → Fructose excreted in urine → → Benedict's test positive and glucostix test negative in urine.

Hereditary Fructose Intolerance due to Aldolase B Deficiency is a Serious Condition

Deficiency of Aldolase B is a serious condition compared to deficiency of fructokinase.

Biochemical Basis

❑ Fructose is converted to fructose-1-phosphate and gets trapped inside the hepatocytes → inorganic phosphate depleted → impaired ATP synthesis → as there is less ATP, gluconeogenesis is impaired (Remember: 6 ATPs are required for synthesis of glucose from 2 molecules of lactate).
❑ Excessive ATP breakdown → ↑ Uric acid production → Gout
❑ In addition, fructose-1-phosphate inhibits glycogenolysis → severe '**post-prandial' hypoglycemia**
❑ Accumulation of charged compound inside the cell causes cellular toxicity → damage to hepatocytes → impaired liver function.

Clinical Features

❑ Infants are normal until the introduction of fructose/sucrose in the diet. Thus, the features begin usually after weaning (withdrawal from breastfeeding).

□ Nausea and vomiting after intake of fructose containing food with signs of hypoglycemia—sweating, tremors and convulsions.
□ Impaired liver function—Raised aminotransferase, hypoalbuminemia, impaired clotting.

Treatment

□ Complete restriction of fructose. (As fructose can be formed from glucose, there is no exogenous requirement for fructose.)
□ In due course of time, the disease becomes milder.

High Fructose Diet Causes Hypertriglyceridemia

□ Fructose enters the glycolysis bypassing the regulatory step catalyzed by PFK-1. Thus, fructolysis is not inhibited even if there is enough ATP. This excess ATP favors lipogenesis in liver.
□ In addition, fructose activates the transcription factors like carbohydrate-response element-binding protein (ChREBP) which are involved in lipogenesis.
□ Therefore **fructose is not an alternative for glucose in diabetics**.
□ This is why sweetening agents like high-fructose corn syrup should be used in moderation.

Inconvertibility of Galactose and Glucose

□ Lactose is the predominant dietary source of galactose.
□ Galactose is exclusively metabolized in the liver.
□ Catabolic pathway of galactose is known as 'Leloir pathway'.
□ The net reaction is conversion of galactose to glucose.
□ At the same time, formation of UDP-galactose is also important. If galactose is not available, UDP-glucose can be converted to UDP-galactose as the epimerase reaction is reversible. So, Galactose is not essential in diet.
□ UDP-galactose is the active donor of galactose.
□ UDP-galactose is essential for the formation of compounds like **lactose, glycosaminoglycans, glycoproteins, galactocerebrosides and other glycolipids (Fig. 3)**

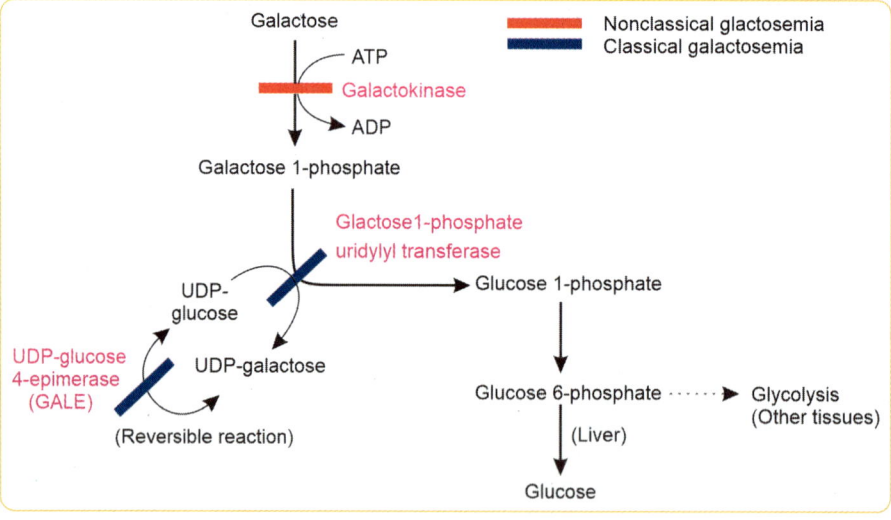

Fig. 3: The net reaction is conversion of galactose to glucose; thus, galactose is not essential in diet

Galactosemia Results from Impaired Galactose Metabolism

Galactosemia is an inborn error of galactose metabolism inherited in autosomal recessive manner.

Enzyme Deficiency

- Galactose-1-phosphate uridyltransferase (Most common)
- Galactokinase
- UDP-Galactose 4'-epimerase (GALE)

Pathophysiology

- Classic galactosemia is due to galactose-1-phosphate uridyltransferase deficiency. Deficiency of other enzymes is relatively benign.
- Galactose derived from milk is accumulated in the form of galactose-1-phosphate.
- This uses all the ATP and depletes inorganic phosphates in the cell.
- Accumulation of charged compound damages the cells of liver, kidney and brain.
- Galactose accumulated in galactosemia enters the lens of the eye and gets converted to osmotically active galactitol (dulcitol) by aldose reductase. This leads to accumulation of water → damage to lens fibers → loss of transparency → cataract.

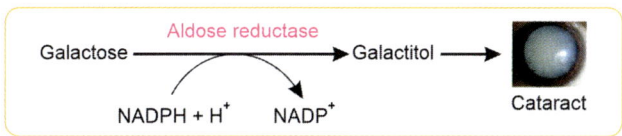

Clinical Features

- Postprandial hypoglycemia—due to depletion of ATPs
- Poor feeding, diarrhea, failure to thrive
- Cataract—due to conversion of galactose to osmotically active galactitol (dulcitol)
- Hepatomegaly, jaundice—due to hepatic cell damage
- Mental retardation—due to defective formation of galactocerebrosides
- *E. coli* sepsis
- Premature ovarian failure in female

Treatment

- Complete restriction of galactose and galactose containing sugars like lactose
- Breastfeeding should be avoided

Key Points

- Glucose can be converted into fructose by polyol pathway.
- Diabetic cataract is due to excessive accumulation of sorbitol (polyol).
- Fructose enters the glycolysis bypassing the regulatory step of glycolysis.
- Essential fructosuria due to fructokinase deficiency is a benign condition.
- Hereditary fructose intolerance due to aldolase B Deficiency is a serious condition.
- High fructose diet cause hypertriglyceridemia.
- Galactose and glucose are interconvertible.
- The classical galactosemia occurs due to deficiency of the enzyme galactose-1-phosphate uridyl transferase.

SELF-ASSESSMENT

Short Answer Questions

1. Write the reactions in which glucose is converted to fructose.

2. Explain the biochemical basis of diabetic cataract.

3. Continued intake of high fructose diet can cause hypertriglyceridemia – How?

4. How is fructose metabolized in our body? Describe the inherited enzymatic defects leading to fructosuria. Explain why fructose challenge in patients with these disorders leads to hypoglycemia?

5. Specify the inherited enzyme defect in galactosemia. Explain how normal body requirement of galactose is met when the patient is kept on galactose-free diet.

6. It is Galactose an essential component of our diet? Justify your answer. Why do children with galactosemia develop cataract?

7. Mention any three clinical features of galactosemia and explain the biochemical basis of those three.

8. Fructosuria is a benign condition whereas hereditary fructose intolerance is a severe condition. Explain.

Multiple Choice Questions

1. Which is the correct sequence of reactions in the polyol pathway?
   a. Glucose fructose sorbitol
   b. Glucose sorbitol fructose
   c. Fructose sorbitol glucose
   d. Fructose glucose sorbitol

2. Which of the following transporter is specific for fructose?
   a. GLUT1   b. GLUT3
   c. GLUT4   d. GLUT5

3. All of the following are true about hereditary fructose intolerance, EXCEPT
   a. Fasting hypoglycemia
   b. Deficiency of Aldolase B
   c. Impaired liver function
   d. Hyperuricemia

4. If galactose is not available in the body, which of the following enzyme can convert UDP-glucose to UDP-galactose?
   a. Galactose-1-phosphate uridyl transferase
   b. Galactokinase
   c. Glucokinase
   d. UDP-Galactose 4-epimerase

ANSWERS

1. b.	2. d.	3. a.	4. d.

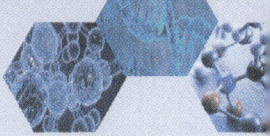

10.7 REGULATION OF BLOOD GLUCOSE

You are aware of the term 'homeostasis, coined by W.B. Cannon. Homeostasis is "the maintenance of a stable internal environment within an organism". Maintaining a stable internal environment in our body includes regulation of blood glucose level also. Glucose is an obligate metabolic fuel for the brain under physiologic conditions and the brain cannot synthesize glucose or store it as glycogen. So, even during fasting, basal level of glucose needs to be maintained. This is achieved by regulation of enzymes through feedback mechanism and neurohormonal control.

NORMAL BLOOD GLUCOSE LEVELS

- ❑ Fasting levels: 70-100 mg/dL
- ❑ Postprandial: up to 140 mg/dL

Optimal glucose concentration is maintained within physiological limits by:
- ❑ The rate of glucose entry into the blood circulation
- ❑ The rate of its removal from the circulation.

Routes of glucose entry into the blood

❑ Fed state:
   ○ Through portal system to the liver and then to systemic circulation
 ❑ Fasting state
   ○ Hepatic glycogenolysis
   ○ Hepatic and renal gluconeogenesis

Glucose removal from the blood

❑ Fed as well as Fasting state:
   ○ Utilization by RBCs and Brain
 ❑ Fed State:
   ○ Conversion to glycogen in liver and muscles
   ○ Conversion to fatty acid in lipogenic tissues
   ○ Conversion into other sugars:
     ➢ To fructose in seminal fluid
     ➢ To lactose in mammary gland
     ➢ To ribose via HMP shunt in actively dividing cells
     ➢ To mannose and galactose for glycoprotein synthesis

Chapter 10 Carbohydrate Metabolism

PHASES OF GLUCOSE HOMEOSTASIS BASED ON THE FAST-FED CYCLE

State	Well-fed state (3-4 hrs. after meal)	Post- Absorptive state (16-18 hrs. after meal)	Fasting(1-1½day)	Prolonged fasting/ Starvation (2-3 days)
Source of glucose	Diet	Hepatic glycogen and Gluconeogenesis	Hepatic and Renal gluconeogenesis	Renal and hepatic gluconeogenesis
Tissues using glucose	All	All tissues, but in Liver, muscle and adipose tissue, the rate of utilization is slowed	Brain and RBCs and cells lacking mitochondria; small amount by muscle	Brain at a slower rate, RBCs normal rate
Major fuel of brain	Glucose	Glucose	Glucose and ketone bodies	Ketone bodies and glucose

We will discuss the fast-fed cycle in Chapter 16 *(Refer to page no. 337)* and we have already discussed the modes of regulation of glycolysis, gluconeogenesis and glycogenolysis. The regulation can be both short-term and long-term.

Short-term Regulation

❑ Allosteric modulators of rate limiting enzymes
❑ Phosphorylation and dephosphorylation mediated by insulin and glucagon

Long-term Regulation

❑ Control of gene expression by insulin and glucagon
In this chapter, we will focus on the role of insulin and glucagon on the regulation of blood glucose and diabetes mellitus.

INSULIN

Insulin is the only Hypoglycemic Hormone in Our Body

Hypoglycemia is more dangerous than hyperglycemia. So, nature has chosen only one hypoglycemic hormone, i.e. Insulin.

On the other hand, there exists a variety of hormones to increase blood glucose. They are glucagon, growth hormone, cortisol, thyroxine and catecholamine. As these hormones counter the action of insulin, these are known as counterregulatory hormones.

Insulin is Released in Response to a High Blood Glucose

Beta cells in the islets of Langerhans of pancreas have low-affinity GLUT2 receptors with high Km value. So, glucose enters through these receptors only when the blood glucose level is high. Entry of glucose triggers the release of insulin in the following way:

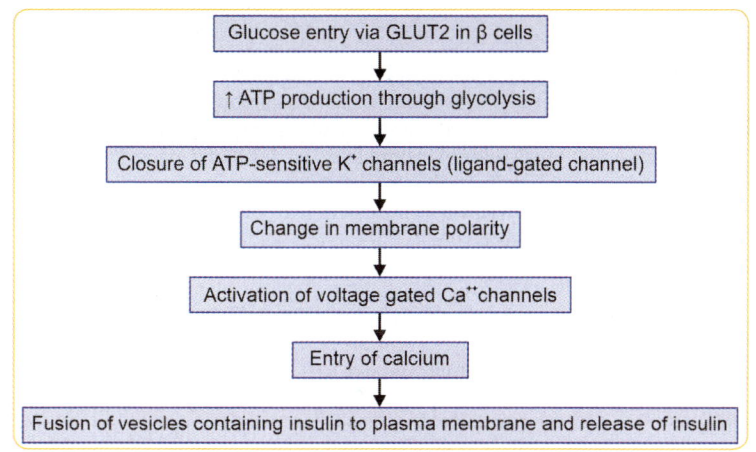

Mechanism of insulin secretion

Phases of Insulin Secretion

Phase	Description
Basal	There is constant basal secretion of insulin which is important to supress undue lipolysis, proteolysis, and glycogenolysis.
First-phase	Release of insulin in response to high-blood glucose begins within 2-minutes and continues for 10 to 15 minutes. Failure of first-phase response in response to high intravenous glucose is seen in people with insulin resistance who may develop type 2 DM in future.
Second-phase	The first-phase is followed by sustained release of insulin independent of high blood glucose. The 2nd phase continues till the restoration of euglycemia.

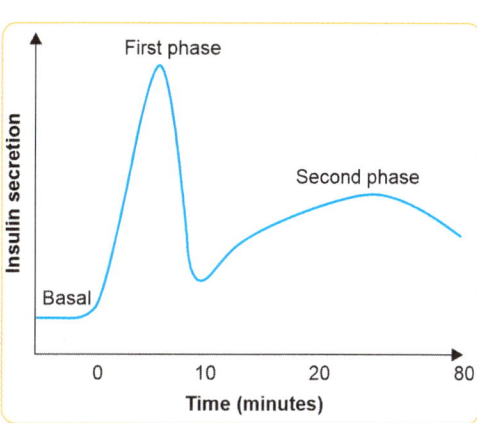

Note: In addition to high glucose, amino acids like arginine also trigger insulin release.

C-Peptide

C-peptide (connecting peptide) is a 31-amino-acid polypeptide that connects insulin's A-chain to its B-chain in the proinsulin molecule. During conversion of proinsulin to insulin, C-peptide is removed. So, equimolar amount of C-peptide and insulin are released in blood. So, measurement of C-peptide has the following diagnostic purposes.

❑ Insulinoma (tumor of beta cells) – high level of insulin and high level of C-peptide
❑ Untreated type I DM – endogenous insulin itself is low and C-peptide level will also be low
❑ Early stages of type II DM - C-peptide level will be normal or high (hyperinsulinemia due to insulin resistance)

Insulin Acts Through Receptor with Tyrosine Kinase Activity

❑ Endogenously produced insulin has a short-half of 4 to 6 minutes. Insulin binds to membrane bound receptors with intrinsic tyrosine kinase activity.
❑ Insulin receptor is a glycoprotein with two alpha and two beta subunits.
❑ Alpha subunits are extracellular and bind to insulin while beta subunits are intracellular and involved in the signal transduction.
❑ Binding of insulin induces a conformational change in the receptor and activates the intrinsic tyrosine kinase activity of beta subunits.
❑ Each beta subunit phosphorylates the tyrosine residue of the other. This is known as cross phosphorylation.

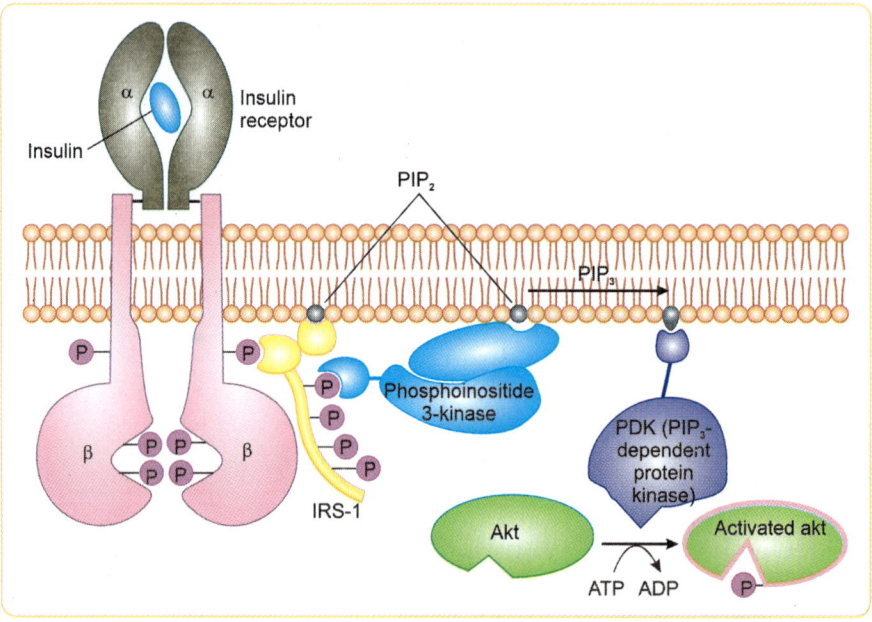

❑ Phosphorylated tyrosine residues are recognized and bound by insulin receptor substrate (IRS) proteins.
❑ Tyrosine kinase activity of insulin receptor phosphorylates the tyrosine residues of IRS also. Phosphotyrosine residues of IRS protein is recognized by Phosphoinositide-3-kinase enzyme.
❑ As the name suggests, PI3 kinase converts phosphatidylinositol 4,5- bisphosphate (PIP 2) into phosphatidylinositol 3,4,5-trisphosphate (PIP 3).
❑ PIP 3 is a second messenger (Chapter 38) *(Refer to page no. 586)*. It activates phosphatidylinositol dependent protein kinase (PDK) which in turn activates many other proteins especially Protein kinase B, also known as Akt.

Akt is Responsible for Many Downstream Effects of Insulin

❑ Akt mediated phosphorylation of cytosolic GLUT4 carrying vesicle → Translocation of GLUT4 to plasma membrane.

- Akt inhibits glycogen synthase kinase by phosphorylation → Glycogen synthesis activated.
- Akt activates ATP-citrate lyase by phosphorylation → Substrates for fatty acid synthesis are made available
- Akt phosphorylates FoxO transcription factors which is needed for the expression of gluconeogenic enzymes → Phosphorylated FoxO is ubiquitinated and destroyed → Gluconeogenesis is inhibited (Fig. 1).

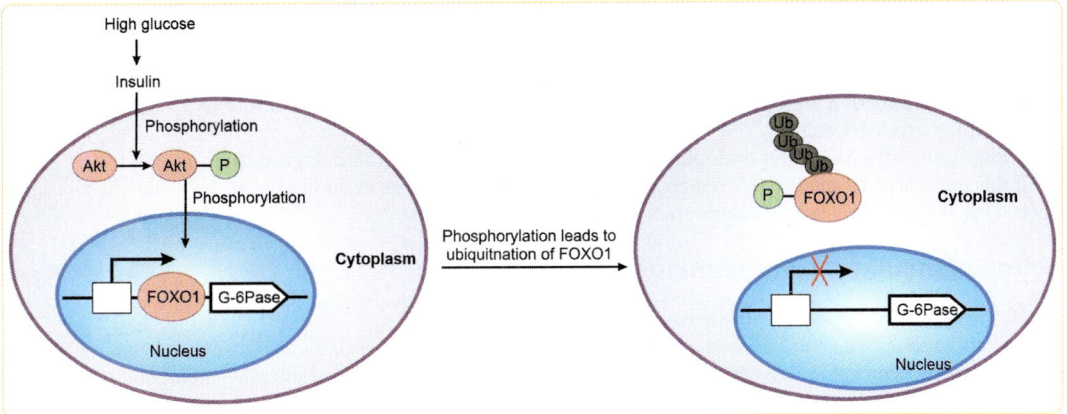

Fig. 1: Insulin-regulated nuclear exclusion of FOXO and its effect on transcription of glucose-6 phosphatase

Insulin Stimulates Phosphatase-1 and Phosphodiesterase 3

- Stimulation of phosphatase-1 leads to dephosphorylation of cAMP-dependent protein kinase A, glycogen synthase, glycogen phosphorylase, phosphorylase kinase, PFK-2 etc → Inhibition of blood glucose increasing pathways
- Stimulation of phosphodiesterase 3 leads to degradation of cAMP to AMP → Inhibition of blood glucose increasing pathways

Note: "Glucagon favours phosphorylation and Insulin favours dephosphorylation" is an oversimplified statement. Insulin favours phosphorylation of certain enzymes through Akt and dephosphorylation of certain enzymes through phosphatase-1.

Anabolic Effect of Insulin is Explained by MAP Kinase Pathway

Most of the metabolic effects of insulin are explained by PI3 kinase pathway whereas the anabolic effects can be explained by MAP kinase pathway.

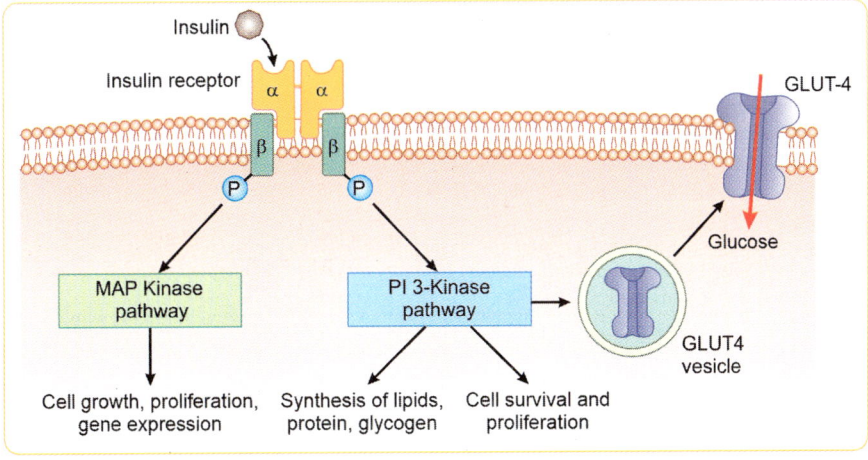

Insulin Resistance is Due to Excess Accumulation of Fatty Acids in Skeletal Muscles

❑ Failure of the cell to respond to insulin is known as insulin resistance. The resistance can be at the level of receptor or at the post-receptor signaling.
❑ Excessive fatty acids inside the cytosol interfere with the insulin signaling.
❑ Skeletal muscles utilize free fatty acids at rest. So, excess dietary fat leads to excess accumulation of fatty acids in the myocytes and insulin resistance in myocytes.
❑ Pancreas tries to compensate the resistance by producing excess insulin.
❑ Excessive production increases the chance of protein misfolding and induces **unfolded protein response** in the endoplasmic reticulum.
❑ Endoplasmic stress activates **endoplasmic reticulum-associated protein degradation** (ERAD).
❑ If this process is not intervened, pancreatic failure and frank diabetes mellitus ensue.
Note: Hyperinsulinemia in type 2 diabetes mellitus is due to insulin resistance.

Action of Insulin is Countered by Many Hormones

❑ Glucagon – regulation of blood glucose **during fasting**
❑ Catecholamines – increased blood glucose **during acute stress** (fight or flight response)
  ○ Hypoglycemia stimulates catecholamine secretion. This explains why there is sweating and palpitations during hypoglycemia.
  ○ Catecholamines works through the second messengers cAMP and calcium. The effects are short-lived.
  ○ **Note:** Catecholamines stimulate glycolysis in muscle but inhibit glycolysis in liver
❑ Glucocorticoids – increased blood glucose **during chronic stress** (chronic infection, persistent psychological stress)
  ○ Glucocorticoids act at the gene level. The effects are **long-lived**.
  ○ Therefore, patients taking glucocorticoids develop glucose intolerance.

Metabolic Action	Insulin	Glucagon
Glycogen synthesis	↑	↓
Glycolysis (energy release)	↑	↓
Lipogenesis	↑	↓
Protein synthesis	↑	↓
Glycogenolysis	↓	↑
Gluconeogenesis	↓	↑
Lipolysis	↓	↑
Ketogenesis	↓	↑

	Insulin	Glucagon
Major role	Decreases the blood glucose and Anabolic role	Maintains the blood glucose level
Produced by	β-cells in islet of Langerhans	α-cells in islet of Langerhans
Receptor	Receptor with intrinsic tyrosine kinase activity	Seven transmembrane serpentine G-protein coupled receptor
Target organ	Liver and extrahepatic tissues	Only Liver

DIABETES MELLITUS

Diabetes mellitus is defined as a heterogeneous group of syndromes characterized by an elevation of fasting blood glucose caused by a relative or absolute deficiency of insulin.

American Diabetic Association criteria for DM		
	Diabetes	Pre-diabetes
HbA$_1$C	≥ 6.5 %	5.7-6.4%
	OR	
Fasting Blood Sugar	≥126 mg/dl*	100-125
	OR	
2-hour plasma glucose during a 75 gm OGTT	≥200 mg/dl*	140-199
	OR	
Random Blood Sugar in a patient with classic symptoms of hyperglycemia† or hyperglycemic crisis	≥200 mg/dl	---

Fasting: no caloric intake for at least 8 hours
*In the absence of unequivocal hyperglycemia, result should be confirmed by repeat testing
†Polyuria, polydipsia & unexplained weight loss

Classification of Diabetes Mellitus

Diabetes mellitus is a phenotype which can be seen in many conditions like Type 1 DM, Type 2 DM, Gestational diabetes and other types.

Common differences between type 1 & type 2 DM		
	Type 1	Type 2
Age of onset	Usually before puberty	Usually after adulthood
Development of symptoms	Rapidly	Gradually
Genetic predisposition	Moderate	Strong
Acute Complications	Diabetic ketoacidosis is more common	Hyperosmolar state
Plasma Insulin level	Low to absent due to beta cell failure	High in the early stage (Hyperinsulinemia) followed by beta cell failure
Response to oral hypoglycemic agents	Absent	Present
Treatment	Insulin is necessary	Life-style modifications (Diet & Exercise) along with drugs. Insulin is required in some patients

Latent Autoimmune Diabetes in Adults (LADA)

❑ LADA, sometimes called as type 1.5 diabetes, is a progressive autoimmune diabetes like type 1 but manifests in the adulthood.
❑ LADA should be differentiated from type 2 DM.
❑ People with LADA require insulin.

Maturity-Onset Diabetes of the Young

❑ Diabetes mellitus is a polygenic disease. But maturity-onset diabetes of the young (MODY) is a monogenic disease.
❑ Monogenic refers to defect in one gene, i.e. In MODY 10, the insulin gene on chromosome 11 is affected.
❑ We have learnt that glucokinase mediated glycolysis and ATP production is necessary for the secretion of insulin from the pancreatic β cells. When the glucokinase gene is defective, insulin secretion is affected.
❑ All types of MODY are inherited in **autosomal dominant** manner.

Type	Defective gene & (Protein)
MODY 1	*HNF4α* (hepatocyte nuclear factor 4α)
MODY2	*GCK* (glucokinase)
MODY 3	*HNF1α* (hepatocyte nuclear factor 1α)
MODY 4	*IPF1* (insulin promoter factor 1)
MODY 5	*HNF1β* (hepatocyte nuclear factor 1β)
MODY 6	*NeuroD1* (neurogenic differentiation factor 1)
MODY 7	*KLF1* (Kruppel-like factor 1)
MODY 8	*CEL* (carboxyl ester lipase)
MODY 9	*PAX4* (paired box transcription factor 4)
MODY 10	*INS* (insulin)
MODY 11	*BLK* (B-lymphocyte-specific tyrosine kinase)

Autosomal dominant Neonatal diabetes is due to mutation of KCNJ11 (inwardly rectifying potassium channel Kir6.2).

Complications of Diabetes Mellitus

Acute	Chronic
❑ Diabetic ketoacidosis (more in type I DM) ❑ Diabetic non-ketotic hyperosmolar coma (more in type II DM) ❑ Hypoglycemia (due to skipping of meal after antidiabetic drugs)	❑ Microvascular (Retinopathy, Neuropathy, Nephropathy) ❑ Macrovascular (Accelerated atherosclerosis)

Advanced Glycation End Products (AGE)

Chronic compilations of Diabetes mellitus are due to advanced glycation end products. Aldehyde group of reducing sugar like glucose nonenzymatically reacts with amino group of proteins to form a reversible **Schiff base**. This reaction is known as **Maillard reaction**. When the Schiff base undergoes irreversible **Amadori rearrangement**, advanced glycation end products are produced, e.g. Carboxymethyl-lysine.

Mechanism of Harmful Effects of AGE

- ❑ AGE forms undesirable protein cross-links and protein adducts leading to protein dysfunction.
- ❑ AGE binds to receptors called RAGE (receptors for AGE) and upregulate the expression of inflammatory genes.
- ❑ Generation of free radicals

$$
\begin{array}{c}
\underset{\text{Reducing sugar}}{\begin{array}{l}
\overset{H}{\underset{C}{}}\diagup\!\!\!\!\diagdown O \\
H-C-OH \\
HO-C-H \\
H-C-OH \\
H-C-OH \\
CH_2OH
\end{array}}
\quad + \underset{\substack{\varepsilon\text{-amino group}\\ \text{of lysine residue}}}{NH_2-Protein}
\rightleftharpoons
\underset{\text{Schiff base}}{\begin{array}{l}
\overset{H}{\underset{C}{}}\diagdown N-Protein \\
H-C-OH \\
HO-C-H \\
H-C-OH \\
H-C-OH \\
CH_2OH
\end{array}}
\rightleftharpoons
\underset{\substack{\text{Ketoamine - Amadori product}\\ \text{(AGE- Advanced glycation end product)}}}{\begin{array}{l}
CH_2-NH-Protein \\
C-O \\
HO-C-H \\
H-C-OH \\
H-C-OH \\
CH_2OH
\end{array}}
\end{array}
$$

Protein – AGE
(Stable protein adduct)

AGE – Protein – AGE
(Stable protein crosslinking)

Accumulation of AGEs in the vessel wall is responsible for the microvascular and macrovascular complications of diabetes. Interestingly, accumulation of AGE is responsible for the ageing of skin.

Diabetic Dyslipidemia

Absolute or relative deficiency of insulin in DM promotes lipolysis in the adipocytes and excess free fatty acids are released in the circulation. High blood glucose and high free fatty acids promote the formation of excess triglycerides in liver and there is excess release of VLDL. Lipoprotein lipase requires insulin. So, in DM, clearance of VLDL is also reduced. So, hypertriglyceridemia is seen in diabetic patients. There is increased small dense LDL and low HDL levels. All of this accelerates the process of atherosclerosis.

Competency

- ❑ BI3.9   Discuss the mechanism and significance of blood glucose regulation in health and disease
- ❑ BI3.10   Interpret the results of blood glucose levels and other laboratory investigations related to disorders of carbohydrate metabolism

Key Points

- ❑ Insulin is the only hypoglycemic hormone in the body.
- ❑ Glucagon, Epinephrine, Cortisol, etc. are the counter regulatory hyperglycemic hormones.
- ❑ Insulin acts through the receptor with intrinsic tyrosine kinase activity.
- ❑ Sulphonyl urea drugs inhibit the K^+ sensitive ATP Channels.
- ❑ Insulin resistance is due to excess accumulation of fatty acids in skeletal muscles.
- ❑ HbA1C >6.5 is diagnostic of Diabetes mellitus.
- ❑ Diabetes mellitus is a polygenic disease. But maturity-onset diabetes of the young (MODY) is a monogenic disease.
- ❑ Accumulation of advanced glycation end products in the vessel wall is responsible for the vascular complications of diabetes.

Chapter 10 Carbohydrate Metabolism

Short Answer Questions

1. Name the hormones that regulate glycolysis and gluconeogenesis and their mechanism of action.

2. What are the diagnostic criteria for diabetes mellitus (DM)? Differentiate type I and type II DM. Discuss the biochemical mechanisms leading to metabolic derangements and acute complications in DM.

3. Discuss with a suitable diagram of how the rise in blood glucose level leads to the secretion of insulin from β-cells of the pancreas. Describe the mechanism of β-cell dysfunction in type 2 diabetes mellitus.

4. Explain the three phases of insulin secretion with the help of a schematic diagram.

5. What is the diagnostic role of C-peptide of insulin?

6. Explain how excessive fatty acids in skeletal muscle cause insulin resistance?

7. Differentiate the metabolic actions of insulin and glucagon in a tabular format.

8. How advanced glycation end products are produced in our body? Why are they detrimental to health?

9. Write a short note on diabetic dyslipidaemia.

10. Differentiate type 1 and type 2 Diabetes Mellitus.

Multiple Choice Questions

1. All of the following are involved in the glucose induced secretion of insulin from the beta cells of pancreas, EXCEPT
   a. Glucose entry via GLUT2 transporter
   b. ↑ATP production through glycolysis
   c. Closure of ATP-sensitive K+ channels
   d. Inhibition of voltage gated Ca++channels

2. What is the nature of insulin receptor?
   a. G-protein coupled receptor
   b. Receptor tyrosine kinase
   c. Non-receptor tyrosine kinase
   d. Receptor with ion channels

3. What is the other name of Akt?
   a. Protein kinase A   b. Protein kinase B
   c. Protein kinase C   d. Protein kinase D

4. Which is the HbA1C cut-off value for diagnosis of diabetes mellitus?
   a. 5.5   b. 6
   c. 6.5   d. 7

5. All of the following are involved in insulin mediated cellular signaling, EXCEPT
   a. Protein kinase B
   b. Phosphoinositide-3-kinase
   c. Phosphatidylinositol dependent protein kinase
   d. Phospholipase

ANSWERS

1. d.	**2.** b.	**3.** b.	**4.** c.	**5.** d.

Amino Acid Metabolism

11.1 DISPOSAL OF AMINO GROUP AND UREA CYCLE

PROTEINS ARE NOT STORED IN THE BODY

In our body, carbohydrates are stored in the form of glycogen and lipids are stored in the form of triglycerides. However, proteins are not stored. Excess dietary proteins are degraded.

In addition to dietary proteins, many proteins present in the body undergo degradation. Inside the cell, there are two protein degradation systems.

Ubiquitin-proteasomal system	Lysosomal protein degradation
ATP-dependent	ATP-independent
Degrades mostly intracellular, short-lived, regulatory proteins	Degrades mostly extracellular, long-lived structural proteins as well as intracellular proteins
Proteolysis is achieved by 20S core subunit of proteasome	Proteolysis is done by acid hydrolases like cathepsins

Ubiquitin proteasomal system is discussed at the end of this chapter.

Proteins are degraded to amino acids. Amino acids are made up of two parts—amino group and carbon skeleton.

- Amino group is converted into urea and excreted in urine.
- Carbon skeleton is glucogenic or ketogenic or both.

We will discuss the fate of amino group in this Chapter and the fate of carbon skeleton in the next chapter.

Overview of Protein Catabolism

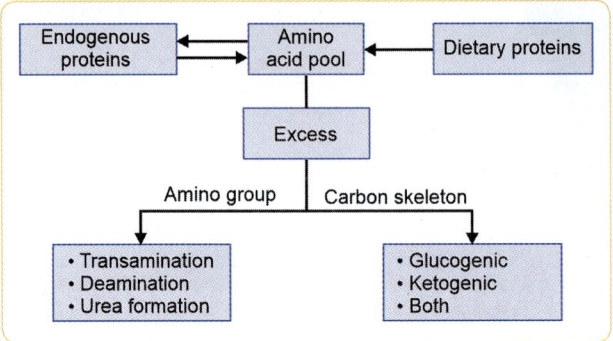

Proteins are Degraded at Many Places but Urea is Produced in Liver Only

Urea is produced from ammonia in the liver. Glutamate is the only amino acid that undergoes significant amount of oxidative deamination to produce ammonia in liver. However, protein degradation takes place at multiple locations of the body. Thus, amino group of all the amino acids must be converted to glutamate.

Amino Group of Most Amino Acids are Incorporated into α-keto Glutarate

Most of amino acids transfer their amino group to the TCA cycle intermediate α-keto glutarate to form glutamate. Glutamate acts as a transport form of NH_3. In liver, glutamate releases free NH_3 in glutamate dehydrogenase catalyzed reaction. This free NH_3 is converted into urea.

In the brain, glutamate combines with NH_3 to produce glutamine, which is a major transport form of NH_3 from brain. Glutaminase enzyme in liver releases free NH_3 for urea synthesis. In kidney, NH_3 released by glutaminase buffers the protons and is excreted as ammonium salts.

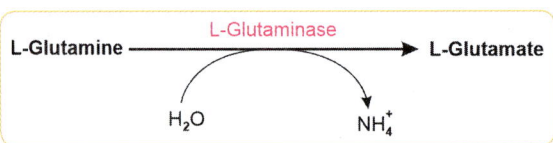

Alanine is the Transport form of Ammonia from Muscles to Liver

Carbon skeleton of branched chain amino acids are preferentially utilized by the skeletal muscle. Amino group of branched chain amino acids is transferred to the glycolytic end product pyruvate to produce alanine. Two molecules of alanine can be converted back to one molecule of glucose in liver through glucose-alanine (Cahill) cycle. (Cahill cycle is discussed in Chapter 10.3) *(Refer to page no. 137)*

Let us discuss the transamination reaction catalyzed by transaminases and oxidative deamination catalyzed by glutamate dehydrogenase.

> **Remember**
>
> **Transport form of NH_3 from Peripheral Tissues to Liver**
> - From all tissues—Glutamate
> - From Brain—Glutamine
> - From muscle—Alanine

TRANSAMINATION REACTIONS

Transamination is the transfer of amino group from an amino acid to a keto acid. Amino acid which gave away the amino group becomes keto acid. Keto acid, which accepted amino group turns into an amino acid. Most of the time the acceptor ketoacid is α-keto glutarate.

All amino acids **except lysine, threonine, proline and hydroxyproline** undergo transamination.

CONCEPT CORNER

Why four amino acids do not undergo transamination?
- Proline is an imino acid. So, it cannot undergo transamination. The same goes for hydroxyproline.
- Transamination of threonine and lysine would produce toxic keto acids.

$$AA_1 + KA_2 \xleftarrow{\text{Transaminase}} KA_1 + AA_2$$

$$\text{Alanine} + \alpha\text{-keto glutarate} \xleftrightarrow[\text{PLP}]{\text{Alanine transminase/SGPT}} \text{Pyruvate} + \text{glutamate}$$

$$\text{Aspartate} + \alpha\text{-keto glutarate} \xleftrightarrow[\text{PLP}]{\text{Aspartate transaminase/SGOT}} \text{Oxaloacetate} + \text{glutamate}$$

All transaminases require pyridoxal phosphate prosthetic group (Vitamin B_6). ALT is present only in the cytosol whereas AST has both cytosolic (20%) and mitochondrial (80%) isoenzymes.

Transamination reactions are example for ping-pong type of bi–bi (2-substrate, 2-product) mechanism of enzyme kinetics (Chapter 5) *(Refer to page no. 63).*

First, there is spontaneous (noncatalyzed) formation of a Schiff base between the carbonyl group of pyridoxal-5'-phosphate and the amino group of the amino acid. Then, the amino acid leaves as ketoacid leaving the amino group in the PLP as pyridoxamine. Finally, the ketoacid comes and there is transfer of amino group to the ketoacid.

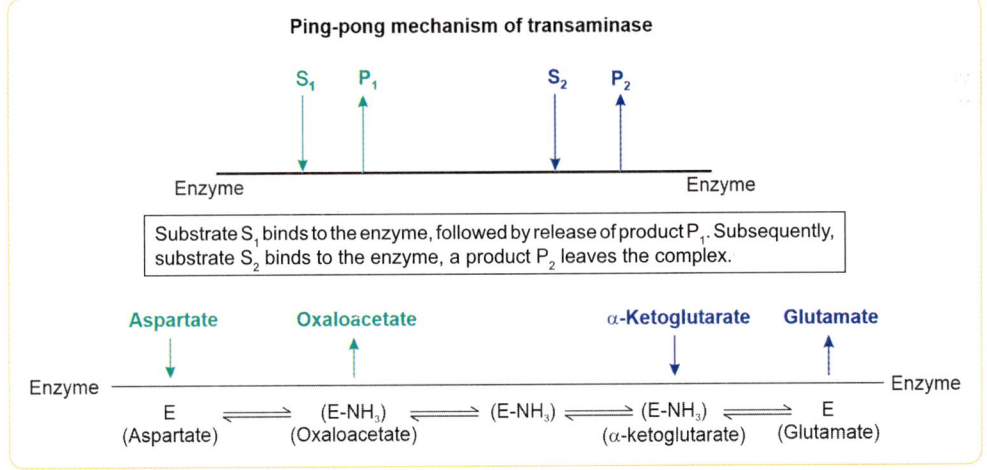

Significance of Transamination Reaction

- Production of glutamate for the subsequent deamination in the liver
- Synthesis of nonessential amino acids like aspartate and alanine
- Pyruvate—alanine transamination is a part of glucose-alanine cycle.

OXIDATIVE DEAMINATION

Glutamate Dehydrogenase Releases Free NH_3 in the Hepatic Mitochondria

In liver, glutamate dehydrogenase catalyzes the **oxidative deamination** of glutamate (oxidation → electrons are transferred to NAD(P); Deamination → ammonia is produced)

Chapter 11 Amino Acid Metabolism

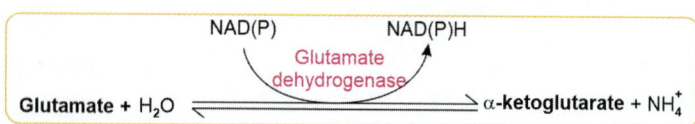

- This is a reversible reaction
- Takes place in mitochondrial matrix
- Enzyme can use **both NAD and NADP** as cofactor
- Allosterically activated by ADP and inhibited by ATP and GTP
- Reverse reaction happens when ammonia is in excess.
- Transamination followed by glutamate dehydrogenase mediated deamination is known as *transdeamination*.

Sources of NH₃ in the Body and its Metabolism

Sources of NH₃

- Glutamate dehydrogenase reaction
- Glutaminase reaction
- Intestinal bacteria acting on nonabsorbed amino acids (This is why local antibiotics are used in hepatic encephalopathy to reduce ammonia production)
- Monoamine oxidase reaction
- Catabolism of purines and pyrimidines.

Metabolic Fate of NH₃

- Conversion into urea is the major fate of NH_3
- Synthesis of glutamine by glutamine synthetase enzyme
- Synthesis of glutamic acid from NH_3 and α ketoglutarate by glutamate dehydrogenase enzyme
- A small amount of NH_3 is excreted in urine as such.

Glutamine is Formed from Glutamate and Ammonia

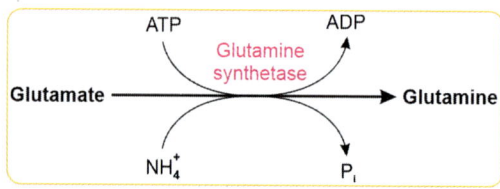

Glutamine is a Transport form as well as a Ready Reservoir of Amino Group in Synthetic Reactions

In addition to the source of ammonia in liver (for urea) and in kidney (for ammonium ions), glutamine is the source of amino group for the following five biosynthetic reactions:
1. N_3 and N_9 of purine ring
2. Carbamoyl phosphate by CPS II → N_3 of pyrimidine ring
3. Amino group of glucosamine-6-phosphate
4. Synthesis of GMP from IMP
5. Conversion of Aspartate to Asparagine

Free NH₃ is Toxic to Brain

- Excess NH_3 shifts the equilibrium of glutamate dehydrogenase reaction towards the formation of glutamate. This depletes α-KG and *impairs TCA cycle*.
- Glutamate binds to NH_3 to produce glutamine by glutamine synthetase. This reaction causes *depletion of ATP*.

Hepatic Encephalopathy

Hepatic encephalopathy is an altered level of consciousness as a result of liver failure. The liver is the site of detoxification of NH_3. When the liver fails, ammonia accumulates. As we have already learnt, NH_3 is toxic to the brain.

Management

- Dietary protein restriction (0.5 g/kg/day)
- Antibiotics (Rifaximin/Neomycin) to minimize the NH_3 production by intestinal flora
- Lactulose – a mild osmotic laxative is metabolised by colonic bacteria to produce short-chain fatty acids resulting in the lowering of local pH and thus converting easily diffusible ammonia to less-diffusible ammonium.
- Hemodialysis
- Liver transplantation for selected patients

UREA CYCLE

Urea Cycle Converts the Toxic Ammonia to Water-soluble Urea

Urea cycle also known as Krebs-Henseleit cycle is the first metabolic cycle to be discovered (Glycolysis is the first metabolic pathway to be discovered).
Organ: Liver
Subcellular location: Cytoplasm and mitochondria; (First 2 steps in mitochondria)
Five enzymes are involved in urea cycle.

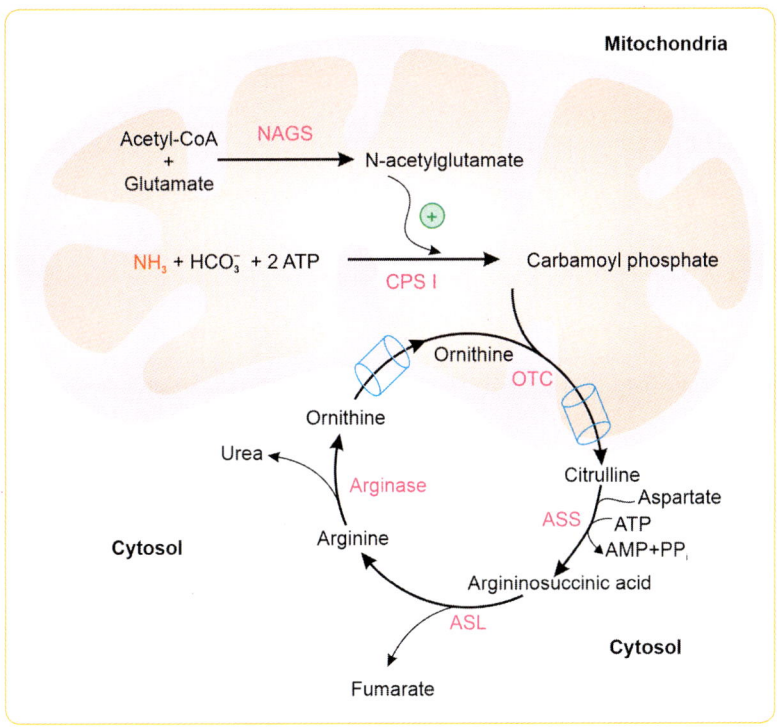

Abbreviations: ASL, argininosuccinate lyase; ASS, argininosuccinate synthetase; CPS I, carbamoyl phosphate synthetase I; OTC, ornithine tanscarbamoylase; NAGS, N-acetyl glutamate synthase

Note: Name of all the three enzymes catalyzing the cytoplasmic reactions begin with 'A'

Reactions of Urea Cycle

- ❑ **Carbamoyl phosphate synthetase-I (CPS-I):** CPS I catalyzes the *first* and *rate-limiting* step of the urea cycle in which Carbamoyl phosphate is formed by the condensation of CO_2, NH_3 and ATP. This reaction takes place in mitochondria and requires the hydrolysis of **2 moles of ATPs.**
- ❑ **Ornithine transcarbamoylase (OTC):** OTC catalyzes the transfer of carbamoyl phosphate to ornithine forming citrulline. Both ornithine and citrulline are non-protein amino acids (Chapter 1) *(Refer to page no. 11)*. This reaction takes place in mitochondria whereas the subsequent steps of the urea cycle take place in the cytoplasm.
- ❑ **Arginino succinate synthetase (ASS):** ASS links the amino group of aspartate to citrulline. Aspartate is the immediate source of 2nd nitrogen of Urea.
  This reaction requires the hydrolysis of ATP to AMP. Therefore, **two high-energy phosphates** are utilized.
- ❑ **Arginino succinate lyase (ASL):** ASL catalyzes the cleavage of Argininosuccinate to Arginine and Fumarate. *We will discuss how fumarate interlinks urea cycle with TCA cycle soon.*
- ❑ **Arginase (ARG):** Arginase is a hydrolase. It hydrolyses arginine into urea and ornithine. Thus, ornithine is reformed at the end of the cycle. This ornithine can enter mitochondria to begin the cycle again.

Note: Compare the oxaloacetate of TCA cycle and ornithine of the urea cycle. Both are reformed at the end.

Carbamoyl Phosphate Synthetase is the First and Rate-limiting Enzyme of Urea Cycle

	CPS I	CPS II
Location	Mitochondria	Cytoplasm
Pathway involved	Urea cycle	Pyrimidine biosynthesis
Source of NH_3	Free NH_3 liberated from glutamate dehydrogenase reaction	Glutamine
Activated by	N-acetyl glutamate	Phosphoribosyl pyrophosphate (PRPP)

Stoichiometry of Urea Cycle

$$CO_2 + NH^{4+} + 3\ ATP + aspartate + 2\ H_2O \rightarrow urea + 2\ ADP + 2\ P_i + AMP + PP_i + fumarate$$

Three moles of ATP, 1 mole each of ammonium ion and of aspartate and five enzymes are needed for the synthesis of 1 mol of urea.

Some authors consider the hydrolysis of PP_i to $2\ P_i$ as another additional high-energy phosphate bond required for urea synthesis. Thus, **four high-energy phosphates** are needed for 1 mole of urea synthesis.

Urea Cycle is Linked to TCA Cycle

Fumarate released from urea cycle is converted into malate and then to oxaloacetate in the TCA cycle. Oxaloacetate can be transaminated to aspartate, which can be reused by the urea cycle. Thus, the TCA cycle and urea cycle are interlinked and hence described as bicycle.

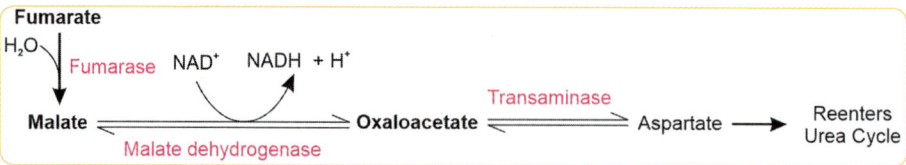

Source of Carbon and Nitrogen of Urea

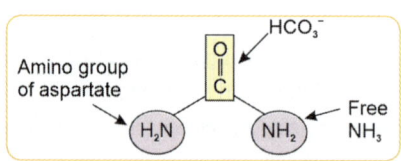

Regulation of Urea Cycle

CPS-I is allosterically stimulated by N-acetyl glutamate which is synthesized by the enzyme NAG synthase.

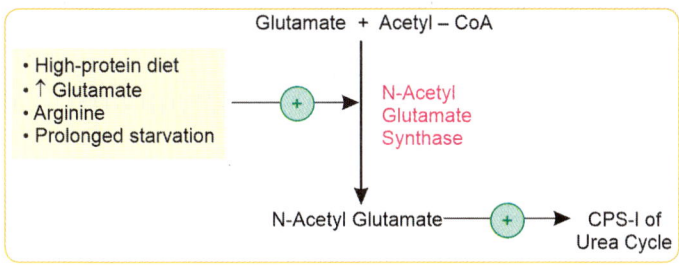

Increased protein catabolism induces glutamate dehydrogenase $\rightarrow\uparrow NH_3 \rightarrow\uparrow$ urea formation

Nonprotein Nitrogenous Compounds (NPN)

NPN are the nitrogenous compounds containing nitrogen other than proteins and peptides.
- ❑ Urea—major nonprotein nitrogenous compound
- ❑ Creatine
- ❑ Uric acid
- ❑ Nucleotides

UREA CYCLE DISORDERS

Disorder	Deficient enzyme	Features
Hyperammonemia type 1	CPS I	Severe form of hyperammonemia; encephalopathy; respiratory alkalosis (stimulation of respiratory center by ammonia)
Hyperammonemia type 2	Ornithine tanscarbamoylase (OTC)	Most common urea cycle disorder; X-linked inheritance Females exhibit protein aversion. ↑ Glutamine, ornithine and NH₃ in blood of newborn orotic aciduria
Citrullinemia	Argininosuccinate synthetase	↑ Citrulline in blood and CSF
Citrullinemia type 2	Mitochondrial aspartate-glutamate carrier AGC2 (citrin)	Recurring episodes of hyperammonemia with associated neuropsychiatric symptoms (*Ref. Harrison 20th Ed., p. 3022*)
Argininosuccinic aciduria	Argininosuccinate lyase	↑Argininosuccinate in blood, urine and CSF Friable, tufted hair (trichorrhexis nodosa)
Argininemia	Arginase	Usually presents in adulthood. ↑Arginine in blood, CSF and urine Arginine competes with lysine and cysteine resulting in their excretion, i.e. Cysteine-Lysinuria.
N-acetyl glutamate synthase (NAGS) deficiency	N-acetyl glutamate synthase (NAGS)	Same features of hyperammonemia type I but responds to treatment with N-acetyl glutamate.
HHH syndrome	Ornithine translocase/permease of mitochondrial membrane	**Hyperornithinemia:** As ornithine cannot enter mitochondria, it accumulates in blood. **Hyperammonemia:** As ornithine is not available, urea cycle cannot operate, i.e. NH₃ cannot be detoxified. **Homocitrullinuria:** As ornithine is not available, ornithine homologue lysine reacts with carbamoyl phosphate to produce homocitrulline. (Difference between lysine and ornithine is just one CH₂ group)

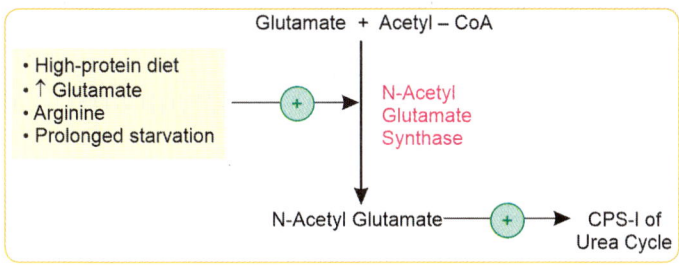

Note: The image contents reference the glutamate/N-acetyl glutamate synthesis pathway diagram.

Glutamate + Acetyl – CoA

- High-protein diet
- ↑ Glutamate
- Arginine
- Prolonged starvation

N-Acetyl Glutamate Synthase

N-Acetyl Glutamate ⟶ CPS-I of Urea Cycle

Chapter 11 **Amino Acid Metabolism**

181

Biochemical Basis of Orotic Aciduria in Urea Cycle Disorders

Mitochondrial ornithine tanscarbamoylase deficiency → ↑carbamoyl phosphate levels → carbamoyl phosphate exits to cytoplasm → ↑pyrimidine synthesis → mild orotic aciduria.

Treatment of Urea Cycle Defects

❑ Protein restriction to 50% of the daily dietary protein allowance and ensuring the optimal growth of the child
❑ Arginine supplementation, except in patients of arginase deficiency
❑ Ammonia scavenging agents—phenylbutyrate and sodium benzoate
❑ In severe hyperammonemia—hemodialysis.

Arginine Stimulates NAG Synthase and thus Ameliorate Hyperammonemia

❑ The semiessential amino acid arginine becomes dietarily essential in urea cycle disorders. Thus, it needs to be supplemented.
❑ Moreover, arginine stimulates the enzyme NAG synthase that produces N-acetyl glutamate which in turn stimulates the CPS-I.
❑ In arginase deficiency, arginine supplementation should not be done.

Phenylbutyrate and Benzoate are Used to Scavenge Ammonia in Urea Cycle Disorders

Phenylbutyrate (BuPhenyl) is a prodrug that undergoes β-oxidation to produce the active drug 'Phenylacetate'. Phenylacetate binds to glutamine to produce phenylacetylglutamine which can be excreted in urine (Fig. 1).

Phenylbutyrate — $(CH_2)_3$-COOH → β-oxidation → **Phenylacetate** — CH_2-COOH

Acetyl-CoA

CH_2-CO-NH COOH \ / CH | $(CH_2)_2$ | CO-NH$_2$

Phenylacetylglutamine

Non-enzymatic reaction

NH$_2$ COOH \ / CH | $(CH_2)_2$ | CO-NH$_2$

Glutamine

Excreted in urine

Benzoate is converted into benzoyl-CoA, which is conjugated with glycine by Phase II biotransformation reaction in liver to produce hippuric acid, which is excreted in urine. Removal of glycine necessitates its synthesis by glycine synthase complex which utilises free ammonia. So, glycine removal leads to decreased blood ammonia.

Phenylbutyrate is superior to benzoate since one molecule of phenylacetate causes excretion of two nitrogens while benzoate causes excretion of only a single nitrogen.

Benzoate can also be used in nonketotic hyperglycemia to remove excess glycine in the body.

Section II Intermediary Metabolism

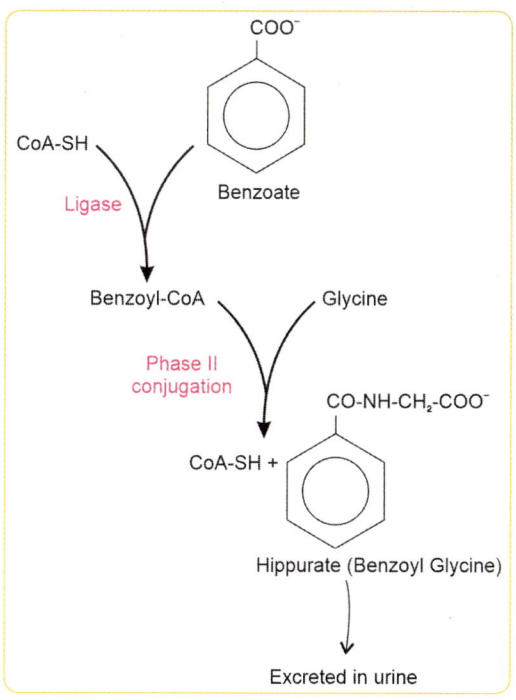

UBIQUITIN-PROTEASOMAL DEGRADATION SYSTEM

Ubiquitin is a small globular protein that is involved in the destruction of many but not all intracellular, short-lived, regulatory proteins.

The last amino acid, i.e. the carboxy terminal amino acid of ubiquitin is glycine.

This glycine forms a peptide bond with the ε amino group of lysine of target protein to be destroyed. This bond is a non α peptide bond or isopeptide bond.

Transfer of ubiquitin to the target protein is mediated by 3 different enzyme systems–E1 (ubiquitin activating enzyme), E2 (ubiquitin conjugating enzyme) and E3 (ubiquitin ligase). This process requires ATP.

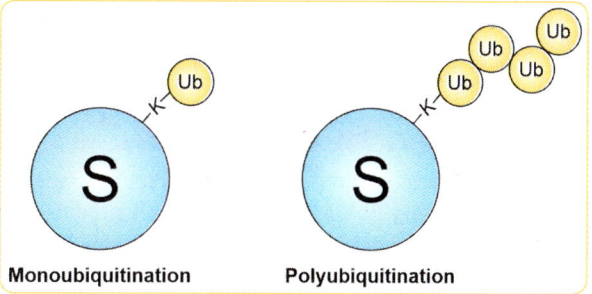

Monoubiquitination   **Polyubiquitination**

Abbreviations: S-Substrate, K-Lysine, Ub-Ubiquitin

Mutation in the E3 ubiquitin ligase will affect the half-life of cellular regulatory proteins and is associated with many cancers.

Ubiquitin itself contains multiple lysine (K) residues to which another ubiquitin can be linked through isopeptide bond (polyubiquitination).

Polyubiquitinated proteins are destroyed in the proteasome. However, monoubiquitination of proteins is involved in intracellular localization and trafficking. **At least four ubiquitin residues** should be added to a protein for its destruction.

Proteasome is a barrel shaped structure with a lid, base and core. Lid is involved in the removal of ubiquitin. The core contains protease activity. Digested proteins are released as amino acids.

Bortezomib, a proteasomal inhibitor is used in the treatment of multiple myeloma.

Chapter 11 Amino Acid Metabolism

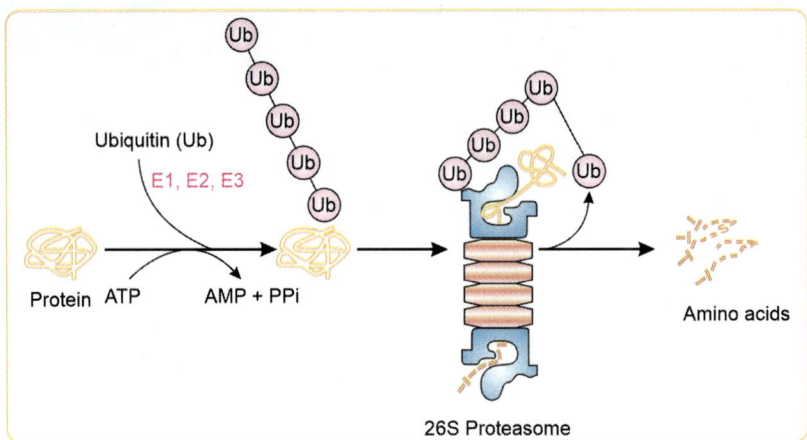

26S Proteasome

Corner

❑ The VHL gene encodes an E3 ubiquitin ligase that regulates expression of hypoxia-inducible factor 1. Loss of VHL is associated with increased expression of vascular endothelial growth factor (VEGF), which induces angiogenesis. (*Ref. Harrison 20th Ed., p. 2754*)

❑ Parkin, which causes autosomal recessive early-onset Parkinson's disease, is an ubiquitin ligase. (*Ref. Harrison 20th Ed., p. 3043*)

❑ Normally, p53 is bound to MDM2, a ubiquitin ligase which targets p53 for degradation in the proteasome. (*Ref. Harrison 20th Ed., p. 462*)

Competency

❑ BI5.4   Describe common disorders associated with protein metabolism

Key Points

❑ There is no storage form of protein. Dietary proteins when taken in excess are degraded.

❑ Bortezomib, drug used in multiple myeloma is an inhibitor of proteasomes.

❑ Glutamate is the only amino acid that undergoes a significant amount of oxidative deamination in liver mitochondria.

❑ Transport form of NH_3 from Peripheral Tissues to Liver:

From all tissues—Glutamate

From Brain—Glutamine

From muscle—Alanine

❑ Synthesis of 1 mole of urea requires 3 moles of ATP.

❑ Fumarate is the link between urea cycle and TCA cycle.

❑ Glutamate dehydrogenase is unique in the way that it can use either NAD^+ or $NADP^+$

❑ Glutamine is a transport form as well as a ready reservoir of nitrogen in synthetic reactions.

❑ Free NH_3 is toxic to brain because it depletes α-ketoglutarate from TCA cycle.

❑ Urea cycle disorders lead to mild orotic aciduria due to excess carbamoyl phosphate in cytoplasm channeled into pyrimidine biosynthesis.

❑ Phenylbutyrate and benzoate are used to scavenge ammonia in urea cycle disorders.

❑ Short-lived, intracellular, regulatory proteins are usually degraded by ubiquitin-proteasomal pathway.

❑ Long-lived, extracellular, structural proteins are usually degraded in lysosome by various enzymes like cathepsins.

Short Answer Questions

1. Justify the statement, "Glutamate serves as the gateway through which all amino acid nitrogen must pass for its release as ammonia."

2. Discuss the role of transamination and oxidative deamination (giving specific biochemical reactions) in the transfer of amino groups from the periphery to the liver.

3. Explain the reasons for the occurrence of encephalopathy in advanced cases of cirrhosis and rationale of the strategy for its treatment.

4. Why hyperammonemia is toxic to the brain?

5. What are transamination reactions? Giving two examples discuss the importance of these reactions.

6. Why the urea cycle is known as 'bicycle'?

7. What is the biochemical basis of arginine supplementation in patients of urea cycle disorders? In which one of the urea cycle disorders, arginine supplementation is not practiced and why?

8. With suitable chemical reactions, explain the biochemical basis of use of phenylbutyrate and benzoate in patients of urea cycle disorders

9. Mention the sources of carbon and nitrogen of a urea molecule produced in the body.

10. Defect in the ubiquitin- proteasomal system leads to various disorders including cancer – Why?

11. Differentiate:
    a. Protein degradation by the ubiquitin-proteasomal system and lysosomes
    b. Carbamoyl Phosphate Synthetase I & Carbamoyl Phosphate Synthetase II

Clinical Case-Based Questions

1. A 45-year-old male with a history of cirrhosis of the liver is brought to the emergency centre by family members for acute mental derangement, disorientation. O/E he is disoriented with icteric sclera and abdominal distension with fluid. Blood ammonia level is elevated.
   a. Explain the basis of the clinical presentation in this case.
   b. What other tests are performed to assess liver function? (Chapter 45)

Multiple Choice Questions

1. Urea cycle converts the toxic ammonia to water-soluble urea. In which of the following organs does urea cycle take place?
   a. Intestine   b. Liver
   c. Kidney   d. Brain

2. Transamination is the first step in the catabolism of many amino acids. All of the following amino acids undergo transamination, EXCEPT:
   a. Alanine   b. Threonine
   c. Glutamate   d. Aspartate

3. Which of the following amino acids undergoes significant amount of oxidative deamination in liver?
   a. Glutamate   b. Aspartate
   c. Alanine   d. Threonine

4. Which of the following amino acid is the transport form of ammonia from brain to liver?
   a. Alanine
   b. Asparagine
   c. Glutamine
   d. Histamine

5. Which of the following urea cycle enzyme is coded by X-chromosome?
   a. Arginase
   b. Argininosuccinate lyase
   c. Argininosuccinate synthase
   d. Ornithine tanscarbamylase

6. Which of the following statements about transamination reactions is correct?
   a. Coupling with ATP hydrolysis is required for transamination
   b. Transamination reactions are irreversible
   c. Transamination reactions require NAD^+ or $NADP^+$
   d. Transamination reactions require pyridoxal-5'-phophate

7. Which of the following statements about the urea cycle is correct?
   a. Carbamoyl phosphate supplies both of the nitrogen atoms of urea in the urea cycle
   b. Argininosuccinate is lyzed to urea and ornithine in the urea cycle
   c. The formation of urea from the urea cycle yields energy
   d. Arginine is hydrolyzed to urea and ornithine in the urea cycle

8. In which of the following urea cycle disorder, arginine supplementation should not be done?
   a. Hyper ammonemia type I
   b. Hyper ammonemia type II
   c. Hyper citrullinemia
   d. Hyperargininemia

9. HHH syndrome is due to defect in:
   a. Ornithine transcarbamoylase
   b. Ornithine permease
   c. Ornithine decarboxylase
   d. Ornithine

10. What is the minimum number of ubiquitin molecules that must be attached to the target protein for its destruction?
    a. 2   b. 4
    c. 8   d. 10

11. To which of the following amino acids of target protein, ubiquitin is added?
    a. Proline   b. Glycine
    c. Serine   d. Lysine

12. What is the nature of bond between ubiquitin and the amino acid of the target protein?
    a. Glycosidic bond
    b. Hydrogen bond
    c. Isopeptide bond
    d. Hydrophobic interaction

13. Which of the following is a proteasomal inhibitor?
    a. Semaxanib
    b. Bortezomib
    c. Alvocidib
    d. Tipifarnib

14. Which of the following amino acids is the source of ammonia in kidney?
    a. Glutamine   b. Alanine
    c. Methionine   d. Glycine

15. In ureotelic animals, carbamoyl group are transferred to:
    a. Urea   b. Uric acid
    c. Ornithine   d. Creatine

16. Hydrolysis occurs at which step of urea cycle?
    a. Cleavage of arginine
    b. Formation of arginosuccinate
    c. Formation of citrulline
    d. Formation of ornithine

17. Which of the following is the allosteric activator of carbamoyl phosphate synthase I ?
    a. Glutamine   b. Oxaloacetate
    c. N-acetyl aspartate d. N-acetyl glutamate

18. Polyubiquitinated proteins are destroyed in:
    a. P-bodies   b. Lysosomes
    c. Proteasomes d. Peroxisomes

19. Which of the following is used as an ammonia scavenger in urea cycle disorders?
    a. Phenylbutyrate   b. Phenyl thiolactone
    c. Phenoxy methane   d. Phenmetformin

20. Which of the following urea cycle reaction takes place in mitochondria?
    a. Arginase
    b. Argininosuccinate lyase
    c. Argininosuccinate synthase
    d. Ornithine tanscarbamylase

ANSWERS					
1. b.	2. b.	3. a.	4. c.	5. d.	6. d.
7. d.	8. d.	9. b.	10. b.	11. d.	12. c.
13. b.	14. a.	15. c.	16. a.	17. d.	18. c.
19. a.	20. d.				

1. **Deficiency of which of the following enzyme will result in similar condition?**
   a. Ornithine transcarbamylase
   b. Argininosuccinate synthase
   c. Argininosuccinate lyase
   d. Arginase

1. **Ans. (c) Argininosuccinate lyase**

Friable and tufted hair (trichorrhexis nodosa) are shown in the image. You can easily guess that this is a case of argininosuccinate lyase deficiency, i.e. argininosuccinic aciduria.

Chapter 11 Amino Acid Metabolism

11.2 SYNTHESIS OF NONESSENTIAL AMINO ACIDS

Chapter Outline

- Dietarily Essential Amino Acids
- Dietarily Nonessential Amino Acids
- Semiessential Amino Acid
- Synthesis of Non-essential Amino Acids
- Nonessential Amino Acids can Become Essential

CLASSIFICATION OF AMINO ACIDS BASED ON THE DIETARY REQUIREMENT

Essential Amino Acids

Phenylalanine, valine, threonine, tryptophan, methionine, leucine, isoleucine, lysine and arginine, histidine are the 10 essential amino acids, which are indispensable and must be taken in diet.

To remember these, you can use the classical mnemonics like AV HILL MP TT or PVT TIM HALL. After learning the entire metabolism, you will learn how nonessential amino acids are synthesized in the body and you will easily remember the essential ones by exclusion.

This is how I used to remember the essential amino acids

- All the 3-branched chain amino acids (Leucine, Isoleucine, Valine)
- All the 3-basic amino acids (Histidine, Arginine, Lysine)
- 2 aromatic amino acids (Phenylalanine, tryptophan)
- Other 2 amino acids that end with '9': Threonine and methionine

Arginine is a Semi Essential Amino Acid

Arginine can be synthesized in the body to some extent. But during increased demand (pregnancy, lactation, growth spurt, convalescence), arginine has to be supplemented in diet. This is why arginine is a semi essential amino acid.

Dietarily Nonessential Amino Acids can be Synthesized in the Body

Alanine, aspartate, glutamate, glutamine, asparagine, serine, glycine, cysteine, tyrosine, proline and selenocysteine are the 11 amino acids which can be synthesized in the body through metabolic processes. Thus, they are dietarily nonessential.

CONCEPT CORNER

Question: "While unicellular bacteria can synthesize all 20 amino acids, Why cannot humans synthesize them?".
Answer: During the course of evolution, genes for synthesis of those amino acids requiring multiple enzymes and more energy were deleted from higher organisms. Life found it economical to obtain them from diet. Synthetic pathways of those amino acids, which can be synthesized in one or two steps were retained.

Nonessential amino acid	Synthesized from
Via transamination of ketoacids	
Alanine	Pyruvate
Aspartate	Oxaloacetate
Glutamate	α-ketoglutarate
Through other intermediates	
Serine	3-phosphoglycerate

Contd…

Section II Intermediary Metabolism

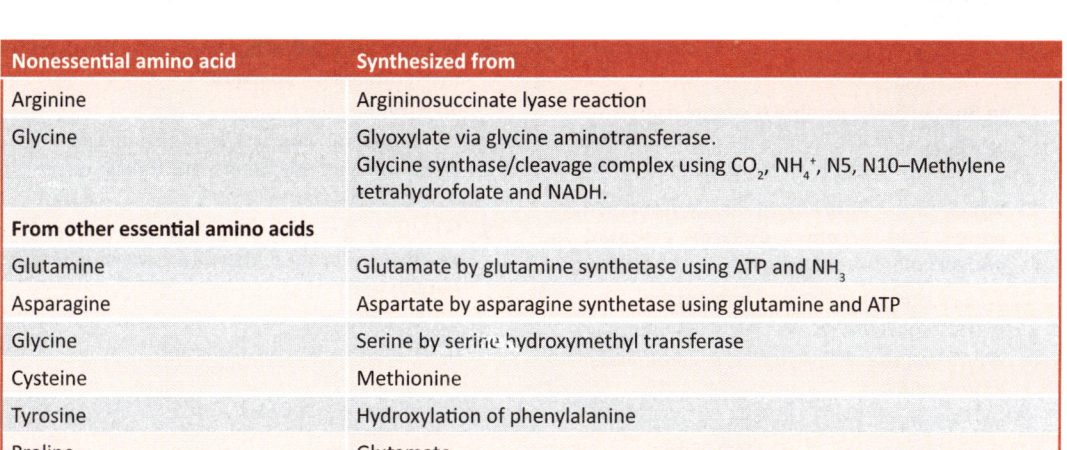

Nonessential amino acid	Synthesized from
Arginine	Argininosuccinate lyase reaction
Glycine	Glyoxylate via glycine aminotransferase. Glycine synthase/cleavage complex using CO_2, NH_4^+, N5, N10–Methylene tetrahydrofolate and NADH.
From other essential amino acids	
Glutamine	Glutamate by glutamine synthetase using ATP and NH_3
Asparagine	Aspartate by asparagine synthetase using glutamine and ATP
Glycine	Serine by serine hydroxymethyl transferase
Cysteine	Methionine
Tyrosine	Hydroxylation of phenylalanine
Proline	Glutamate
Selenocysteine	Serine

Synthesis of Asparagine is Intrinsically Linked to Glutamine

Glutamine synthetase uses free NH_3 for amide bond formation. However, asparagine synthetase uses amide group of glutamine. Thus, nonessential amino acid glutamine is essential for the synthesis of another nonessential amino acid, i.e. asparagine.

Nonessential Amino Acids can Become Essential Under Some Circumstances

❑ In phenylketonuria, phenylalanine hydroxylase is deficient and thus tyrosine becomes dietarily essential.
❑ In homocystinuria, conversion of methionine to cysteine is affected and cysteine becomes dietarily essential.

Key Points

❑ Unicellular organisms can synthesize all the 20(+1) standard amino acids whereas humans cannot.
❑ Ten amino acids cannot be synthesised in our body and must be taken in diet. These are the dietarily essential amino acids.
❑ Dietarily non-essential aminoacids can be synthesized in the body.
❑ Arginine is a semi essential amino acid – the synthesis cannot meet the need during the periods of high-demand like pregnancy, lactation, child growth, recovery from illness (convalescence).
❑ Nonessential amino acids can become essential under some circumstances.

SELF-ASSESSMENT

Short Answer Questions

1. Name the dietarily essential amino acids and why they are essential in diet?

2. Name two nutritionally non-essential amino acids which can be synthesized in the body from other essential amino acids. Give the reaction in each case.

3. Name two nutritionally non-essential amino acids which can be synthesized from TCA cycle and glycolytic intermediates. Give the reaction for any one of them.

4. With suitable examples, explain how dietarily nonessential amino acids can become essential under some circumstances.

Multiple Choice Questions

1. **In human body, proline is synthesized from:**
   a. Glutamate
   b. Glycine
   c. Lysine
   d. Valine

2. **Which of the following dietarily nonessential amino acid becomes dietarily essential in patients of homocystinuria?**
   a. Cysteine
   b. Methionine
   c. Homocysteine
   d. Phenylalanine

3. **Which of the following is a semi essential amino acid?**
   a. Arginine
   b. Alanine
   c. Serine
   d. Methionine

4. **Which of the following enzyme is involved in the synthesis of aspartate?**
   a. Alkaline phosphatase
   b. Serum glutamic pyruvic transaminase
   c. Serum glutamic oxaloacetic transaminase
   d. γ glutamyl transferase

5. **Which of the following amino acid can be synthesized from a glycolytic intermediate in human body?**
   a. Aspartate
   b. Glutamate
   c. Histidine
   d. Serine

11.3 FATE OF CARBON SKELETON

Chapter Outline

❏ Glucogenic and Ketogenic Amino Acids
❏ Important Figure Showing the Entry of Amino Acids into the TCA Cycle

In the previous chapter, we discussed the fate of amino group. Here, we will study about the fate of carbon skeleton.

GLUCOGENIC AND KETOGENIC AMINO ACIDS

Amino acids that produce pyruvate or TCA cycle intermediates are glucogenic

Transamination of α-amino acids produce α-keto acids which could be pyruvate or intermediates of TCA cycle like oxaloacetate and α-ketoglutarate. We know that pyruvate and all the intermediates of TCA cycle are glucogenic. Thus, those amino acids, which can produce pyruvate and intermediates of TCA cycle are glucogenic.

Amino Acids that Produce Acetoacetyl-CoA or Acetyl-CoA or Both are Ketogenic

Catabolism of some amino acids produce acetoacetyl-CoA or acetyl-CoA or both. Acetoacetyl-CoA is converted into acetyl-CoA which enters TCA cycle and gets completely oxidized. Acetyl-CoA is not a substrate for gluconeogenesis. Thus, those amino acids that produce acetoacetyl-CoA or acetyl-CoA are ketogenic.
Leucine is the purely ketogenic amino acid.

Five Amino Acids are Both Ketogenic and Glucogenic

Carbon skeleton of five amino acids are both glucogenic and ketogenic. They are the 3 aromatic amino acids, lysine and isoleucine.

Purely ketogenic amino acid	Both glucogenic and ketogenic amino acid	Purely glucogenic amino acid
Leucine	Phenylalanine, Tyrosine, Tryptophan, Lysine, Isoleucine	Other 14 amino acids

Chapter 11 Amino Acid Metabolism

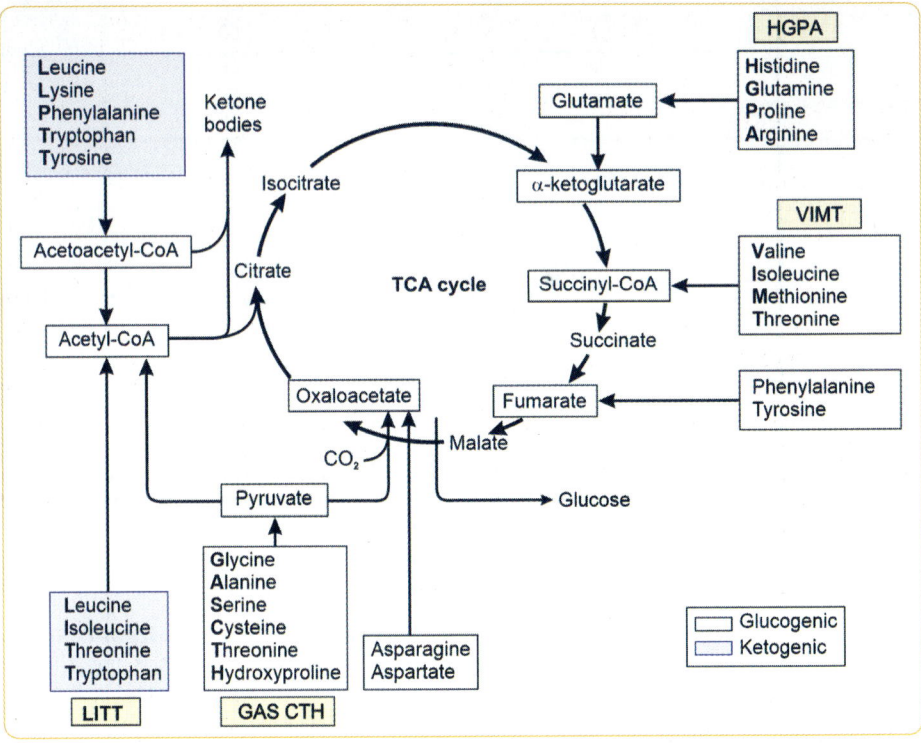

It is very important to understand the above illustration. In transamination reactions (Chapter 11.1) *(Refer to page no. 176)*, we learnt that transamination of aspartate produces oxaloacetate which is a glycolytic intermediate. Therefore, oxaloacetate is glucogenic. In the same way, alanine produces pyruvate and is glucogenic. Try to correlate the fate of other amino acids also. We will discuss the fate of the carbon skeleton of many of the amino acids shown in the illustration in Chapter 11.6 *(Refer to page no. 207)*.

Key Points

- ❑ Amino acids that produce pyruvate or any TCA cycle intermediates are glucogenic.
- ❑ Amino acids that produce acetoacetyl-CoA or acetyl-coA or both are ketogenic.
- ❑ Leucine is the purely ketogenic aminoacid.
- ❑ Phenylalanine, tyrosine, tryptophan, lysine, isoleucine are the five amino acids are both ketogenic and glucogenic

Short Answer Questions

1. Classify the amino acids based on their fate of carbon skeleton. Give an example for each.
2. Why tyrosine is both glucogenic and ketogenic?
3. Explain why alanine is a purely glucogenic amino acid.

Multiple Choice Questions

1. Which one of the following amino acids does not produce an intermediate of TCA cycle?
   a. Leucine
   b. Isoleucine
   c. Valine
   d. Proline

2. All of the following amino acids are converted into succinyl CoA, EXCEPT:
   a. Methionine
   b. Isoleucine
   c. Valine
   d. Histidine

3. All are ketogenic and glucogenic amino acids EXCEPT:
   a. Tyrosine
   b. Isoleucine
   c. Phenylalanine
   d. Serine

4. Catabolism of which of the following amino acid produces fumarate?
   a. Tryptophan
   b. Tyrosine
   c. Threonine
   d. Valine

5. Which of the following amino acid can produces oxaloacetate directly in a single reaction?
   a. Alanine
   b. Cysteine
   c. Threonine
   d. Aspartate

ANSWERS

1. a.	2. d.	3. d.	4. b.	5. d.

11.4 SPECIAL PRODUCTS

Amino acids are the monomeric units of proteins. In addition to protein formation, certain amino acids produce biologically important special products which we will discuss in this chapter.

First, let us have an overview of the amino acids and the special products produced by them. Then, we will discuss all the important special products briefly.

Amino acid	Special products
Glycine	Creatine, Glutathione, Heme, Purine, Sarcosine
Arginine	Creatine, Nitric oxide (NO), Polyamines
Cysteine	Coenzyme A, Taurine
Tryptophan	Serotonin (5-hydroxytryptamine), Melatonin, Niacin
Tyrosine	Melanin, Catecholamines (Dopamine, Norepinephrine, Epinephrine), Thyroxine
Glutamate	Gama Amino Butyric Acid (GABA)
Histidine	Histamine, Homocarnosine
β-alanine	Coenzyme A, Carnosine, Anserine

CREATINE

- Creatine formation requires 3 amino acids—Glycine, Arginine and Methionine
- Creatine metabolism takes place in 3 organs—Kidney, Liver and Muscle
- Glycine, formamide (not guanidino) group of arginine and CH_3 group from S-adenosyl methionine are incorporated into creatine.

Creatine phosphate is a phosphagen (*see* Chapter 4) *(Refer to page no. 47).*

Creatinine is used to Estimate GFR

- Creatinine is an anhydrous metabolite of creatine.
- 1–2% of creatine phosphate is spontaneously and irreversibly converted into its anhydride waste product, creatinine daily in the body, i.e. amount of creatinine produced each day is constant.
- Amount of creatinine produced each day is dependent upon the muscle mass.
- Creatinine is freely filtered by the glomeruli and secreted into the tubules.
- Creatinine clearance is an indicator of glomerular filtration rate (*Chapter 45*) *(Refer to page no. 649).*

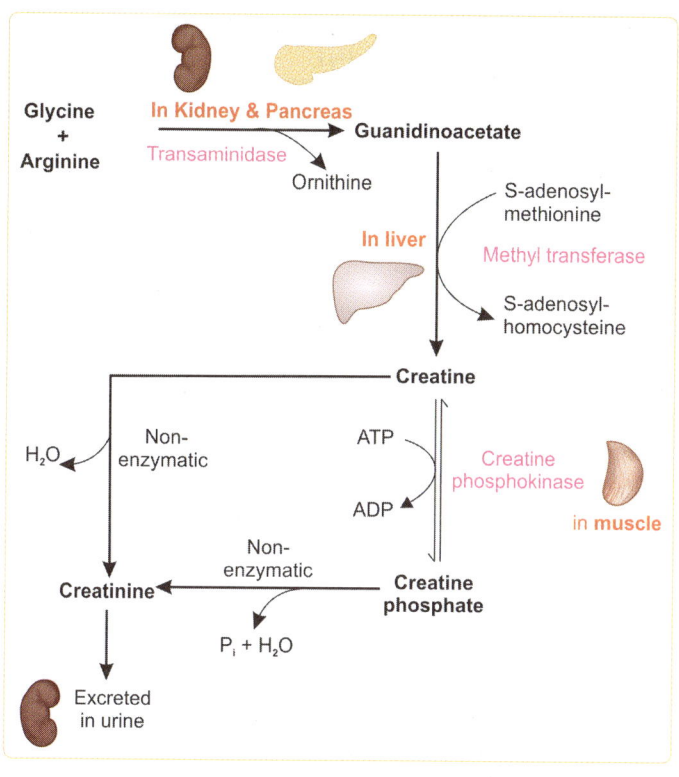

GLUTATHIONE IS THE CELLULAR REDUCING AGENT

- ❏ Glutathione is γ-glutamyl cysteinyl glycine (tripeptide)
- ❏ The bond between glutamate and cysteine is an *isopeptide* bond.

CONCEPT CORNER

Isopeptide bond is an amide bond in which one of the contributors to the peptide bond is non-alpha amino group or non-alpha carboxyl group.

For more: ▶ https://www.youtube.com/watch?v=GOYLvRcvj7Q or scan QR code

Role of Glutathione

- ❏ Major intracellular reducing agent: Helps in maintenance of sulfhydryl/thiol (–SH) groups of the enzymes and proteins in reduced state
- ❏ Maintenance of integrity of RBC membrane
- ❏ Detoxification of H_2O_2 by the glutathione peroxidase catalyzed reaction
- ❏ Transport of neutral amino acids across the membrane of intestine and renal tubules (γ-glutamyl cycle a.k.a. *meister* cycle)
- ❏ Synthesis of leukotrienes C_4, D_4 and E_4 from A_4
- ❏ Detoxification of xenobiotics by glutathione conjugation (Phase II reaction)
- ❏ Coenzyme for prostaglandin H synthase
- ❏ Coenzyme for maleylacetoacetate cis, trans isomerase reaction (in tyrosine catabolism)

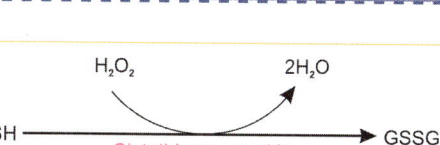

NITRIC OXIDE IS A FREE RADICAL AS WELL AS A GASEOUS SIGNALLING MOLECULE PRODUCED FROM ARGININE

Dibasic amino acid arginine is acted upon by the heme containing cytosolic NO synthase (NOS) to produce NO and citrulline.

$$\text{Arginine} \xrightarrow{\text{NO Synthase}} \text{NO + Citrulline}$$

The NOS enzyme is unique. It requires five cofactors—NADPH, FAD, FMN, heme and tetrahydrobiopterin (BH4). There are three isoforms of the enzyme.

Isoform	Location	Function
eNOS—endothelial (Ca^{++} dependent)	Endothelial cells, platelets, myocardium, endocardium	Arterial relaxation (NO is endothelial derived relaxation factor)
nNOS—neuronal (Ca^{++} dependent)	Neurons	Neurotransmitter involved in learning
iNOS—inducible (Ca^{++} independent)	Macrophages	Killing of bacteria by reactive nitrogen species

Mechanism of action: NO production in endothelium → diffuses into vascular smooth muscle → Activates 'soluble' guanylyl cyclase→ ↑ cGMP → Activation of protein kinase G → Relaxation of vascular smooth muscle.

Termination of action:
❑ cGMP is degraded by phosphodiesterase-5
❑ Nitric oxide itself is an unstable free radical. Therefore, it undergoes decomposition spontaneously.

Drugs Acting through NO

Drug	Mechanism of action	Use
Nitroglycerin	Produces NO	Used in *angina pectoris* to relieve coronary spasm
Sildenafil (Viagra)	Inhibits phosphodiesterase-5 which degrades cGMP →↑ cGMP →prolongation of action of NO	Used in *erectile dysfunction* to dilate the blood vessels of corpora cavernosa.

PUTRESCINE, SPERMIDINE AND SPERMINE ARE THE POLYAMINES DERIVED FROM ORNITHINE

Polyamines are polycations synthesized from nonprotein amino acid ornithine. (Ornithine, itself is a derivative of arginine).

Putrescine (diamine), spermidine (triamine) and spermine (tetramine) are the main polyamines.

Rate-limiting enzyme: Ornithine decarboxylase

Source of ornithine: Arginase mediated cleavage of arginine. Arginase isoenzyme is present in some nonhepatic tissues also.

Peculiar Points to Note

❑ Propylamine group is transferred to putrescine and spermidine to produce spermidine and spermine respectively; not just methyl group.
❑ Propylamine group donor in polyamine synthesis is not SAM, it is decarboxy SAM.
❑ Ornithine decarboxylase enzyme is regulated by ubiquitin-independent proteasomal degradation.

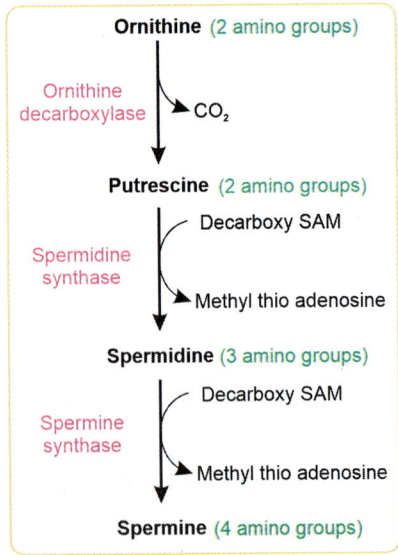

Role of Polyamines

Presence of regularly spaced multiple positive charges is the reason behind the functional capability of polyamines.
- Role in DNA stabilization, signal transduction, cell cycle regulation, chromatin modulation, cell migration.
- Used as a growth factor in cell culture technique.

Polyamine Synthesis Inhibitor

Eflornithine or difluoromethyl ornithine (DFMO), suicidal inhibitor of ornithine decarboxylase enzyme is used in:
- Sleeping sickness—injection form
- Facial hirsutism—topical application (*Ref. Harrison 20th Ed., p. 1559 & 1563*)

TAURINE IS A CYSTEINE DERIVATIVE USED IN CONJUGATION OF BILE SALTS

Taurine is derived from oxidation and decarboxylation of cysteine. Taurine is conjugated to primary bile acids cholic acid and chenodeoxycholic acid to form bile salts-taurocholate, glycocholate, taurochenodeoxycholate and glycochenodeoxycholate.

Taurine conjugation enhances the solubility of bile salts at the pH of body fluids.

CYSTEINE AND β-ALANINE CONTRIBUTE TO THE STRUCTURE OF COENZYME A

Coenzyme A is a thioalcohol (SH) group containing nucleotide cofactor involved in the transfer of acyl groups in metabolic reactions, e.g. acetyl group transfer in PDH catalyzed reaction.

Structure of Coenzyme A

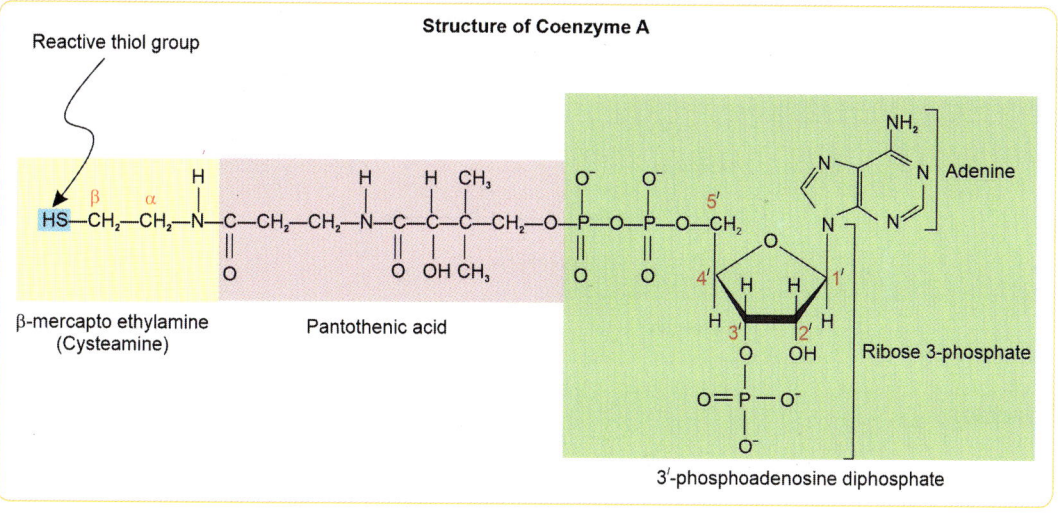

3′-phosphoadenosine diphosphate

Look at the structure of coenzyme A. Can you notice the SH group? This is the reason why we always write coenzyme A as CoA-SH when it is not esterified. (Glutathione also contains SH group and we write it as G-SH)

$$CoA\text{-}SH + Fatty\ acid\text{-}COOH \rightarrow Fatty\ acyl\text{-}CoA + H_2O$$
$$CoA\text{-}SH + Succinate \rightarrow Succinyl\text{-}CoA + H_2O$$

Acyl-CoA contains a thioester bond. This is a high-energy bond. That is why succinyl-CoA to succinate conversion generates ATP/GTP at the substrate-level.

Remember

- Cysteine provides the thioethanolamine portion of coenzyme A
- β-alanine is a component pantothenic acid (β-alanine + Pantoic acid = Pantothenic acid)

Chapter 11 **Amino Acid Metabolism**

197

BIOGENIC AMINES ARE PRODUCED FROM PLP MEDIATED DECARBOXYLATION OF AMINO ACIDS

Decarboxylation of amino acids produces biogenic amines. Pyridoxal phosphate serves as a cofactor in the breaking of carbon-carbon bond. This reaction does not require ATP. All the decarboxylases are lyases (EC4).

Amino acid	Biogenic amine
Glutamate	Gamma-amino butyric acid (GABA)
Histidine	Histamine
5-hydroxy tryptophan	Serotonin (5-hydroxytryptamine)
Tyrosine	Tyramine
DOPA	Dopamine → Norepinephrine and epinephrine
Ornithine	Putrescine
Lysine	Cadaverine

Let us discuss the role of GABA and histamine.

GABA IS AN INHIBITORY NEUROTRANSMITTER

Decarboxylation of α-**carboxyl group** of glutamate by the enzyme *glutamate decarboxylase* produces gamma-amino butyric acid (GABA) which is one of the non-α amino acids. *Pyridoxal phosphate* and magnesium are the cofactors needed for the reaction.

Produced by: Mainly neurons of CNS.

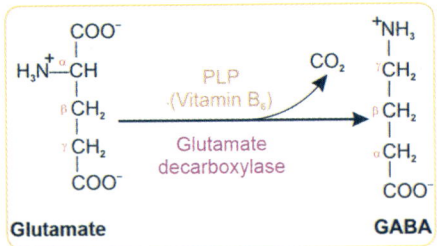

Function

❑ *Inhibitory* neurotransmitter-acts on 2 types of receptors. (GABA$_A$— ligand-gated chloride channel; GABA$_B$— G protein-coupled receptors)
❑ Homocarnosine is a major antioxidant in brain. It is a dipeptide of GABA and histidine.

Clinical Aspects

❑ GABA receptor agonists (e.g. Gabapentin) are used in various conditions like epilepsy, neuropathic pain, spasticity.
❑ Antibodies to glutamic acid decarboxylase, the biosynthetic enzyme for GABA leads to *stiff person syndrome*. (*Ref. Harrison 20th ed, p. 669*)

HISTAMINE

Histamine is the 1° aromatic amine derived from removal of CO_2 (decarboxylation) from histidine by the enzyme histidine decarboxylase. Pyridoxal phosphate (B_6) and magnesium are the cofactors needed for the reaction.
Produced by: Basophils, mast cells, enterochromaffin-like cells of gastric mucosa and some parts of brain.

Function

Acts through 4 types of receptors (H_1 to H_4). All are G protein coupled/7-transmembrane receptors.
❑ Vasodilation and bronchoconstriction via H_1 receptor—mediates *type I immediate hypersensitivity* (anaphylaxis).
❑ *Secretion of HCl* via H_2 receptor in parietal cells

Clinical Aspects

❑ H_1 receptor blockers (cetirizine) are antiallergic drugs with sedation as a side effect.
❑ H_2 receptor blockers (ranitidine) are used to ↓HCl in peptic ulcer disease.

SEROTONIN, MELATONIN AND NIACIN ARE DERIVED FROM TRYPTOPHAN

❑ Hydroxylation and subsequent decarboxylation of tryptophan forms serotonin (5-hydroxytryptamine) which is a potent vasoconstrictor and stimulator of smooth muscle contraction.

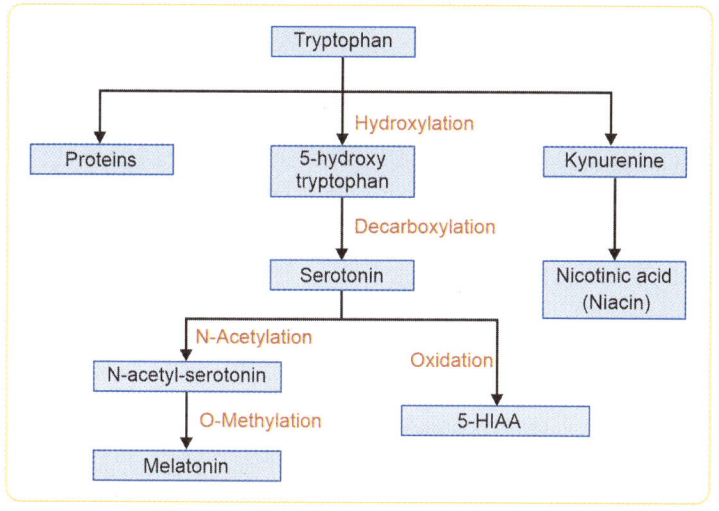

Abbreviation: 5-HIAA, 5-hydroxyindoleacetic acid

❑ N-acetylation of serotonin, followed by its O-methylation in the pineal body, forms melatonin.
❑ In human body, niacin is synthesized from tryptophan with the help of PLP. (60 mg tryptophan = 1 mg niacin)

Carcinoid Syndrome Leads to Niacin Deficiency

❑ Carcinoid syndrome leads to overproduction of serotonin, i.e. excessive conversion of tryptophan into serotonin and thus tryptophan to niacin conversion is affected.
❑ Degradation of serotonin by monoamine oxygenase produces 5-hydroxyindoleacetic acid (5-HIAA), urinary excretion of which is increased in carcinoid syndrome.

CATECHOLAMINE, MELANIN AND THYROXINE ARE DERIVED FROM TYROSINE

Formation of Catecholamines

Catecholamines are synthesized from the aromatic amino acid tyrosine.

Rate-limiting step is catalyzed by *tyrosine hydroxylase* enzyme. It requires *iron (Fe²⁺)* and *tetrahydrobiopterin* as cofactors. Tyrosine hydroxylase is a monooxygenase as it incorporates single atom of oxygen to substrate and reduces the other to water.

Vitamins needed: Vitamin C and B_6

Minerals needed: Iron and copper

Note: Loss of dopaminergic neurons in the substantia nigra leads to Parkinson's disease.

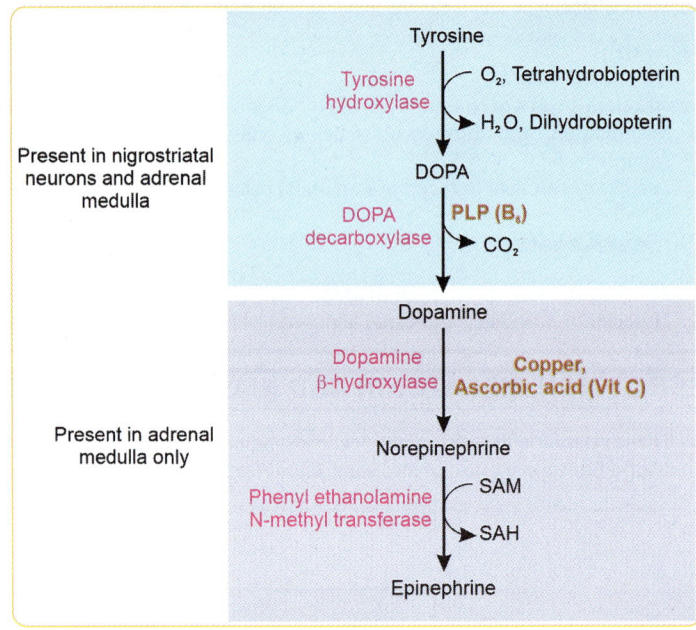

Formation of Melanin

Melanins are the pigments which give color to the skin, hair, iris and retinal pigment epithelium.

❑ Pheomelanin—yellow to red color

❑ Eumelanin—brown to black color

Melanin protects the skin from UV ray damage. Melanin in the iris prevents excessive stimulation of photoreceptors.

Melanin is synthesized from tyrosine. Copper containing enzyme *tyrosinase* converts tyrosine to DOPA and then to dopaquinone. Oxidation of dopaquinone produces dopachrome. Further oxidation produces products, which nonenzymatically polymerise to form melanin.

Albinism is due to Partial or Complete Deficiency of tyrosinase Enzyme

Types of Albinism

❑ Oculocutaneous albinism—Autosomal recessive inheritance (involvement of both skin and eyes)

❑ Ocular albinism—X-linked recessive.

Clinical Features

❑ Absence of pigmentation in skin, hair and iris

❑ Photophobia and reduced vision

❑ Susceptibility to skin cancer

Know the Difference

Tyrosinase is involved in melanin synthesis; Tyrosine hydroxylase is involved in catecholamine synthesis.

REVIEW–SPECIAL PRODUCTS

Special product		Constituents
Creatine (Methyl guanido acetic acid)		**G**lycine, **A**rginine, **M**ethionine (Entire glycine + formamide group of arginine + CH_3 group of SAM)
Carnitine		Lysine and methionine
Glutathione		γ-**G**lutamyl cysteinyl **G**lycine (Peptide bond between glutamate and cysteine is an *isopeptide* bond)
Sarcosine		N-methyl glycine
β-Alanyl dipeptides	Carnosine	β-alanyl histidine
	Anserine	N-methyl carnosine
Homocarnosine		GABA + histidine
Tryptophan derivatives	Serotonin	5-hydroxytryptamine
	Melatonin	N-acetyl-5-methoxytryptamine
	Niacin	60 mg tryptophan = 1 mg niacin

REVIEW–METABOLIC ROLE OF GLYCINE

1. **Creatine** synthesis
2. **Glutathione** synthesis
3. Entire glycine molecule is incorporated as C_4, C_5 and N_7 of purine ring
4. One carbon donor - N^5, N^{10} **methylene THFA** is produced from glycine cleavage system
5. **Heme** synthesis (Glycine + succinyl CoA → δ ALA)
6. **Detoxification of benzoyl CoA** by glycine conjugation to produce **hippuric acid**
7. **Glycine conjugation to bile acids** → Glycocholic acid and glycochenodeoxycholic acid
8. **Every 3rd amino acid of collagen—Glycine is the smallest amino acid** that can be accommodated in the narrow space of intermingling of triple helix
9. **Glycine** itself is an inhibitory neurotransmitter.
10. Sarcosine is N-methyl Glycine

Key Points

- ❑ Glutathione is γ-glutamyl cysteinyl glycine.
- ❑ Creatine is derived from glycine, arginine and methionine (GAM).
- ❑ Carnitine is synthesized from lysine and methionine.
- ❑ Polyamines (putrescine, spermidine, spermine) are synthesised from ornithine.
- ❑ Nitric oxide is a free radical as well as a gaseous signaling molecule produced from arginine.
- ❑ Asymmetric dimethyl arginine (ADMA) interferes with nitric oxide production.
- ❑ Taurine is a cysteine derivative used in conjugation of bile salts.
- ❑ Cysteine and β-alanine contribute to the structure of coenzyme A.
- ❑ Biogenic amines are produced from PLP mediated decarboxylation of amino acids.
- ❑ Serotonin, melatonin and niacin are derived from tryptophan.
- ❑ Antibodies to glutamic acid decarboxylase, the biosynthetic enzyme for GABA leads to stiff person syndrome.
- ❑ Antibodies against glutamic acid decarboxylase are found in type I diabetes mellitus.
- ❑ Tyrosinase is involved in melanin synthesis; tyrosine hydroxylase is involved in catecholamine synthesis.
- ❑ 60 mg tryptophan is equivalent to 1 mg niacin.

Chapter 11 Amino Acid Metabolism

Short Answer Questions

1. Describe the biosynthesis of creatine.

2. Name any two hormones derived from amino acids and also indicate the respective precursor amino acids and the reaction.

3. Name the amino acids involved in the biosynthesis of glutathione. Enumerate the biological role of glutathione.

4. How Nitric oxide is synthesized in the body? What is the physiological role of nitric oxide in blood vessel? Explain the mechanism of action. Mention the ways in which the action of nitric oxide is terminated/regulated. Name any two drugs acting through the production of nitric oxide and their therapeutic use.

5. What is active methionine? Enumerate 4 reactions involving this compound.

6. Name the three polyamines. Mention two physiological functions of polyamines?

7. Which amino acid serves as the source of polyamines? Which enzyme catalyzes the rate-limiting step of polyamine synthesis? Name a therapeutic drug, which inhibits the parasitic polyamine synthesis.

8. What is the amino acid precursor of taurine? What is the metabolic role of taurine in our body?

9. Name the components of coenzyme A.

10. Mention the mechanism by which biogenic amines are produced in our body. Name any 4 amino acids and one biogenic amine produced from each.

11. Differentiate the role of tyrosinase and tyrosine hydroxylase.

12. Enumerate the metabolic role of glycine.

13. Explain why infants with pyridoxine deficiency are prone to develop convulsions.

14. Name any two special products derived from the following amino acids:
    a. Glycine
    b. Tyrosine
    c. Tryptophan

Multiple Choice Questions

1. **All of the following are tyrosine derivatives, EXCEPT:**
   a. Serotonin   b. Melanin
   c. Epinephrine   d. Thyroxine

2. **In context to the amino acid and the special product, which one of the following is a wrongly matched pair?**
   a. Phenylalanine—niacin
   b. Tryptophan—serotonin
   c. Phenylalanine—melanin
   d. Tyrosine—epinephrine

3. **Glycine is useful in all of the following, EXCEPT:**
   a. Purine synthesis
   b. Creatine synthesis
   c. Spermine synthesis
   d. Heme synthesis

4. **Decarboxylation of which of the following amino acids yields a potent vasodilator?**
   a. Aspartate
   b. Arginine
   c. Histidine
   d. Serine

5. **Which of the following chemical reaction is involved in the conversion of noradrenaline to adrenaline?**
   a. Hydroxylation
   b. Carboxylation
   c. Methylation
   d. Dehydrogenation

ANSWERS

| 1. a. | 2. a. | 3. c. | 4. c. | 5. c. |

11.5 ONE-CARBON METABOLISM

Chapter Outline

- One-carbon Groups and Carriers
- Generation of One-carbon Groups
- Utilization of One-carbon Group
- Methyl Folate Trap

One carbon metabolism refers to a group of reactions in which one-carbon groups are generated and utilized in metabolic reactions. As these one-carbon groups are highly volatile in nature, they need specialized carriers for the transfer reactions.

ONE-CARBON GROUPS AND CARRIERS

Structure	Name	Oxidation state	Carriers
$-CH_3$	Methyl	Most reduced	Tetrahydrofolate (THFA) Vitamin B_{12} Activated methionine/S-adenosyl methionine (SAM)
$-CH_2-$	Methylene	Oxidized	
$-CHO$	Formyl	Most oxidized	Tetrahydrofolate
$-CHNH$	Formimino		
$-CH=$	Methenyl		

Amino Acids that Donate One-carbon Groups to the One Carbon Pool

Amino acid	One carbon group produced
Glycine	Methylene group through glycine cleavage complex in the liver mitochondria
Serine	Methylene group via serine hydroxymethyltransferase
Tryptophan	Formate through N-Formylkynurenine
Histidine	Formimino group through FIGLU
Methionine	Methyl group via S-Adenosyl methionine (SAM)

Amino acids producing one-carbon group: GST–HM (Goods and Service Tax approved by Honorable Minister)

GENERATION OF ONE-CARBON GROUPS

1. Glycine producing methylene group via glycine cleavage system:

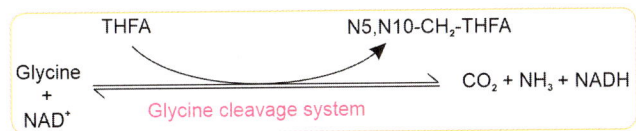

2. Serine producing methylene group via serine hydroxymethyl transferase:

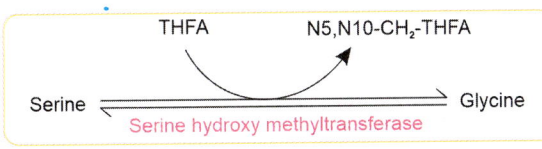

3. Histidine producing formimino group:

Histidine $\xrightarrow{\text{Histidase}}$ Urocanic acid $\longrightarrow$ N-Formimino glutamate (FIGLU) $\xrightarrow[\text{THFA}\quad\text{5-Formimino}]{\text{Glutamate formimino transferase}}$ Glutamate

THFA   5-Formimino THFA

The above reaction is the basis of *Histidine loading test.* Since the conversion of FIGlu to glutamate requires folic acid, excretion of FIGlu in response to a histidine load is diagnostic of folate deficiency.

4. Tryptophan producing formate:

Tryptophan $\xrightarrow{\text{Pyrrolase}}$ N-formyl kynurenine $\xrightarrow[\text{H}_2\text{O}\quad\text{Formate}]{\text{Formylase}}$ Kynurenine

Kynurenine is further degraded by hydroxylation and B_6 dependent kynureninase mediated hydrolysis. Deficiency of B_6 leads to conversion of the hydroxykynurenine into xanthurenate.

This is the biochemical basis of *tryptophan loading test.* i.e. excretion of xanthurenate (**xanthurenic aciduria**) in response to a tryptophan load is diagnostic of vitamin B_6 deficiency.

Utilization of One-carbon Group

1. Conversion of dUMP to dTMP (difference between uracil and thymine is a methyl group at 5th position)
2. Synthesis of serine from glycine by the reversible conversion
3. C_2 and C_8 of purine ring
4. Conversion of homocysteine to methionine by methionine synthase reaction

Conversion of dUMP to dTMP

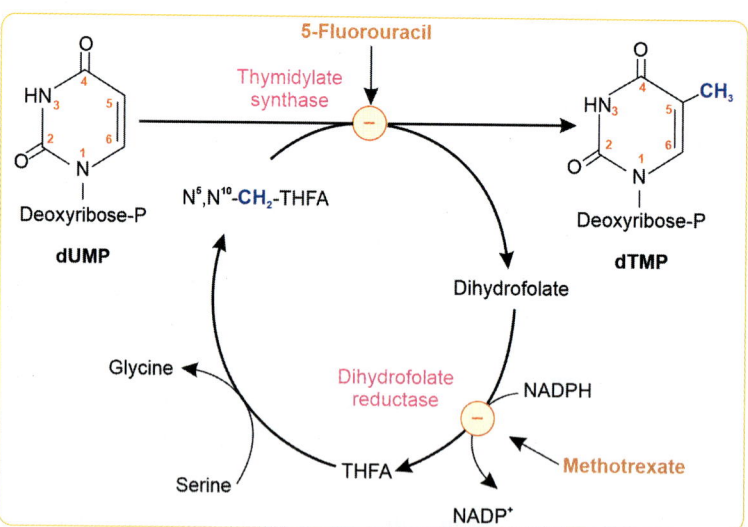

In this reaction, in addition to the release of one-carbon group, oxidation of one carbon carrier THFA is happening i.e. THFA is oxidized to DHFA. Reduction of DHFA to THFA is done by the enzyme dihydrofolate reductase which is the target of methotrexate. Thymidylate synthase is inhibited by the **suicidal** inhibitor 5-fluorouracil.

Overview of One-carbon Metabolism

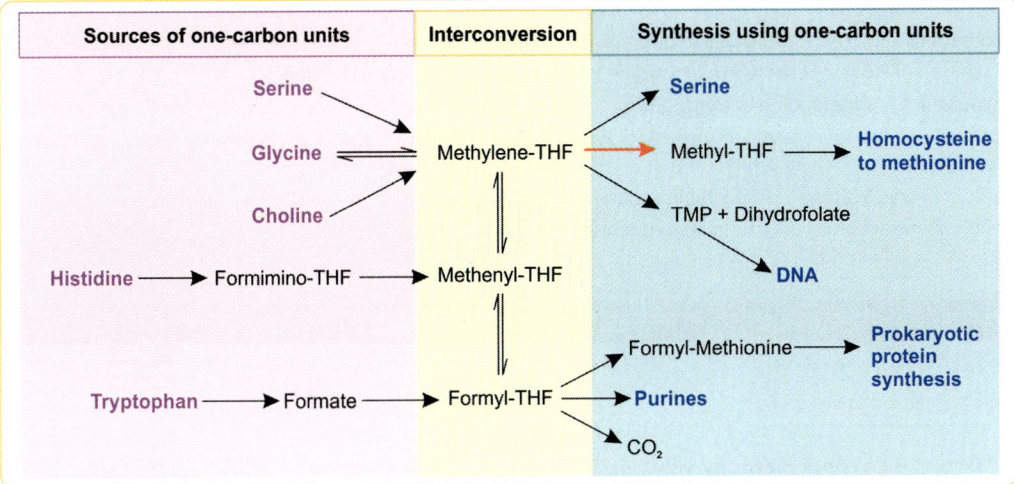

Fig. 1: Sources and utilization of one-carbon groups are illustrated. See the interconvertible nature of formyl, methenyl and methylene THFA. Also, note that production of methyl-THFA is an irreversible reaction. The importance of this will be explained in folate trap.

METHYLFOLATE TRAP

Deficiency of Vitamin B₁₂ Leads to Trapping of Folate in the form of N5-methyl THFA

CH_3-THFA is the most reduced form of THFA, and it is not interconvertible to other forms like methylene and formyl THFA. Moreover, production of CH_3-THFA is an irreversible reaction. The only way to regenerate active forms of THFA is the following B_{12} catalysed reaction. Thus, in deficiency of methyl cobalamine, all the folate in the body will be trapped as methyl-THFA. This is known as *'folate trap'* or *'methyl folate trap'* (*Chapter 33*) (*Refer to page no. 523*).

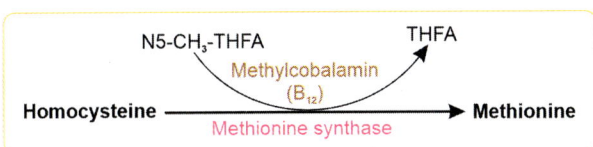

Note: Conversion of homocysteine to methionine does not mean that methionine can be synthesized in the body. This is actually a regeneration of methionine. The homocysteine itself came from methionine.

Key Points

❑ Glycine, serine, tryptophan, histidine, methionine are the five amino acids that act as one-carbon donors.
❑ Formimino glutamate (FIGlu) is a metabolite of histidine.
❑ Excretion of FIGlu in response to a histidine load is diagnostic of folate deficiency.
❑ Kynurenine is a metabolite of tryptophan.
❑ Excretion of xanthurenate (xanthurenic aciduria) in response to a tryptophan load is diagnostic of vitamin B_6 deficiency.
❑ 5-flouorouracill is the suicidal inhibitor of thymidylate synthase.
❑ Methotrexate inhibits dihydrofolate reductase.

SELF-ASSESSMENT

Short Answer Questions

1. Discuss the various means by which one-carbon moieties are formed and utilized in the body.

2. Name two amino acids, which have an important contribution in the synthesis of N5, N10-methylene THFA, specifying the reaction in each case.

3. Name any four amino acids that act as one-carbon donors. Write the chemical reaction of any two of them.

4. Which vitamin acts as a carrier of one-carbon carbon? Name two biochemical reactions in which one-carbon groups are utilised.

5. Which one-carbon group is involved in the conversion of dUMP to dTMP? Name the enzyme catalysing this reaction? Which anticancer drug inhibits this enzyme?

6. Explain the basis of 'folate trap' with suitable biochemical reaction. What is the clinical significance of this?

7. Describe the role of FIGLU excretion test

8. Deficiency of which vitamin can be detected using tryptophan loading test? Explain the biochemical basis.

Multiple Choice Questions

1. **Kynurenine is formed from the metabolism of:**
   a. Phenylalanine
   b. Tryptophan
   c. Tyrosine
   d. Histidine

2. **Metabolism of which of the following amino acids produces FIGlu as an intermediate?**
   a. Histidine   b. Arginine
   c. Cystine   d. Methionine

3. **All of the following amino acids are one-carbon donors, EXCEPT:**
   a. Glycine   b. Serine
   c. Histidine   d. Threonine

4. **All of the following reactions utilize one-carbon groups, EXCEPT:**
   a. *De novo* purine synthesis
   b. dTMP to dUMP synthesis
   c. Regeneration of methionine
   d. Synthesis of serine

5. **Which of the following vitamin deficiency is detected by tryptophan loading test?**
   a. B_1 b. B_2
   c. B_9 d. B_6

ANSWERS

1. b.	2. a.	3. d.	4. b.	5. d.

11.6 INBORN ERRORS OF AMINO ACID METABOLISM

Inborn errors of metabolism are group of disorders in which a single gene defect causes a clinically significant block in a metabolic pathway resulting either in accumulation of substrate behind the block or deficiency of the product.

The term "Inborn errors of metabolism" was coined by Edward Garrod. He described the features of *Albinism, Alkaptonuria, Cystinuria and Pentosuria*, which are collectively known as *Garrod's tetrad*.

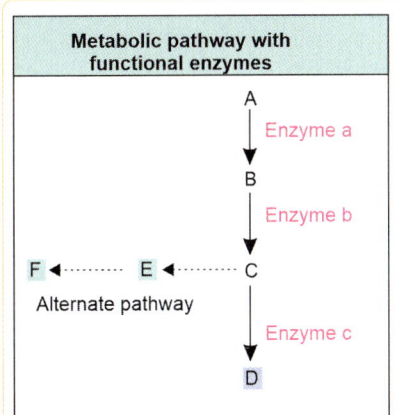
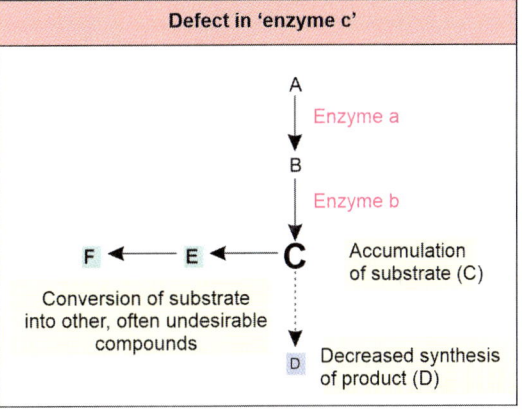

Fig. 1: Showing a metabolic pathway involving 3 enzymes a, b and c. Defect in enzyme c leads to accumulation of substrate C and reduced production of D. In the absence of enzyme c, substrate C undergoes alternate pathways often leading to undesirable products (E and F).

Let us extrapolate the figure with certain diseases which we have already discussed and some others are going to be discussed in this chapter.

	G6PD deficiency	Galactosemia	Phenylketonuria	Homocystinuria
Enzyme deficient	Glucose-6-phosphate dehydrogenase	Galactose-1-phosphate uridyl transferase	Phenylalanine hydroxylase	Cystathionine β synthase
Accumulated substrate	–	Galactose-1-phosphate	Phenylalanine	Homocysteine

Contd...

Chapter 11 Amino Acid Metabolism

	G6PD deficiency	Galactosemia	Phenylketonuria	Homocystinuria
Important metabolite deficient	NADPH	–	Tyrosine	Cysteine
Undesirable compounds	–	Galactitol	Phenylpyruvate, phenyllactate	Homocysteine, Homocysteine adducts with proteins

In this chapter, we will limit our discussion to the inborn errors of amino acid metabolism. We will also discuss the relevant normal metabolism of the amino acid whenever it is necessary for the understanding of the disorder.

GLYCINE

GLYCINE DONATES ONE-CARBON GROUP IN GLYCINE CLEAVAGE SYSTEM CATALYZED REACTION

Glycine is a simple amino acid with 2 carbons and one nitrogen. Cleavage of glycine by glycine cleavage complex of hepatic mitochondria releases ammonium ion, CO_2.

Importance of Glycine Cleavage System

❑ In this process, one carbon group is transferred to tetrahydrofolate to produce N^5, N^{10} methylene THFA.
❑ The pathway is reversible. Thus, glycine can also be synthesized by this pathway.

Defective Glycine Cleavage Complex System Leads to Nonketotic Hyperglycinemia

❑ **Nonketotic hyperglycinemia** is a condition with increased glycine levels due to *defect in glycine cleavage complex.*
❑ Glycine is a neurotransmitter; excitatory in central nervous system and inhibitory in peripheral nervous system.
❑ *Excess glycine* results in *mental retardation* and other neuronal symptoms.
❑ Sodium benzoate (C_6H_5COONa) is used to remove the glycine in this condition because it undergoes phase II conjugation reaction with glycine in liver to produce water-soluble **Hippuric acid**.

Deamination of Glycine Produces Glyoxylate

Oxidative deamination of glycine produces glyoxylate which is further metabolized in the body.
Defect in the metabolism of glyoxylate leads to conversion into oxalic acid and *primary hyperoxaluria*.

Type I primary hyperoxaluria is due to defect in alanine: glyoxylate aminotransferase. Vitamin B_6 may be beneficial in selected patients with type 1 primary hyperoxaluria. (*Ref. Harrison 20th Ed., p. 3016*)

SULPHUR CONTAINING AMINO ACIDS

METHIONINE IS PURELY GLUCOGENIC SINCE IT PRODUCES SUCCINYL-COA

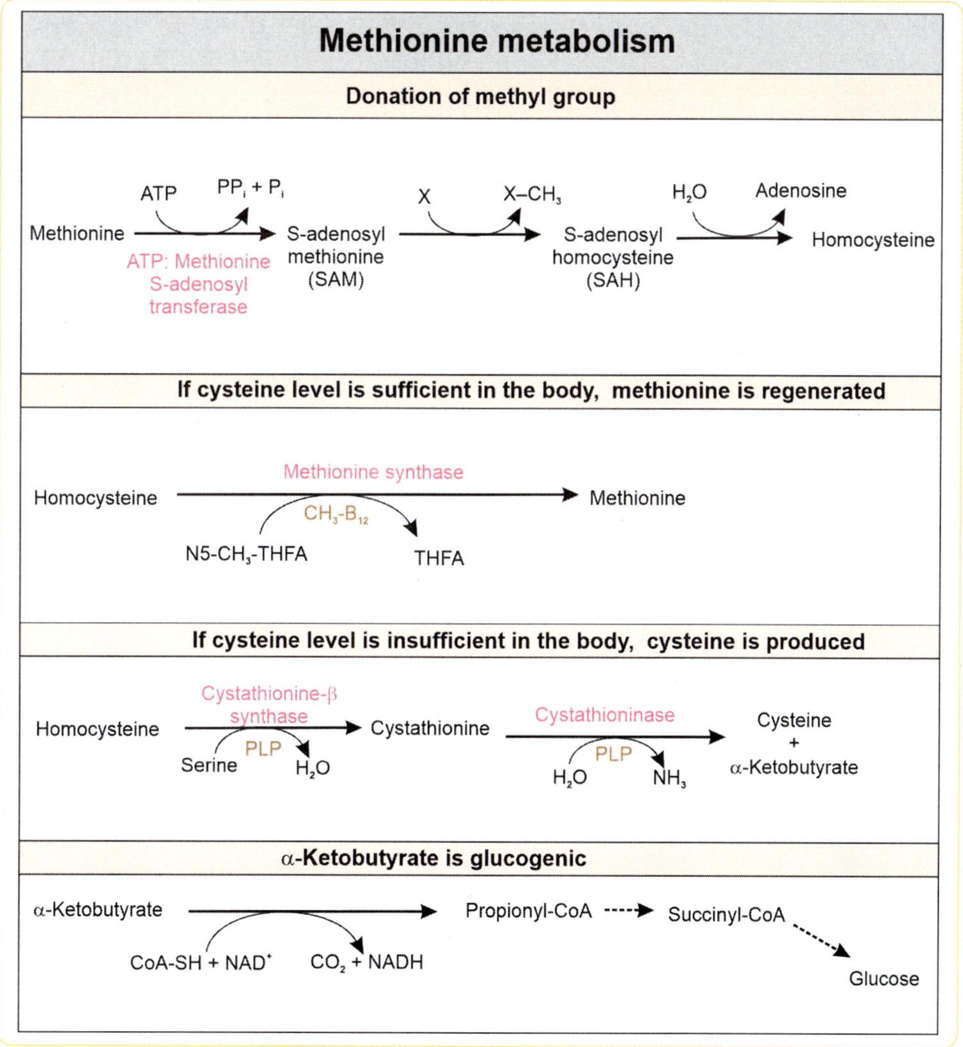

Methionine metabolism

Donation of methyl group

Methionine $\xrightarrow[\text{ATP: Methionine S-adenosyl transferase}]{\text{ATP} \quad \text{PP}_i + \text{P}_i}$ S-adenosyl methionine (SAM) $\xrightarrow{\text{X} \quad \text{X–CH}_3}$ S-adenosyl homocysteine (SAH) $\xrightarrow{\text{H}_2\text{O} \quad \text{Adenosine}}$ Homocysteine

---

If cysteine level is sufficient in the body, methionine is regenerated

Homocysteine $\xrightarrow[\text{N5-CH}_3\text{-THFA} \quad \text{THFA}]{\text{Methionine synthase} \quad \text{CH}_3\text{-B}_{12}}$ Methionine

---

If cysteine level is insufficient in the body, cysteine is produced

Homocysteine $\xrightarrow[\text{Serine} \quad \text{H}_2\text{O}]{\text{Cystathionine-β synthase} \quad \text{PLP}}$ Cystathionine $\xrightarrow[\text{H}_2\text{O} \quad \text{NH}_3]{\text{Cystathioninase} \quad \text{PLP}}$ Cysteine + α-Ketobutyrate

---

α-Ketobutyrate is glucogenic

α-Ketobutyrate $\xrightarrow[\text{CoA-SH} + \text{NAD}^+ \quad \text{CO}_2 + \text{NADH}]{}$ Propionyl-CoA ----▸ Succinyl-CoA ------▸ Glucose

Points to note from this pathway:
- ❏ Cysteine can be produced from methionine.
- ❏ If cysteine is adequate, methionine is regenerated.
- ❏ Vitamin B_6, folate and vitamin B_{12} are involved.

Active Methionine is Involved in Transmethylation Reactions

❑ S-adenosyl methionine (SAM) or AdoMet is the active methionine.
❑ The enzyme ATP: L-methionine S-adenosyl transferase catalyzes the formation of SAM.
❑ This reaction is unique and energy of all the 3 high-energy phosphate bonds of ATP are utilized.
❑ SAM is the main compound involved in transmethylation reactions.
❑ N5 methyl THFA is involved in methylation of homocysteine to regenerate methionine.

Transmethylation reactions	
Substrate	**Product**
SAM mediated transmethylation	
Noradrenaline	Adrenaline
N-acetyl serotonin	Melatonin
Guanidino acetate	Creatine
Phosphatidylethanolamine	Phosphatidylcholine (lecithin)
Cytosine of DNA	5-methyl cytosine
Lysine of histone proteins	Methylated histone
RNA	5' capping
N⁵ methyl THFA mediated transmethylation	
Homocysteine	Methionine

Methionine to Cysteine Conversion is Affected in Homocystinuria

Homocystinuria is an autosomal recessive disease due to defect in methionine catabolism and there will be increased concentration of the sulfur-containing amino acid **homocystine** (Disulphide form) in blood and urine. Classical homocystinuria is due to *cystathionine β synthase* deficiency.

Other forms of homocystinuria are due to defect in 5,10-CH2-THFA reductase, e and vitamin B_{12} lysosomal efflux and metabolism.

Clinical features: Four organ systems are affected mainly.

Clinical features of Homocystinuria

Eye	CVS	Skeletal system	Brain
Ectopia lentis, High myopia	Thromboembolic disorders	Chest wall deformities, Mrfanoid habitus, Osteoporosis	Mental retardation

Laboratory Findings

❑ **Blood:**
  ○ Elevated level of free homocystine and methionine
  ○ Reduced levels of cystathionine, cysteine or cystine
  ○ Analysis is done by high-performance liquid chromatography (HPLC).
❑ **Urine:** ↑ excretion of homocystine.

Treatment

- ☐ High dose vitamin B_6
- ☐ Normal dose folic acid and vitamin B_{12}
- ☐ Supplementation of cysteine in diet since cysteine becomes dietary essential in patients of homocystinuria.
- ☐ Methionine restriction in diet combined with betaine (N,N,N-trimethyl glycine). Betaine is a methyl donor that provides an alternate pathway for conversion of homocysteine back to methionine. This methionine can be incorporated into proteins. The need for methionine by the body will determine the efficacy of betaine therapy. That is why methionine restriction is done.

"Hyperhomocysteinemia refers to increased total plasma concentration of homocysteine with or without an increase in free homocystine (disulfide form)" (*Ref. Harrison 20th Ed., p. 3018*)

Biochemical Basis of Thromboembolic Risk and Atherosclerosis in Hyperhomocysteinemia

- ☐ Homocysteine forms thiolactone adducts with proteins
- ☐ Homocysteine inhibits copper dependent *lysyl oxidase* which is important for cross linking of both collagen and elastin → dysregulation of extracellular matrix → endothelial dysfunction

BRANCHED CHAIN AMINO ACIDS

BRANCHED CHAIN AMINO ACIDS ARE OXIDIZED IN A SIMILAR MANNER TO THAT OF FATTY ACIDS

Branched chain amino acids are not degraded in the liver. The skeletal muscle preferentially utilized them.

First step in the metabolism of branched chain amino acids is the transamination to produce branched chain α-keto acids.

Next, they undergo, oxidative decarboxylation by branched chain ketoacid dehydrogenase (BCKD) enzyme complex, which is analogous to PDH complex and requires the same 5 cofactors (Thiamine pyrophosphate, Lipoic acid, coenzyme A, FAD and NAD).

Look at the end products of the degradation of branched chain amino acids:

- ☐ Leucine produces acetoacetate and acetyl-CoA → purely ketogenic
- ☐ Valine produces succinyl-CoA → purely glucogenic
- ☐ Isoleucine produces acetyl-CoA and succinyl-CoA → Both glucogenic and ketogenic

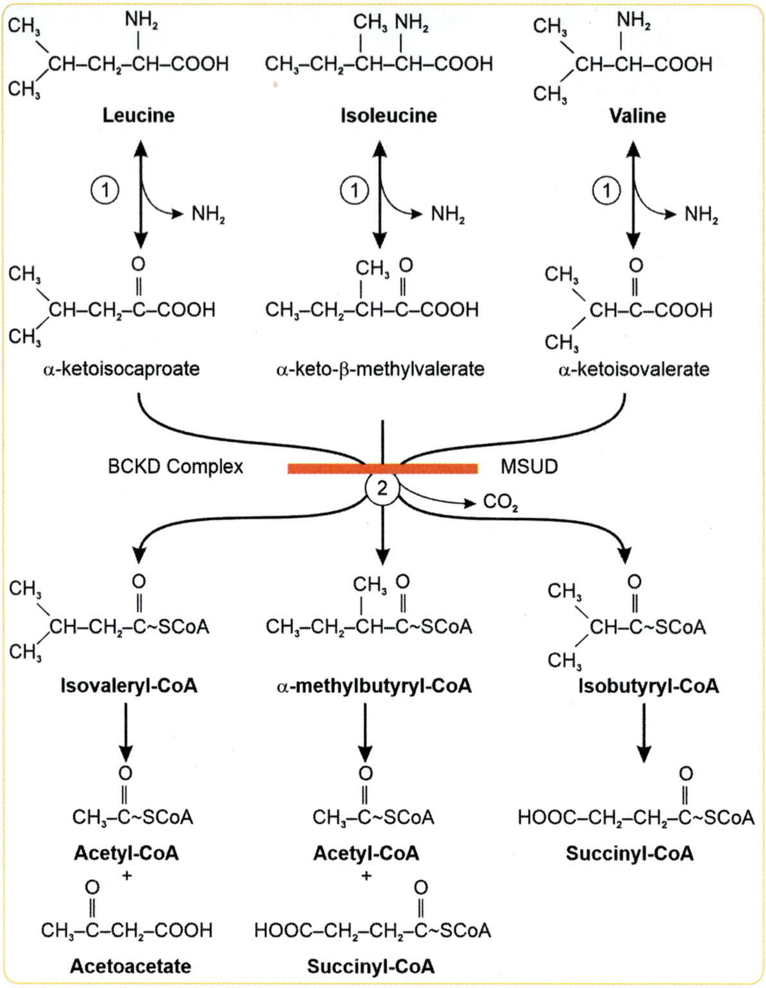

Fig. 2: 1—Transaminase 2—Branched chain ketoacid dehydrogenase (BCKD)

Defect in BCKD Complex Leads to Maple Syrup Urine Disease (MSUD)

Clinical Presentation

Autosomal recessive inheritance; disease manifests in the first week of life with vomiting, hypoglycemia and *neurological impairment.* Characteristic maple syrup/*burnt sugar odor* (not taste) of urine and sweat is due to accumulation of branched chain ketoacids.

Diagnostic Findings

- ❑ ↑Branched chain keto acids and branched chain amino acids in blood and urine is confirmatory.
- ❑ Presence of alloisoleucine (stereoisomeric metabolite of isoleucine) is characteristic.
- ❑ Enzyme defect can be demonstrated in cultured fibroblasts.

Treatment

- ❑ *Restriction of branched chain amino acids,* low enough to prevent mental retardation but adequate enough to allow normal growth
- ❑ *Thiamine supplementation* in cases of thiamine responsive MSUD

AROMATIC AMINO ACIDS

PHENYLALANINE/TYROSINE IS DEGRADED TO FUMARATE AND ACETO-ACETYL-COA

Phenylalanine is a 9-carbon compound (Benzene ring – 6 C and Alanine – 3C). This 9-carbon compound is degraded to two 4-carbon compounds—Fumarate and acetoacetyl-CoA. One carbon is oxidized as CO_2 by the dioxygenase step.

Here, I want you to understand the role of dioxygenase in the opening of aromatic ring. Look at the homogentisate oxidase (dioxygenase) catalyzed reaction in which aromatic ring is opened.

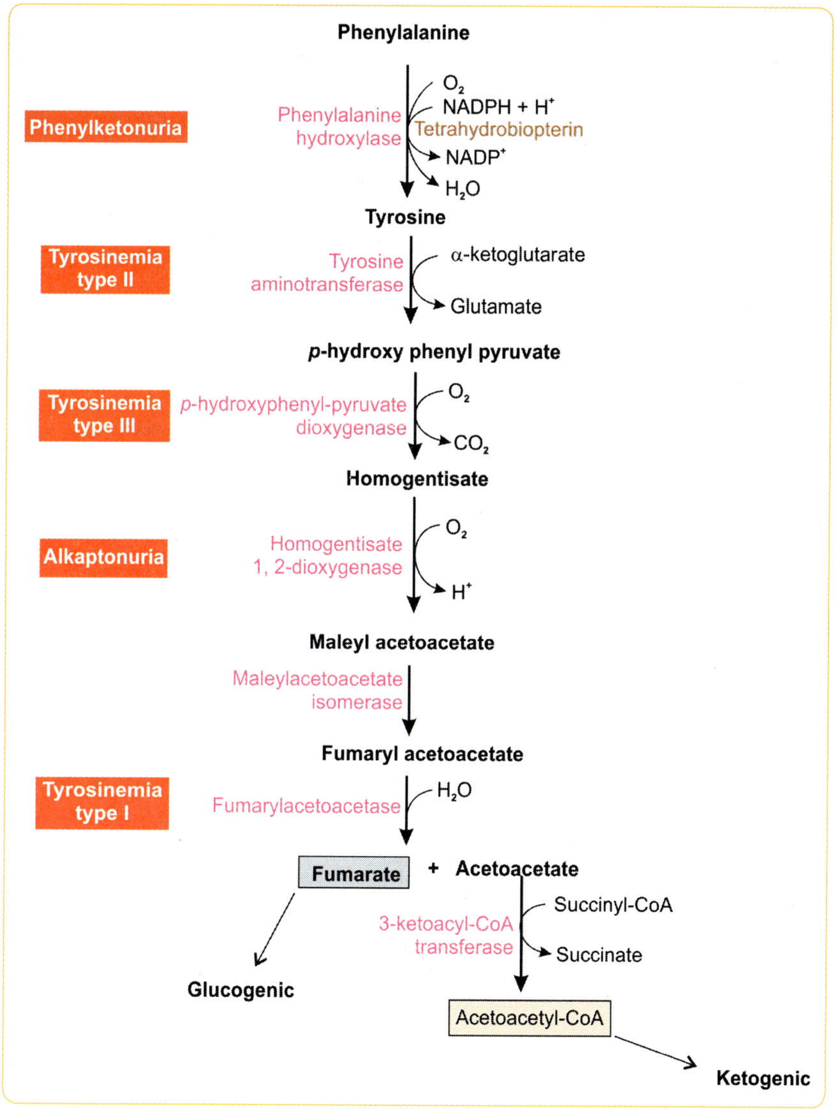

Phenylalanine

Phenylketonuria

Phenylalanine hydroxylase — O_2, NADPH + H$^+$, Tetrahydrobiopterin → NADP$^+$, H_2O

Tyrosine

Tyrosinemia type II

Tyrosine aminotransferase — α-ketoglutarate → Glutamate

***p*-hydroxy phenyl pyruvate**

Tyrosinemia type III

p-hydroxyphenyl-pyruvate dioxygenase — O_2 → CO_2

Homogentisate

Alkaptonuria

Homogentisate 1, 2-dioxygenase — O_2 → H$^+$

Maleyl acetoacetate

Maleylacetoacetate isomerase

Fumaryl acetoacetate

Tyrosinemia type I

Fumarylacetoacetase — H_2O

Fumarate + **Acetoacetate**

3-ketoacyl-CoA transferase — Succinyl-CoA → Succinate

Glucogenic

Acetoacetyl-CoA

Ketogenic

Impaired Conversion of Phenylalanine to Tyrosine Leads to Hyperphenylalaninemias

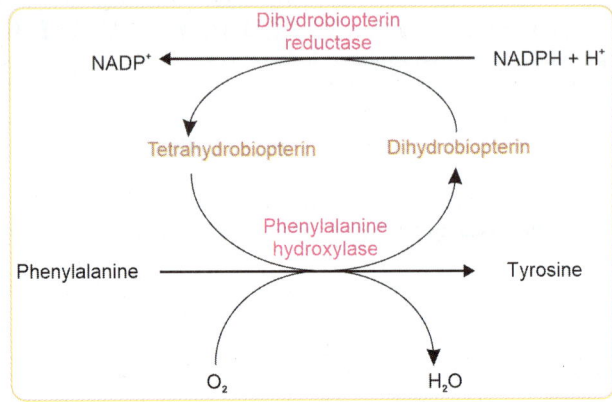

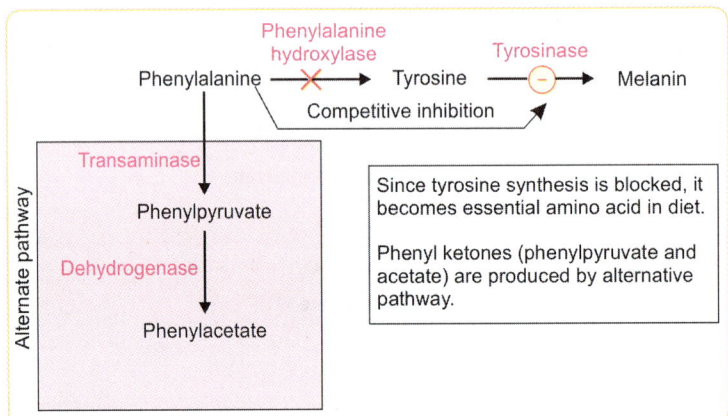

Since tyrosine synthesis is blocked, it becomes essential amino acid in diet.

Phenyl ketones (phenylpyruvate and acetate) are produced by alternative pathway.

Types	Defect
I (Classic PKU)	Phenylalanine hydroxylase
II and III	Dihydrobiopterin reductase
IV and V	Dihydrobiopterin biosynthesis

Clinical Features

Child is usually normal at birth. Breastfeeding introduces phenylalanine and symptoms begin to appear. Vomiting, irritability, convulsions, tremors are the usual presentation. Mousy odor is due to phenylacetate. Child is fair colored and eczematous.

Biochemical Basis of Clinical Features

❑ Tyrosine cannot be synthesized → **Tyrosine becomes dietarily essential**
❑ Increased phenylalanine inhibits the transport of other amino acids required for neurotransmitter synthesis, reduces synthesis and increases degradation of myelin → **Mental retardation**
❑ Phenylalanine is a competitive inhibitor of tyrosinase, the rate-limiting enzyme of melanin synthesis → **Hypopigmentation**

Screening and Diagnosis

- Screening is done by Guthrie test which is a microbiological assay. Urine $FeCl_3$ test for the detection of phenylpyruvate is not specific and no longer used.
- Mass spectrometry based screening has the advantage that it allows the detection of multiple disorders.
- Prenatal diagnosis can be done by PCR based genetic techniques.

Guthrie Test is Used in the Screening of Phenylketonuria

Guthrie test is based on the inhibition of growth of a particular strain of *Bacillus subtilis* (ATCC 6051) by beta-2-thienylalanine and this inhibition can be relieved by phenylalanine, phenyl pyruvic acid and phenyl lactic acid. *Bacillus subtilis* is grown in the presence of inhibitor.
- If the blood from normal infant is added, there will be no growth as there is no phenylalanine to relieve the inhibition.
- When the blood from phenylketonuric infant is added, bacteria will proliferate since unmetabolized phenylalanine and derivatives are abundant in blood.

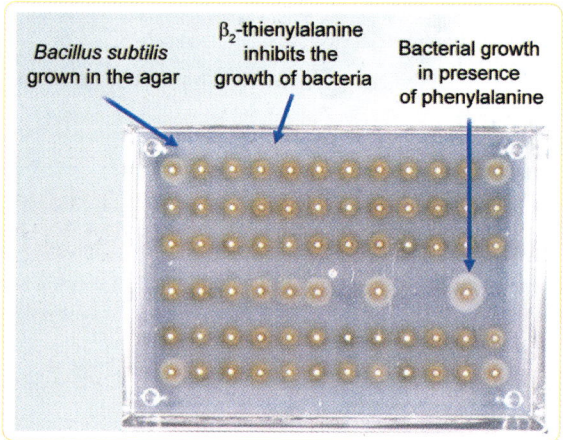

Treatment

- Diagnosis and management should be done before 2 weeks of age to prevent mental retardation.
- Limiting the substrate of the deficient enzyme, i.e. phenylalanine, is the first-line of management. So, breastfeeding should be avoided.
- Initiation of early phenylalanine-free diet can prevent mental retardation and dietary restriction is continued throughout the life according to recent guidelines.
- Carrier mothers should take phenylalanine-free diet during pregnancy because high phenylalanine will affect the growth of a normal fetus also (teratogenic effect).
- Phenylalanine containing artificial sweetener **aspartame** (Methyl ester of aspartate + phenylalanine) should be avoided.

HOMOGENTISATE OXIDASE DEFICIENCY LEADS TO ALKAPTONURIA

- Alkaptonuria is the 1st IEM to be described by Garrod in 1908.
- Autosomal recessive disorder with defect in tyrosine catabolism.
- Deficiency of *homogentisic acid oxidase*/homogentisate 1,2-dioxygenase results in accumulation of homogentisate.
- Urine becomes *black* on standing due to oxidation, (Standing means keeping the urine in the stand/rack for some time!).
- Benedict's test will be positive with the urine since homogentisate is a reducing substance.

❑ Oxidized black pigments (*benzoquinone acetic acid*) gets deposited in connective tissues like vertebra. This is known as *ochronosis*. This causes inflammation of joints (arthritis). Pigmentation can be noted in cartilage of ear.

Treatment

❑ *Nitisinone* (same drug used in type I tyrosinemia)—inhibitor of the enzyme p-hydroxyphenylpyruvate dioxygenase, reduces the urinary excretion of homogentisic acid.
❑ Antioxidant vitamin *ascorbic acid* is given to prevent the oxidation of homogentisic acid to benzoquinone acetic acid.
❑ Low protein diet

TYROSINEMIA

Type	Defect	Involvement
Type I tyrosinemia (tyrosinosis)	Fumarylacetoacetate hydrolase	Hepatorenal
Type II tyrosinemia (Richner-Hanhart syndrome)	Tyrosine aminotransferase	Oculocutaneous
Tyrosinemia type III	p-hydroxyphenylpyruvate hydroxylase	Normal liver function

DISORDERS OF AMINO ACID TRANSPORT

Disease	Defect	Amino acid excreted in urine	Features
Hartnup disease	Neutral amino acid transporter (SLC6A19) in renal tubules and jejunum	Tryptophan	Excretion of tryptophan in urine → ↓synthesis of niacin → Pellagra
Cystinuria	Dibasic amino acid transporter (SLC3A1, SLC7A9) in renal tubules and jejunum	Cystine, Ornithine, Lysine, Arginine (COLA)	Cystine nephrolithiasis
Lysinuric protein intolerance	Dibasic transporter SLC7A7	Lysine, arginine and ornithine	Protein intolerance, hepatosplenomegaly, failure to thrive, muscle weakness, impaired immune function, osteoporosis, renal failure
Dicarboxylic aciduria	Dicarboxylic amino acid transporter	Glutamic acid, aspartic acid	None
Iminoglycinuria	Imino acid transporter	Glycine, proline, hydroxyproline	None
Oasthouse syndrome	Methionine absorption	Methionine	Diarrhea, mental retardation, convulsions, blue eyes and white hair
Drummond syndrome	Tryptophan absorption	Tryptophan	Hypercalcemia with nephrocalcinosis and indicanuria; Urinary tryptophan is broken down by bacteria to indole which is oxidized to indigo blue → this is the reason behind the name *Blue diaper syndrome*

Medium chain acyl-CoA dehydrogenase deficiency produces dicarboxylic aciduria in which dicarboxylic acids of 6 and 8 carbons are formed.

ODOR OF URINE IN VARIOUS DISORDERS

Odor	Disease
Musty or Mousy	Phenylketonuria
Fruity	Diabetic ketoacidosis
Cabbage/Rancid	Tyrosinemia
Boiled cabbage	Hypermethioninemia
Swimming pool	Hawkinsinuria
Burnt sugar (maple syrup)	Maple syrup urine disease
Foul smelling	Urinary tract infections
Sweaty feet	Isovaleric acidemia
Rotten fish	Trimethylaminuria

OTHER INBORN ERRORS OF METABOLISM OF IMPORTANCE

Disorder	Deficient enzyme
Hawkinsinuria	4-Hydroxyphenylpyruvate dioxygenase
DOPA-responsive dystonia	Tyrosine hydroxylase
Histidinemia	Histidine-ammonia lyase
Cystinosis	Lysosomal membrane protein cystinosin involved in efflux
Sulfocysteinuria	Sulfate oxidase or molybdenum cofactor deficiency
Glutaric acidemia type I	Glutaryl-CoA dehydrogenase
Gyrate atrophy of the choroid and retina	Ornithine-δ-aminotransferase
Isovaleric acidemia	Isovaleryl-CoA dehydrogenase
Propionic acidemia	Propionyl-CoA carboxylase
Multiple carboxylase/biotinidase deficiency	Holocarboxylase synthase or biotinidase
Methylmalonic acidemia	Methylmalonyl-CoA mutase/racemase or cobalamin reductase/adenosyltransferase

Competency

- ☐ BI11.5   Describe screening of urine for inborn errors
- ☐ BI5.4   Describe common disorders associated with protein metabolism

Chapter 11   Amino Acid Metabolism

Key Points

- ❑ Type I primary hyperoxaluria is due to defect in alanine: Glyoxylate aminotransferase.
- ❑ S-adenosyl methionine (AdoMet/SAM) is known as active methionine.
- ❑ Tyrosine produces fumarate and acetoacetate. Hence, it is both glucogenic and ketogenic.
- ❑ Maple syrup urine disease is due to defect in Branched chain keto acid dehydrogenase.
- ❑ Tyrosine becomes an essential amino acid in phenylketonuria.
- ❑ Cysteine becomes an essential amino acid in Homocysteinemia.
- ❑ Amino acids excreted in cystinuria are Cystine, Ornithine, Lysine, Arginine
- ❑ COLA) – defect in dibasic amino acid transporter.
- ❑ Benzoquinone acetate, oxidized product of homogentisic acid accumulates in the connective tissue in alkaptonuria (ochronosis). This oxidation can be prevented by Vitamin C.
- ❑ Nitisinone (inhibitor of 4-hydroxyphenylpyruvate dioxygenase) is used in alkaptonuria and type I tyrosinemia.
- ❑ Benedict's test will be positive in alkaptonuria.
- ❑ Limiting the substrate for deficient enzyme (phenylalanine) is the first line therapy in phenylketonuria.
- ❑ Mousy order in phenylketonuria is due to phenyl acetate.
- ❑ Type I tyrosinemia (tyrosinosis) is due to defect in fumarylacetoacetate hydrolase.

SELF-ASSESSMENT

Short Answer Questions

1. What is the enzymatic defect in nonketotic hyperglycinemia? How will you treat this patient taking advantage of a type II biotransformation reaction?

2. What is 'transmethylation'? Give two examples

3. Elevated plasma level of homocysteine is a cardiovascular risk factor – How?

4. Name two inborn errors of amino acid metabolism in which dietary non-essential amino acids become essential. Mention the defective enzymes.

5. What is the enzyme defect in Maple syrup urine disease? Mention the coenzymes required for the activity of these enzymes. Name the amino acids whose metabolism is affected in this condition.

6. Elevated plasma concentration of which amino acid have been implicated as a cardiovascular risk factor? What is the plausible mechanism behind this? Explain with reasons, which vitamin(s) are supplemented with the therapeutic aim to lower the levels of this amino acid?

7. What is the inherited defect in Hartnup's disease? Explain why these patients often present with pellagra?

8. Mention the inherited defect in i) phenyl ketonuria ii) homogentisic aciduria iii) maple syrup urine & iv) homocystinuria.

9. Describe the metabolism of any one aromatic amino acid. Mention the metabolic defects.

Multiple Choice Questions

1. **Which of the following is not a component of Garrod's tetrad?**
   a. Phenylketonuria   b. Cystinuria
   c. Alkaptonuria   d. Pentosuria

2. **Transamination of which of the following amino acids produce glycine?**
   a. Alanine   b. Aspartate
   c. Glutamate   d. Glyoxylate

3. **The vitamin cofactor required for the cystathioninase reaction is:**
   a. B_1 b. B_2
   c. B_5 d. B_6

4. **Which of the following enzyme is deficient in maple syrup urine disease?**
   a. Branched chain amino acid aminotransferase
   b. Branched chain amino acid dehydrogenase
   c. Branched chain keto acid dehydrogenase
   d. Branched chain keto acid oxidase

5. **Phenylalanine is degraded into:**
   a. Fumarate and succinate
   b. Fumarate and acetoacetate
   c. Fumarate and malate
   d. Fumarate and pyruvate

6. **Tetrahydrobiopterin is required for the metabolism of:**
   a. Alanine
   b. Lysine
   c. Phenylalanine
   d. Serine

7. **All of the following are characteristics of phenylalanine hydroxylase, EXCEPT:**
   a. Mixed function oxidase
   b. Tetrahydrobiopterin is a cofactor
   c. NADPH provides the reducing power
   d. Vitamin C is a cofactor

8. **Mousy odor of urine in phenylketonuria is due to:**
   a. Phenylalanine
   b. Phenylpyruvate
   c. Phenyl acetate
   d. Phenyl lactate

9. **In phenylketonuria, the first line therapy is:**
   a. Replacement of the defective enzyme
   b. Replacement of the deficient product
   c. Limiting the substrate for deficient enzyme
   d. Giving the missing amino acid by diet

10. **Aspartame is made up of:**
    a. Aspartic acid and methionine
    b. Asparagine and methionine
    c. Asparagine and phenylalanine
    d. Aspartate and phenylalanine

11. **Richner-Hanhart syndrome is due to defect in:**
    a. Homogentisate oxidase
    b. Partial defect in HGPRTase
    c. Fumarylacetoacetate hydrolase
    d. Tyrosine aminotransferase

12. **A boy was brought with complaints of red, scaly rash and mild cerebellar ataxia by his mother with the history of same symptoms in her older daughter. It was clear that the boy did not have the usual dietary-deficiency form of pellagra. The boy was found to excrete excessive free amino acids in the urine. Which of the following amino acid transporter could be defective in this boy?**
    a. Dibasic amino acid transporter
    b. Dicarboxylic amino acid transporter
    c. Imino acid transporter
    d. Neutral amino acid transporter

13. **Which of the following transporter is defective in Hartnup disease?**
    a SLC3A1   b. SLC6A19
    c. SLC7A7   d. SLC7A9

14. **In which of the following diseases swimming pool odor of urine is seen?**
    a. Hawkinsinuria   b. Isovaleric acidemia
    c. Trimethylaminuria d. Tyrosinemia

15. **The protein defective in cystinosis is responsible for:**
    a. Absorption of cysteine from intestine
    b. Absorption of cysteine from renal tubules
    c. Efflux cysteine from endoplasmic reticulum
    d. Efflux of cysteine from lysosome

16. **Homocystinuria is due to abnormal metabolism of:**
    a. Methionine
    b. Valine
    c. Cysteine
    d. Leucine

17. **Oasthouse syndrome is due to malabsorption of:**
    a. Methionine
    b. Valine
    c. Cysteine
    d. Leucine

Chapter 11 Amino Acid Metabolism

ANSWERS

1. a.	2. d.	3. d.	4. c.	5. b.	6. c.
7. d.	8. c.	9. c.	10. d.	11. d.	12. d.
13. b.	14. a.	15. d.	16. a.	17. a.	

IMAGE-BASED QUESTION

1. **Name the enzyme that is deficient in this disease:**
   a. Tyrosine aminotransferase
   b. Homogentisate oxidase
   c. Fumarylacetoacetate hydrolase
   d. p-hydroxyphenylpyruvate hydroxylase

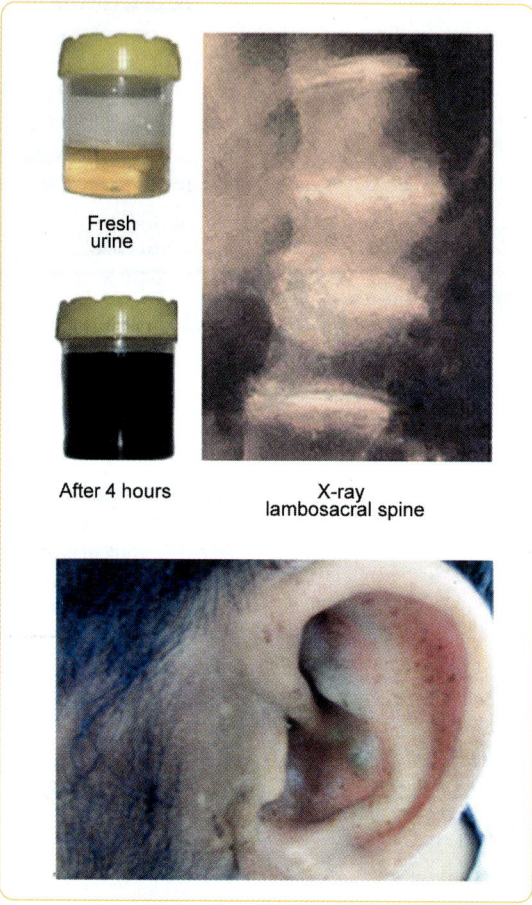

Fresh
urine

After 4 hours

X-ray
lambosacral spine

1. **Ans. (b) Homogentisate oxidase**

The image shows darkening of urine on standing, inflammatory changes in the vertebra and bluish green pigmentation in the ear. All this point out to the diagnosis of alkaptonuria. Thus, the answer is B. All other options are related to tyrosinemia.

11.7 HEME METABOLISM

STRUCTURE OF HEME

- Heme is an iron-porphyrin IX.
- Porphyrins are cyclic tetrapyrrole structure linked by four methyne (=HC—) bridges.

Note: Methyne group is also known as methyne or methene.

 High Yield

Heme Containing Enzymes

- Catalase
- Peroxidases
- Soluble guanylyl cyclase
- NADPH oxidase
- Nitric oxide synthase
- Tryptophan pyrrolase/2,3 dioxygenase
- Cytochrome C oxidase
- Cytochrome P450 (mixed function oxidase)

Heme Containing Nonenzyme Proteins

- Oxygen transporter—Hemoglobin
- Oxygen reservoir—Myoglobin, cytoglobin, neuroglobin
- Carrier of electrons - Cytochrome C

SYNTHESIS OF HEME

First and last three reactions take place inside the mitochondria and the rest of the reactions take place in cytoplasm.

Chapter 11 Amino Acid Metabolism

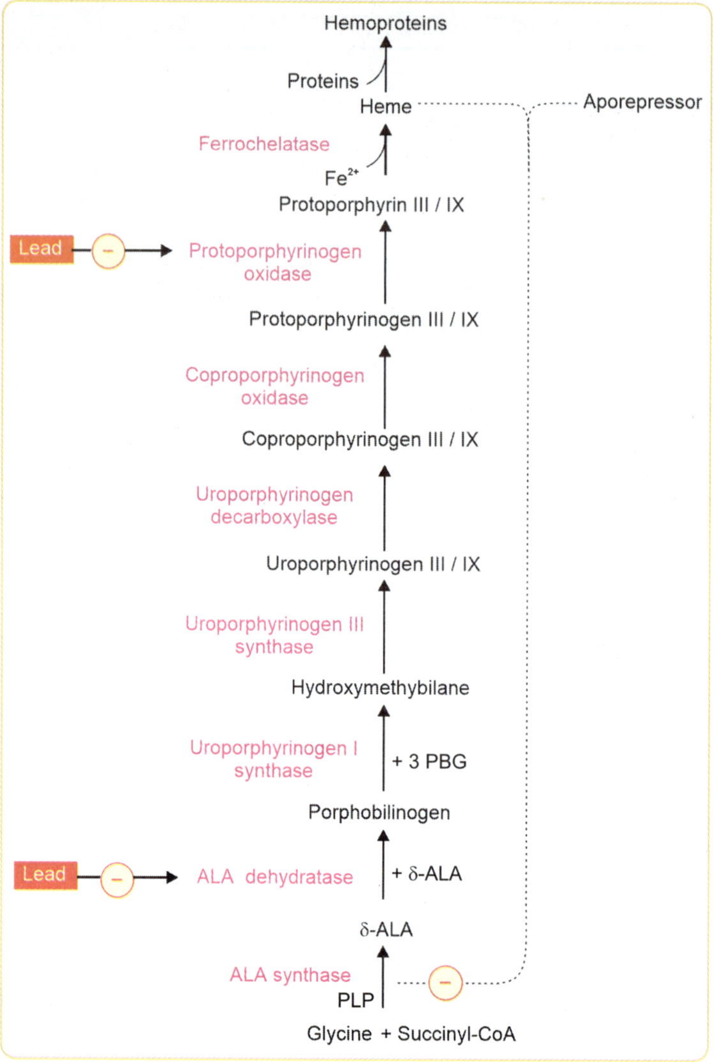

Hemoproteins

Proteins

Heme ⋯⋯⋯⋯⋯⋯⋯⋯ Aporepressor

Ferrochelatase

Fe^{2+}

Protoporphyrin III / IX

Lead ─ ⊖ ⟶ Protoporphyrinogen oxidase

Protoporphyrinogen III / IX

Coproporphyrinogen oxidase

Coproporphyrinogen III / IX

Uroporphyrinogen decarboxylase

Uroporphyrinogen III / IX

Uroporphyrinogen III synthase

Hydroxymethybilane

Uroporphyrinogen I synthase | + 3 PBG

Porphobilinogen

Lead ─ ⊖ ⟶ ALA dehydratase | + δ-ALA

δ-ALA

ALA synthase ⋯ ⊖ ⋯
PLP

Glycine + Succinyl-CoA

ALA Synthase is the Rate-limiting Enzyme of Heme Synthesis

There are 2 isoforms of ALA synthase.

ALA synthase 1	ALA synthase 2
Housekeeping enzyme present in all nucleated cells.	Erythroid tissue-specific enzyme
Inhibited by heme	Not inhibited by heme

Vitamin B₆ Deficiency Leads to Anemia

Pyridoxal phosphate (Vitamin B_6) is the cofactor for the δ-amino levulinic acid (ALA) synthase enzyme.

$$\text{Glycine + succinyl CoA} \xrightarrow[\text{ALA synthase}]{\text{PLP (B}_6\text{)}} \delta \text{ amino levulinic acid}$$

Thus, B_6 deficiency leads to reduced formation of heme and microcytic hypochromic anemia.

PORPHYRIAS ARE DUE TO DEFECT IN HEME SYNTHESIS

  High Yield

Type	Enzyme defect	Clinical features
Hepatic porphyria		
Acute intermittent porphyria	❑ Porphobilinogen deaminase a.k.a, Hydroxymethylbilane synthase a.k.a, Uroporphyrinogen I synthase	Abdominal pain, neuropsychiatric symptoms
Porphyria cutanea tarda	Uroporphyrinogen decarboxylase	Photosensitivity; *most common porphyria*
Hereditary coproporphyria	Coproporphyrinogen oxidase	Abdominal pain, neuropsychiatric symptoms and photosensitivity
Variegate porphyria	Protoporphyrinogen oxidase	
Erythropoietic porphyrias		
Congenital erythropoietic porphyria	Uroporphyrinogen III synthase	Photosensitivity
Protoporphyria	Ferrochelatase	
X-linked protoporphyria (XLP)	ALA synthase 2	

Harrison's Corner

Even though Harper says "Individuals with low ALAS2 activity develop anemia, not porphyria", Harrison mention it as porphyria. (*Ref. Harrison 20th Ed., p.2985*).

Inheritance of Porphyria

All porphyrias are inherited in autosomal dominant manner except:
- ❑ Congenital erythropoietic porphyria → autosomal recessive inheritance
- ❑ X-linked protoporphyria (XLP) → X-linked recessive inheritance

Biochemical Basis of Photosensitivity in Porphyria

Conjugated double bonds joining the pyrrole rings in the porphyrins absorb light.

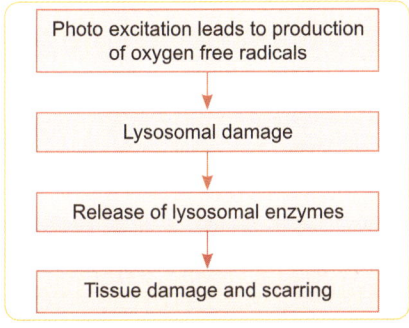

Photo excitation leads to production of oxygen free radicals

↓

Lysosomal damage

↓

Release of lysosomal enzymes

↓

Tissue damage and scarring

Soret Peak

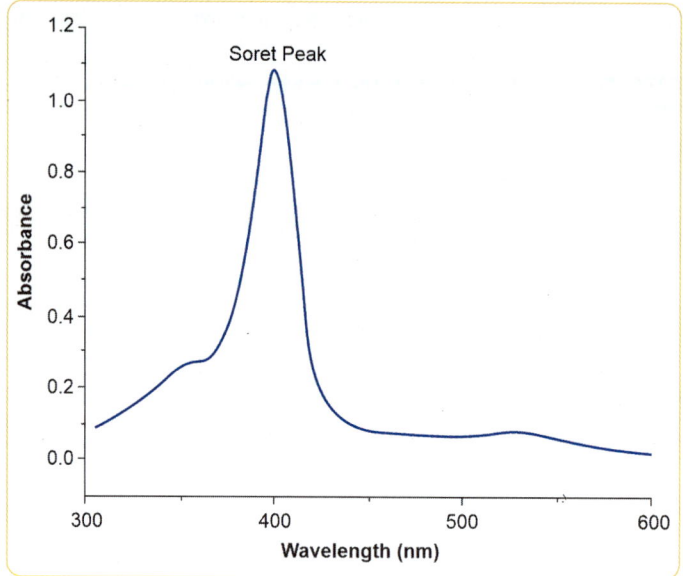

During spectrophotometry, absorption curve for porphyrins in 5% solution of hydrochloric acid peaks around 410 nm. This peak is known as soret band.

ACUTE INTERMITTENT PORPHYRIA

Due to defect in porphobilinogen deaminase/hydroxymethylbilane synthase/Uroporphyrinogen I synthase. Porphobilinogen cannot be deaminated so it accumulates and δ ALA also accumulates.

Clinical Features

- Usually expressed after puberty.
- Usually presents with abdominal pain. Patient may undergo unnecessary abdominal surgery.
- Presents with neuropsychiatric manifestations. Patient may be wrongly diagnosed as a case of schizophrenia.
- Urine gets darkened on exposure to air due to oxidation of porphobilinogen to porphobilin.
- There is **no photosensitivity** since the metabolic block in AIP is earlier to the production of porphyrins. Porphobilinogen does not contain any conjugated double bonds like porphyrins and it cannot absorb light.

Treatment of Acute Intermittent Porphyria

- Avoiding fasting, alcohol and Cyt P450 inducers
- Excess intake of carbohydrates (Glucose loading)
- Administration of hematin (hydroxide of heme) to repress ALA synthase

 Clinical Correlation

Phenobarbitone may precipitate acute porphyria in susceptible individuals.

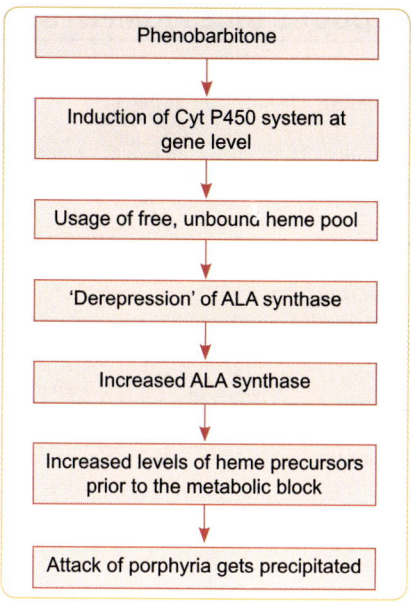

```
┌─────────────────────────────┐
│       Phenobarbitone        │
└─────────────────────────────┘
              │
              ▼
┌─────────────────────────────┐
│ Induction of Cyt P450 system│
│       at gene level         │
└─────────────────────────────┘
              │
              ▼
┌─────────────────────────────┐
│ Usage of free, unbound heme │
│           pool              │
└─────────────────────────────┘
              │
              ▼
┌─────────────────────────────┐
│ 'Derepression' of ALA synthase│
└─────────────────────────────┘
              │
              ▼
┌─────────────────────────────┐
│   Increased ALA synthase    │
└─────────────────────────────┘
              │
              ▼
┌─────────────────────────────┐
│ Increased levels of heme    │
│ precursors prior to the     │
│      metabolic block        │
└─────────────────────────────┘
              │
              ▼
┌─────────────────────────────┐
│Attack of porphyria gets     │
│        precipitated         │
└─────────────────────────────┘
```

 Clinical Correlation

Role of porphyrins in cancer photodynamic therapy

Porphyrins absorb light and get excited to produce free radicals. This property is used in photodynamic therapy. Porphyrin analogues are injected into the patient. Cancer cells due to their high demand for heme takes up more porphyrins. When a LASER light (blue or red) is passed, the porphyrins get excited and produce free radical damage to cancer cells.

PLUMBOPORPHYRIA

❑ Lead (Pb) causes acquired porphyria, which is known as 'Plumboporphyria.'
❑ Lead inhibits SH groups (thiol) of the enzymes ferrochelatase and ALA dehydratase.

Laboratory Findings

❑ Anemia with basophilic stippling
❑ ↑ RBC free protoporphyrinogen levels
❑ ↑ Excretion of δ-ALA and coproporphyrin in urine

**arrison's** Corner

Tyrosinemia Leads to Porphyria Like Features

Succinylacetone accumulates in hereditary tyrosinemia type I (fumarylacetoacetate hydrolase deficiency). Succinylacetone is structurally similar to ALA and inhibits ALA dehydratase leading to porphyria like features. *(Ref. Harrison 20th Ed., p.2990)*

HEME DEGRADATION PRODUCES BILE PIGMENTS AND CARBON MONOXIDE

- ❑ Senescent RBCs are destroyed in reticuloendothelial system releasing hemoglobin.
- ❑ Globin part enters the amino acid pool.
- ❑ Heme iron is usually oxidized to ferric iron forming hemin.
- ❑ Hemin is the substrate of microsomal heme oxygenase system.
- ❑ α-methenyl bridge between pyrrole ring I and II is broken and is released as CO.
- ❑ Cyclic ring becomes linear to produce biliverdin.
- ❑ In birds, biliverdin is excreted as such.
- ❑ In humans, biliverdin is further reduced by NADPH dependant biliverdin reductase enzyme to produce bilirubin.

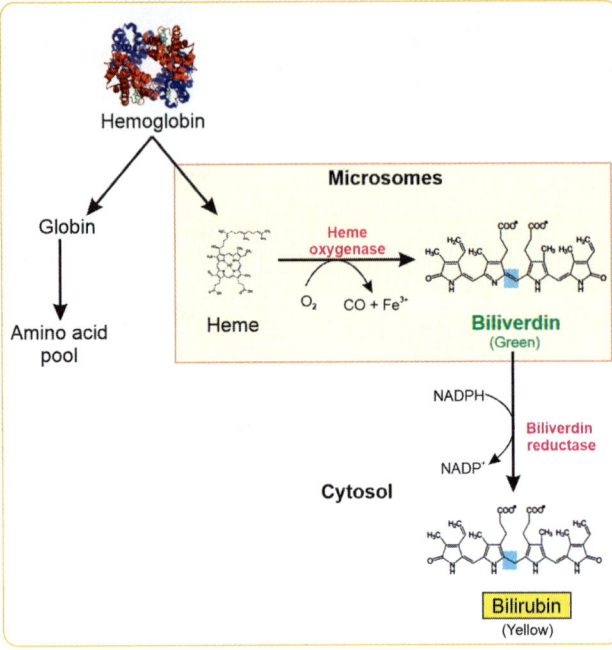

Bilirubin is Transported to Liver by Albumin

Bilirubin is sparingly water-soluble. It is transported by albumin in plasma for further metabolism in the liver. One molecule of albumin carries 2 molecules of bilirubin, one in high affinity site and another in low affinity site. Drugs like sulfonamides and salicylates can easily displce bilirubin in low affinity site.

Bilirubin Metabolism in Liver

Uptake	Bilirubin is removed from albumin and taken up by a *high capacity*, saturable facilitated transport system. Bilirubin binds to *ligandin* inside the cell.
Conjugation	❑ Nonpolar bilirubin is conjugated with 2 molecules of glucuronic acid by *UDP-glucuronosyl transferase* of the endoplasmic reticulum in 2 steps. ❑ The enzyme is inducible by drugs like phenobarbitone.
Biliary excretion	❑ Multispecific organic anion transporter (**MOAT**), a member of ATP-binding Cassette (**ABC**) transporter located in plasma membrane of biliary canaliculi excretes bilirubin diglucuronide into bile by active transport. ❑ Inducible by drugs like phenobarbitone. ❑ Rate-limiting step of bilirubin metabolism.

Difference Between Conjugated and Unconjugated Bilirubin

Property	Conjugated	Unconjugated
Water solubility	+	Low
Affinity for lipids	−	+
Bound to serum Albumin	+	+++
Renal excretion	+	−
Van den Bergh reaction	Direct	Indirect
Lipid membrane permeability	−	+

Van den Bergh's Test

Vandenberg reaction is a method for the detection and differentiation of hyperbilirubinemia. This reaction is the historical basis of calling conjugated and unconjugated bilirubin as direct and indirect bilirubin respectively.

Principle and Procedure

Bilirubin in the serum reacts with diazotised sulfanilic acid to produce purple colored azobilirubin. Diazo reagent is prepared freshly by mixing sulfanilic acid in dilute HCl and sodium nitrite.

$$Sulfanilic\ acid + HCl + NaNO_2 \rightarrow diazotised\ sulfanilic\ acid$$
$$Diazotised\ sulfanilic\ acid + Bilirubin \rightarrow azobilirubin\ (purple)$$

The color developed can be measured by colorimetry or spectrophotometry.

Interpretation

Observation	Inference	Interpretation
Purple color develops after addition of diazo reagent within 30 seconds.	Direct +Ve test	Conjugated bilirubin
Color develops after addition of methanol (methanol makes unconjugated bilirubin soluble)	Indirect +Ve test	Unconjugated bilirubin
The color develops initially, deepens after addition of CH$_3$OH.	Biphasic response	Both conjugated and unconjugated

Note: Modified Jendrassik's method using caffeine benzoate as an accelerator is used in most of the clinical laboratories for the estimation of bilirubin.

Urobilinogens are Formed in Intestine from Conjugated Bilirubin

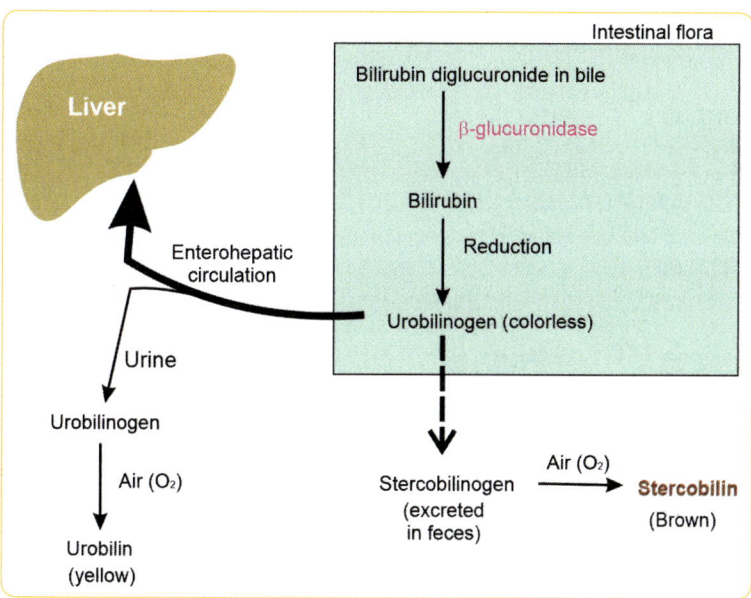

Jaundice is a Symptom of Many Different Diseases

Jaundice also known as icterus is yellowish pigmentation of the skin, sclera, and other mucous membranes due to high blood bilirubin levels.

Hyperbilirubinemia is not Synonymous with Jaundice

❑ Normal bilirubin: 0.1 to 1.2 mg/dL
❑ Hyperbilirubinemia: >1 mg/dL
❑ Clinical jaundice: >2 mg/dL

Causes of Jaundice

Hemolytic (Prehepatic) jaundice—Unconjugated hyperbilirubinemia:
❑ Intravascular and extravascular hemolysis
❑ Ineffective erythropoiesis in bone marrow
Hepatocellular (Hepatic)—Mixed hyperbilirubinemia
❑ Cirrhosis of liver due to various causes
❑ Hepatocellular carcinoma
Obstructive (posthepatic)—Conjugated hyperbilirubinemia
❑ Gallstones
❑ Cancer of pancreas

Choluric versus Acholuric Jaundice

❑ As water-soluble conjugated bilirubin is excreted in the urine of patients of hepatic and posthepatic (obstructive) jaundice, these 2 conditions are known as choluric jaundice.
❑ Hemolytic jaundice (excess unconjugated bilirubin) is known as acholuric jaundice.

Laboratory Findings in Various Types of Jaundice

Laboratory findings	Hemolytic (Prehepatic)	Hepatocellular (Hepatic)	Obstructive (post-hepatic)
Urinary findings			
Urobilinogen	↑	Normal or ↓	Absent
Bile salts	Absent	+	+++
Bile pigments	Absent (Acholuric)	++	+++
Stool			
Stercobilinogen	↑	↓	Absent; clay colored stool
Nature of stool	Normal	Oily and foul smelling (Steatorrhea)	
Liver function test			
Clotting time and PT	Normal	Increased in patients of chronic conditions.	
Enzymes raised	LDH	ALT, AST rise is more than ALP	1. Alkaline Phosphatase 2. 5'-nucleotidase 3. Gamma glutamyl transferase
Others			
Vandenberg reaction	Indirect	Biphasic	Direct
Reticulocyte count	↑	Normal	

Patients with Cholelithiasis Often Pass Clay-colored Stools

Characteristic color of stool is due to stercobilinogen, which is formed from reduction of urobilinogen. Urobilinogen is formed from degradation of bile pigments by bacterial action.

As there is obstruction of bile flow into intestine in cholelithiasis, there is *no formation of stercobilinogen* and patient passes clay colored stools.

PHYSIOLOGICAL JAUNDICE OF NEWBORN

Bilirubin production is higher in infants than adults due to *accelerated hemolysis and immature hepatic conjugating system. Unconjugated* bilirubin level rises. Jaundice usually appears on the 3rd or 4th day of life.

Treatment

- ❑ Sun exposure and phototherapy with 450 nm light
- ❑ Phenobarbital to induce conjugation systems
- ❑ Albumin infusion in severe cases to prevent kernicterus.

Role of Phototherapy in Physiological Jaundice

Infant is exposed to 450 nm blue light exposing all the body, covering eyes and male genitals.

Photoisomerization

- ❑ ZZ, ZE, EE and EZ are the four structural isomers of bilirubin.
- ❑ ZZ is the stable, more insoluble form. Other forms are relatively soluble and are known as lumirubins. Phototherapy converts the ZZ form into lumibirubins. Monoglucuronylated lumirubins are easily excreted.

Photooxidation converts bilirubin to colorless, water soluble forms.

Precautions during phototherapy: Phototherapy may cause dehydration. Frequent breastfeeding should be ensured.

Role of Albumin Infusion in Physiological Jaundice

Albumin has one high affinity binding site and one low affinity binding site for unconjugated bilirubin. So, to prevent the excess unconjugated bilirubin from crossing the blood brain barrier of neonates, albumin infusion can be done.

Premature Infants are more Prone to Kernicterus

Physiological jaundice is exaggerated in premature infants. If the bilirubin level is more than 20–25 mg/dL, albumin binding becomes saturated. Lipophilic unconjugated bilirubin can easily enter through the underdeveloped blood brain barrier and gets deposited in areas like **basal ganglia**. This leads to damage of neurons. This clinical condition is known as *kernicterus*.

 Clinical Correlation

> **Drugs should be used cautiously in infants to avoid kernicterus.**
> Drugs like *aspirin and sulfonamides* should be used cautiously in infants, especially in premature infants, as they displace bilirubin from albumin and increase the risk of kernicterus.

INHERITED CAUSES OF HYPERBILIRUBINEMIA

- ❑ *Unconjugated:* Crigler-Najjar syndrome type I and II, Gilbert syndrome
- ❑ *Conjugated:* Dubin Johnson and Rotor syndrome

Crigler-Najjar Type I versus Crigler-Najjar II

	Crigler-Najjar type I	Crigler-Najjar type II	Gilbert syndrome
Type of defect in UDP-glucuronosyl transferase	*Complete* absence of activity due to a **nonsense** mutation	*Partial* defect in enzyme activity due to a **missense** mutation	*Promoter* mutation resulting in *variable rate of transcription* of normal enzyme
Clinical course	Fatal within first 15 months	Benign	Completely benign
Response to phenobarbitone	No response as there is no enzyme at all	Respond	Responds well

Dubin Johnson and Rotor

Dubin Johnson (Black liver disease)	Rotor
Defective efflux due to *ABC2* mutation	Defective reuptake due to organic anion transporter mutation
Nonvisualization of gallbladder with bromsulphthalein (BSP)	Visualized

DELTA BILIRUBIN (BILIPROTEIN)

Chromatographic fractionation of serum bilirubin	
α bilirubin	Unconjugated bilirubin
β bilirubin	Bilirubin monoglucuronide
γ bilirubin	Bilirubin diglucuronide
δ bilirubin	Albumin bound conjugated bilirubin

❏ Delta bilirubin or biliprotein is the fraction of conjugated bilirubin that is covalently bound to albumin.
❏ Half-life of bilirubin is 4 hours but the half-life of delta bilirubin is 12–14 days (half-life of albumin)
❏ This is responsible for the lab finding of:
  ○ Absence of bilirubinuria in some patients with conjugated hyperbilirubinemia as the delta bilirubin is not filtered by renal glomeruli.
  ○ Persistent hyperbilirubinemia even after the disappearance of jaundice in cases of obstructive jaundice.

Competency

❏ BI6.11   Describe the functions of haem in the body and describe the processes involved in its metabolism and describe porphyrin metabolism.
❏ PA25.1   Describe bilirubin metabolism, enumerate the etiology and pathogenesis of jaundice, distinguish between direct and indirect hyperbilirubinemia

Key Points

❏ Heme is an iron containing protoporphyrin IX.
❏ All nucleated cells can synthesize heme.
❏ Heme synthesis involves both the cytoplasm and mitochondria.
❏ ALA synthase is the rate-limiting enzyme of heme synthesis.
❏ There is no photosensitivity in acute intermittent porphyria.
❏ Lead causes acquired porphyria by inhibiting ALA dehydratase and Ferrochelatase.
❏ Porphyria cutanea tarda is the most common porphyria.
❏ Bilirubin is the end product of heme metabolism in humans. Birds excrete biliverdin.
❏ Carbon monoxide is produced during the heme oxygenase mediated degradation of heme.
❏ Delta bilirubin is the conjugated bilirubin that is covalently bound to albumin.
❏ Alkaline phosphatase, 5'nucleotidase and, gamma glutamyl transferase are raised in obstructive jaundice.

Short Answer Questions

1. Explain why in acute intermittent porphyria photosensitivity is absent, though it is a prominent feature of other porphyrias?

2. Mention two causes of neonatal jaundice. What is the rationale behind the administration of phenobarbital and albumin in severe neonatal jaundice?

3. Explain why vitamin K injection is often administered to patients with chronic obstructive jaundice. *(Chapter 33)*

4. Mention two conditions associated with unconjugated hyperbilirubinemia. "Albumin infusion and phototherapy can reduce the chance of kernicterus in neonatal jaundice" – Explain how?

5. Explain how hematin administration relieves the symptoms of porphyria whereas alcohol ingestion exacerbates them?

6. Mention the property of porphyrins which help in their detection in body fluids. Explain the biochemical basis of the use of porphyrins in cancer photodynamic therapy.

7. How can choluric and acholuric jaundice be differentiated by using liver function tests performed in a biochemistry laboratory?

8. How does phototherapy affect the solubility of unconjugated bilirubin?

9. Mention the substrates required for heme biosynthesis. How is the biosynthesis of heme regulated?

10. What do you understand by the term Plumboporphyria? What is the biochemical basis of this? What are the urinary findings in this condition?

11. Which heavy metal toxicity is associated with porphyria? Explain the biochemical basis.

12. Discuss the biochemical basis of photosensitivity in porphyrias. Mention the enzymatic defect in acute intermittent porphyria and protoporphyria.

13. Explain how phenobarbitone therapy may precipitate acute porphyria in susceptible cases.

14. What is the biochemical basis of use of hematin and β-carotene in treatment of porphyria patients?

15. Briefly describe the process of the conjugation of bilirubin.

16. Explain the biochemical findings in obstructive jaundice.

17. Explain how biliary obstruction could impair blood coagulation?

18. Explain why bilirubin is present in the urine of infective hepatitis but absent in the urine of a newborn with Rh incompatibility?

Multiple Choice Questions

1. **True about bilirubin is:** **(JIPMER May 2015)**
   a. Conjugated with glycine
   b. Facilitates absorption of carbohydrate from the gut
   c. Is a steroid
   d. Bound to albumin in circulation

2. **Conjugated hyperbilirubinemia is seen in:**
   a. Breast milk jaundice **(AIIMS May 2010)**
   b. Crigler-Najjar syndrome
   c. Dubin Johnson syndrome
   d. Gilbert's syndrome

3. **True about Gilbert syndrome is:**
   a. Autosomal dominant **(AIIMS May 2010)**
   b. Unconjugated hyperbilirubinemia
   c. Normal life span
   d. All are true

4. **The test used to diagnose Dubin Johnson syndrome is:** **(AIPG 2009)**
   a. Serum transaminases
   b. BSP test
   c. Hippurate test
   d. Gamma glutamyltransferase level

5. **Which is the best biochemical marker of intrahepatic cholestasis of pregnancy (ICP)?**
   a. Alkaline phosphatase **(AIPG 2011)**
   b. Bilirubin
   c. Bile acids
   d. Bile salts

Chemistry of Lipids

DEFINITION OF LIPIDS

Carbohydrates and proteins are defined by their chemical nature. Unlike them, lipids are defined by their physical properties.

"Lipids are a heterogeneous group of compounds, including fats, oils, steroids, waxes, and related compounds that are insoluble in water and soluble in organic solvents".

CLASSIFICATION OF LIPIDS

Type	Description	Example
Simple lipids	Esters of fatty acids and alcohols (glycerol or long chain alcohols)	Triglyceride (neutral fat) and waxes (fatty acids esterified with long chain alcohol)
Compound/Complex lipids	Esters of fatty acids containing groups in addition to an alcohol (glycerol or sphingosine)	Phospholipids, aminolipids, glycolipids, sulfolipids

Contd...

Type	Description	Example
Precursor and derived lipids	Derivatives or precursors of the simple and complex lipids	Fatty acids, glycerol, steroids, other alcohols, fatty aldehydes, ketone bodies, fat soluble vitamins, steroid hormones

Oils are fats which are in liquid state at room temperature.

We will discuss each of the lipid type one by one. Let us begin with fatty acid.

Note: Fat refers to triglyceride/triacylglycerol.

FATTY ACIDS ARE ALIPHATIC CARBOXYLIC ACIDS

Fatty acids are aliphatic carboxylic acids with the general formula $R-(CH_2)_n-COOH$. Fatty acids in our body are found usually esterified with the alcohol groups of glycerol, sphingosine and cholesterol. *Free fatty acids are bound to albumin* in the plasma.

CLASSIFICATION OF FATTY ACIDS

Based on the chain length	❑ Short chain—2–6 carbons ❑ Medium chain—8–12 carbons ❑ Long chain—14–18 carbons ❑ Very long chain—More than 20 carbons

Based on the number of double bonds	❑ Saturated—No double bonds
	❑ Unsaturated
	  ○ Monounsaturated—Presence of single double bond
	  ○ Polyunsaturated—More than one double bond

Based on the branching	Branched chain fatty acids, e.g. Phytanic acid
Odd or even number of chain length	❑ Odd chain fatty acids—Present in ruminant fat (e.g. cattle milk) which too is of microbial origin; not found in humans ❑ Even chain fatty acids

Saturated Fatty Acids

Short chain	Medium chain	Long chain	Very long chain
Acetic acid—2C Butyric acid—4C Caproic acid—6C	Caprylic—8C Capric—10C Lauric—12C	Myristic—14C Palmitic—16C Stearic—18C	Behenic acid—22C Lignoceric acid—23C

Unsaturated Fatty Acids

Common name	Chain length, number and location of double bonds	Family
Monounsaturated fatty acids (MUFA)—One double bond		
Palmitoleic acid	16:1; Δ 9	ω^7
Oleic acid	18:1; Δ 9; cis	ω^9
Elaidic acid	18:1; Δ 9; trans	ω^9
Polyunsaturated fatty acids (PUFA)		
Dienoic acids (2 double bonds)		
Linoleic acid	18:2; Δ 9, 12	ω^6
Trienoic acids (3 double bonds)		

Contd...

Common name	Chain length, number and location of double bonds	Family
Gamma linolenic acid	18:3, Δ 6, 9, 12	ω^6
Alpha linolenic acid	18:3, Δ 9, 12, 15	ω^3
Tetraenoic acid (4 double bonds)		
Arachidonic acid	20:4, Δ 5, 8, 11, 14	ω^6
Pentaenoic acid (5 double bonds)		
Eicosapentaenoic (Timnodonic) acid	20:5, Δ 5, 8, 11, 14, 17	ω^3
Hexaenoic acid (6 double bonds)		
Docosahexaenoic (cervonic) acid	22:6, Δ 4, 7, 10, 13, 16, 19	ω^3

Know the difference

Phyto means plant. Both phytanic and phytic acids are of plant origin.

❑ Phytanic acid—branched chain fatty acid found in dairy and animal meat.

❑ Phytic acid (Phytate) is nothing but inositol polyphosphate. It is an antinutrient since it interferes with the absorption of calcium, iron, copper etc.

Note: Defect in the oxidation of phytanic acid leads to Refsum disease (Chapter 13.1) *(Refer to page no. 254)*.

α and ω Nomenclature

The carbon to which –COOH group is attached is the α carbon. Omega carbon is the one at the distal end of alpha carbon *(ω is the last letter of the Greek alphabet)*.

Some fatty acid structures are given to understand the omega nomenclature. Can you write the number of carbons present in the fatty acids shown in the image below without looking at the table above? Give it a try.

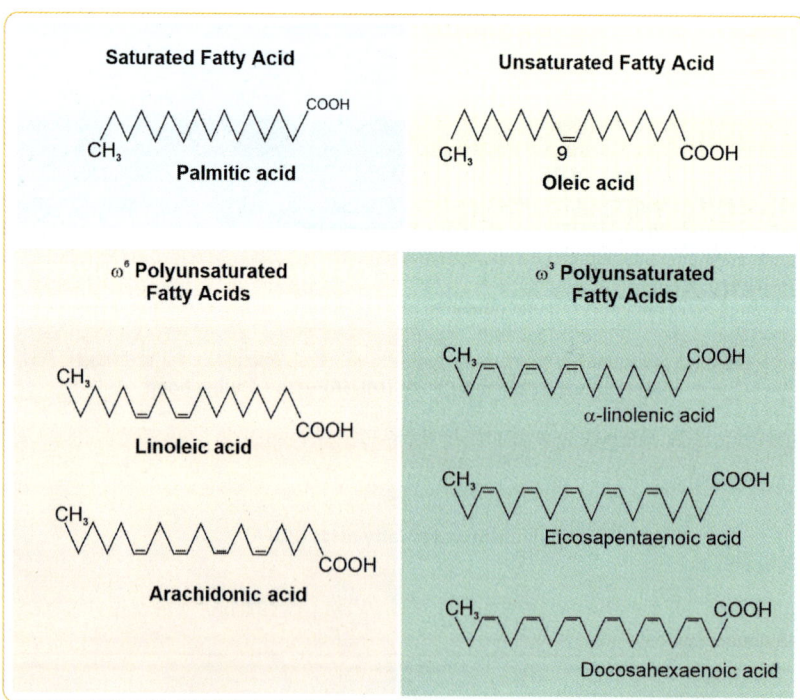

Number and the Nature of Double Bonds Significantly Alter the Property of the Fatty Acids

Melting temperature of a fatty acid is directly proportional to the chain length but inversely proportional to the degree of unsaturation, i.e. number of double bonds.

Double bonds can be of either *cis* or *trans* in nature. All the naturally occurring fatty acids in the body are of *cis* type.

cis double bond

trans double bond

Look at the bend produced in the hydrocarbon chain because of the cis double bond. We will learn the importance of this in *Chapter 36* *(Refer to page no. 564)*.

Polyunsaturated Fatty Acids (PUFA)

The PUFA are the fatty acids containing ≥ 2 double bonds. For example, arachidonic acid is a 20-carbon fatty acid with 4 double bonds at positions 5, 8, 11 and 14. It is a ω^6 fatty acid ($20-14 = 6$). Some PUFA can be synthesized in our body by desaturation and elongation of existing fatty acids.

Importance of PUFA

- ❑ All the essential fatty acids are polyunsaturated fatty acids.
- ❑ They are responsible for the fluidity of the membrane.
- ❑ Arachidonic acid is important for synthesis of eicosanoids.
- ❑ High intake of PUFA is considered to be beneficial in preventing cardiovascular disease.
- ❑ PUFA are the target of free radicals and undergo lipid peroxidation. Thus, PUFA are like double-edged sword.

CONCEPT CORNER

How PUFA can prevent cardiovascular diseases?
Compared to saturated fatty acids, PUFA are preferentially incorporated into cholesterol and help in the excretion of cholesteryl esters in the bile and thus reduce the level of circulating cholesterol.

Essential Fatty Acids cannot be Synthesized in the Body and have to be Taken in the Diet

Essential fatty acids are the two polyunsaturated fatty acids, which must be taken in diet as there is no desaturase enzyme in human body that can introduce double bonds beyond 9th carbon atom.
- ❑ Linoleic acid 18:2, Δ 9, 12 (ω^6 fatty acid)
- ❑ α-linolenic acid 18:3, Δ 9, 12, 15 (ω^3 fatty acid)

Arachidonic acid can be synthesized from linoleic acid. Therefore, it is not considered as 'purely' essential fatty acid and linoleic acid is considered as the most important PUFA.

Deficiency: Phrynoderma—follicular hyperkeratosis due to EFA and vitamin A deficiency.

Chapter 12 Chemistry of Lipids

ω³ Fatty Acids

ω³ fatty acids are those that contain a double bond in the ω³ position.

Fatty acid	Number of carbons	Number of double bonds	Position of double bonds	Dietary source
α-linolenic acid 18:3 *cis*-($\Delta^{9, 12, 15}$)	18	3	9, 12, 15	Flaxseed oil, soya bean oil
Eicosapentaenoic acid (EPA) (timnodonic acid)	20 (eicosa)	5 (penta)	5, 8, 11, 14, 17	Fish oil
Docosahexaenoic acid (DHA) (cervonic acid)	22 (docosa)	6 (hexa)	4, 7, 10, 13, 16, 19	Fish oil, breast milk

Beneficial Effects of ω³ Fatty Acids

Infantile Brain Growth and Development

DHA (cervonic acid) is important for brain and retinal development of infants and children. This shows the *importance of breastfeeding* since breast milk contains DHA.

Cardio Protective Role

Cyclooxygenase enzyme produces prostaglandins. When it acts on arachidonic acid, it produces thromboxane A_2 and prostacyclin I_2 which are strong platelet-aggregator and weak platelet antiaggregator respectively. When it acts on EPA, TxA_3 and PGI_3 are produced which are weak platelet-aggregator and strong antiaggregator respectively. Therefore, ω³ fatty acids shift the equilibrium towards the beneficial antiatherogenic state (*explained in Chapter 13.7*) *(Refer to page no. 302)*.

So, it is recommended to take adequate ω3 fatty acids. The ideal ω6 fatty acid to ω3 fatty acids ratio in diet is 1:1. The optimal ratio is 4:1.

Iodine Number and Saponification Number are used for Characterization of Fatty Acids

	Iodine number	Saponification number
Definition	Number of grams of iodine needed to saturate 100 grams of fatty acid	Number of milligrams of KOH required to saponify 1g of fat
Importance	Determines the *degree of unsaturation* of a fatty acid. Higher the number higher the unsaturation.	Measure of *chain length* of a fatty acid. Higher the number lower the chain length.

Hydrogenation of Fatty Acids Increases the Shelf-life but Introduces *trans*-double Bonds

Hydrogenation of vegetable oils (oils are nothing but fat in liquid state) is done to increase the shelf-life of the fatty acids.

Hydrogenated fatty acids become solid and are known as margarine/Vanaspati/Plant ghee/Dalda.
However, *partial hydrogenation* introduces *trans* double bonds in the fatty acids.
Deep-frying of foods also introduces *trans* double bonds in fatty acids.

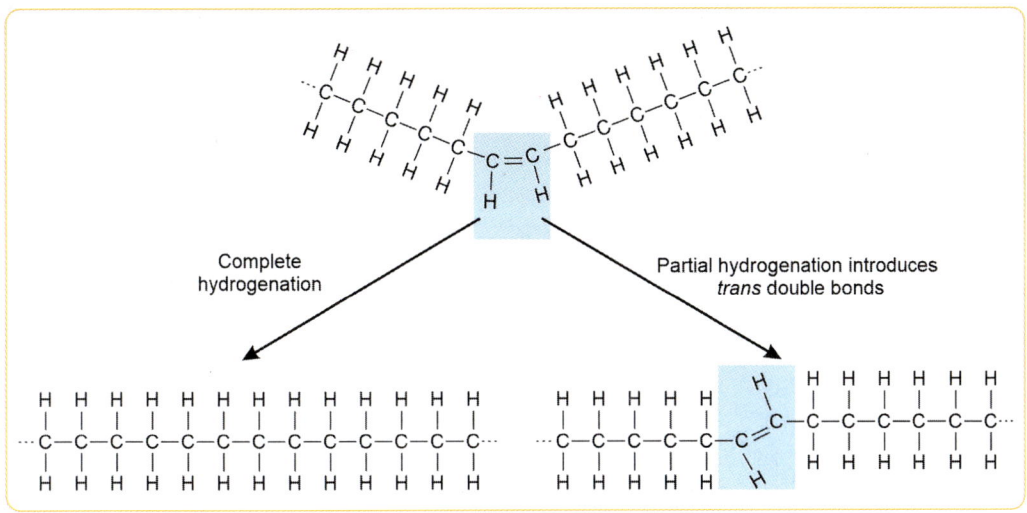

Complete hydrogenation

Partial hydrogenation introduces *trans* double bonds

TRANS FATTY ACIDS ARE DETRIMENTAL TO HEALTH

All fatty acids in the human body are *cis* fatty acids. *Trans* fatty acids are derived from ruminant fat and hydrogenated vegetable oils.

"*Trans* fatty acids compete with essential fatty acids and may exacerbate essential fatty acid deficiency. Moreover, they are structurally similar to saturated fatty acids and have comparable effects in the promotion of hypercholesterolemia and atherosclerosis." (*Ref. Harper 30th Ed., p. 242*).

LIPID PEROXIDATION LEADS TO RANCIDITY OF FATTY ACIDS

Fatty acids especially polyunsaturated fatty acids on exposure to repeated heat, light and oxidizing agents produce dicarboxylic acids, aldehydes and ketones. These degradative products give colour change and unpleasant taste to the rancid oil. Food additives like tocopherol, butylated hydroxy anisole/toluene (BHA/BAT) are added to oils to prevent oxidation.

ACYLGLYCEROLS ARE FATTY ACIDS ESTERIFIED WITH GLYCEROL

When OH group of the glycerol is esterified with COOH group of fatty acids, acyl glycerol is produced.
- Monoacylglycerol → esterified with one fatty acid
- Diacylglycerol → esterified with two fatty acids; usually different
- Triacylglycerol → esterified with three fatty acids; usually different, triglycerides are the storage form of lipids in the adipose tissue.

Chapter 12 Chemistry of Lipids

Simple TAG

Mixed TAG

Based on the fatty acid component, triacylglycerol (TAG) are divided into simple and mixed TAG. In simple TAG, the three fatty acids esterified are same. As shown in the image, a mixed TAG contains two or three different fatty acids. Mixed TAG is more abundant in our body than simple TAG.

CHOLESTEROL

Cholesterol is a 27-carbon compound. It is made up of a 19-carbon cyclopentano perhydro phenanthrene (CPP) ring system and an 8-carbon side chain. The 3rd carbon of the CPP ring is hydroxylated. No enzyme in human body is capable of degrading CPP ring.

Cholesterol is an isoprene (5 C) derivative. Other isoprene derivatives (Polyprenoids) found in human body are: Ubiquinone (Coenzyme Q), Dolichol, all fat-soluble vitamins, steroid hormones.

Cholesterol is amphipathic because of the presence of a hydroxyl group at the 3rd carbon. When this OH group is esterified to the COOH group of a fatty acid, cholesteryl esters are produced. Cholesteryl esters are more hydrophobic compared to cholesterol. **Cholesterol is found only in animals. For all practical purposes, cholesterol is not found in plants.**

All sterols contain CPP ring system. Sitosterol is a plant (phyto) sterol. Ergosterol is a fungal (myco) sterol. Cholesterol is an animal (zoo) sterol. Cholesterol is not found in plants. Vegetable oils do not contain cholesterol.

Role of Cholesterol in Membrane

Cholesterol is a normal component of cell membrane. It modulates the membrane fluidity.

Rigid steroid ring system interferes with the lateral movement of fatty acid chains.

- At lower temperatures: ↑ fluidity
- At higher temperatures: ↓ fluidity

It is abundant in specialized portion of membrane known as lipid rafts, which are involved in signaling process.

Derivatives of Cholesterol

Cholesterol is the precursor of:
❑ Bile acids
❑ Steroid hormones (Mineralocorticoids, glucocorticoids and sex steroids)
❑ Vitamin D

COMPLEX LIPIDS

Phospholipids

❑ Phospholipids are *complex lipids.*
❑ Phospholipids are made up of a nitrogen containing alcohol, esterified to the phosphatidic acid (*see the image*) or sphingosine.
❑ Because of presence of both polar and nonpolar molecules, phospholipids are amphipathic in nature.
❑ Phosphate and the group attached to it is hydrophilic and is termed as head group.
❑ Rest of the hydrophobic portion of the molecule is called tail.

Depending upon the alcohol backbone, there are two types of phospholipids.
❑ **Glycerophospholipids:** Phosphatidylcholine (Lecithin), phosphatidylinositol, phosphatidylserine, phosphatidylethanolamine (cephalin) and plasmalogens etc.
❑ **Sphingophospholipids:** Sphingomyelin only.

Glycerophospholipids

❑ Phosphatidic acid is the basic structure present in all the glycerophospholipids.
❑ Phosphatidic acid is 1, 2 diacylglycerol-3-phosphate.
❑ Group in the position X will decide the name of the phospholipid.

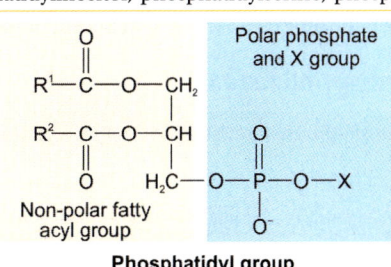

Phosphatidyl group

Group in position X	Name of phospholipid
Myo-inositol	Phosphatidylinositol
Serine	Phosphatidylserine
Glycerol	Phosphatidylglycerol
Ethanolamine	Phosphatidylethanolamine (cephalin)
Choline (trimethyl ethanolamine)	Phosphatidylcholine (lecithin)
Phosphatidylglycerol	Cardiolipin (diphosphatidylglycerol)

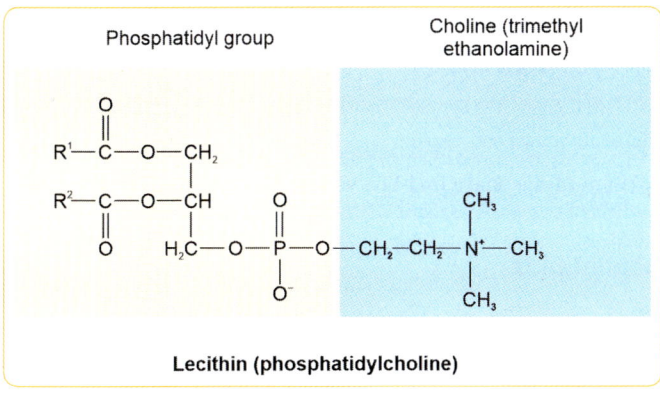

Lecithin (phosphatidylcholine)

Sphingophospholipid

❑ Sphingomyelin is the only sphingophospholipid.
❑ Sphingosine is an 18-carbon amino alcohol with two hydroxyl groups; sphingosine biosynthesis requires palmitoyl-CoA and serine.
❑ When a fatty acid combines with the amino group of sphingosine, ceramide is produced.

Fatty acid-COOH + $_2$HN-Sphingosine → Ceramide

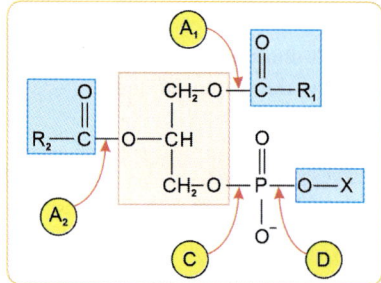

Ceramide

❑ Alcohol group of sphingosine of ceramide reacts with phosphoryl choline to produce sphingomyelin

Ceramide-OH + phosphoryl choline → Sphingomyelin

Except for sphingomyelin, all other sphingolipids are not phospholipids. We will discuss the other sphingolipids after finishing the discussion of phospholipids.

Phospholipases

Phospholipases are the enzymes that catalyse the hydrolysis of phospholipids into fatty acids and other components.

Fig. 1: Site of cleavage of various types of phospholipases

Phospholipase	Action upon lecithin produces
A$_1$	Acyl glycerophosphorylcholine and fatty acid
A$_2$	Lysolecithin + fatty acid
B	Both PLA$_1$ and PLA$_2$ activities
C	1,2 diacylglycerol and phosphoryl choline
D	Phosphatidic acid and choline

❑ Phospholipase A$_2$ present in the snake and bee venom releases lysolecithin, which is responsible for the hemotoxic effects.

Functions of Phospholipids

❑ **Components of cell membrane:** Inner and outer membrane contains different phospholipids (asymmetry of membranes). Outer membrane contains lecithin and sphingomyelin. Inner membrane contains cephalin and phosphatidylserine.

- **Micelles:** Phospholipids form micelles and help in transport of fat and fat-soluble vitamins.
- **Lung surfactant:** *Dipalmitoyl lecithin* is secreted by type II pneumocytes—lines the inner membrane of alveoli and prevents the collapse of alveoli during expiration.
- **Cell signaling:** Phosphatidylinositol 4,5-bisphosphate (PIP$_2$) on cleavage by phospholipase C yields diacylglycerol (DAG) and inositol trisphosphate (IP$_3$) which are 2nd messengers. Diacylglycerol activates protein kinase C. IP$_3$ opens endoplasmic reticulum calcium channels.
- **Apoptosis:** Flipping of phosphatidylserine toward outer leaflet of cell membrane leads to apoptosis.
- Eicosanoid synthesis – Plasma membrane phospholipids are acted upon by phospholipase A2 and give arachidonic acid for the synthesis of eicosanoids.

Cardiolipin or Diphosphatidylglycerol is Predominantly Found in Inner Mitochondrial Membrane

Etymology: Initially discovered in heart; hence the name.

Cardiolipin is *diphosphatidylglycerol,* i.e. two phosphatidic acid moieties attached to one glycerol molecule.

Cardiolipin (Diphosphatidyl glycerol)

It is *predominantly found in the inner mitochondrial membrane.*

Physiological Role of Cardiolipin

- ↑ the efficiency of the oxidative phosphorylation by acting as proton trap.
- Involved in signaling of apoptosis (programmed cell death).
- Has anticoagulant activity.

Clinical Correlation

Pathological Role of Cardiolipin

- Barth syndrome—defective synthesis of cardiolipin; X-linked recessive inheritance; mutation of tafazzin gene; associated with 3-methylglutaconic aciduria and dilated cardiomyopathy.
- Antiphospholipid syndrome—thrombotic condition leading to abortion. Anticardiolipin antibodies are found in SLE and antiphospholipid antibody syndrome.

Pulmonary Surfactant

Nature: Lung surfactant is a complex of lipids and proteins. Dipalmitoyl phosphatidylcholine (DPPC) is the major lipid component. SP-A, B, C and D are the surfactant-associated proteins. SP-A and D are noncollagen collagens. They too contain Gly-X-Y sequence. However, they form non-helical globular regions.

Secreted by: Type II alveolar cells

Function: Surfactant lines the inner membrane of alveoli and prevents the collapse of alveoli during expiration. They improve the lung compliance.

Deficiency: Premature infants produce less surfactant and are more prone for respiratory distress syndrome.

Chapter 12 Chemistry of Lipids

Plasmalogen and Platelet Activating Factor are Ether Lipids

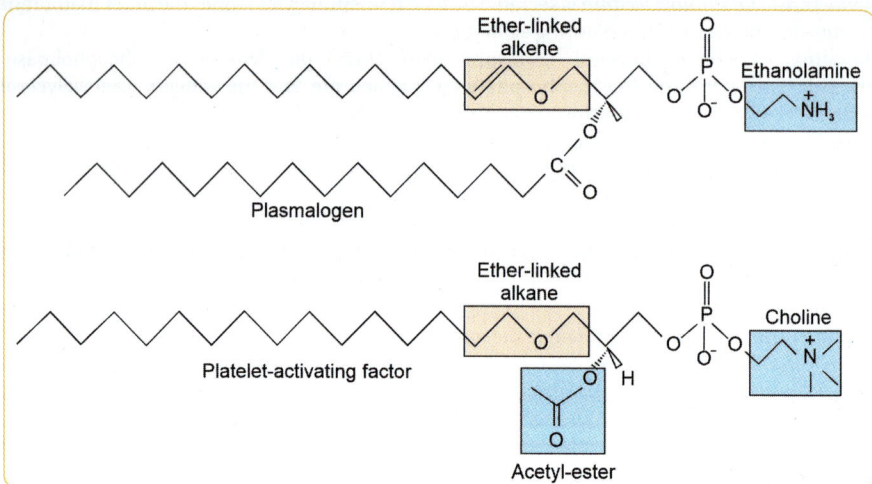

Fig. 2: The structure is given for the understanding purpose. No need to memorize.

Plasmalogen

- ❑ Plasmalogens are phospholipids with an ether linkage in Sn1 carbon
- ❑ Abundant in heart and brain
- ❑ Plasmalogens are precursor to platelet activating factor

Platelet Activating Factor

- ❑ PAF is also an ether lipid
- ❑ Signaling molecule with important role in inflammation and anaphylaxis
- ❑ Potent inflammatory molecule—stimulates platelet aggregation and the release of serotonin (a vasoconstrictor) and causes bronchoconstriction.

GLYCOLIPIDS

Cerebrosides

You know ceramide is sphingosine + fatty acid. When the OH group of ceramide forms a glycosidic bond with glucose or galactose, cerebrosides are produced.

Glucose + ceramide = Glucocerebroside
Galactose + ceramide = Galactocerebroside

Gangliosides

- ❑ Gangliosides are sialic acid containing glycosphingolipids.
- ❑ Sialic acid (N-acetylneuraminic acid) is a 9-cabon monosaccharide derived from pyruvate and mannose (N-acetylmannosamine)

- Based on the number of sialic acids, gangliosides are classified into GM, GD, GT, GQ. G stands for ganglioside and M, D, T, Q stands for mono, di, tri and quad/tetra respectively.
- Based on the relative electrophoretic mobility each type is subclassified. e.g. GM1, GM2, GM3.
- **GM1 ganglioside is the receptor for cholera toxin.**

Easy Way to Remember the Components of Various Sphingolipids

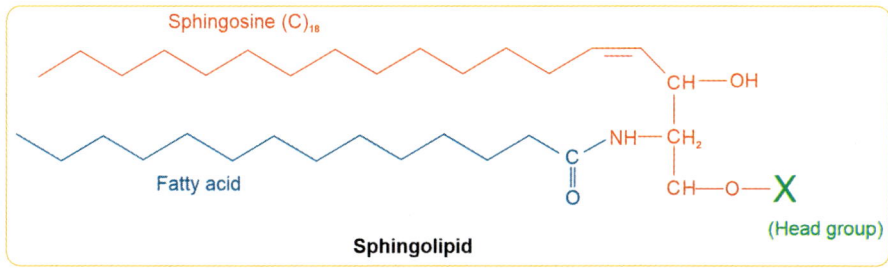

Sphingosine (C)₁₈

Fatty acid

CH—OH

NH—CH₂

CH—O—X

(Head group)

Sphingolipid

Sphingolipid	X (as shown in the image)
Ceramide	H
Sphingomyelin	Phosphoryl choline
Glucosylcerebroside	—Glc (in nonneural tissue)
Galactocerebroside	—Gal (in neural tissue)
Globosides	Several neutral sugars (Glc, Gal, GalNAc)
Gangliosides	Complex carbohydrates with sialic acid

Cerebroside versus Ganglioside

	Cerebroside	Ganglioside
Sugar	Contains a single sugar—either glucose or galactose	Contains up to 7 monosaccharides
Sialic acid	Absent	Present—one or multiple
Charge	Neutral	Negatively charged because of the sialic acid

STRUCTURES THAT CAN BE FORMED BY PHOSPHOLIPIDS/ AMPHIPATHIC LIPIDS

Amphipathic lipids have both polar and non-polar parts. So, in aqueous medium they can form the following:
- Bilayer membrane—*we will discuss this extensively in Chapter 36 (Refer to page no. 564).*
- Micelle—micelles in water are organized in such a way that the hydrophilic head faces the solvent and the hydrophobic parts are directed toward the interior. Micelles can carry lipid soluble compounds like fat-soluble vitamins in the core.
- Liposomes have been discussed further as follows:

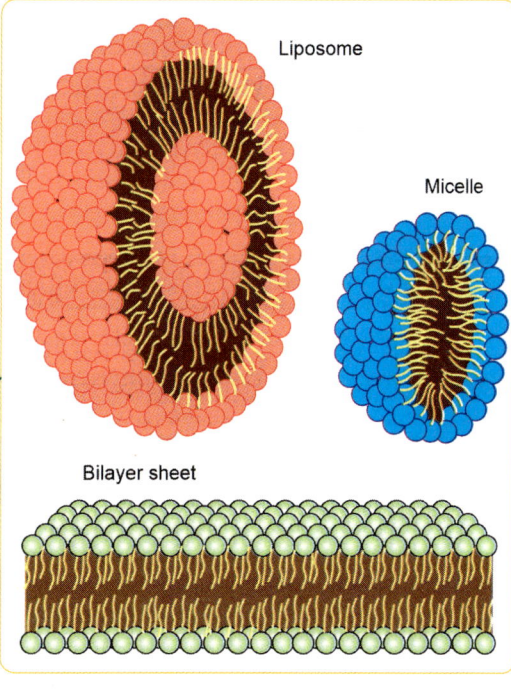

Liposome

Micelle

Bilayer sheet

LIPOSOMES

❏ Liposomes are formed when amphipathic lipids like phospholipids are sonicated in an aqueous medium.
❏ They form a vesicle carrying the aqueous medium inside their core.
❏ Unlike micelles, which can carry only nonpolar substances, liposomes can carry both polar and nonpolar substances.

Uses

❏ Study of membrane transport
❏ **Drug delivery:**
  ○ Liposomes combined with antibodies against tissue/cancer specific antigen can deliver the drug exactly to the target tissues.
  ○ Delivery of drugs like amphotericin B to the target sites, which otherwise produce serious effects if administered in other forms.
❏ **Gene therapy:** Liposomes can be used to carry DNA for gene therapy.

LIPID PEROXIDATION

Lipid peroxidation is reactive oxygen species mediated nonenzymatic autoxidation of lipids.

Harmful Effect

❏ *In vitro*, lipid peroxidation is the cause of rancidity of edible oils on repeated heating.
❏ *In vivo*, peroxidation of membrane lipids is implicated in aging, atherosclerosis, inflammatory diseases and cancer.

Marker

❏ Malondialdehyde (MDA) and hydroxynonenal (HNE) are the end products of lipid peroxidation. MDA assay is used to measure oxidative stress in the body.
❏ Lipofuscin also known as "age pigment" accumulation is a result of membrane lipid peroxidation.

Food Preservatives

Propyl gallate, butylated hydroxyanisole and butylated hydroxytoluene are antioxidants used to prevent lipid peroxidation.

Competency

- ☐ BI4.1 Describe and discuss main classes of lipids (Essential/non-essential fatty acids, cholesterol and hormonal steroids, triglycerides, major phospholipids and sphingolipids) relevant to human system and their major functions
- ☐ BI11.24 Enumerate advantages and/or disadvantages of use of unsaturated, saturated and trans fats in food

Key Points

- ☐ α-linolenic acid, eicosapentaenoic acid (timnodonic acid) and docosahexaenoic acid (cervonic acid) are the ω³ fatty acids.
- ☐ Linoleic and linolenic acid are the only two essential fatty acids. Arachidonic acid can be synthesized from linoleic acids
- ☐ Phytanic acid is a branched chain fatty acid.
- ☐ Sphingosine is an 18-carbon amino alcohol derived from palmitoyl-CoA and serine.
- ☐ Sphingomyelin is the only sphingophospholipid.
- ☐ Ceramide is an amide formed through the reaction of amino group of sphingosine and carboxyl group of a fatty acid. Ceramide = Sphingosine + fatty acid
- ☐ Glucose + ceramide = Glucocerebroside
- ☐ Gangliosides are complex sphingolipids with sialic acid (derivatives of NANA) residues.
- ☐ GM1 ganglioside is the receptor for cholera toxin in human intestine.
- ☐ Lipoic acid (thioctic acid) is a sulfur containing fatty acid.
- ☐ Lignoceric acid is a 23-carbon, saturated fatty acid present in cerebrosides.
- ☐ Cardiolipin is diphosphatidyl glycerol; exclusively seen in inner mitochondrial membrane.
- ☐ Plasmalogen is an ether lipid synthesized in peroxisomes.
- ☐ Platelet-activating factor is an ether-linked phospholipid. It causes pulmonary vasoconstriction, bronchoconstriction, systemic vasodilation, increased capillary permeability.
- ☐ Separation of a particular fatty acid from a mixture of lipids is done by Gas-liquid chromatography.

SELF-ASSESSMENT

Short Answer Questions

1. Classify lipids and give an example for each type.

2. Why certain polyunsaturated fatty acids are essential in diet? Name the dietarily essential fatty acids for humans. Indicate for each, the number of carbon atoms present and the number and position of double bonds. Mention the important functions served by these in our body.

3. Why arachidonic acid is not an essential fatty acid in humans? Mention the deficiency manifestations of essential fatty acids.

4. Premature baby tends to develop respiratory distress syndrome – Why?

5. Define polyunsaturated fatty acids (PUFA). Name any two. Enumerate the important roles of PUFA.

6. What is the role of cholesterol in the cell membrane? Mention any two cholesterol derivatives.

7. Name the ω³ fatty acids and mention their dietary source and beneficial effects.

8. Define iodine number and saponification number. Mention their applications.

9. What is the result of partial hydrogenation of fatty acids? What advise will you give regarding the dietary intake of partially hydrogenated fatty acids to a patient with cardiovascular disease? Justify your answer.

10. Why trans-fatty acids are considered detrimental to health?

11. Classify complex lipids with examples for each type.

12. Differentiate cerebroside and ganglioside.

13. Name the three structures that can be formed by phospholipids or amphipathic Lipids? Mention their physiological and therapeutic roles?

14. Explain the biochemical basis which enables liposomes to be used in anticancer therapy and gene therapy?

15. What are all the markers of lipid peroxidation? How can you prevent peroxidation of edible oils?

16. Explain why babies born prematurely are prone to Acute Respiratory Distress Syndrome (ARDS)?

17. Name any four important phospholipids. Write their functions and location in the cell.

18. What is the rationale behind the addition of vitamin E to the commercially available vegetable oils?

Multiple Choice Questions

1. Which of the following is a derived lipid?
   a. Fatty acids
   b. Glycolipid
   c. Phospholipid
   d. Triglyceride

2. Free fatty acids in the blood are carried by:
   a. Albumin
   b. HDL
   c. LDL
   d. VLDL

3. Which of the following is a branched chain fatty acid?
   a. Cervonic acid
   b. Timnodonic acid
   c. Phytic acid
   d. Phytanic acid

4. Arachidonic acid has 20 carbon atoms with:
   a. 2 double bonds
   b. 3 double bonds
   c. 4 double bonds
   d. 5 double bonds

5. All of the following are ω^3 fatty acids, except:
   a. α-linolenic acid
   b. Eicosapentaenoic acid
   c. Docosahexaenoic acid
   d. γ-linolenic acid

6. Which of the fatty acid is present exclusively in breast milk?
   a. Arachidonic acid
   b. Docosahexaenoic acid
   c. Linoleic acid
   d. Linolenic acid

7. Snake venom contains:
   a. Phospholipase A_1
   b. Phospholipase A_2
   c. Phospholipase C
   d. Phospholipase D

8. Phospholipid contains:
   a. Hydrophilic heads and hydrophobic tails
   b. Long water-soluble carbon chains
   c. Positively charged functional groups
   d. Both (b) and (c)

9. Which of the following phospholipid is predominantly found in the inner mitochondrial membrane?
   a. Cardiolipin
   b. Cephalin
   c. Lecithin
   d. Sphingomyelin

10. Cervonic acid is present in:
    a. Fish oil
    b. Coconut oil
    c. Soya bean oil
    d. Ghee

11. Which of the following lipids is responsible for stimulation of hypersensitivity, activation of clotting?
    a. Cardiolipin
    b. Platelet activating factor
    c. Phosphatidylethanolamine
    d. Lecithin

12. Arachidonic acid can be synthesized from:
    a. Linoleic acid
    b. α-linolenic acid
    c. Palmitic acid
    d. Stearic acid

13. All of the following is saturated fatty acids, except:
    a. Linoleic acid
    b. Oleic acid
    c. Palmitic acid
    d. Stearic acid

14. Which of the following phospholipid is associated with apoptosis?
    a. Phosphatidylcholine
    b. Dipalmitoylphosphatidylcholine
    c. Phosphatidylserine
    d. Phosphatidylinositol 4,5-bisphosphate

15. All of the following are true about sphingolipids, except:
    a. Amphipathic in nature
    b. Two hydrocarbon tails are formed by fatty acids
    c. Sphingosine is an 18 carbon amino alcohol
    d. Sphingomyelin is a phospholipid

16. Defect in which of the following lipid is associated with Barth syndrome?
    a. Sphingomyelin
    b. Phosphatidylethanolamine
    c. Phosphatidylserine
    d. Diphosphatidylglycerol

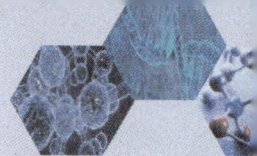

17. **All of the following are phospholipids, except:**
    a. Cardiolipin
    b. Cerebroside
    c. Sphingomyelin
    d. Surfactant lipid

18. **Which of the following lipid does not contain glycerol?**
    a. Lecithin
    b. Sphingomyelin
    c. Cardiolipin
    d. Ceramide

19. **Respiratory distress syndrome in premature infants is due to inadequate secretion of which one of the following lipids?**
    a. Dipalmitoylphosphatidylcholine
    b. Sphingomyelin
    c. Cholesterol
    d. Phosphatidylinositol

20. **All of the following are isoprene derivatives (polyprenoids), except:**
    a. Ubiquitin
    b. Dolichol
    c. β carotene
    d. α tocopherol

21. **Which of the following is true about cholesterol?**
    a. Made up of 25 carbons
    b. There is a hydroxyl group at 5th carbon
    c. Contain the cyclic tetra perhydro phenanthrene ring
    d. Amphipathic in nature

22. **Which of the following is a phytosterol?**
    a. Sitosterol
    b. Cholesterol
    c. Ergosterol
    d. Calcitriol

ANSWERS

1. a.	2. a.	3. d.	4. c.	5. d.	6. b.
7. b.	8. a.	9. a.	10. a.	11. b.	12. a.
13. a.	14. c.	15. b.	16. d.	17. b.	18. b.
19. a.	20. a.	21. d.	22. a.		

13 CHAPTER

Lipid Metabolism

13.1 FATTY ACID OXIDATION

FREE FATTY ACIDS ARE NOT FREE IN THE BLOOD

The word 'free' in the free fatty acids refers to the free carboxylic acid groups.

Free fatty acids are non-esterified fatty acids (NEFA). In blood, they are carried by albumin. The hydrophobic clefts in the albumin bind to fatty acids with high affinity.

FATTY ACIDS RELEASE MORE ENERGY COMPARED TO CARBOHYDRATES

Fatty acids are more reduced (electron dense) as compared to carbohydrates. Thus, oxidation of fatty acids releases more electrons compared to oxidation of glucose, which is used to generate more ATPs in the electron transport chain. This explains why the calorific value of fats (9 kcal/g) is more than double of carbohydrates (4 kcal/g).

METABOLIC FATE OF FATTY ACIDS DIFFERS DEPENDING UPON THE CHAIN-LENGTH AND DEGREE OF SATURATION

SHORT-CHAIN FATTY (C_2-C_6) ACIDS ARE PRODUCED IN THE BODY FROM MICROBIAL FERMENTATION OF DIETARY FIBERS

- Intestinal flora ferments the dietary fiber into short-chain fatty acids especially butyrate (4-carbon).
- Only 5% of the short-chain fatty acids produced in the colon are excreted in stool while 95% is absorbed and used by the colonic epithelial cells for their energy needs.
- Butyrate has tumor suppressor/anti-proliferative activity.

MEDIUM-CHAIN FATTY ACIDS (C_8-C_{14}) ARE TRANSPORTED AS SUCH IN THE PORTAL BLOOD

❑ Unlike long-chain fatty acids, medium chain fatty acids do not require incorporation into chylomicrons. They also do not require carnitine for their transport into the mitochondria.

❑ Medium chain triglycerides require neither pancreatic lipase nor micelle formation.

❑ Medium chain fatty acids are used in the therapy of CPT-I deficiency (discussed below).

❑ Medium-chain triglycerides are present in coconut oil in large amount.

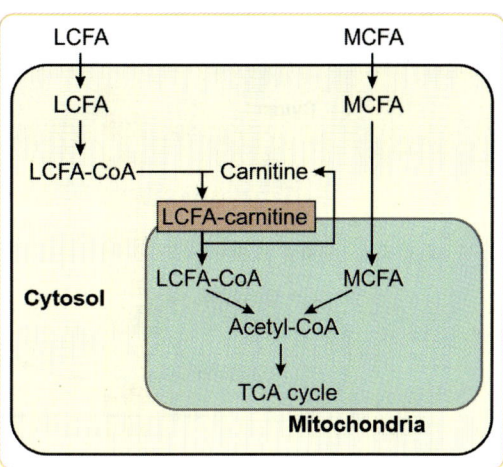

OXIDATION OF LONG-CHAIN FATTY ACIDS

Fatty Acid Must be Activated Before Oxidation

Like any other molecule we studied so far, fatty acids also need to be activated before oxidation.

$$R-\underset{\substack{\|\\O}}{C}-OH + HS-CoA \xrightarrow[\text{ATP} \quad \text{AMP + PPi}]{\text{Acyl-CoA synthetase}} R-\underset{\substack{\|\\O}}{C}\sim S-CoA + H_2O$$

Fatty acid　　Coenzyme A　　　　　　　　　Fatty acyl-CoA

Two high-energy phosphate bonds of ATP are utilised in this reaction.

Carnitine Shuttle - Long-chain Fatty Acyl-CoA Requires Carnitine for the Transport into Mitochondria

❑ Carnitine is β-hydroxy-γ-trimethyl ammonium butyrate. It is synthesised from Lysine and Methionine with the help of Vit. C for hydroxylation.

❑ It is Zwitterionic in nature; helps in the transport of long-chain fatty acyl groups across the mitochondrial membrane since the inner mitochondrial membrane is impermeable to long-chain acyl CoA.

❑ Carnitine is not required for the transport of small and medium chain fatty acids.

❑ Carnitine palmitoyltransferase-I, is an enzyme in the outer mitochondrial membrane. It catalyses the transfer of fatty acyl-CoA to carnitine producing acyl-carnitine.

❑ Acyl-carnitine is translocated into mitochondrial matrix through the inner mitochondrial membrane translocase.

❑ Acyl-carnitine is transferred to CoA-SH to produce fatty acyl-CoA and Carnitine by CPT II of the inner membrane. Carnitine is sent back to intermembranous space via the same translocase and it is reutilized.

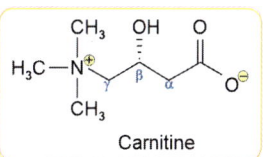

Fig. 1: α-carbon is the carbon to which the functional group, (here, COOH) is attached. Notice the hydroxyl group at β position and trimethyl ammonium $(CH_3)_3N^+$ group at γ position. Also notice the zwitterionic nature – presence of positive and negative charges

❑ Since carnitine shuttles back and forth of inner membrane, this process is known as 'Carnitine Shuttle'

❑ CPT I is inhibited by malonyl-CoA, which is an intermediate of fatty acid synthesis. This is how fatty acid synthesis and oxidation are reciprocally regulated.

Chapter 13 Lipid Metabolism

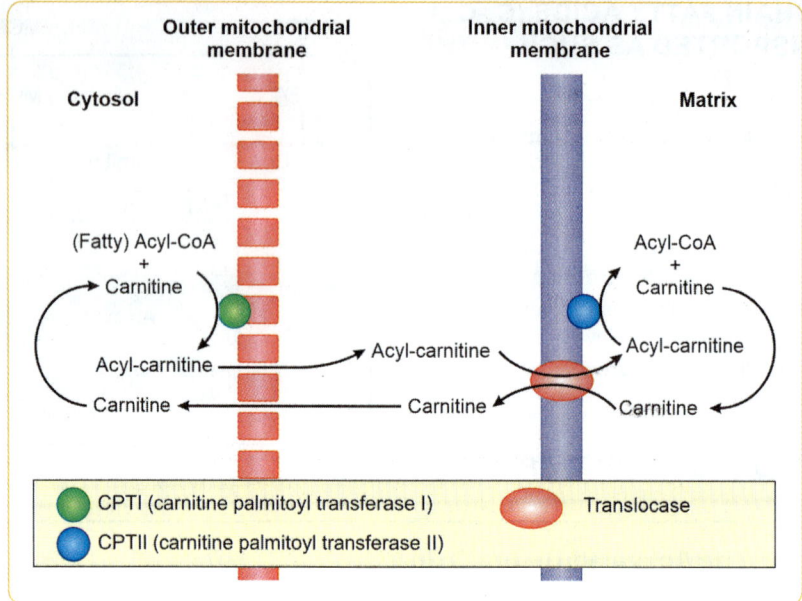

Fig. 2: Carnitine shuttle – notice that carnitine can be reutilised.
By Original image: Slagtvectorization: Own work - File:Acyl-CoA from cytosol to the mitochondrial matrix.gif, CC0

β-Oxidation of Long-chain Saturated Fatty Acids Follow the same Recurrent Theme we have Studied in TCA Cycle

Four repetitive steps of fatty acid oxidation can be easily remembered as:
1. FAD dependent oxidation
2. Hydration
3. NAD+ dependent oxidation
4. Thiolysis

FAD Dependent Oxidation

- ❑ The enzyme acyl-CoA dehydrogenase removes hydrogen from both α and β carbons and transfer it to FAD.
- ❑ Dehydrogenation produces a double bond between α and β carbons.
- ❑ The product is called enoyl-CoA (en – double bond; oyl - fatty acyl)
- ❑ The FADH$_2$ produced is carried by **electron transferring flavoprotein** (ETF) to the electron transport chain.

Hydration

Hydration is catalyzed by enoyl-coA hydratase. Hydration facilitates the dehydrogenation going to happen in the next step.

NAD+ Dependent Oxidation

NAD+ dependent oxidation is catalyzed by hydroxyacyl-CoA dehydrogenase. β carbon is oxidized and now it contains keto group. As the oxidation involves the beta carbon, the pathway is named beta oxidation.

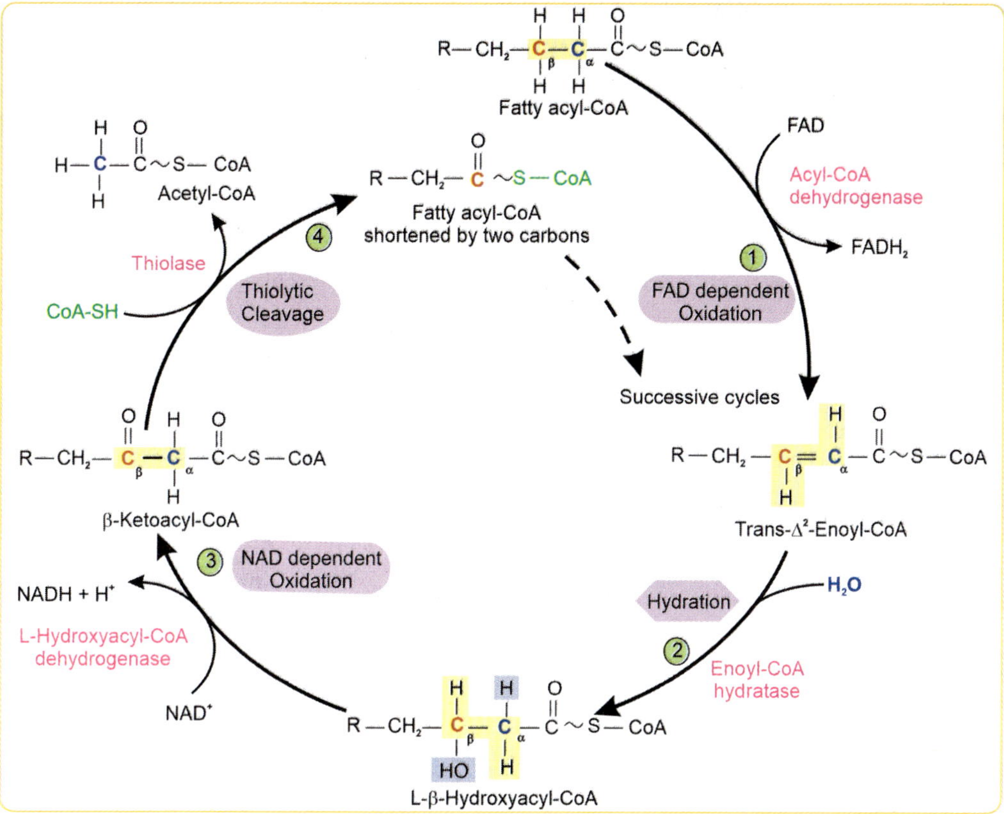

R—CH$_2$—C—C—C~S—CoA
Fatty acyl-CoA

H—C—C~S— CoA
Acetyl-CoA

Thiolase

R—CH$_2$—C ~S—CoA
Fatty acyl-CoA
shortened by two carbons

CoA-SH

④ Thiolytic Cleavage

FAD

Acyl-CoA dehydrogenase

① FAD dependent Oxidation

FADH$_2$

Successive cycles

R—CH$_2$— C —C —C~S—CoA
β-Ketoacyl-CoA

R—CH$_2$— C = C —C~S—CoA
Trans-Δ²-Enoyl-CoA

③ NAD dependent Oxidation

NADH + H⁺

L-Hydroxyacyl-CoA dehydrogenase

NAD⁺

Hydration

—H$_2$O

② Enoyl-CoA hydratase

R —CH$_2$— C — C —C~S— CoA
HO   H
L-β-Hydroxyacyl-CoA

Thiolysis

Thiolysis is catalyzed by the enzyme thiolase, which is a transferase (EC2). Thiolase transfers the CoA-SH to the beta carbon and thus releases acetyl-CoA and a fatty acyl-CoA that is two carbons shorter, which undergoes further rounds of β oxidation.

End Products

Thus, each round of β oxidation produces, 1 FADH$_2$, 1 NADH and 1 acetyl-CoA. If the fatty acid is even numbered, 2 acetyl-CoA molecules are produced in the last round. If the fatty acid is odd numbered, 1 acetyl-CoA and 1 propionyl-CoA are produced in the last round.

Fate of Products of Fatty Acid Oxidation

❑ The FADH$_2$ and NADH donates their electrons to the electron transport chain.
❑ Acetyl-CoA is completely oxidized in the TCA cycle and produces FADH$_2$ and three NADH and one ATP/GTP.
❑ When the TCA cycle is not operational in liver, acetyl-CoA is diverted to ketogenesis.

Energy Yield from β-oxidation of Palmitic Acid

Palmitic acid is a 16C fatty acid. It will undergo seven rounds of β oxidation.

Energy Yield from β-oxidation of Stearic Acid

Source	ATP yield	Total ATP
7 FADH2	x 1.5 ATP	10.5
7 NADH	x 2.5 ATP	17.5
8 acetyl CoA	x 10 ATP	80
Activation		-2
Net yield		106

Stearic acid contains 18C. Thus, there will be one extra round of oxidation and one more acetyl-CoA produced. Therefore, add 14 ATP (2.5 + 1.5 + 10) to the ATP yield of palmitic acid, you will get 120 ATPs.

Fate of Odd-chain Fatty Acids

Intake of odd chain fatty acids is minimal unless a person eats a lot of ruminant fat. Moreover, only one propionyl-CoA is produced in the last cycle of β-oxidation. The fate of propionyl-CoA is usually conversion to succinyl-CoA and to enter the TCA cycle. All the TCA cycle intermediates are glucogenic. Thus, propionyl-CoA is glucogenic.

Propionyl-CoA Derived From Odd Chain Fatty Oxidation Is Glucogenic

Propionyl-CoA + HCO₃⁻ — Carboxylase / Biotin, ATP → ADP + Pᵢ → D-Methyl malonyl-CoA ⇌ Racemase ⇌ L-Methyl malonyl-CoA → B₁₂ Mutase → Succinyl-CoA

Fig. 3: Enzymes involved (in order): Biotin dependent Propionyl-CoA carboxylase, Methylmalonyl-CoA racemase/epimerase and B12 dependent Methylmalonyl-CoA mutase

Vitamin B₁₂ Deficiency Leads to Methylmalonic Aciduria

Methyl malonyl-CoA mutase enzyme requires **adenosyl cobalamine (B₁₂)** cofactor. That is why B₁₂ deficiency leads to methylmalonic aciduria.

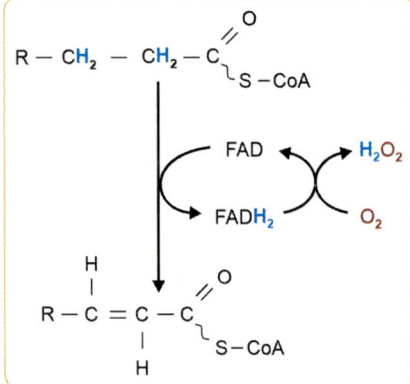

β Oxidation of Very Long Chain Fatty Acids Takes Place in Peroxisomes

- Fatty acids with more than or equal to 22 carbons (e.g. lignoceric acid) first undergo β-oxidation in the peroxisome for some cycles till **Octanoyl-CoA** is produced.
- Unlike mitochondria, there is no electron transport chain and electron transferring flavoprotein in peroxisomes. So, peroxisomal FAD-dependent acyl-CoA dehydrogenase a.k.a. acyl-CoA oxidase transfers the electrons from FADH₂ to molecular oxygen producing **H₂O₂**. Catalase present in the peroxisome detoxifies H₂O₂.
- The other steps of peroxisomal beta oxidation are similar to mitochondrial β-oxidation.
- Octanoyl-CoA, Acetyl-CoA and NADH used are utilized by mitochondria.

Branched Chain Fatty Acids Undergo α Oxidation

We have seen how both the hydrogen atoms of β carbon are removed in β oxidation. Branched chain fatty acids like phytanic acid contain a methyl group at beta carbon. This does not allow the β carbon to be oxidized. Therefore, they undergo alpha oxidation in the peroxisomes.

Oxidation of Unsaturated Fatty Acids Require Isomerase and Reductase

- Oxidation of double bonds positioned at the odd numbered carbon as in palmitoleic acid (16:Δ9) require only an isomerase enzyme to shift the position of double bonds to enable the oxidation.
- Oxidation of double bonds positioned at the even numbered carbon require a reductase in addition to isomerase.

Regulation of fatty acid oxidation is discussed in Ketone bodies portion.

Role of fatty acid oxidation inhibitors in angina is discussed in Chapter 16 (Refer to page no. 335).

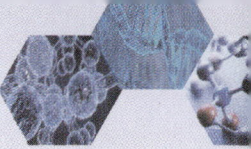

Review of Various Types of Fatty Acid Oxidation

Type	Importance	Subcellular location	Products
Mitochondrial β oxidation of even chain fatty acids	Major mode	Mitochondrial matrix	Molecules of Acetyl CoA, $FADH_2$ and NADH
Mitochondrial β oxidation of odd chain fatty acids	Propionyl-CoA produced is glucogenic	Mitochondrial matrix	Molecules of Acetyl CoA, Only one propionyl-CoA, $FADH_2$ and NADH
Peroxisomal β oxidation	Trimming of very long chain fatty acids (VLCFA).	Peroxisome	Acetyl-CoA, Octanoyl-CoA, H_2O_2
α oxidation	Phytanic acid (branched chain) oxidation in brain and liver. Defect leads to Refsum disease.	Peroxisome	Formyl-CoA, Propionyl-CoA, Acetyl-CoA, Isobutryl-CoA, No ATPs are produced.
ω oxidation	Minor mode. Important in conditions of impaired β oxidation.	Microsome (ER)	Dicarboxylic acids

DISORDERS OF FATTY ACID OXIDATION

Carnitine Deficiency and Fatty Acid Oxidation Defects Lead to Hypoglycemia

❑ Carnitine deficiency leads to defective fatty acid oxidation and hypoglycemia.
❑ Hypoglycin present in unripe ackee fruit is responsible for Jamaican Vomiting Sickness. Hypoglycin inhibits medium and short chain acyl-CoA dehydrogenase and thus inhibits β-oxidation. Unripe **lychee fruit** is thought to produce acute toxic encephalopathy in pediatric population in a similar manner to that of ackee fruit.

CONCEPT CORNER

Even chain fatty acids cannot be converted to glucose. Then, how come defect in fatty acid oxidation leads to hypoglycemia?
❑ Fats have sparing action on glucose. If fatty acids are not oxidized, glucose will be utilized in place of fatty acids.
❑ Energy required for the gluconeogenesis comes from fatty acid oxidation. So, defect in fatty acid oxidation leads to hypoketotic (nonketotic) hypoglycemia.
Why non-ketotic?
❑ Ketone bodies are produced from fatty acid oxidation. If fatty acids cannot be oxidised, ketone bodies cannot be produced, right?

▶ goo.gl/JvpMYt or scan QR code

Peroxisomal Disorders

❑ Peroxisomes are the subcellular organelles present in all nucleated cells. They are abundant in liver and kidney. Peroxisomes are involved in both catabolic and anabolic reactions.
❑ **Catabolic reactions:** β-oxidation of very long chain fatty acids, α-oxidation of phytanic and many dicarboxylic acids; detoxification of hydrogen peroxide by catalase.
❑ **Anabolic reactions:** Synthesis of bile acids and plasmalogens (ether lipids). It should be noted that plasmalogens are important components of myelin sheath.
❑ Peroxisomal disorders are a group of inborn errors of metabolism due to impairment in peroxisome function.
❑ Both defects in the biogenesis of peroxisomes and defect in a single peroxisomal enzymes have been documented.

Defect in Peroxisomal Biogenesis

Peroxin proteins required for peroxisomal biogenesis are coded by *PEX* group of genes. Defect in these leads to the following *clinical continuum*.

Chapter 13 Lipid Metabolism

253

Disorder	Severity
Zellweger syndrome (cerebrohepatorenal syndrome)	Most severe
Neonatal adrenoleukodystrophy	Intermediate
Infantile Refsum disease	Mildest

Peroxisomal Disorders due to Single Enzyme Defect

Disorder	Defect	Consequent pathogenies
Refsum disease	Phytanoyl-CoA hydroxylase a.k.a phytanic acid oxidase	See below
Hyperoxaluria type 1	Alanine: glyoxylate aminotransferase	Failure of glyoxylate to glycine conversion → accumulation
X-linked adrenoleukodystrophy	*ABCD1* gene mutation	Failure of transport of VLCFA into peroxisomes → accumulation

Refsum's Disease

- **Inheritance:** Autosomal recessive
- **Defect: Peroxisomal α-oxidation of phytanic acid**
- **Clinical Features:** Retinitis pigmentosa, cerebellar ataxia, Sensory neural hearing loss, Neuropathy
- **Lab findings:** ↑protein level in CSF, Abnormal bile acid pattern, ↓ plasmalogen
- **Treatment:** Phytanic acid restriction, i.e. restriction of fats from ruminating animals (Cow and sheep are ruminants. So, dairy products and sheep meat are restricted.)

Medium-chain Acyl-CoA Dehydrogenase Deficiency is Common and Fatal

- Medium-chain acyl-CoA dehydrogenase (MCAD) acts on C_4-C_{12} fatty acyl-CoA.
- MCAD deficiency is the most common mitochondrial β-oxidation defect.
- Symptoms are precipitated by condition in which glucose is not available and fats need to be utilised like infection and fasting.
- Present with hypoketotic hypoglycemia and sudden infant death syndrome
- Presence of unmetabolised octanoylcarnitine and C_6-C_{10} dicarboxylic acids is diagnostic of the disorder by tandem mass spectrometry.
- **Treatment:**
  - The primary goal of treatment should be to provide adequate calories, avoiding fasting, rigorous treatment of infections
  - Intravenous glucose to treat acute episodes
  - Oral carnitine to clear the toxic metabolites

 *Revise*

Three types of fatty acids			
	Short-chain	Medium-chain	Long-chain
Number of carbons	<8	8-12	>12
Presence in diet	Absent	Less	Large amount
Presence in feces	Substantial amount	Absent	Minimal amount
Site of absorption	Colon	Small Intestine	Small intestine
The requirement of pancreatic lipase and micelle formation for digestion	No	No	Yes

Competency

❏ BI4.2   Describe the processes involved in digestion and absorption of dietary lipids and also the key features of their metabolism

Key Points

❏ Free fatty acids are bound to albumin in the blood.
❏ Oxidation of fatty acids requires them to be linked to Coenzyme A. Two high-energy phosphate bonds are needed in this step.
❏ Carnitine is synthesized from lysine and methionine with the help of vitamin C.
❏ Carnitine mediated fatty acid transport is an example of carrier mediated transport.
❏ Fatty acyl-CoA dehydrogenase produces $FADH_2$.
❏ Complete oxidation of palmitate (C16) produces 106 ATPs.
❏ Oxidation of odd chain fatty acids produce acetyl-CoA and propionyl CoA.
❏ Cerebro-hepato-renal syndrome of Zellweger results from the absence of peroxisomes. Therefore, there will be accumulation of very-long-chain fatty acids, especially in the brain. The gene mutated are of PEX family.
❏ Oxidation of odd numbered double bonds in unsaturated fatty acids require isomerase only.
❏ Oxidation of even numbered double bonds requires reductase and isomerase.

SELF-ASSESSMENT

Short Answer Questions

1. What does the word 'free' in free fatty acids refer to? How free fatty acids are transported in the blood?

2. During catabolism, fatty acids release more energy compared to carbohydrates. Why?

3. Describe the role of bile salts in lipid metabolism.

4. What is the source of short-chain fatty acids in our body?

5. What is the biochemical basis of the use of coconut oil in patients with CPT-I deficiency?

6. What is the role of carnitine in cellular metabolism? Explain why carnitine deficiency in infants is associated with non-ketotic hypoglycemia?

7. Name the four repetitive steps of β-oxidation of fatty acids. How many rounds of β-oxidation will palmitic acid undergo? Name the catabolic products derived from β-oxidation of palmitic acid and their numbers. Calculate the ATP yield.

8. How many propionyl-CoA molecules are derived from the oxidation of a saturated fatty acid with 17-carbons? Mention the fate of propionyl-CoA

9. In which subcellular organelle does the β-oxidation of very-long-chain fatty acids take place? What is the end-product?

10. How branched-chain fatty acids are oxidized in the body?

11. In a tabular format, compare the various types of oxidation of fatty acids.

12. What is the metabolic defect in Refsum's disease? Mention the clinical features, laboratory findings and the treatment.

13. Name any two disorders of peroxisomal biogenesis and two disorders due to defect in the single peroxisomal enzyme.

14. What is the common clinical presentation of Medium-chain Acyl-CoA Dehydrogenase Deficiency? How will you treat a patient with this condition?

Multiple Choice Questions

1. **Free fatty acids in blood are carried by**
   a. VLDL   b. LDL
   c. Albumin   d. Chylomicron

2. **Which of the following fatty acids is produced by fermentation of dietary fiber by colonic flora?**
   a. Palmitate
   b. Butyrate
   c. Oleate
   d. Linoleate

3. **What is essential for transfer of fatty acid across mitochondrial membrane?**
   a. Creatine   b. Protein X
   c. Carnitine   d. Ligandin

4. **Carnitine helps in**
   a. Transport of fatty acids from mitochondria to cytosol
   b. Transport of fatty acids from cytosol to mitochondria
   c. Transport of pyruvate in to mitochondria
   d. Transport of Malate in Malate shuttle

5. **Carnitine synthesis requires**
   a. Leucine and Methionine
   b. Lysine and Methionine
   c. Glycine and Lycine
   d. Glycine and Methionine

6. **Which of the following is NOT a step of β-oxidation?**
   a. NADP dependent oxidation
   b. FAD dependent oxidation
   c. Thiolysis
   d. Hydration

7. **Methylmalonyl-CoA mutase reaction requires**
   a. Vitamin B_1 b. Vitamin B_2
   c. Vitamin B_{12} d. Biotin

8. **Alpha-oxidation takes place in**
   a. Mitochondria   b. Peroxisomes
   c. Cytosol   d. Golgi apparatus

9. **In Zellweger syndrome, there is**
   a. Accumulation of long fatty acid
   b. Accumulation of short chain fatty acids
   c. Accumulation of very long chain fatty acids
   d. Accumulation of medium chain fatty acids

10. **Trimming of very long chain fatty acids takes place by**
    a. Mitochondrial α oxidation
    b. Mitochondrial β oxidation
    c. Peroxisomal α oxidation
    d. Peroxisomal β oxidation

11. **What are the products formed after thiolytic cleaves of fatty acyl-CoA in final step of fatty acid oxidation?**
    a. Fatty acyl-CoA and acetyl-CoA
    b. Fatty acyl-CoA and acetate
    c. Fatty acyl-CoA and acetoacetyl-CoA
    d. Fatty acid and acetyl-CoA

12. **All of the following are produced by peroxisomal β-oxidation of very long chain fatty acids, EXCEPT**
    a. Acetyl-CoA
    b. Malonyl-CoA
    c. Octanoyl-CoA
    d. $FADH_2$

13. **Which one of the following disease is associated with C6-C10 dicarboxylic aciduria?**
    a. Refsum disease
    b. Medium chain acyl-CoA dehydrogenase deficiency
    c. Zellweger disease
    d. Tangier disease

ANSWERS

1. c.	2. b.	3. c.	4. b.	5. b.	6. a.
7. c.	8. b.	9. c.	10. d.	11. a.	12. b.
13. b.					

13.2 KETOGENESIS AND UTILIZATION OF KETONE BODIES

Chapter Outline

Let's begin with the question: Fats give us more energy. Still, why should the body produce ketone bodies?

Brain does not have any storage form of energy. Maintenance of optimal blood glucose level is of utmost important for the proper functioning of brain.

Brain cannot utilize fatty acids, since,
- ❏ It is hypothesized that albumin bound fatty acids cannot cross the blood brain barrier
- ❏ Absence of certain enzymes of fatty acid oxidation in the brain

When glucose is not available, brain needs an alternative fuel. This alternative fuel is ketone bodies, which are nothing but water-soluble fatty acid oxidation products.

KETOGENESIS

Ketone bodies are produced exclusively by the liver from fatty acids in conditions when glucose is not available (e.g. fasting, starvation, intake of ketogenic diet or in the absence of insulin like Diabetes mellitus.

Remember the adage 'Fats are burnt in the flame of carbohydrates'. Oxidation of acetyl-CoA in TCA cycle requires oxaloacetate, derived from carbohydrates.

During starvation and uncontrolled DM, fatty acids are mobilized from adipose tissue and are oxidized by liver to acetyl-CoA. Oxaloacetate cannot be generated when carbohydrates are less. Without oxaloacetate, acetyl-CoA cannot be oxidized in the TCA cycle. So, the accumulated acetyl-CoA is diverted to ketogenesis.

Following are the steps of ketone body formation:
1. Condensation of 2 molecules of acetyl-CoA
2. Formation of HMG-CoA
3. Breakdown of HMG-CoA
4. Reduction of acetoacetate to beta-hydroxy butyrate
5. Spontaneous decarboxylation of acetoacetate to acetone

All the steps of ketogenesis takes place in the hepatic mitochondria.

Step 1: Thiolase catalyzes the condensation of two molecules of acetyl-CoA to acetoacetyl-CoA. This is the reversal of final step of β-oxidation of fatty acids.

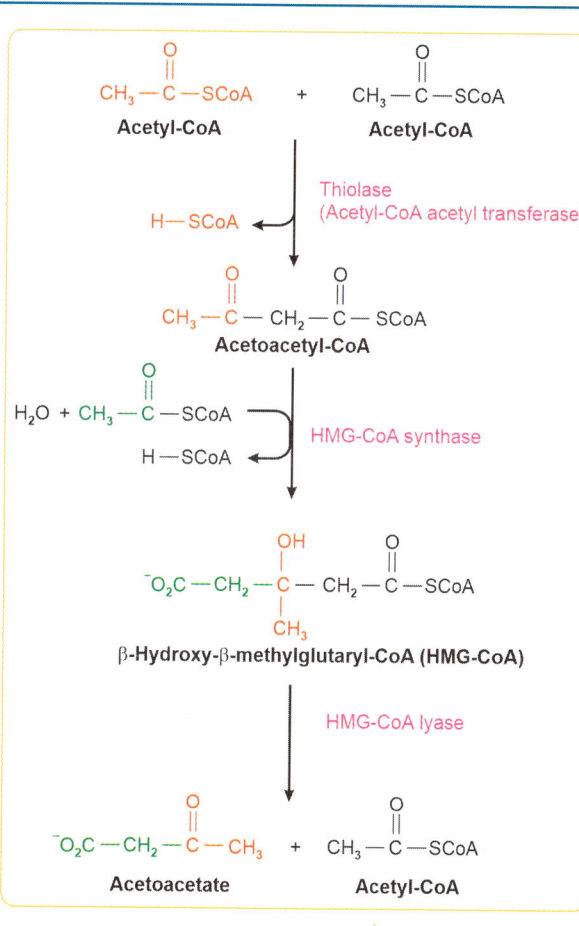

Chapter 13 Lipid Metabolism

257

Step 2: Another acetyl-CoA is condensed to the acetoacetyl-CoA to produce β-Hydroxy-β-Methyl Glutaryl-CoA a.k.a HMG-CoA. This reaction is catalyzed by mitochondrial HMG-CoA synthase. (*Cytosolic HMG-CoA synthase isoenzyme is involved in cholesterol synthesis*)

Step 3: HMG-CoA is broken down to acetoacetate and acetyl-CoA by HMG-CoA lyase, the rate-limiting enzyme of ketone body formation. Acetoacetate is the first ketone body formed. So, it is known as **primary ketone body.**

Step 4: Aceto acetate is enzymatically reduced with NADH to produce β-hydroxy butyrate.

Aceto acetate can also undergo spontaneous decarboxylation to produce acetone, which is exhaled out and gives fruity odor to the breath.

Since β-hydroxy butyrate and acetone are produced from acetoacetate, they are known as **2° ketone bodies.**

β-hydroxy Butyrate is NOT a Ketone

Look at the structure of β-hydroxy butyrate. It doesn't contain a keto group. It is a secondary alcohol. Therefore, it won't give a positive result with Rothera's test/Gerhardt's test for ketones.

UTILIZATION OF KETONE BODIES

❑ Ketone bodies are water-soluble and are freely transported in the blood; used by extra-hepatic tissues.
❑ During starvation,
  ○ Skeletal and cardiac muscle, and renal cortex preferentially use ketone bodies.
  ○ Up to 20% of brain's energy requirement can be met by ketone bodies. The remainder must be supplied by glucose.
❑ In extra-hepatic tissues, aceto acetate and β-hydroxy butyrate are converted to acetyl-CoA and this acetyl-CoA is oxidized in the TCA cycle to release energy. *Remember: Ketone bodies are produced in hepatic mitochondira since acetyl-CoA cannot enter the TCA cycle. However, in extrahepatic tissues, acetyl-CoA enters the TCA cycle!*
❑ Acetoacetate needs to be activated by transfer of CoA-SH from succinyl-CoA to produce acetoacetyl-CoA. *This reaction compromises the GTP/ATP produced by the succinate thiokinase reaction. When, we calculate the energy-yield of acetoacetate we will reduce this.*
❑ The enzyme transferring CoA-SH to acetoacetate is known as Succinyl CoA-acetoacetate CoA transferase or Thiophorase.
❑ Thiophorase enzyme is absent in liver. So, liver cannot utilise ketone bodies. *Liver produces ketone bodies for the utilization of extra-hepatic tissues.*
❑ Finally, thiolase transfers one more CoA-SH molecule to acetoacetyl-CoA to break it into two molecules of acetyl-CoA.
❑ Acetyl-CoA enters the TCA cycle.

Section II Intermediary Metabolism

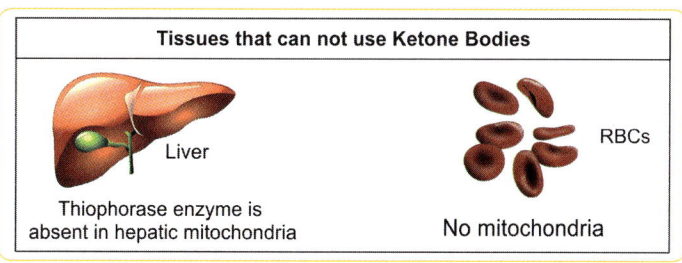

β-Hydroxybutyrate

CH$_3$ — C — CH$_2$ — C (OH, H) with O, O$^-$

D-β-Hydroxybutyrate dehydrogenase

NAD$^+$
NADH + H$^+$

Acetoacetate

CH$_3$ — C — CH$_2$ — C with O (double bond), O, O$^-$

β-ketoacyl-CoA transferase (Thiophorase)

Succinyl-CoA
Succinate

Acetoacetyl-CoA

CH$_3$ — C — CH$_2$ — C with O, O, S-CoA

Thiolase

CoA-SH

CH$_3$ — C (O, S-CoA) + CH$_3$ — C (O, S-CoA)

2 Acetyl-CoA

Note: β-ketoacyl-CoA transferase is also known as Succinyl CoA-acetoacetate CoA transferase and Thiophorase.

Tissues that cannot use Ketone Bodies

Tissues that can not use Ketone Bodies	
Liver	RBCs
Thiophorase enzyme is absent in hepatic mitochondria	No mitochondria

ENERGY AUDIT

I want you to know the difference between "Energy Yield from ketone bodies in extra-hepatic tissues" and "Energy-Yield from β-oxidation of fatty acids when acetyl-CoA is converted to ketone bodies in the liver"

Energy Yield from Ketone Bodies in Extra-hepatic Tissues

From Acetoacetate

Two acetyl-CoA molecules produced by acetoacetyl-CoA enters the TCA cycle to produce 20 *(2 X 10)* ATP equivalents. Reduce one ATP equivalent compromised by CoA-SH transfer from succinyl-CoA. You will get net yield of **19 ATP equivalents.**

From β-hydroxybutyrate

One NADH is produced during the oxidation of β-hydroxybutyrate to acetoacetate. If we add the P:O ratio of NADH, i.e. 2.5 ATP to the above, we will get **21.5 ATP equivalents**.

Energy-Yield from β-oxidation of Fatty Acids When Acetyl-CoA is Converted to Ketone Bodies in the Liver

- ❑ 7 cycles of beta oxidation of palmitate form 8 acetyl CoA, producing 4 moleules of acetoacetate. *This acetoacetate cannot be converted to acetyl-CoA in the liver.*
- ❑ One molecule of NADH and $FADH_2$ are produced in one cycle of beta oxidation. Thus, 7 cycles will produce 28 ATP.
- ❑ Reduce the 2 ATPs utilized in the activation of fatty acids, you will get **26 ATPs** when one molecule of palmitate is oxidized to 4 molecules of acetoacetate.
- ❑ One molecule of NADH is utilized for the reduction of acetoacetate to β-hydroxy butyrate. Thus, you reduce 10 (4 × 2.5) ATPs from the above 26. Therefore, you get **16 ATPs** when one molecule of palmitate is oxidised to 4 molecules of β-hydroxy butyrate.

In liver, one molecule of Palmitate produces 106 ATPs. Although one molecule of palmitate produces only 16 ATPs, when it is oxidized to ketone bodies in liver. This is very important when many molecule of fatty acids need to be converted to ketone bodies in the liver within the constraints of the limited capacity of electron transport chain.

KETOACIDOSIS

Ketoacidosis refers to the presence of increased ketoacids in blood and urine (Ketonemia + Ketonuria).

Uncontrolled diabetes mellitus and prolonged starvation are the two important causes of ketoacidosis.

Pathogenesis

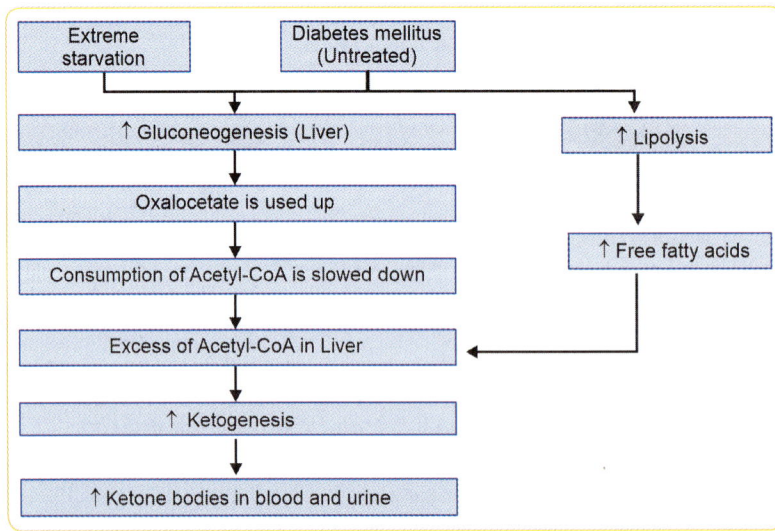

Section II Intermediary Metabolism

260

Excessive Production of Ketone Bodies is Harmful

- ❑ Ketone bodies accumulate in the body when the production exceeds that of the utilization.
- ❑ Ketonemia is accumulation of ketone bodies in blood. Ketonuria is excretion of ketone bodies in urine *(ketone bodies are water-soluble)*. Ketosis is the combination of both.
- ❑ Acetoacetic acid and β-hydroxy butyric acid are organic acids with **pKa of 3.6 and 4.7** respectively.
- ❑ At the physiological pH of the blood (7.35 to 7.45), they **dissociate** to produce acetoacetate and β-hydroxy butyrate with concomitant **release of proton**.
- ❑ These protons need to be **buffered by the alkali reserve of the body** i.e. the bicarbonate (HCO_3^-) buffer.
- ❑ Excessive generation of ketone bodies causes **depletion of alkali reserve** of the body → **Metabolic acidosis with high anion gap and Hyperkalemia.**
- ❑ Metabolic acidosis results in derangement of enzymes and ion pumps of the body → Homeostasis cannot be maintained → dangerous to the body.

β-hydroxy Butyrate is the Major Ketone Body in Diabetic Ketoacidosis and Starvation

Diabetes and starvation leads to raised NADH/NAD ratio. This raised levels of NADH converts acetoacetate to β-hydroxy butyrate. This explains the possibility of Rothera's nitroprusside test being negative in a patient of diabetic ketoacidosis.

β-hydroxybutyrate is converted to acetoacetate in response to treatment with insulin. Ketone body levels may appear to increase if measured by laboratory assays that use the nitroprusside reaction. *(Ref. Harrison 20th Ed., p. 2870).*

Ketogenic Diets are Prescribed for Refractory Epilepsy

Ketogenic diet is a high-fat, adequate-protein, low-carbohydrate diet, which was initially used to control refractory epilepsy, is now being tried in many weight-reduction regimens like Atkin's diet. The exact mechanism of anticonvulsant action of ketone bodies is yet to be identified.

REGULATION OF KETOGENESIS

To understand the regulation of ketogenesis, let's recall the metabolic state that drives ketogenesis. "Acetyl-CoA produced from β-oxidation of adipocyte derived free fatty acids cannot enter the TCA cycle because of lack of oxaloacetate. So, acetyl-CoA is converted to ketones"
Thus, ketogenesis is regulated at three levels:

Level	Regulation
1. Mobilization of free fatty acids from adipocytes	❑ Glucagon and epinephrine promotes the mobilization of fatty acid while insulin inhibits. ❑ ↓ [Insulin]/[Glucagon] promotes ketogenesis.
2. Entry of fatty acids into β-oxidation in hepatic mitochondria	❑ CPT-I is the rate-limiting step for the entry of fatty acids. ❑ Malonyl-CoA inhibits CPT-I. Malonyl-CoA is produced by acetyl-CoA carboxylase. Acetyl-CoA carboxylase activity is reduced when insulin/glucagon ratio is low. Therefore, there is no malonyl-CoA to inhibit CPT-I when insulin is low and glucagon is high.
3. Entry of acetyl-CoA into TCA cycle in hepatic mitochondria	❑ Availability of oxaloacetate determines the fate of acetyl-CoA. ❑ Normally, Acetyl-CoA activates pyruvate carboxylase to produce oxaloacetate. ❑ During starvation, all the available pyruvate/oxaloacetate is channeled into gluconeogenesis.

Chapter 13 Lipid Metabolism

Competency

- ❏ BI4.2   Describe the processes involved in digestion and absorption of dietary lipids and also the key features of their metabolism

Key Points

- ❏ Ketogenesis is the process of diverting the acetyl-CoA, derived primarily during high rates of fatty acid oxidation in the liver, into the synthesis of ketone bodies.
- ❏ Acetoacetate is the primary ketone body. Beta-hydroxy butyrate and Acetone are the secondary ketone bodies.
- ❏ β-hydroxy butyrate doesn't contain the keto group; produces negative result with Rothera's test and Gerhardt's test.
- ❏ Brain uses ketone bodies during starvation.
- ❏ HMG-CoA lyase catalyse the rate-limiting step of ketogenesis.
- ❏ Liver cannot utilize ketone bodies because of absence of thiophorase.
- ❏ RBCs cannot utilize ketone bodies because of absence of mitochondria.

SELF-ASSESSMENT

Short Answer Questions

1. Name the substrate and rate-limiting enzyme of ketogenesis. Name the primary and secondary ketone bodies. Explain why the liver cannot utilize ketone bodies.

2. Name the tissues that cannot utilize ketone bodies and explain why?

3. Explain the pathogenesis of diabetic ketoacidosis.

4. Explain the regulation of ketogenesis.

5. Describe the pathway of synthesis and utilization of Ketone bodies and the regulation of ketogenesis.

6. How and in which organ are ketone bodies formed? Indicate their harmful effect and useful function in the body.

7. Why is a small amount of dietary carbohydrate essential to prevent ketosis?

Multiple Choice Questions

1. **Why brain cannot use free fatty acids?**
   a. Free fatty acids are toxic to brain
   b. Albumin bound free fatty acid cannot enter brain
   c. Brain has enough storage of energy
   d. Free fatty acids provide the least amount of energy

2. **The major ketogenic organ is**
   a. Liver
   b. Kidney
   c. Intestine
   d. Muscle

3. **Ketone bodies are usually produced from the oxidation of**
   a. Glucose
   b. Glutamate
   c. Glycine
   d. Palmitate

4. **Ketogenesis takes place in**
   a. Hepatic mitochondria
   b. Hepatic cytoplasm
   c. Hepatic microsome
   d. Hepatic lysosome

5. **Which of the following is known as primary ketone body?**
   a. Acetoacetate
   b. β-hydroxy butyrate
   c. Acetone
   d. All of the above

6. **Which of the following does not contain a keto group?**
   a. Acetoacetate
   b. β-hydroxy butyrate
   c. Acetone
   d. All of the above

7. **Which of the following cannot utilize ketone bodies?**
   a. RBCs
   b. Skeletal muscle
   c. Cardiac muscle
   d. Brain

8. **Liver cannot utilize ketone bodies because of absence of**
   a. Succinyl-CoA synthetase
   b. Succinyl CoA: Acetoacetate CoA transferase
   c. Thiolase
   d. Thioesterase

9. **Which of the following reaction is common to both ketogenesis and cholesterol synthesis**
   a. Formation of HMG-CoA
   b. Reduction of HMG-CoA
   c. Carboxylation of HMG-CoA
   d. Oxygenation of HMG-CoA

10. **Atkin's diet is**
    a. Protein restricted low-calorie diet
    b. Carbohydrate restricted low-calorie diet
    c. Fat restricted low-calorie diet
    d. Mineral restricted low-calorie diet

11. **The rate of ketone body production is mainly determined by the relative rate of which of the following key metabolic pathways?**
    a. Proteolysis
    b. Glycogenolysis in muscle
    c. Glycolysis
    d. Lipolysis

ANSWERS

1. b.	**2.** a.	**3.** d.	**4.** a.	**5.** a.	**6.** b.
7. a.	**8.** b.	**9.** a.	**10.** b.	**11.** d.	

13.3 FATTY ACID SYNTHESIS

Fatty acid synthesis is a.k.a **Lynen's spiral** (Feodor Lynen was awarded the Nobel Prize for his studies on fatty acid synthesis). Liver and adipocyte are the major site of fatty acid synthesis.

ACETYL-COA AND NADPH ARE REQUIRED FOR FATTY ACID SYNTHESIS

Two carbon acetyl-CoA is the building block for fatty acid synthesis. Reducing power is derived from NADPH.
❑ Acetyl-CoA is derived mainly from dietary carbohydrates.
❑ Reducing power is derived mainly from HMP shunt.
The pyruvate dehydrogenase reaction producing acetyl-CoA takes place inside the mitochondria. But the fatty acid synthesis takes place in the cytosol. Acetyl-CoA can not be directly transported out of mitochondria. So, how is acetyl-CoA made available for fatty acid synthesis? Let us see.

EXCESS CALORIC STATE PROMOTES FATTY ACID SYNTHESIS

❑ Excess dietary carbohydrates promote the synthesis of fatty acids. The oxidation of excess dietary carbohydrates lead to excess of NADH and ATPs. (Insulin is produced in response to glucose. Insulin is lipogenic. It upregulates the level of enzymes involved in lipogenesis)
❑ High NADH inhibits the isocitrate dehydrogenase of TCA cycle and leads to accumulation of citrate, which comes out of mitochondria and provides one molecule of acetyl-coA and one molecule of NADPH.
❑ Citrate also inhibits the PFK-1, rate-limiting enzyme of PFK-1 and thus diverts glucose to HMP shunt to generate the NADPH required for lipogenesis.

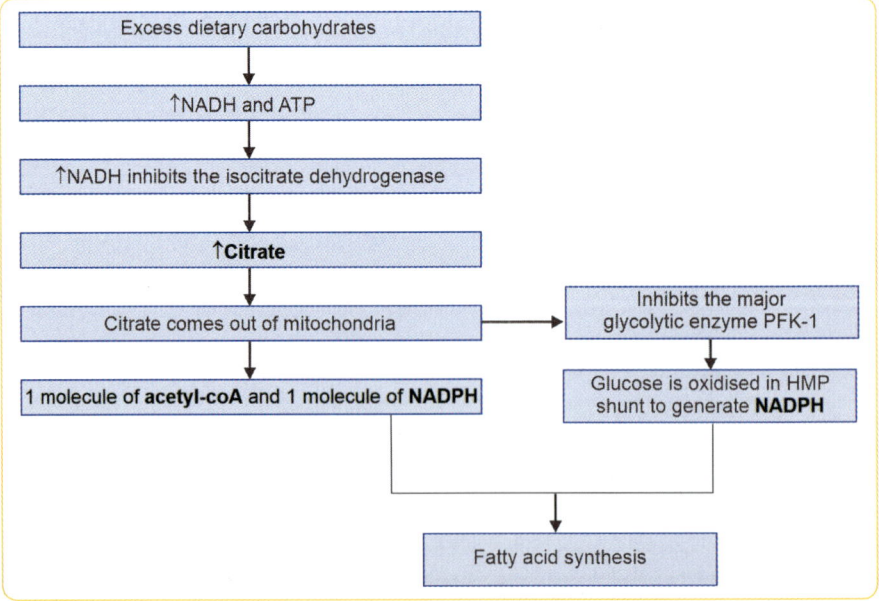

Citrate Provides the Acetyl-CoA Required for Fatty Acid Synthesis

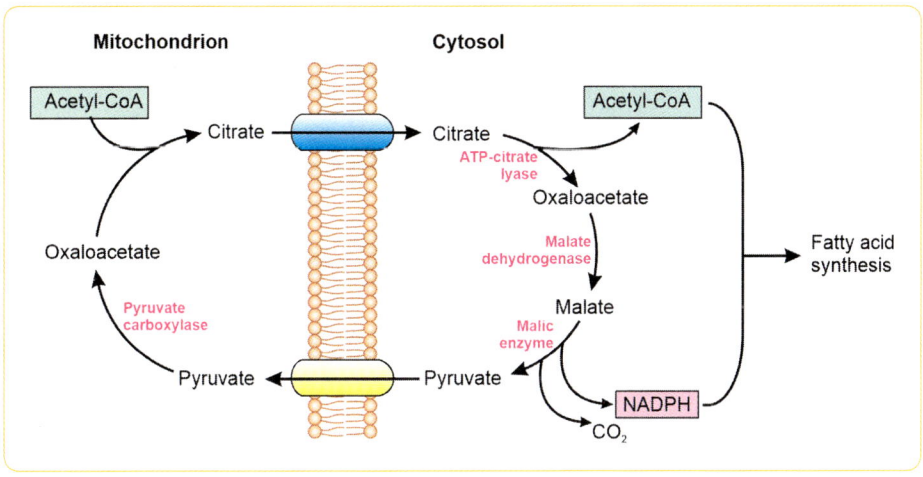

Fatty Acid Synthesis is NOT the Reversal of Oxidation

In carbohydrate metabolism, we have studied that gluconeogenesis is not a complete reversal of glycolysis since only the reversible/near-equilibrium reactions are common to both glycolysis and gluconeogenesis.

Fatty acid synthesis and oxidation are entirely different reactions taking place in two different cellular compartments. The former takes place in cytosol while the later takes place inside mitochondrial matrix.

	Fatty acid oxidation	Fatty acid synthesis
Subcellular Location	Mitochondria	Cytoplasm
Coenzymes utilized	NAD+, FAD, CoA-SH	Pantothenate, NADPH, Biotin (Memory Aid: Punjab National Bank)
Enzymes	Fatty acyl-CoA dehydrogenase and trifunctional protein	Single multifunctional polypeptide enzyme which exist as a dimer
Acyl group linked to SH group of	Coenzyme A	Acyl-Carrier Protein (ACP)

Acetyl-CoA Carboxylase Catalyzes the Rate-limiting Step of Fatty Acid Synthesis

Acetyl-CoA carboxylase, a multifunctional enzyme mediates the ligation (joining) of CO_2 to acetyl-CoA. This is one of the 4 carboxylases that require biotin.

Remember

Only 4 carboxylases require Biotin.
1. Acetyl-CoA carboxylase
2. Pyruvate carboxylase
3. Propionyl-CoA carboxylase
4. Methyl-crotonyl-CoA carboxylase.

Chapter 13 **Lipid Metabolism**

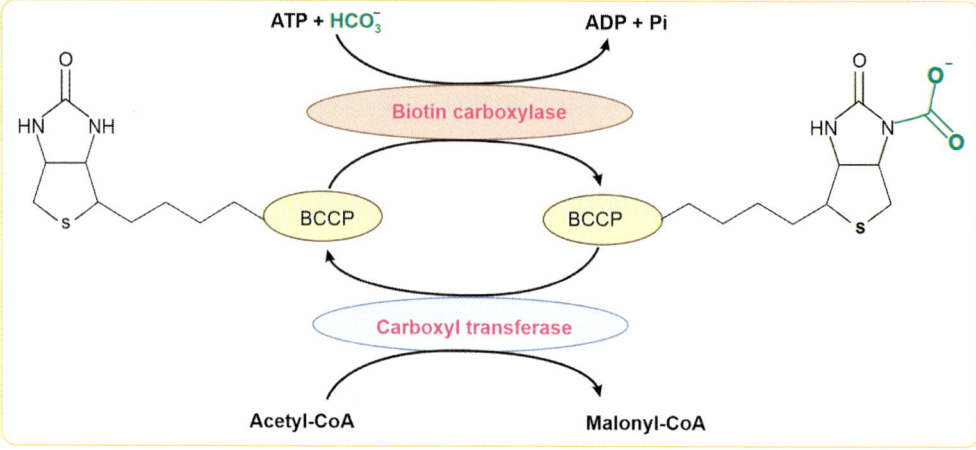

Fig. 1: Look at the two enzyme activities-carboxyltransferase and biotin carboxylase present in the same polypeptide. First, biotin is carboxylated with the help of ATP. BCCP is the carrier for both biotin and carboxy biotin. Carboxyl transferase activity transfers the CO_2 from carboxy biotin to acetyl-CoA.

The product of acetyl-CoA carboxylase i.e. malonyl-CoA enters the fatty acid synthase complex. *Regulation of this enzyme will be discussed at the completion of the topic.*

Fatty Acid Synthase Complex is a Homodimer of 6 Enzymes and Acyl Carrier Protein

Fatty acid synthase is a multifunctional enzyme. Six enzymatic activities and acyl carrier protein (ACP) are present in one single polypeptide. Single polypeptide is functionally inactive. Two polypeptides (homodimer) are needed for the activity.

Reactions of Fatty Acid Synthesis

1. Attachment to Acyl-Carrier protein
2. Condensation by (Acyl-malonyl-ACP) condensing enzyme
3. Reduction by β-Ketoacyl-ACP reductase
4. Dehydration by 3-Hydroxyacyl-ACP dehydratase
5. Reduction by Enoyl-ACP reductase
6. Thioesterase reaction

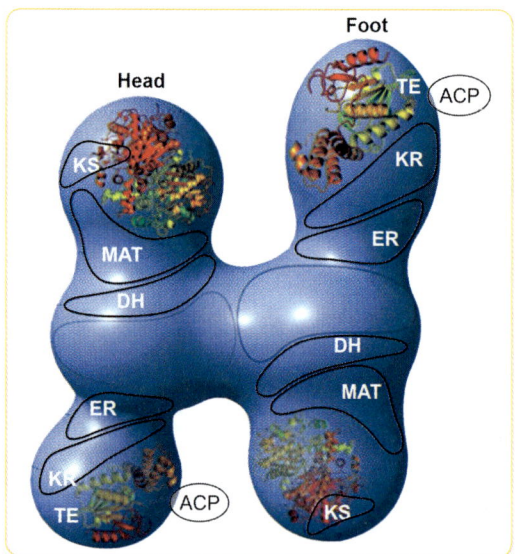

Fig. 2: The figure shows the fatty acid synthase homodimer which is arranged in head to foot manner. Note the 6 enzymatic activities. Growing fatty acid chain is linked to acyl carrier protein throughout the synthesis till it is released finally by the thioesterase

Reaction 1: Attachment to ACP

❑ First, an acetyl-CoA molecule binds to the SH group of pantothenic acid of ACP.
❑ Then, this is transferred to the SH group of cysteine of ketoacyl synthase. Thus, the SH group of ACP is vacant and binds to a malonyl-CoA.
❑ These reactions are catalyzed by Malonyl/acetyl transacylase.

Fig. 3: Reactions of fatty acid synthesis

Reaction 2: Condensation

Beta-ketoacyl synthase (KS) catalyzes the condensation of acetyl-CoA and malonyl-CoA releasing CO_2. This results in the formation of the acetoacetyl group attached to ACP. SH group gets free.

Reactions 3, 4, 5

The acetoacetyl group attached to ACP is reduced, dehydrated, and reduced again to produce butyryl group attached to ACP. The butyryl group is then transferred to the SH group of cysteine of KS. The vacant ACP-SH binds to new malonyl group. These reactions are repeated, resulting in the formation of the palmitoyl group attached to ACP.

CONCEPT CORNER

Fatty acid synthase enzyme complex can incorporate only two carbons to the growing fatty acid synthesis. So, only even chain fatty acids are found in animals.

Comparison of Fatty Acid Oxidation and Synthesis

Oxidation	Synthesis

$R-CH_2-CH_2-CH_2-C-SCoA$ (with O double bond, γ β α carbons)

Oxidation → FAD / FADH$_2$ — Acyl-CoA dehydrogenase

$R-CH_2-C=C-C-SCoA$ *Trans–Δ^2–enoyl-CoA*

Hydration → H$_2$O — Enoyl-CoA hydratase

$R-CH_2-C-C-C-SCoA$ **3-Hydroxyacyl-CoA**

Oxidation → NAD$^+$ / NADH + H$^+$ — Hydroxyacyl-CoA dehydrogenase

$R-CH_2-C-CH_2-C-SCoA$ **3-Ketoacyl-CoA**

Thiolysis → CoA-SH — β-Ketothiolase

$R-CH_2-C-SCoA + H_3C-C-SCoA$

Acyl-CoA shortened by two carbon atoms   **Acetyl-CoA**

Synthesis side:

$H_3C-C-ACP + {}^-OOC-CH_2-C-ACP$
Acetyl-ACP   **Malonyl-ACP**

Condensation → ACP + CO$_2$ — β–Ketoacyl-ACP synthase

$H_3C-C-CH_2-C-S-ACP$ **Acetoacetyl-ACP**

Reduction → NADPH + H$^+$ / NADP$^+$ — β-Ketoacyl-ACP reductase

$H_3C-CH-CH_2-C-S-ACP$ (with OH) **3-Hydroxy-butyryl-ACP**

Dehydration → H$_2$O — 3-Hydroxyacyl-ACP dehydratase

$H_3C-C=C-C-S-ACP$ **Crotonyl-ACP**

Reduction → NADPH + H$^+$ / NADP$^+$ — Enoyl-ACP reductase

$H_3C-CH_2-CH_2-C-S-ACP$ **Butyryl-ACP**

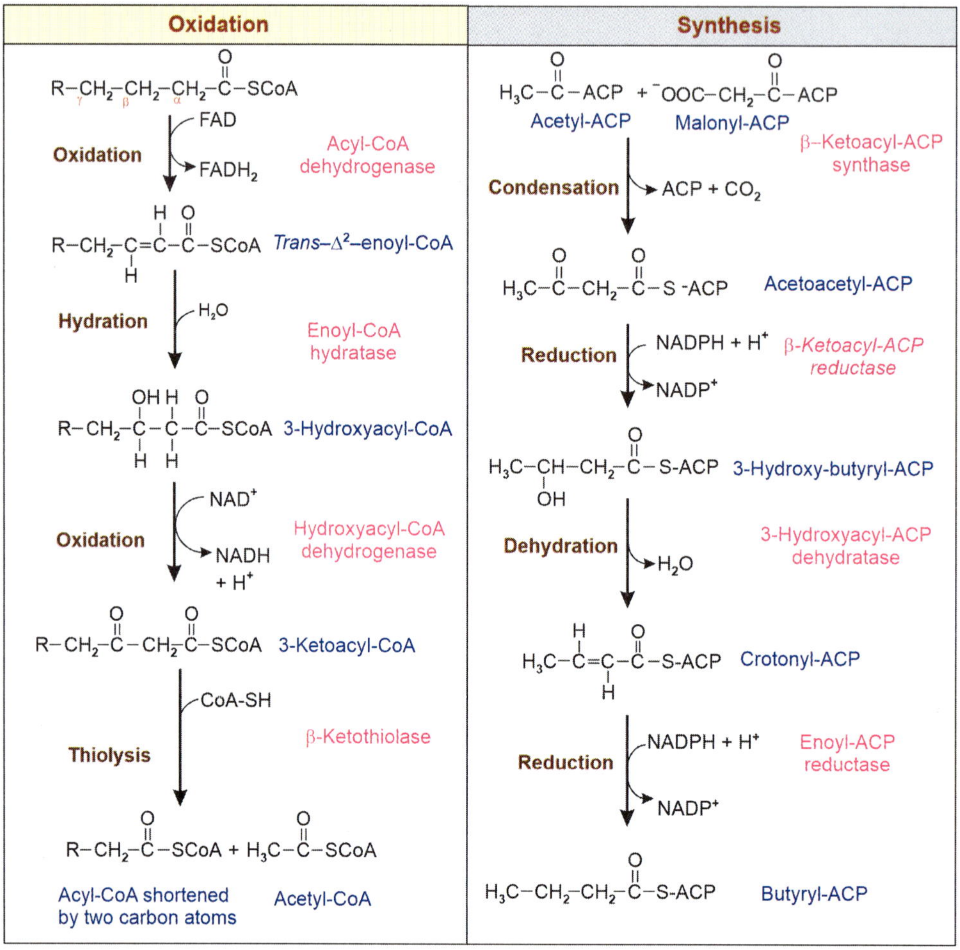

Fig. 4: Compare the fatty acid oxidation and synthesis. The steps are reverse opposite

Step in fatty acid oxidation	Reverse step in fatty acid synthesis
Thiolysis (final step)	Condensation (first step)
NAD$^+$ dependent oxidation	NADPH mediated reduction
Hydration	Dehydration
FAD dependent oxidation (first step)	NADPH mediated Reduction

Malonyl-CoA is used to add 2-Carbon Units

Other than the first acetyl-coA, Malonyl-CoA is the source of addition of 2 carbons in fatty acid synthesis.

Acetyl-CoA is carboxylated to malonyl-CoA utilising ATP, then malonyl-CoA is decarboxylated during condensation. Why should nature spend the energy to carboxylate?

There are two advantages of using malonyl-CoA:
1. Release of CO$_2$ from malonyl-CoA releases substantial amount of energy, which was trapped during the condensation of acetyl-CoA and CO$_2$ with the help of ATP. This drives the reaction forward.
2. Malonyl-CoA inhibits CPT-1 and prevents the simultaneous operation of fatty acid oxidation during fatty acid synthesis.

Palmitate is the End Product of Fatty Acid Synthase

Thioesterase component of fatty acid synthase acts as a chain-length ruler. When the growing fatty acid chain reaches the length of 16 carbons, thioesterase removes it from acyl carrier protein.

The net reaction is,

8 Acetyl-CoA + 7 ATP +14 NADPH + 14 H$^+$ → Palmitate + 14 NADP$^+$ + 8 CoA +6 H$_2$O + 7 ADP + 7 Pi

Elongation of Fatty Acids

Higher length fatty acids are produced by elongation systems in endoplasmic reticulum and to some extent in mitochondria.

Microsomal fatty acid elongase system elongates both saturated and unsaturated fatty acids. This uses NADPH as reducing agent, Malonyl-CoA as two-carbon donor. (Sometimes NADH can also be used by the reductase)

Microsomal elongase system	Mitochondrial elongase system
Most active	Less active
Malonyl-CoA is the two-carbon donor	Acetyl-CoA is the two-carbon donor
Elongates C$_{10}$ to C$_{22}$ and C$_{24}$ fatty acids	Elongates only medium-chain and short-chain fatty acids

Polyunsaturated Fatty Acids are Derived By Desaturation and Elongation

❏ Introduction of double bonds is done by microsomal cytochrome b5 desaturases using NADPH
❏ Linoleic acid is converted to arachidonic acid in which γ-linolenic acid is an intermediate. This is why arachidonic acid is not essential provided adequate linoleic acid is taken in diet.
❏ Desaturase enzymes cannot introduce double bonds beyond the Δ9 position. This is why certain polyunsaturated fatty acids are essential and must be taken in diet.

REGULATION OF FATTY ACID SYNTHESIS

Acetyl-CoA carboxylase is the rate-limiting enzyme of fatty acid synthesis.
It is regulated by allosteric regulation, covalent modification and induction/repression.

Allosteric Regulation

❏ Acetyl-CoA carboxylase is active when it is polymerized. Citrate promotes the polymerization. Citrate is an indicator of high-energy status. Palmitoyl-CoA inhibits the polymerization.
❏ Citrate can even partially relieve the inhibition done by phosphorylation.

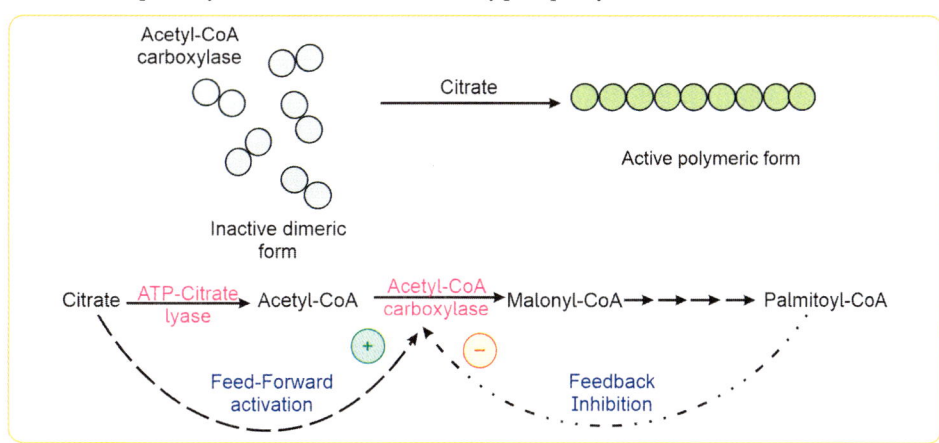

Covalent Modification

Acetyl-CoA carboxylase is active in dephosphorylated state and inactive in phosphorylated state.

❑ Phosphorylation is done by AMP activated protein kinase. **AMP is the indicator of low-energy state**. Anabolic reactions are inhibited during low-energy state.

❑ Dephosphorylation and activation of acetyl-CoA carboxylase is done by insulin activated protein phosphatase 2A *(lipogenic action of insulin)*.

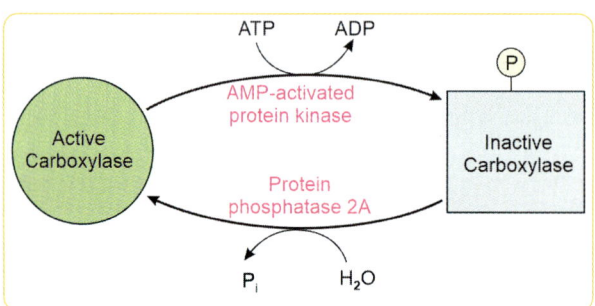

Insulin is Lipogenic (Induction)

❑ Insulin induces the production of ATP-citrate lyase and acetyl-CoA carboxylase at gene level.

Excess Fatty Acyl-CoA Inhibits Fatty Acid Synthesis

Fatty acyl-CoA is usually esterified into triacylglycerol in liver and adipose tissue. Non-esterified fatty acyl-CoA inhibits the synthesis of the fatty acids by inhibiting,

❑ Acetyl-CoA carboxylase
❑ Tricarboxylate transporter (citrate transporter)
❑ Mitochondrial ATP-ADP exchange transporter

Competency

❑ BI4.2 Describe the processes involved in digestion and absorption of dietary lipids and also the key features of their metabolism

Key Points

❑ Fatty acid synthesis is not the reversal of fatty acid oxidation.

❑ ATP, Mg^{2+}/Mn^{2+}, biotin, NADPH and pantothenic acid are needed for fatty acid synthesis from acetyl-CoA.

❑ For fatty acid synthesis, acetyl-CoA is transported from mitochondria to cytosol in the form of citrate.

❑ Fatty acid synthase complex is a homodimer of 2 polypeptide chains containing 6 enzyme activities and acyl carrier protein.

❑ Acyl-CoA inhibits the Carnitine palmitoyl transferase-I (CPT-I), acetyl-CoA carboxylase, and tricarboxylate transporter, this is why fatty acid synthesis and oxidation does not take place simultaneously.

❑ Biotin dependent Acetyl-CoA carboxylase, rate limiting enzyme of fatty acid biosynthesis is made up of multiple subunits; Citrate is the allosteric activator that helps in polymerization of subunits.

❑ Acetyl-CoA carboxylase is active in dephosphorylated state and inactive in phosphorylated state; Insulin mediates dephosphorylation while glucagon and epinephrine mediates phosphorylation.

❑ NADPH for fatty acid synthesis is derived mainly from HMP shunt.

❑ 16-carbon fatty acid palmitate is the end product of fatty acid synthase complex in most of the tissues.

❑ Elongation of fatty acid occurs in endoplasmic reticulum and to some extent in mitochondria.

❑ Polyunsaturated fatty acids are synthesized by desaturase and elongase enzymes in the endoplasmic reticulum.

❑ Heme containing, NADPH/NADH requiring Cytochrome b5 is the fatty acid desaturase in the endoplasmic reticulum.

❑ Desaturase enzymes cannot introduce double bonds beyond the Δ9 position. This is why certain fatty acids are essential and must be taken in diet.

SELF-ASSESSMENT

Short Answer Questions

1. Explain how excess caloric state promotes fatty acid synthesis?

2. Mention the rate-limiting step of fatty acid synthesis and fatty acid oxidation. How are fatty acid synthesis and fatty acid oxidation reciprocally regulated? Give the ATP yield of complete oxidation of palmitic acid (16C).

3. Which intermediate of the TCA cycle provides acetyl-CoA and NADPH required for the synthesis of fatty acids? Illustrate with a schematic diagram.

4. Explain why fatty acid synthesis is not the reversal of fatty acid oxidation.

5. Write the rate-limiting reaction of fatty acid synthesis and mention the vitamin cofactor. Explain how this reaction is regulated by various means?

6. Other than the first acetyl-CoA, Malonyl-CoA is the source of addition of 2 carbons in fatty acid synthesis. What is the advantage of using malonyl-CoA compared to acetyl-CoA?

Multiple Choice Questions

1. The enzyme catalyzing the rate-limiting step of fatty acid synthesis:
   a. Acetoacetate synthetase
   b. Acetyl-CoA carboxylase
   c. Thiophorase
   d. Thioesterase

2. Which of the following is the allosteric activator of Acetyl-CoA carboxylase?
   a. Malonyl-CoA   b. Acetyl-CoA
   c. Citrate   d. Biotin

3. Which of the following TCA cycle intermediate comes out of mitochondria to provide the acetyl-CoA required for lipogenesis?
   a. Citrate   b. Malate
   c. Oxaloacetate   d. Isocitrate

4. Acetyl-CoA carboxylase activity requires:
   a. Pantothenate
   b. Biotin
   c. Thiamine pyrophosphate
   d. Riboflavin

5. **Final step of fatty acid synthesis is catalyzed by:**
   a. Thiolase
   b. Thiophorase
   c. Thioesterase
   d. Hydroxylase

6. **Major source of NADPH for fatty acid synthesis is:**
   a. HMP shunt
   b. TCA cycle
   c. Glycolysis
   d. Uronic acid pathway

7. **Lipogenesis occurs in:**
   a. Liver
   b. Skeletal muscles
   c. Myocardium
   d. Lungs

8. **End product of cytosol fatty acid synthetase in humans is:**
   a. Oleic acid
   b. Arachidonic acid
   c. Linoleic acid
   d. Palmitic acid

9. **Role of citrate in fatty acid synthesis is:**
   a. Carrier of one carbon unit from mitochondria to cytosol
   b. Provides ATP
   c. Activates acetyl-CoA carboxylase
   d. Provide NADH for fatty acid synthesis

10. **NOT true about fatty acid synthase:**
    a. Six enzyme activities
    b. Present in cytosol
    c. All the activities are present in one polypeptide in eukaryotes
    d. All the activities are present in one polypeptide in prokaryotes

11. **Acetyl-CoA is transported out of the mitochondria in order to serve as a substrate for fatty acid or cholesterol synthesis. Which of the following enzymes used in this transport process provides NADPH required for these reductive biosynthesis reactions?**
    a. ATP-citrate lyase
    b. Citrate synthase
    c. Malate dehydrogenase
    d. Malic enzyme

12. **During fatty acid synthesis, growing fatty acyl group is covalently linked to:**
    a. SH group of Coenzyme A
    b. SH group of acyl-carrier protein
    c. COOH group of Coenzyme A
    d. COOH group of acyl-carrier protein

13. **Insulin causes increased lipogenesis by all of following except:**
    a. Inhibition of pyruvate dehydrogenase
    b. Increasing activity of acetyl-CoA carboxylase
    c. Decrease of cAMP in cell
    d. Transport of glucose inside adipocytes via GLUT4
    Acetyl-CoA is needed for lipogenesis. Pyruvate dehydrogenase produces Acetyl-CoA. So, inhibition of PDH does not favor lipogenesis.

ANSWERS

1. b.	**2.** c.	**3.** a.	**4.** b.	**5.** c.	**6.** a.
7. a.	**8.** d.	**9.** c.	**10.** d.	**11.** d.	**12.** b.
13. a.					

13.4 METABOLISM OF CHOLESTEROL

Sterols are characterized by the presence of cyclopentanoperhydrophenanthrene nucleus. Based on their source, sterols are classified into three types.
1. Phytosterol – present in plants. They are actively excreted even if they are absorbed from the diet. Campesterol, Sitosterol, and Stigmasterol are some phytosterols.
2. Ergosterol – present in fungi
3. Zoosterol, e.g. Cholesterol is present only in animals, not in plants.

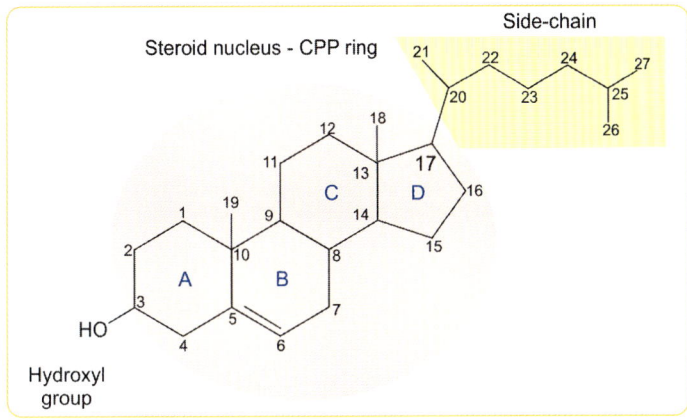

Fig. 1: Structure of Cholesterol

The structure of cholesterol has been already discussed in the Chapter 12 *(Refer to page no. 238)*.

ROLE OF CHOLESTEROL

❑ In plasma membrane:
   ○ Regulator of membrane fluidity *(Chapter 36)* *(Refer to page no. 564)*
   ○ Component of lipid rafts *(Chapter 36)* *(Refer to page no. 565)*
❑ Precursor of:
   ○ Steroid hormones
   ○ Vitamin D
   ○ Bile Acids
❑ Nerve conduction
❑ Signal transduction e.g. cholesterol is important for Hedgehog signaling.

CHOLESTEROL SYNTHESIS

Site: All nucleated cells can synthesize cholesterol
Subcellular location: Cytoplasm
Precursor: Acetyl-CoA

Chapter 13 Lipid Metabolism

273

Rate-limiting enzyme: HMG-CoA reductase (Smooth Endoplasmic membrane bound)

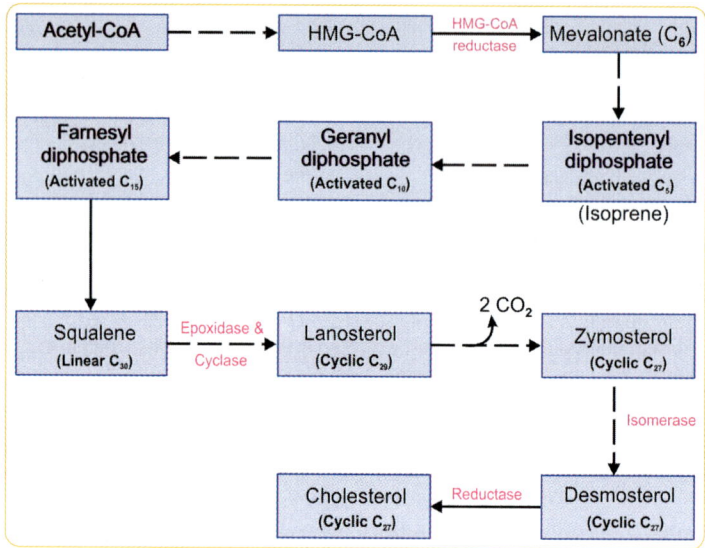

Look at the simplified schematic presentation of the cholesterol synthesis. Acetyl-coA is the precursor of cholesterol. Formation of HMG-CoA is catalyzed by HMG-CoA synthase. This is similar to ketogenesis (Chapter 13.2) *(Refer to page no. 257)*. However, this reaction is catalyzed by cytosolic isoenzyme of HMG-CoA synthase. As we have already discussed in Chapter 12 *(Refer to page no. 257)*, cholesterol is an isoprene derivative.

Statins Inhibit HMG-CoA Reductase

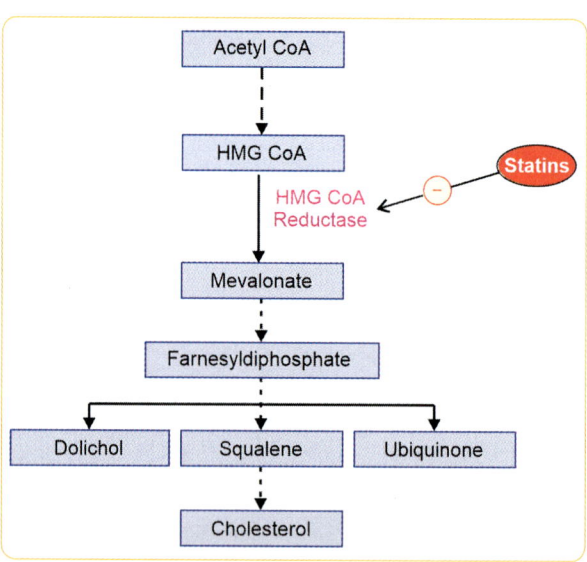

Statins (Atorvastatin, Lovastatin, etc.) are **competitive inhibitors** of the rate-limiting enzyme, HMG-CoA reductase.

Cholesterol synthetic pathway is also involved in the synthesis of many other important compounds like farnesyl pyrophosphate, Dolichol and coenzyme Q10 (ubiquinone).

- ❑ Farnesyl pyrophosphate and geranyl pyrophosphate are involved in prenylation of proteins.
- ❑ Dolichol is involved in N-glycosylation of proteins in endoplasmic reticulum.
- ❑ Ubiquinone (Q10) is the electron carrier in the mitochondrial respiratory chain.
  Inhibition of ubiquinone synthesis is one of the reasons for statin-induced myopathy.

Once heart failure is well-established, statin therapy may not be as beneficial and theoretically could even be detrimental by depleting ubiquinone in the electron transport chain. *(Ref. Harrison 20th Ed., p. 1775).*

REGULATION OF CHOLESTEROL SYNTHESIS

The rate-limiting enzyme of cholesterol synthesis is HMG-CoA reductase. We will study the regulation under the usual headings – Allosteric, Covalent and Induction/repression

Allosteric Regulation

Mevalonate the product of HMG-CoA reductase and cholesterol the ultimate product of the pathway both allosterically inhibit the enzyme (feedback inhibition).

Covalent Modification

HMG-CoA reductase is inactive in phosphorylated form. Phosphorylation is done by AMP activated protein kinase (AMPK) previously known as HMG-CoA reductase kinase.

Dephosphorylation is done by phosphatase. AMPK and phosphatase are also regulated by covalent modification.

How to Remember HMG-CoA Reductase is Inactive in Phosphorylated State?

As we have already learnt in Chapter 5 *(Refer to page no. 51)*, most catabolic enzymes are active in the phosphorylated state and most anabolic enzymes are active in the dephosphorylated state. HMG-CoA reductase is an enzyme of anabolic state. So, it will be active in dephosphorylated state, right?

Why AMP should Inactivate HMG-CoA Reductase?

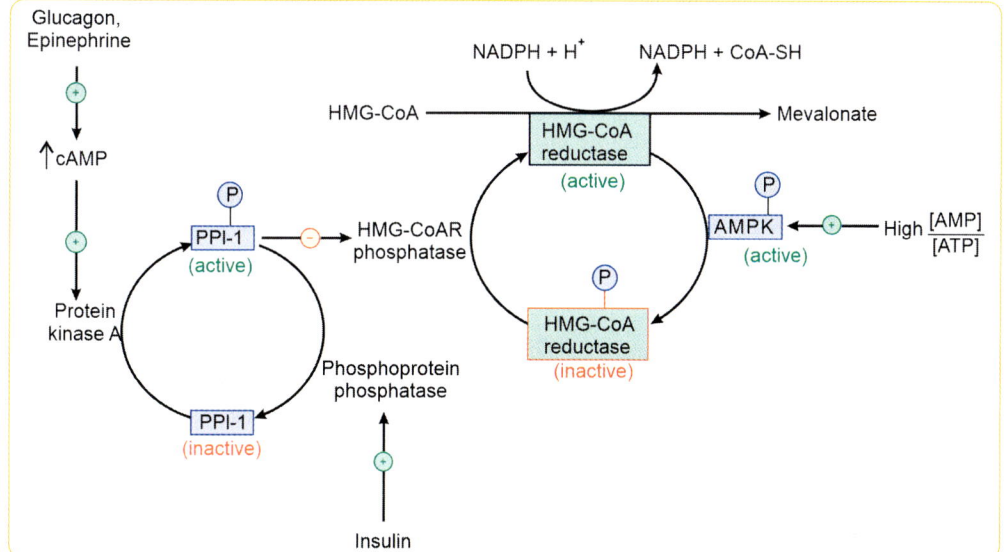

Fig. 2: Regulation of cholesterol synthesis; PKA - protein kinase A; PPI-1: phosphoprotein phosphatase inhibitor-1; HMG-CoAR – HMG-CoA reductase; AMPK: AMP activated protein kinase. Glucagon and epinephrine inhibit HMG-CoA reductase. Insulin and thyroxin activate the enzyme

Chapter 13 Lipid Metabolism

The major source of AMP is from the adenylate kinase reaction (Chapter 10) *(Refer to page no. 126)*. When ATP is low in the cell, adenylate kinase converts two molecules of ADP to one molecule of ATP and AMP. Thus, AMP level rises during low energy status inside the cell. We donot need cholesterol to be synthesized in such condition. That is why AMP activates AMPK/HMG-CoA reductase kinase.

Induction and Repression

SREBP-SCAP-Insig interaction: Transcription of HMG-CoA reductase gene is regulated by a transcription factor known as sterol regulatory element binding protein (SREBP). Binding of this protein to DNA sequences known as *sterol regulatory element* (SRE) increases the transcription of HMG-CoA reductase gene.

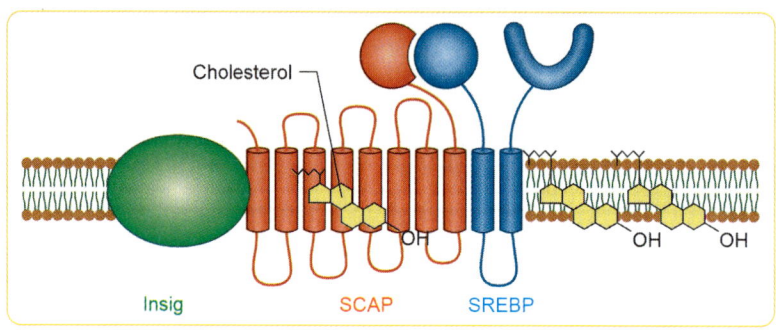

SREBP is a Positive Regulator

❑ When the amount of cellular cholesterol is adequate, SREBP is not required in the nucleus. Therefore, it is kept inactive by binding to a protein known as SCAP (SREB cleavage activating protein).
❑ SCAP is a cholesterol sensor. When the cellular cholesterol levels fall, SCAP escorts SREBP to the Golgi complex.
❑ In Golgi, SREBP is partially cleaved by proteases.
❑ The partial cleavage exposes the DNA-binding domain of the SREBP
❑ SREBP translocates to the nucleus and increases the transcription of HMG-CoA reductase.

Insig is a Negative Regulator by Two Ways

❑ First, when the cellular cholesterol level is adequate, Insig-1 (insulin-induced gene) binds to SCAP and prevent it from escorting SREBP to Golgi.
❑ Second, Insig-1 mediates the degradation of HMG-CoA reductase enzyme.
❑ *In Chapter 11.1 (Refer to page no. 175), we have seen that regulatory proteins are destroyed by ubiquitin proteasomal system.*
❑ In the presence of excess cholesterol, Insig-1 binds to the sterol-sensing domain of HMG-CoA-reductase and mediates the degradation through ubiquitin-proteasomal system. (Remember: Both HMG-CoA reductase and Insig are located on the ER membrane).

Note: Although Insig is insulin-induced gene, it is a negative regulator of cholesterol since its actions is **sterol-dependent**.

Review of Regulation of Cholesterol

Regulator	Mechanism	Effect on cholesterol synthesis
SREBP	Increases the transcription of HMG-CoA reductase (HMG-CoAR)	↑
SCAP	Escorts SREBP to Golgi for the proteolytic activation	↑
Mevalonate	Allosteric inhibitor of HMG-CoAR	↓
AMP activated protein kinase	Phosphorylation of HMG-CoAR	↓
Insig	Retains scap in ER membrane Mediates degradation of HMG-CoAR	↓

FATE OF CHOLESTEROL

In all the cells	In specific type of cells
❏ Incorporation into membrane ❏ Converted to cholesteryl esters for storage	❏ Excreted as bile acids (In hepatocytes) ❏ Conversion to steroid hormones (In steroidogenic tissues)

STEROID HORMONES

❏ Mitochondrial cytochrome P450 side chain cleavage enzyme (P450scc) a.k.a desmolase converts cholesterol to pregnenolone and isocaproaldehyde.

❏ Isocaproaldehyde is converted to propionyl-CoA and thus side chain of cholesterol is glucogenic.

❏ **Pregnenolone is the precursor of all the steroid hormones.**

❏ Steroidogenic Acute Regulatory (StAR) protein is essential for the transport of cholesterol to the inner mitochondrial membrane.

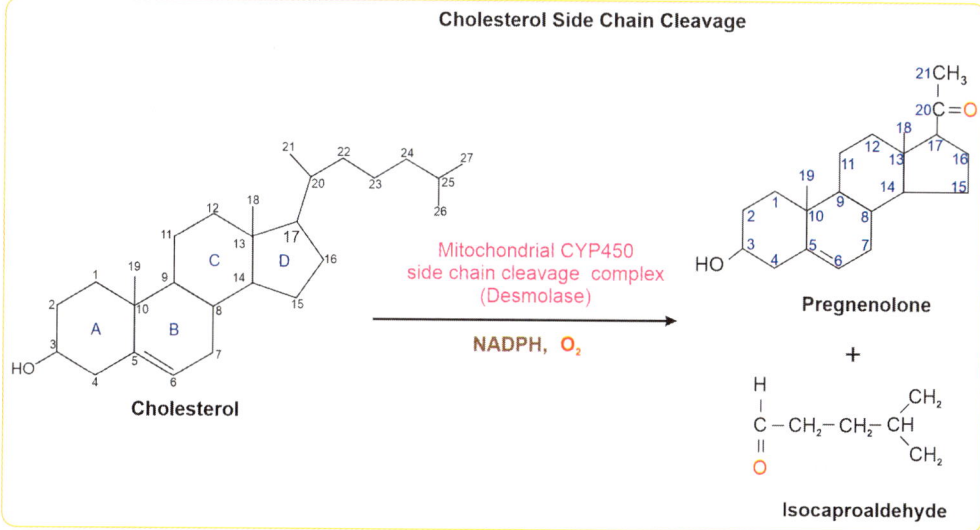

Fig. 3: ACTH – Adreno Cortico Tropic Hormone, P450scc – Cytochrome P450 side chain cleavage, Isocaproaldehyde produced by the cleavage is converted to propionyl-CoA

BILE ACIDS

❏ Bile acids are 24-carbon compounds derived from 27-carbon cholesterol in the endoplasmic reticulum of hepatocytes.

❏ 7 alpha-hydroxylase (CYP7A1), cytochrome P450 enzyme catalyzes the rate-limiting step; requires vitamin C and NADPH.

❏ 1° bile acids are synthesized in the hepatic smooth endoplasmic reticulum whereas 2° bile acids are produced from the chemical modification of primary bile acids by the **intestinal bacteria**.

❏ Conjugation of glycine or taurine (a cysteine derivative) with bile acids produces bile salts.

Chapter 13 Lipid Metabolism

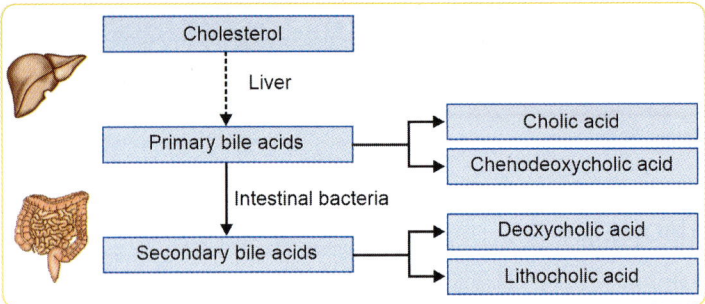

Farnesoid X Receptor (FXR) is a Bile Acid Sensor

❑ FXR belongs to nuclear receptor superfamily. Raised bile acids, especially Chenodeoxy cholic acid, binds to FXR and leads to suppression of cholesterol 7 alpha-hydroxylase (CYP7A1), the rate-limiting enzyme of bile acid synthesis.

❑ FXR agonists like obeticholic acid can be used in primary biliary cholangitis, formerly known as primary biliary cirrhosis

Clinical Correlation

Antibiotic associated Diarrhea

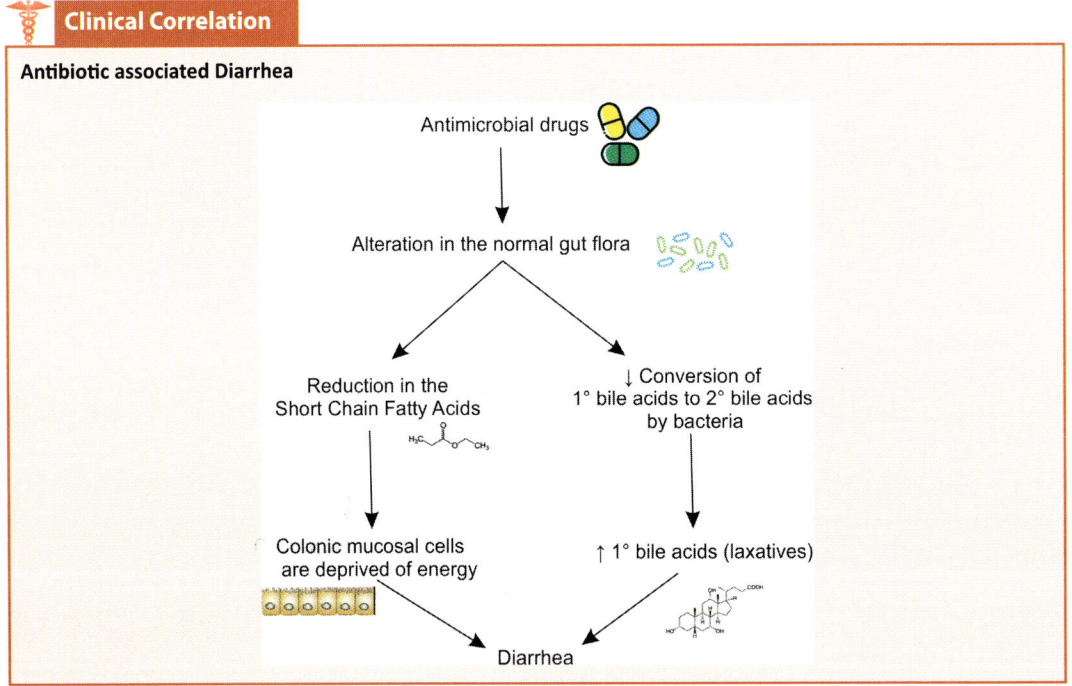

Competency

❑ BI4.2	Describe the processes involved in digestion and absorption of dietary lipids and also the key features of their metabolism

- All nucleated cells can synthesize cholesterol.
- HMG-CoA reductase catalyzes the rate-limiting step of cholesterol synthesis; active in dephosphorylated state; inhibited by statins.
- Sterol regulatory element binding protein (SREBP) is a positive regulator whereas Insig is a negative regulator of cholesterol biosynthesis.
- 7-α hydroxylase catalyzed reaction in the endoplasmic reticulum is the committed, rate-limiting step of bile acid synthesis; requires NADPH and Vitamin C.
- Cholesterol is the precursor of Bile acids, Steroid hormones and Vitamin D.
- Cholesterol, Ubiquinone, Dolichol and Vitamins A, D, E, and K, and β-carotene (provitamin A) are isoprene derivatives.

SELF-ASSESSMENT

Short Answer Questions

1. Describe the three major mechanisms of cholesterol removal in the liver.
2. State the biological functions of cholesterol.
3. How is cholesterol absorbed from the intestine, transported in blood between tissues and eliminated from our body?
4. Explain how homeostasis of the cellular cholesterol content is maintained. Name one drug that lowers plasma cholesterol.
5. Which is the rate-limiting step of cholesterol biosynthetic pathway? Mention a therapeutic drug that inhibits this enzyme. Explain how the enzyme is regulated by allosteric modulators, covalent modification and induction/repression?
6. Discuss the role of SREBP (sterol regulatory element binding protein) in the regulation of intracellular cholesterol homeostasis.
7. Name the primary and secondary bile acids. How secondary bile acids are produced from primary bile acids? With reference to bile acid synthesis, Mention the following
   a. Rate-limiting enzyme
   b. Subcellular organelle in which the reaction takes place.
   c. Vitamin cofactor required

Multiple Choice Questions

1. **All of the following are the dietary sources of cholesterol, Except:**
   a. Animal meat   b. Egg yolk
   c. Liver   d. Vegetable oil

2. **Cholesterol is the precursor of all of the following, Except:**
   a. Chenodeoxy cholic acid
   b. Steroid hormones
   c. Taurine   d. Vitamin D

3. **Which of the following is not synthesized from mevalonate?**
   a. Farnesyl pyrophosphate
   b. Geranyl pyrophosphate
   c. Dolichol   d. Ubiquitin

4. **Which of the following is NOT true for cholesterol metabolism?**
   a. HMG-CoA reductase is the key regulator of cholesterol biosynthesis
   b. Biosynthesis takes place in the cytoplasm
   c. Reduction reactions use NADH as cofactor
   d. Cholesterol is transported by LDL in plasma

5. **All of the following are true statements regarding the regulation of cholesterol synthesis, Except:**
   a. Binding of SREBP to SRE enhances the transcription of HMG-CoA reductase
   b. HMG-CoA reductase is active in phosphorylated form
   c. Insig1 is involved in the degradation of HMG-CoA reductase
   d. Mevalonate inhibits HMG-CoA reductase

6. **Which of the following is the rate-limiting enzyme of bile acid synthesis?**
   a. Microsomal 7-α hydroxylase
   b. Mitochondrial 7-α hydroxylase
   c. Microsomal 17-α hydroxylase
   d. Mitochondrial 17-α hydroxylase

Chapter 13   Lipid Metabolism

7. **The rate limiting enzyme of bile acid synthesis is a**
   a. Monooxygenase   b. Dioxygenase
   c. Reductase   d. Dehydrogenase

8. **Which of the following is the natural ligand for farnesoid X receptor?**
   a. Chenodeoxycholic acid
   b. Cholesterol
   c. Estrogen
   d. HMG-CoA

9. **Which of the following is the common precursor in the synthesis of all steroid hormones?**
   a. Pregnenolone
   b. 17-OH pregnenolone
   c. Isocaproaldehyde
   d. Androstenedione

ANSWERS

1. d.	**2.** c.	**3.** d.	**4.** c.	**5.** b.	**6.** a.
7. a.	**8.** a.	**9.** a.			

13.5 LIPOPROTEINS

Skeletal muscles and heart utilise free fatty acids at resting state. Lipids are hydrophobic and water-insoluble. How does our body make the dietary and liver-produced lipids soluble in blood? This is achieved by combining the hydrophobic lipids with amphipathic molecules like proteins and phospholipids. This transport form is known as lipoproteins.

Lipids alone – insoluble
Lipids + proteins (lipoprotein)- soluble

Based on their density, lipoproteins are classified into four major classes – namely chylomicron, very low-density lipoprotein (VLDL), low-density lipoproteins (LDL) and high-density lipoproteins (HDL). Chylomicron is the least dense whereas HDL is the densest. Albumin bound free (non-esterified) fatty acids are not usually considered as lipoprotein. Electrophoretic separation of lipoproteins is slightly different compared to the density-based separation *(shown in the illustration)*.

On serum electrophoresis, HDL, VLDL and LDL separates at alpha, pre-beta and beta regions respectively. So, HDL, VLDL and LDL are known as alpha-lipoprotein, pre-beta lipoprotein and beta lipoprotein respectively. Chylomicrons do not move from the point of sample application/origin (mentioned as immobile fraction in the illustration).

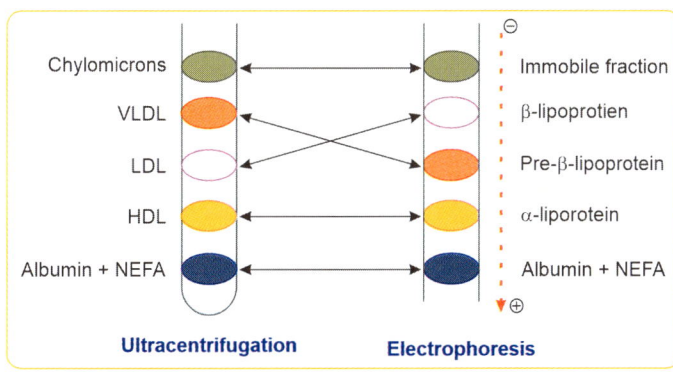

Ultracentrifugation   **Electrophoresis**

Types of lipoproteins and their role

Lipoprotein	Role
Chylomicron	Transport of dietary (exogenous) triacylglycerol from intestine through lymphatic duct to peripheral tissues.
VLDL (Very Low-density lipoprotein)	Transport of endogenous triacylglycerol from liver to peripheral tissues.
LDL (Low density lipoprotein)	Transport of cholesterol from liver to peripheral tissues.
HDL (High density lipoprotein)	Transport of cholesterol from peripheral tissues to liver (Reverse cholesterol transport).

General structure of lipoprotein

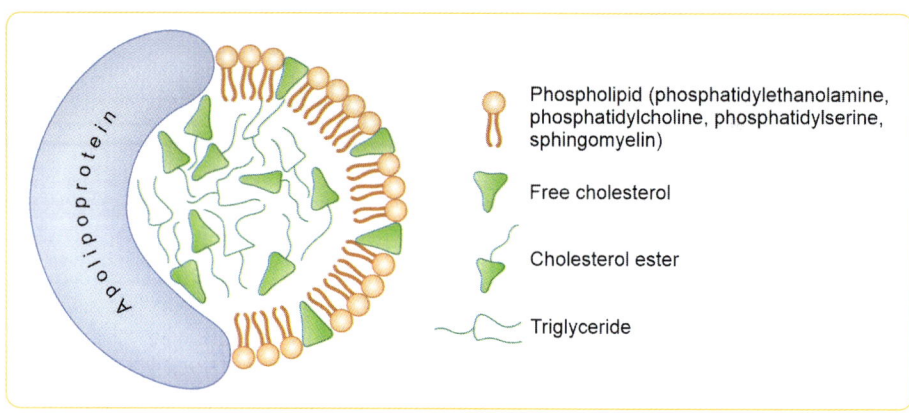

As we have already seen, lipoproteins are made of lipids (triacylglycerol, cholesterol and cholesteryl ester), amphipathic proteins and amphipathic phospholipids. In the illustration, notice the location of apolipoprotein and phospholipids. They are in the periphery, right? Amphipathic molecules have both hydrophobic and hydrophilic regions. Hydrophilic portion of the phospholipids and proteins interact with the aqueous phase of the blood. Look at the location of free cholesterol (Cholesterol contains a free -OH at 3rd position). Therefore, it is also amphipathic and located in the outer part of the lipoprotein. Look at the location of triglycerides (TAG) and cholesteryl esters (CE). They are more hydrophobic. Therefore, they can be accommodated only in the inner core of the lipoprotein molecule.

Phospholipids are very important for the formation of lipoproteins. Their hydrophobic tails interact with TAG and CE. Therefore, any defect in hepatic synthesis of phospholipids leads to accumulation of lipids in the liver (**fatty liver**). Another important thing to keep in mind is that lipoproteins are dynamic in nature. Lipids and proteins of lipoproteins constantly undergo degradation, exchange and modification, which we will study in detail further.

Remember

❑ Lipoprotein with highest triacylglycerol – chylomicrons
❑ Lipoprotein with highest cholesterol - LDL
❑ Lipoprotein with highest protein and phospholipid – HDL

APOLIPOPROTEINS

Apolipoproteins are the Protein Components of Lipoproteins

Apoprotein + Lipids = Lipoproteins

❑ Most of the apolipoproteins are synthesized in liver. They serve both structural and functional roles in the metabolism of lipoproteins.

- Apolipoproteins are named based on the electrophoretic mobility of lipoproteins.
  - Apoproteins of β-lipoprotein (LDL) and pre β-lipoproteins (VLDL) are named ApoB.
  - Apoproteins of α lipoprotein (HDL) are named ApoA.
  - Subtypes are indicated with Roman numerals, e.g. Apo AI and Apo AII whereas the nomenclature of apoB48 and B100 is a special case *(the reason will be explained in Chapter 23)* *(Refer to page no. 408)*.
- Defect in the synthesis of apolipoproteins leads to accumulation of lipids in the liver (fatty liver).
- Defect in one type of apoprotein leads to defective metabolism of specific type(s) of lipoprotein.
- LDL is the only lipoprotein with a single type and single number of apoprotein, i.e. apoB100.
- All other lipoproteins contain more than one type and many numbers of apoproteins. Some are assembled when they are made, and others are acquired during their circulation in blood.
- **HDL acts as a reservoir of apoproteins** for other lipoproteins.

Apoprotein	Component of	Function
Apo A-I	HDL	Activates the enzyme Lecithin: Cholesterol acyl transferase (LCAT)
Apo B-48	Chylomicron	Structural component
Apo B-100	LDL, VLDL, IDL	Ligand for LDL receptor Structural component of VLDL, IDL and LDL
Apo C-II	Chylomicron and VLDL	**Activates** lipoprotein lipase (LPL)
Apo C-III	VLDL	**Inhibits** lipoprotein lipase
Apo D	HDL	Different from other apoproteins. Synthesised in brain and testes. Associated with neurological dysfunction and is a biomarker of androgen insensitivity syndrome.
Apo E	VLDL, chylomicron remnants, and a subset of HDL	Ligand for hepatic remnant receptor Helps in hepatic uptake of VLDL and Chylomicron APO E*4 allele is a risk factor for Alzheimer's disease.

CHYLOMICRONS

Chylomicron Transports the Dietary Triacylglycerols and Cholesterol

- Dietary lipids are packaged into chylomicrons (CM) in the intestinal smooth endoplasmic reticulum and secreted into lacteals (lymphatic vessels).
- These lacteals form thoracic duct, which drains into the systemic circulation at the level of left subclavian vein. Ninety percent of chylomicrons is triglycerides. Chylomicrons also transport fat-soluble vitamins.
- **That is why any defect in intestinal digestion (e.g. biliary obstruction) of fat leads to deficiency of fat-soluble vitamins**.
- ApoB48 is the structural protein of chylomicron. Microsomal triglyceride transfer protein (MTP) is required for the assembly of chylomicron and other apoB containing lipoproteins.
- Newly secreted chylomicrons from intestine are known as 'nascent chylomicrons'. Only apoB48 and apo A are present in the nascent chylomicron.
- Apo CII and apo E are acquired from HDL molecule.
- Chylomicron provides fatty acids for adipocytes, skeletal and cardiac muscles in the post absorptive state.
- Fatty acids from triglycerides are released by the enzyme lipoprotein lipase (LPL) located in the capillary endothelium of these organs.
- LPL is anchored to the capillary endothelium via heparan sulphate. Apo CII is the cofactor required for the enzyme.
- The fate of free fatty acids released by the LPL depends upon the tissue.
  - In muscles, fatty acids are used for energy production (muscle at rest utilize free fatty acids as fuel).
  - In adipocytes, fatty acids are reesterified and stored as TAG for future need.

After the action of LPL, the amount of TAG in chylomicron is decreased and it becomes 'chylomicron remnant'. These remnants are taken up by hepatic remnant receptors (LDL receptor related protein and LDL receptors) for which ApoE act as a ligand. Regulation of LPL activity can be easily remembered in this way: Chylomicrons

Chapter 13   Lipid Metabolism

are produced in the fed state. Insulin is also produced in fed state. Therefore, insulin enhances the LPL activity. Phospholipids and Apo CII are activators of LPL whereas apo AII and apo CIII are inhibitors.

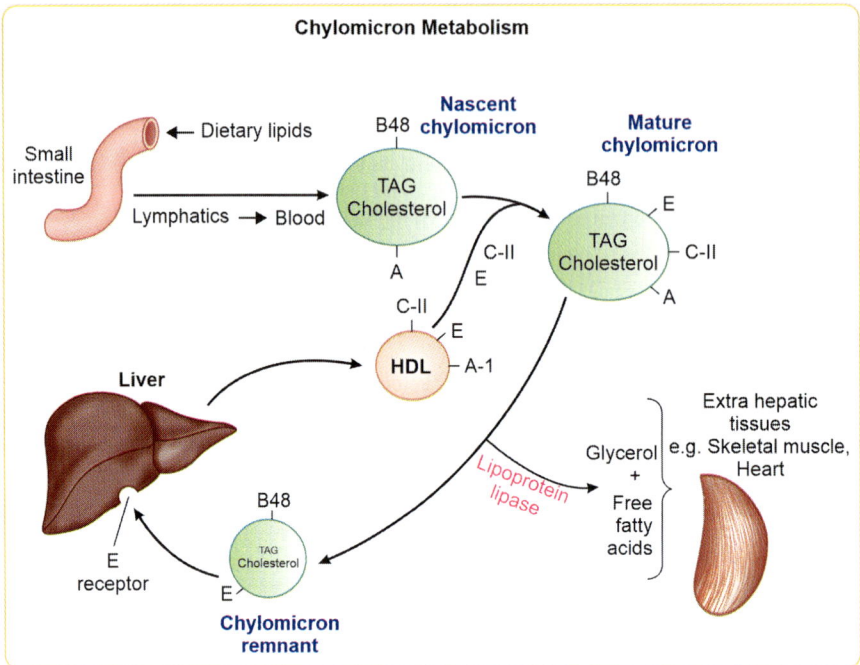

Chylomicron is a short-lived molecule. It is cleared from the circulation within an hour of its production. In the blood, chylomicron stays for just 5 to 10 minutes. **This is why chylomicrons are not seen on serum electrophoresis of a fasting person.**

Defect in Lipoprotein Lipase or apo CII Leads to Familial Hyperchylomicronemia

- ❑ LPL is required for the breakdown of chylomicrons.
- ❑ Any defect in lipoprotein lipase or its cofactor apo CII leads to defective clearance of chylomicrons and accumulation of chylomicrons and triglycerides.
- ❑ LPL deficiency manifests in the first decade of life whereas apo CII defect manifests a little later.

Three Major Features

Clinical feature	Biochemical basis
Abdominal pain due to Pancreatitis	Capillary obstruction by accumulated chylomicrons
Lipemia retinalis	In the absence of LPL activity, chylomicrons persisting in the **systemic** circulation are visualized in fundus examination.
Eruptive xanthoma (xantho-yellow)	Deposition of triglycerides in the skin and subcutaneous tissues leads to **painless**, yellow papules known as xanthomas. LPL deficiency leads to sudden appearance of these kind of papules and hence the name eruptive xanthoma. Eruption occurs when the TAG level exceeds 2000 mg/dL.

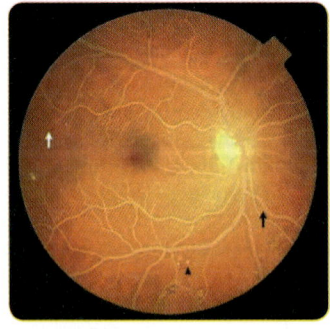

Fig. 1: Lipemia retinalis

Note: LPL defect per se does not increase the risk of cardiovascular disease.

VLDL

VLDL is the Transporter of Endogenously Produced Triglycerides and Cholesterol

❑ Fatty acids synthesized in liver are esterified with glycerol to produce triglycerides and packed with apoB100, phospholipids and cholesteryl esters and exported in the form of VLDL.

❑ As VLDL is rich in TAG, the density is very low. (Chylomicrons have the highest amount of TAG and hence the lowest density).

❑ Similar to chylomicron formation, VLDL formation also requires Microsomal triglyceride transfer protein (MTP). So, defect in MTP leads to absence of all apoB containing lipoproteins (abetalipoproteinemia) which we will discuss soon.

❑ Nascent VLDL has only one molecule of apoB100. Apo CII and apo E are obtained from HDL.

❑ In circulation, VLDL loses TAGs in two ways:

1. Similar to chylomicrons, VLDL is acted upon by the LPL enzyme.
2. VLDL gives its TAG and takes cholesteryl ester from HDL in the reaction catalyzed by cholesteryl ester transfer protein (CETP) present on the HDL molecule.

We will discuss CETP during the discussion of HDL.

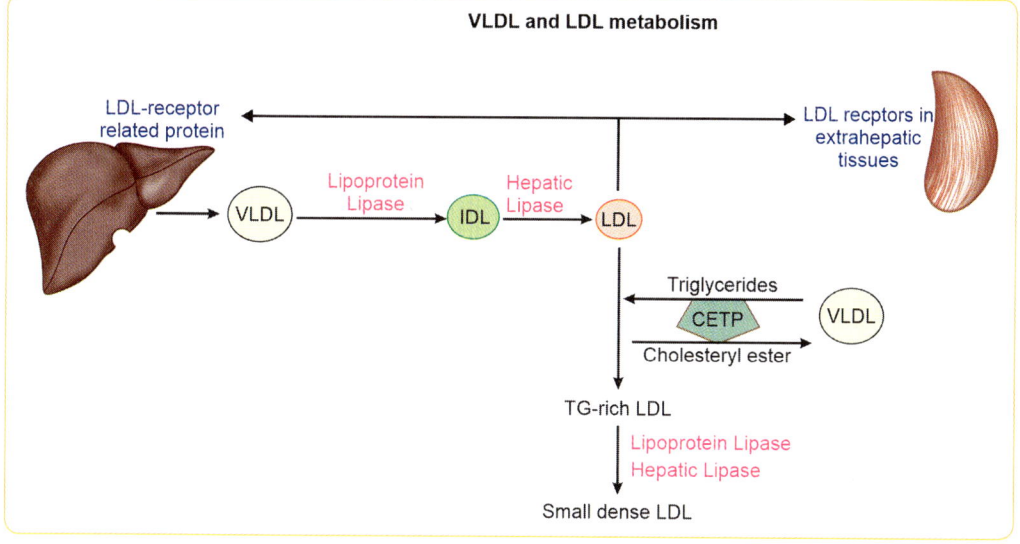

VLDL and LDL metabolism

❑ Thus, TAG content of the VLDL is reduced and the cholesteryl ester content is increased. This molecule is known as Intermediate Density Lipoprotein (IDL).

❑ In liver, IDL molecule is acted upon by hepatic lipase and TAG content is further reduced to produce cholesterol-rich low-density lipoprotein. (Hepatic lipase converts IDL to LDL)

❑ Not all LDL molecules are derived from IDL. Some amount of LDL molecules can be directly produced from liver.

At this junction, it is important to differentiate Lipoprotein lipase and hepatic lipase.

	Lipoprotein lipase	Hepatic lipase
Location	Adipocytes Striated muscles	Liver
Cofactor needed	Phospholipids and Apo CII	None
Lipoprotein substrate	Chylomicron, VLDL	VLDL, HDL
Role	Breakdown of triglycerides in chylomicron and VLDL	❑ Conversion of IDL to LDL ❑ Conversion of HDL₂ to HDL₃ i.e. *post-prandial* triglyceride-rich HDL to the *post-absorptive* triglyceride-poor HDL.

LDL

LDL is the Transporter of Cholesterol from Liver to Peripheral Tissues

We have studied how LDL is formed from IDL and learnt that some LDL can be formed directly.
- ❑ ApoB100 of LDL is unique - there is only one apoB100 in an LDL molecule and it is not exchangeable with other lipoproteins.
- ❑ **LDL is the lipoprotein with highest amount of cholesterol**.
- ❑ LDL is the source of cholesterol for extrahepatic tissues. *(All nucleated cells can produce some amount of cholesterol)*
- ❑ LDL is the major source of cholesterol for steroidogenic tissues like testis, ovary and adrenal gland.
- ❑ LDL molecule is taken up by the LDL receptors (ApoB100/ApoE receptor) located in liver and extra-hepatic (peripheral) tissues.
- ❑ ApoB100 of LDL serves as a ligand. **Only 30% of the LDL molecule is taken up by peripheral tissues, 70% is taken back by the liver.**

Intake of LDL Through LDL Receptor is a Regulated Process

- ❑ LDL particle is taken up by LDL receptor through receptor-mediated endocytosis.
- ❑ LDL receptor is a glycoprotein receptor encoded by chromosome 19. It is located in the clathrin coated pits of the cell membrane.
- ❑ LDL receptor is found in all nucleated cells, major amount is seen in liver, which is responsible for clearance of 70% of LDL from the circulation.
- ❑ Endocytosed LDL fuses with lysosome and the LDL receptor dissociates due to lower pH. After dissociation, the **receptor is recycled to the cell surface when PCSK9 is not present.**
- ❑ PCSK9 (proprotein convertase subtilisin kexin 9) a protease which binds to LDL receptor and degrades the LDL receptor inside the lysosomes. This prevents the recycling of LDL receptor.
- ❑ Certain virus (hepatitis C virus, vesicular stomatitis virus) can associate themselves with LDL molecules and enter the cell through LDL-receptor.
- ❑ Goldstein and Brown were awarded the 1985 Nobel Prize in Physiology or Medicine for their identification of LDL receptor gene and its relation to familial hypercholesterolemia.

Intake of Oxidized LDL Through Scavenger Receptor is Unregulated and Accelerates Atherosclerosis

- ❑ The half-life of LDL particle is around 3 days. This considerable time in the circulation increases the chance of oxidation of lipids and protein (apoB100) of LDL.
- ❑ HDL *(which we will discuss shortly)* prevents the oxidation of LDL.
  - ○ HDL associated enzymes like **Paraoxonase** prevent lipid peroxidation.
  - ○ Thus, **reduction in the level of HDL increases the chance of LDL getting oxidized**.
- ❑ **Oxidized LDL is a ligand for scavenger receptors (SR-A1)** on macrophages. This uptake is not a regulated process and thus oxidized LDL accumulates in macrophages to produce lipid-laden, foamy macrophages (foam cells).
- ❑ Foam cells secrete inflammatory cytokines and promotes the formation of atherosclerotic plaque.

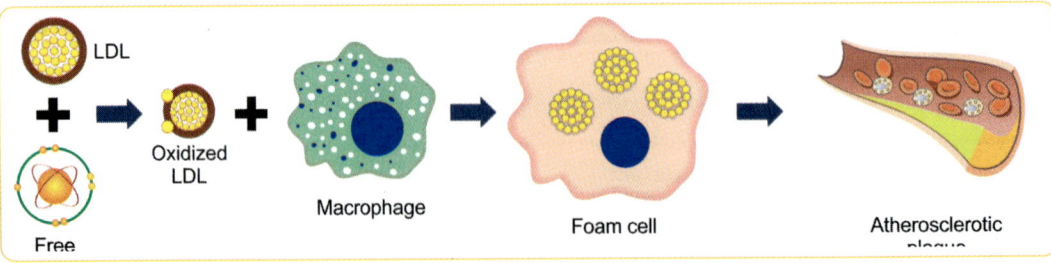

LDL | Oxidized LDL | Macrophage | Foam cell | Atherosclerotic plaque
Free

Defective LDL Clearance Leads to Familial Hypercholesterolemia (FH)

❑ Familial hypercholesterolemia (FH) is an **autosomal dominant** disease due to defect in either the LDL receptors or the ligand-binding region of apo B-100. This leads to accumulation of LDL molecules.
❑ **Clinical Features:**
  ○ Highly elevated level of cholesterol increases the risk of coronary heart disease.
  ○ Tendinous xanthoma: Xanthomas develop in the tendon especially the Achilles tendon.
  ○ Xanthelasma: Yellowish deposit of cholesterol underneath the skin, usually on or around the eyelids).
❑ Manifestation of the disease in homozygous mutation is severe and affected individuals succumb to the disease unless rigorous treatment is administered.

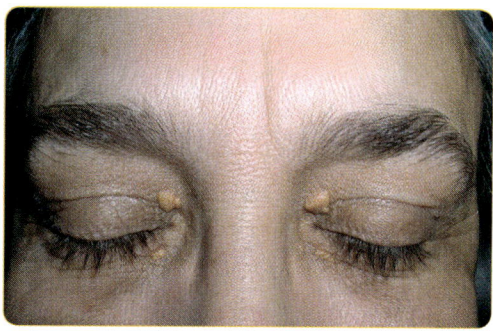

Xanthelasma palpebrarum
Fig. 2: By Klaus D. Peter, Gummersbach, Germany [CC BY 3.0 de (https://creativecommons.org/licenses/by/3.0/de/deed.en)], from Wikimedia Commons

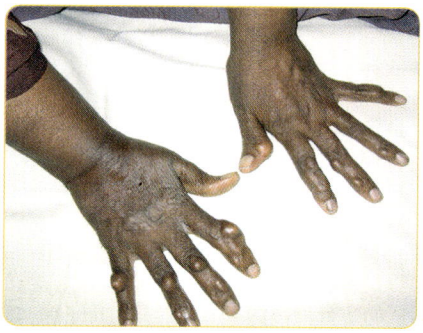

Tendon Xanthoma
Fig. 3: By Anita A Kumar, et al (CC BY 2.0), via Wikimedia Commons

In addition to lifestyle changes, familial hypercholesterolemia requires rigorous drug treatment. Biochemical basis of lipid-lowering therapy is given in the table as follows:

Hypolipidemic drug	Mechanism of action
Statins	Competitive inhibitor of **HMG-CoA reductase**
Ezetimibe	Inhibits intestinal cholesterol absorption **(inhibits Niemann-Pick C1–like 1 -NPC1L1)**
Cholestyramine (Bile acid binding resin)	Excretion of bile acids in the stool
Fibrates (e.g. Fenofibrate)	PPARα activator ($\downarrow$TAG synthesis in liver)
Niacin	$\downarrow$ lipolysis in adipocytes; $\downarrow$ VLDL; $\uparrow$HDL
Evolocumab	Monoclonal antibody against PCSK9
Mipomersen	Antisense RNA for apolipoprotein B-100 mRNA
Lomitapide	Microsomal triglyceride transfer protein (MTP) inhibitor

Microsomal Triglyceride Transfer Protein Defect Leads to Abetalipoproteinemia

❑ Microsomal triglyceride transfer protein (MTP) is needed for packaging of triglycerides (TAG) into apo B containing lipoproteins.
❑ Absence of MTP leads to absence of all beta-lipoproteins (Chylomicron, VLDL and LDL) in the blood, i.e. abetalipoproteinemia a.k.a Bassen-Kornzweig syndrome.
❑ Blood cholesterol level is low, and TAGs get accumulated in the liver and intestine.
❑ As chylomicrons are absent, there will be severe fat-soluble vitamin deficiency.
❑ **Clinical Presentation:**
  ○ Accumulation of TAG in the intestine leads to malabsorption.
  ○ Alteration in membrane lipids lead to RBCs with spine like projections – acanthocytosis.

○ Neurological symptoms include ataxia and retinitis pigmentosa
○ Fat-soluble vitamins should be supplemented to prevent early death.

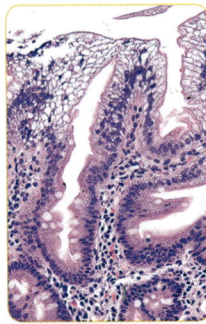

Fig. 4: Duodenal biopsy showing enterocytes with a clear cytoplasm (due to lipid accumulation) characteristic of abetalipoproteinemia

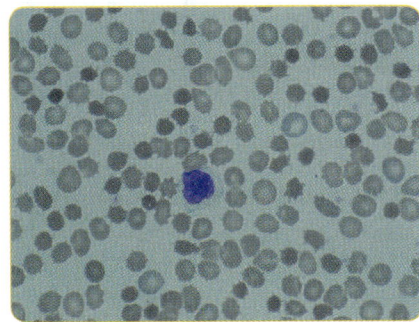

Fig. 5: Acanthocytes

HDL

HDL is the Transporter of Cholesterol from Peripheral Tissue to Liver

❏ HDL molecule is produced by the liver and intestine. ApoAI is the structural protein of HDL. It also contains apoC and apoE.

❏ Nascent HDL produced from intestine does not contain apo C and apo E whereas nascent HDL produced from liver contains both. Hepatic nascent HDL acts as a reservoir of apolipoproteins for intestinal nascent HDL and other lipoproteins.

❏ Nascent HDL, a.k.a discoidal HDL is nothing but an Apo AI containing phospholipid bilayer with some free cholesterol.

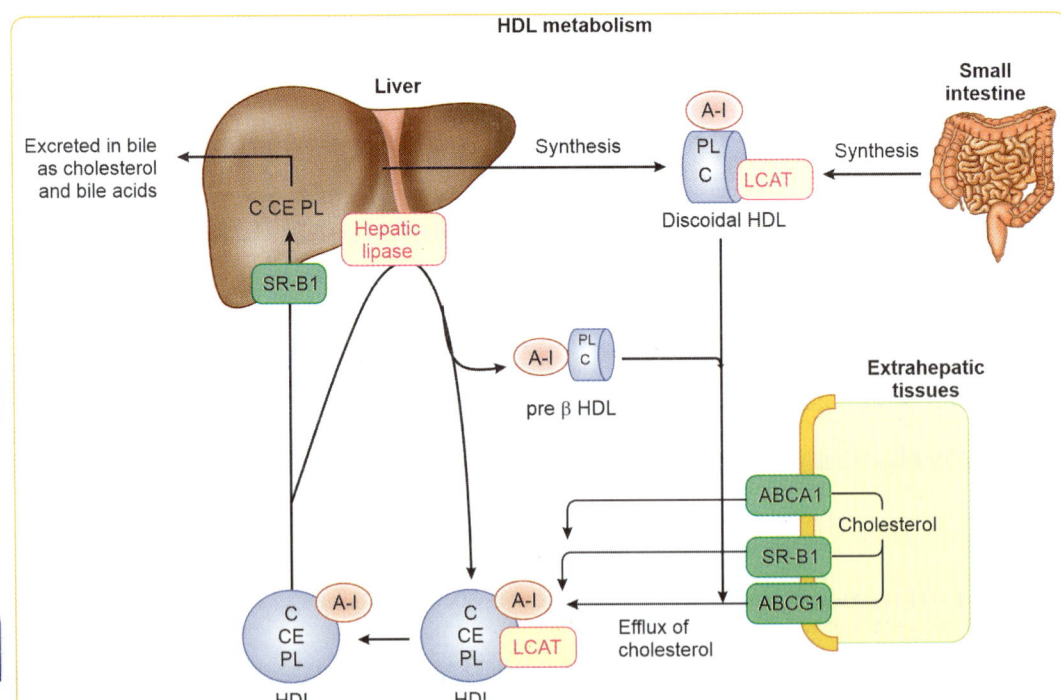

Role of LCAT

- Free cholesterol is amphipathic. Therefore, it occupies the outer portion of HDL.
- To move it into the interior of HDL molecule, it needs to be made hydrophobic by adding a fatty acid to the hydroxyl group of cholesterol.
- Lecithin Cholesterol acyltransferase (LCAT) is an enzyme produced by liver which binds to HDL in a reversible manner.
- LCAT transfers the fatty acid from lecithin (phospholipid present in HDL) to cholesterol and thus produces cholesteryl ester and lysolecithin.

Cholesterol + Lecithin → Cholesteryl Ester + Lysolecithin

- Cholesteryl esters are more hydrophobic than cholesterol and move inside the core of HDL. Thus, LCAT action converts the discoidal HDL to spherical HDL (a.k.a HDL_3) with a non-polar core.
- Accumulation of cholesteryl ester inside the HDL can cause feedback inhibition of LCAT.

Role of CETP

- The above-mentioned feedback inhibition is relieved by CETP protein.
- Cholesteryl ester transfer protein catalyzes the transfer of cholesteryl ester from HDL to VLDL, IDL and LDL in exchange for triacylglycerol.
- Therefore, Cholesteryl ester content in HDL is reduced and thus product inhibition of the LCAT enzyme is relieved.
- Inhibition of CETP was considered as a therapeutic way to increase the level of HDL. However, the development of CETP inhibitor drugs was stopped because of premature death in study subjects caused by the drug.

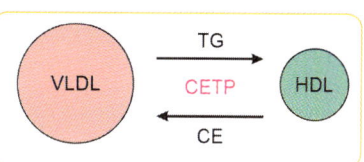

Fig. 6: Action of Cholesteryl ester transfer protein. CE- cholesteryl ester, TG-triglyceride.

LCAT Deficiency

We have studied that LCAT is required for esterification of free cholesterol in the discoidal HDL. Deficiency of LCAT leads to accumulation of discoidal HDL and accumulation of free cholesterol. Accumulation of LCAT deficiency can be partial or complete.

	Complete LCAT deficiency	Partial LCAT deficiency
Also known as	Familial LCAT deficiency or **Norum's disease**	Fish eye disease
Corneal opacification	+	-
Hemolytic anemia	+	-
End-stage renal disease	+	-

Subtypes of HDL

Based on the composition, HDL is divided into 3 subtypes – HDL_1, HDL_2 and HDL_3.

Subtype	Description
HDL_3	Spherical HDL formed from nascent discoidal HDL by the action of LCAT
	A-I / PL / C Nascent discoidal HDL → LCAT → HDL_3 → LCAT → HDL_2
HDL_2	Spherical, Less dense, smaller than HDL_3. Formed from HDL_3 by the action of LCAT, more variable component of total HDL, responsible for the anti-atherogenic effect
HDL_1	Minor fraction of total HDL with apoE as the major component. *Don't confuse HDL1 with nascent HDL. Both are different.*

In our body, there is a constant conversion of HDL_3 to HDL_2 and vice versa mediated by LCAT, hepatic lipase and CETP. This is known as **HDL cycle**.

Reverse Cholesterol Transport

LDL mediates the transport of cholesterol from liver to peripheral tissues. HDL mediated return of cholesterol from extrahepatic tissues to the liver is known as reverse cholesterol transport.

Major proteins and Enzymes involved in reverse cholesterol transport:
- ABCA1 and ABCG1 *(ABC stands for ATP-binding Cassette Protein)*
- Lecithin: cholesterol acyltransferase (LCAT)
- SR-B1

Receptor	Ligand
LDL receptor	apo B-100 and apo E
LDL-receptor-related protein-1 (LRP-1), a.k.a. remnant receptor	apo E
HDL receptor	apo A-I
Scavenger receptor A1	oxidised LDL

LIPOPROTEIN (a)

- Lipoprotein (a) is made up of a specific apoprotein apo (a) which is linked to ApoB of LDL particle via disulphide bond.

$$\text{i.e. } Lp(a) = LDL + apo\,(a)$$

- Apo(a) contains kringle domains that are present in plasminogen also.

  Because of the presence of Apo(a), Lp(a) interferes with the activity of plasminogen and hence it is thrombogenic.

  Number of kringle domains are determined by genetic polymorphism.

  Video lecture on Lp(a): ▶ youtu.be/VTqOg_Sr7rc or scan QR code

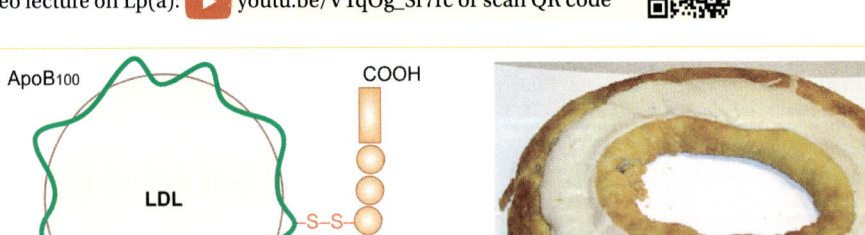

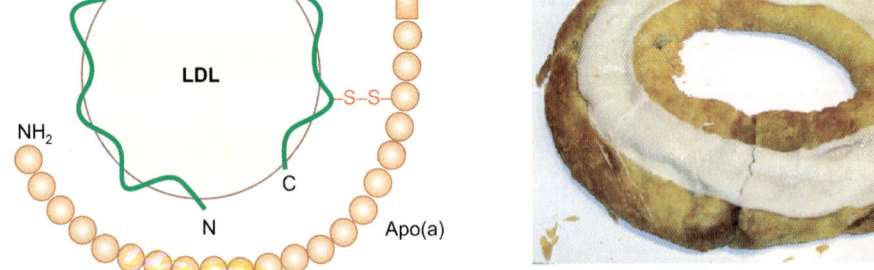

Kringle pastry

Fig. 7: Structure of lipoprotein (a) – Apo (a) is linked to ApoB via disulphide linkages

LIPOPROTEIN X

Lipoprotein-X (Lp-X) is an abnormal low-density lipoprotein found in patients with **cholestasis**. It is rich in phospholipids and poor in proteins.

HYPERLIPOPROTEINEMIAS

❏ Hyperlipoproteinemias can be primary (inherited defect in lipoprotein metabolism) or secondary due to other conditions.

Primary Hyperlipoproteinemias

❏ Frederickson's classification is a **phenotypic classification** based on the appearance of fasting plasma sample after keeping the sample standing for 12 hours at 4°C and the analysis of lipid profile of the same sample.

In the table below, Frederickson's type is mentioned in Roman numeral and the genetic nomenclature is given within bracket.

Type	Lipoprotein, elevated	Plasma appearance	Molecular defects	Clinical feature
I (Familial chylomicronaemia syndrome)	Chylomicrons	Lactescent (milk-like)	LPL and Apo C-II	Eruptive xanthoma, pancreatitis, lipemia retinalis, No risk of CAD
IIa (Autosomal dominant Familial hypercholesterolemia and others)	LDL	Clear	LDL receptor	Tendon xanthoma, coronary artery disease (CAD), premature death
IIb (Familial combined hyperlipidemia)	LDL and VLDL	Clear	Apo C-III	Associated with other risk factors like obesity, glucose intolerance, insulin resistance, and hypertension. Risk of premature CAD
III (Familial dysbetalipoproteinemia or Broad beta disease)	Chylomicron and VLDL remnants	Turbid	Apo E	Palmar, Tuberoeruptive xanthomas, risk of coronary artery disease
IV (Familial hypertriglyceridemia)	VLDL	Turbid	Apo A-V	No xanthomas CAD risks +/−
V (Familial hypertriglyceridemia, a.k.a Mixed hyperlipidemia)	Chylomicrons and VLDL	Lactescent	Apo A-V and GPIHBP1	Eruptive xanthoma, Pancreatitis, CAD risk +/−

Secondary forms of Hyperlipidemia

In addition to the defects in genes directly involved in lipoprotein metabolism, hyperlipidemia can also be caused by some other life-style related conditions and pathological disorders.

Conditions	Major lipid elevated
Obesity, Diabetes mellitus type 2, Von-Gierke disease, Cushing syndrome	VLDL
Hypothyroidism, Nephrotic syndrome, Cholestasis	LDL
Multiple myeloma	IDL
Alcohol*, Exercise, Estrogen therapy	HDL

*Alcohol is a teratogen and carcinogen.

HYPOLIPOPROTEINEMIAS

❏ Abetalipoproteinemia *(explained earlier)*
❏ Familial hypobetalipoproteinaemia – due to apoB gene mutation
❏ Tangier's disease (Familial alpha lipoprotein deficiency)

Tangiers Disease (Fish Eye Disease)

- ❑ Autosomal recessive disorder due to mutations in *ABCA1* gene.
- ❑ Extremely marked reduction in HDL-Cholesterol and apoA-I
- ❑ Markedly accelerated catabolism of apoA-I and apo A-II
- ❑ C/F: Enlarged orange tonsils (due to dietary carotenoid deposition), Hepatosplenomegaly and Peripheral neuropathy

LIPOTROPIC FACTORS

Liver is not the primary organ for storage of fat. Accumulation of fat (TAG) in liver is known as fatty liver. Lipotropic factors prevent the development and progression of fatty liver. Choline, Methionine, Betaine, Vitamin E and Selenium are the lipotropic factors.

Lipotropic Factor	Role
Choline (Trimethyl ethanolamine)	Synthesis of phosphatidyl choline (lecithin) → lipoprotein assembly and removal of triglycerides
Methionine	S-Adenosyl Methionine is the methyl donor for the synthesis of choline
Selenium and Vitamin E	Prevents lipid peroxidation

Lipotropic factors - CBSE-Modi - Choline, Betaine, Selenium, Vit E and Methionine

LABORATORY ESTIMATION OF LIPID PROFILE

Lipid profile is a panel of blood tests done to assess the risk of a person developing coronary artery disease.

Precaution for Sample Collection

- ❑ Blood sample should be collected after an overnight fast of at least 12 hours.
- ❑ Triglyceride value is the one, which is affected by non-fasting conditions. As triglyceride value is used to indirectly calculate the LDL value, this will influence the LDL value also.
- ❑ Patient should be on a stable diet for the past 2-3 weeks.
- ❑ There should not be any major illness or surgery in the past 2 weeks, as low values are obtained in these conditions.
- ❑ Freshly separated plasma or serum should be used if possible.

In most clinical laboratories, the total cholesterol and TAGs in the plasma are measured using enzymatic methods, and then the cholesterol in the supernatant is measured after precipitation of apoB-containing lipoproteins to determine the HDL-Cholesterol.

In most of the laboratories, LDL is not directly estimated. It is calculated using the **Friedwald formula.**

$$\text{LDL Cholesterol} = (\text{Total Cholesterol}) - \{(\text{HDL Cholesterol}) + (\text{TAG}/5)\}$$

This formula assumes that 20% of total triglycerides is carried by VLDL. But this formula **cannot be used when TAG level exceeds 400 mg/dL.**

Optimum Lipid Profile

Parameter	Desirable range
Total Cholesterol	<200 mg/dL
Triglycerides	<150 mg/dL
HDL-Cholesterol	>40 mg/dL
LDL-Cholesterol	<100 mg/dL

Competency

Key Points

- Triglyceride content is maximum in chylomicron.
- Chylomicron is the transporter of dietary (exogenous) triglycerides; It is the lipoprotein with least electrophoretic mobility.
- Lipoprotein lipase is attached to the capillary endothelium of adipose tissue, heart, skeletal muscle and lactating mammary gland by heparan sulfate. Injection of heparin causes release of this enzyme from heparan sulfate.
- VLDL is the transporter of endogenous triglycerides from liver to peripheral tissues.
- HDL is the mediator of reverse cholesterol transport from peripheral tissues to liver.
- LDL receptor is a glycoprotein receptor coated with clathrin.
- Molecular mechanism behind the generation of apo B-48 (in intestine) and apo B-100 (in liver) from same ApoB gene is RNA editing (through deamination of cytosine to uracil of mRNA).
- ABCA1 and ABCG1, Lecithin: cholesterol acyltransferase (LCAT) and SR-B1 are involved in reverse cholesterol transport.
- Lipoprotein-X (Lp-X) is an abnormal low-density lipoprotein found in cholestasis.
- Apo B of Lp(a) has structural similarity with plasminogen i.e. both contain kringle domain.
- Excess triglycerides produce soft, eruptive xanthomas whereas excess cholesterol produces hard, tendon xanthomas.
- APOE*4 allele is a risk factor for Alzheimer's disease.

SELF-ASSESSMENT

Short Answer Questions

1. Classify lipoproteins and write the function of each.

2. Discuss the metabolism of plasma lipoprotein which has a pathogenic role in the development of atherosclerosis.

3. Draw a schematic diagram of the structure of Lp(a). What does 'a' in Lp(a) denote? Why persons with a high level of Lp(a) are prone to coronary artery disease?

4. Justify the statement – "HDL cholesterol has a preventive role in atherosclerosis"

5. What are lipotropic factors? Name any two.

6. What is reverse cholesterol transport? Draw a schematic diagram and name the enzymes and proteins involved in this process.

7. Discuss how the triacylglycerols derived from the diet are transported in the blood and taken up by the tissues.

8. Name any four hypolipidemic drugs and mention the mechanism of action of each.

9. What is the role of cholesteryl ester transfer protein (CETP) in the lipoprotein metabolism?

10. What is the defect in abetalipoproteinemia? Mention the clinical presentation.

11. Name the major proteins and enzymes involved in reverse cholesterol transport and discuss the role of each.

12. Name any four causes of secondary forms of hyperlipidemia.

13. Name a disease that presents with hypolipidemia and orange tonsils. What is the enzymatic defect?

Multiple Choice Questions

1. During lipoprotein fractionation using ultracentrifugation, the uppermost layer will consist of
   a. HDL
   b. LDL
   c. VLDL
   d. Chylomicron

2. ApoB48 is the apoprotein present in
   a. Chylomicron
   b. VLDL
   c. LDL
   d. HDL

3. All of the following is True about apolipoproteins EXCEPT:
   a. Apo A-I is the major protein component of HDL
   b. Constitute the peripheral region of plasma lipoproteins
   c. Divided into A, B and C only
   d. Apo A, B and C are divided into further subtypes

4. Which of the following is NOT critical for optimal HDL function?
   a. LCAT
   b. Lipoprotein lipase
   c. CETP
   d. Apolipoprotein A

5. Lipoprotein with least electrophoretic mobility is
   a. Chylomicrons
   b. VLDL
   c. LDL
   d. HDL

6. The maximum triglyceride content is found in
   a. Chylomicrons
   b. VLDL
   c. LDL
   d. HDL

7. HDL is rich in
   a. Triglycerides
   b. Cholesterol
   c. Apolipoproteins
   d. Phospholipids

8. Apoprotein(s) present in a significant amount in nascent chylomicron is/are
   a. apo B48
   b. apo CII
   c. apo E
   d. All of the above

9. The source of apo E and apo CII for chylomicrons is
   a. VLDL
   b. IDL
   c. LDL
   d. HDL

10. All of the following statements are true, EXCEPT
    a. Chylomicrons are involved in the transport of exogenous triglycerides
    b. VLDL is involved in the transport of endogenous triglycerides
    c. LDL is involved in the transport of dietary triglycerides
    d. HDL is involved in the reverse cholesterol transport

11. Which of the following is NOT involved in reverse cholesterol transport?
    a. LCAT
    b. SRB1
    c. ABCA1
    d. Lipoprotein lipase

12. Which of the following lipase is regulated by glucagon?
    a. Lipoprotein lipase
    b. Pancreatic lipase
    c. Gastric lipase
    d. Hormone sensitive lipase

13. A person on fat free, carbohydrate rich diet continues to get obese. Which of the following lipoprotein is likely to be elevated in his blood?
    a. Chylomicron
    b. VLDL
    c. LDL
    d. HDL

14. All of the following are true, EXCEPT
    a. Fish-eye disease is due to partial deficiency of LCAT
    b. Norum disease is due to complete LCAT deficiency
    c. Tangier disease is due to ATP binding cassette transporter 1 deficiency
    d. Barth syndrome is due to accumulation of cardiolipin

15. Lecithin cholesterol acyl transferase is activated by
    a. Apo AI
    b. Apo B48
    c. Apo B100
    d. Apo C

16. Familial hypercholesterolemia is due to
    a. Defective lipoprotein lipase
    b. Defective LDL receptors
    c. Defective scavenger receptors
    d. Defective VLDL receptors

17. The enzyme mediating the conversion of Intermediate Density Lipoprotein to Low Density Lipoprotein is
    a. Lecithin Cholesterol Acyl Transferase
    b. Lipoprotein lipase
    c. Cholesteryl ester transfer protein
    d. Hepatic Lipase

18. All of the following are true about Cholesteryl Ester Transfer Protein, EXCEPT
    a. Associated with HDL
    b. Facilitates the transfer of cholesteryl ester from HDL to LDL
    c. Facilitates the transfer of triacylglycerol from LDL to HDL
    d. Facilitates the transfer of triacylglycerol from HDL to LDL

19. **The lipid lowering drug ezetimibe acts by**
    a. Inhibition of HMG coA reductase
    b. Inhibition of Niemann–Pick C1-Like 1
    c. Activation of PPAR
    d. Sequestering bile acids

20. **The lipid lowering drug lomitapide acts by**
    a. Inhibiting HMG coA reductase
    b. Inhibiting microsomal triglyceride transfer protein
    c. Activating PPAR
    d. Activating LCAT

21. **All of the following are true, EXCEPT**
    a. Apo AI is the cofactor for LCAT
    b. Apo CII is the cofactor for lipoprotein lipase
    c. Apo CIII inhibits lipoprotein lipase
    d. Apo AII activates LCAT

22. **All of the following are true about lipoprotein X, except**
    a. Lipoprotein X is an abnormal fraction of LDL
    b. Found in patients of familial lecithin: cholesterol acyltransferase deficiency
    c. Found in patients of cholestasis
    d. Associated with neurological disorders

23. **Apo D is associated with**
    a. Hepatobiliary diseases
    b. Neurodegenerative disorders
    c. Hematological diseases
    d. Musculoskeletal diseases

24. **Which of the following isoforms of ApoE confers the greatest risk of developing Alzheimer's disease?**
    a. Apo E1   b. Apo E2
    c. Apo E3   d. Apo E4

25. **Which of the following is NOT true for cholesterol metabolism?**
    a. HMG-CoA reductase is the key regulator of cholesterol biosynthesis
    b. Biosynthesis takes place in the cytoplasm
    c. Reduction reactions use NADH as cofactor
    d. Cholesterol is transported by LDL in plasma

26. **Friedwald formula cannot be used when triglyceride levels are**
    a. >350 mg/dL
    b. >400 mg/dL
    c. >200 mg/dL
    d. >300 mg/dL

27. **An infant presented with diarrhea. Blood examination revealed acanthocytes, undetectable cholesterol and triglyceride. VLDL and LDL fractions were absent on lipoprotein electrophoresis. Oral vitamin A 25 000 IU daily, 25-hydroxy vitamin D 10 000 IU daily, and vitamin K 5 mg daily were prescribed. Vitamin E was added later on. Sequencing for which of the following gene could have been done to confirm the diagnosis?**
    a. Lipoprotein lipase
    b. Microsomal triglyceride transfer protein
    c. Cholesterol Ester transfer protein
    d. Lecithin Cholesterol Acyl transferase

28. **Apo(a) of Lp(a) is similar to**
    a. Factor VIII   b. Fibrin
    c. Plasminogen   d. Thrombin

29. **Orlistat inhibits**
    a. Pancreatic lipase
    b. Gastric lipase
    c. Phospholipase A2
    d. All of the above

30. **Broad beta band in electrophoretic pattern of lipoproteins present in:**
    a. Type I hyperlipoproteinemia
    b. Type IIa hyperlipoproteinemia
    c. Type III hyperlipoproteinemia
    d. Type IV hyperlipoproteinemia

31. **Total cholesterol – 300 mg/dL; HDL-cholesterol – 40 mg/dL, TAG – 200 mg/dL. Calculate the LDL-cholesterol.**
    a. 90 mg/dL
    b. 120 mg/dL
    c. 220 mg/dL
    d. 260 mg/dL

ANSWERS

1. d.	**2.** a.	**3.** c.	**4.** b.	**5.** a.	**6.** a.
7. c.	**8.** a.	**9.** d.	**10.** c.	**11.** d.	**12.** d.
13. b.	**14.** d.	**15.** a.	**16.** b.	**17.** d.	**18.** d.
19. b.	**20.** b.	**21.** d.	**22.** d.	**23.** b.	**24.** d.
25. c.	**26.** b.	**27.** b.	**28.** c.	**29.** a.	**30.** c.
31. c.					

13.6 ACYLGLYCEROLS AND PHOSPHOLIPIDS METABOLISM

Triglycerides are the storage form of fatty acids and it is the best storage form of energy in the body because one gram provides 9 kilocalories. Moreover, triglycerides don't need water for their storage while glycogen requires water for storage. Adipocytes are involved in both synthesis and storage of triglycerides. Liver is also a major site of triacylglycerol synthesis but it is not involved in storage. Triglyceride synthesis takes place in the smooth endoplasmic reticulum.

Triglycerides are made of glycerol and fatty acids. Fatty acids are acquired from the chylomicrons by the action of lipoprotein lipase located on the capillary walls of adipocytes.

WHITE ADIPOSE TISSUE LACK GLYCEROL KINASE

Phosphatidic acid is the precursor of triglycerides as well as glycerophospholipids. Glycerol-3-phosphate is needed for the synthesis of phosphatidic acid. However, white adipocytes lack the enzyme glycerol kinase (Liver and brown adipocytes have glycerol kinase). Then, how glycerol-3-phophate is produced in adipocytes?

Glycerol-3-Phosphate

Fatty acyl-CoA — Glycerol-3-phosphate acyltransferase **committed step**

Lysophosphatidic acid

Fatty acyl-CoA — Lysophosphatidate acyltransferase

Phosphatidic acid

Phosphatase

Diacylglycerol

Fatty acyl-CoA — Diacylglycerol acyltransferase

Triacylglycerol

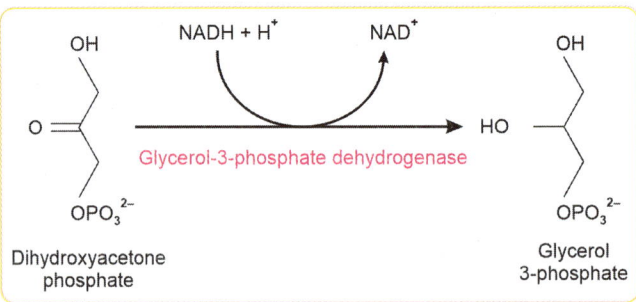

Dihydroxyacetone phosphate — Glycerol-3-phosphate dehydrogenase (NADH + H⁺ → NAD⁺) → Glycerol 3-phosphate

Dihydroxyacetone phosphate, intermediate of glycolysis is reduced to glycerol-3-phosphate by the enzyme glycerol-3-phosphate dehydrogenase. This is why **minimal amount of glycolysis is necessary to synthesize triglycerides in adipocytes**. Glucose enters adipocytes through insulin mediated GLUT-4.

Glyceroneogenesis Provides Glycerol-3-phosphate

The synthesis of glycerol-3-phosphate from pyruvate in adipocytes is known as glyceroneogenesis. This is a shortened version of gluconeogenesis where pyruvate is converted to DHAP and then to glycerol-3-phosphate.

Glycerol-3-phosphate is Esterified with Fatty Acyl-CoA

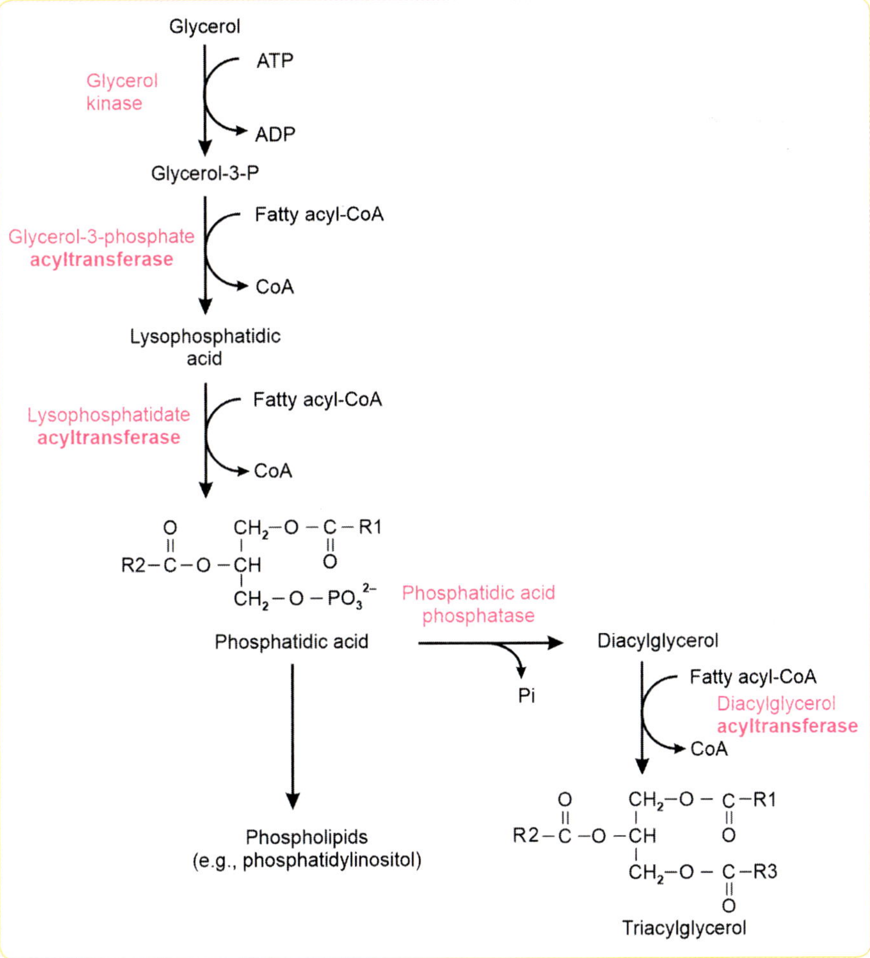

In glycerol-3-phosphate, two hydroxyl groups are free. Esterification of one fatty acyl-CoA (activated form of fatty acid) produces lysophosphatidic acid and the CoA is removed. Esterification of another fatty acid produces phosphatidic acid. Finally, the phosphoryl group is removed and another fatty acid is added to produce triacylglycerol.

PHOSPHATIDIC ACID IS THE PRECURSOR OF BOTH TRIGLYCERIDES AND GLYCEROPHOSPHOLIPIDS

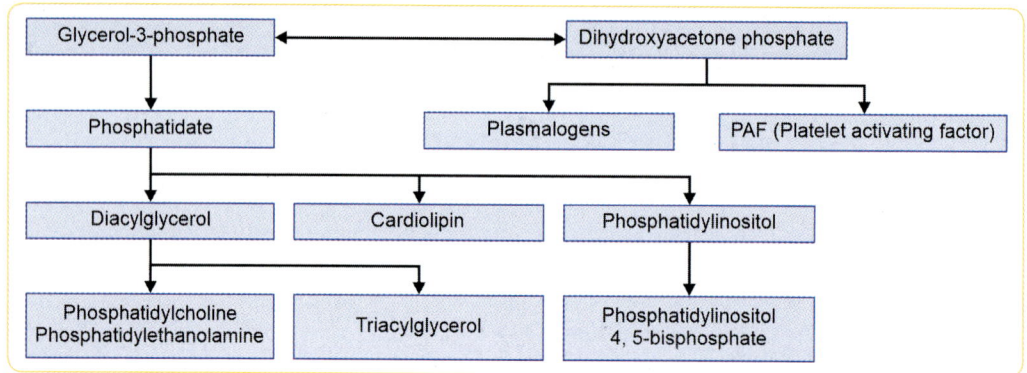

Hormone Sensitive Lipase Mediates the Breakdown of Fat in Adipocytes

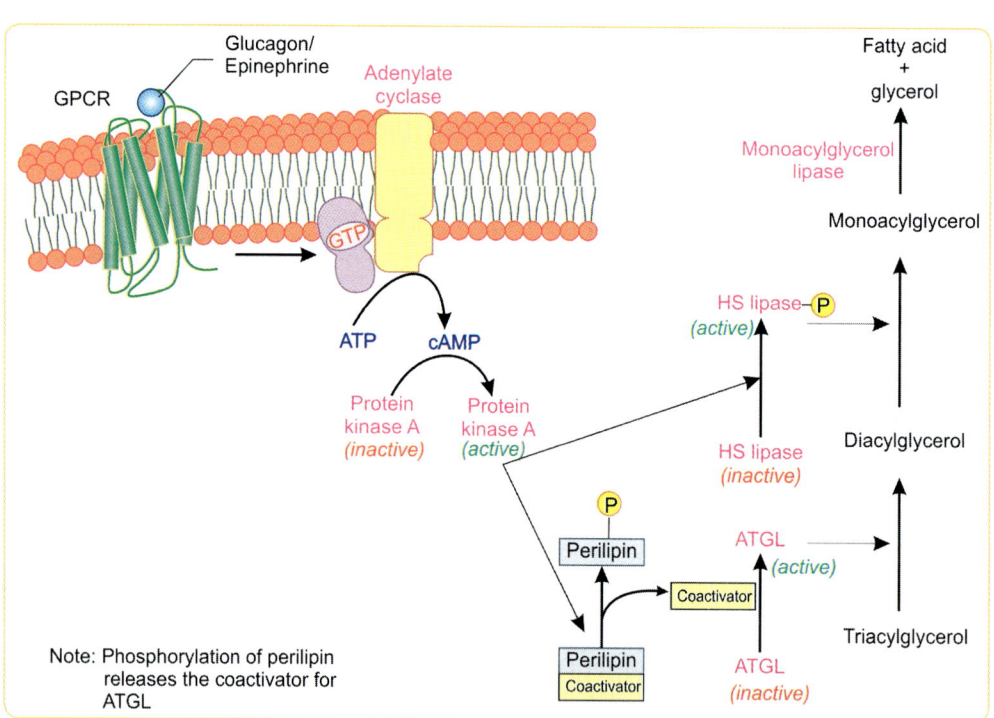

Fig. 1: 7TM receptor – 7 transmembrane receptor, ATGL - Adipose triglyceride lipase, HS lipase – Hormone sensitive lipase, TAG – Triacylglycerol, DAG -Diacylglycerol, MAG - Monoacylglycerol

During fasting, glucagon and epinephrine activate hormone sensitive lipase (HSL) through the second messenger cAMP. HSL is active in phosphorylated form.

❑ Triacylglycerol is surrounded by a protein known as perilipin.

❑ Hormone mediated phosphorylation of perilipin makes TAG accessible to Adipose triglyceride lipase (ATGL) which produces diacylglycerol from TAG.

❑ HSL breaks diacylglycerols into monoacyl glycerol and free fatty acids.

Free Fatty Acids are Not Free

- ❑ Nonesterified/unesterified fatty acids mobilized from adipocytes are not free in plasma. They are carried by albumin. Inside the cells, they are transported by fatty acid binding proteins.
- ❑ Fatty acid binding proteins are emerging biomarkers in various disease conditions like cardiac injury and acute kidney injury.

LYSOSOMAL STORAGE DISORDERS

Lysosomal storage disorders are rare inborn errors due to defect in the synthesis and transport of enzymes and proteins, required for the normal lysosomal functions. Absence of enzymes lead to accumulation of the substance, which is otherwise degraded.

Classification of Lysosomal Storage Disorders

- ❑ Mucopolysaccharidoses (Chapter 41) *(Refer to page no. 610)*
- ❑ GM2 Gangliosidoses
- ❑ Neutral Glycosphingolipidoses
- ❑ Glycoproteinoses (Chapter 41) *(Refer to page no. 603)*
- ❑ Mucolipidoses e.g. I-cell disease (type II) and Pseudo-Hurler polydystrophy (type III)
- ❑ Leukodystrophies
- ❑ Disorders of Neutral Lipid metabolism
- ❑ Disorders of Glycogen metabolism (Chapter 10.4) *(Refer to page no. 147)*

Sphingolipidoses

- ❑ Due to genetic defect in the lysosomal degradation of sphingolipids.
- ❑ All are inherited in autosomal recessive manner except Fabry's disease, which is X-linked.

Disease	Deficient enzyme	Lipid accumulating	Clinical features
Tay-Sachs disease	Hexosaminidase A	GM2 ganglioside	Mental retardation, blindness, muscular weakness, Cherry-red spot in retina
Sandhoff disease	β-Hexosaminidases A and B		Macrocephaly; Hyperacusis, Cherry-red spot
Fabry's disease	α-Galactosidase	Globotriaosylceramide	X-linked recessive, Renal failure, Angiokeratomas
Gaucher's disease	Acid β-Glucosidase	Glucosylceramide	Hepatosplenomegaly, Osteoporosis, Mental retardation
Niemann-Pick disease	Sphingomyelinase	Sphingomyelin	Mental retardation, Hepatosplenomegaly, death during infancy
Metachromatic leukodystrophy	Arylsulfatase A	3-Sulfogalactosylceramide	Mental retardationand features of demyelination
Krabbe's disease	β-Galactosidase	Galactosylceramide	Mental retardation. Absence of myelin.
Multiple sulfatase deficiency	Active site cysteine to Cα-formylglycine converting enzyme	Sulfatides; mucopolysaccharides	Mental retardation, Retinal degeneration Vacuolated and granulated cells
Farber's disease	Ceramidase	Ceramide	Skeletal deformation, mental retardation, death during infancy

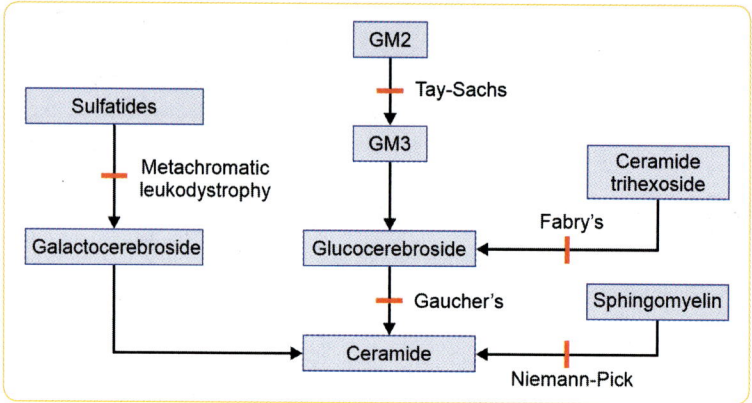

Treatment

❑ Enzyme replacement therapy and bone marrow transplantation are used in Gaucher's disease. Gene therapy is a promising treatment option.
❑ Substrate reduction therapy is a novel technique of inhibiting the formation of the substrate, which is been accumulated, e.g., Miglustat is used for Gaucher disease, eliglustat for both Gaucher and Fabry disease.

Competency

❑ BI4.2   Describe the processes involved in digestion and absorption of dietary lipids and also the key features of their metabolism

Key Points

❑ Adipose tissue lack glycerol kinase.
❑ Phosphatidic acid is the precursor of triglycerides as well as glycerophospholipids.
❑ Hormone sensitive lipase mediates the breakdown of fat in adipocytes.
❑ The enzymes deficient in Tay-Sachs, Fabry's and Niemann-Pick disease are Hexosaminidase A, α-Galactosidase and Sphingomyelinase respectively.
❑ The enzymes deficient in metachromatic leukodystrophy, Krabbe's and Farber's disease are Arylsulfatase A, β-Galactosidase and Ceramidase respectively

Short Answer Questions

1. A minimal amount of glycolysis is required in adipocytes for the triacylglycerol synthesis – Justify.

2. Name any two sphingolipidoses and the enzymatic defect. Mention the treatment approaches to this condition.

Multiple Choice Questions

1. All of the following tissues contain glycerol kinase, EXCEPT
   a. Liver
   b. Mammary gland
   c. Kidney
   d. White adipose tissue

2. Which of the following cant NOT be a component of triglycerides?
   a. Glycerol
   b. Linoleic acid
   c. Palmitic acid
   d. Sphingosine

3. Which of the following is a GM_2 gangliosidosis?
   a. Sandhoff's disease
   b. Niemann-Pick disease
   c. Gaucher's disease
   d. Fabry's disease

4. All of the following disorders are inherited in autosomal recessive mode, EXCEPT
   a. Sandhoff's disease
   b. Niemann-Pick disease
   c. Gaucher's disease
   d. Fabry's disease

5. Miglustat is an example of
   a. Augmentation therapy
   b. Substrate reduction therapy
   c. Enzyme replacement therapy
   d. Gene therapy

ANSWERS

1. d.	**2.** d.	**3.** a.	**4.** d.	**5.** b.

13.7 EICOSANOIDS

Eicosanoids are autacoids (local hormones) produced usually by the oxidation of 20-carbon arachidonic acid (Greek: *Eicosi – twenty*). Their actions are short-lived. Arachidonic acid is derived from membrane phospholipid by the action of phospholipase A_2.

Eicosanoids comprise of,

- ❑ Prostanoids, leukotrienes (LTs) and lipoxins (LXs) are the eicosanoids.
- ❑ Prostanoids include prostaglandins (PGs), prostacyclins (PGIs), and thromboxane (TXs).

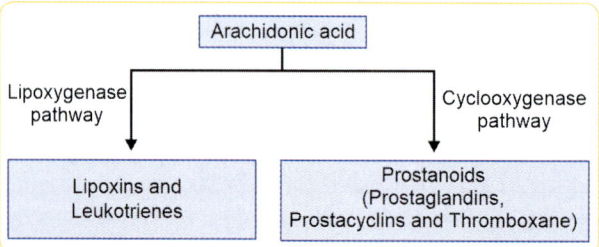

SYNTHESIS OF PROSTAGLANDINS

- ❑ Prostaglandins (PGs) are classified into series 1, 2 and 3 based on the number of double bonds.
- ❑ The precursor of series 1, 2 and 3 PGs are linoleic, arachidonic, and eicosapentaenoic acids respectively.
- ❑ Series 2 PGs are the most common ones.
- ❑ Membrane phospholipids are the source of arachidonic acid. Phospholipase A2 cleaves arachidonic acid from phospholipids.
- ❑ Hormones like epinephrine activate Phospholipase A2.
- ❑ Arachidonic acid is oxidized by prostaglandin-endoperoxide synthase 2 (prostaglandin G/H synthase and cyclooxygenase).
- ❑ PGH2 is the precursor of various other prostaglandins (illustrated in the image).

Note: Prostaglandins are not stored like the other autacoids, i.e. Histamine and Serotonin.

CYCLOOXYGENASE IS A SUICIDE ENZYME

Cyclooxygenase can switch off the synthesis of prostaglandins by undergoing self-catalyzed destruction. That is why this enzyme is known as suicidal enzyme. (*Do not confuse this with suicidal inhibition, Aspirin is a suicidal inhibitor of cyclooxygeanse*).

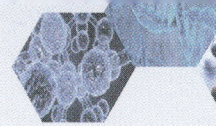

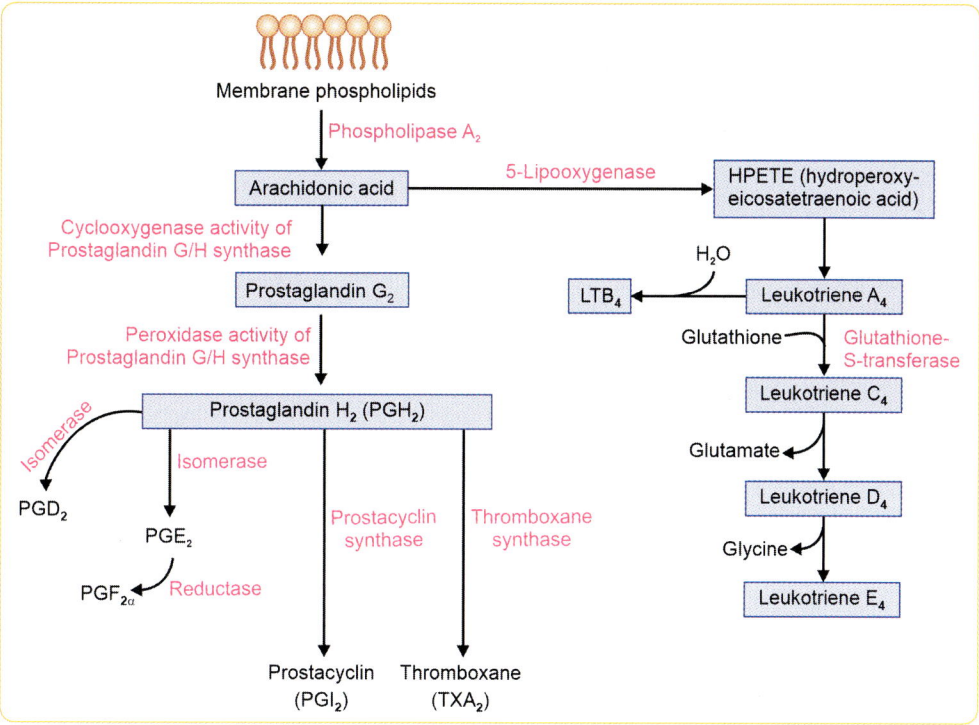

Fig. 1: https://upload.wikimedia.org/wikipedia/commons/4/40/Eicosanoid_synthesis.svg PGH2 synthase is a bifunctional enzyme. Note the role of glutathione. Steroids inhibit Phospholipase A2. NSAIDs inhibit Cyclooxygenase. Leukotriene C4,D4 and E4 are known as slow-releasing substance of anaphylaxis.

BIOLOGICAL FUNCTIONS OF PROSTAGLANDINS

- Mediators of inflammation
- Pyrogens – Cause fever during inflammation. Fever is beneficial to some extent that it disallows the growth of pathogens. PGE_2 reset the thermostat in the hypothalamus
- Algogen – Pain producing. PGE_2 enhances another Algogen bradykinin
- PGE_2 and $PGF_{2\alpha}$ are important for child birth (parturition).
- Patency of the ductus arteriosus is maintained by prostacyclin (PGI_2) in fetal life.
- Gastric intestinal mucosal integrity is maintained by prostaglandins. That is why $NSAID_S$ cause gastric ulcer.
- Thromboxane A$_2$ causes platelet aggregation. Prostacyclin (PGI_2) acts as a vasodilator and inhibitor of platelet aggregation. Normally, their production is in a balanced state.

 Clinical Correlation

In patent ductus arteriousus, we give Indomethacin (an NSAID) to close the duct.
On the other hand, it is important to keep the duct open in duct-dependent congenital heart diseases like transposition of great arteries (TGA). Therefore, we use prostaglandin E1 (PGE1), also known as alprostadil.

Chapter 13 **Lipid Metabolism**

Isoenzymes of Cyclooxygenase

- ❏ COX1 and 2 are the two isoenzymes of cyclooxygenase.
- ❏ COX1 is constitutionally expressed and is involved in the production of cytoprotective prostaglandins that are needed for the regular bodily functions like renal function, maintenance of gastric mucosal integrity.
- ❏ COX2 is inducible and is involved in the production of inflammatory cytokines.

Inhibitors of Prostaglandin Synthesis

- ❏ Steroids like corticosterone inhibits the release of arachidonic acid from membrane phospholipids.
- ❏ Non-steroidal anti-inflammatory drugs like aspirin, indomethacin, ibuprofen and diclofenac inhibits the cyclooxygenase 1 and 2.
- ❏ Cyclooxygenase 2 specific inhibitors like celecoxib were developed in the aim of preventing the inhibition of cytoprotective prostaglandin synthesis. Moreover, celecoxib is found to be associated with early cardiovascular mortality.

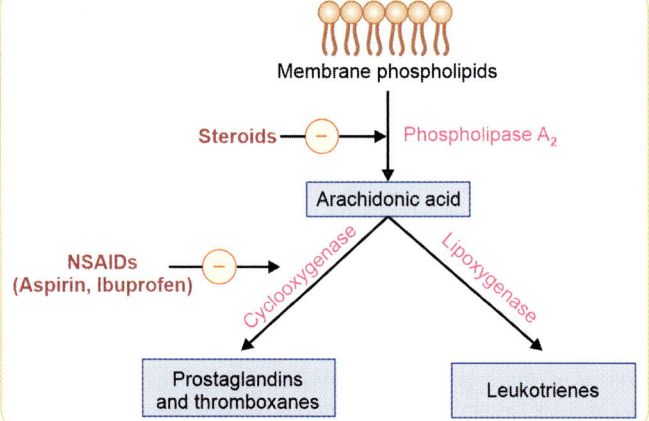

Snake Venom Contains Phospholipase A$_2$

Snake venom contains various types of phospholipase A$_2$. Lysolecithin produced by PLA$_2$ is responsible for the hemorrhagic manifestations.

Therapeutic uses of Prostaglandins

Some of the prostaglandins listed here are synthetic.

Prostaglandin	Formulation	Use
Misoprostol (PGE$_1$)	Tablet	Peptic ulcer and Medical termination of pregnancy
Dinoprostone (PGE$_2$)	Vaginal gel	Induction of labor
Carboprost (PGF$_{2\alpha}$)	Injection	Postpartum hemorrhage control
Epoprostenol (PGI$_2$)	Injection	Pulmonary hypertension
Lantanoprost (PGF$_{2\alpha}$)	Eye drops	Glaucoma
Alprostadil (PGE$_1$)	IV injection	To maintain the patency of the ductus arteriosus.
	Intra cavernous injection	Erectile dysfunction

Low Dose Aspirin is used in the Prophylaxis of Coronary Artery Disease

Thromboxane A_2 promotes platelet aggregation and prostacyclin (PGI_2) inhibits platelet aggregation. Aspirin inhibits thromboxane A_2 formation in platelets and prostacyclin formation in endothelial cells. Platelets are non-nucleated. So, once inhibited the enzyme is inhibited for the life-time of the platelet. But, endothelial cells can produce new prostacyclin since they are nucleated. Thus, the overall effect is inhibition of platelet aggregation in the blood vessels.

Biochemical basis of Aspirin Induced Asthma

Aspirin inhibits cyclooxygenase. Because of this, all the available arachidonic acid is **diverted into lipoxygenase pathway** to produce leukotrienes which are potent bronchoconstrictors. This is how aspirin induces asthma in susceptible persons.

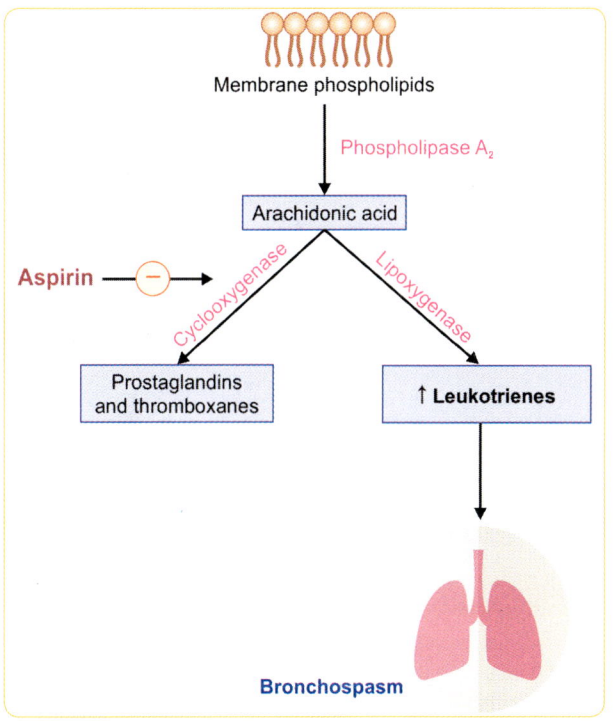

Cardioprotective Role of ω^3 Fatty Acids can be Explained by the Prostanoids Produced from Them

Series 3 prostanoids (3 double bonds) are produced from ω^3 fatty acids.

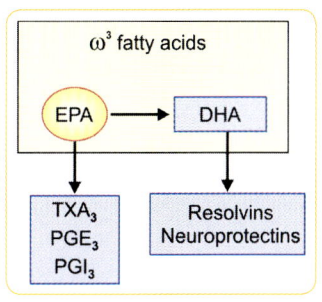

Thromboxane A$_3$ is a weak platelet aggregator compared to TXA$_2$.

PGI$_3$ is a potent platelet disaggregator compared to PGI$_2$.

Thus, the net effect of intake of ω³ fatty acids is the promotion of platelet disaggregation. Moreover, resolvins and protectins produced from docosahexaenoic acid are involved in resolution of inflammation.

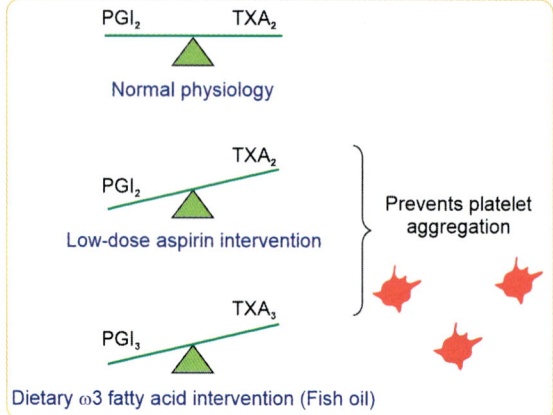

Fish Oil is Rich in ω³ Fatty Acids

Fish Oil is Used in

❑ Familial chylomicronemia syndrome
❑ Familial hypertriglyceridemia
❑ Dysmenorrhea

Fish Oil is Contraindicated in

❑ Type IIa hyperlipoproteinemia
❑ Patients with bleeding tendency

arrison's
Corner

"Fish oil or concentrated omega-3 fatty acid supplements impair platelet function. They alter platelet biochemistry to produce more PGI3, a more potent platelet inhibitor than prostacyclin (PGI2), and more thromboxane A3, a less potent platelet activator than thromboxane A2. Diets rich in omega-3 fatty acids can result in a prolonged bleeding time and abnormal platelet aggregation". (*Ref. Harrison 20th Ed., p.407*)

Competency

❑ BI4.6   Describe the therapeutic uses of prostaglandins and inhibitors of eicosanoid synthesis

Key Points

❑ Arachidonic acid is the source of eicosanoids.

❑ Thrombaxane A_2 causes platelet aggregation whereas Prostacyclin (PGI2) inhibits.

❑ Cyclooxygenase is a suicide enzyme.

❑ Aspirin is a suicidal inhibitor of cyclooxygenase.

❑ Snake venom contains phospholipase A2

❑ Aspirin can induce asthma in susceptible individuals.

❑ Thromboxane A3 is a weak platelet aggregator compared to TXA2 whereas PGI3 is a potent platelet disaggregator compared to PGI2.

❑ Fish oil is contraindicated in type IIa hyperlipoproteinemia

SELF-ASSESSMENT

Short Answer Questions

1. How is the precursor of prostaglandins made available for their synthesis? Name a drug, which inhibits this step.

2. Mention any two biological roles of prostaglandins.

3. Explain how ω³ Fatty Acids shifts the equilibrium of eicosanoids towards a cardioprotective state.

4. What is the biochemical basis of the following?

   a. Aspirin induces asthma in susceptible individuals

   b. Low dose aspirin is used in people at risk of coronary artery disease

   c. Cardioprotective role of ω3 Fatty Acids

Multiple Choice Questions

1. Which of the following enzymes is known as 'suicide enzyme'?

   a. Cyclooxygenase   b. Lipoxygenase

   c. Phospholipase C   d. Xanthine oxidase

2. Which of the following is required for the synthesis of Leukotrienes C4, D4 and E4 from Leukotriene A4?

   a. Anserine   b. Carnitine

   c. Carnosine   c. Glutathione

3. Eicosanoids are usually derived from

   a. Oleic acid   b. Linoleic acid

   c. Linolenic acid   d. Arachidonic acid

4. Which of the following are component of slow-releasing substance of anaphylaxis?

   c. LT $A_4B_4C_4$

   c. LT $B_4C_4D_4$

   c. LT $C_4D_4E_4$

   c. LT $A_4E_4F_4$

5. Which of the following fatty acid present in fish-oil is known for its cardio-protective action?

   a. Arachidonic aid

   b. Eicosapentaenoic acid

   c. Linoleic acid

   d. Palmitic acid

6. Snake venom contains

   a. Phospholipase A_1

   b. Phospholipase A_2

   c. Phospholipase C

   d. Phospholipase D

7. Which of the following inhibits platelet aggregation?

   a. TXA2

   b. PGI2

   c. LTA3

   d. PGE2

Chapter 13 Lipid Metabolism

ANSWERS

| 1. a. | 2. d. | 3. d. | 4. c. | 5. b. | 6. b. |
| 7. b. | | | | | |

Metabolism of Alcohol

ABSORPTION OF ALCOHOL

Ethanol (C_2H_5OH) is completely miscible with water because it can make hydrogen bonds with water molecules. Major site of alcohol absorption is the *upper small intestine*.
❑ Stomach—20% absorption
❑ Small intestine—80% absorption
❑ Mouth—readily absorbed
Rate of absorption is increased when alcohol is taken on empty stomach. Food intake reduces the rate of absorption by delaying the gastric emptying. Alcohol can freely diffuse across the cells. Metabolism of alcohol and its effect are influenced by gender, body weight, and genotype.

METABOLISM OF ALCOHOL

Alcohol is metabolized in liver by three different systems:
1. Cytosolic alcohol dehydrogenase—major pathway
2. Microsomal ethanol oxidizing system (MEOS)—induced by chronic alcohol ingestion
3. Peroxisomal catalase (minor pathway)

Cytosolic Alcohol Dehydrogenase Pathway

❑ Alcohol dehydrogenase is the enzyme with EC number 1.1.1.1
❑ This is a NAD^+ dependent dehydrogenase.
❑ NAD^+ binding domain of this and some other dehydrogenases is known as **Rossmann fold**.
❑ Converts ethanol to acetaldehyde and produces $NADH + H^+$.

Excess Alcohol Induces Microsomal Ethanol Oxidizing System

❑ Microsomal ethanol oxidizing system (MEOS) is an alternate pathway of ethanol metabolism.
❑ MEOS activity increases after chronic alcohol consumption.
❑ *CYP2E1* is the predominant enzyme that converts ethanol to acetaldehyde.

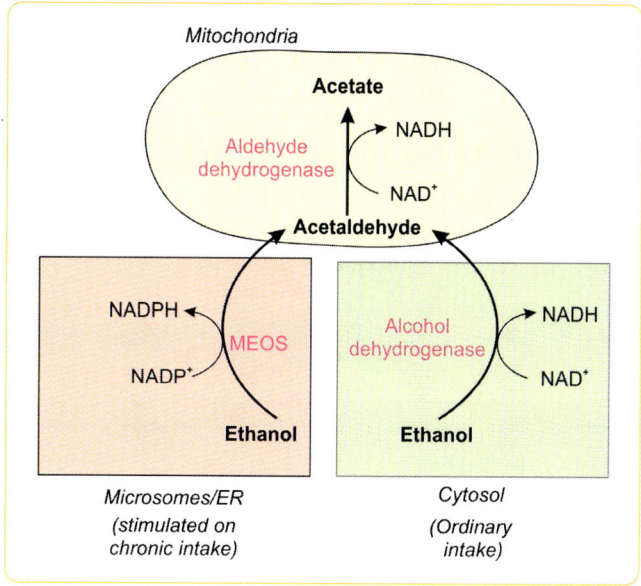

Zero-Order Kinetics

Kinetics of alcohol elimination is said to be a zero-order process. This means that rate of removal of alcohol from the body is constant irrespective of the amount of alcohol. This is because of the saturation of the alcohol dehydrogenase with even low concentrations of alcohol.

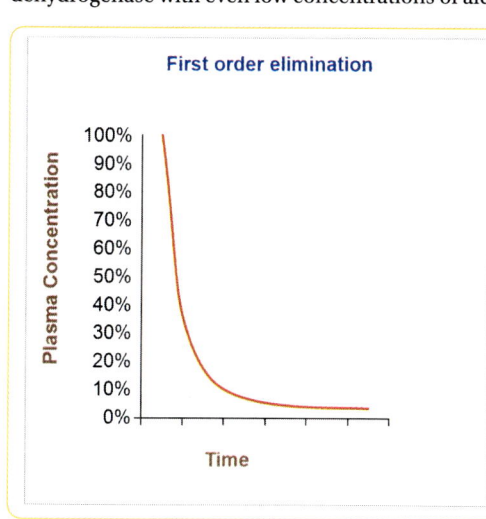

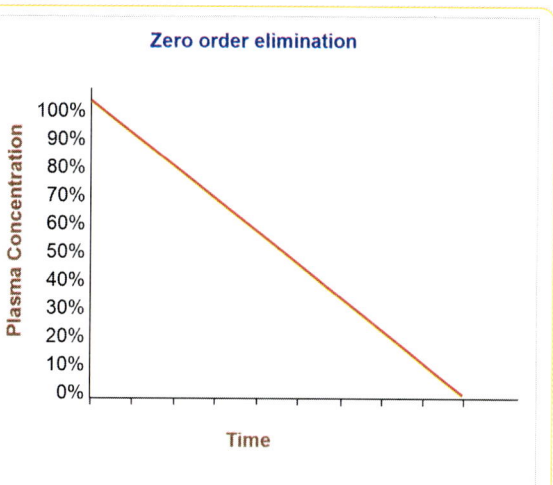

Methanol and ethylene glycol are also metabolized by the same system that metabolizes the ethanol.

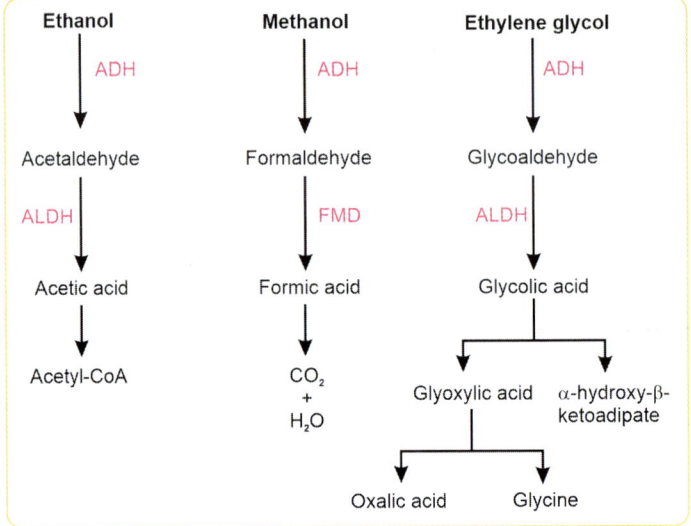

Ingestion of illicit liquor and methylated sprit leads to methanol poisoning. Formaldehyde produced by methanol is more toxic than acetaldehyde. It causes toxic amblyopia and blindness.

Ethylene glycol is a component of antifreeze. Both methanol and ethylene glycol poisoning cause raised anion gap metabolic acidosis.

Increased NADH/NAD Ratio is the Biochemical Basis of Metabolic Derangements in Alcoholism

Hypoglycemia	↑ NADH promotes the conversion of pyruvate to lactate and oxaloacetate to malate. Thus, it depletes the glucogenic substrates.
Lactic acidosis	↑ NADH causes the conversion of pyruvate to lactate
Gout	❑ **Overproduction:** Alcohol increases urate synthesis by enhancing the turnover of adenine nucleotides ❑ **Under excretion:** Lactic acid competes with uric acid for excretion in the urinary tubules
Fatty liver	↑ NADH inhibits the isocitrate dehydrogenase. Citrate comes out of mitochondria and fatty acid synthesis is promoted.
Liver damage	Acetaldehyde forms adduct with proteins and toxic to hepatocytes

1 gram of alcohol provides 7 kcal. However, these calories are known as empty calories since they are not associated with nutrients like vitamins and minerals.

Reasons for Thiamine Deficiency in Alcoholism

❑ Alcoholics do not take food properly. Therefore, the chance of dietary thiamine deficiency is more common.
❑ Moreover, alcohol inhibits thiamine absorption.

Biochemical Basis of Fetal Alcohol Syndrome: Disruption of Retinoic Acid Signaling

❑ Retinol, a form of vitamin A is an alcohol.
❑ Alcohol dehydrogenase is also involved in the conversion of retinol to retinoic acid, an important molecule needed for growth and development.

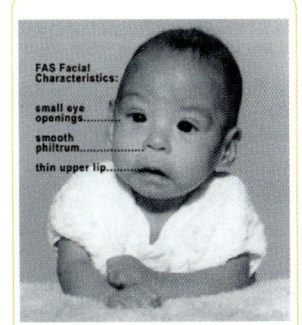

FAS-Fetal Alchol syndrome

 **Mnemonic**

Alcohol is a 7-letter word. Calorific value of alcohol is 7kcal/gram.

❑ Excess ethanol competes with the retinol for the conversion by alcohol dehydrogenase and thus affects the retinoic acid synthesis and thus disrupts the retinoic acid signaling pathway.

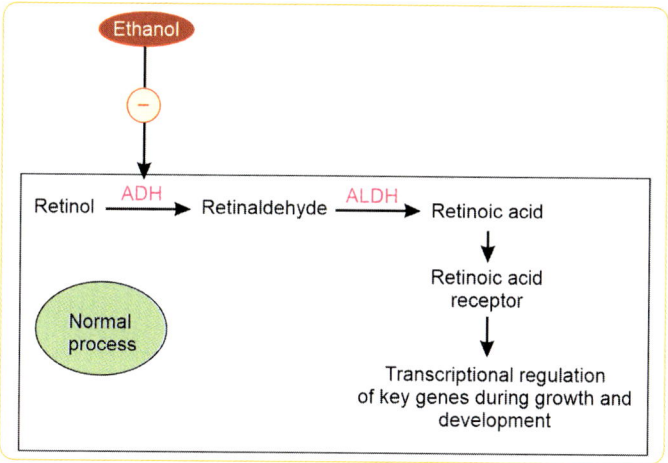

Methanol Poisoning is Treated by Administration of Ethanol

Ethanol competes with methanol and ethylene glycol for the active site of the alcohol dehydrogenase enzyme. This is the biochemical basis of use of ethanol in these poisoning conditions.

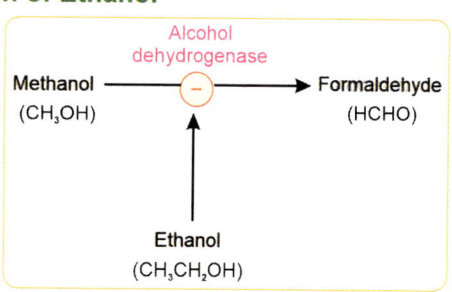

Carbohydrate-Deficient Transferrin is a Marker of Chronic Alcoholism

❑ Transferrin is a glycoprotein.
❑ Alcohol inhibits the glycosylation of several glycoproteins, including transferrin.
❑ Chronic alcoholism leads to transferrin deficient in four to five sialic acid residues.
❑ Consumption of >80 g of alcohol/day leads to an increase in the plasma carbohydrate-deficient transferrin (CDT) concentration.
❑ Other marker of alcoholic liver disease: γ-glutamyl transpeptidase.

Drugs Inhibiting Alcohol Metabolism

Drug	Target Enzyme	Use
Disulfiram	Aldehyde dehydrogenase	Aversion therapy
Fomepizole	Alcohol dehydrogenase	Ethylene glycol poisoning

Chapter 14 Metabolism of Alcohol

Competency

❑ PA12.1   Enumerate and describe the pathogenesis of disorders caused by alcohol
❑ PH1.20   Describe the effects of acute and chronic ethanol intake
❑ PH1.21   Describe the symptoms and management of methanol and ethanol poisonings

Key Points

- ❑ Major site of absorption of alcohol is the upper small intestine.
- ❑ Calorific value of alcohol is 7 kcal/g.
- ❑ Respiratory quotient of alcohol is 0.6.
- ❑ Chronic consumption of ethanol induces microsomal CYP2E1.
- ❑ Carbohydrate-deficient transferrin is a marker of chronic alcoholism.
- ❑ γ-glutamyl transferase is elevated in various liver diseases including alcoholic liver diseases.
- ❑ Moderate alcohol consumption increases the synthesis of apoA-I; increases the level of HDL; lowers the incidence of coronary heart disease.
- ❑ Moderate alcohol use increases the risk of breast cancer, hypertension, and stroke in women.

SELF-ASSESSMENT

Short Answer Questions

1. Explain the biochemical basis of hypoglycemia, lactic acidosis, fatty degeneration of liver seen in alcoholic patients.

2. Name two drugs acting on the metabolism of alcohol. Mention their enzyme targets.

3. What is the biochemical basis of the following?
   a. Use of Ethanol to treat methanol poisoning
   b. Fetal alcohol syndrome

Multiple Choice Questions

1. A 3-year-old girl was brought into the emergency room. She was cold and clammy and breathing rapidly. She was confused and lethargic. Her mother indicated that she had accidentally ingested automobile antifreeze (ethylene glycol) while playing in the garage. Following gastrointestinal lavage and activated charcoal administration, a nasogastric tube for ethanol was administered. How will ethanol help in relieving the symptoms?
   a. Conjugate with ethylene glycol to form a soluble compound
   b. Induce the alcohol dehydrogenase enzyme
   c. Competitively inhibit the metabolism of ethylene glycol
   d. Promote the excretion of metabolite of ethylene glycol

2. Enzyme system in which of the following organelles is induced on chronic ingestion of alcohol?
   a. Cytosol   b. Mitochondria
   c. Microsome   d. Lysosome

3. Which of the following drug inhibits the enzyme aldehyde dehydrogenase?
   a. Flumazenil   b. Fomepizole
   c. Disulfiram   d. Ethanol

4. The conventional treatment for methanol toxicity is to administer ethanol. Which of the following explains the basis of this treatment?
   a. Ethanol acts as a competitive inhibitor to methanol
   b. Ethanol acts as a non-competitive inhibitor to methanol
   c. Ethanol destroys the enzymatic activity of alcohol dehydrogenase
   d. Ethanol blocks the entry of methanol within the cells

ANSWERS

1. c.	2. c.	3. c.	4. a.

15 CHAPTER

Nucleotide Chemistry and Metabolism

NUCLEOSIDE, NUCLEOTIDE AND NUCLEIC ACID

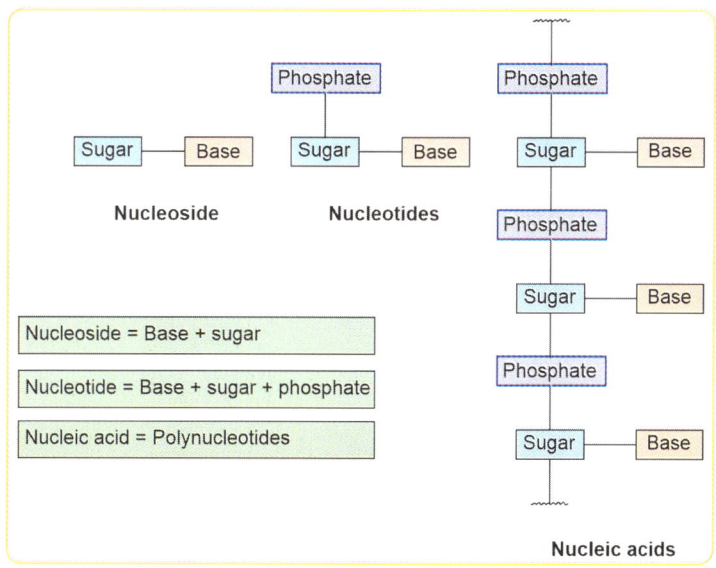

Nucleoside = Base + sugar

Nucleotide = Base + sugar + phosphate

Nucleic acid = Polynucleotides

Nitrogenous Bases

Bases can be either purines or pyrimidines.

Purine Bases

Adenine
(6-aminopurine)

Guanine
(2-amino-6-oxypurine)

Hypoxanthine
(6-oxypurine)

Xanthine
(2,6-dioxypurine)

Uric acid
(2,6,8-trioxypurine)

By just looking at the structures, you can easily understand the following facts:
- Deamination of adenine produces hypoxanthine.
- Deamination of guanine produces xanthine.
- Oxidation of hypoxanthine produces xanthine.
- Oxidation of xanthine produces uric acid.
- Uric acid is the most oxidized purine.

Pyrimidine Bases

Pyrimidine

Cytosine
(2-oxy-4-aminopyrimidine)

Uracil
(2,4-dioxy pyrimidine)

Thymine
(5-methyl uracil)

Points to Ponder

- Cytosine is found in both DNA and RNA. Uracil is found only in RNA.
- Deoxyribothymidine (Deoxyribose + thymine) is found only in DNA.
- Ribothymidine (Ribose + thymine) is found in tRNA.

Thymine is Synthesized from Methylation of dUMP in the Body

- Thymine is nothing but 5-methyl uracil. So, conversion of dUMP to TMP by the enzyme thymidylate synthase requires methyl (one-carbon) donor.

- N^5, N^{10}-methylene-THFA is the donor of one carbon group in this reaction. One peculiar thing about this reaction is that *one-carbon carrier THFA gets oxidized to DHFA during the reaction.* This DHFA is reduced back by the enzyme dihydrofolate reductase. The anticancer drug, methotrexate, is a competitive inhibitor of this enzyme.
- 5-fluorouracil is a prodrug that is converted into fluoro deoxy uridylic acid (F-dUMP) which is a potent inhibitor of thymidylate synthase (suicidal inhibition).

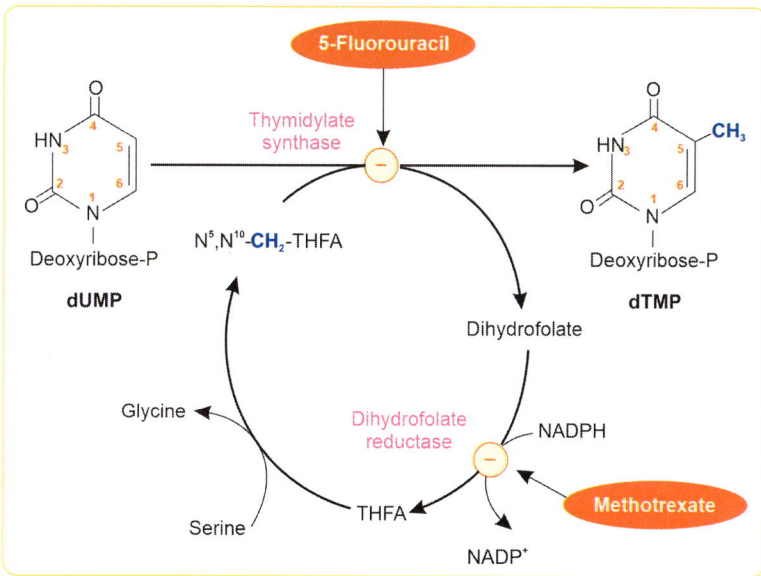

Fig. 1: Note the drugs acting as inhibitors. Deoxy TMP is usually written as TMP instead of dTMP

Deamination of Methylated Cytosine Produces Thymine

Look at the structure of cytosine, uracil and thymine.
- Thymine is nothing but 5-methyl uracil.
- Deamination of cytosine produces uracil.
- Thus, deamination of methylated cytosine will produce thymine.

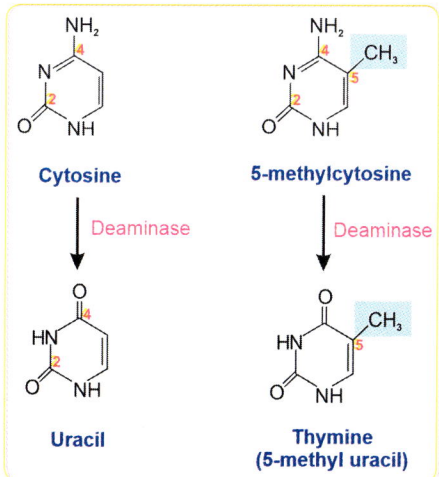

Nucleosides (Base + Sugar)

Nucleosides are produced when N_9 of purine or N_1 of pyrimidine forms a β-N-glycosidic bond with C_1 OH group of ribose or deoxyribose.
- (deoxy) Adenosine = Adenine + (deoxy) ribose
- Guanosine = Guanine + ribose
- Inosine = Hypoxanthine + ribose
- Uridine = Uracil + ribose
- Thymidine = Thymine + ribose
- Cytidine = Cytosine + ribose

Nucleotides (Base + Sugar + Phosphate)

Nucleotides are formed when 5′ OH group of nucleosides is phosphorylated.
Examples:
- AMP = 5′ adenosine monophosphate = adenylic acid
- CDP = 5′ cytidine diphosphate
- dGTP = 5′ deoxy guanosine triphosphate
- dTTP = 5′ deoxy thymidine triphosphate (more commonly designated TTP)
- cAMP = 3′–5′ cyclic adenosine monophosphate

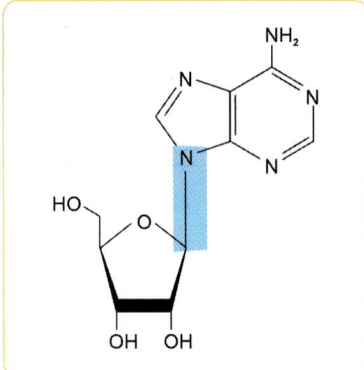

Fig. 2: The structure of adenosine is shown here. The bond between base and ribose is β-N-Glycosidic bond of nucleotides. It is always in anti-conformation in nucleic acids to prevent steric hindrance.

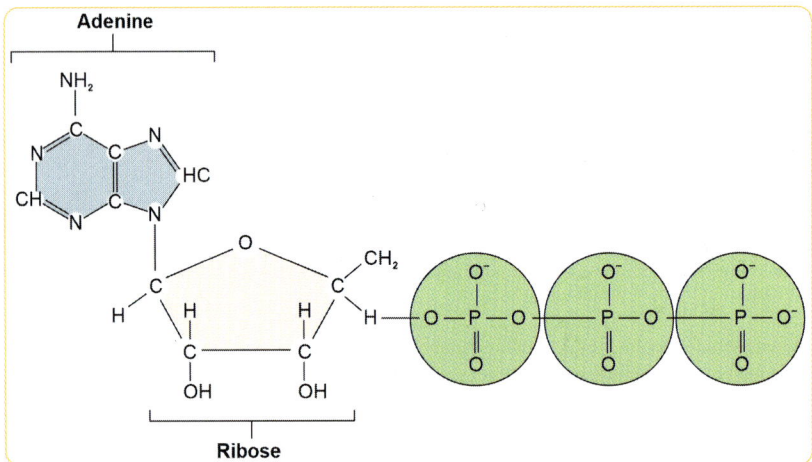

Fig. 3: Structure of 5° Adenosine triphosphate

Cyclic Nucleotides are Produced by Cyclase Enzymes

 High Yield

Cyclic nucleotides are nucleoside monophosphates in which the bond between the sugar and phosphate is cyclic in nature. Cyclic AMP and cyclic GMP are the two cyclic nucleotides in our body. They are involved in cell signalling (Chapter 38) *(Refer to page no. 583)*. Cyclic nucleotides are produced by cyclase enzymes. Adenylyl cyclase and guanylyl cyclase are the enzymes, which convert 5′ ATP to 3′, 5′ cAMP and 5′ GTP to 3′, 5′ cGMP respectively. Cyclase enzymes exist in membrane-bound and cytosolic (soluble) forms.

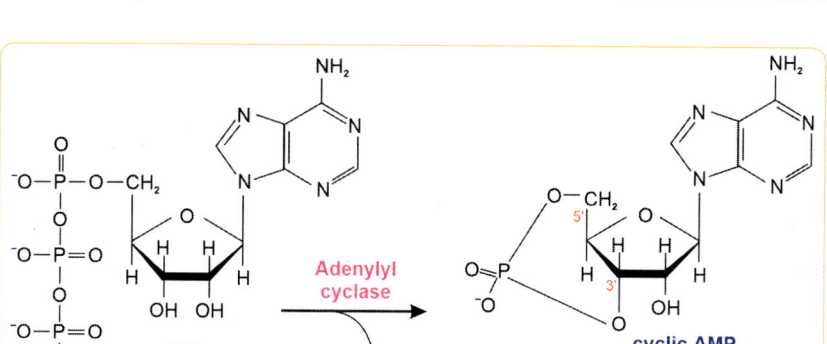

Biological Role of Nucleotides

- ❏ Adenine nucleotides are components of the coenzymes, NAD^+, $NADP^+$, FAD and CoA
- ❏ Monomers of nucleic acids—DNA and RNA
- ❏ Used to activate substrates for biosynthetic reactions
  - ○ UDP-glucose → Glycogen, glycoproteins
  - ○ UDP-glucuronic acid → Type II biotransformation (conjugation) reactions
  - ○ CDP-diacylglycerol → Phosphatidyl inositol synthesis
  - ○ CDP-ethanolamine → Phosphatidyl ethanolamine synthesis
  - ○ S-adenosylmethionine → Methyl donor
  - ○ GDP-L-fucose → Glycoproteins
- ❏ ATP is the universal currency of energy. Thermodynamically unfavored reactions are made favorable by coupling with ATP hydrolysis.
- ❏ Cyclic AMP and cyclic guanosine monophosphate (GMP) are 2nd messengers in signal transduction
- ❏ ATP, ADP and AMP are allosteric regulators for many enzymes
- ❏ ATP dependent phosphorylation regulates the action of enzymes and membrane transporters
- ❏ ADP-ribosylation is a covalent modification having multiple roles *(Chapter 38, end portion) (Refer to page no. 588).*

DE NOVO PURINE BIOSYNTHESIS

In our body, purine nucleotides are produced by de novo (from simple molecules) or by salvage (recycling) pathways. We will first look at the de *novo* pathway.

SOURCES OF CARBON AND NITROGEN ATOMS OF PURINE RING

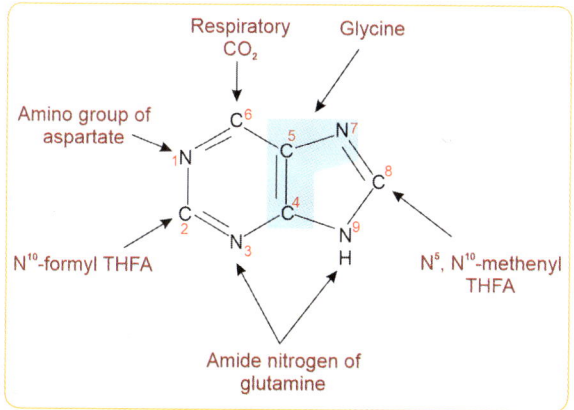

Note: Standard textbooks mention N^{10}-formyl THFA as the sources of both C_2 and C_8. But in Harper textbook, it is given that N^5, N^{10}-methynyl THFA is the source of C_8.

KEY POINTS IN *DE NOVO* PURINE BIOSYNTHESIS

- ❏ The purine ring system is assembled on phosphoribosyl pyrophosphate (PRPP)
- ❏ 11 enzymatic reactions are involved and some of the enzymes are multifunctional.
- ❏ Successive steps of activation by phosphorylation followed by displacement assemble the purine ring.
- ❏ 3 amino acids are involved in purine synthesis—glycine, glutamine and aspartate
  - ○ All the carbons and nitrogen of glycine are incorporated into the purine ring.
  - ○ Glutamine and aspartate donate only amino group.
- ❏ One-carbon groups for purine ring are carried by folate.
- ❏ IMP is the first purine nucleotide formed. AMP and GMP are formed from IMP.
- ❏ AMP synthesis requires GTP; GMP synthesis requires ATP. This is a wonderful cross-regulation!
- ❏ 6 ATPs are required for the assembly of a purine ring (Inosine monophosphate).

PRPP SYNTHESIS IS THE FIRST STEP OF PURINE SYNTHESIS

- ❏ Ribose-5-phosphate produced by the HMP shunt is the substrate for the enzyme PRPP synthetase.
- ❏ In coupled reactions, when there is conversion of ATP to AMP, PP_i is hydrolyzed and makes the reaction irreversible.
- ❏ However, in this reaction, ATP → AMP does not mean PP_i is hydrolyzed. Here, pyrophosphate is transferred to the 1st carbon of ribose. (It will be hydrolyzed in the next step!)
- ❏ Since PRPP can go for many pathways like *de novo* and salvage synthesis of purine and pyrimidine nucleotides, synthesis of NAD^+, and $NADP^+$, this step is not a committed step.

Glutamine Phosphoribosyl Pyrophosphate Amidotransferase Catalyzes the Committed Step of *de novo* Purine Synthesis

- ❏ PRPP is the activated form of ribose-5-phosphate.
- ❏ Amino group of glutamine is transferred to the 1st carbon of PRPP; pyrophosphate is hydrolyzed in this step and makes this reaction irreversible. (So, ultimately 2 high-energy phosphate bonds are utilized)

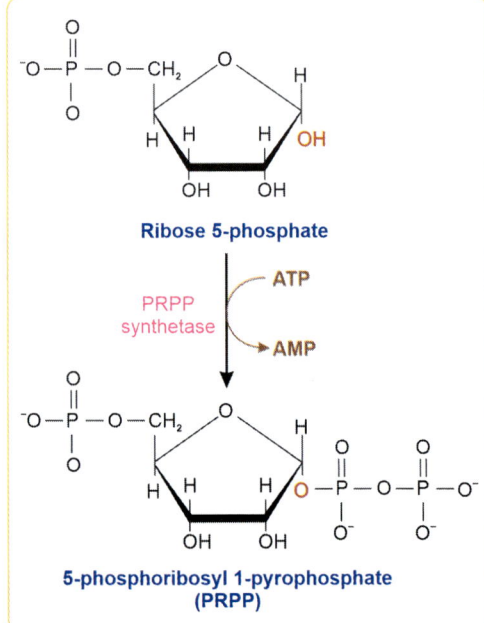

Ribose 5-phosphate

PRPP synthetase

ATP

AMP

5-phosphoribosyl 1-pyrophosphate (PRPP)

□ 5-phosphoribosyl 1-amine has only one fate, i.e. to enter *de novo* purine synthesis. So, this is the committed step.

□ The only limiting factor for this enzyme is the availability of PRPP. i.e. in addition to the committed step, PRPP synthetase reaction is also regulated.

Energy Cost

Can you observe that 6 ATP equivalents are needed for the assembly of purine ring? You should consider the hydrolysis of PP_i as 2 ATP equivalents to get the calculation right. Can you recall that 6 ATP equivalents are needed for the synthesis of 1 molecule of glucose from 2 molecules of pyruvate/lactate?

AMP and GMP are Synthesized from IMP

□ IMP is the first purine nucleotide synthesized
□ AMP and GMP are derived from IMP in a two-step reaction
□ AMP synthesis requires GTP; GMP synthesis requires ATP
□ Adenine is 6-amino purine. The amino group is given by aspartate.
□ Guanine is 2-amino 6-oxy (keto) purine. The amino group is given by glutamine and the keto group is introduced by dehydrogenase reaction.

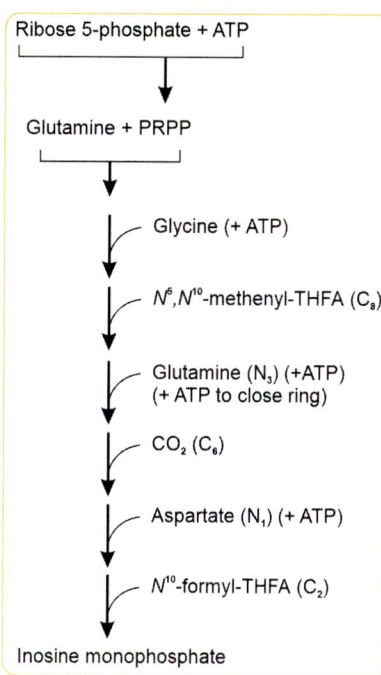

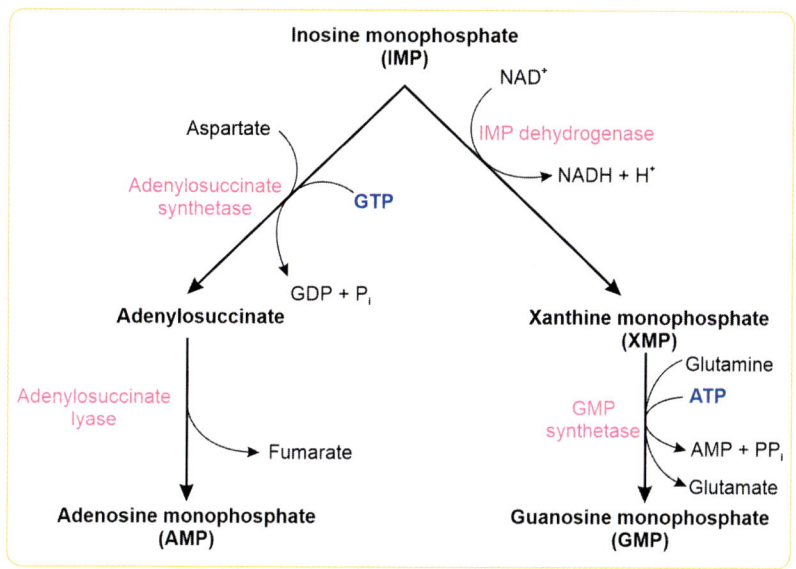

Note: Let us revise the enzyme nomenclature using these reactions. Synthetases require ATP/GTP. They belong to ligases. Lyases mediate the nonhydrolytic cleavage of bonds. Dehydrogenases remove hydrogen using NAD, FAD or NADP.

Regulation of *de novo* Purine Synthesis

Enzyme	Inhibited by
PRPP synthetase	ADP and GDP
Glutamine phosphoribosyl pyrophosphate amidotransferase	AMP, GMP, ADP, GDP, IMP
IMP dehydrogenase	GMP
Adenylosuccinate synthetase	AMP

PYRIMIDINE BIOSYNTHESIS

SOURCES OF PYRIMIDINE RING ATOMS

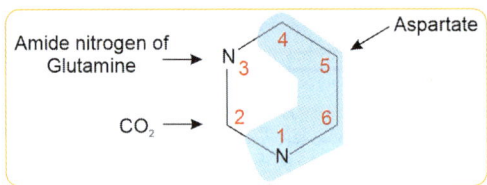

Fig. 4: Glutamine and CO_2 produce carbamoyl phosphate

Note: Thus, we can consider carbamoyl phosphate as the source of C_2 and N_3.

KEY POINTS IN *DE NOVO* PYRIMIDINE BIOSYNTHESIS

- ❑ *De novo* pyrimidine biosynthesis takes place in *both cytosol and mitochondria.*
- ❑ Carbamoyl phosphate synthetase II is the rate-limiting enzyme.
- ❑ Base is synthesized first and then joined to PRPP.
- ❑ OMP is the first pyrimidine nucleotide synthesized.
- ❑ Uridine monophosphate (UMP) is derived from OMP.
- ❑ dTMP is derived by methylation of dUMP by the enzyme thymidylate synthase.
- ❑ Cytidine triphosphate (CTP) is formed by amination of UTP.
- ❑ Nucleotide mono-, di-, and triphosphates are interconvertible.

CPS II CATALYZES THE RATE-LIMITING REACTION OF *DE NOVO* PYRIMIDINE SYNTHESIS

- ❑ Rate-limiting step of *mammalian de novo* pyrimidine biosynthesis is CPS II.
- ❑ **Aspartate transcarbamoylase is the rate limiting enzyme of *de novo* pyrimidine synthesis in bacteria.**

	CPS I		CPS II
Location	Mitochondria		Cytoplasm
Pathway involved	Urea cycle		Pyrimidine biosynthesis
Source of NH_3	Free NH_3 liberated from glutamate dehydrogenase reaction		Glutamine
Allosteric activator	N-acetyl glutamate		Phosphoribosyl pyrophosphate
Inhibitor	–		UTP

Know the Difference

Purine synthesis	Pyrimidine synthesis
All *de novo* purine synthesis enzymes are cytosolic	Except for mitochondrial dihydro orotate dehydrogenase, all enzymes of pyrimidine synthesis are cytosolic.
Purine ring is built upon the phosphoribosyl pyrophosphate (PRPP)	PRPP is added at later stage

SALVAGE PATHWAYS

De novo synthesis of purine and pyrimidine bases is an energy expensive process hence, all the cells recycle purine and pyrimidine bases by converting them into nucleotides. Salvage pathway takes place primarily in extrahepatic tissues, whereas de novo pathway takes place primarily in the liver.

The RBC and brain are entirely dependent upon purine salvage pathway. Brain lacks the PRPP-amido transferase enzyme.

Enzymes involved in purine salvage pathway:
❏ Adenine phosphoribosyltransferase (APRT)
❏ Hypoxanthine-guanine phosphoribosyltransferase (HGPRT)

$$\text{Hypoxanthine + PRPP} \xrightarrow{\text{HGPRTase}} \text{IMP + PP}_i$$

$$\text{Guanine + PRPP} \xrightarrow{\text{HGPRTase}} \text{GMP + PP}_i$$

$$\text{Adenine + PRPP} \xrightarrow{\text{APRTase}} \text{AMP + PP}_i$$

Ribose-5-phosphate is added to the free base to produce mononucleotide. Phosphoribosyl pyrophosphate (PRPP) acts as the source of ribose-5-phosphate.

Subsequent hydrolysis of inorganic pyrophosphate by pyrophosphatase ($\Delta G^{0'}$ = –19.2 kJ/mol) makes this reaction essentially irreversible.
❏ Hypoxanthine guanine phosphoribosyltransferase enzyme can act on both hypoxanthine and guanine to produce IMP and GMP respectively.
❏ Adenine phosphoribosyltransferase acts in the similar manner to produce AMP.

Regulation

❏ Feedback inhibition by the end products, i.e. IMP and GMP
❏ Lack of bases for salvage → ↑PRPP → stimulation of PRPP amidotransferase → *increased de novo* synthesis.

NUCLEOSIDE DIPHOSPHATES ARE REDUCED TO DEOXYNUCLEOSIDE DIPHOSPHATE

❏ Ribonucleotide reductase also known as ribonucleoside diphosphate reductase reduces the 2' OH group of the nucleoside diphosphate with the help of thioredoxin (a protein).

- Thioredoxin gets oxidized in this process which is reduced by the enzyme thioredoxin reductase using NADPH.
- Thioredoxin reductase is a selenocysteine containing enzyme (other selenocysteine containing enzymes: Glutathione peroxidase, 5′ deiodinase)

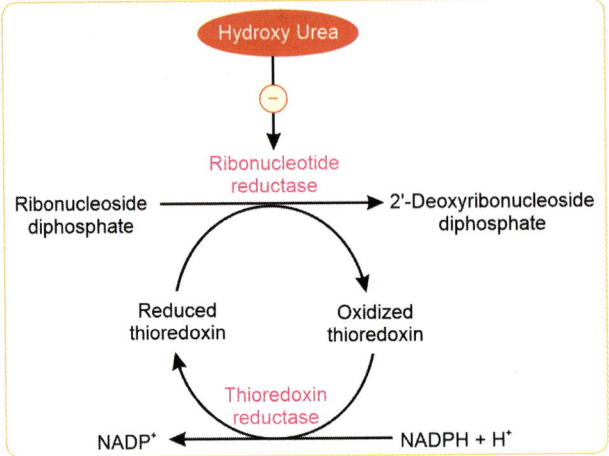

Note: Only nucleoside diphosphates can act as a substrate for ribonucleotide reductase

Hydroxy Urea Inhibits Ribonucleotide Reductase

- Inhibition of ribonucleotide reductase by hydroxyurea results in impaired deoxy ribonucleotide synthesis → Impaired DNA synthesis.
- Used in chronic myeloid leukemia.
- Hydroxyurea increases the level of fetal hemoglobin (HbF). HbF interferes in the sickling process → useful in sickle cell disease.

CATABOLIC END-PRODUCTS

Uric Acid is the End Product of Purine Catabolism in Primates

Xanthine oxidase catalyzes the formation of uric acid. This is a *molybdenum* containing *flavoenzyme*. This is one of the few reactions in our body that generates H_2O_2.

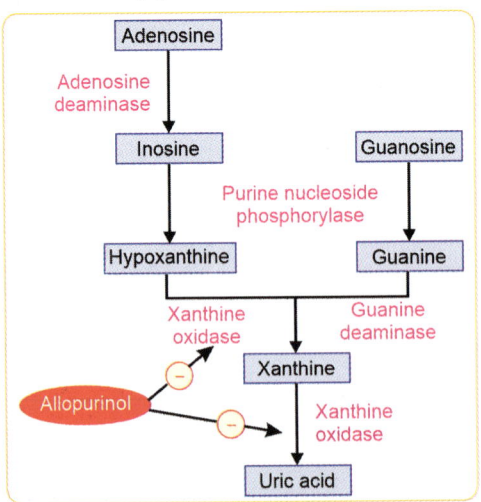

In Nonprimate Mammals, Uric Acid is Converted into Allantoin

- ❑ In nonprimate mammals, uricase also known as urate oxidase converts uric acid to water-soluble allantoin.
- ❑ Primates including humans lack uricase.
- ❑ Uric acid is an antioxidant. This can also be a reason for prolonged life span of primates compared to nonprimates.

$$\text{Uric acid} + H_2O + O_2 \xrightarrow{\text{Uricase}} \text{Allantoin} + H_2O_2 + CO_2$$
(Low solubility) → (High solubility)

Products of Pyrimidine Catabolism are Water-soluble

- ❑ CO_2, NH_3, β-alanine, and β-aminoisobutyrate are the end products of pyrimidine catabolism.
- ❑ β-alanine is produced from cytosine/uracil catabolism.
- ❑ β-aminoisobutyrate produced from thymine catabolism.

DISORDERS OF NUCLEOTIDE METABOLISM

ADENOSINE DEAMINASE (ADA) DEFICIENCY IS ONE OF THE CAUSES OF SEVERE COMBINED IMMUNODEFICIENCY (SCID)

- ❑ When ADA is deficient, adenosine takes an alternate route for degradation (Adenosine → AMP → Inosine).
- ❑ But deoxy AMP has no alternate route of degradation → Accumulation of dAMP
- ❑ dATP is produced from dAMP → ↑ dATP inhibits ribonucleotide reductase
- ❑ DNA synthesis is halted → Rapidly proliferating cells like lymphocytes are more sensitive to the inhibition → Immunodeficiency

HYPERURICEMIA CAN LEAD TO GOUT

- ❑ Gout is a clinical syndrome characterized by hyperuricemia and recurrent acute arthritis due to deposition of monosodium urate (MSU) crystals.
- ❑ Even though uric acid is not toxic, it has temperature dependent limited solubility.
- ❑ As the pK_1 of uric acid is 5.8, uric acid exists as monosodium urate at the alkaline pH of plasma and as uric acid in the acidic pH of urine.
- ❑ Maximum solubility of urate at 37°C is 7 mg/dL.
- ❑ Solubility decreases in the cooler and more acidic parts of the body like metatarsophalangeal joint and urate is deposited in the form of monosodium urate monohydrate crystals, which is known as *tophi*.

Pathology of Gouty Arthritis

Uric acid remains protonated in acidic pH of urine and it is ten times less soluble compared to urate. This is the basis of formation of renal stones in hyperuricemia.

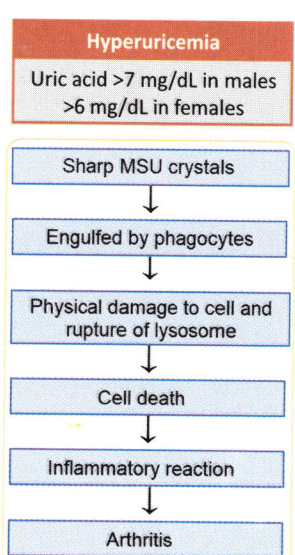

Causes of hyperuricemia	
1° hyperuricemia	**2° hyperuricemia**
❑ Over activity of PRPP synthetase ❑ Partial and complete deficiency of HGPRT	❑ Increased cell turn-over: Leukemia, psoriasis, polycythemia vera ❑ Hemolytic anemia ❑ Radiotherapy and chemotherapy leads to tumor lysis syndrome which causes urate nephropathy ❑ Hereditary fructose intolerance ❑ Von-Gierke's disease ❑ Salicylates (underexcretion) ❑ Preeclampsia

Another way of classification of hyperuricemia is the causes of overproduction and underexcretion.

Overproduction	Underexcretion
Idiopathic Over activity of PRPP synthetase Deficiency of HGPRT Malignancies Hemolytic anemia Psoriasis	Renal failure Acidosis Diuretics Hypothyroidism Preeclampsia

arrison's
Corner

Hyperuricemia due to both overproduction and underexcretion:
❑ Von-Gierke disease (Gluose-6-phosphatase deficiency)
❑ Hereditary fructose intolerance Aldolase B (Frucose-1-phosphate aldolase deficiency)
❑ Alcohol *(Chapter 14) (Refer to page no. 308)*
❑ Shock   *(Ref. Harrison 19th Ed., p.431e-2; Table 431e-2)*

Diagnosis

❑ Needle aspiration of affected joint or tophi and microscopic examination is mandatory for confirming the diagnosis.

Treatment

Treatment modality is different for an acute attack and maintenance therapy.

Acute attack: Acute attack is treated by drugs to relieve pain (analgesics) and drugs that inhibit inflammation (anti-inflammatory).

❑ **Colchicine** → inhibits microtubule polymerization by binding to tubulin → inhibits the movement of phagocytes (*Colchicine has no effect on metabolism or excretion of urate*).

❑ Nonsteroidal anti-inflammatory drugs (NSAIDs)

Maintenance therapy: Aim is to keep the level of uric acid low (5.0–6.0 mg/dL). Dietary restriction of purine rich foods does not have that much role. Avoiding alcohol is important.

Drugs used are:

Uricosuric agents—Drugs that enhance the excretion of uric acid	
Probenecid, sulfinpyrazone, benzbromarone	Inhibits the tubular reabsorption
Xanthine oxidase inhibitors—decrease the formation	
Allopurinol	Suicidal inhibitor
Febuxostat	Noncompetitive inhibitor

Contd...

Recombinant urate oxidase—used in special situations	
Rasburicase	Converts uric acid to more soluble allantoin; used to prevent uric acid nephropathy in patients of tumor lysis syndrome

arrison's
Corner

Serum uric acid levels can be normal or low at the time of an acute attack of gout, as inflammatory cytokines can be uricosuric and effective initiation of hypouricemic therapy can precipitate attacks. *(Ref. Harrison 19th Ed., p.2234)*

Know the Difference

	Gout	Pseudogout
Accumulation of	Mono sodium urate (MSU) crystals	Calcium pyrophosphate dihydrate (CPPD)
Etiopathogenesis	Overproduction or underexcretion of uric acid	Excessive breakdown of ATP → pyrophosphate → bind with calcium → Hyaline cartilages are affected most commonly
Joint involved	1st metatarso phalangeal	Knee
Shape of crystal	Needle	Rhomboid
Appearance of crystal on polarized light	Negative birefringence	+ve birefringence

LESCH-NYHAN SYNDROME

X-linked recessive disorder with *complete absence of HGPRT* enzyme of purine salvage pathway. Enzymes of *de novo* synthesis of purines are absent in brain. Therefore, brain is dependent entirely upon salvage pathway.

Reason for Hyperuricemia

❑ HGPRTase enzyme reutilizes (salvages) purine bases. When this enzyme is deficient, guanine, xanthine or hypoxanthine are not salvaged and therefore degraded to uric acid.
❑ Increased PRPP stimulates the *de novo* purine synthesis and excess purine nucleotides are degraded to uric acid.

Clinical Features

❑ Affected children are usually males as the disorder is X-linked.
❑ Hyperuricemia
❑ Choreoathetosis
❑ Spasticity
❑ Mental retardation
❑ Self-mutilation—biting tongue, fingers, chewing lips—striking feature of this disease.

Treatment

❑ No curative treatment is available. Removal of deciduous teeth is done to avoid biting. Other management is symptomatic.
❑ Genetic counseling and carrier testing for prevention.

☺ LESCH ☺

Lips and fingers are bitten
Excessive uric acid (Gout)
Self-mutilation
Choreoathetosis
HGPRTase deficiency

*Partial deficiency of HGPRTase results in the **Kelley-Seegmiller
syndrome**. It is associated with hyperuricemia but there are no
central nervous system manifestations.*

OROTIC ACIDURIA

Orotic aciduria is a *disorder of pyrimidine synthesis* associated with
excessive excretion of orotic acid in urine.

The UMP synthase is a bifunctional enzyme with two enzymatic
activities within the same polypeptide - Orotate phosphoribosyl
transferase and OMP decarboxylase

The UMP synthase defect leads to orotic aciduria. So, UMP
cannot be synthesized. This leads to failure of DNA synthesis
and megaloblastic anemia. If you give *uridine* to the patient orally, the metabolic block is surpassed and other
pyrimidines can be synthesized.

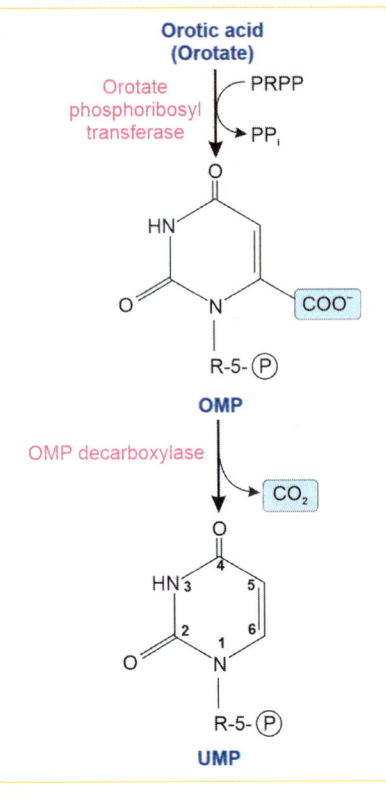

Etiological types of orotic aciduria	
Hereditary	**Disorders of pyrimidine synthesis:** ❑ **Type I** orotic aciduria—Orotate phosphoribosyltransferase and OMP decarboxylase (*both* the activities of bifunctional enzyme) ❑ **Type II** orotic aciduria—OMP *decarboxylase* defect
	Urea cycle disorders: ❑ Ornithine transcarbamoylase (OTC) deficiency
Acquired	❑ **Drugs:** 　○ Allopurinol 　○ 5-fluorouracil ❑ Other disorders causing mitochondrial dysfunction: 　○ Reye syndrome (due to aspirin intake during viral illness)

Biochemical Basis of Orotic Aciduria Urea-cycle Disorders

Mitochondrial ornithine transcarbamoylase deficiency → ↑carbamoyl phosphate levels → enters cytoplasm →
↑pyrimidine synthesis → mild orotic aciduria

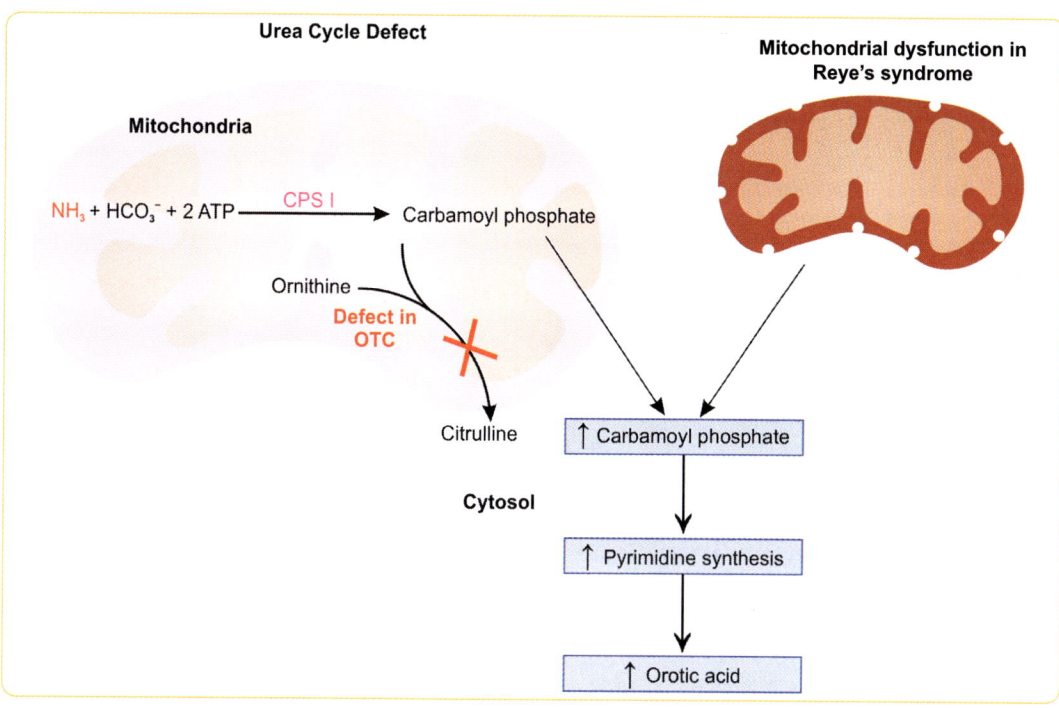

Biochemical Basis of Orotic Aciduria in Reye's Syndrome

Mitochondrial dysfunction → Carbamoyl phosphate cannot be utilized and leaks to cytoplasm → ↑pyrimidine synthesis → mild orotic aciduria

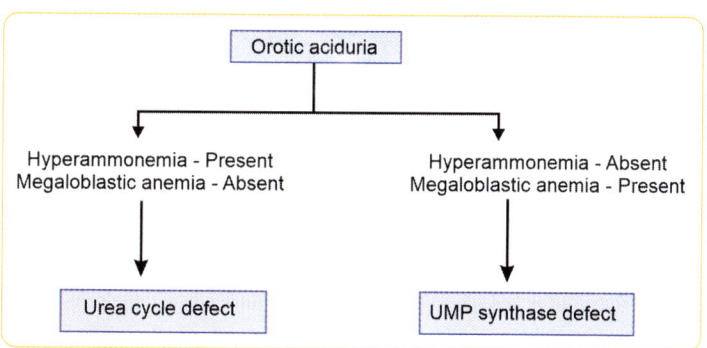

 **Important** *Revise*

Disorders of purine and pyrimidine metabolism

Defect	Disease
PRPP synthetase over activity	Gout
Partial deficiency of HGPRTase	Kelley-Seegmiller syndrome
Complete deficiency of HGPRTase	Lesch-Nyhan syndrome

327

Contd...

 **Important** ***Revise***

Disorders of purine and pyrimidine metabolism

Defect	Disease
Adenosine deaminase deficiency	Autosomal recessive severe combined immunodeficiency (SCID)
Purine nucleoside phosphorylase	SCID
Orotate phosphoribosyltransferase and orotidylic acid decarboxylase (Bifunctional enzyme)	Orotic acid aciduria type 1
Orotidylic acid decarboxylase activity only	Orotic acid aciduria type 2

INHIBITORS OF PYRIMIDINE BIOSYNTHESIS

Drug	Enzyme inhibited	Use
Leflunomide	Dihydroorotate dehydrogenase	Disease-modifying antirheumatic drug (DMARD)
Methotrexate	Mammalian dihydrofolate reductase	Anticancer
Trimethoprim	Prokaryotic dihydrofolate reductase	Antibiotic
Proguanil	Protozoal dihydrofolate reductase	Antimalarial
5-fluorouracil (5-FU)	Thymidylate synthase	Anticancer

Hydroxyurea inhibits the formation of both purine and pyrimidine deoxyribonucleotides.

COFFEE, TEA AND COCOA CONTAIN METHYL XANTHINES

Xanthine (2, 6-dioxy purine) is a purine base. Methylxanthines are present in coffee, tea and cocoa.
- Coffee contains caffeine.
- Tea contains theophylline.
- Cocoa contains theobromine.

Methylxanthines inhibit the phosphodiesterase enzyme and increase the level of cyclic AMP.

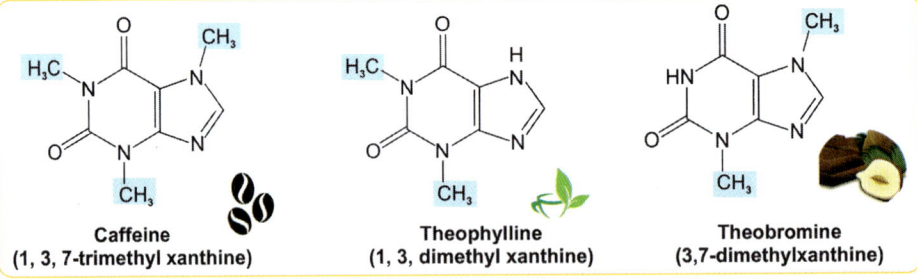

Fig. 5: Structure of methylxanthines. They differ in the number of methyl groups

MODIFIED BASES ARE FOUND IN TRANSFER RNA

In addition to the ordinary bases, transfer RNA contains lot of modified bases like 5-methylcytosine, 5-hydroxymethylcytosine, and pseudouridine (ψ). *We will discuss this in Section III.*

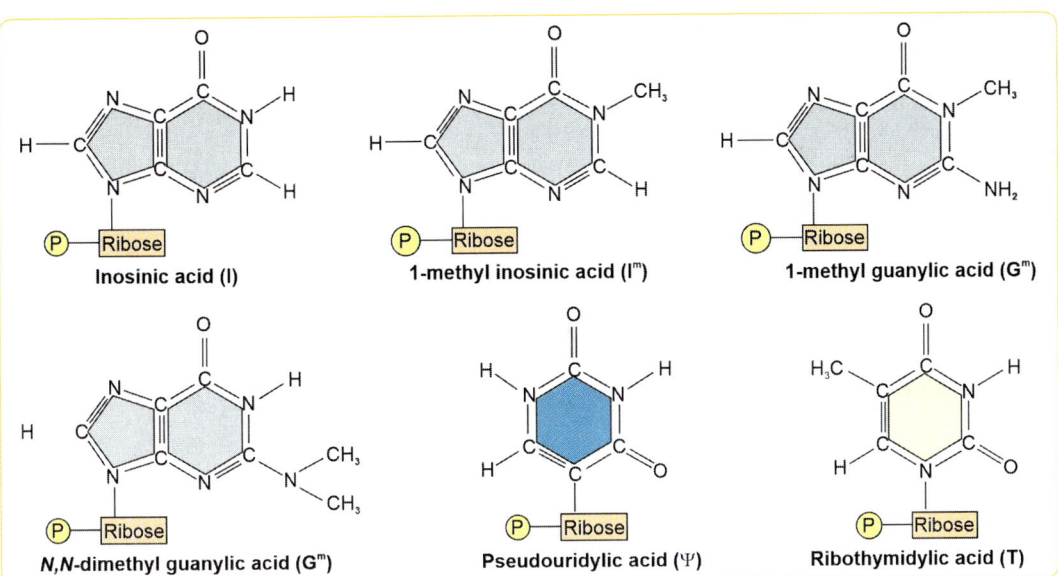

Inosinic acid (I) | 1-methyl inosinic acid (Iᵐ) | 1-methyl guanylic acid (Gᵐ)

N,N-dimethyl guanylic acid (Gᵐ) | Pseudouridylic acid (Ψ) | Ribothymidylic acid (T)

SYNTHETIC ANALOGUES ARE USED IN RESEARCH AND THERAPEUTICS

Base Analogues

5-bromouracil (5-BrU) is a thymine analogue. It is incorporated into DNA instead of thymine. However, in further replication cycle ionized form of bromouracil base pairs with guanine resulting in (transition) mutation. 5-BrU is used as an experimental mutagen in research.

5-BrU (Keto)　　Adenine　　5-BrU (Enol)　　Guanine

Fig. 6: 5-Bromouracil causes transition mutation (purine replaced by another purine)

Nucleoside Analogues

❑ Nucleoside analogues inhibit the DNA synthesis. So, they are used as anti-viral and chemotherapeutic agents.
❑ Dideoxynucleotides does not contain 3′ OH group in the ribose. So, they cannot be further elongated to produce 3′→ 5′ phosphodiester bond. So, DNA chain is terminated.
❑ Cytarabine is cytosine arabinoside. This contains arabinose sugar instead of ribose.

	Example and uses
Deoxyadenosine analogues	Didanosine (ddI) (HIV)
Deoxycytidine analogues:	Cytarabine (chemotherapy) Zalcitabine/dideoxycytidine (HIV)
Guanosine analogue	Abacavir (HIV) Acyclovir (Herpes virus)
Thymidine analogue	Zidovudine also known as azidothymidine (HIV)

Chapter 15 **Nucleotide Chemistry and Metabolism**

329

MECHANISM OF ACTION OF ACYCLOVIR

❑ Acyclovir is an antiviral drug used mainly for herpes zoster infection.
❑ Acyclovir is selectively acted upon by the viral thymidine kinase.
❑ The enzyme phosphorylates acyclovir to a monophosphate.
❑ The acyclovir monophosphate (acyclo-GMP) is subsequently converted into acyclovir triphosphate (acyclo-GTP) by human enzymes.
❑ Acyclo-GTP persists is a potent inhibitor viral DNA polymerase compared to human DNA polymerases. Thus, viral replication is selectively inhibited in the infected cells.

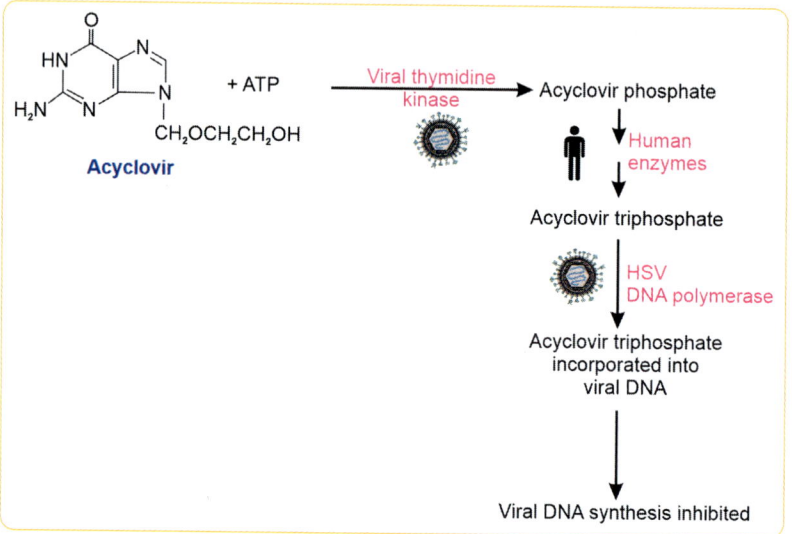

https://youtu.be/uaViOLPKW8E or scan QR code

Competency

❑	BI6.2	Describe and discuss the metabolic processes in which nucleotides are involved
❑	BI6.3	Describe the common disorders associated with nucleotide metabolism
❑	BI6.4	Discuss the laboratory results of analytes associated with gout & Lesch Nyhan syndrome

Key Points

- Nucleotides are dietarily not essential since both purine and pyrimidine nucleotides can be synthesized by *de novo* or salvage pathways.
- Nucleotides in the diet are usually degraded to uric acid.
- Phosphoribosyl pyrophosphate (PRPP) is required for the *de novo* and salvage synthesis of purine and pyrimidine nucleotides, NAD$^+$, and NADP$^+$.
- Glutamine phosphoribosyl Pyrophosphate amidotransferase (Amido phosphoribosyl transferase) catalyzes the committed step of *de novo* purine synthesis.
- Inosine monophosphate (IMP) is the first purine nucleotide to be synthesized and orotidine monophosphate (OMP) is the first pyrimidine nucleotide to be synthesized.
- 6 ATP equivalents are required for the assembly of a purine ring.
- Thymine is synthesized from methylation of dUMP.
- Deamination of methylated cytosine produces thymine
- RBCs and brain lack the capacity of *de novo* purine biosynthesis. They are dependent on salvage pathway.
- Uric acid is the end product of purine metabolism in primates. Allantoin is the end product of purine metabolism in nonprimate mammals.
- Uricase converts uric acid to allantoin; used to prevent urate nephropathy in tumor lysis syndrome.
- Xanthine oxidase is a molybdenum containing flavoprotein.
- Complete deficiency of HGPRTase leads to Lesch-Nyhan syndrome (X-linked disorder) whereas partial deficiency results in Kelley-Seegmiller syndrome.
- CPS II catalyze the rate-limiting step of *de novo* pyrimidine synthesis.
- Hydroxy urea inhibits ribonucleotide reductase.
- β-alanine is the catabolic end product of cytosine/uracil.
- β-aminoisobutyrate is the catabolic end product of thymine.

SELF-ASSESSMENT

Short Answer Questions

1. Enumerate any four biological roles of nucleotides.

2. A boy with ornithine transcarbamoylase (OTC) deficiency was found to have an increased level of orotic acid in his urine. What is the biochemical basis of this?

3. Explain why adenosine deaminase deficiency leads to Severe Combined Immunodeficiency (SCID).

4. Explain why rasburicase is given to cancer patients undergoing chemotherapy

5. Explain the biochemical basis of hyperuricemia in Von Gierke's disease and Lesch-Nyhan syndrome.

6. Name two causes of hyperuricemia caused by both overproduction and underexcretion of uric acid.

7. Name any four inherited enzyme defects associated with hyperuricemia.

8. Define Gout. Explain the biochemical basis of secondary gout. Add a note on clinical features and management.

9. Differentiate Gout and Pseudogout.

10. Name a synthetic nucleotide analog, which is used in the treatment of gout. Mention its mechanism of action.

11. Name the diseases that are associated with defects in purine degradation pathway along with their symptoms, molecular basis, and management strategies.

12. Describe Lesch-Nyhan Syndrome under the following headings – Mode of Inheritance, Deficient enzyme, Clinical features, Management

13. Draw a diagram to show the sources of nitrogen and carbon atoms of the purine ring when it is produced by *de novo* synthetic pathway.

14. Name the end products of purine and pyrimidine catabolism.

15. What is the enzyme target of hydroxyurea? In which disease this drug is used.

16. What are nucleotide analogues? Explain their role in medicine.

17. Indicate the general mechanism of action of synthetic nucleotide analogues. Name one example each of an antiviral, anticancer and immunosuppressant drug of this group and specify their mechanism of action.

18. What is the mechanism of action of acyclovir?

Clinical Case-Based Questions

1. A 40-year-old man presented with severe pain and swelling in his right great toe. Past medical history revealed that he is diabetic and hypertensive. He is on thiazides and metformin. Personal history revealed that he is a chronic alcoholic. On local examination, there was swelling, erythema, warmth and tenderness of the affected area. Synovial fluid was aspirated and revealed needle-shaped crystals that were negatively birefringent under polarizing microscopy.

    a. What is the likely diagnosis? Mention the biochemical basis of this disorder.
    b. Name any hereditary conditions that can cause this disorder.
    c. What is the procedure that is needed for the definitive diagnosis of this disorder?

Multiple Choice Questions

1. **Nucleotides serve all of the following roles, EXCEPT:**
    a. Monomeric units of nucleic acids
    b. Mediators in cellular signaling
    c. Source of energy
    d. Structural component of membrane

2. **Which of the following amino acids is directly involved in purine biosynthesis?**
    a. Glycine
    b. Tryptophan
    c. Ornithine
    d. Alanine

3. **Which of the following base is usually not present in DNA?**
    a. Adenine
    b. Guanine
    c. Cytosine
    d. Uracil

4. **End product of purine catabolism is:**
    a. Creatine
    b. Urea
    c. Uric acid
    d. Water + ammonia

5. **End product of purine catabolism in nonprimate mammals is:**
    a. Allantoin
    b. Inosine
    c. Uric acid
    d. Xanthine

6. **Beta-alanine is derived from:**
    a. Adenine
    b. Guanine
    c. Adenosine
    d. Uracil

7. **Nitrogen atoms in purines are derived from all, EXCEPT:**
    a. Aspartate
    b. Glutamine
    c. Glycine
    d. THFA

8. **Lesch Nyhan syndrome is due to:**
    a. Total deficiency of HGPRT
    b. Partial deficiency of HGPRT
    c. Total deficiency of PRPP
    d. Partial deficiency of PRPP

9. **In humans, rate limiting step of *de novo* pyrimidine synthesis is:**
    a. Aspartate transcarbamoylase
    b. Carbamoyl phosphate synthase-II
    c. Dihydroorotate dehydrogenase
    d. Ornithine transcarbamoylase

10. **Orotic aciduria is a disorder of:**
    a. Purine synthesis
    b. Pyrimidine synthesis
    c. Purine catabolism
    d. Pyrimidine catabolism

11. **All of the following are correct statements regarding sources of purine ring, EXCEPT:**
    a. C_2 is from formyl THFA
    b. C_4, C_5 and C_7 are from glycine
    c. C_6 is from respiratory CO_2
    d. N_3 and N_9 are from amide nitrogen of glutamine

12. **Which of the following condition causes hyperuricemia due to both increased production and decreased excretion?**
    a. PRPP synthetase over activity
    b. Glucose-6-phosphatase deficiency
    c. HGPRT deficiency
    d. Renal failure

13. **Which of the following is the substrate for the enzyme ribonucleotide reductase?**
    a. Nucleoside monophosphate
    b. Nucleoside diphosphate
    c. Nucleoside triphosphate
    d. All of the above

14. **Inhibition of which of the following enzyme is responsible for the anticancer action of 5-fluorouracil?**
    a. Dihydrofolate reductase
    b. Thymidylate kinase
    c. Thymidylate reductase
    d. Thymidylate synthase

15. **All of the following are true about xanthine oxidase, EXCEPT:**
    a. Contains iron
    b. Contains molybdenum
    c. Flavoprotein
    d. Produces H_2O

16. **A 1-year-old female patient is lethargic, weak, and anemic. Her height and weight are low for her age. Her urine contains an elevated level of orotic acid. The activity of uridine monophosphate synthase is low. Administration of which of the following is most likely to alleviate her symptoms?**
    a. Adenine
    b. Guanine
    c. Thymidine
    d. Uridine

17. **All of the following are true regarding acute gouty arthritis, EXCEPT:**
    a. Allopurinol is effective to treat the acute attack
    b. MSU crystals are needle shaped and negatively birefringent
    c. Serum uric acid levels can be normal or low at the time of an acute attack
    d. Tophi are made up of monosodium urate crystals (MSU)

18. **A male child presents with delayed development and scarring of the lips and hands. His parents have restrained him because he obsessively chews on his lips and fingers, which of the following is likely to occur in this child?**
    a. Increased levels of 5-phosphoribosyl-1 pyrophosphate (PRPP)
    b. Decreased purine synthesis
    c. Decreased levels of uric acid
    d. Glycogen storage disorder

19. **The rate of DNA synthesis in a culture of cells could be most accurately determined by measuring the incorporation of which of the following radioactively labeled compounds?**
    a. Adenine   b. Guanine
    c. Phosphate   d. Thymidine

20. **BrdU is an analogue of:**
    a. Adenosine   b. Cytidine
    c. Guanosine   d. Thymidine

21. **Which of the following causes hyperuricemia?**
    a. HGPRT overactivity
    b. PRPP synthetase deficiency
    c. Glucose 6-phosphatase deficiency
    d. Glucose phosphate dehydrogenase deficiency

22. **Deficiency of all of the following enzymes lead to hyperuricemia, EXCEPT:**
    a. HGPRT   b. Aldolase B
    c. Adenosine deaminase
    d. Glucose-6-phosphatase

23. **Acyclovir inhibits:**
    a. Viral adenylate kinase
    b. Viral ribonucleotide reductase
    c. Viral thymidylate kinase
    d. Viral thymidylate synthase

ANSWERS

1.	d.	2.	a.	3.	d.	4.	c.	5.	a.	6.	d.
7.	d.	8.	a.	9.	b.	10.	b.	11.	b.	12.	b.
13.	b.	14.	d.	15.	d.	16.	d.	17.	a.	18.	a.
19.	d	20.	d.	21.	c.	22.	c.	23.	c.		

Chapter 15 Nucleotide Chemistry and Metabolism

16 CHAPTER

Integration of Metabolism

Integration of metabolism ensures that all the organs get the appropriate metabolic fuels all the time. *So far, we have studied the metabolism of carbohydrate, proteins and lipids along with their regulation. In this chapter, we will first review what we have studied and then discuss how our body integrates the metabolism.*

TYPES OF METABOLIC REACTIONS WE HAVE STUDIED SO FAR

Type of reaction	Reaction	Enzyme catalyzing
Oxidation of substrate	Ethanol to acetaldehyde	Alcohol dehydrogenase
Reduction of substrate	HMG CoA to Mevalonate	HMG-CoA reductase
Carboxylation	Pyruvate to oxaloacetate	Pyruvate carboxylase
Oxidative decarboxylation (Removal of CO_2 and electrons)	Pyruvate dehydrogenase	Pyruvate to Acetyl-CoA
Oxidative deamination (Removal of ammonia and electrons)	Glutamate to ammonia and NAD(P)H	Glutamate dehydrogenase
Hydrolysis	Sucrose to glucose and fructose	Sucrase
Hydration	Fumarate to Malate	Fumarase
Dehydration	2-phosphoglycerate to phosphoenolpyruvate	Enolase
Condensation by dehydration	Peptide bond formation	Peptidyl transferase
Group-transfer	Transfer of phosphoryl group from ATP to Glucose	Hexokinase or Glucokinase
Isomerization	Glucose-6-phosphate to fructose-6-phosphate	Phosphohexose isomerase

STORAGE FORM OF ENERGY-RELEASING MACROMOLECULES

Carbohydrates, lipids and proteins are the energy-producing macromolecules. Carbohydrate is stored as glycogen. Lipids are stored as triacylglycerol in adipocytes. The storage capacity for glycogen is limited, whereas fat storage capacity is unlimited. Proteins have no storage form. Body proteins are broken down during starvation.

METABOLIC PROFILE OF ORGANS

Brain

- ❏ There is no stored fuel in brain and it uses glucose as the fuel.
- ❏ Brain requires continuous supply of glucose (120 g of glucose per day).
- ❏ **Brain cannot use free fatty acids** (Chapter 13.1) *(Refer to page no. 248)*.
- ❏ Hypoglycemia leads to neuroglycopenic symptoms.
- ❏ **Glucose enters the brain through GLUT1 and GLUT3**.
- ❏ **During starvation, brain can use ketone bodies produced by the liver.**
- ❏ **Ammonia is detoxified into glutamine.**
- ❏ De novo purine synthesis enzymes are absent; brain is dependent on purine salvage pathway.

Heart Primarily Uses Fatty Acids Aerobically

- ❏ Cardiac myocytes are rich in mitochondria and their metabolism is purely aerobic under normal conditions.
- ❏ The main fuels are free fatty acids and ketone bodies.
- ❏ Lipoprotein lipase present in the cardiac capillary endothelium releases free fatty acids from the chylomicron and VLDL.
- ❏ Only under hypoxic conditions, glucose is fermented to lactic acid.
- ❏ Glucose transport takes place via both GLUT1 (insulin independent) and GLUT4 (insulin dependent).

Biochemical Basis of Ischemia-Reperfusion Injury

Biochemical Changes in Ischemia

- ❏ Under ischemic condition, heart switches to the anaerobic metabolism. You know that glucose is the only fuel that can be utilized anaerobically. Anaerobic glycolysis leads to accumulation of lactate → intracellular acidification.
- ❏ As there is no oxygen to accept the electrons in mitochondrial electron transport chain, the direction of proton pump gets reversed. i.e. F_0F_1ATP pump becomes F_0F_1 ATPase and consumes ATP to pump protons from the matrix into the cytosol. Surprising, isn't it? This leads to intracellular acidification.
- ❏ Cell tries to get rid of the protons by activating Na^+H^+ exchanger. H^+ goes out and Na^+ comes in. This Na^+ cannot be pumped out because of failure of ATP requiring $Na^+ K^+$ ATPase. So, $Na^+ Ca^{++}$ exchanger is activated resulting in calcium overload. Raised intracellular calcium activates many hydrolases that can cause cell damage.
- ❏ As there is shortage of ATP during ischemia, available ADP is converted to ATP by adenylate kinase/Myokinase enzyme.

$$\text{ADP} + \text{ADP} \underset{}{\overset{\text{Myokinase}}{\rightleftharpoons}} \text{ATP} + \text{AMP}$$

Biochemical Changes During Reperfusion

- ❏ When oxygen is restored, AMP is degraded to uric acid by xanthine oxidase reaction. We have studied that xanthine oxidase produces H_2O_2 – reactive oxygen species.
- ❏ When the oxygen is restored, there is a sudden increase in fatty acid oxidation and an increased NADH production.
- ❏ Abrupt change in pH and calcium leads to opening of mitochondrial permeability transition pores (mPTP).
- ❏ This is why partial fatty oxidation inhibitors have been tried in some types of angina.

Chapter 16 Integration of Metabolism

RBCs Lack Mitochondria and Convert Glucose to Lactate

❏ Glucose is the only fuel that can be utilized in the absence of oxygen. So, anaerobic glycolysis is the only source of energy for RBCs.
❏ 30% of glucose is metabolized in HMP shunt to produce NADPH.
❏ As there are no mitochondria, ketone bodies cannot be used.
❏ Rapaport-Luebering cycle operating helps in the unloading of oxygen but compromises the ATP production by 1,3 BPG kinase step.
❏ Pyruvate kinase deficiency compromises the substrate level phosphorylation → No ATP production → Inability to maintain $Na^+K^+ATPase$ pumps → RBC lysis → Hemolytic anemia

At Resting State, Skeletal Muscle uses free Fatty Acids

❏ Skeletal muscle uses free fatty acids at rest, glucose while exercising, amino acids and ketone bodies during fasting and starvation.
❏ In fed state, entry of glucose is through insulin dependent GLUT-4; Glucose is converted to glycogen in muscle.
❏ After an overnight fasting, the level of glucose transporters is reduced in skeletal muscle.
❏ Skeletal muscle does not contain glucagon receptor.
❏ Catecholamines stimulate glycogenolysis in skeletal muscle.
❏ Muscle glycogenolysis can't increase the blood glucose because of absence of glucose-6-phosphatase (Chapter 10.4) *(Refer to page no. 144)*.
❏ Creatine phosphate is a phosphagen (*Chapter 4*) *(Refer to page no. 47)* in muscles.
❏ Lactate produced in anaerobic glycolysis of muscle is utilized by liver for gluconeogenesis (Cori cycle Chapter 10.3) *(Refer to page no. 134)*

Source of ATP for skeletal muscle during vigorous physical activity	
Source of ATP in a 100-m sprint	**Source of ATP in a marathon (Aerobic metabolism)**
❏ First 4 to 5 seconds – Stored ATP, Creatine phosphate ❏ After 5 seconds – Anaerobic glycolysis	❏ Till 4 minutes – Blood glucose ❏ 5 to 18 minutes – Hepatic glycogen ❏ 19 to 70 minutes – Muscle glycogen ❏ Till 4000 minutes – Triacylglycerol

Liver is a Versatile Organ

❏ Liver preferentially utilizes fatty acids and carbon skeleton of amino acids.
❏ In fed state, glucose enters the liver through high capacity, low affinity glucose transporter GLUT2.
❏ Liver stores glycogen and provides glucose during fasting.
❏ Liver converts fatty acids to ketone bodies during starvation.
❏ Liver cannot utilize the ketone bodies because of absence of the thiophorase (β-ketoacyl-CoA transferase) enzyme.
❏ Kidney is an important gluconeogenic organ next to liver.

Fuel Preference of Organs

Organ	Fed state	Fasting	Starvation
Brain		Glucose	Glucose and Ketone bodies
Muscles	Glucose, Fatty acids		Fatty acid
Liver	Glucose	Fatty acid	Amino Acid
Adipose tissue	Glucose		Fatty acids
Heart	Fatty acid		Fatty acid and Ketone bodies

In fed state, adipocytes take up fatty acids through lipoprotein lipase for storage purpose, not as a fuel.

Section II Intermediary Metabolism

FAST-FEED CYCLE

Let us discuss the metabolic changes under the following conditions:
- ❑ Absorptive (Fed) state (0 to 4 hours after meal)
- ❑ Post-absorptive (Fasting) state
- ❑ Starvation state

Fed or Absorptive State

- ❑ The absorptive state is the initial 3 to 4 hours period after ingestion of a normal meal.
- ❑ Nutrient-rich blood from intestine reaches the liver via portal system.
  - ○ Excess glucose is converted into glycogen by the liver.
  - ○ Amino acids are utilized for protein synthesis.
- ❑ There is a transient increase in plasma glucose, amino acids and triacylglycerols occurs. Elevated glucose and amino acids lead to increased secretion of insulin.
- ❑ Triglycerides are packaged into chylomicrons and enter the systemic circulation via thoracic duct (lymphatic system).
  - ○ Free fatty acids from chylomicrons are used by adipocytes for TAG synthesis.
  - ○ Free fatty acids from chylomicrons are used to produce energy by β oxidation in cardiac and skeletal muscles.
- ❑ Insulin is the major hormone regulating the metabolism in fed state.
  - ○ Promotes glycogen synthesis in liver and skeletal muscle.
  - ○ Promotes lipid synthesis in adipocytes.
  - ○ Promotes the uptake of branched chain amino acids by skeletal muscle.
- ❑ During the absorptive period, most of the tissues use glucose as a fuel (Fig. 1).

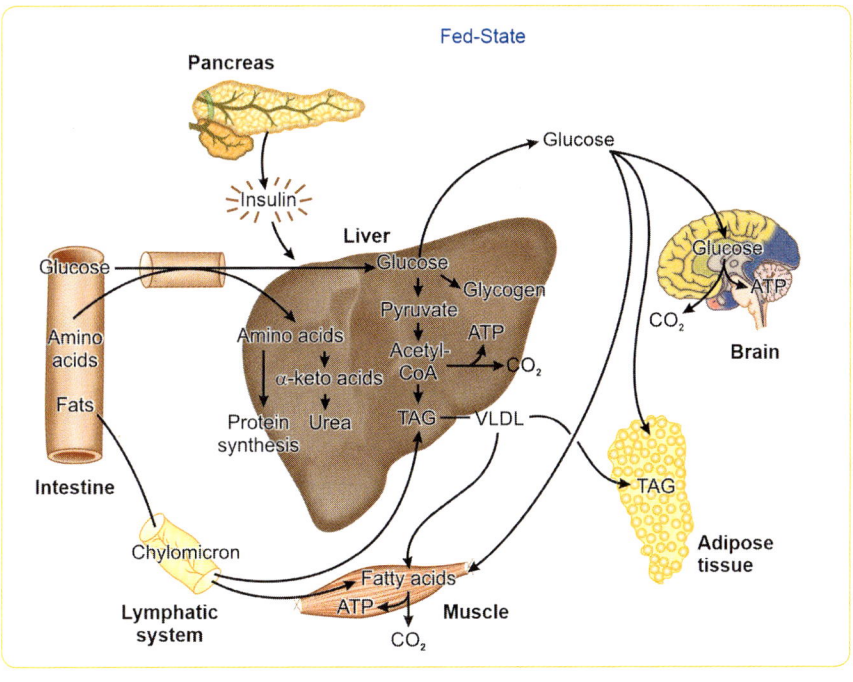

Fig. 1: Events during fed state

Post-Absorptive (Fasting) State

- ❑ Blood glucose level decreases after a meal. Insulin level decreases and the glucagon level increases.
- ❑ Glucagon is produced from the alpha cells of pancreas.

- ❑ Glucagon
    - ○ Increases the level of cAMP and activates protein kinase A.
    - ○ Enhances glycogenolysis in liver and inhibits glycogen synthesis.
    - ○ Stimulates gluconeogenesis in liver.
    - ○ Mobilizes fatty acids from adipocytes.
- ❑ Fatty acids have sparing action on glucose, i.e. mobilized fatty acids are preferentially utilized by liver and muscle and thus sparing the glucose to be used by brain and RBCs.

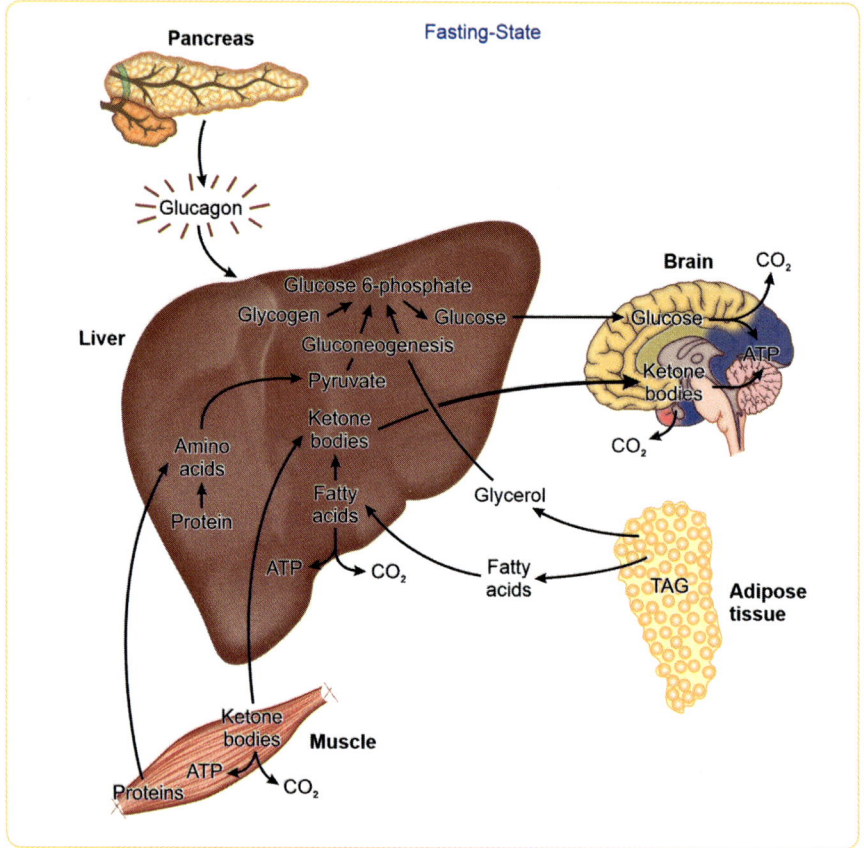

Fig. 2: Events during fasting state

STARVATION STATE

- ❑ When fasting is prolonged to the state of starvation, the highest priority is to maintain the optimum glucose level for brain.
- ❑ Gluconeogenic substrates like pyruvate and lactate are low. So, amino acids are broken down.
- ❑ But, break down of proteins is detrimental. So, the body minimizes protein break down after few days of starvation.
- ❑ Conversion of fatty acids to ketone bodies by liver becomes significant after the 3rd day of starvation.
- ❑ Ketone bodies supply 30% of brain's energy demand after 3rd day of starvation.
- ❑ Skeletal and cardiac muscle can also use ketone bodies.
- ❑ Excess ketone bodies lead to ketoacidosis.

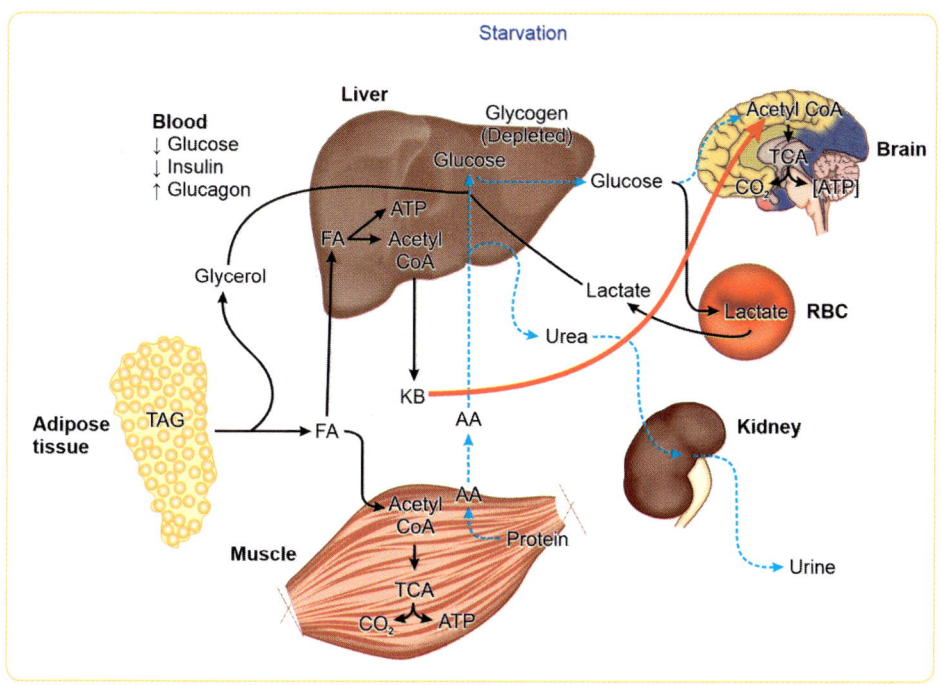

REFEEDING AFTER STARVATION

❑ Refeeding after prolonged starvation leads to potentially fatal fluid and electrolyte abnormality known as refeeding syndrome.

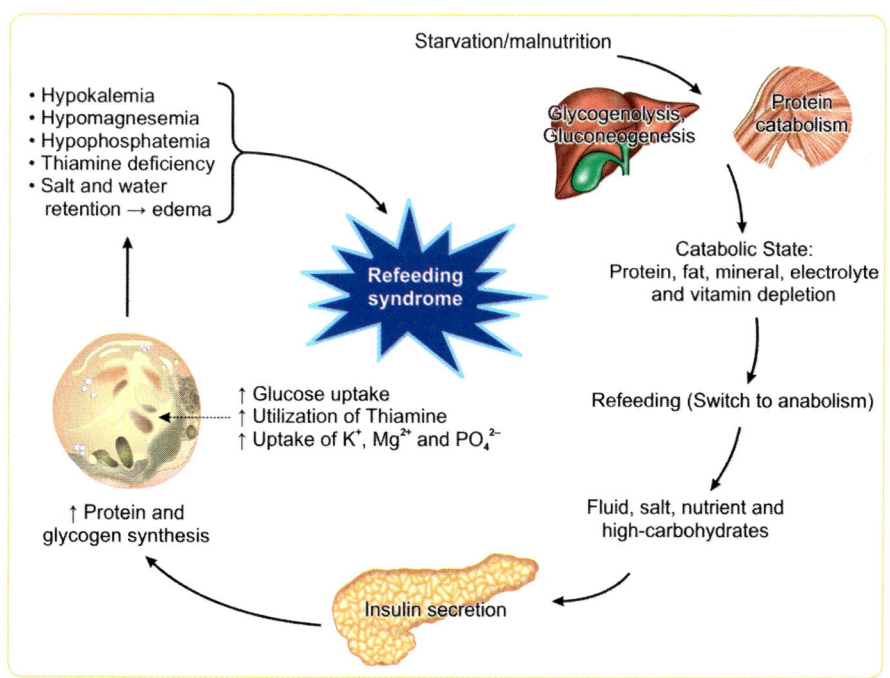

- During refeeding, insulin level increases. Insulin enhances the **intracellular shift of potassium** leading to hypokalemia.
- **Magnesium and phosphate are also taken up inside the cells** leading to decreased level in blood.
- It is always better to administer thiamine before refeeding (Chapter 33) *(Refer to page no. 520)*

SOURCES AND FATE OF METABOLIC FUELS

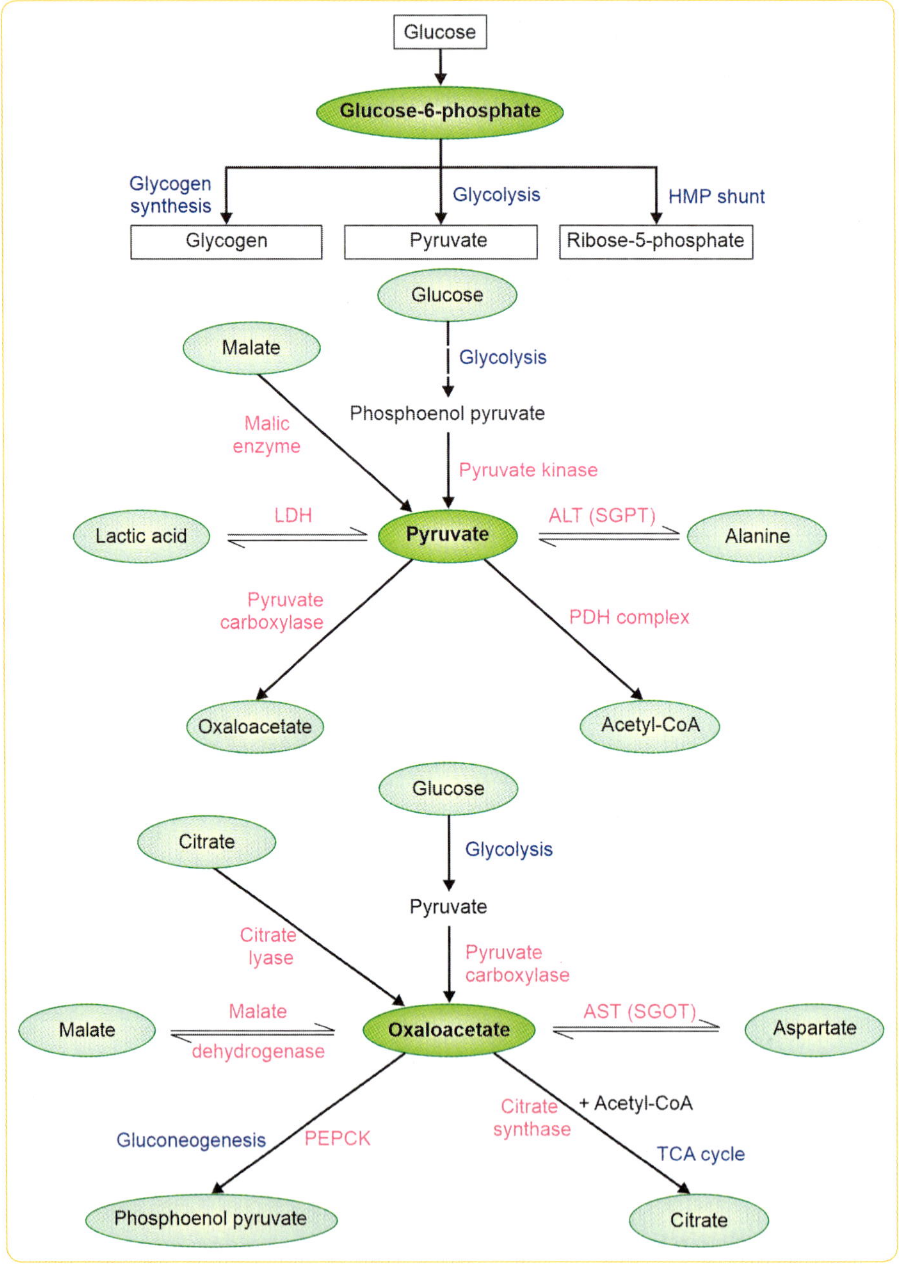

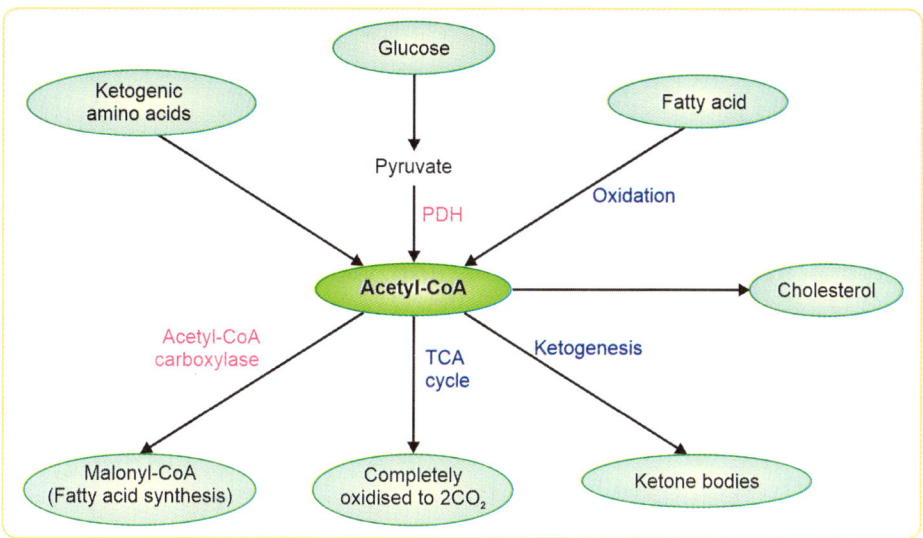

CALORIC HOMEOSTASIS, OBESITY AND METABOLIC SYNDROME

Molecules Involved in the Control of Appetite

During exam preparation, you may feel that eating thrice a day is a waste of time. But it is crucial for an organism to take adequate macromolecules and micronutrients in diet for the sake of survival.

The appetite stimulating molecules are known as orexigens whereas appetite supressing (or satiety inducing) molecules are known as anorexigens (Anorexia is loss of appetite).

The appetite is regulated by neurohormonal system. **Ghrelin** produced from the stomach is the start-signal for eating. Ghrelin induces appetite by stimulating Neuropeptide Y (NPY)/Agouti-related peptide (AgRP) neurons in hypothalamus.

When the food travels through the intestine, various anorexigenic molecules like Insulin, amylin, cholecystokinin, gastric inhibitory polypeptide (GIP) and glucagon-like peptide-1 are released. Anorexigenic molecules induce satiety.

Peptide YY produced from the ileum in response to fat-rich food is a stop-signal for eating. Peptide YY mediated inhibition of gastric emptying is known as **ileal brake phenomenon**. Storage of fat in adipocytes induces the production of leptin which is another anorexigenic molecule.

Remember the orexigenic peptides: Ghrelin, Neuropeptide Y (NPY) and Agouti related peptide (AgRP)

Leptin

Adipocytes are not just storage depot for fat. They do secrete many cytokines and hormones. **Adipocyte secreted hormones are known as adipokines**. Leptin is one among them. The quantity of leptin secretion is directly proportional to the adipocyte mass. Leptin binds to the membrane bound leptin receptors on the hypothalamus and **induces satiety** through JAK/STAT pathway. Leptin deficiency or leptin resistance leads to obesity. Because of its association with obesity, leptin gene is named as OB gene. In addition to caloric homeostasis, leptin is also important for puberty. **Leptin shows permissive effect on gonadotrophin releasing hormone during puberty.**

Obesity is a Chronic Inflammatory State

Obesity is a type of malnutrition. It has been found that obesity leads to a low-grade chronic inflammatory state in the body thus increasing the risk of diabetes mellitus, cardiovascular diseases and cancer.

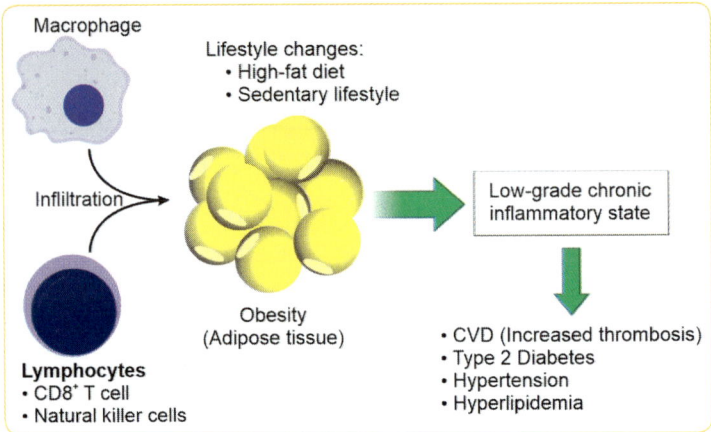

NCEP ATP III Criteria for Metabolic Syndrome

Metabolic syndrome (syndrome X) is diagnosed if any of the three or more is present.

❑ Central obesity	**Waist circumference:** ❑ >40 inches in males ❑ >35 inches in female
❑ Elevated Triglyceride	≥150 mg/dL or on hypolipidemic treatment
❑ Reduced HDL cholesterol	<40 mg/dL in males; <50 mg/dL in females on hypolipidemic treatment
❑ High blood pressure	Systolic ≥130 mm Hg; diastolic ≥85 mm Hg on anti-hypertensive treatment
❑ Impaired glucose tolerance	Fasting glucose ≥110 mg/dL or on treatment for Diabetes mellitus

❑ Syndrome X – metabolic syndrome
❑ Syndrome Y – Coronary slow-flow phenomenon
❑ Syndrome Z – Syndrome X + Obstructive sleep apnea

Competency

❑ BI6.1   Discuss the metabolic processes that take place in specific organs in the body in the fed and fasting states.
❑ PY3.11   Explain energy source and muscle metabolism

Section II Intermediary Metabolism

Key Points

❑ At resting state, skeletal muscle uses fatty acids as a fuel.

❑ For a 100 m sprinter, creatine phosphate is the principal source of ATP in the initial 5 seconds followed by anaerobic glycolysis.

❑ In a marathon runner, aerobic metabolism is the principal source of ATP. Blood glucose and free fatty acids derived from the breakdown of triacylglycerols in adipose tissue are the fuel sources.

❑ Refeeding after prolonged starvation leads to potentially fatal fluid and electrolyte abnormality known as refeeding syndrome.

❑ Peptide YY mediated inhibition of gastric emptying is known as ileal brake phenomenon.

❑ Ghrelin, Neuropeptide Y (NPY) and Agouti-related peptide (AgRP) are appetite-inducing (orexigenic) peptides.

❑ Leptin, an adipokine secreted by adipose tissues induces satiety.

❑ Obesity is a chronic inflammatory state - leads to cardiovascular disease and cancer.

SELF-ASSESSMENT

Short Answer Questions

1. Enumerate the metabolic profile of the following organs:
   a. RBC
   b. Skeletal muscle
   c. Liver

2. List the biochemical alterations at the end of 100 m sprint in muscle and blood.

3. Caloric homeostasis – short term and long-term

4. How does leptin help in the long-term control of caloric homeostasis? Which tissue/organ produces leptin in our body?

5. With a schematic illustration, show the sources and metabolic fate of:
   a. Pyruvate
   b. Acetyl-CoA
   c. Oxaloacetate

6. Explain why acetyl-CoA is said to be at the crossroad of metabolism.

7. Explain the biochemical basis of 'refeeding syndrome'.

Multiple Choice Questions

1. All of the following reactions are examples of oxidative decarboxylation, EXCEPT:
   a. Isocitrate dehydrogenase
   b. Pyruvate dehydrogenase
   c. Alpha ketoglutarate dehydrogenase
   d. Glutamate dehydrogenase

2. Which of the following glucose transporters are located in the brain?
   a. GLUT 1 and GLUT 2
   b. GLUT 1 and GLUT 3
   c. GLUT 2 and GLUT 3
   d. GLUT 2 and GLUT 4

3. All of the following are true about metabolic profile of brain, EXCEPT:
   a. Brain can't use free fatty acids
   b. Glucose enters the brain through GLUT1 and GLUT3.
   c. During starvation, brain produces ketone bodies.
   d. Ammonia is detoxified into glutamine

4. Which of the following is the major source of energy for cardiac myocytes?
   a. Amino acid   b. Fatty acid
   c. Glucose   d. Glycogen

5. Which of the following is the source of ATP for skeletal muscle in the second hour of marathon?
   a. Creatine phosphate
   b. Conversion of glucose to lactate
   c. Complete oxidation of glucose
   d. Oxidation of fatty acid

6. Which type of metabolic fuel is utilized for generating glucose by liver under conditions of severe starvation?
   a. Glycogen
   b. Fatty acids
   c. Starch
   d. Amino acid

7. **All of the following statements about metabolic profile of muscles are correct, EXCEPT:**
   a. After an overnight fasting, the level of glucose transporters is reduced.
   b. Glucagon and epinephrine stimulate glycogenolysis
   c. Muscle glycogenolysis can't increase the blood glucose
   d. Creatine phosphate is a phosphagen in vertebrate muscle

8. **All of the following are true about leptin, EXCEPT:**
   a. It is an adipokine
   b. It limits food intake
   c. Act through cyclic AMP
   d. It is a peptide hormone

9. **Which of the following molecule mediates ileal brake phenomenon?**
   a. Glucagon
   b. Peptide YY
   c. Ghrelin
   d. Leptin

10. **Which of the following increases appetite?**
    a. α-melanocyte-stimulating hormone
    b. Melanin-concentrating hormone (MCH)
    c. Glucagon-related peptide-1
    d. Serotonin

11. **The combination of metabolic syndrome (syndrome X) and which of the following is known as syndrome Z?**
    a. Diabetes mellitus
    b. Obstructive sleep apnea
    c. Cardiovascular disease
    d. Hypertension

12. **Which of the following hormone shows permissive effect during puberty?**
    a. Leptin
    b. Growth hormone
    c. Insulin
    d. GnRH

ANSWERS

| 1. d. | 2. b. | 3. c. | 4. b. | 5. d. | 6. d. |
| 7. b. | 8. c. | 9. b. | 10. b. | 11. b. | 12. a. |

Metabolism of Xenobiotics/ Biotransformation Reactions

Xenobiotics are chemical compounds that are foreign to the human body such as drugs, food additives and pollutants.

XENOBIOTICS ARE METABOLIZED PREDOMINANTLY BY LIVER IN 2 PHASES

Xenobiotics are metabolized in *2 phases* predominantly in *liver*. Usually, phase I reactions are followed by phase II reactions. But, drugs already possessing a polar group like –OH, $-NH_2$, or –COOH may enter phase II directly and become conjugated without prior phase I metabolism, e.g. isoniazid.

	Phase I	Phase II
Type of reaction	Hydroxylation (major)	Conjugation
Mechanism	Introduction or exposing reactive, polar group	Addition of polar compounds
Increase in hydrophilicity	Small	Large
Consequence	May result in metabolic activation or inactivation	Facilitates excretion

Biotransformation May Lead to Bioactivation or Detoxification

- Metabolism of xenobiotics *does not always lead to detoxification* but sometimes results in conversion of a procarcinogen to carcinogen. So, the term *biotransformation* of xenobiotics is preferred over detoxification of xenobiotics.
- Biotransformation reactions are not limited to xenobiotics. Endogenous compounds (endobiotics) like bilirubin and steroids also undergo biotransformation reactions. Thus, biotransformation reactions are involved in metabolism of both xenobiotics and endobiotics (entoxification).

Phase I reactions	Phase II reactions
Addition of polar group or unmasking of polar group	Conjugation reactions to make water soluble
❑ *Hydroxylation* by CYP450 (major) ❑ Deamination ❑ Dehalogenation ❑ Desulfuration ❑ Epoxidation ❑ Reduction ❑ Peroxygenation ❑ Hydrolysis	❑ Glucuronide conjugation ❑ Glutathione conjugation ❑ Sulfation ❑ Acetylation ❑ Methylation ❑ Glycine conjugation

PHASE I REACTIONS

Phase I reactions are mediated mainly by the Cytochrome P450 superfamily enzymes. For easy understanding, I am giving the properties of the enzyme under the following headings:

P450: On spectrophotometry, maximum absorption is seen at *450 nm*, when the reduced (ferrous) form of cytochrome enzymes bind to carbon monoxide, hence the name P450.

Nomenclature: CYP superfamily consists of more than 150 isoforms. Each member is given a unique name, e.g. CYP2E6 (2 → family E → Subfamily 6 → member).

Location: Liver, small intestine, brain and lung. Major amount is present in liver.

Subcellular location: Membrane bound in *smooth endoplasmic reticulum* (microsomes) and inner mitochondrial membrane.

Prosthetic group: FMN and heme

Function: Involved in *phase I* metabolism of endogenous and exogenous substances and *steroid synthesis*.

Type of reaction: Monooxygenase also known as mixed-function oxidases since they incorporate one atom from O_2 into substrate and reduce the other atom to water. Reactive free-radical is produced as the intermediate of the reaction.

$$RH + O_2 + NADPH + H^+ \rightarrow R\text{-}OH + H_2O + NADP^+ \text{ (NADPH is the electron donor)}$$

Substrate specificity: Broad; catalyzes around 60 types of reactions

Inducibility:

Cytochrome P450 enzymes are highly inducible, i.e. certain substances can increase the transcription of specific CYP isoforms. The following are some of the CYP isoforms and their inducers.

❑ CYP1A—Polyaromatic hydrocarbons
❑ CYP2B—Phenobarbitone
❑ CYP2E—Ethanol
❑ CYP3A—Anticonvulsants, rifampicin

Inhibition: Drugs like cimetidine, ketoconazole and ciprofloxacin compete with CYP450 enzyme and result in reversible competitive inhibition.

Genetic polymorphism: Certain CYP450 enzymes exist in polymorphic forms among population. This explains the varied drug response to same drug in different people. SNP arrays are available in the market to predict the pharmacogenetic effect of a drug in a person.

Drug Interaction

Induction or inhibition of a CYP enzyme by a drug alters the metabolism of the substrate of the enzyme.
e.g. Phenobarbitone → ↑ CYP2C9 → ↑ metabolism of warfarin → Warfarin therapy failure

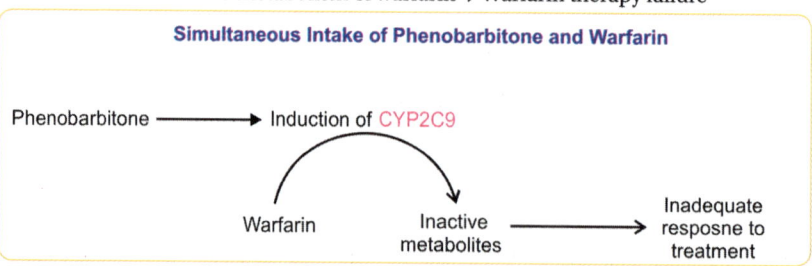

Procarcinogen to Carcinogen Conversion

- Aflatoxin B1 is hepatotoxin derived from the fungus aspergillus flavus.
- Aflatoxin is converted to reactive *epoxide* form by CYP450 system.
- Aflatoxin epoxide causes G → T mutation in p53 gene and covalent adducts with the lysyl residues of proteins.

Chronic Smoking and Alcoholism: A Deadly Combination

Chronic alcoholism leads to induction of the CYP2E1 microsomal enzyme. This leads to increased conversion of procarcinogens present in smoke to carcinogens.

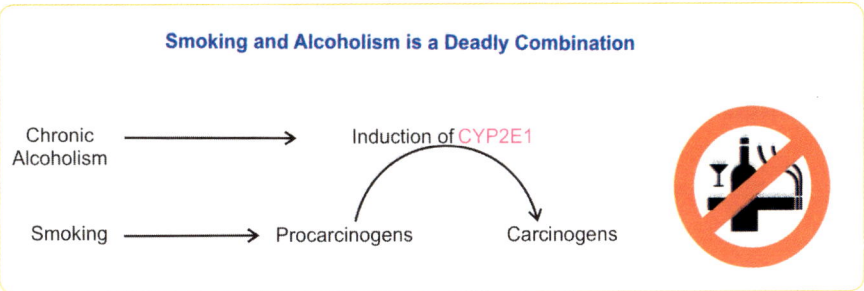

Phase II Reactions (Conjugation)

Conjugation reactions are Phase II reactions which usually follow Phase I reactions. There are certain substances like *isoniazid which directly enter phase II reactions without phase I*. Products of conjugation reactions have *more molecular weight and less active* than their substrates.

Type of conjugation	Donor	Enzyme	Substrate that is conjugated
Glucuronide conjugation	UDP-glucuronic acid	UDP-glucuronosyl-transferase	□ Bilirubin, cortisol □ Benzoic acid
Glutathione conjugation	γ-Glutamyl cysteinyl glycine (G-SH)	Glutathione-S-transferase	Electrophilic substrates
Sulfation	PAPS (3′-phosphoadenosine-5′-phosphosulfate)	Sulfotransferase	Steroid hormones, chloramphenicol
Methylation	SAM/AdoMet (S-adenosyl methionine)	Methyltransferase	□ Histamine → N-methyl histidine (inactivation) □ Noradrenaline → adrenaline (potentiation)
Acetylation	Acetyl-CoA	N-acetyltransferase (polymorphism results in fast and slow acetylators)	Isoniazid (directly enters phase II)
Glycine conjugation	Glycine	Glycine-N-acyltransferase	Glycine + Benzoyl CoA → Hippuric acid

Fast and Slow Acetylators

Isoniazid is metabolized by acetylation. Genetic polymorphisms leads to individual with fast and slow acetylation capacity. Fast acetylation leads to rapid elimination of drug and failure of treatment. Slow acetylation leads to accumulaiton of drug and toxicity.

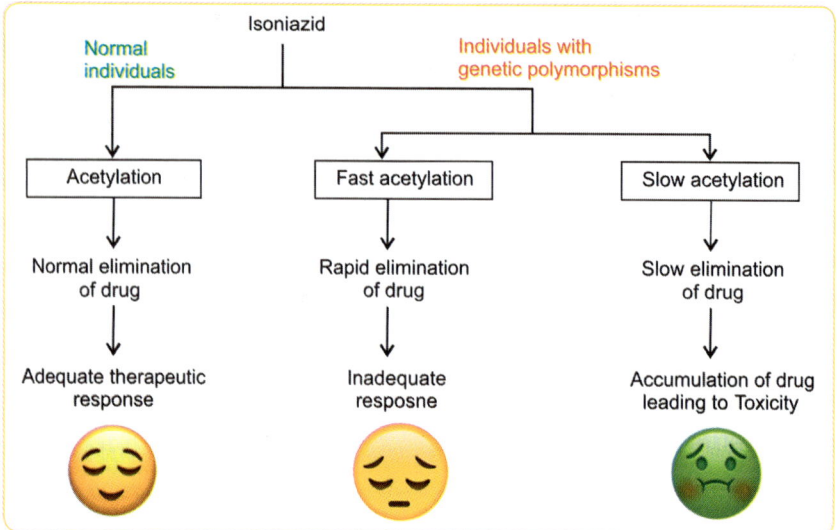

 Clinical Correlation

"Paracetamol Poisoning"   ▶ youtu.be/-dzliufA4g4 or scan QR code

"Acetaminophen is metabolized predominantly by a phase II reaction to innocuous sulfate and glucuronide metabolites; however, a small proportion of acetaminophen is metabolized by a phase I reaction to a hepatotoxic metabolite formed from the parent compound by the CYP2E1. This metabolite, N-acetyl-p-benzoquinoneimine (NAPQI), is detoxified by binding to "hepatoprotective" glutathione to become harmless, water-soluble mercapturic acid, which undergoes renal excretion."

❑ N-acetyl-p-benzoquinoneimine (NAPQI) depletes the glutathione levels.
❑ N-acetyl cysteine (specific antidote for paracetamol poisoning) replenishes the glutathione levels.

CONCEPT CORNER

Urinary cortisol measurement is done in patients suspected with Cushing's syndrome. You may wonder how the lipid soluble cortisol comes in urine. Apply your biochemistry knowledge here. Cortisol is conjugated with glucuronic acid and excreted in urine!

Competency

❑ BI7.5   Describe the role of xenobiotics in disease

Key Points

❑ Phase I hydroxylation reactions are catalyzed by cytochrome P450 mixed function monooxygenase.
❑ Conjugation of glycine to Benzoyl-CoA produces water-soluble hippuric acid.
❑ N-acetyl-p-benzoquinone imine (NAPQI) is the toxic intermediate produced by paracetamol which is conjugated with glutathione and excreted safely.
❑ N-acetyl cysteine is given to replenish glutathione levels in paracetamol poisoning.
❑ Isoniazid enters the phase II reaction directly.
❑ 3'-phosphoadenosine-5'-phosphosulfate (PAPS) is the sulfate donor for Phase II sulfation reactions.

SELF-ASSESSMENT

Short Answer Questions

1. Explain why cytochrome P450 is called as monooxygenase?

2. Explain detoxification by conjugation with two examples.

3. Smoking and alcoholism together are a deadly combination of increased risk of cancer – Explain.

4. Tabulate the differences between phase I and phase II biotransformation reactions.

5. With a suitable example, explain how the CYP450 system is involved in the conversion of a procarcinogen to a carcinogen.

6. How is acetaminophen metabolised in the body? Which is the antidote for paracetamol poisoning? Explain the mechanism of action.

7. A 37-year old female was put on warfarin following valve replacement. Following an episode of convulsions, she was prescribed phenytoin. After a week, it was noticed that the prothrombin time (PT) was markedly reduced. The clinician then had to increase the dose of warfarin to maintain PT at the optimal level.

   a. What is the biochemical basis of this drug interaction?

   b. What is the mechanism of action of warfarin? (Chapter 33)

Multiple Choice Questions

1. All of the following are phase-I biotransformation reactions, EXCEPT:
   a. Acetylation
   b. Hydrolysis
   c. Oxidation
   d. Reduction

2. The major group of enzyme catalyzing the phase-I biotransformation reaction is referred as:
   a. Mono-oxygenases
   b. Mixed function oxidases
   c. Cytochrome P450 system
   d. All of the above

3. Which of the following compound undergoes glucuronidation?
   a. Aspirin
   b. Methanol
   c. Bilirubin
   d. Phenyl acetate

4. Which of the following enzymes is involved in the Glucuronidation reactions?
   a. UDP-Glucose pyrophosphorylase
   b. UDP-Glucuronosyl transferase
   c. Glucan transferase
   d. Glucuronidase

5. Which of the following is the sulfur donor for sulfation reactions?
   a. Phospho pantothenic acid
   b. Taurine
   c. Phosphoadenosine phosphosulfate
   d. Cysteine

6. 450 in CYP450 denotes
   a. 45% of cytochrome CYP1000
   b. 450th member in the family
   c. Molecular weight is 450 kilodalton
   d. Absorption peak at 450 nm

7. Which of the following drug undergoes phase II biotransformation before entering phase I biotransformation reaction?
   a. Isoniazid
   b. Ethambutol
   c. Chloramphenicol
   d. Erythromycin

8. Which of the following isoform of CYP is involved in metabolism of ethanol?
   a. CYP1A
   b. CYP2B
   c. CYP2E
   d. CYP3A

9. Which of the following is used as the specific antidote for acetaminophen poisoning?
   a. N-acetyl cysteine
   b. Cysteine
   c. Activated charcoal
   d. Egg white

10. Hippuric acid is produced from the conjugation of Benzoyl-CoA with:
    a. Glutathione
    b. Glycine
    c. Glutamate
    d. Cysteine

ANSWERS

| 1. a. | 2. d. | 3. c. | 4. b. | 5. c. | 6. d. |
| 7. a. | 8. c. | 9. a. | 10. b. | | |

III

SECTION

Molecular Biology

Section Outline

Nucleic Acids: Introduction

NUCLEIC ACIDS ARE POLYNUCLEOTIDES (NUCLEOTIDE = NITROGENOUS BASE + PENTOSE SUGAR + PHOSPHATE)

- Nucleic acids containing ribose sugar are ribonucleic acids, i.e. RNA.
- Nucleic acids containing deoxy ribose sugar are deoxyribonucleic acids, i.e. DNA.
- In addition to the difference in sugars, DNA does not contain uracil.

CONCEPT CORNER

Advantage of Thymine Over Uracil in DNA

Spontaneous deamination of cytosine in DNA produces uracil which is corrected back to cytosine by the DNA repair enzymes normally. If uracil were present in DNA, the repair enzymes would not be able to distinguish between the uracil which was originally present in DNA from deaminated cytosine. *This is explained in detail in Chapter 21 (Refer to page no. 381).*

DNA versus RNA

		DNA	RNA
1°structure	Sugar	2' deoxy ribose	Ribose
	Bases	Adenine, Guanine, Thymine, Cytosine	Uracil instead of thymine
2° structure		Double-stranded helix	Single-stranded – can form random coil and stem loops
Stability		Alkali resistant due to 2' deoxy ribose	Less stable
Location (Eukaryotes)		Nuclear DNA is found in nucleus	Synthesized in nucleus to function mostly in cytoplasm

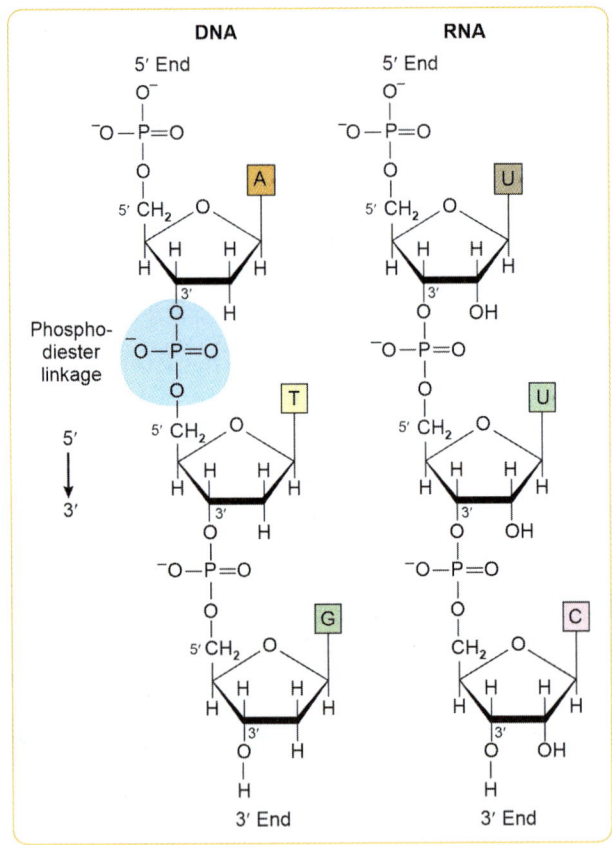

β-D-Ribose   β-D-Deoxyribose

DNA is More Stable than RNA

❑ RNA undergoes spontaneous alkaline hydrolysis (without any enzyme) because of the 2′ hydroxyl group.
❑ Absence of oxygen in the 2′ carbon makes deoxyribose of DNA stable.
❑ This is why nature might have chosen DNA as the genetic material.
❑ This is also the reason why RNA isolation (separation of RNA from cells/tissues) experiment is done at slightly acidic pH whereas DNA isolation is done under slightly alkaline pH.

PROPERTIES OF NUCLEIC ACIDS

- ❏ Polarity/Directionality
- ❏ Negatively charged
- ❏ Maximum UV absorbance (λ_{max} = 260 nm)

NUCLEIC ACIDS ARE DIRECTIONAL MOLECULES

Nucleic acids have polarity, i.e. they are directional molecules.
- ❏ 5' end is the end with free (unesterified) phosphate group.
- ❏ 3' end is the end with free OH group.
- ❏ Nucleic acid sequences are always written in 5'→3' direction. GAATC should be understood as 5'GAATC3'.
- ❏ During the synthesis of nucleic acids, 3' OH group of existing nucleotide is esterified to the 5' phosphoryl group of the incoming nucleotide via **3'→5' phosphodiester bond**.

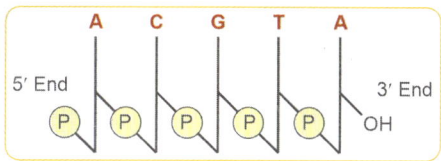

NUCLEIC ACIDS CARRY NEGATIVE CHARGE AT PHYSIOLOGICAL pH - DNA ELECTROPHORESIS

- ❏ Due to the presence of the phosphodiester linkages, nucleic acids carry significant negative charge at physiologic pH.
- ❏ This is the basis of separation of DNA or RNA by electrophoresis.
- ❏ As the DNA or RNA is negatively charged at alkaline pH, it will move towards the anode. Electrophoretic movement is based on the mass/charge ratio. So, shorter DNA molecules will move faster than the longer molecules.

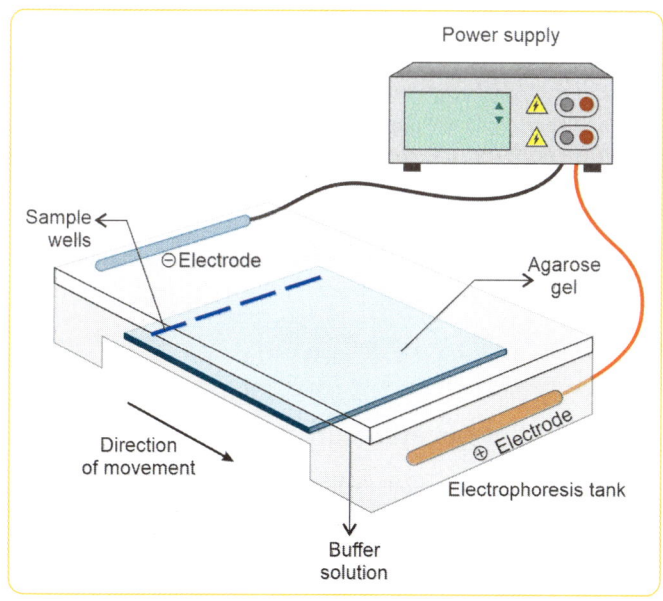

DNA samples are loaded into the wells made on an agarose or polyacrylamide gel. DNA is colorless. Then, how will we track the movement of DNA during electrophoresis and how will we visualize the DNA after electrophoresis?

- ❏ During sample loading, we mix DNA with colored dyes like bromophenol blue. Movement of bromophenol blue helps in tracking the movement of DNA.
- ❏ During gel preparation, we mix ethidium bromide to the agarose. Ethidium bromide intercalates into the spaces between the stacked base pairs of the DNA. It fluoresces under UV light when it intercalates with DNA. So, after electrophoresis, we visualize the gels under UV light to look for the orange color fluorescence (Fig. 1).

Fig. 1: Visualization of agarose gel under UV light

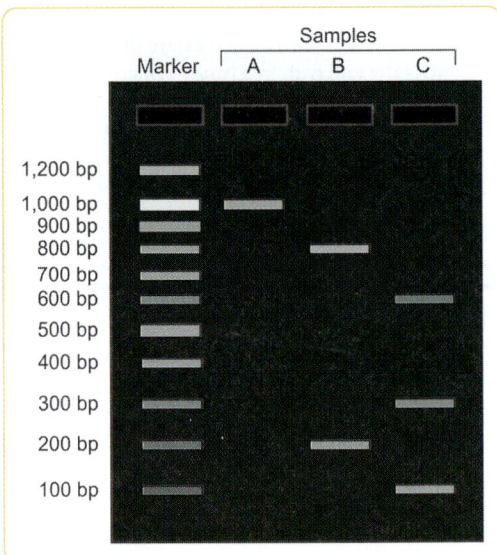

Fig. 2: Schematics of DNA electrophoresis of marker and three samples

Marker is also known as DNA ladder. This is a mixture of DNA fragments of known size. This is used to find out the size of sample DNA by comparison (Fig. 2).

NUCLEIC ACIDS ABSORB UV LIGHT

- ❏ The conjugate double bonds present in the purine and pyrimidine bases maximally absorb ultraviolet light at the wavelength of 260 nm.

T_m - Melting Temperature of DNA

T_m is the temperature at which half of the DNA is converted to single-stranded/random coil state.

Determinants

- ❏ GC content (3 hydrogen bonds), Salts (Na^+, Mg^{2+}) – increase T_m
- ❏ Formamide – decrease T_m (Formamide disrupts the secondary structures)

Hyperchromicity of DNA After Denaturation

Increased UV absorption of DNA after denaturation (dsDNA to ssDNA) is known as hyperchromicity. We know that it is the nitrogenous base of DNA that actually absorb the UV light not sugar or phosphates.

- ❏ In double-stranded DNA, bases are stacked over each other. This limits the absorbance of UV light by bases.
- ❏ In single-stranded DNA, bases are free to absorb more light.

In other words, denatured single-stranded DNA have increased UV absorption by 30–40% because **non-stacked bases can absorb more UV light** than the stacked bases (Fig. 3).

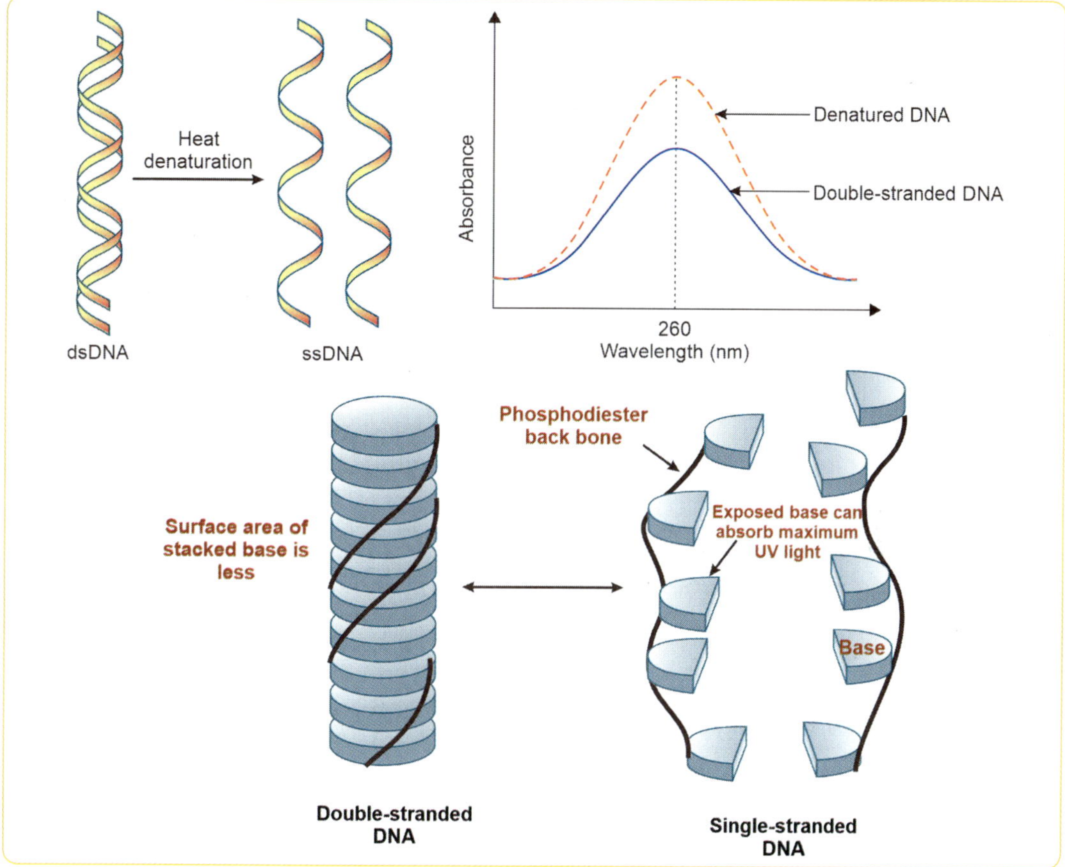

Fig. 3: Stacking of bases limits the UV absorption in double-stranded DNA

260/280 Ratio is used to Assess the Purity of DNA and RNA

❏ Nucleic acids have absorbance maxima at 260 nm.
❏ It is found that the absorbance of UV light by DNA/RNA at 260 nm is double than that of the absorbance at 280 nm.
❏ 260/280 ratio of pure DNA/RNA is in the range of 1.8 to 2.

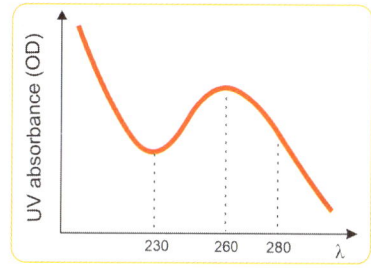

Nucleic acids also absorb UV light at 230 nm. 260/230 absorbance ratio is higher than that of 260/280 ratio. A low 260/230 ratio indicates contamination with chemicals used during nucleic acids isolation.

CHARGAFF'S RULE

Erwin Chargaff and his colleagues analyzed the DNA from various organisms and came to the following conclusion.

In a double-stranded DNA

Number of purines = Number of pyrimidine

$[A + G] = [T + C]$

$[A] = [T]$

$[G] = [C]$

$[A + T] \neq [G + C]$

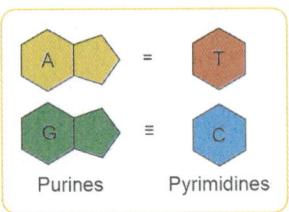

Note: Chargaff's rule is NOT applicable to RNA and ssDNA.

Later, Watson and Crick proposed their base-paring rule based on Chargaff's rule. But, unfortunately Chargaff's contribution was not recognized by the Nobel committee.

Why Purines Pair with only Pyrimidines? Why Purine-purine and Pyrimidine-Pyrimidine Base Pairing is not Possible?

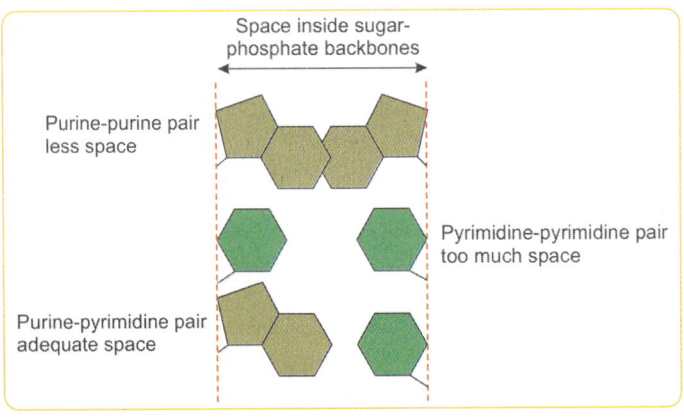

STRUCTURE OF DNA

Let us begin our discussion on the structure of DNA. Chargaff, Avery and Rosalind Franklin were the pioneers in the field of study of DNA.

❏ The famous Avery–MacLeod–McCarty experiment demonstrated that DNA is the substance that causes bacterial transformation.

❏ Chargaff's rule has been already discussed.

❏ Rosalind Elsie Franklin was an X-ray crystallographer. She and Maurice Wilkins extensively studied the molecular structures of DNA and RNA (Figs 4 and 5). Watson and Crick's DNA model is based on her findings only. Franklin was not given Noble Prize since she passed away due to ovarian cancer at the age of 37.

❏ The Nobel Prize in physiology or medicine of 1962 was awarded to Francis Crick, James Watson and Maurice Wilkins with this brief history, let us move to the structure of DNA.

Section III Molecular Biology

DIFFERENT FORMS OF DNA

	A	B	Z (Zig zag)
Size	Broad	Medium	Compact
Helix sense	Right-handed	Right-handed	Left-handed
Base pair per turn	11	10.5	12
Glycosidic bond angle	Anti	Anti	Pyrimidines: anti Purines: syn
Seen in	❑ Dehydrated DNA ❑ DNA-RNA hybrid ❑ Double-stranded regions of RNA	Physiological form described by Watson and Crick in 1953	Occurs naturally in regions of DNA with alternating purines and pyrimidines (e.g. poly GC)

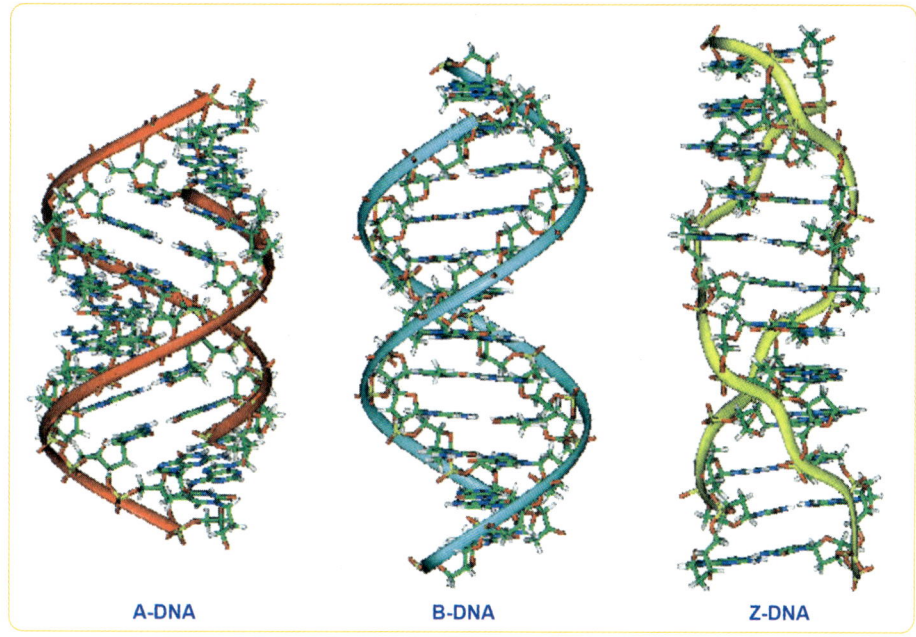

A-DNA   B-DNA   Z-DNA

Fig. 4: A, B and Z DNA; (Can you see that Z DNA is left handed?)

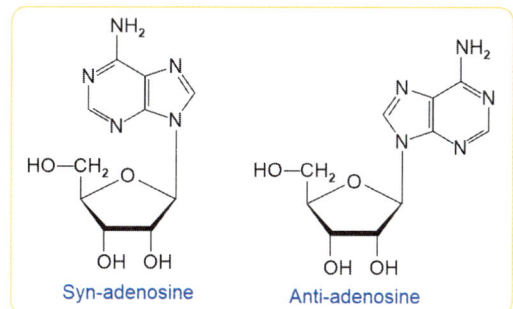

Syn-adenosine   Anti-adenosine

Fig. 5: Beta N-glycosidic bonds of nucleic acids in our body is usually of anti-conformation. This avoids the steric hindrance

FEATURES OF B-DNA

Watson and Crick's double helical model of DNA is based on B-DNA.

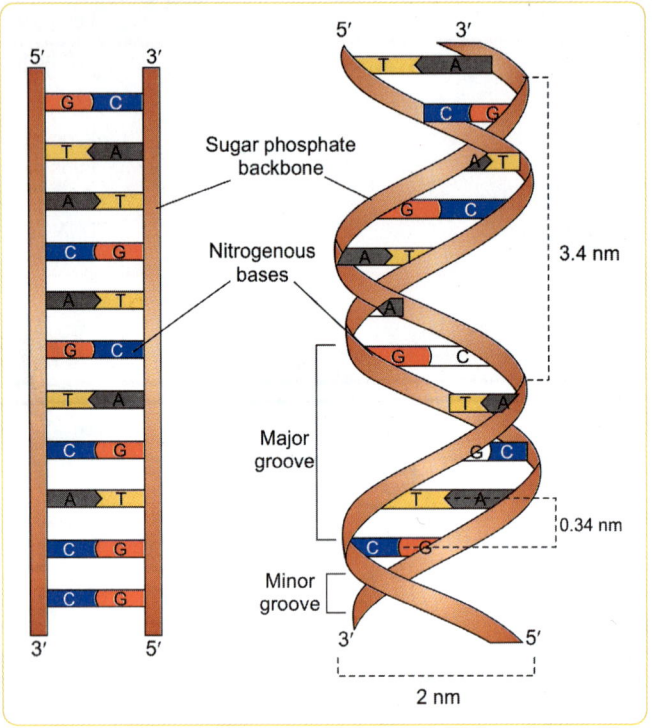

Double helix	DNA is 2 polynucleotide strands coiled around a central axis
Dimensions	Diameter of helix → 2 nm Pitch of helix → 3.4 nm No of base pairs per turn → 10 So, distance between 2 adjacent bases → 0.34 nm
Anti-parallel	Two strands are anti-parallel to each other.
Complementary nature	Bases in the two strands are complementary to each other
Base pairing via hydrogen bond	A = T; G ≡ C. As there are 3 hydrogen bonds, GC interaction is stronger.
Grooves	Because of coiling around the central axis, 2 grooves are formed—major and minor. These grooves are the **sites for interaction with proteins**.

BONDS IN DNA

- ❑ **Hydrogen bond**—between the bases of opposite strand
- ❑ **Phosphodiester bond**—between two nucleotides of the same strand
- ❑ **β-N-Glycosidic bond**—between base and deoxyribose
- ❑ **Stacking forces (Hydrophobic interactions)**—bases are stacked over each other.

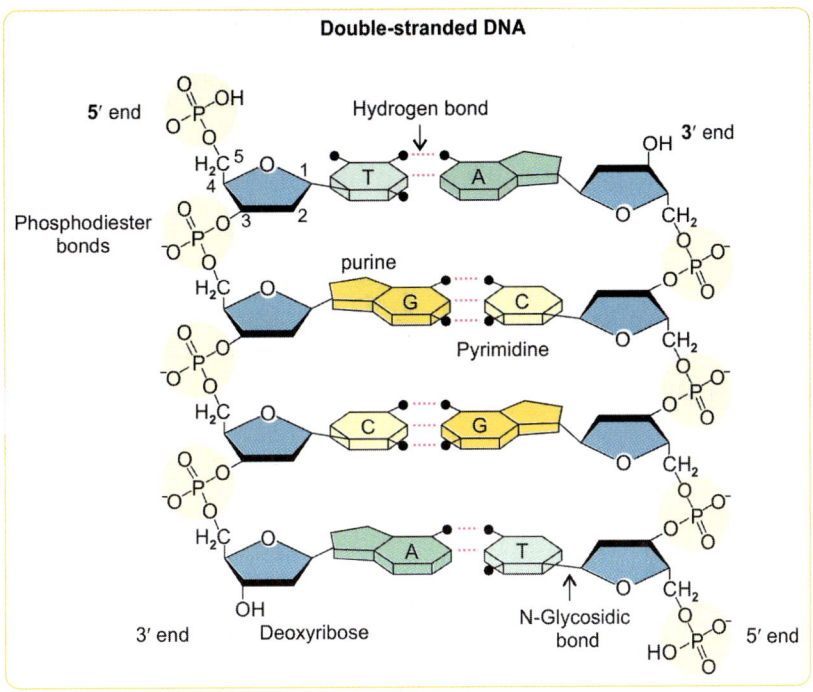

Fig. 6: Appreciate the bonds in DNA and the antiparallel nature of strands

Hoogsteen Base Pairing is a non-Watson-Crick base Pairing

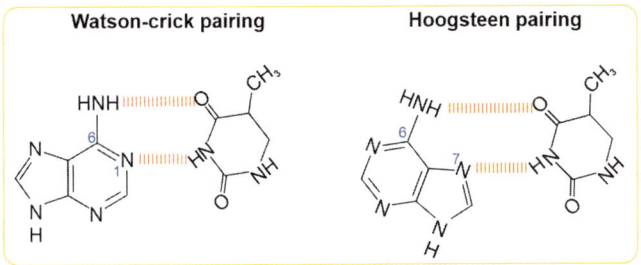

A-T refers to Watson-crick base pairing between A and T whereas:

- ❑ A*T refers to Hoogsteen base pairing.
- ❑ A*T involves N7 of adenine hydrogen bonding to C6 of T. In A-T, N7 of adenine is not involved in bonding.
- ❑ **Triplex DNA** formation is due to Hoogsteen base pairing.
- ❑ Certain modified bases in tRNA can be involved in Hoogsteen base pairing.

STRUCTURE OF RNA

Primary structure of RNA is nothing but the sequence of bases in 5'→3' direction linked by 3'→5' phosphodiester bond. The sugar in RNA is always ribose.

RNA contains uracil instead of thymine. But, transfer RNA contains thymine and many unusual and modified bases.

RNA is usually single-stranded. Complementary regions of the single-stranded RNA base-pair with each other to form secondary structures.

RNA CAN FORM EXTENSIVE SECONDARY STRUCTURES

RNA can form various secondary structures. The most familiar one to you could be stem-loop. Look at the image. Stem is formed by regions complementary to each other and the non-complementary regions loop out.

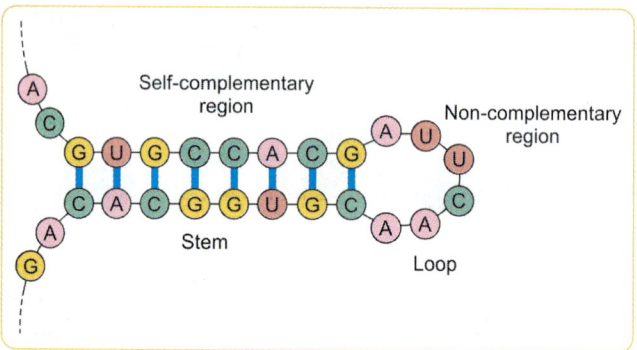

 High Yield

Stem-loop Structure Plays Many Roles

❑ Prokaryotic rho-independent transcription termination involves stem-loop formation.
❑ Histone mRNA contains a stem-loop structure in the 3' end instead of poly A tail. This stem-loop structure provides stability to histone mRNA.
❑ Selenocysteine is coded by the stop-codon UGA. Stem-loop structure in the mRNA directs the formation of selenocysteine.
❑ Stem-loop structures present in the 3' and 5' untranslated regions of mRNAs are involved in the regulation of gene expression.
❑ Stems and loops of tRNA serve important functions (we will discuss this in Chapter 25)
This completes the introduction of nucleic acids.

 *Extra Mile*

Aptamers

❑ Because of their extensive secondary structure, nucleic acids can form motifs that can bind to various molecules.
❑ Aptamers are short, single-stranded oligonucleotides that can bind to various molecules like proteins with high affinity and specificity similar to antibodies.
❑ Pegaptanib (Macugen) is the first therapeutic aptamer approved by the US FDA in 2004 as an anti-angiogenic medicine for the treatment of neovascular (wet) age-related macular degeneration (AMD). Pegaptanib binds to vascular endothelial growth factor (VEGF) and prevent the formation of new blood vessels.

Competency

❑ BI7.1   Describe the structure and functions of DNA and RNA

Key Points

❏ Nucleic acids are polynucleotides.
❏ DNA doesn't contain uracil.
❏ RNA undergoes spontaneous alkaline hydrolysis.
❏ DNA is more stable than RNA.
❏ DNA is negatively charged.
❏ Nucleic acids absorb UV light maximally at 260 nm.
❏ Dehydrated DNA, DNA-RNA hybrid and double-stranded regions of RNA exist in A form.
❏ Triplex DNA formation is due to Hoogsteen base pairing.
❏ The grooves in the DNA are the sites for interaction with proteins.

SELF-ASSESSMENT

Short Answer Questions

1. Tabulate the differences between DNA and RNA.

2. What do you understand by the term 'hyperchromicity of DNA after denaturation'? Explain the basis of this phenomenon.

3. Explain the principle behind assessing the purity of DNA using spectrophotometry.

4. In a sample of DNA, if adenine content is 23%, what will be the amount of guanine?

5. Enumerate the salient features of Watson and Crick's double-helical DNA-model.

Multiple Choice Questions

1. **Which one of the following bonds links two nucleotides in a nucleic acid?**
   a. $3' \rightarrow 3'$ phosphodiester bond
   b. $3' \rightarrow 5'$ phosphodiester bond
   c. $5' \rightarrow 5'$ phosphodiester bond
   d. $5' \rightarrow 5'$ phosphotriester bond

2. **Nucleic acids absorb UV light maximally at the wavelength of:**
   a. 260 nm
   c. 410 nm
   b. 280 nm
   d. 320 nm

3. **Bacteria X has an adenine content of 33% in its genome. What is the expected thymine content in the genome of X?**
   a. 22%
   c. 44%
   b. 33%
   d. 67%

4. **In a sample of DNA, if adenine content is 23%, what will be the amount of guanine?**
   a. 23%
   c. 54%
   b. 46%
   d. 27%

5. **At physiologic pH, DNA is:**
   a. Positively charged
   b. Negatively charged
   c. Amphoteric
   d. Uncharged

6. **DNA-RNA hybrid and dehydrated DNA exist in:**
   a. A form
   c. C form
   b. B form
   d. Z form

7. **Approximate number of bases in one turn of in B-DNA helix is:**
   a. 9.5
   c. 11.5
   b. 10.5
   d. 12

8. **Triplex DNA is formed because of:**
   a. Guanosine repeats
   b. Hoogsteen pairing
   c. Palindromic sequences
   d. Polypyrimidine sequences

ANSWERS

1. b.	**2.** a.	**3.** b.	**4.** d.	**5.** b.	**6.** a.
7. b.	**8.** b.				

19
CHAPTER

Organization of DNA

A diploid human cell contains ~6 billion base pairs distributed among 46 chromosomes. In Chapter 18 *(Refer to page no. 360)*, we have studied that the distance between each nucleotide in B-DNA is 0.34 nm. So, each diploid cell contains ~2 meters of DNA (6 billion × 0.34 nm = 2.04 metres). The diameter of a normal cell nucleus is 10 μm.

How can a 10 μm nucleus contain 2 meters of DNA?

This is achieved by the compact packaging of DNA with histone and other proteins.

Let us differentiate chromosome and chromatin (Figs 1 and 2).

❑ Chromosome is a more tightly-packed DNA found only during metaphase. Chromosomes are made of chromatids.

❑ Chromatin is a less tightly-packed DNA found throughout the interphase (the phase between two mitoses).

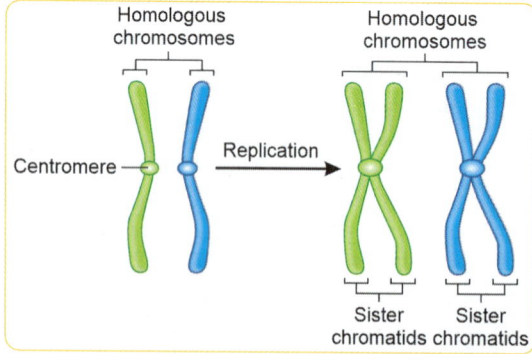

Fig. 1: Highest level of condensation of DNA is seen in chromosomes

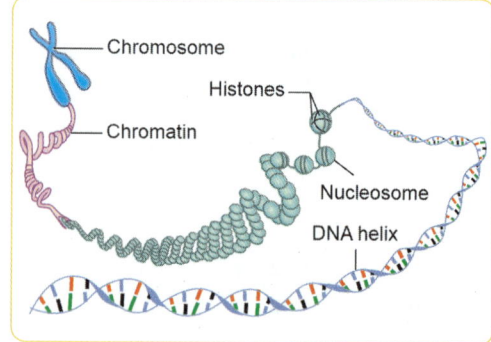

Fig. 2: Chromatin is less condensed than the chromosome

KARYOTYPING IS USED TO VISUALIZE CHROMOSOMES

Karyotyping is a chromosome visualization technique in which chromosome pairs are arranged in decreasing order of size except chromosome 21 and 22. For karyotyping, peripheral blood is collected in a heparin-containing vial. Phytohemagglutinin (mitogen for T-cells) is added, and the cells are cultured at 37°C for three days. Then colchicine is used to arrest the growing cells in metaphase, cells are placed in hypotonic saline to rupture them, followed by spreading them on a glass slide for fixation. Giemsa, one of the many stains, is used to stain the condensed chromosomes (G-banding). Dark bands appear when Giemsa is used hence the name G-banding. Karyotyping technique is used to detect the presence of aneuploidy (an extra copy or absence of a chromosome), euploidy, deletion, or translocation of a portion of a chromosome. **Karyotyping can't detect microdeletions (deletion involving lesser than 5 Mb)**.

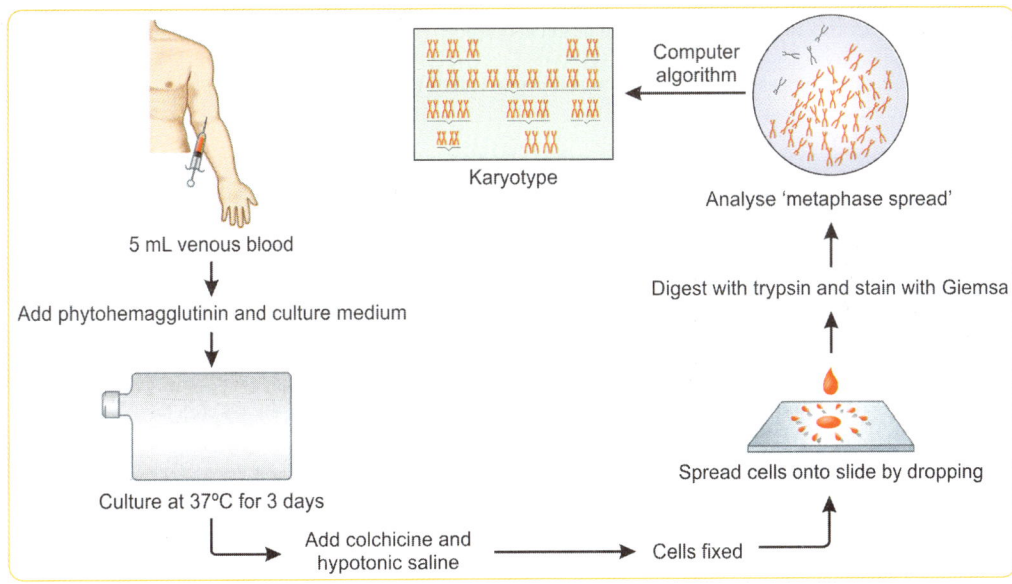

Fig. 3: Karyotyping procedure

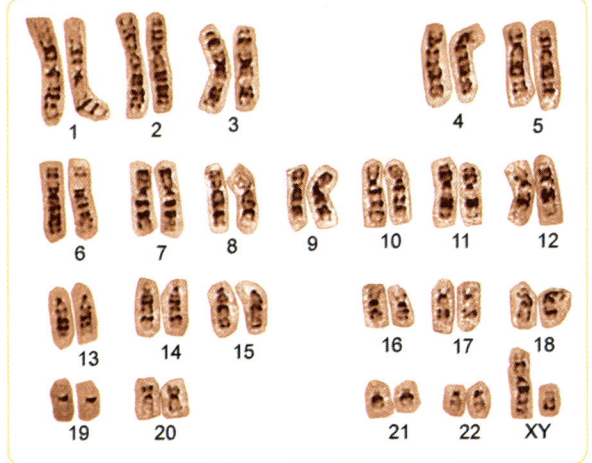

Fig. 4: Karyotyping by G-banding

The chromosomes are arranged in the decreasing order of size. The largest chromosome is the chromosome 1. The smallest chromosome is the Y chromosome. The smallest autosome is the chromosome 21, not 22 (Figs 3 and 4).

Chapter 19 **Organization of DNA**

FLUORESCENCE IN SITU HYBRIDIZATION (FISH) IS SUPERIOR TO KARYOTYPING

Fluorescence in situ Hybridization (FISH) is a molecular cytogenetic technique that is used to identify chromosomal abnormalities that cannot be detected by karyotyping. Let us analyze each word in the term Fluorescence In Situ Hybridization (Figs 5 and 6).

❑ 'Fluorescence' refers to the use of **fluorescent labelled** single-stranded DNA probes which are complementary to a specific segment of a chromosome.

❑ 'In situ' means original place. In FISH, the chromosomes are intact and are adhered to a glass slide.

❑ 'Hybridization' refers to the binding of a single-stranded DNA probe to its complementary region on the chromosome.

FISH is performed by denaturing the double-stranded DNA in the fixed chromosomes on a microscope slide. Once DNA is denatured, two different fluorescently labelled DNA probes are used. The first probe serves as a control and hybridizes with DNA to a region other than the target sequence. The second probe (test probe) hybridizes to the target sequence.

❑ In trisomy, an extra fluorescent spot with the test probe will be seen.

❑ In translocation, fusion band is seen

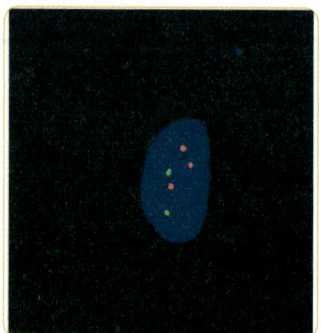

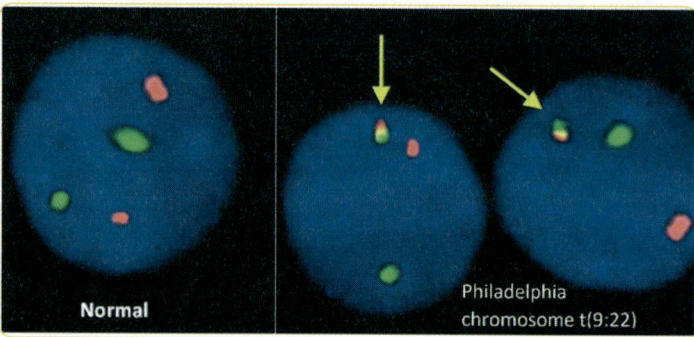

Normal

Philadelphia chromosome t(9:22)

Fig. 5: Trisomy 21- Probe for chr. 13 is labelled with green fluorescence which acts as a control. Probe for chr.21 is labelled with red fluorescence – showing three signals, so trisomy.

Fig. 6: Normal FISH pattern – Chr.9 (red) and chr.22 (green) Philadelphia chromosome – t(9:22) – fusion of green and red indicates translocation

The Advantage of FISH over Karyotyping

❑ Cells at any phase of the cell-cycle can be analyzed. So, there is no need to culture the cells; results can be obtained in a shorter duration.

❑ Microdeletions (deletion involving lesser than 5 Mb) can also be detected.

DNA PACKAGING

DNA is negatively charged because of the phosphodiester backbone. Negatively charged DNA is packed with positively charged histone proteins. Histone proteins are rich in positively charged amino acids, lysine and arginine. There are five histone proteins: H1, H2A, H2B, H3, and H4. Proteins other than the histones in the nucleus are known as non-histone nuclear proteins.

Instead of histones, protamines (arginine-rich proteins) bind to the DNA during spermatogenesis.

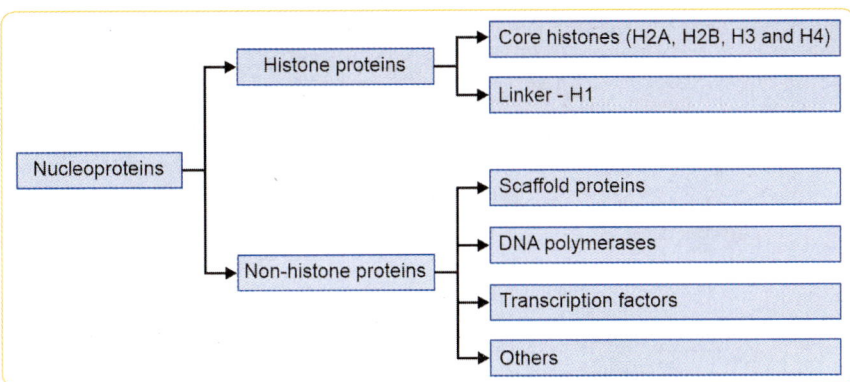

Nucleosomes (10 nm Fiber) are the 1st Level of DNA Packaging in Eukaryotes

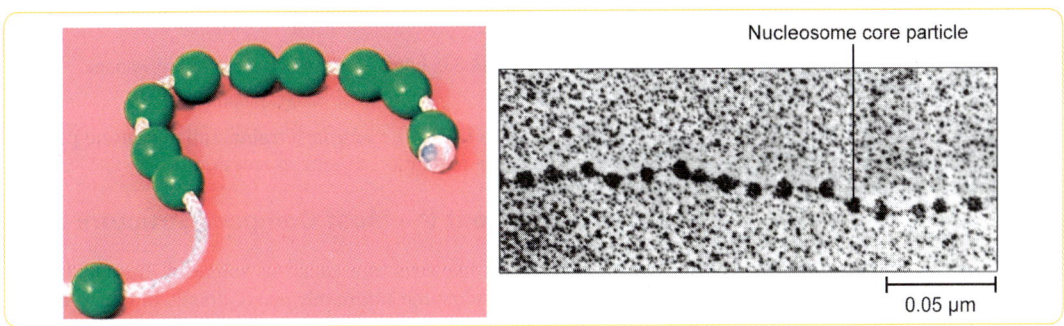

Nucleosome core particle

0.05 µm

Fig. 7: Electron microscopic view of decondensed chromatin showing beads on a string appearance

Beads on a String

When decondensed chromatin is viewed under an electron microscope, it looks like the beads on a string *(compare the images) (Fig. 7)*. This beaded structure is the nucleosomes. Nucleosome is nothing but histones complexed to DNA. **Nucleosome (10 nm fiber) is the basic unit of chromatin.**

Two molecules of each of H2A, H2B, H3, and H4 (Core histones) form a histone octamer. 1 and ¾ turns (~146 bp) DNA double helix is wounded around the histone octamer in a **left-handed** fashion. Nucleosome makes the DNA 7-fold compact. The DNA between two nucleosomes is called as the **linker DNA** and is **bound by H1 histone**.

Structure of Nucleosome (10 nm Fiber)

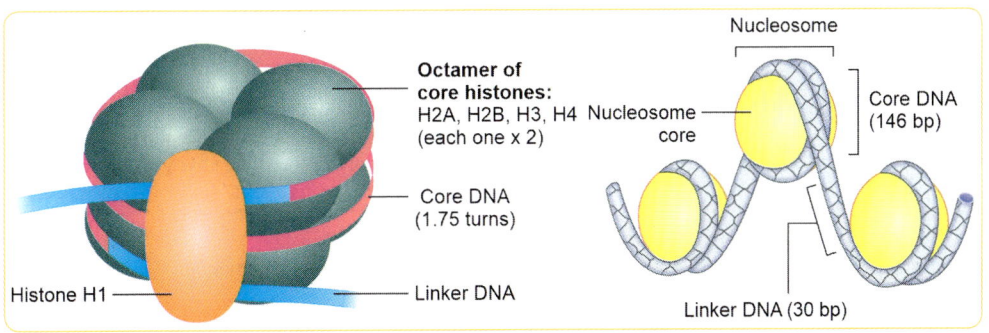

Octamer of
core histones:
H2A, H2B, H3, H4
(each one x 2)

Core DNA
(1.75 turns)

Histone H1

Linker DNA

Nucleosome
core

Nucleosome

Core DNA
(146 bp)

Linker DNA (30 bp)

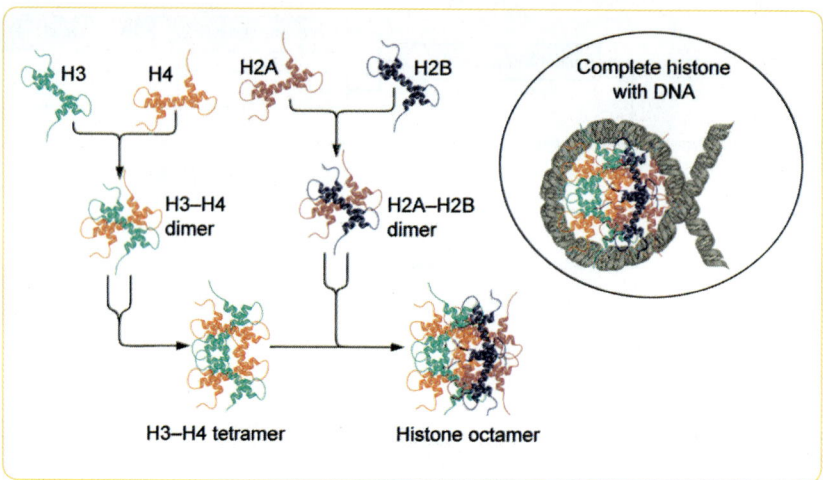

Note: The amino-terminal ends of histones protrude through DNA. The amino-terminal tail of histone proteins undergoes various modifications which play a role in the regulation of gene expression.

Further Packaging of Nucleosomes Brought out the Most Compact Structure

10 nm nucleosome is folded and compacted to 30 nm helix-like structure called us solenoid. This structure is further compacted manifolds to form the chromatin or chromosome ultimately, exact details of which are not known.

DNA → Nucleosome → Solenoid → Chromatin fibre → Chromatid → Chromosome

 High Yield

Histone protein and Histone mRNA are unique in many ways:

❑ Histones contain a high number of amino acids with positively charged side chains. So, during electrophoresis, they move towards cathode even at alkaline pH.
❑ Highly conserved during evolution – an amino acid sequence of histones of various species shows similarity. Core histones are more conserved than H1 histone.
❑ Histone mRNA does not contain polyA tail.
❑ Histone genes do not contain introns.

DNA-HISTONE INTERACTION PLAYS A VITAL ROLE IN THE ORGANIZATION OF DNA STRUCTURE

Histone proteins undergo various modifications (mentioned in the table). Histone modification is an epigenetic change. **Epigenetics is the study of heritable changes in gene expression without any change in DNA nucleotide sequence.** Histone modifications and DNA methylation are the two major epigenetic changes. *We will discuss the histone modifications in this chapter. DNA methylation will be discussed in Chapter 27 (Refer to page no. 455).*

Histone modification	Amino acid involved	Effect
Acetylation (done by histone acetylase)	Lysine	Usually euchromatin formation
Deacetylation (done by histone deacetylase)	Lysine	Usually heterochromatin formation.

Contd...

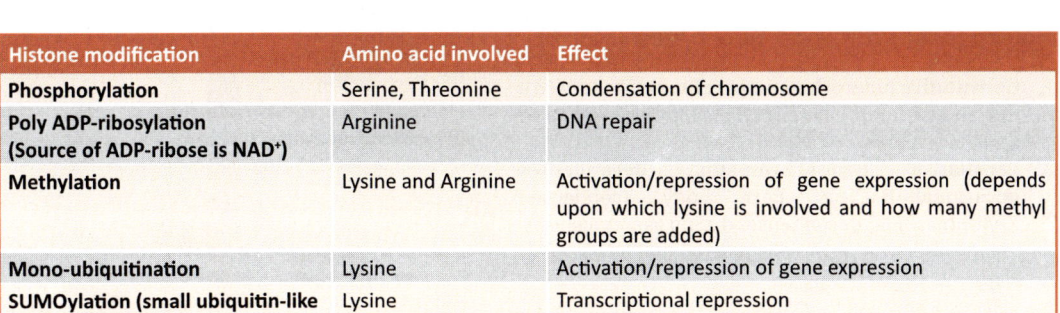

Histone modification	Amino acid involved	Effect
Phosphorylation	Serine, Threonine	Condensation of chromosome
Poly ADP-ribosylation (Source of ADP-ribose is NAD⁺)	Arginine	DNA repair
Methylation	Lysine and Arginine	Activation/repression of gene expression (depends upon which lysine is involved and how many methyl groups are added)
Mono-ubiquitination	Lysine	Activation/repression of gene expression
SUMOylation (small ubiquitin-like modifier addition)	Lysine	Transcriptional repression

Histone Code can be Written, Read and Erased

Let us consider the example of acetylation.
- ❏ Histone acetylase enzymes acetylate the $\in$-amino group of lysine of histone tail – Code writers
- ❏ Bromodomain containing proteins recognize this acetyl lysine group – Code readers
- ❏ Histone deacetylase enzymes remove the acetyl group – Code erasers

Histone Post-translational Modification can Lead to Chromatin Remodeling (Fig. 8)

- ❏ Chromatin remodeling is the dynamic process that changes the chromatin architecture. The position of a nucleosome is altered in the chromatin by ATP-dependent chromatin remodeling proteins.
- ❏ Chromatin remodeling proteins recognize modified histone residues.

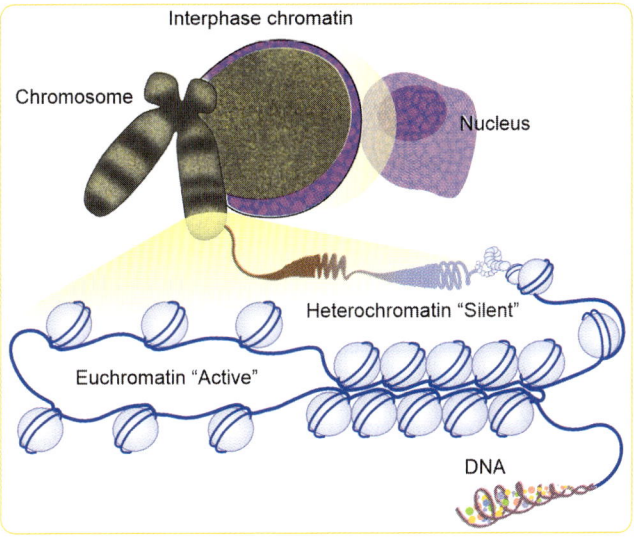

Fig. 8: Chromatin remodeling converts euchromatin to heterochromatin and vice versa

Euchromatin Versus Heterochromatin

	Euchromatin	Heterochromatin
On G-Banding	Lightly stained	Darkly stained
Transcriptionally	Active	Inactive
Evolutionary significance	Found both in prokaryotes and eukaryotes	Found only in eukaryotes
Epigenetic change associated	Histone acetylation	Histone deacetylation
Sensitivity to DNAse digestion	Sensitive	Resistant

Heterochromatin is of Two Types (Fig. 9)

1. **Constitutive heterochromatin (always condensed):** Centromeric and Telomeric DNA is always condensed and inactive.
2. **Facultative heterochromatin (decondenses occasionally):** e.g. Barr body (condensed X-chromosome) decondenses during gametogenesis.

Barr body formation (Lyon's hypothesis) will be discussed in Chapter 30 *(Refer to page no. 481).*

TOPOLOGICAL PROPERTIES OF DNA

Topological properties are those which are not changed by DNA deformations and remains constant as long as both the DNA strands are intact, e.g., linking number.

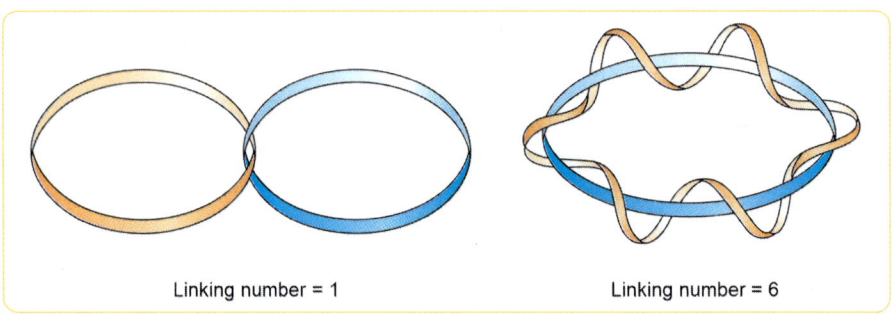

Fig. 9: Centromere and telomere form the constitutive heterochromatin

Linking number is the number of times one strand of DNA has to be passed through the other strand so as to separate both the strands completely, in other words, Linking number is the number of times one strand winds around the other. It is an integer for a closed circular DNA. For example, linking number of a closed circular DNA with 1050 bp is 100, assuming there is 10.5 bp per turn.

Components of linking number: Linking number = Twist + Writhe

   Note: The details of twist and writhe are not necessary for the exam purpose.

Linking number = 1   Linking number = 6

SUPERCOILING OF DNA

We know that DNA is a double helical coil. Coiling of the DNA double helical coil is known as DNA supercoiling. There are two types of supercoiling:
1. Positive supercoiling
2. Negative supercoiling
   - Negative supercoils have right-handed supercoil turns. **Negative supercoiling leads to underwinding of DNA.**
   - Positive supercoils have left-handed supercoil turns. **Positive supercoiling leads to overwinding of DNA.**

DNA is Usually Negatively Supercoiled

DNA in our body is negatively supercoiled. Negatively supercoiled DNA has a lesser linking number than the positively supercoiled DNA. Negative-supercoiling makes the separation of DNA energetically favourable during replication and transcription.

Topoisomerases

Topoisomers are DNA forms that differ in their linking number. Topoisomerases are the enzymes which can change the linking number of DNA.

Replication and transcription induce supercoiling in the DNA. This problem of replication and transcription-induced supercoiling is dealt by topoisomerase enzymes. There are two types of topoisomerases *(discussed in Chapter 21)* *(Refer to page no. 384).*

Competency

❏ AN73.2   Describe technique of karyotyping with its applications

Key Points

❏ The Y chromosome is the smallest of all the chromosomes. Chromosome 21 is the smallest autosome.
❏ Microdeletions (deletion involving lesser than 5 Mb) can be detected by FISH but not by karyotyping.
❏ Nucleosomes (10 nm Fiber) are the first Level of DNA Packaging in Eukaryotes.
❏ The DNA between two nucleosomes is the linker DNA; bound by H1 histone.
❏ In a nucleosome, ~146 bp length of DNA wraps 1.75 times around core histones.
❏ Centromeric and Telomeric DNA are highly repetitive.
❏ Centromeric and telomeric DNA is in the form of constitutive heterochromatin, i.e. always condensed and inactive.
❏ Barr body is an example for facultative heterochromatin.
❏ Acetylation of epsilon-amino group of lysine of core histone proteins lead to euchromatin formation or facilitates gene expression.
❏ In our body, DNA is usually present in a negatively super-coiled form.
❏ Topoisomers are DNA forms that differ in their linking number.

SELF-ASSESSMENT

Short Answer Questions

1. Differentiate Chromatin and Chromosome.

2. With a suitable diagram describe the structure and composition of a nucleosome.

3. What is linker DNA? Which histone is bound to it?

4. The functional status of a region of the genome depends on the pattern of specific chromatin modifications. Explain.

5. What is the advantage of 'Fluorescence In Situ Hybridization' over Karyotyping?

6. Name the histone code writer, reader and eraser.

7. Tabulate the differences between euchromatin and heterochromatin.

8. What are the two types of heterochromatin? Give an example for each.

Chapter 19   Organization of DNA

Multiple Choice Questions

1. **What is the approximate number of base pairs associated with a single nucleosome?**
   a. 146
   b. 292
   c. 73
   d. 1460

2. **Histone proteins are rich in which of the following amino acids?**
   a. Histidine and lysine
   b. Lysine and arginine
   c. Arginine and histidine
   d. Histidine and valine

3. **Which of the following is an example for facultative heterochromatin?**
   a. Centromere
   b. Telomere
   c. Barr body
   d. All of the above

4. **Which of the following is the donor of ADP-ribose for ADP-ribosylation reactions?**
   a. Adenosine triphosphate
   b. Adenosine monophosphate
   c. Nicotinamide adenine dinucleotide
   d. Flavin mononucleotide

5. **Which of the following is the smallest autosome?**
   a. Chromosome 1
   b. Y chromosome
   c. Chromosome 21
   d. Chromosome 22

6. **Linker DNA is bound to which of the following histone?**
   a. H1
   b. H2A
   c. H2B
   d. H3

7. **Nucleosome is also called as:**
   a. 10 nm fiber
   b. 30 nm fiber
   c. 50 nm fiber
   d. 70 nm loop

8. **Topoisomers are DNA forms that differ in:**
   a. GC content
   b. melting temperature
   c. Coding region
   d. Linking number

9. **DNase hypersensitive regions are:**
   a. Methylated DNA
   b. Heterochromatin region
   c. Transcriptionally active chromatin
   d. RNA bound DNA region

ANSWERS

1. a.	2. b.	3. c.	4. c.	5. c.	6. a.
7. a.	8. d.	9. c.			

20
CHAPTER

Human Genome

Gene is defined as "a nucleotide sequence in a DNA molecule that acts as a functional unit for the production of a protein, a structural RNA, or a catalytic or regulatory RNA molecule."

HUMAN GENOME PROJECT

A genome of an organism is the complete set of nucleotide sequences including all of its genes. Frederick Sanger pioneered the technique of DNA sequencing which earned him the Nobel Prize for the second time. DNA of bacteriophage (φ X174) is the first sequenced DNA. *Haemophilus influenza* is the first free-living organism to have its genome completely sequenced.

The human genome project (HGP) was initiated by National Institute of Health (NIH) and Department of Energy (DOE) with the aim to sequence the entire human genome. Co-operation of Celera Genomics, a private company, accelerated the speed of sequencing. The first draft of HGP was published in 2001 in the journal "Nature". Bacterial Artificial Chromosome (BAC) and Yeast Artificial Chromosome (YAC) were the vectors widely used in HGP. The sequence of the entire human genome was completed in 2003. According to Human Genome Project, genetic similarity between all humans is 99.9%.

Key Findings of Human Genome Project

- ❏ Size of haploid genome: 3 billion base pairs
- ❏ Number of protein-coding genes: <25,000
- ❏ At least 30% of the genome consists of repetitive sequences
- ❏ Alu family constitutes ~10% of human genome
- ❏ Largest gene: *DMD* gene coding for dystrophin protein
- ❏ Only around 1.5 % of human genome is made of protein-coding (exonic) DNA

Human Genome is yet to be Annotated

Annotation is the process of associating function to the DNA elements of the genome. Human genome sequencing is complete, but annotation is yet to be done. ENCODE project (Encyclopaedia of DNA Elements) attempts annotation of human genome.

JUNK DNA IS NOT JUNK!

- ❑ Only around 1.5% of human genome is made of protein-coding (exonic) DNA. This does not mean that the rest of the DNA is transcriptionally inactive and junk.
- ❑ **ENCODE** project estimated that roughly 80% of human genome is transcriptionally active at some point of time in one or more cell types. (Ref. Harper 30th Ed., p.102) *(These regions are involved in the production of non-coding RNAs)*

Coding Regions are Interrupted by Intervening Sequences

In the human genome, the sequence which codes for mature RNA molecules are interrupted by non-coding regions. The regions that appear in the mature RNA are known as exons, whereas intervening non-coding regions are known as introns. The size of the intron is highly variable. Histone gene is devoid of introns while the DMD gene has the highest number of introns.

Extra edge: *Introns and exons are present in tRNA also. So, all exons are not protein-coding.*

The Largest Gene is DMD, but the Largest Protein is Titin

Why is the largest gene not producing largest protein? It is because of the size of introns.

Dystrophin	Titin
Largest **gene** in human	Largest **protein** in human
DMD Gene size: 2.22 MB	*TTN* Gene size: 0.28 MB
mRNA size: 14.1 KB	mRNA size:101.5 KB

At Least 30% of Human Genome Consists of Repetitive Sequences

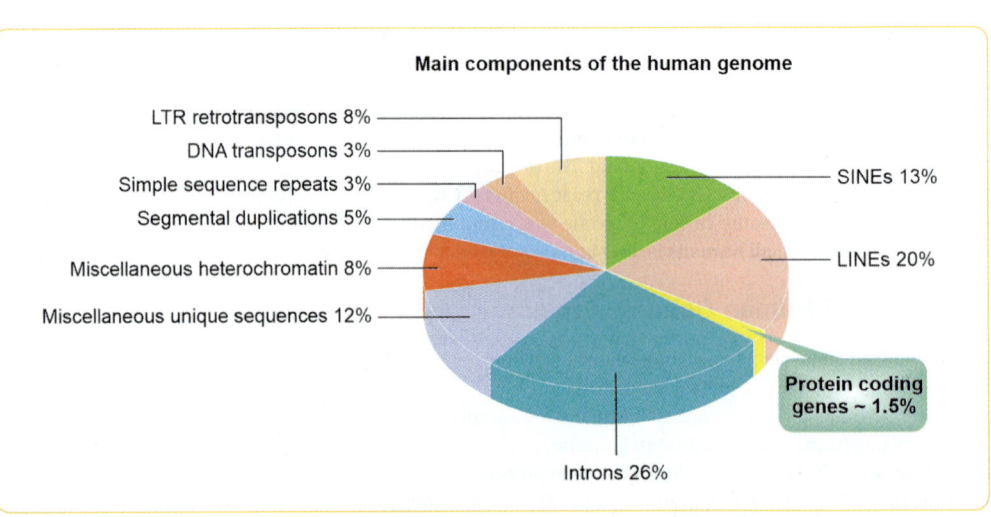

Main components of the human genome

LTR retrotransposons 8%
DNA transposons 3%
Simple sequence repeats 3%
Segmental duplications 5%
Miscellaneous heterochromatin 8%
Miscellaneous unique sequences 12%
SINEs 13%
LINEs 20%
Protein coding genes ~ 1.5%
Introns 26%

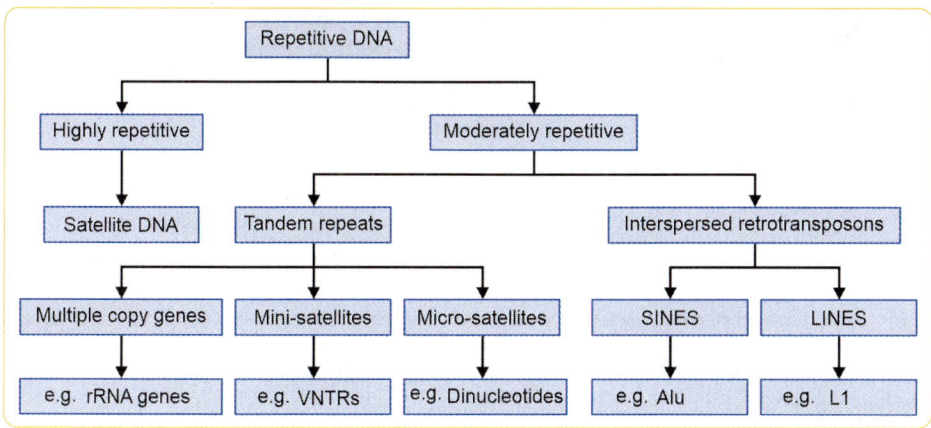

Centromeric and Telomeric DNA Constitutes the Satellite DNA

The centromere is rich in adenine and thymine. Human telomere contains the repeat sequence of TTAGGG. On density gradient centrifugation, centromeric and telomeric DNA form separate satellite bands because of their unique base composition.

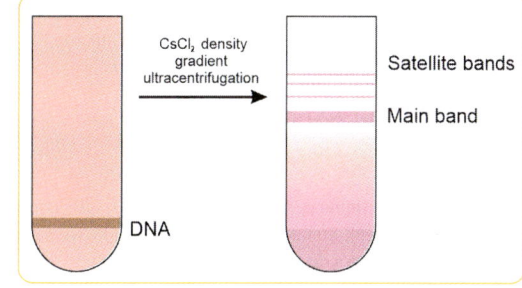

Tandem Repeats versus Interspersed Repeats

Tandem repeats are repetitive DNA sequences which lie back to back while interspersed repeats are intervened by other sequences.

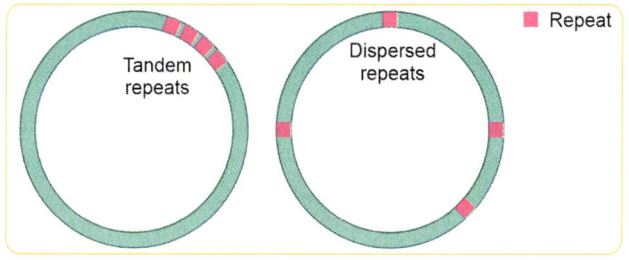

rRNA Genes are Found in Multiple Numbers

In the human genome, only two copies (maternal and paternal allele) are present for most of the genes. But, hundreds of ribosomal RNA genes are found.

What could be the reason?

rRNA is needed for protein synthesis. Multiple copies of rRNA are needed for the enormous amount of proteins produced in the cells.

Microsatellites are 2 to 6 Nucleotides Repeated up to 50 Times

Microsatellite repeats/short tandem repeats are also known as simple sequence repeats since the repeat of two nucleotides (dinucleotide repeat) is the most commonly repeated element.

Chapter 20 Human Genome

Microsatellite repeats are superior markers for DNA fingerprinting compared to VNTR analysis.

- ❑ At any given locus, the number of repeats may vary between two individuals—**polymorphic**
- ❑ Number of repeats varies between 2 chromosomes—**usually Heterozygous**
- ❑ Multiple allelic forms are possible since the repeat size is up to 50 times—**multiallelic**
- ❑ Codominantly inherited from father and mother—**heritable**
- ❑ Can be **easily detected** by PCR
- ❑ Used in linkage analysis

Increase in Trinucleotide Repeats Leads to Various Disorders

In addition to the dinucleotides, trinucleotide repeats are also found both in the coding and the non-coding regions of DNA. Increased number of trinucleotide repeats leads to disorders. In these disorders, the number of repeats increases with successive generation and thus the age of onset decreases and severity of the disease increases. This is known as **anticipation**. The molecular basis of this is **replication slippage** *(details of which is beyond the scope of this book)*.

Disease	Repeat	Inheritance
Fragile X syndrome	CGG [remember egg (female) for X]	X-linked
Friedreich's ataxia	GAA	Autosomal Recessive
Myotonic dystrophy	CTG	Autosomal dominant with variable penetrance
Spinobulbar muscular atrophy (Kennedy disease), Spinocerebellar ataxia, Huntington's disease	CAG; CAG codes for glutamine (Q). Thus, these diseases are known as poly Q disorders.	Autosomal dominant except Kennedy disease which is X-linked.

The mutation in nucleotide repeat disorders can be either **Premutation** or **Full mutation.** Let's take the example of Fragile X syndrome which is due to the excess of CGG repeats in the FMR1 gene.

- ❑ Normal number of CGG repeats in *FMR1* gene: 6 to 55
- ❑ Premutation: 55–200 repeats
- ❑ Disease: >200 repeats

- ❑ Myotonic dystrophy type 2 is a **tetranucleotide** (CCTG) repeat disorder.
- ❑ Spinocerebellar ataxia type 10 and 31 are due to the increased **pentanucleotide** repeats.

Variable Number Tandem Repeats (VNTR) are Repetitive Elements with Longer Repeats

- ❑ Repeat size is 10 to 100 bp in VNTR
- ❑ Similar to STR, VNTR regions are also polymorphic, i.e. the number of repeats varies between the individuals in a population.
- ❑ RFLP-PCR technique can identify the number of VNTR repeats.
- ❑ STR is superior to VNTR in identification of a person by DNA fingerprinting. That is why, nowadays, STR is used instead of VNTR (Fig. 1).

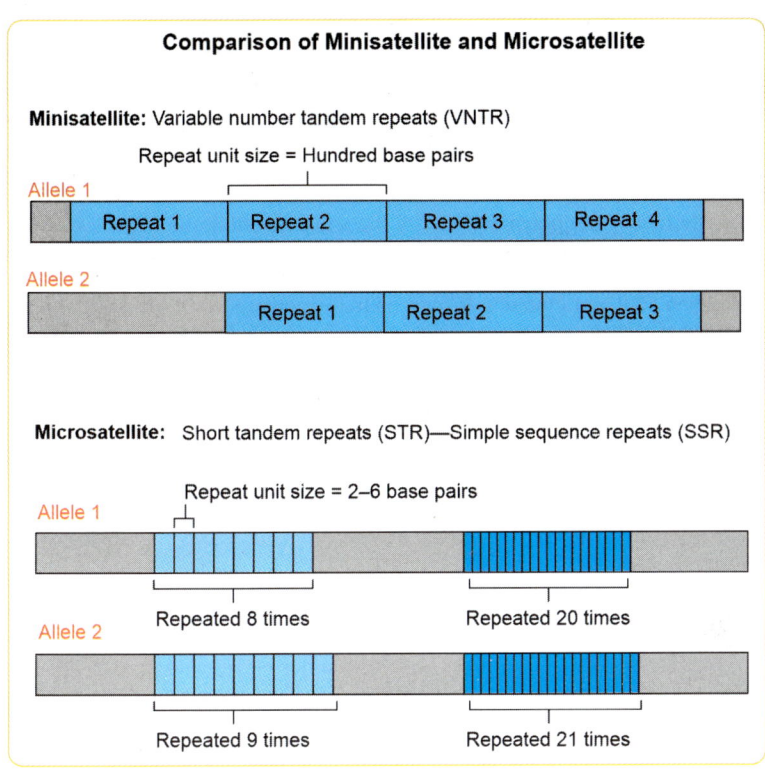

Fig. 1: Comparison of VNTR and microsatellites

The Combined DNA Index System (CODIS) is the United States national DNA database of convicts created and maintained by the FBI. A biological sample obtained from crime scene is compared with the CODIS database to find out the criminals. CODIS uses twenty microsatellite (STR) markers. In our country, Centre for DNA Fingerprinting and Diagnostics (CDFD) has the access to CODIS.

Transposons

Transposons are the DNA sequences that can move from one location of the genome to the other.
Transposons are of two types:
1. DNA transposons
2. RNA transposon (retrotransposons)
RNA transposons are also known as retrotransposons since they require conversion into DNA before insertion into genome. During evolution, most of the transposons were inactivated by methylation and heterochromatin formation. In humans, only few retrotransposons are active, e.g. Alu element.

Alu Elements are Actively Transcribed in the Body

- ❑ Alu sequence was discovered by using restriction enzyme AluI derived from *Arthrobacter lutues*.
- ❑ Alu belongs to the Short Interspersed Nuclear Element (SINE) family.
- ❑ Alu RNA resembles 7S RNA (component of signal recognition particle) and 4.5S RNA.
- ❑ Alu constitutes ~10% of the human genome.
- ❑ Insertion of Alu into a normal gene disrupts the gene and results in disease.

SINGLE NUCLEOTIDE POLYMORPHISMS

DNA sequencing of various humans revealed that on average, 1 out of every 1000 bp varies from one person to the next. This single nucleotide base difference that exists among the population is known as single nucleotide polymorphism (SNP, *pronounced as Snip*). As 1 out of 1000 bp varies, the human genome contains 14 million SNPs.

❏ Most of these SNPs are harmless. But a few are disease-associated and are capable of altering the responsiveness to drugs (pharmacogenomics).

❏ For example, a single nucleotide polymorphism of the gene encoding the hepatic cholesterol transporter ABCG5/G8 has been found in 21% of patients with gallstones, but only in 9% of the general population. (Ref. Harrison 19th Ed., p.2077)

❏ The difference between SNP and mutation is that the frequency of SNP in the population is >1% while the frequency of mutation is <1%.

Techniques to Detect SNP

❏ DNA sequencing – Gold standard
❏ RFLP-PCR
❏ Single-strand conformational polymorphism (electrophoresis)
❏ Denaturing high-performance liquid chromatography.

MITOCHONDRIAL GENOME

Mitochondrial DNA is unique in many ways.

Features of Mitochondrial DNA

❏ Maternally inherited since the paternal mitochondrial DNA are destroyed upon fertilization
❏ Constitutes 1% of total human DNA (Size 16,569 bp)
❏ Closed Circular, double-stranded (Heavy and light strand)
❏ Multiple copies (up to 10 copies) are present in a single cell (polyploidy).
❏ Contains 37 genes
  ○ 13 protein-coding genes for the components of electron transport chain
  ○ 24 non-protein coding genes
    ➢ 22 tRNA gene
    ➢ 2 rRNA genes (an intron is present between these 2 rRNA genes)
❏ Evidence for mitochondria being prokaryotic in origin:
  ○ Circular double-stranded DNA with only one intron
  ○ Rate of mutation is high
  ○ Genetic code is different from the universal genetic-code
  ○ Can synthesize some proteins with own ribosomes *(Most other proteins are imported from cytosol which are coded by nuclear genome.)*

	Nuclear DNA	Mitochondrial DNA
Shape	Linear	Circular
Ploidy	Diploid in somatic cells, haploid in germ cells	Polyploid
Inheritance	Inherited from both parents	Maternal
Histone proteins	Present	Absent
Number of genes	Not yet fully estimated (~25000)	37
Proofreading and repair activity	Present	Absent
Recombination	+	-
Protein coding regions	1 to 2%	93%
Number of stop codons	3	4

Key Points

- ❑ Human haploid genome is made up of 3 billion (10^9) base pairs.
- ❑ In the human genome, only around 1.5% of DNA is protein-coding. But, more than 90% of genome is transcriptionally active.
- ❑ Number of protein-coding human genes is around 25,000. Number of proteins formed from 25,000 genes is >2.5 lakh.
- ❑ DMD gene coding for dystrophin protein is the largest human gene.
- ❑ Microsatellites are two to six nucleotides repeated up to 50 times.
- ❑ Premutation is seen in patients of trinucleotide repeat disorders like Fragile X syndrome.
- ❑ Transposons are the DNA sequences that can move from one location of the genome to the other.
- ❑ Alu belongs to the Short Interspersed Nuclear Element (SINE) family.
- ❑ A single-nucleotide polymorphism (SNP) is a single base variation in a position of the DNA between members of a species or paired chromosomes in an individual.
- ❑ Mitochondrial DNA is double-stranded, circular, maternally inherited; contains 37 genes.

SELF-ASSESSMENT

Short Answer Questions

1. Enumerate the key findings of the human genome project.
2. What do you understand by the term 'annotation'? Is the human genome fully annotated?
3. Do you agree with the term 'Junk DNA'? Justify your answer.
4. The largest gene in humans is DMD, but the largest protein is Titin. Explain why the largest gene is not producing the largest protein.
5. What are microsatellite repeat sequences? Why are they extensively used as genetic markers? Mention one example of microsatellite instability leading to disease.
6. What do you understand by the term 'single nucleotide polymorphisms (SNP)'? Discuss their medical importance. Name any two techniques for the detection of SNPs.
7. Enumerate the salient features of the mitochondrial genome.
8. Tabulate the differences between nuclear DNA and mitochondrial DNA.

Multiple Choice Questions

1. **True about microsatellite sequence is:**
   a. Small satellite
   b. Extrachromosomal DNA
   c. Short sequence (2-6) repeat DNA
   d. Looped-DNA

2. **All of the following reactions take place in mitochondria, EXCEPT**
   a. DNA synthesis
   b. Protein synthesis
   c. Fatty acid synthesis
   d. β-oxidation of fatty acids

3. **Mitochondrial DNA is:**
   a. Open circular   b. Closed circular
   c. Nicked circular   d. Linear

4. **In the entire genome, the exonic DNA constitutes:**
   a. 1.5%
   b. 0.5%
   c. 0.1%
   d. 0.02%

5. **Alu element is an example of:**
   a. Minisatellite
   b. Microsatellite
   c. LINE
   d. SINE

6. **VNTR sequences are example of:**
   a. Minisatellite
   b. Microsatellite
   c. LINE
   d. SINE

7. **Number of protein-coding genes present in mitochondrial genome is:**
   a. 37
   b. 22
   c. 13
   d. 2

8. Which of the following trinucleotide repeat disorder and the repeat is wrongly paired?

   a. Kennedy disease - CAG
   b. Friedreich's Ataxia - CGG
   c. Myotonic dystrophy – CTG
   d. Huntington's chorea - CAG

9. All of the following diseases are due to excessive number of glutamine residues in proteins, except:

   a. Huntington's chorea
   b. Spinocerebellar ataxia
   c. Myotonic dystrophy
   d. Spinobulbar muscular atrophy

10. Which of the following disease is X-linked?

   a. Huntington's chorea
   b. Spinocerebellar ataxia
   c. Myotonic dystrophy
   d. Spinobulbar muscular atrophy

ANSWERS

1. c.	2. c.	3. b.	4. a.	5. d.	6. a.
7. c.	8. b.	9. c.	10. d.		

DNA Replication and Repair

DNA replication is a process of making another exact copy of parent DNA. It is necessary for the survival of species. DNA replication takes place during the S phase (synthesis phase) of the cell cycle.

DNA REPLICATION IS SEMI-CONSERVATIVE IN ALL DOMAINS OF LIFE

 High Yield

Features of DNA Replication

- ❑ Semiconservative – each parent strand acts as a template for a new daughter strand
- ❑ Template dependent
- ❑ Begins at specific origin(s) of replication
- ❑ Cannot start from scratch; requires a primer
- ❑ The direction of synthesis is always 5′ to 3′

- ❑ Bi-directional and antiparallel to parent strand
- ❑ Semi-discontinuous (continuous in leading strand; discontinuous in lagging strand)
- ❑ DNA is hemimethylated during replication
- ❑ Error-free – information is faithfully copied
- ❑ Post-replication error is repaired

During DNA replication, the DNA double helix is separated. Each strand acts as a template for the synthesis of new, complementary strand.

Each of the two DNA molecules produced at the end of replication contains one parent strand and one daughter strand. *Half of the parent strand is conserved, hence the name 'semiconservative'.*

Matthew Meselson and Franklin Stahl proved the semiconservative nature of replication by their experiments using density gradient centrifugation with N^{15} and N^{14} nitrogen isotopes (Fig. 1).

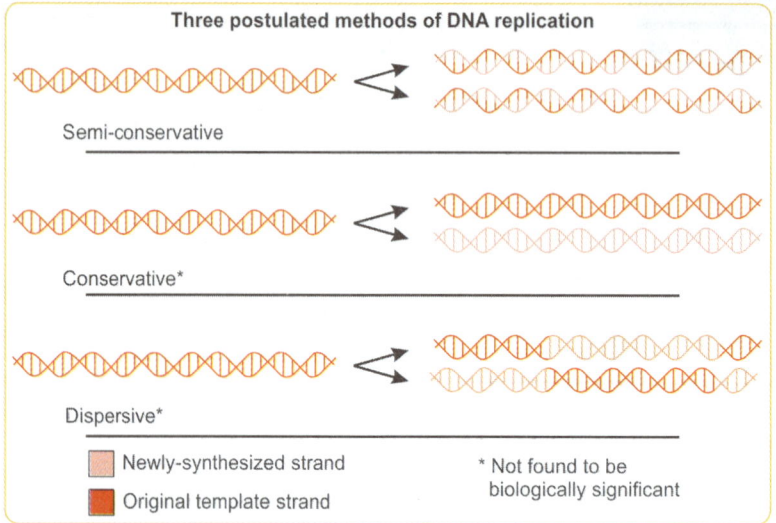

Fig. 1: Illustration of three suggested models for replication. Out of these semi-conservative model has been experimentally proved

Source: Adenosine, CC BY-SA 2.5, https://commons.wikimedia.org/w/index.php?curid=2644927

Most of the research activities in DNA replication have been carried out in prokaryotes. So, we will first discuss the prokaryotic DNA replication in detail followed by eukaryotic DNA replication.

PROKARYOTIC DNA REPLICATION

DNA REPLICATION BEGINS AT SPECIFIC REGION OF DNA

Replication of DNA cannot start from a random site. It starts from a particular region of DNA known as the origin of replication or oriC. Replication proceeds bidirectionally from oriC. The size of E. coli genome is 4.6 Mb (million base pairs), and the whole genome is replicated in about 20 minutes.

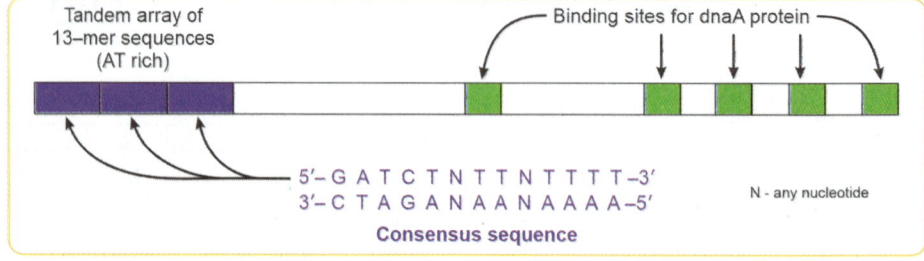

OriC contains repeats of five 9 bp sequence and three 13 bp AT-rich sequences. *As you already know, there are only two hydrogen bonds between adenine and thymine. So, DNA easily gets separated in this region.*

DnaA PROTEIN UNWINDS DNA AT *oriC*

DnaA is a bacterial protein that binds to the 9 bp sequences of *oriC*. Binding of around eight molecules of DnaA to 9bp sequence causes ATP-independent unwinding of AT-rich region.

FURTHER UNWINDING IS DONE BY ATP-DEPENDANT DnaB HELICASE

Further separation of DNA strands is done by a hexameric DnaB protein. As DnaB unwinds the DNA helix, this protein is called as DNA helicase *(c.f. sucrase breaks sucrose).*

DnaB utilizes ATP hydrolysis to open the double helix, and DnaC protein helps in the loading of DnaB (Fig. 2).

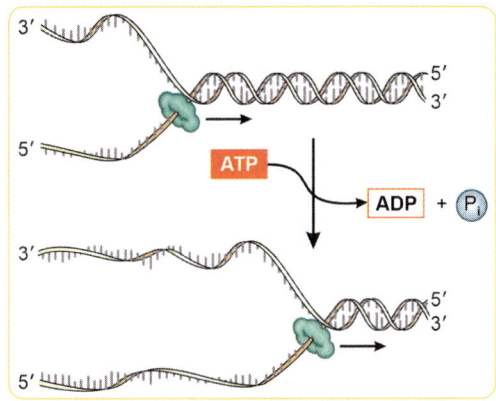

Fig. 2: DnaB in action – Look at the hexameric nature

ssDNA-BINDING PROTEINS (SSBs) STABILIZE THE ssDNA

❑ DNA strands separated by the helicase can reassociate.
❑ This is prevented by the binding of multiple single-stranded DNA-binding proteins (SSBs) to the separated DNA.
❑ ssDNA-binding proteins (SSBs) show co-operative binding, i.e. binding of one SSB facilitates the binding of another SSB.

DnaG MAKES RNA PRIMER

DNA polymerase enzymes cannot begin the replication from scratch. They need an existing free 3'-OH group to which they can add nucleotide. This free 3'-OH group is provided by an RNA primer. This RNA primer is synthesized by DnaG which is a DNA directed RNA polymerase. DnaG is known as primase.

REPLICATION BUBBLE, REPLICATION FORK AND REPLISOME

The region of DNA opened by helicase looks like a bubble. Enzymes of DNA replication act on this bubble region. Helicase acts on both sides of this bubble. So, the bubble progressively enlarges as the new DNA is being synthesized. As you can see in the image, one replication bubble consists of two replication forks. Replisome refers to all the proteins that work at the replication fork.

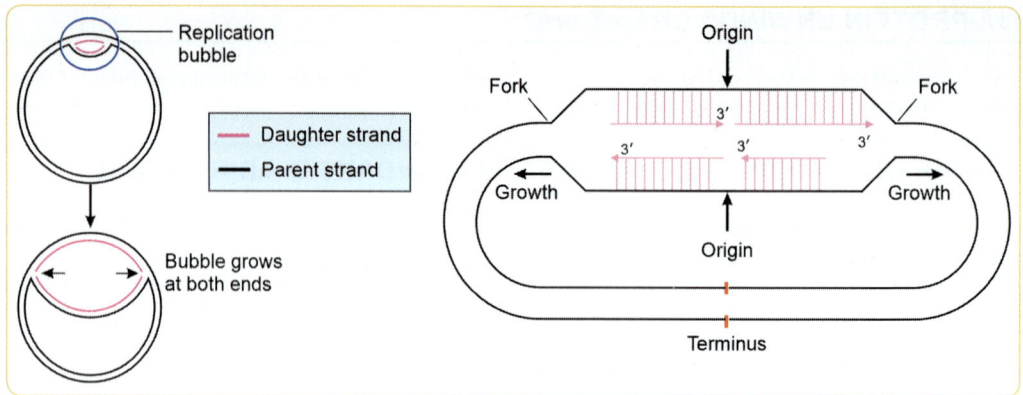

CHALLENGES IN DNA REPLICATION ARE SOLVED BY INGENIOUS SOLUTIONS

DNA replication encounters many challenges. We will discuss only two challenges.

First, the formation of replication bubble induces supercoiling in the non-replicated regions. Topoisomerase solves this problem.

Topoisomers are forms of DNA which differ in their linking number. Topoisomerases are enzymes which can alter the linking number. There are two types of topoisomerases.

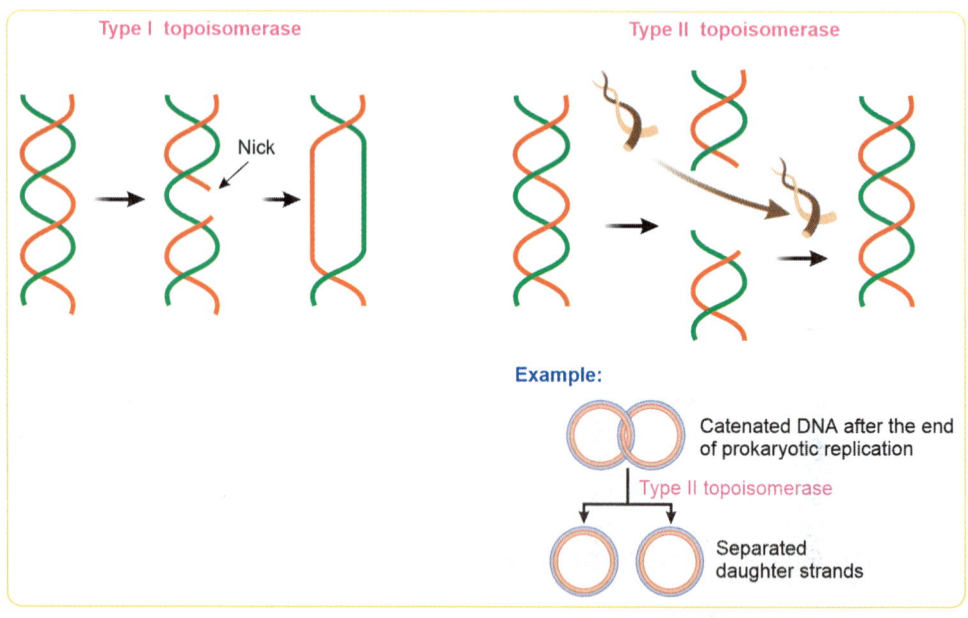

	Type I	Type II
Mechanism of action	Makes a transient cut on one of the strands and passes the other strand through → 360° rotation → sealing of the cut	Cuts both strands → passage of another double helix → sealing
Change in linking number	Increments of 1	Increments of 2

Contd...

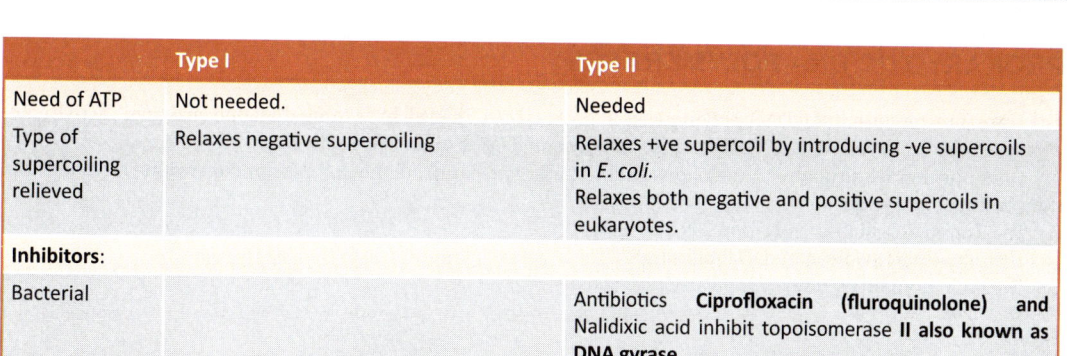

	Type I	Type II
Need of ATP	Not needed.	Needed
Type of supercoiling relieved	Relaxes negative supercoiling	Relaxes +ve supercoil by introducing -ve supercoils in *E. coli*. Relaxes both negative and positive supercoils in eukaryotes.
Inhibitors:		
Bacterial		Antibiotics **Ciprofloxacin (fluroquinolone) and** Nalidixic acid inhibit topoisomerase **II also known as DNA gyrase**
Human	**Camptothecin** (anti-cancer drug)	**Doxorubicin, Etopo**side (anti-cancer drug)

The second problem is because of the antiparallel nature of the DNA. The two DNA strands are of opposite polarity, and DNA polymerases only synthesize DNA in 5' to 3' direction. So, DNA is made in opposite directions on each template. Let us have a close-up view of replication fork to understand this further.

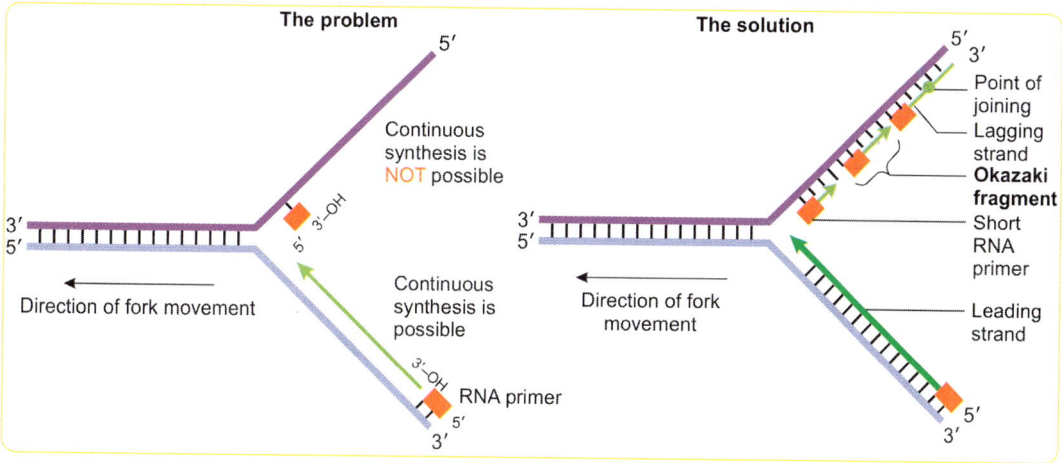

There are two replication forks. We are looking at one. DNA can be synthesized only in 5' to 3' direction. In one strand, the direction of movement of helicase and direction of movement of DNA polymerase are the same. So, DNA can be synthesized continuously (marked by green arrow). This strand is known as leading strand.

On the other strand, which is known as lagging strand, the direction of movement of helicase and DNA polymerase are opposite. So, DNA polymerase cannot synthesize DNA continuously. This problem is solved by synthesizing DNA in small stretches and joining them. Thus, multiple RNA primers are also needed. **Short stretches (100–1000 bp) of DNA synthesized on the lagging strand are known as Okazaki fragments.**

In the end, the RNA primers need to be removed and replaced with DNA. Removal of RNA primers is done by DNA polymerase I; joining of the Okazaki fragments is done by DNA ligase. This is why DNA replication is said to be **semi-discontinuous**; continuous in leading strand and discontinuous in lagging strand.

LEADING STRAND VERSUS LAGGING STRAND

	Leading strand	Lagging strand
Also known as	Forward strand	Retrograde strand
Nature of DNA synthesis	Continuous synthesis toward the replication fork	Discontinuous synthesis away from the replication fork
Number of RNA Primers required	Single	Multiple

Now, let us turn our attention to DNA polymerases.

PROKARYOTIC DNA POLYMERASES

There are three predominant DNA polymerases in prokaryotes. They all have more than one enzymatic activity, i.e. All these polymerases have 5'→3' polymerase activity and 3'→5' exonuclease activity.

DNA polymerase structure is analogous to a human right-hand. So, the domains in the enzyme are named as thumb, finger and palm.

❑ The thumb domain is involved in DNA binding
❑ The fingers are involved in dNTP binding
❑ The palm domain contains the polymerase activity.

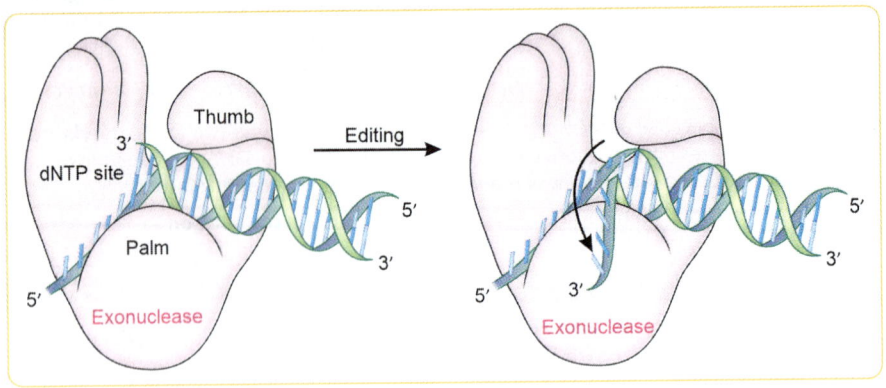

DNA Polymerases are Template Dependant

❑ DNA polymerases make a new strand of DNA from the template strand with strict adherence to the Watson-Crick base pairing rule.
❑ If the sequence of the parent strand is 5'ATGCGTA3', the sequence of the daughter strand will be 5'TACGCAT3' *(DNA strands are antiparallel).*

Incorrect Base-Pairing is Removed by Proof-Reading

DNA polymerase requires magnesium for the correct orientation of nucleotides in the active site. Catalysis starts only after the correct Watson-Crick base pairing of incoming nucleotide with the template DNA.

When the incoming nucleotide base pairs correctly, 3' OH group of the existing nucleotide makes a nucleophilic attack on the α-phosphoryl group of the incoming nucleotide resulting in the release of pyrophosphate. The hydrolysis of pyrophosphate ($\Delta G^{0'}$ = -19.2 kJ/mol) drives the reaction forward. The bond created is 3'→ 5' phosphodiester bond.

Sometimes, mistakenly wrong nucleotides can be added. This slows down the enzymatic catalysis and prevents the addition of next nucleotide. Slowed catalysis triggers the DNA strand with the wrong nucleotide to be transferred to the 3'→5' exonuclease active site. This site cleaves the 3'→5' phosphodiester bond of the mispaired nucleotide, and the DNA returns to the polymerase site. That is why **3'→5' exonuclease activity is known as proof-reading activity.**

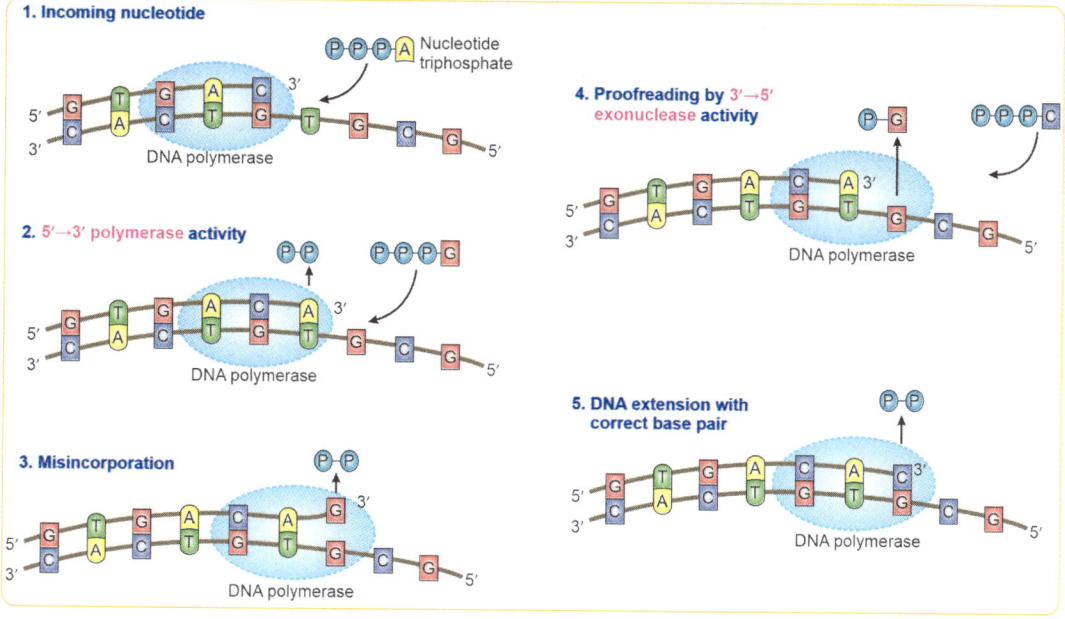

All the prokaryotic DNA polymerases have 5'→3' polymerase and 3'→5' exonuclease activity. Prokaryotic DNA polymerase I has additional 5'→3' exonuclease activity. This activity is important for removal of the RNA primers *(shown in the image below)* and DNA repair *(to be discussed)*.

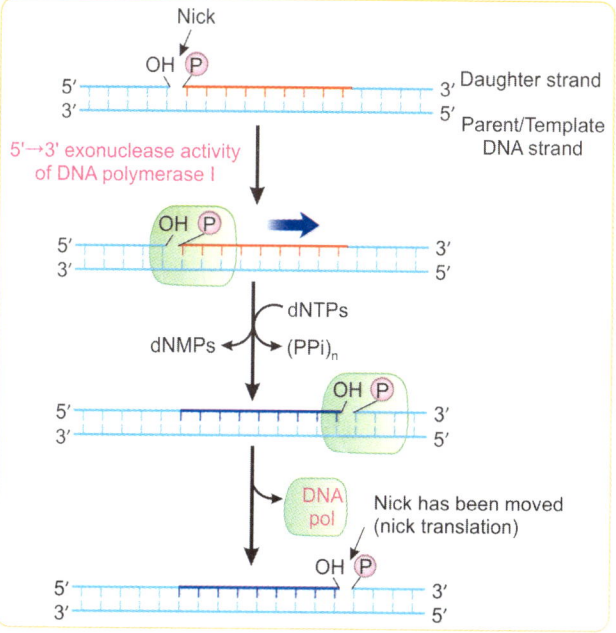

Fig. 3: Illustration of 5'→3' exonuclease activity. Nick lacks the phosphodiester bond. DNA polymerase I cleaves the nucleotides from the nick. Because of the cleavage, nick is moving from one position to another position. This is why it is known as nick-translation. DNA ligase seals Nick. DNA polymerase I is used in in-vitro incorporation of radiolabelled probes using nick-translation technique

DNA Polymerases can Continuously Add Nucleotides

Usually, enzymes bind to the substrate and catalyze the conversion of substrate to product, and the product will be released from the enzyme. The enzyme will bind to new substrate again. But, DNA polymerases progressively add on nucleotides without falling off from the DNA template. **The number of nucleotides that can be added by the polymerase continuously is known as processivity**. Among the three, DNA **polymerase III is the most processive** enzyme.

Processivity of DNA pol III is due to a protein known as β-**clamp or sliding clamp.** This clamp prevents the falling off of DNA polymerase. There is a clamp loader protein which helps in the loading of the clamp to DNA (Fig. 4).

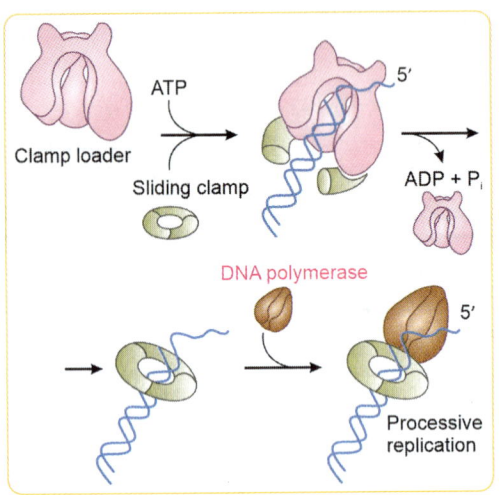

Fig. 4: Role of sliding clamp as the processivity factor

Let us review the following three types of prokaryotic DNA polymerases.

I	First DNA polymerase discovered by Arthur Kornberg, hence known as Kornberg's enzyme The only polymerase with 5'→3' exonuclease activity. Three activities: 5' → 3' polymerase activity - Polymerization 3' → 5' exonuclease activity – Proofreading **5' → 3' exonuclease activity** - Removal of RNA primer, DNA repair, nick-translation
II	❑ Contains 5' → 3' polymerase and 3' → 5' exonuclease activity ❑ Proofreading activity is low, also known as. Error-prone polymerase.
III	❑ **Major enzyme of replication** having 5' → 3' polymerase and 3' → 5' exonuclease activity ❑ Biggest enzyme with >10 subunits ❑ **Highly Processive** enzyme involved in both leading and lagging strand synthesis. ❑ Involved in DNA mismatch repair.

RNA Primers are Removed by DNA Pol I

We have seen that RNA primer is needed for DNA polymerase to act. There are multiple RNA primers in the lagging strand. They need to be removed, and the Okazaki fragments need to be joined. DNA polymerase I removes the RNA primers with its 5' → 3' exonuclease activity. It fills the gap with its polymerization activity. DNA ligase seals the nick.

RNase H can also Remove RNA Primers

RNase H can also remove RNA primers. This is a unique type of ribonuclease. It can cleave only the RNA located in the DNA-RNA hybrid. It can't cleave free RNA.

 Video lecture on RNAseH:
https://youtu.be/KVBE1DqQ_ro or scan QR code

Type II Topoisomerase is Needed for the Separation of Replicated DNA in E. coli

In E. coli, the genome is circular. So, we get two catenated circular DNA after the completion of replication. Type II topoisomerases separate these catenated DNA strands. (*Remember: type II topoisomerases can make double-stranded cuts and pass the other strand through the gap and then seal the cut*).

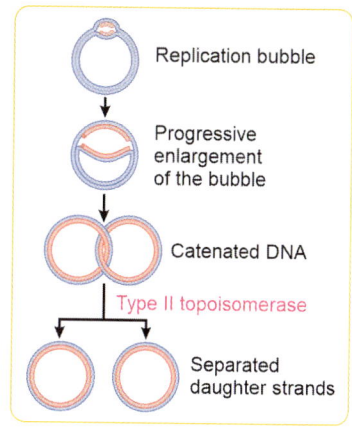

DNA is Hemimethylated During Replication

In DNA, specific sequences are methylated. For example, Adenine of 5'GATC3' sequence is always methylated by methylase enzymes. During replication, the daughter strand is unmethylated. This is known as hemimethylation of DNA during replication. *Importance of methylation will be clear when we study mismatch repair (Fig. 5).*

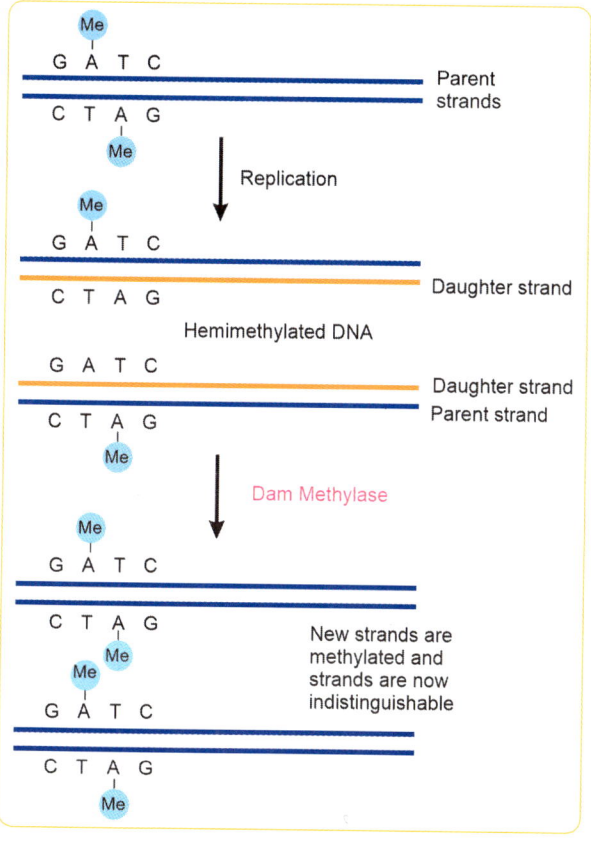

Fig. 5: Dam – Deoxyadenosine methylase

Review: Proteins and Enzymes of Prokaryotic DNA Replication

Protein	Role
DnaA	Recognizes origin of replication
Helicase (DnaB)	Unwinding
SSBPs	Prevents the hybridization of unwound DNA
Topoisomerases	Relaxes supercoiling
Primase (DnaG)	RNA primer synthesis
DNA Polymerases	Template-directed DNA synthesis
RNase H	Removal of RNA Primers
Ligase	Nick sealing
Dam methylase	Methylation of adenine residues of GATC sites

EUKARYOTIC DNA REPLICATION

Basic theme of DNA replication is same as in prokaryotes. The differences are discussed below.

Multiple origin of replication: As the size of the genome is bigger, multiple origins of replication are required. The origin is recognized by complex of 6 proteins known as origin recognition complex (ORC).

EUKARYOTIC DNA POLYMERASES

There are more DNA polymerases in eukaryotes compared to prokaryotes.

α	**Primase** activity (synthesis of RNA primer) and polymerase activity without proof-reading; less processive
β	Base-excision repair
γ	**Mitochondrial DNA** replication
δ	**Lagging strand synthesis;** nucleotide and Base-excision repair
ε	**Leading strand synthesis;** nucleotide and Base-excision repair
θ	DNA repair

POLYMERASE SWITCH AND SLIDING CLAMP

Polymerase with low processivity initiates the replication. Later, polymerase with high processivity takes over. This handover is known as polymerase switching. As we have already seen, DNA Pol III is the most processive prokaryotic polymerase. In eukaryotes, DNA pol δ and ε are processive. Proliferating cell nuclear antigen (PCNA) acts as the sliding clamp (processivity factor) in eukaryotes.

	Switching
Prokaryotes	DNA pol I → III
Eukaryotes	DNA pol α → δ or ε

END REPLICATION PROBLEM IN EUKARYOTES

It is important to understand the structure of telomere first.

Telomere

Ends of the chromosomes are called as telomere (Tele – end). Telomeres contain the hexameric sequence 5'TTAGGG3' repeated up to 2500 times. 3' ending strand is rich in Guanosine and is lengthier than the other strand. So, it is folded back to form a loop with the help of various proteins. Moreover, unusual Guanine-Guanine hydrogen bonding is seen in the looped region (Fig. 6).

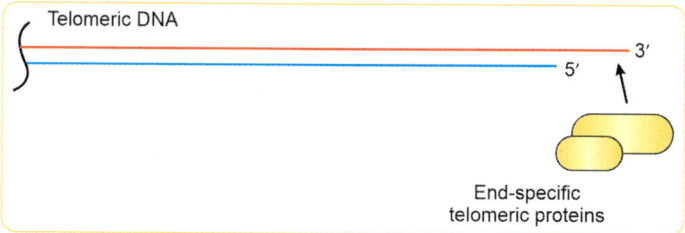

Fig. 6: Simplified structure of telomere

The ends of linear eukaryotic DNA cannot be replicated entirely during **lagging strand** synthesis.

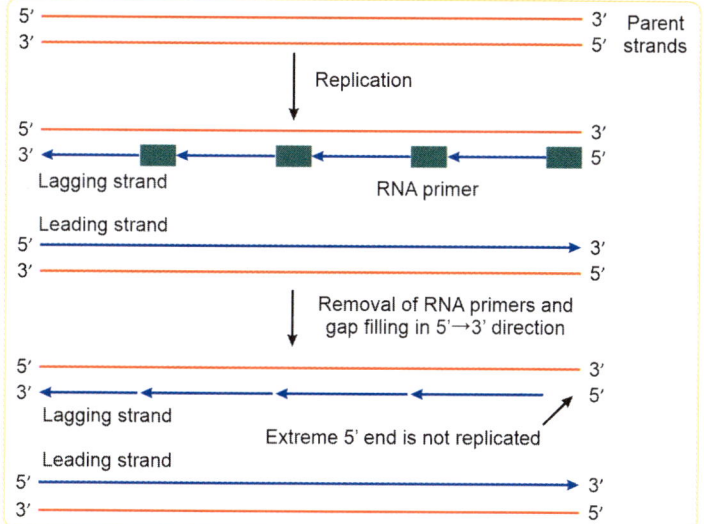

This is due to the removal of the final RNA primer which leaves a gap (shown in the above image with black arrow). With successive replications, there is progressive shortening of telomeric DNA. This is known as end-replication problem. This is the main reason for ageing. As the prokaryotic DNA is circular, there is no end replication problem.

Telomerase Solves End Replication Problem in Eukaryotes

Telomerase is a **reverse transcriptase** carrying its own RNA template. Telomerase elongates the G-rich strand of telomeric DNA.

Telomerase acts in three repetitive steps:

1. **Base pairing:** RNA template of the telomerase is complementary to the hexameric repeating sequence of G-rich strand. So, telomerase could base pair.
2. **Elongation:** Telomerase acts as a reverse transcriptase by elongating the G rich strand using its own RNA template.
3. **Translocation:** Once the elongation is over, telomerase change its location so that it can base pair again to start another round of elongation.

Once the G rich strand is elongated to a particular length, synthesis of the other strand can be initiated with the help of normal cellular DNA polymerases.

Importance

❑ Telomerase is expressed predominantly in stem cells and cancer cells making these cells immortal (Fig. 7).
❑ The absence of telomerase in somatic cells explains the replicative senescence and aging.

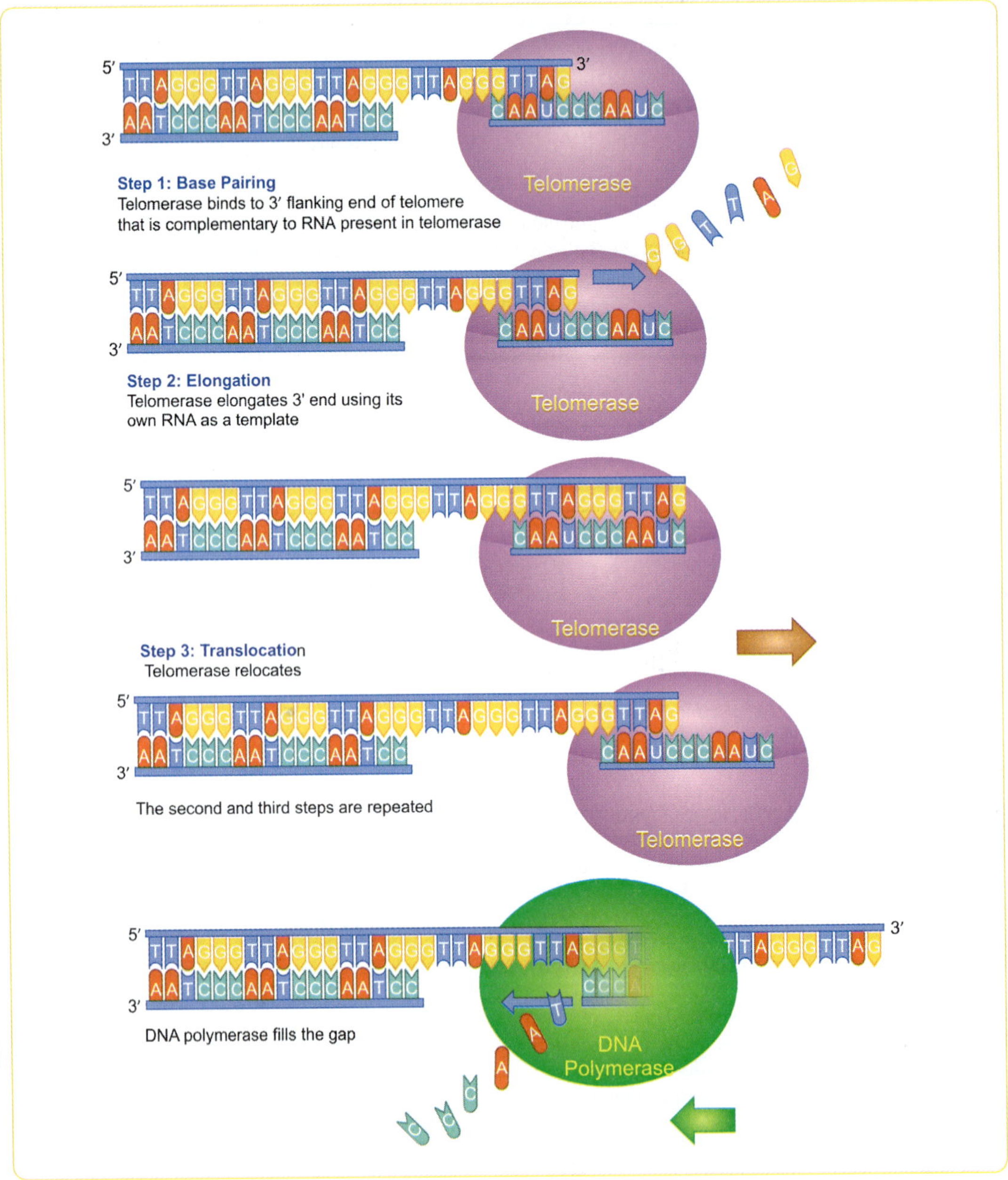

Step 1: Base Pairing
Telomerase binds to 3′ flanking end of telomere that is complementary to RNA present in telomerase

Step 2: Elongation
Telomerase elongates 3′ end using its own RNA as a template

Step 3: Translocation
Telomerase relocates

The second and third steps are repeated

DNA polymerase fills the gap

Fig. 7: Simplified illustration of action of the telomerase

Source: Uzbas, F, CC BY-SA 3.0, https://commons.wikimedia.org/w/index.php?curid=17635942

Replication in Prokaryotes versus Eukaryotes

	Prokaryotic	Eukaryotic
Site	Cytoplasm	Nucleus
Origin of replication	Single	Multiple
Replication bubbles	Single	Multiple
Rate of polymerization	1000 bp/second	50 base/second
End replication problem	Doesn't happen as the DNA is circular	Happens as the DNA is linear
Length of Okazaki fragments	1000 -2000	100-200
Cofactor for DNA ligase	NAD$^+$	ATP
Sliding Clamp	β clamp	PCNA
RNA primer is removed by	DNA Polymerase I	Flap endonuclease-1 RNase H

Salient Features of DNA Replication – Review

Feature	Description
Semiconservative	DNA replication produces two DNA copies, and each contains one of the parent strands and one daughter strand.
Template-dependent	DNA polymerase can produce new DNA by using the complementary base pairing rule.
Begins at origin(s) of replication	DNA replication begins at a single specific origin of replication in prokaryotes and multiple origins in eukaryotes.
Requires RNA primer	Free 3' OH group required for the DNA polymerase is provided by a short piece of RNA which is removed later.
Direction of synthesis is always 5' to 3'	Incoming nucleotides are added to the 3' OH group of existing nucleotide via 3' → 5' phosphodiester bond.
Bi-directional	From the origin of replication, the replication forks move in both directions.
Semi-discontinuous	DNA synthesis is continuous in leading strand but discontinuous in lagging strand.
Errors are corrected	Due to complementary base-pairing rule and proof-reading activity of DNA polymerase.
Hemimethylated	During replication, parent strand of DNA is methylated whereas the daughter strand is not. Immediately after replication, daughter strand also methylated.
Post-replication error is repaired	DNA repair machinery repairs any DNA damage that arises after replication.

DNA REPAIR

DNA IS THE ONLY MACROMOLECULE THAT IS REPAIRED

Macromolecules like proteins, if damaged, will be destroyed in our body. But, DNA even if it is damaged, requires to be preserved and repaired because of its vital function.

Integrity of DNA during and after replication is maintained by:
- ❑ Geometric symmetry during complementary base-pairing
- ❑ Proof-reading activity of DNA polymerase
- ❑ DNA repair mechanisms

❑ Various checkpoint mechanisms that check the DNA for any errors during cell cycle (Chapter 31) *(Refer to page no. 485)*

	Error rate (rate of wrong base incorporated)
Polymerase activity alone	10^{-3} to 10^{-4}
Proofreading	10^{-6} to 10^{-7}
Repair mechanisms	10^{-9} to 10^{-10}

Here is the tabulation of various types of damages and how the damage is repaired and the diseases due to the defects in repair.

Type of damage	Repair mechanism	Disease due to failure
Base mismatch during replication	Mismatch repair	Hereditary nonpolyposis colon cancer
UV induced thymine dimer	Nucleotide excision repair	Xeroderma pigmentosum Cockayne syndrome Trichothiodystrophy
Damage to single base	Base excision repair	MUTYH-associated polyposis
Double-stranded breaks		**End-joining repair**
	Homologous end joining repair	Bloom syndrome Werner syndrome
	Non-homologous end joining repair	Severe combined immuno deficiency disease

*The above repair mechanisms are seen in both prokaryotes and eukaryotes. But **direct repair by photolyase** is present only in prokaryotes.*

METHYL-DIRECTED MISMATCH REPAIR (PROKARYOTIC)

There could be a base mismatch during replication which escaped the proofreading activity of DNA polymerase.

Whenever there is a mismatch, it is necessary to differentiate the parent strand from the newly-synthesized daughter strand. This is achieved by methylation of adenine of GATC sequence of parent strands.

MutS and MutL proteins recognize the mismatch.

MutH is an endonuclease that makes a nick in the daughter strand which contains **unmethylated** GATC region.

This nick is extended by the combined action of DNA helicase II (UvrD), exonucleases (Exo) and Single-Strand binding proteins (SSB).

The gap is filled by DNA polymerase III (not Pol I), and DNA ligase seals the nick.

Finally, daughter strand is also methylated on the adenine of GATC sequences.

Eukaryotic analogues of MutS and MutL are MSH2 (MutS homolog), MLH1, MSH3 and MSH6. Defect in these proteins leads to **Hereditary Nonpolyposis Carcinoma of the Colon** (HNPCC).

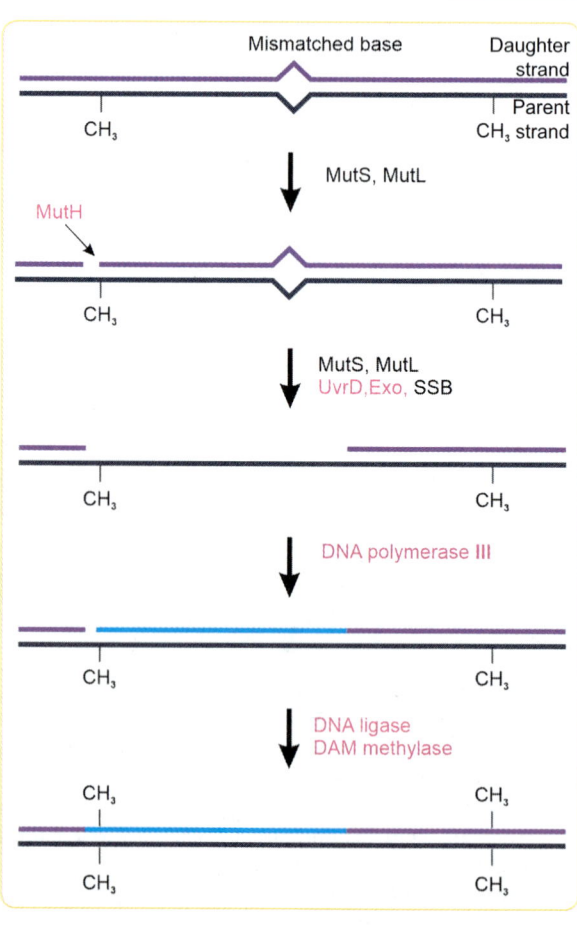

Section III Molecular Biology

BASE EXCISION REPAIR (PROKARYOTIC)

Nitrogenous bases of DNA can be altered by either spontaneous deamination or chemical agents. For example, cytosine spontaneously gets deaminated to uracil. Uracil is not supposed to be present in the DNA. This will alter the genetic code if uncorrected.

DNA glycosylase (e.g. Uracil N-glycosylase) removes any uracil present in the DNA. This enzyme breaks the β-N-glycosidic bond between uracil and the sugar-phosphate backbone. This creates an abasic site (apyrimidinic site). This site is recognized by AP endonuclease makes a nick (small cut) in the phosphodiester backbone of the AP site. DNA polymerase I removes the nucleotides from the nick by 5′ → 3′ exonuclease activity and replaces the gap with new nucleotides. DNA ligase seals the final nick (Fig. 8).

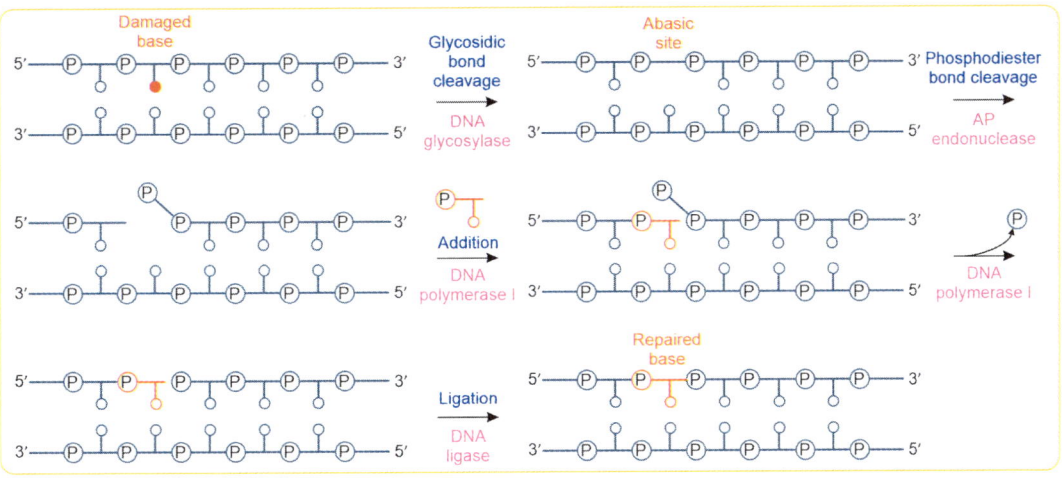

Fig. 8: Base excision repair

NUCLEOTIDE EXCISION REPAIR (PROKARYOTIC)

Ultra-violet rays cause pyrimidine (thymine) dimer formation in the same strand of DNA. This pyrimidine dimer interferes with the replication and transcription, if not repaired.

Nucleotide excision repair is complex in humans. At least 18 proteins are involved in this process. A defect in nucleotide excision repair leads to xeroderma pigmentosum.

Xeroderma Pigmentosum

Xeroderma pigmentosum is a **familial cancer syndrome** with **autosomal recessive** inheritance. **Genetic defect:** Defect could be in one of the multiple genes (XP-A through G) involved in **nucleotide excision repair** (NER). As a result, ultraviolet rays induced **cyclobutane thymidine dimers** are not repaired, and mutations accumulate in the affected individuals.

Xeroderma pigmentosum variant (XP-V) is due to defect in DNA polymerase eta (Pol η).

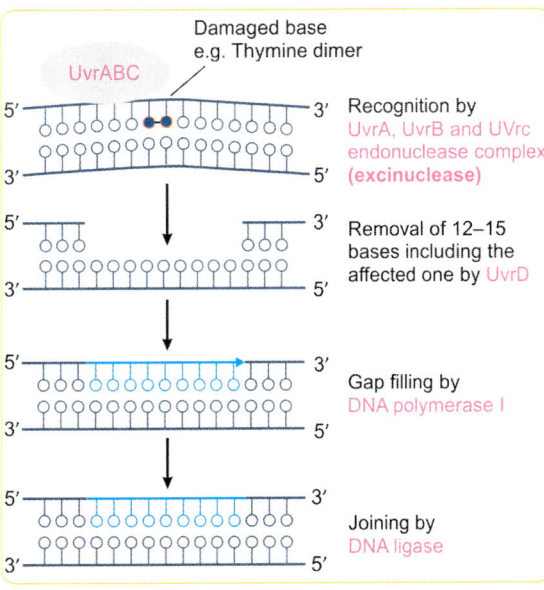

Clinical Features

Affected persons are extremely sensitive to any source of UV light, especially sunlight. The exposed skin develops excessive dryness, pigmentation, atrophy, and hyperkeratosis (thickened precancerous outgrowths of the dermis). Due to unknown mechanisms, neurodegeneration also happens.

Cancer Proneness

As the DNA repair mechanism is defective, people with XP are highly susceptible to developing basal cell carcinoma, squamous cell carcinoma and malignant melanomas.

Treatment

There is no definitive treatment. Avoidance of all sources of UV light is the prevention to be taken. Precancerous lesions need to be removed by surgery. Most of the patients succumb to metastasis in their forties.

Cockayne Syndrome

❑ Due to defect in transcription coupled nucleotide excision repair.
❑ Clinical features: failure to thrive, microcephaly, hearing loss, nervous system anomalies, photosensitivity
❑ **Patients are not predisposed to cancer**.

DOUBLE-STRAND BREAK REPAIR

Double-stranded breaks (DSB) in DNA are caused by ionizing radiation, toxic chemicals, or ultraviolet light (UV). DSB can be repaired by,
❑ Nonhomologous end joining (NHEJ)
❑ Homologous recombination (HR)
For easy understanding, let us differentiate the two.

	Nonhomologous end joining	Homologous recombination
Phase of cell cycle	Takes place during G_0 and G_1 phase of cell-cycle	Takes place during S, G2, and M phase
Description	Broken ends are re-joined without much processing	Homologous sequences in the sister chromatid is used as a template to repair the break
Loss of genetic material	Less efficient; There is always a loss of genetic material	More efficient; Chance of loss of genetic material is less
Diseases due to defect	Severe combined immunodeficiency disease (SCID)	Bloom syndrome Werner syndrome AT-like disorder Nijmegen breakage syndrome Rothmund-Thomson syndrome Breast cancer susceptibility 1 and 2 (BRCA1, BRCA2)

 High Yield

CRISPR-Cas9 system is a powerful tool for genome editing. Cas9 is an endonuclease that makes double stranded break in the DNA. This can be used either for gene disruption or gene replacement.
❑ If no homologous DNA strand is provided, the cell will repair the double stranded break by NHEJ leading to gene knockout.
❑ If a homologous DNA strand is provided, the cell will repair the double stranded break by homology-directed recombination leading to gene replacement. This is used to replace defective genes with functional genes.

REVIEW–PROTEINS AND ENZYMES INVOLVED IN DNA REPAIR

Type	Proteins and Enzymes involved
Mismatch repair	Dam methylase MutH, MutL, MutS proteins DNA helicase II **DNA polymerase III** DNA ligase
Base-excision repair	DNA glycosylases AP endonucleases DNA polymerase I DNA ligase
Nucleotide-excision repair	ABC excinuclease Helicase DNA polymerase I DNA ligase
Direct repair (photoreactivation)	DNA photolyase

- ❏ Terminal deoxynucleotidyl transferase (TdT) is a **template-independent DNA polymerase** involved in the VDJ recombination of gene segments (generation of antibody diversity).
- ❏ Klenow fragment is DNA polymerase I without 5′ → 3′ exonuclease activity.

Competency

- ❏ BI7.2   Describe the processes involved in replication and repair of DNA

Key Points

- ❏ DNA replication is semiconservative; bidirectional; discontinuous
- ❏ DNA polymerase always requires a primer with free 3′OH group.
- ❏ Direction of DNA and RNA synthesis is 5′ to 3′.
- ❏ Okazaki fragments are short stretches of DNA synthesized discontinuously on the lagging strand.
- ❏ DNA polymerase III is involved in mismatch repair.
- ❏ CRISPR-Cas9 system is a genome-editing tool being tried in many genetic disorders.
- ❏ Klenow fragment is DNA polymerase I without 5′→ 3′exonuclease activity.
- ❏ Telomerase is an RNA-dependent DNA polymerase (reverse transcriptase) carrying its own RNA template.
- ❏ Helicases unwind the double helical DNA, whereas topoisomerases uncoil the supercoiled DNA.
- ❏ Bloom syndrome occurs due to defective homologous repair of double-stranded breaks.
- ❏ Xeroderma pigmentosum and Hereditary Non-Polyposis Colon Cancer occur due to defect in Nucleotide excision repair and mismatch repair respectively.

Chapter 21   DNA Replication and Repair

SELF-ASSESSMENT

Short Answer Questions

1. Write the functional significance of 5'→3' and 3'→5' exonuclease activities of DNA polymerase.

2. How DNA gyrase changes the topological property of DNA? Name an antibiotic that inhibits gyrase activity.

3. What is the end replication problem in eukaryotes? Why E. coli bacteria do not face this problem? How do rapidly dividing human cells overcome this problem?

4. Draw a schematic diagram to depict the main steps of nucleotide excision repair (NER) in eukaryotes and name the enzymes involved. Name one human disease associated with a defect in NER.

5. Why activation of telomerase is necessary for a cell to become cancerous, though it is not active in most human cells? Explain the mechanism of action of telomerase.

6. What are the different types of double-stranded break repair in eukaryotes? Name any one disease each that arise due to defect in these repair mechanisms.

7. Explain why uracil is present in RNA and not in DNA.

8. (i) List the commonest DNA defect that occurs due to UV ray exposure. (ii) Name DNA repair defect in:
   a. Xeroderma Pigmentosum
   b. Hereditary nonpolyposis colorectal cancer
   (iii) Write the mechanism of the anticancer effect of Adriamycin.

Clinical Case-based Question

1. An 8-year-old boy was presented at dermatology clinic with skin tumour on right cheek and scattered areas of hyperpigmentation. He also avoided exposure to sunlight because it caused blister on his skin. There was no similar history in the family.
   a. What is the provisional diagnosis?
   b. Mention the genetic cause that leads to this disorder.

Long Answer Question

1. How is the fidelity of DNA replication maintained during cell division? Enumerate the different cellular DNA repair mechanisms for repair of damaged DNA. Discuss any one of them with a schematic diagram. Add a note on diseases associated with defective DNA repair.

Multiple Choice Questions

1. In which phase of the cell cycle does DNA replication occur?
   a. S phase
   b. G1 phase
   c. G2 phase
   d. M phase

2. DNA replication follows which of the following model(s)?
   a. Conservative
   b. Semiconservative
   c. Dispersive
   d. All of the above

3. All of the following are features of DNA replication, except:
   a. Semiconservative
   b. Semidiscontinous
   c. Template independent
   d. Requires RNA primer

4. Which is the only prokaryotic DNA polymerase with 5'→3' exonuclease activity?
   a. DNA Polymerase I
   b. DNA Polymerase II
   c. DNA Polymerase III
   d. DNA Polymerase IV

5. Which of the following is the cofactor for Prokaryotic DNA ligase?
   a. Tetrahydrobiopterin
   b. ATP
   c. NAD
   d. FAD

6. Which one of the following protein is involved in methyl-directed mismatch repair in humans?
   a. ATM
   b. BRCA1
   c. MSH6
   d. Rad51

7. **Telomerase is a:**
   a. RNA dependent RNA polymerase
   b. DNA dependent RNA polymerase
   c. DNA dependent DNA polymerase
   d. RNA dependent DNA polymerase

8. **DNA gyrase is:**
   a. Eukaryotic DNA topoisomerase I
   b. Eukaryotic DNA topoisomerase II
   c. Prokaryotic DNA topoisomerase I
   d. Prokaryotic DNA topoisomerase II

9. **All of the following are true about telomerase, except:**
   a. Ribonucleoprotein
   b. Reverse transcriptase
   c. RNA containing ribozyme
   d. Active in cancer cells and germ cells

10. **DNA glycosylase is involved in:**
    a. Mismatch repair
    b. Base excision repair
    c. Nucleotide excision repair
    d. Direct repair

11. **A cell is placed in a medium containing radioactively labelled thymidine. After the cells undergo replication three times, what percentage of the cells will have both strands of DNA labelled?**
    a. 25%   b. 50%
    c. 75%   d. 100%

12. **All of the following are true about Okazaki fragments, EXCEPT:**
    a. DNA segments
    b. synthesized by DNA polymerase
    c. Discontinuous
    d. Located in leading strand

13. **All of the following are true statements about Cockayne syndrome, except:**
    a. Defective transcription-coupled repair
    b. Photosensitivity
    c. Increased risk of cancer
    d. Mental retardation

14. **OriC is a:**
    a. Gene   b. Protein
    c. Enzyme   d. DNA sequence

ANSWERS

1. a.	**2.** b.	**3.** c.	**4.** a.	**5.** c.	**6.** c.
7. d.	**8.** d.	**9.** c.	**10.** b.	**11.** c.	**12.** d.
13. c.	**14.** d.				

22
CHAPTER

Polymerase Chain Reaction

Chapter Outline

- Components of PCR
- PCR Reaction Profile
- Phases of PCR
- Reverse Transcription PCR (RT- PCR)
- Real-time PCR (q PCR)
- Variants of PCR

Polymerase chain reaction (PCR) is the in vitro, enzymatic, DNA amplification technique to amplify a specific DNA segment of interest.

Kary B Mullis was awarded the Nobel Prize in the year 1993 "for his invention of the PCR method".

In PCR, repetitive rounds of amplification of DNA is done using template DNA, specific primers and DNA dependent DNA polymerase enzyme. Products of the previous round act as template for the subsequent rounds, hence chain reaction.

Let us compare in vivo DNA replication and in vitro PCR reaction.

	DNA replication	PCR
Nature of primer	RNA	DNA
Separation the two strands of DNA	Helicase	Heat
Name of enzyme that elongates the new strand of DNA	Thermolabile DNA polymerases (usually)	Thermostable DNA polymerases
Proof reading	+	±

COMPONENTS OF PCR

- Template DNA
- Pair of DNA Primers
- Thermostable DNA polymerase
- dNTPs
- Buffer with divalent cation

Let us discuss every component briefly.

Template DNA

- 50–100 ng of pure DNA is needed.
- Should be free of proteins and lipids.

- DNA can be isolated from blood, tissues, cultured cells, hair follicle, etc.
- Purity of the DNA is assessed by 260/280 ratio (OD value of 1.8 to 2)

Pair of Primers

Unlike RNA polymerase, DNA polymerase requires free 3' OH group to begin the polymerization reaction. Primers are designed in such a way that they will anneal/hybridize to the region flanking the DNA sequence to be amplified.

Points to be Considered while Designing Primer

- **Length:** 18–25 nucleotides – If the primer is too short, there will be non-specific hybridization. If the primer is too long, the melting temperature and annealing temperature will be very high.
- **Melting temperature:** 55–65° C; $T_m = 4(G + C) + 2(A + T)$
- **Annealing temperature:** Lower than the melting temperature
- **GC Content:** 40–60%
- **Primer secondary structures due to self-complementarity:** Avoided
- **Repeating nucleotide sequences:** Avoided

Thermostable DNA Polymerases

Before the discovery of thermostable polymerases, PCR was done using the Klenow fragment (DNA polymerase I without 5'→ 3' exonuclease activity) of E. coli DNA polymerase. The problem with Klenow fragment is that it gets inactivated by high temperature.

Thermostable polymerases are derived from bacteria living in hot springs. Some of the thermostable polymerases and their origin are as follows:
- *Taq (Thermus aquaticus)*
- *Pfu (Pyrococcus furiosus)*
- *Vent (Thermococcus litoralis)*
- *Tth (Thermococcus thermophiles)*

Taq Polymerase

- DNA dependent DNA polymerase derived from *Thermus aquaticus*
- Half-life of 45 minutes at 95°C.
- **Extension rate:** 2 to 5 kb/min
- **Processivity:** 50–60 bases
- The reason behind the thermostability of Taq polymerase is the increased hydrophobicity of the core of the enzyme and stabilization by electrostatic forces

dNTPs (Deoxynucleotide Triphosphates)

- Equimolar concentrations of dATP, dCTP, dGTP, dTTP are used.
- ddNTP (dideoxynucleotide triphosphates) are NOT used in PCR; ddNTPs are used in sanger's chain-termination method of DNA sequencing.

Buffer

- Buffer is used to maintain the optimum pH (pH 8) for the activity of Taq polymerase.
- Buffer also contains divalent cations like Mg^{2+} which is a cofactor for Taq polymerase.

PCR REACTION

All these components (reaction mixture) are taken in a PCR tube and placed in the thermocycler (also known as PCR machine). Thermocycler can change the temperature of the reaction mixture rapidly.

When the reaction mixture is heated at around 94°C, the hydrogen bond between the DNA strands break. When the thermocycler rapidly cools down to annealing temperature, the primers anneal to the template DNA.

On raising the temperature to the optimum temperature of *taq* polymerase, i.e. 72°, elongation of the primer takes place. The cycle is repeated again and again up to 30 to 40 times (Fig. 1).

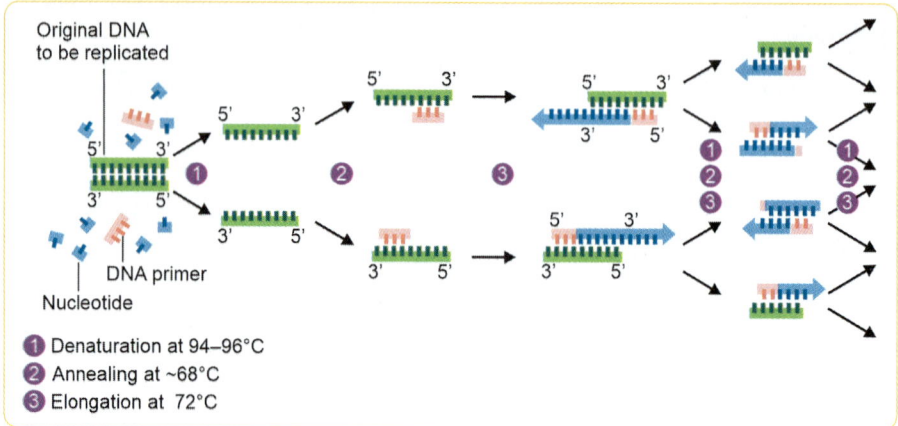

Fig. 1: Schematic drawing of a complete PCR cycle – notice the exponential amplification.
Source: Enzoklop - Own work, CC BY-SA 3.0, https://commons.wikimedia.org/w/index.php?curid=32003643

Phases of PCR

A PCR reaction consists of the following three phases:
1. **Exponential:** Exact doubling of product is accumulating at every cycle (assuming 100% reaction efficiency). The reaction is very specific and precise.
2. **Linear:** The reaction components are being consumed, the reaction is slowing and components are starting to degrade.
3. **Plateau:** In due course of PCR reaction, the components get depleted and polymerase efficiency decreases. So, the reaction ceases and no more products are being made. If left long enough, the PCR products will begin to degrade.

Visualization of PCR Amplicons

The products produced by PCR amplification are known as amplicons. These are of a specific size based on the primer we have designed. So, we can do a gel electrophoresis to visualize the presence of a specific product size.

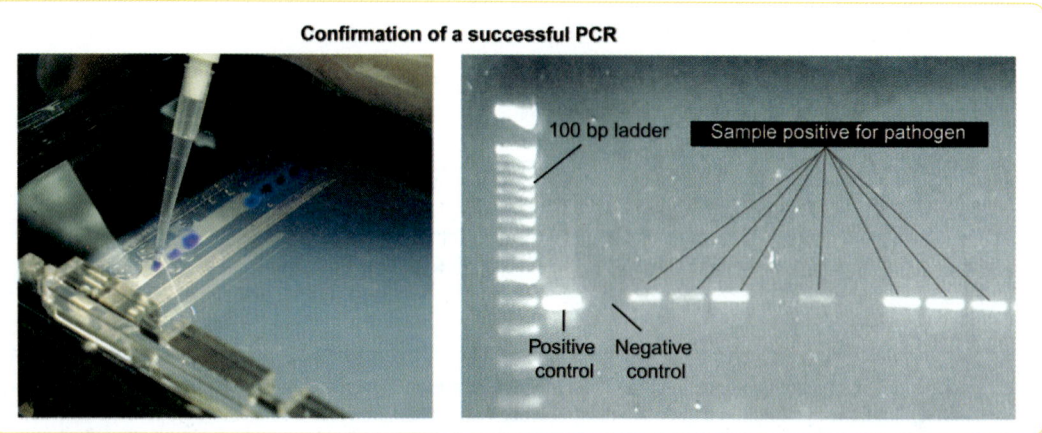

Advantages of PCR

❑ PCR is exquisitely sensitive. A single DNA molecule can be amplified and detected.
❑ PCR can be made highly specific using the stringency of hybridization at relatively high temperature.
❑ The sequence of the target need not be known. Only the flanking sequence needs to be known.
❑ Even targets larger than 10 kb can also be amplified by PCR using specific methods.

REVERSE TRANSCRIPTION PCR (RT-PCR)

❑ RT-PCR is a 2-step process – reverse transcription followed by PCR.
❑ The initial sample we take is RNA.
❑ Reverse transcription is the RNA directed DNA synthesis.
❑ Reverse transcriptases present in retroviruses like HIV convert RNA to DNA (known as complementary DNA/ cDNA).
❑ RT enzymes used for reverse transcription is usually derived from MMLV/AMV (Moloney strain of murine leukaemia virus reverse transcriptase/ Avian myeloblastosis virus)
❑ Reverse transcription is done at 37-42° using DNA primers.

Three Types of DNA Primers used for Reverse Transcription/cDNA Synthesis

1. Oligo dT primers
2. Random primers (random hexamers or random decamers)
3. Sequence-specific primers

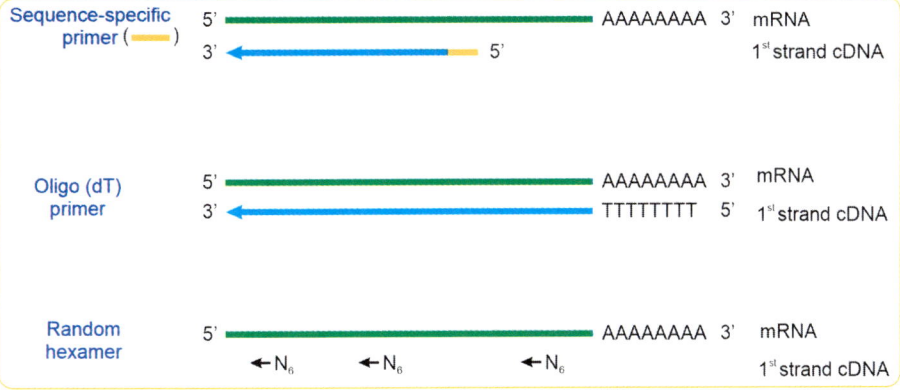

Oligo dT Primers

❑ Oligo dT primers bind to the poly-A tail of mRNA.
❑ Only mRNAs can be converted to cDNA.
❑ mRNAs not having poly A tail-like histone mRNA can't be converted to cDNA by this method.
❑ The efficiency of reverse transcription by oligo dT priming is less when the mRNA is longer.

Random Primers (Random Hexamer or Decamer)

❑ All types of mRNAs can be converted to cDNA.
❑ Longer RNAs can also be efficiently converted to cDNA.

Sequence-specific Primers

When there is a need to produce the cDNA of a specific gene (e.g. cDNA of insulin), sequence-specific primers are used.

cDNA Synthesis Reaction

All the components of reverse transcription are kept in a water bath or dry heat block at 42° C for forty minutes. RT enzyme mixture present in the cDNA can interfere in the downstream processes. So, it is important to inactivate the enzyme by heating the reaction mixture to 92°C before storage.

cDNA produced by reverse transcription is the first strand of cDNA. That is why the commercially available kits for cDNA synthesis are known as "1st strand cDNA Synthesis Kit". When the first strand of cDNA is amplified by PCR, double-stranded cDNA is generated. Amplified DNA is detected by electrophoresis.

Applications of RT-PCR

❑ Used in detection of RNA viruses
❑ To study mRNA expression levels in cells, tissues in a **semiquantitative** manner.
❑ To study for the presence of active infection, e.g. TB

REAL-TIME PCR (qPCR)

Real-time PCR also known as quantitative PCR is used to quantify the amount of initial DNA in the reaction. Traditional PCR detects the DNA by electrophoresis at the end-point of the reaction. Real-time PCR generally uses a fluorescent dye to measure the amount of product as it is generated. This allows for data collection after each cycle of PCR instead of only at the end of the 20 to 40 cycles.

Three General Methods used for the qPCR

1. DNA-binding dyes: (SYBR Green, Syto 9, Eva green)
2. Hydrolysis probes: (Taq Man, Beacons, Scorpions)
3. Hybridization probes

We will limit our discussion only to the dye-binding method. In this method, the PCR mixture is prepared along with a double-stranded DNA binding dye (e.g. SYBR green). The dye shows fluorescence only when it binds to dsDNA. The dye shows increased fluorescence upon binding to the double-stranded DNA.   After each cycle, the level of fluorescence is measured. Fluorescence detectors capture the signals and convert them to graphical representation on the screen.

Role of Housekeeping Genes

During cDNA synthesis, we take an equal amount of mRNA for all the samples. But the efficiency of reverse transcriptase can vary from sample to sample. To normalize this, we should also study the expression of a housekeeping gene in addition to the gene of interest. The expression of housekeeping genes is constant over time.

Reference gene used for normalization	Physiological role
GAPDH (Glyceraldehyde 3 phosphate dehydrogenase)	Glycolytic enzyme with moonlighting functions
28S ribosomal protein	Subunit of the ribosome with peptidyl transferase activity
TFRC (Transferrin receptor)	Intracellular iron uptake
HPRT (Hypoxanthine-guanine phosphoribosyltransferase)	Purine salvage pathway
TBP (Tata box binding protein)	Transcription factor

Analysis of Real-time Data

It is important to understand the following three terms used in qPCR.

1. **The baseline** is defined as PCR cycles in which a reporter fluorescent signal is accumulating but is beneath the limits of detection of the instrument.

2. **Threshold**: is an arbitrary level of fluorescence chosen on the basis of the baseline variability. A signal that is detected above the threshold is considered a real signal that can be used to define the threshold cycle (CT) for a sample. The threshold can be adjusted for each experiment so that it is in the region of exponential amplification across all plots.
3. **The threshold cycle (Ct)** is defined as the fractional PCR cycle number at which the reporter fluorescence is greater than the threshold. The CT is a basic principle of real-time PCR and is an essential component in producing accurate and reproducible data.

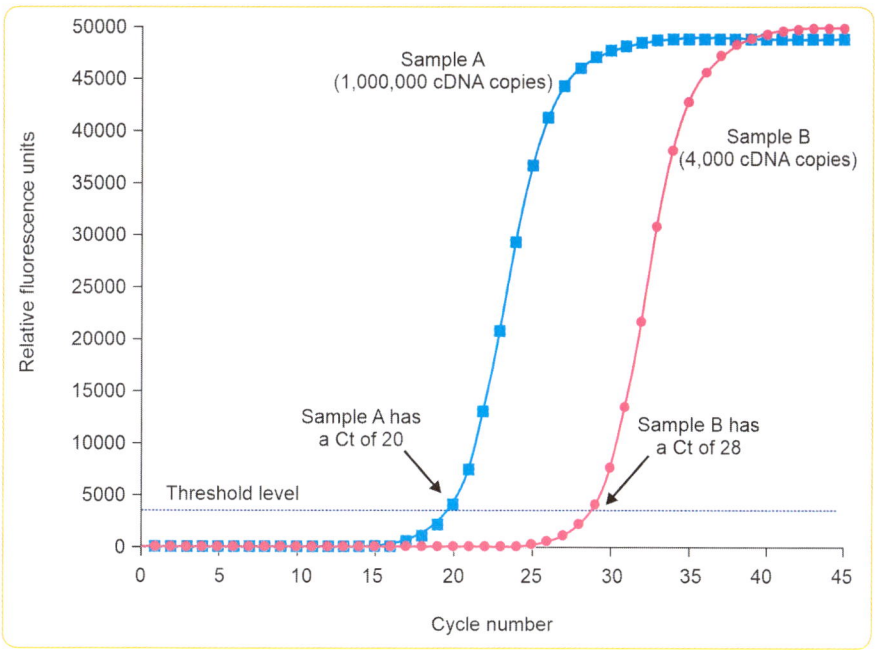

Fig. 2: The original amount of cDNA (RNA) was higher in sample A compared to sample B. Sample A contains 2^8-fold higher amount of RNA

If the initial amount of cDNA (in turn mRNA) is less, it will take a long time to achieve the threshold fluorescence, and thus the Ct value will be higher. The initial amount of cDNA/mRNA is inversely proportional to the Ct value. For example, when the threshold is set for a Real-time PCR and sample A has a Ct value of 20 and sample B has a Ct value of 28, it means that sample A has 2^(Ct of B -Ct of A), that is 2^8 times, more RNA than sample B (Fig. 2).

Fold gene expression is usually calculated using the following method:

$$\Delta Ct = Ct \text{ (gene of interest)} - Ct \text{ (housekeeping gene)}$$
$$\Delta\Delta Ct = \Delta Ct \text{ (treated sample/case)} - \Delta Ct \text{ (untreated sample/control)}$$
$$\text{Fold gene expression} = 2^{-(\Delta\Delta Ct)}$$

Advantages of Real-Time PCR

❑ Amplification can be monitored in real time.
❑ Less than two-fold change can also be detected.
❑ No post-run processing like electrophoresis is required.

Variants of PCR

As PCR is versatile, there are so many variations of PCR available. Interested readers can further study these techniques in the literature.

- ❏ Multiplex PCR
- ❏ Nested PCR
- ❏ Colony PCR
- ❏ Digital PCR
- ❏ Suicide PCR
- ❏ Methylation-specific PCR (MS-PCR)
- ❏ Amplification-refractory mutation system (ARMS)-PCR

▶ **Video lecture of this chapter:** https://youtu.be/TL_bRtjZteQ or scan QR code

Key Points

- ❏ Polymerase chain reaction is an in vitro, enzymatic DNA amplification technique.
- ❏ Template DNA, Pair of DNA Primers, Thermostable DNA polymerase, dNTPs, Buffer with divalent cation are the components of PCR.
- ❏ Taq polymerase is a thermostable DNA polymerase.
- ❏ ddNTPs are used in Sanger's chain-termination method of DNA sequencing. ddNTPs are not used in PCR.
- ❏ Reverse transcriptase is an RNA-dependant DNA polymerase.
- ❏ SYBR green dye is used in real-time PCR.
- ❏ Oligo dT primers, Random primers (random hexamers or decamers) and Sequence-specific primers are used for cDNA synthesis.

SELF-ASSESSMENT

Short Answer Questions

1. Name any four components of polymerase chain reaction and their roles.

2. Why is only a specific DNA sequence amplified by PCR though the reaction cocktail contains the whole genome? Explain the reason for the extreme sensitivity of PCR. What is RT-PCR?

3. What are all the points to be kept in mind while designing a primer for PCR?

4. Name three priming techniques for cDNA synthesis. Which one will you use for the synthesis of cDNA of histone gene?

5. What is the advantage of realtime PCR over conventional PCR?

6. Enumerate the medical applications of PCR and real-time PCR.

Multiple Choice Questions

1. **PCR is a:**
   a. DNA degradation technique
   b. DNA amplification technique
   c. DNA sequencing technique
   d. all of these

2. **All of the following are used in a PCR reaction, except:**
   a. Buffer
   b. ddNTPs
   c. Oligonucleotide primer pair
   d. Template DNA

3. **All of the following are used in a PCR reaction, except:**
   a. Taq polymerase
   b. dNTP
   c. Primer
   d. Radiolabelled DNA probe

4. **Enzyme used for cDNA synthesis is**
   a. DNA dependent RNA polymerase
   b. DNA dependent DNA polymerase
   c. RNA dependent RNA polymerase
   d. RNA dependent DNA polymerase

5. **In PCR, deoxy nucleotide triphosphates (dNTPs) are added to the growing DNA strand during the ___ phase.**
   a. Extension
   b. Annealing
   c. Denaturation
   d. None

6. **A dideoxynucleotide does not contain**
   a. 2′ and 3′ OH group
   b. 3′ and 4′ OH group
   c. 4′ and 5′ OH group
   d. 2′ and 5′ OH group

7. **All of the following are priming strategies for cDNA synthesis, except:**
   a. Random hexamer
   b. Random decamer
   c. Gene-specific primer
   d. oligo d(A) primer

8. **The technique for accurate quantification of gene expression is:**
   a. Northern blot   b. PCR
   c. Real-Time Reverse Transcriptase PCR
   d. Reverse Transcriptase PCR

9. **All of the following are used in cDNA synthesis, except:**
   a. MMLV-reverse transcriptase
   b. Genomic DNA
   c. deoxynucleotide triphosphate
   d. Oligo d(T) primer

ANSWERS

| 1. b. | 2. b. | 3. d. | 4. d. | 5. a. | 6. a. |
| 7. d. | 8. c. | 9. b. | | | |

Transcription

Chapter Outline

"DNA → RNA → PROTEIN" IS THE CENTRAL DOGMA OF LIFE

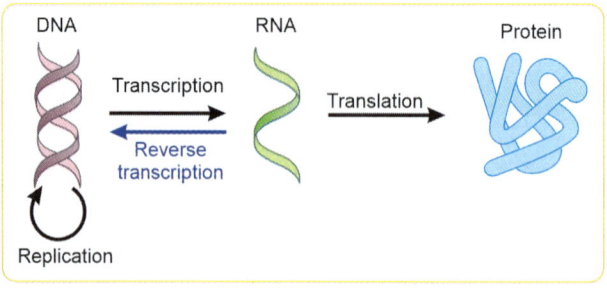

The central dogma of life suggested the flow of information as DNA → RNA → protein

- DNA replication—DNA directed DNA synthesis
- Transcription—DNA directed RNA synthesis
- Translation—RNA directed protein synthesis

Exceptions to Central Dogma

Discovery of retrovirus revealed that the flow of information is not always unidirectional.
- Reverse transcription—RNA directed DNA synthesis; seen in retrovirus
- RNA replication—RNA directed RNA synthesis; seen in RNA virus

VARIOUS TYPES OF RNAs

Type	Heterogeneity	Relative abundance	Function
Protein coding RNA			
Messenger RNA (mRNA)	The most heterogeneous RNA type with one lakh species	2–5% of total RNA	Carries the information for protein synthesis
Non-protein coding RNAs			
Transfer RNA (tRNA)	40 to 50 different species	10–15%	Adaptor molecule in translation
Ribosomal RNA (rRNA)		The most abundant RNA (80% of total RNA)	Combine with proteins to form ribosomes 28S rRNA is a ribozyme having peptidyl transferase activity
Small nuclear RNA (snRNA)	~30 species		Spliceosome-mediated splicing of heteronuclear RNA
Small nucleolar RNA (snoRNA)	>3000 species	<1% of total	Processing of ribosomal RNA
Micro RNA (miRNA) (usually endogenous) **Small interference RNA) (siRNA) usually exogenous)**	>14,000 speices	<1% of total	Size: 20–22 nucleotides Regulation of gene expression by RNA interference/Post-transcriptional gene silencing (PTGS)
Long noncoding RNAs (lncRNAs)	~30,000 species		Size: >200 nucleotides □ Contributes to structural aspects of chromatin □ Regulation of gene expression □ X-chromosome inactivation (Barr body formation)

Non-coding RNA refers to RNAs that do not code for a protein. It is estimated that **more than 80% of human genome is transcriptionally active** but may not produce protein. Much of this non-protein coding transcriptionally active regions are involved in the regulation of gene expression.

Before beginning the discussion of the transcription process, let us compare replication and transcription.

Replication	Transcription
Takes place once during the cell-cycle	Frequent process
Whole DNA is replicated	Only a segment of DNA is transcribed
Primer is needed	Not needed
dNTPs are needed	NTPs are needed
Efficient proof-reading and repair	No repair and less efficient proof-reading

RNA POLYMERASE

In prokaryotes, single RNA polymerase synthesizes all types of RNAs, whereas there are 3 types of polymerases in eukaryotes.

The Similarity Between DNA Polymerase and RNA Polymerase

- ❑ Template-dependent
- ❑ Requires Magnesium
- ❑ Processive in nature
- ❑ The direction of synthesis is 5′ to 3′
- ❑ Watson-Crick base-pairing rule is followed

Difference Between DNA Polymerase and RNA Polymerase

Unlike DNA polymerases, RNA polymerases:
- ❑ Do not require any primer
- ❑ Use ribonucleotide triphosphates instead of deoxy NTPs
- ❑ Incorporate Uracil instead of Thymine
- ❑ Don't have any repair activity

TEMPLATE STRAND VERSUS CODING STRAND

It is important to understand the following terminologies:
- ❑ **Template strand is the strand which is being acted upon by RNA polymerase.**
- ❑ The mRNA produced resembles the sequence of the opposite strand (except U in place of T). This is why the opposite strand is known as coding, sense or plus strand, i.e. by seeing the sequence of coding DNA strand, we can find out the RNA sequence easily.
- ❑ Gene sequences given in scientific literature are always of the coding strand.

A A C G U A G G G U C A C A U C . . .	RNA transcript
. . . G C A T A C A A C A C A C C T T G C A T C C C A G T G T A G . . .	Template, or antisense (-), strand
. . . C G T A T G T T G T G T G G A A C G T A G G G T C A C A T C . . .	Coding, or sense (+), strand

PROKARYOTIC TRANSCRIPTION

Prokaryotic RNA polymerase core is made up of 5 subunits - 2 α subunits, β, β′ and ω. β and β′ subunits are involved in the catalysis.

PROMOTER

In DNA replication, we have seen that OriC is the origin of replication. How do RNA polymerases know where to begin the transcription? DNA sequences known as promoters help in this. Promoter sequences are not transcribed by RNA polymerases (with exceptions). They lie upstream to the transcription initiation site.

RNA polymerase can't bind to the promoter. It requires the help of another protein known as σ factor. As there are various promoters, there exist various types of σ factors. σ factors act catalytically, i.e. after the release, they assist initiation of another core polymerase.

RNA polymerase holoenzyme = Core polymerase ($\alpha_2\beta\ \beta'\omega$) + σ factor

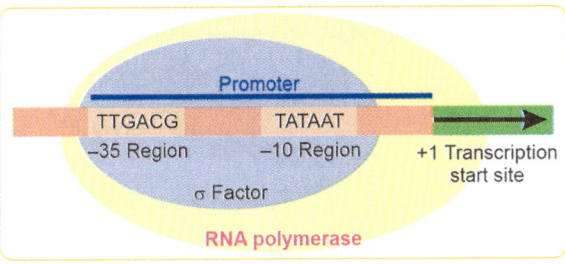

Bacterial promoters are simple. There are two promoters, –35 sequence and –10 sequence as shown in the image. +1 indicates the transcription start site (TSS). The nucleotide 5′ to the TSS is not transcribed and is known as –1. There is no zero site.

–10 element (TATAAT) is known as Pribnow box. This is not TATA box.

Efficiency of the Promoter

TATAAT is the consensus sequence of -10 region. Those promoters having the same six nucleotide sequence are termed as "strong" promoters. They are transcribed frequently compared to weak promoters which differ from the consensus sequence.

THREE PHASES OF TRANSCRIPTION

Similar to replication, transcription can also be described in 3 phases – Initiation, Elongation, and Termination.

Initiation

As we have seen, σ factor helps in the promoter recognition. RNA polymerase slides along the DNA instead of repeatedly binding and dissociating from it. σ subunit decreases the affinity of RNA polymerase for general regions of DNA by a factor of 10^4.

When the RNA polymerase holoenzyme binds to the DNA, both –35 sequence and –10 sequence are in the double-helical state. This is known as closed promoter complex. Then, –35 sequence remains in double-stranded state while the **Pribnow box (–10 sequence)** is unwound by the RNA polymerase. This is known as open promoter complex. This completes the initiation.

Elongation

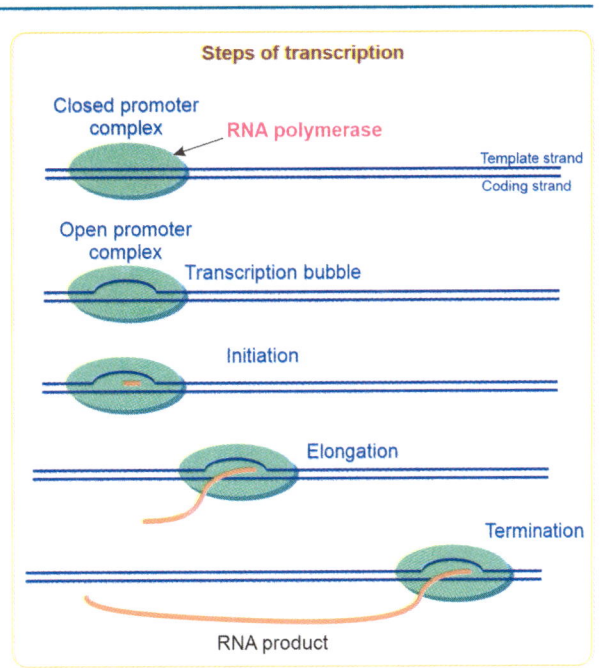

RNA polymerase does not require a primer. It can add nucleotides from scratch. DNA-RNA hybrid formed initially is unstable. RNA polymerase transcribes a few nucleotides and then releases the RNA and start afresh. This is known as transient excursion of RNA polymerase. **Once, about 10 nucleotides have been added**, RNA polymerase moves away from the promoter sequence. This is known as promoter clearance or escape.

Promoter clearance marks the beginning of elongation phase.

Termination

Termination of transcription can be either ρ-dependent or ρ independent.

ρ-Dependent Termination

ρ-protein is a hexameric ATP-dependant helicase. It binds to the nascent RNA and pulls it away from RNA polymerase and template DNA.

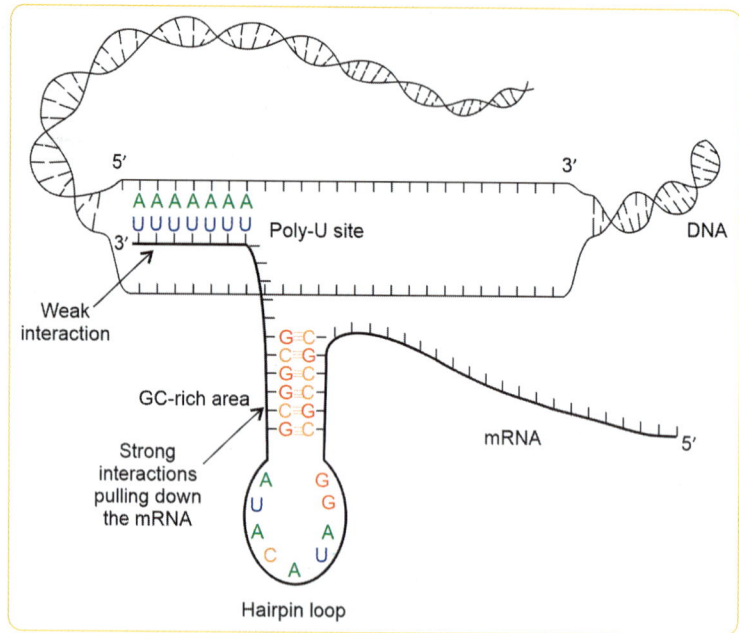

Hairpin loop

ρ-Independent Termination

DNA template contains the signal for termination of transcription. Transcription of a GC-rich palindromic DNA sequence forms a stem-loop in the mRNA followed by poly U nucleotides. The interaction between A of DNA and U of RNA is weak. So, the nascent RNA dissociates from the DNA template and enzyme.

TRANSCRIPTION AND TRANSLATION ARE COUPLED IN PROKARYOTES

❑ In prokaryotes, the mRNA produced is polycistronic, i.e. multiple start and stop codons are present in the same RNA; multiple proteins can be produced from single mRNA.
❑ As there is no clear cytoplasmic-nuclear demarcation in prokaryotes, translation begins even before the completion of transcription. This knowledge will be useful to understand the phenomenon of attenuation (Chapter 27) *(Refer to page no. 451)*.

POST-TRANSCRIPTIONAL PROCESSING OF RNA

In prokaryotes, mRNA does not undergo much post-transcriptional processing while rRNA and tRNA undergo modifications.

Modifications in tRNA:
❑ Ribonuclease P, a ribozyme cleaves the 5′ end.
❑ CCA trinucleotide is added to the 3′ end by a transferase.
❑ Bases are modified by various enzymes. tRNA is the RNA with high proportion of modified bases.

Modifications in rRNA:
- ❑ Certain bases and ribose are methylated.
- ❑ Pseudouridine and dihydrouridine are formed.

TRANSCRIPTION IN EUKARYOTES

Transcription in eukaryotes is different in the following ways:

	Prokaryotic	Eukaryotic
Location	Cytoplasm	Nucleus
Transcription-translation coupling	Before completion of transcription, translation begins	Not possible because of compartmentalization
Number of RNA polymerase	Single RNA polymerase synthesizes all types of RNA	3 types of polymerases, one for each type of RNA
Promoters	Promoters are simple	Promoters are complex
Initiation factors	Initiation factors not needed	Necessary
Recognition of promoter	σ factor	Tata binding protein part of Transcription Factor II D
Operons	The transcriptional unit is polycistronic **There is an operator region downstream of a promoter in operons**	Monocistronic; No operons; No operators
Post-transcriptional modification of mRNA	Nil	Extensive

EUKARYOTIC RNA POLYMERASES

There are three eukaryotic RNA polymerases. They are involved in the transcription of following types of RNAs.

Type	Location	RNA transcribed	Effect of α-amanitin
I	Nucleolus	18, 5.8 and 28S rRNA	Insensitive
II	Nucleoplasm	mRNA, miRNA and snRNA	Highly sensitive (1 µg/mL)
III	Nucleoplasm	tRNA, 5S rRNA and U6 snRNA	Less sensitive (10 µg/mL)

EUKARYOTIC TRANSCRIPTIONAL REGULATION COMPLEX

The following four DNA elements control transcription:
1. Core promoters – TATA box (also known as Hogness box), Inr (initiator) element, Downstream promoter element (DPE)
2. Promoter-proximal elements – CAAT box and GC box
3. Enhancer and repressor elements
4. Other regulatory elements

The first two are required for basal gene expression whereas the last two are involved in regulated gene expression (Fig. 1).

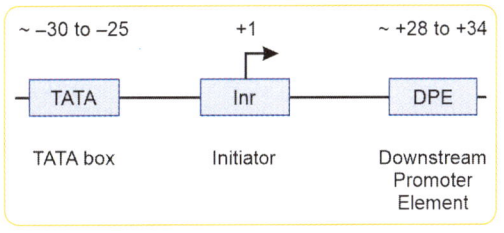

Fig. 1: Core promoters of Drosophila

TRANSCRIPTION FACTORS

Transcription factors are proteins involved in transcription. There are two types.
1. **General transcription factors**—bind to the core promoters and are required for the basal transcription, e.g. TFIID binds to TATA box as well as Inr.

2. **Specific transcription factors**—bind to the proximal promoter elements or enhancer/repressor, e.g. CAAT box transcription factor binds to CAAT box, and Sp1 transcription factor binds to GC box.

General Transcription Factors

General transcription factors are classified based on the RNA polymerases. TF I, II and III are those needed for RNA polymerase I, II and III respectively. Individual transcription factors are named alphabetically. We will focus on RNA polymerase II only.

Recognition of promoter is mostly done by TATA-binding protein which is a part of TFIID.

TFIID is a multisubunit protein made up of:
- ❑ TATA box binding protein (TBP) – binds to the minor groove of DNA and recognizes the core promoter, i.e. TATA box
- ❑ Many (10–12) subunits of TBP associated factors (TAF) that bind to non-TATA promoters.

Transcription factor	Role
TFIID	Binds to TATA box
TFIIA	Stabilize the pre-initiation complex by binding to TBP of TFIID
TFIIB	Bridges RNA polymerase and TFIID
TFIIF	Recruitment of RNA polymerase to promoter site
TFIIE	Recruits TFIIH
TFIIH	❑ Helicase activity – unwinds the promoter region ❑ Kinase activity – phosphorylates the C-terminal domain (CTD) of RNA Polymerase

CARBOXY-TERMINAL DOMAIN (CTD) OF RNA POLYMERASE II IS UNIQUE

Carboxy-terminal of eukaryotic RNA Pol II contains a serine-rich sequence (Tyr-Ser-Pro-Thr-Ser-Pro-Ser) repeated multiple times. TFIIH phosphorylates these serine residues. This recruits the enzymes involved in RNA processing (capping, splicing and polyadenylation). **Phosphorylation of CTD marks the beginning of elongation phase of transcription.**

Very less is known about the termination of transcription in eukaryotes.

POST-TRANSCRIPTIONAL MODIFICATIONS

Newly-synthesized heterogenous nuclear RNA (hnRNA) undergoes certain post-transcriptional modifications to become functional mRNA. These are 5′ capping, 3′ tailing, splicing and RNA editing.

5′ 7-mG CAPPING

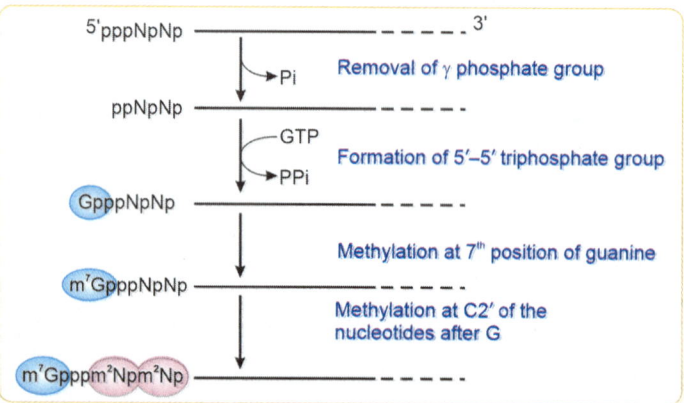

Addition of 7-methylguanosine to the 5′ end is via 5′-5′ triphosphate linkage rather than the usual 3′-5′ phosphodiester bond. This cap is known as cap 0. S-Adenosyl methionine is the methyl group donor. Addition of methyl groups to 2′OH group of ribose molecules of the 2nd and 3rd nucleotides produces cap 1 and 2 respectively. Capping occurs prior to the splicing. To be precise, capping is a co-transcriptional modification. Capping is done during the elongation phase itself.

Role of Capping

- Nuclear export of RNA
- Protection from 5′→3′ exonuclease
- Ribosomal recognition and circularization of mRNA during translation

REMOVAL OF INTRONS (SPLICING)

Introns are the intervening non-protein coding regions between exons. Introns are removed, and exons are joined together in a process known as splicing.

Splicing is of two types—spliceosome mediated and self-splicing (ribozyme activity of introns). Self-splicing is seen in both prokaryotes and eukaryotes, whereas spliceosome mediated splicing is seen only in eukaryotes.

The sequence of the intron is unique. It always starts with GU and ends with AG, and there is a branch site containing A in the middle.

Splicing Involves Two Transesterification Reactions (*Shown in the Image as 1 and 2*) and the Removal of Intron as a Lariat

First, 2′-OH group of Adenylate residue of branch site attacks the 5′ end of the intron. This creates an atypical 2′→5′ phosphodiester bond in the intron. This is the 1st transesterification reaction (transfer of phosphodiester bond). Next, 3′-OH group of the exon1 attacks the 3′ end of the intron. Thus, a 3′→5′ phosphodiester bond is made between exon 1 and 2 (2nd transesterification reaction). The intron is released as a lariat-shaped structure.

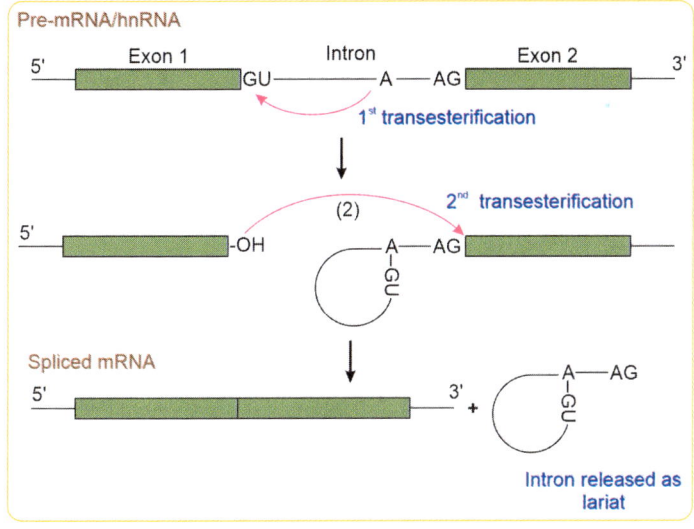

Spliceosome-mediated Splicing

In due course of evolution, ribonucleoproteins evolved to assist the process of splicing. RNA components of these ribonucleoproteins are uracil-rich small nuclear RNAs (snRNA). So, these ribonucleoproteins are known as snRNPs *(pronounced as 'snurps')* and are named as U1 through U6 (there is no U3). They all assemble together to form a supramolecular assembly known as spliceosome. Spliceosomes are found in the specialized region of nucleus known as nucleolus and are found only in eukaryotes. Other than snRNPs, splicing requires many other proteins known as splicing factors.

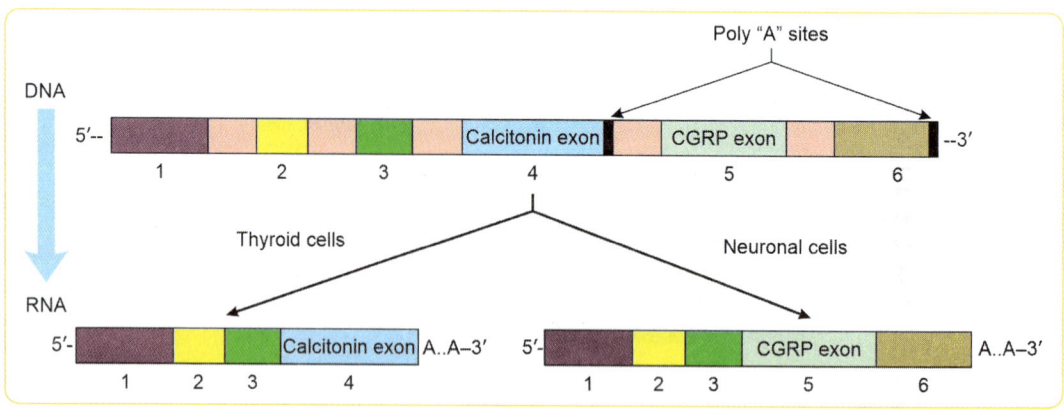

❑ U1—base pairs to the 5′ splice site containing GU.
❑ U2—base pairs to the branch point.
❑ U5—binds to the 5′ splice site and then the 3′ splice site
❑ U4—base-pairs with the U6 and prevents premature activity
❑ U6—catalyzes the splicing

Splice site mutations alter the size of the protein and lead to diseases like β-thalassemia. About 15% of human genetic diseases occur due to defective splicing.

Alternative Splicing is One Way to Expand the Number of Proteins Produced from a Gene

CALCA gene is made up of six exons. In thyroid cells, first four exons are spliced together to produce calcitonin protein. But, in the neurons, exons 1,2,3,5 and 6 are spliced together to form calcitonin gene-related peptide. Thus, one mRNA can produce many proteins by alternative splicing.

ADDITION OF POLY A TAIL AT 3′ END

AAUAAA is the signal for polyadenylation located in the 3′ end. Polyadenylation factor recognizes this and endonuclease cleaves 20 nucleotides downstream from this signal. Then, the polyA polymerase adds up to 250 adenylate residues. Length of the tail may vary depending on the mRNA. Not all mRNAs contain poly A tail. Histone mRNA contains a stem-loop structure instead of poly A tail.

Role of polyA tail:

❑ Protects from 3'→5' exonuclease; enhances the stability of mRNA. Thus, lengthier the polyA tail, more stable is the mRNA.

❑ Plays a role in circularization of mRNA during translation.

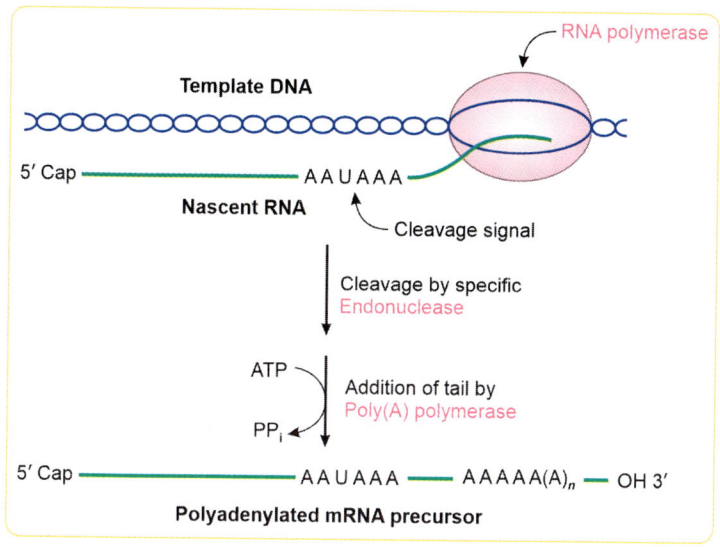

Polyadenylated mRNA precursor

There can be many polyadenylation signals in certain genes. This will lead to mRNAs of different length by alternate polyadenylation.

RNA EDITING

RNA editing is a post-transcriptional modification that changes the mRNA sequence and as a result, alters the protein produced by that mRNA.

RNA editing has been observed in all types of RNA – mRNA, tRNA and rRNA.

Types of RNA editing:

❑ Insertion or deletion of nucleotide(s)

❑ Base modification by deaminase enzymes

Example

Apo B_{100} of VLDL and ApoB$_{48}$ of chylomicron are products of the same ApoB gene. But the RNA is processed differently. In liver, apo B mRNA is 100% translated to produce apoB$_{100}$ protein. In the intestine, same apoB mRNA is acted upon by the cytosine deaminase enzyme that results in a premature stop codon and only 48%(2152/4563) of mRNA is translated. This produces apoB48 protein which lacks the hepatic receptor binding domain.

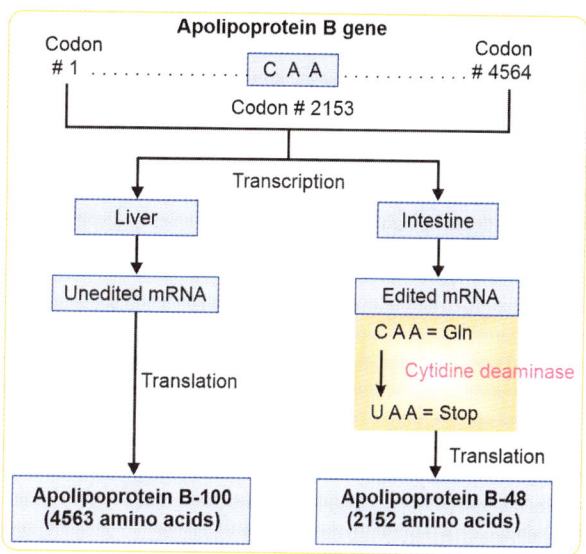

Non Protein-coding Sequences will be Present in mRNA Even After the Post-transcriptional Modifications

It is important to know that there are untranslated regions (UTRs) in the mature processed mRNA. They are important for ribosomal recognition, determining the half-life of mRNA, etc. The length of the UTR varies in different genes (Fig. 2).

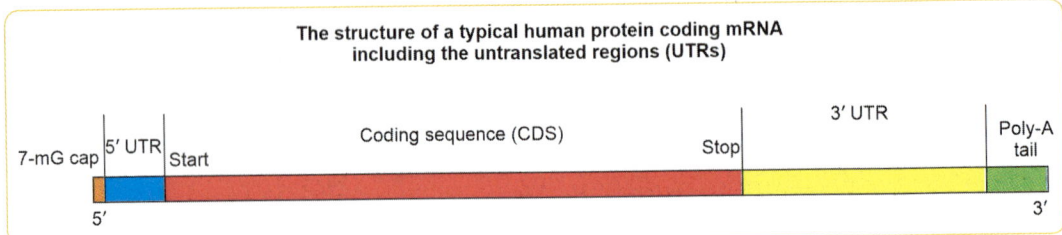

Fig. 2: UTR – Untranslated regions; RNA secondary structures like stem-loop are usually seen in UTR

High Yield

- Shine-Dalgarno sequence is present in the 5' UTR of all prokaryotic mRNAs (Chapter 25) *(Refer to page no. 428)*
- Histone mRNA does not contain a polyA tail. It contains a stem-loop structure in the 3' UTR which protects the mRNA from the exonuclease attack.
- Micro RNAs bind to the 3' UTR and repress the gene expression (Chapter 27) *(Refer to page no. 447)*.
- Ferritin mRNA contains stem-loop structures in the 5' UTR. Transferrin receptor mRNA contains stem-loop structures in the 3' UTR (Chapter 27) *(Refer to page no. 447)*.
- Riboswitches are the regulatory RNA elements found in the untranslated regions of mRNA. Riboswitches bind to metabolites and alter the translation of the mRNA (Chapter 27) *(Refer to page no. 447)*.

INHIBITORS OF TRANSCRIPTION

IN PROKARYOTES

- **Rifampicin** (anti-tubercular drug) – binds to β subunit of prokaryotic RNA polymerase → prevents the promoter clearance step and inhibits the RNA production. Rifampicin resistance in tubercular bacilli is due to mutation of β subunit.
- Actinomycin also known as Dactinomycin: Insert itself in between the bases of DNA helix (Intercalation) and thus inhibits transcription.

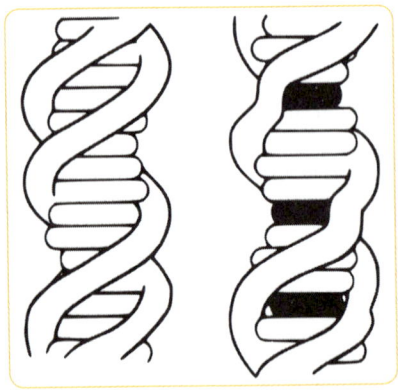

IN EUKARYOTES

❏ α-amanitin is a toxin produced by the poisonous mushroom *Amanita phalloides* (death cap); inhibits RNA polymerase II → production of mRNA is inhibited.

❏ Promoters for RNA polymerase III are usually located in the DNA sequences within the transcribed section of the gene (intragenic promoter).
❏ Poly A polymerase is a template-independent RNA polymerase.
❏ The nucleolus is formed by looping out of rRNA coding DNA segments.
❏ Heparin is an inhibitor of free RNA polymerase. This property is utilized in research.
❏ The transcriptome is the total of all the RNA molecules, expressed from the genes of an organism.
❏ Epitranscriptome refers to changes in gene expression without changes in RNA sequences. Post-transcriptional modifications of RNA species mediate epitranscriptomic changes.

Amanita phalloides

Competency

❏ BI7.2   Describe the processes involved in transcription

Key Points

- ❏ Transcription is the DNA directed RNA synthesis. Reverse transcription is the RNA directed DNA synthesis.
- ❏ There is only one type of RNA polymerase in prokaryotes to synthesize all types of RNA.
- ❏ RNA polymerase is processive; template dependent; doesn't require primer; requires magnesium/zinc; direction of synthesis is 5′ → 3′.
- ❏ Where to begin the transcription is determined by promoter sequences, and the frequency of transcription is determined by the efficiency of the promoter.
- ❏ rRNA is the most abundant cellular RNA.
- ❏ tRNA contains the maximum number of modified bases.
- ❏ Rifampicin inhibits the β-subunit of prokaryotic RNA polymerase.
- ❏ Alpha-amanitin inhibits RNA polymerase II.
- ❏ In eukaryotes, the promoter is recognized by transcription factor II D (TFIID).
- ❏ Carboxy-terminal domain of RNA polymerase II is phosphorylated by Transcription Factor II H (TFIIH).
- ❏ Operator region is not found in eukaryotes.
- ❏ Only mRNA codes for the protein. All other RNAs are non-protein coding.
- ❏ The sequence of intron always starts with GU and ends with AG.
- ❏ Multiple proteins from a single gene may be generated by differential RNA processing (alternative splicing, alternate promoter utilization etc.)
- ❏ Mechanism by which ~25 thousand protein-coding genes produce more than 2.5 lakh peptide sequences - Differential RNA processing
- ❏ The molecular mechanism behind the generation of Apo B-48 (in the intestine) and Apo B-100 (in the liver) from same Apo B gene is RNA editing (through deamination of cytosine to uracil of mRNA).

SELF-ASSESSMENT

Short Answer Questions

1. Name the different types of RNA. Write their salient features. Mention their functions.

2. Resistance to rifampicin arises from mutations in the β-subunit of RNA polymerase that alter residues of the rifampicin binding site on RNA polymerase, resulting in decreased affinity for rifampicin. Explain.

3. Write the functions of small nuclear ribonucleoproteins (snRNAs- U1, U2, U4, U5 and U6) in spliceosome- catalyzed splicing.

4. Explain the mechanism of action of antitubercular drug rifampicin. What is the biochemical basis of resistance of M. tuberculosis to rifampicin?

5. In eukaryotic mRNAs, write the nucleotide which is attached to the 5′ end, and the bond involved for the formation of the cap. Explain the importance of the cap in mRNA function. What is the difference between cap 0 and cap 1?

6. Explain how Rho-independent termination of transcription takes place in prokaryotes.

7. Write the two transesterifications by which introns are removed and exons are spliced together in eukaryotes. Mention one disease caused by defective splicing.

8. What is RNA editing? Explain with an example.

Multiple Choice Questions

1. **Which of the following is the most abundant RNA species in the cell?**
   a. Micro RNA   b. Messenger RNA
   c. Ribosomal RNA   d. Transfer RNA

2. **Sigma factor is a component of:**
   a. 3′ to 5′ endonuclease
   b. DNA ligase
   c. DNA polymerase
   d. RNA polymerase

3. **Which of the following eukaryotic transcription factor recognizes the promoter?**
   a. Rho   b. Pribnow box
   c. Sigma   d. TFIID

4. All of the following are non-coding RNAs, except:
   a. mRNA   b. tRNA
   c. rRNA   d. lncRNA

5. In Eukaryotes, mRNA is produced by:
   a. RNA Polymerase I
   b. RNA Polymerase II
   c. RNA Polymerase III
   d. RNA Polymerase IV

6. The DNA segment from which the primary mRNA transcript is copied or transcribed is called:
   a. Coding strand
   b. Initiator methionine domain
   c. Template strand
   d. Translation unit

7. The bond present in cap-0 of mRNA is:
   a. 7-methyl guanosine attached by 5'→5' diphosphate bridge
   b. 7-methyl guanosine attached by 3'→5' triphosphate bridge
   c. 7-methyl guanosine attached by 3'→5' phosphodiester bridge
   d. 7-methyl guanosine attached by 5'→5' triphosphate bridge

8. Which of the following is the equivalent of pribnow box in eukaryotes?
   a. ACAT box   b. CAAT box
   c. GC box   d. Hogness box

9. A mutation to this sequence in eukaryotic messenger RNA will affect the process by which the 3'-end poly-A tail is added to the mRNA?
   a. AAUAAA   b. CAAT
   c. TATAAA   d. GU... A ... AG

10. Poly A tail is translated into:
    a. Polylysine   b. Polyglutamate
    c. Polyproline   d. None

11. Which of the following transcription factor is involved in the phosphorylation of Carboxy terminal domain (CTD) of RNA polymerase II?
    a. TFIID   b. TFIIA
    c. TFIIB   d. TFIIH

12. To which of following DNA elements does special transcription factors bind?
    a. TATA box   b. Inr element
    c. DPE   d. CAAT box

13. Sequence of an intron always starts and ends with:
    a. GU,AG   b. AG,GU
    c. GA,GU   d. UG,AG

14. 5S rRNA is transcribed by:
    a. RNA polymerase I
    b. RNA polymerase II
    c. RNA polymerase III
    d. RNA polymerase IV

15. All of the following are true about splicing, except:
    a. Intron is removed in the form of a lariat
    b. Introns begin with GU and ends with AG
    c. Spliceosomes are seen in E. coli
    d. Two transesterification reactions are involved

16. apoB48 and apoB100 are the two proteins with different length produced from apoB gene by:
    a. Alternative polyadenylation
    b. Alternative splicing
    c. mRNA editing
    d. mRNA silencing

17. The part of the gene which is retained in the mature mRNA is:
    a. Exon   b. Intron
    c. hnRNA   d. miRNA

ANSWERS

1. c.	2. d.	3. d.	4. a.	5. b.	6. c.
7. d.	8. d.	9. a.	10. d.	11. d.	12. d.
13. a.	14. c.	15. c.	16. c.	17. a.	

Chapter 23 Transcription

24

CHAPTER

Genetic Code

Chapter Outline

- Deciphering the Genetic Code
- Initiation and Stop Codons
- Features of Genetic Code
- Wobble Hypothesis
- Human Mitochondrial Genetic Code
- Mutations and Various Types

Genetic code is the relationship between the linear sequence of 4-letter nucleotide alphabet of DNA and the 20-letter amino acid language of proteins.

DECIPHERING THE GENETIC CODE

- The genetic code (Fig. 1) was deciphered through the study of protein synthesis in cell-free extracts, *in vitro*.
- Synthetic polyribonucleotide poly-U was found to code only for polyphenylalanine.
- Synthetic polyribonucleotide poly-A was found to code only for polylysine.
- Synthetic polyribonucleotide poly-C was found to code only for polyproline.

First letter	Second letter: U	Second letter: C	Second letter: A	Second letter: G	Third letter
U	UUU, UUC Phe UUA, UUG Leu	UCU, UCC, UCA, UCG Ser	UAU, UAC Tyr UAA Stop UAG Stop	UGU, UGC Cys UGA Stop UGG Trp	U C A G
C	CUU, CUC, CUA, CUG Leu	CCU, CCC, CCA, CCG Pro	CAU, CAC His CAA, CAG Gln	CGU, CGC, CGA, CGG Arg	U C A G
A	AUU, AUC, AUA Ile AUG Met	ACU, ACC, ACA, ACG Thr	AAU, AAC Asn AAA, AAG Lys	AGU, AGC Ser AGA, AGG Arg	U C A G
G	GUU, GUC, GUA, GUG Val	GCU, GCC, GCA, GCG Ala	GAU, GAC Asp GAA, GAG Glu	GGU, GGC, GGA, GGG Gly	U C A G

Fig. 1: The human genetic code

INITIATION CODON

Initiator codon or start codon is the first codon of a mRNA translated by a ribosome. AUG is the most common start codon.

AUG codes for N-formyl Methionine in prokaryotes and Methionine in eukaryotes.

TERMINATION CODONS/STOP/NONSENSE CODONS

Stop codons do not code for any particular amino acid. Presence of any of the following codons signal the end of polypeptide synthesis:

- UAA (ochre)
- UGA (opal)
- UAG (amber)

U Are Away (UAA), U Gone Away (UGA), U Are Gone (UAG)

Properties of genetic code:

- Unambiguous
- Degenerate/Redundant
- Nearly universal
- Nonoverlapping (the triplet codons are read in only one order. E.g AUGGTGA can not be read as UGGTGA)
- Not punctuated (the triplet codons are continous. No base serves as a comma)
- Collinear (sequence of RNA corresponds to the linear sequence of amino acids)

Let us discuss the properties one by one.

- **Unambiguous:** Each codon specifies only one amino acid. There is no ambiguity in this.
- **Degeneracy/Redundancy:** One amino acid can be coded by multiple codons. This is known as degeneracy of genetic code.

Exception to degeneracy of genetic code:

- Methionine and tryptophan are the only amino acids coded by single codon.
- Methionine is coded only by AUG and Tryptophan is coded only by UGG.

Synonymous codons are those codons that specifies the same amino acids, e.g. GGA, GGG, GGU, and GGC codes for glycine and all these 4 codons can be said to be synonymous. Degeneracy of genetic code can be explained by wobbling phenomenon.

Wobble Hypothesis

- The degeneracy arises from imprecision (wobble) in the pairing of the 3rd base of the codon with the 1st base of the anticodon. (shown in the image).
- If U and G are at the wobble position of the codon, there will be no wobbling (shown in the table).
- Wobble hypothesis was proposed by Francis Crick.

Base-pairing Combinations	
Base at 5' end of anticodon	Base at 3' end of codon
I*	A, C, or U
G	C or U
U	A or G
A	U
C	G

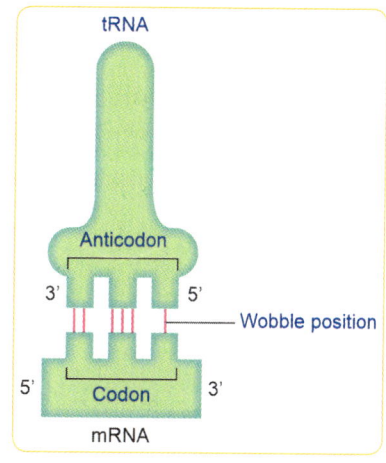

	3 2 1	3 2 1	3 2 1
Anticodon	(3') G–C–I	G–C–I	G–C–I (5')
Codon	(5') C–G–A	C–G–U	C–G–C (3')
	1 2 3	1 2 3	1 2 3

You can notice that when inosine is the first base of anticodon, it can base pair with A, U or C of the codon's 3rd base

Area Under the Curve

The advantage of wobbling is that even if there is a mutation in the 3rd base of the codon, there is a chance that there is no change in the amino acid encoded.

There are 64 codons. But, the number of tRNA are lesser.

Nearly Universal

Let it be an E. coli or elephant, Genetic code is same in all organisms with some exceptions.

Exception to the universal nature of genetic code is found in mitochondrial genome of yeast, fruit fly and vertebrates.

Human Mitochondrial Genetic Code

Codon	Should code for	Actually codes for
UGA	Stop	Tryptophan
AGA, AGG	Arginine	Stop
AUA	Isoleucine	Methionine

MUTATIONS

Mutation can be defined as a sudden change in genetic sequence of an individual which may or may not lead to variation of phenotype.

Mutations can be classified in many ways. Let us discuss them in detail:

MUTATION AT GENE LEVEL

❑ **Point mutation:** Point mutation is a change in single base. This could be either transition or transversion.
  ○ **Transition** is substitution of a purine by a purine or a pyrimidine by a pyrimidine.
  ○ **Transversion** is substitution of a purine by a pyrimidine and vice versa.
  According to the effect, point mutation can be further divided into:
  ○ Silent
  ○ Missense – acceptable, partially acceptable, non-acceptable
  ○ Non-sense
  Silent mutations produce no change in amino acid. This is because of the degeneracy of the genetic code. GAA to GAG mutation is an example of silent transition. (GAA, GAG both codes glutamate).
  Missense mutations incorporate wrong amino acids. These are of three types. Hemoglobin variants are given here for example.

Acceptable (normal O$_2$ carrying)	Hb Bristol	β67 Val → Asp
	Hb Hikari	β61 Lys → Asn
	Hb Sidney	β67 Val → Ala
Partially Acceptable	HbS (Sickle cell Hb)	β6 Glu → Val
	HbC	β6 Glu → Lys
Unacceptable (No O$_2$ binding)	Hb (Hb Boston) (Methemoglobinemia)	His E7 → tyrosine

Nonsense mutation produces a premature stop codon, results in premature termination of protein synthesis, e.g. Crigler Najjar syndrome type I is due to a nonsense codon producing truncated UDP-glucuronosyl transferase.

Q. Mutation in sickle cell anemia is CAG to CUG. How will you describe this?
A. Partially acceptable, missense, transverse, point mutation.

❏ **Frameshift mutation:** These are the mutations resulting in change of reading frame. This is due to insertion and deletion of bases not in multiples of three.

Frame shift mutations alter the reading frame. Deletion of 3 bases won't affect the reading frame but it will produce a protein which is less by one amino acid.

For example, type O blood group has a frame shift mutation in the gene encoding the terminal glycosyltransferase that results in the production of an inactive protein. *(Ref. Harper 30th Ed., p.698)*

MUTATIONS AT CHROMOSOMAL LEVELS – CHROMOSOMAL ABERRATIONS

❏ Inversion
❏ Translocation
❏ Deletion
❏ Duplication

BASED ON THE CELL INVOLVED

❏ Somatic mutations: Non-hereditary
❏ Germline mutations: Hereditary, e.g. Li fraumeni syndrome is due to germline p53 mutations.

BASED ON THE EFFECT IN GENE FUNCTION

❏ Gain of function mutation
❏ Loss of function mutation

Examples of gain of Function Mutations

❏ FGFR3 mutation in achondroplasia *(Ref. Harrison 20th Ed., p. 2965)*
❏ BCR-ABL fusion tyrosine kinase in Chronic myeloid leukemia
❏ Protein aggregation (malfunction) in trinucleotide repeat disorders
❏ Point mutation in *RAS* protooncogene leading to continuous activity

Mutations of DNA other than Exons

❏ Splice site mutations: Lead to faulty splicing, e.g. some cases of β-thalassemia
❏ Promoter mutations: Alter the transcriptional activity, e.g. Gilbert syndrome occurs due to mutation in the UDP-glucuronosyltransferase promoter

Dynamic Mutations

Trinucleotide expansion diseases are dynamic in nature, i.e. the number of repeats increases with successive generation. (The number of repeats will be higher in son compared to the father).

Beneficial mutations: People with homozygous CCR5Δ32 mutation are resistant to infection with M-tropic strains of HIV-1.

DRIVER MUTATIONS AND PASSENGER MUTATIONS

❑ Mutations that cause or accelerate the process of carcinogenesis are known as driver mutations.
❑ Mutations found in cancer cells are not involved in carcinogenesis are passenger mutations.

▶ Video lecture of Genetic code: https://youtu.be/E_jXEwLvgX4 or scan QR code

Competency

❑ BI7.3   Describe gene mutations

Key Points

❑ Degeneracy of genetic code is explained by wobbling of codon anticodon base pairing.
❑ Synonymous codons are the different codons coding for the same amino acid.
❑ Isoaccepting tRNAs are the different tRNAs which carry the same amino acid.
❑ UAA (ochre), UGA (opal) and UAG (amber) are the stop codons.
❑ Methionine (AUG) and tryptophan (UGG) are the only amino acids coded by a single codon.
❑ AUG is the initiator codon that codes for methionine in eukaryotes and N-formyl methionine in prokaryotes.
❑ Poly U codes for phenylalanine. Poly A codes for polylysine.

SELF-ASSESSMENT

Short Answer Questions

1. Enumerate any four features of genetic code and explain each.

2. What do you understand by the term 'degeneracy of genetic code'? Explain the molecular basis.

3. Classify various types of DNA mutations. Name a beneficial DNA mutation that gives protection against HIV-1.

4. Give an example of a partially acceptable missense mutation. Explain why nature has selected this mutation in coastal regions of the world.

5. **Differentiate**
   a. Transition and Transversion
   b. Driver mutation and passenger mutation
   c. Missense and nonsense mutation

Multiple Choice Questions

1. **Assume 4 nucleotides code for an amino acid. What is the number of amino acids coded possible?**

   a. 4   b. 16

   c. 64   d. 256

2. **Multiple codons coding for the same amino acid is known as ....**
   a. Degeneracy
   b. Transcription
   c. Mutation
   d. Frame shift mutation

3. **All of the following are synonymous codon pair, EXCEPT:**
   a. CAU and CAC
   b. AUU and AUC
   c. AUG and AUA
   d. AAU and AAC

4. **Anticodon 5′-I-G-C-3′ can recognize all of the following codons EXCEPT:**
   a. GCA
   b. GCC
   c. GCG
   d. GCU

5. **Poly A codes for:**
   a. Poly lysine
   b. Poly glycine
   c. Poly phenylalanine
   d. Poly valine

6. **Amber codon is:**
   a. UAA
   b. UAG
   c. UGA
   d. UGG

7. **Degeneracy of the genetic code can be explained by:**
   a. Wobble hypothesis
   b. Watson-Crick base pairing
   c. Chargaff rule
   d. Translational recoding

8. **The initiator AUG in prokaryotes codes for:**
   a. Methionine
   b. N-Formyl methionine
   c. Phenyl alanine
   d. Tyrosine

9. **Of the ____ different possible codons, ____ specify amino acids and ____ signal stop.**
   a. 20, 17, 3
   b. 180, 20, 60
   c. 64, 61, 3
   d. 61, 60, 1

10. **What sequence on the template strand of DNA corresponds to the first amino acid inserted into a protein?**
    a. TAC
    b. UAA
    c. UAG
    d. AUG

11. **The change (denoted in capital letter) in 5′ mRNA sequence of a gene from transcription start site from cauuagc-Ucaugauucagc-caugaag... to cauuagc-Acaugauucagccaugaag is called a _____ mutation.**
    a. Missense
    b. Frameshift
    c. Silent
    d. Non-sense

ANSWERS

1. d.	2. a.	3. c.	4. c.	5. a.	6. b.
7. a.	8. b.	9. c.	10. a.	11. c.	

25
CHAPTER

Translation

Translation is the RNA mediated synthesis of proteins. During translation, the four-letter language of mRNA is translated to the 20-letter language of proteins.

The process of translation requires:
- tRNAs
- Aminoacyl tRNA synthetases and all the 20 amino acids
- Ribosomes
- mRNA
- Various protein factors

Let us discuss each component in detail.

TRANSFER RNA

Transfer RNA acts as an adaptor during protein synthesis. It is the carrier of amino acids. Anticodon-codon interaction ensures the incorporation of correct amino acids *(Chapter 24)* *(Refer to page no. 422)*. The number of tRNA in each organism is variable. In humans, there are about 41tRNAs in cytoplasm and 21tRNAs in mitochondria.

Remember

Mitochondria has its own protein-synthesizing machinery

1° Structure of tRNA

- As we know, the primary structure of RNA is the sequence of nucleotides from 5' to 3' end.
- tRNA is made up of 73–93 nucleotides.

- Nucleotides in tRNA sequence contain many unusual and modified bases, e.g. dihydrouridine, pseudouridine, inosine, etc.
- 5′ end is always base paired, and 3′ end is free or attached to an amino acid.
- 3′ end of all the tRNA ends with the same sequence, i.e. CCA.
- This trinucleotide sequence is post-transcriptionally added to the tRNA by the enzyme CCA tRNA nucleotidyltransferase.

2° Structure of tRNA

- 2° structure of tRNA is clover-leaf shaped.
- Base-paired, double-stranded regions form stems, whereas the non- base-paired regions form loops.

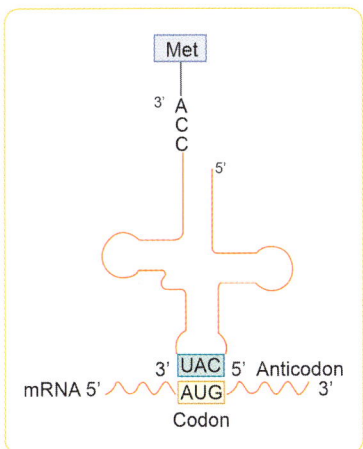

Fig. 1: tRNA as an adapter

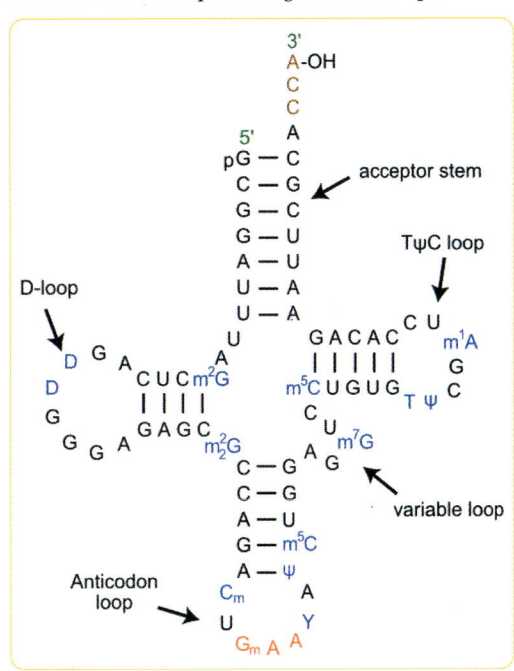

Fig. 2: The secondary structure of tRNA Phe from the *Sources:* Yikrazuul - Own work, CC BY-SA 3.0, https://commons.wikimedia.org/w/index.php?curid=10126790

3° Structure of tRNA

- 3° structure of tRNA is inverted, L-shaped.
- Each tRNA has unique 3° structure
- Stem and loop together form an arm. For example, D-arm is formed by D-stem and the D-loop.

Chapter 25 Translation

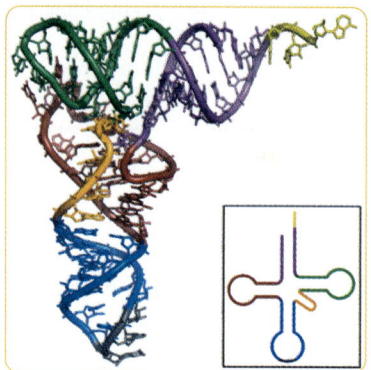

Fig. 3: 3° structure of tRNA - CCA tail (yellow), Acceptor stem (purple), Variable loop (orange), D arm (red), Anticodon arm (blue) with Anticodon (black), T arm (green)

Functions of the Parts of tRNA

Component	Function
Acceptor stem	Hydroxyl group of adenosyl nucleotide of CCA forms an ester bond with the carboxyl group of aminoacid.
DHU arm (DiHydroUridine) or D arm	Helps in recognition of tRNA by specific aminoacyl tRNA synthetase
Anticodon arm	Base pairs with the codon of mRNA
TψC arm (RiboThymidine, Pseudouridine (ψ), Cytidine)	Helps in ribosomal recognition
Variable loop	This contains unique sequences that are specific for that particular tRNA.

AMINOACYL tRNA SYNTHETASE

Aminoacyl tRNA synthetase is a ligase *(EC 6)* that adds a specific amino acid to the 3'OH group of adenosyl residue of tRNA *(we know that all tRNAs end with CCA)*. Carboxyl group of the amino acid reacts with OH group of adenosine to form an ester bond.

In humans, there are at least 20 aminoacyl tRNA synthetases specific for each amino acid. The number of cytosolic tRNA is higher than 40. Thus, many tRNAs exist to carry the same amino acid (degeneracy of genetic code). Different tRNAs carrying the same amino acid are known as **Isoaccepting tRNA**.

The process of adding amino acid to the tRNA is known as **charging** of tRNA, and for this, the amino acid is **activated** using the hydrolysis of ATP to AMP. As you know, ATP to AMP hydrolysis is considered as the expense of two ATP equivalents/two high-energy phosphate bonds.

Acceptor stem and anticodon arm of tRNA are crucial for the aminoacyl tRNA synthetase reaction.

Amino Acid + ATP → Aminoacyl-AMP + PPi	(Reaction 1 in the image)
Aminoacyl-AMP + tRNA → Aminoacyl-tRNA + AMP	(Reaction 2 in the image)

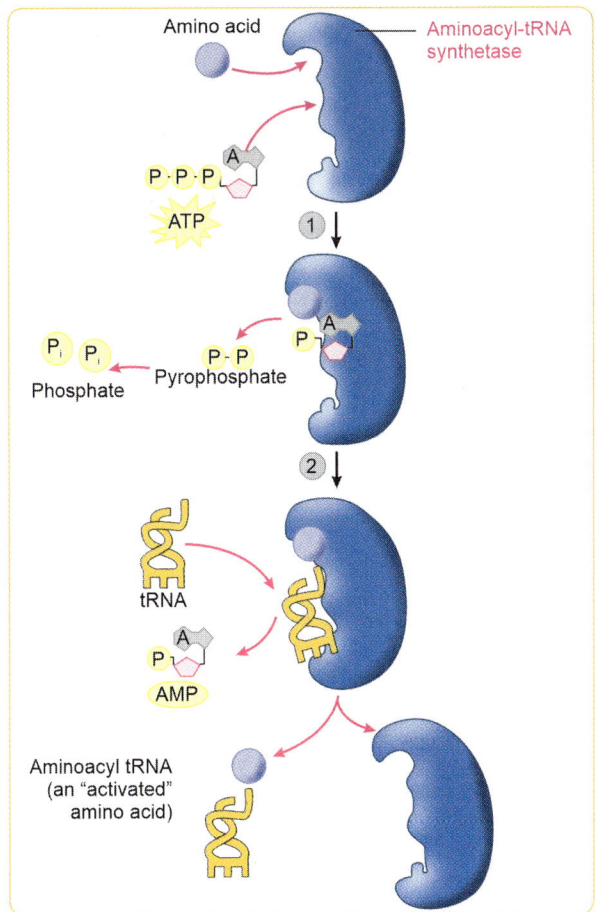

Aminoacyl tRNA synthetases have proof-reading activity. It recognizes the anticodon arm and other parts of tRNA. If a wrong amino acid is incorporated, the CCA bound amino acid is shifted to an editing site of the enzyme and removed. Thus, incorporation of correct amino acid into the protein is ensured. This is why some authors describe tRNA-aminoacyl tRNA synthetase interaction as the **second genetic code**.

MESSENGER RNA

- ❏ mRNA carries the information for protein synthesis.
- ❏ Triplet codons specify for amino acids (Chapter 24) *(Refer to page no. 422)*.
- ❏ The entire mRNA is not translated, there are untranslated regions (UTR) in the 5' and 3' end (Chapter 23) *(Refer to page no. 408)*.
- ❏ mRNA is translated in 5'→ 3' direction.
- ❏ Start codon AUG marks the end of the 5' untranslated region

RIBOSOMES

Venkatraman Ramakrishnan, Indian-American and British citizen was the recipient of 2009 Noble Prize **for studies of the structure and function of the ribosome**.

Ribosomes are ribonucleoprotein supramolecular assemblies. mRNA and tRNA interact with ribosomes to produce the polypeptide. At a given point, single mRNA can be acted upon by multiple ribosomes. This is known as **polysomes**.

Two-third of ribosome is RNA. Ribosomal RNA and proteins assemble to form ribosomes. Based on the sedimentation coefficient, prokaryotic and eukaryotic ribosomes are classified as shown in the image.

The ribosome is made up of small and large subunits. Small subunit is involved in recognition of mRNA, whereas large subunit is involved in the catalysis (peptidyl transferase activity).

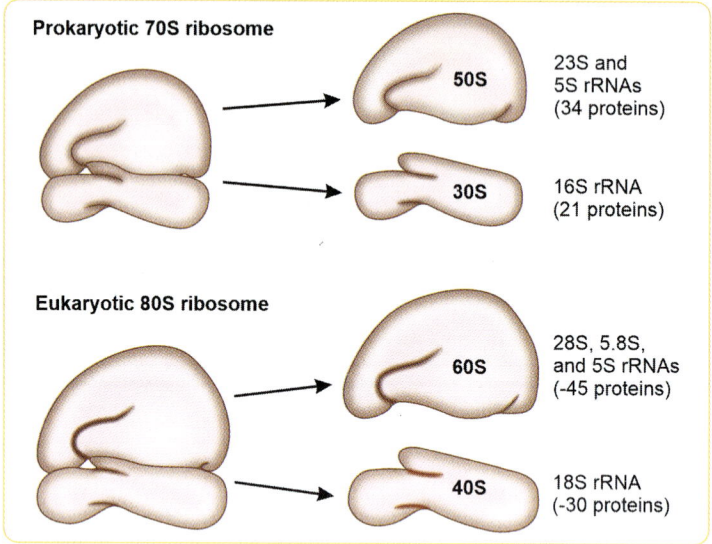

Ribosomes Contain Three Binding Sites for tRNA

Ribosomal proteins provide the scaffold for protein synthesis. There are three tRNA binding sites in ribosome formed by small and large subunit together.
- ❑ **A site:** Aminoacyl tRNA entry site
- ❑ **P site:** Peptidyl site where the growing peptide is located
- ❑ **E site:** Exit site where the empty tRNA leaves

TRANSLATION REQUIRES MANY PROTEIN FACTORS

In addition to ribosomal proteins, translation requires many other proteins. Translation is divided into three phases – initiation, elongation and termination. Protein factors involved in the first two steps are named by adding the suffix 'factor' to these steps. Proteins required for termination are known as 'release factors'.

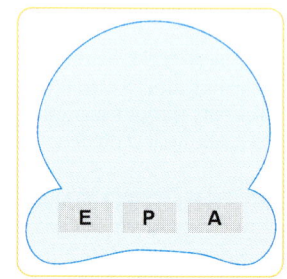

PROKARYOTIC TRANSLATION

Initiation

Initiation phase involves the recognition of the start codon and bringing the 1st amino acyl tRNA to the ribosome. As already mentioned, the small ribosomal subunit (30S) is involved in the recognition. The large subunit is involved in the catalysis. So, only the small subunit should first bind to mRNA. Binding of large subunit must be prevented. We will discuss how these events are achieved.

1. Recognizing the Start Codon

All prokaryotic mRNAs contain a unique sequence in the 5′ untranslated region, known as Shine-Dalgarno (SD) sequence. 16S rRNA of 30S ribosomal subunit contains the complementary sequence to SD sequence. 16S rRNA base pairs with the mRNA in such a way that the correct reading frame is established.

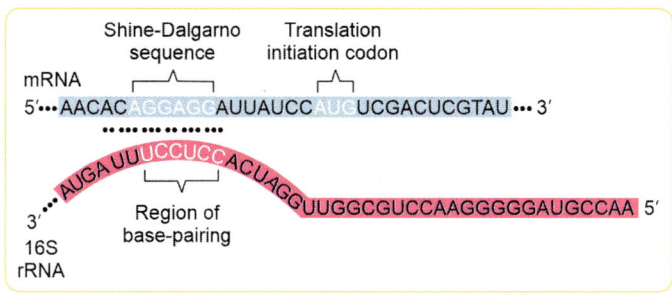

2. Preventing the Premature Binding of 50S Subunit

❑ Initiation factor 1 (IF1) binds to the A site and blocks the entry of any aminoacyl tRNA.

❑ IF 3 binds to the 30S subunit and prevents the premature binding of the 50S subunit.

3. Bringing the 1ˢᵗ Amino Acid and Formation of 70S Initiation Complex

❑ Once the position of AUG is established, the 1ˢᵗ amino acyl tRNA (initiator tRNA), i.e. N-formyl Methionyl tRNA (fMet-tRNA) needs to be brought to the ribosome.

❑ Initiation factor-2 (IF-2) brings the fMet-tRNA to the P-site.

❑ IF-2 is a G-protein evolutionarily related to heterotrimeric G-proteins and monomeric Ras G-protein.

❑ IF-2-GTP-initiator tRNA ternary complex binds to the ribosome. Once correct base-pairing between the anticodon of the initiator tRNA and the AUG codon of mRNA happens, GTP is hydrolyzed to GDP. GDP bound IF-2 cannot bind to tRNA and is released.

❑ IF-2 stimulates the association of large (50S) subunit.

❑ Thus, 70S preinitiation complex is formed, and the stage is set for elongation.

Remember

❑ IF-1 prevents the premature association of 50S, IF-2 brings the initiator tRNA to P site and IF-3 blocks the A site.

❑ GTP hydrolysis is a common theme in all the steps of translation.

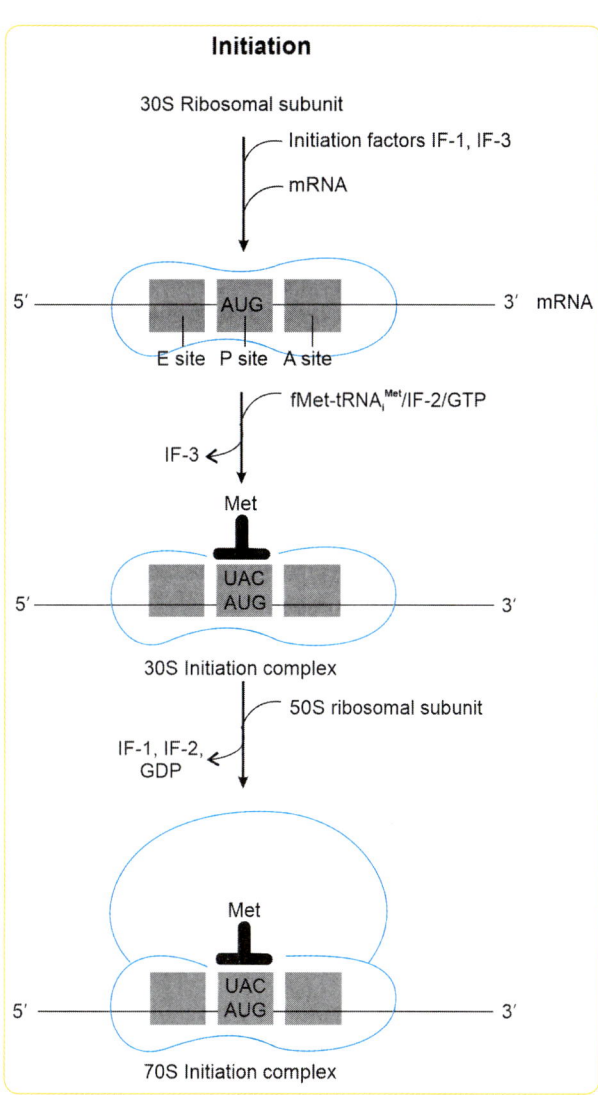

Chapter 25 Translation

Elongation

N-formyl methionyl tRNA is the only tRNA that can enter P-site. All other aminoacyl tRNAs can enter only A site. They are brought by a G-protein EF-Tu (Elongation factor-Tu). GTP bound EF-Tu binds to the aminoacyl tRNA and brings it to the A site. Once correct base-pairing between the anticodon of the aminoacyl tRNA and the codon of mRNA happens, GTP is hydrolyzed to GDP. GDP bound EF-Tu cannot bind to tRNA and is released. To be active again, GDP of EF-Tu needs to be replaced with GTP. EF-Ts, a guanine nucleotide exchange factor helps in the GTP-GDP exchange of EF-Tu.

Differentiation of Start Codon AUG and Internal AUG

❑ Internal AUG refers to the AUG codons located after the start codon. Why does translation usually begin at the start codon AUG, not at the internal AUGs?
❑ In prokaryotes, the Shine-Dalgarno sequence helps to recognize the start codon AUG. Initiation is restricted to the AUG located on the 3′ side of the SD sequence.
❑ IF-2, not EF-Tu binds to fMet-tRNA. So, fMet-tRNA is prevented from binding to internal AUG.

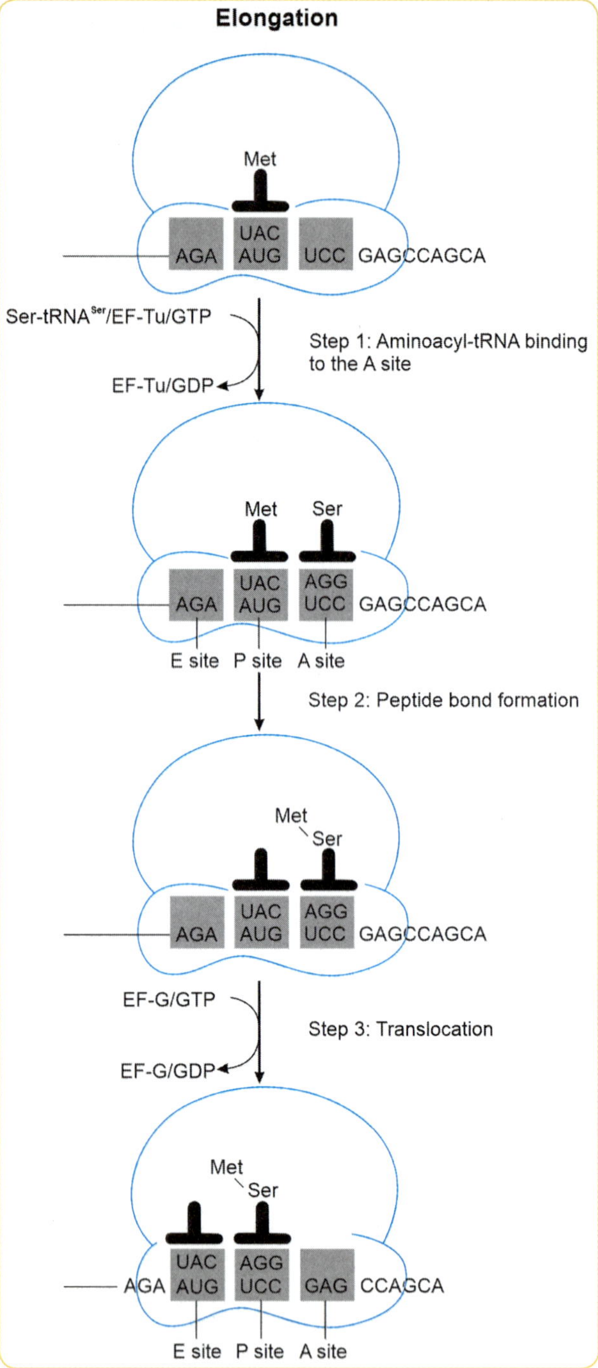

Peptidyl Transferase is a Ribozyme

The fMet-tRNA is in the P-site, and the aminoacyl tRNA is in the A site. 23S rRNA of the 50S subunit catalyzes the peptide bond formation between the carboxyl group of methionine and the amino group of the next amino acid (serine in the illustration).

Note the following important points:
❑ After the peptide formation, the tRNA in the P-site becomes empty. But, the amino acyl residue resides in the P-site only.
❑ **Peptidyl transferase reaction is a thermodynamically spontaneous reaction** achieved due to catalysis by proximity and orientation. *(No ATP is needed for this particular step but energy is needed overall)*
❑ Even though methionine is the first amino acid to be added during the protein synthesis, it may be removed later.
❑ Proteins are synthesized from N-terminal to C-terminal.

Elongation

Ser-tRNA^Ser/EF-Tu/GTP

EF-Tu/GDP

Step 1: Aminoacyl-tRNA binding to the A site

E site P site A site

Step 2: Peptide bond formation

EF-G/GTP

EF-G/GDP

Step 3: Translocation

E site P site A site

Section III | Molecular Biology

434

Translocation of mRNA and tRNA is Required for the Translation to Continue

As the P-site tRNA is now empty (does not carry any aminoacid), it should leave. The peptide-linked tRNA of the A site should move to the P-site so that the next aminoacyl tRNA can come to the A-site. mRNA should move by a distance of 3 nucleotides, i.e. one codon. This translocation is achieved by elongation factor G (EF-G, also known as translocase) and conformational changes of the ribosome. Translocation requires hydrolysis of GTP to GDP.

Remember

EF-Tu brings the aminoacyl tRNA to A site, EF-Ts helps in GDP-GTP exchange and EF-G is the translocase.

Termination

Once the stop codon is in the A site, no aminoacyl tRNA comes to A site since stop codons cannot be recognized by any tRNA. The stop codons are recognized by release factors (RFs). Release factors induces the peptidyl transferase to add water to the growing peptide, i.e. hydrolysis of peptide from the tRNA. There are three RFs.

- ❑ **RF1:** Recognizes UAA or UAG
- ❑ **RF2:** Recognizes UAA or UGA
- ❑ **RF3:** GTPase that removes the RF1 and RF2 on completion of their role.

At the end, the polypeptide leaves the P site. Ribosomes dissociate and can start a new cycle of translation.

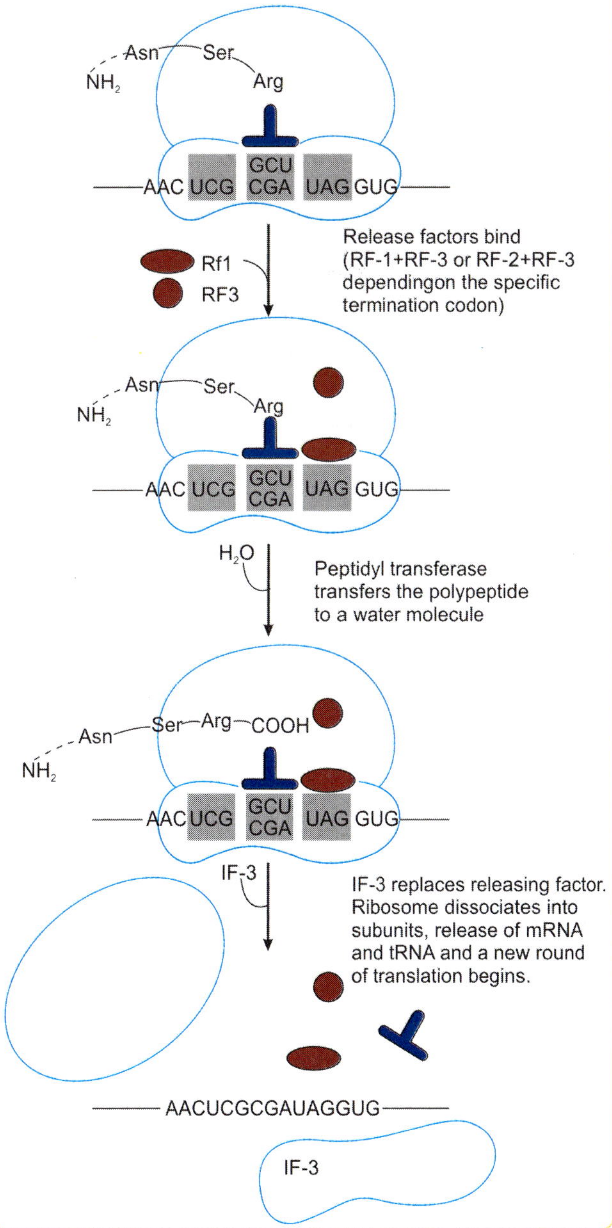

Energy Expenditure

- ❑ Even though peptidyl transferase reaction is a thermodynamically spontaneous reaction, energy input is needed for activation of amino acids (ATP → AMP), entry of amino acid to A site (GTP → GDP) and translocation of tRNA and mRNA (GTP → GDP).
- ❑ Thus, the **formation of one peptide bond requires 4 ATP equivalents or 4 high-energy phosphate bonds.**

Termination

Release factors bind (RF-1+RF-3 or RF-2+RF-3 dependingon the specific termination codon)

Peptidyl transferase transfers the polypeptide to a water molecule

IF-3 replaces releasing factor. Ribosome dissociates into subunits, release of mRNA and tRNA and a new round of translation begins.

— AACUCGCGAUAGGUG —

IF-3

EUKARYOTIC TRANSLATION

Translation in eukaryotes is similar to prokaryotes. Initiation, Elongation and Termination are the three steps. Let us discuss the differences.

Chapter 25 Translation

Ribosomes: Eukaryotic ribosome is 80S particle made up of 60S large subunit and 40S small subunits. *See the diagram under the heading ribosomes.*

Initiator tRNA: Initiator tRNA is methionyl tRNA instead of N-formyl methionyl tRNA.

Initiation: There is no Shine-Dalgarno sequence in eukaryotic mRNA. The 1st AUG on the 5' end of the mRNA is usually the start codon. Kozak sequence, the eukaryotic equivalent of Shine-Dalgarno sequence is present in some mRNAs. Analogue of IF-2 in eukaryotes is eIF-2 which brings the initiator tRNA, i.e. methionyl tRNA.

The following are the events in eukaryotic initiation:

❑ Binding of a ternary complex consisting of the initiator methionyl-tRNA, (met-tRNA$_i$), GTP, and eIF-2 to the 40S ribosome forms **43S preinitiation complex** *(memory aid: 40S + ternary = 43S).*

❑ **Circularization of mRNA:** It is important to note that almost all mRNAs are circularized before translation. Poly A tail binding protein (PABP) bind to the 3' poly A tail. Eukaryotic initiation factor eIF4F which is made up of 4A,4G and 4E subunits binds to the 7-methyl guanosine cap. 4E binds to the cap. 4G binds to the PABP. 4A acts as a mediator. It also acts as RNA helicase.

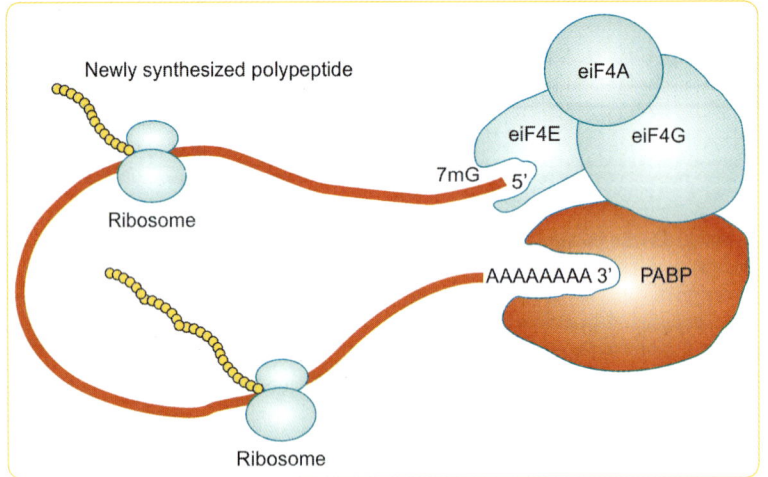

CONCEPT CORNER

What is the advantage of circularization of mRNA?
Circularization ensures that only a complete mRNA is translated. Otherwise, there is a possibility of translating incompletely synthesized or partially degraded mRNAs. You know that formation of peptide bond requires 4 ATP equivalents. Translating incomplete mRNAs means waste of energy. To avoid this, circularization might have evolved.

❑ Binding of mRNA to the 43S preinitiation complex forms **48S initiation complex**.
❑ Combination of the 48S initiation complex with the 60S ribosomal subunit forms the **80S initiation complex** and results in positioning of Met-tRNA$_i$ at the P site. This marks the commencement of elongation phase.

Elongation

At the beginning of elongation, met-tRNA$_i$ occupies the P site, and the A site is ready to receive an aminoacyl-tRNA. Elongation involves the following steps:

❑ Positioning the correct aminoacyl-tRNA in the A site of the ribosome
❑ Formation of the peptide bond
❑ Shifting the mRNA by one codon relative to the ribosome

Elongation phase requires three elongation factors which are similar to prokaryotic ones. Let us compare the prokaryotic and eukaryotic elongation factors for easy understanding.

23S rRNA is the peptidyl transferase in prokaryotes whereas 28S rRNA of the 60S subunit is the peptidyl transferase in eukaryotes.

Eukaryotic elongation factor	Prokaryotic counterpart	Function
eEF-1 α	EF-Tu	Delivery of aminoacyl tRNA to ribosome
eEF-1 β γ	EF-Ts	Guanine nucleotide exchange factor
eEF-1 δ	EF-G	Translocation

Termination

The mechanism of termination is similar to that of prokaryotes. There are two release factors in humans compared to the three releasing factors of prokaryotes.
- ❑ eRF-1 → recognizes all the stop codons
- ❑ eRF-3 → assist eRF-1 in terminating translation.

BIOSYNTHESIS OF SELENOCYSTEINE

Selenocysteine is the 21st amino acid coded by the stop codon **UGA (stop codon reassignment)**. A **stem loop structure** present in the 3' untranslated region of mRNA direct the synthesis of selenocysteine. This process is known as **translational recoding. Serine** is the precursor of selenocysteine. Selenocysteine is not a result of post-translational modification. It is added in a **cotranslational manner**.

▶ Video lecture on selenocysteine: https://youtu.be/4-RgA3ZcGmc or scan QR code

INHIBITORS OF TRANSLATION

	Inhibitor
Eukaryotic	❑ **Cycloheximide** – inhibits peptidyl transferase activity of 60S subunit ❑ **Ricin and Diphtheria toxin**
Prokaryotic	**Antibiotics:** ❑ **30S inhibitors:** Aminoglycosides (Streptomycin), Tetracycline ❑ **50S inhibitors:** Chloramphenicol, Erythromycin, Linezolid Lincomycin **Memory aid:** Buy AT 30 CELL at 50
Both	**Puromycin:** Tyrosyl tRNA analogue

Mechanism of Action of Tetracycline

Tetracycline binds to the A site of 30S ribosomal subunit and blocks the incoming aminoacyl tRNA from binding to the mRNA-ribosome complex.

Ricin is an Extremely Toxic Plant Glycosidase

- ❑ Ricin is the toxic protein present in the seeds of castor beans (*Ricinus communis*)
- ❑ Ricin is a **glycosidase.** It cleaves an adenine residue from the 28S rRNA (peptidyl transferase)
- ❑ As low as 500 μg is lethal for a human.

Diphtheria Toxin Causes ADP-Ribosylation of Elongation Factor-2

- ❑ Mammalian elongation factor-2 (EF-2) contains a unique amino acid called **diphthamide** (histidine derivative).

Fig. 4: Castor beans

- Diphtheria toxin and exotoxin A of pseudomonas aeruginosa mediate the transfer of ADP-ribose from NAD to the diphthamide residue (ADP-ribosylation) of EF-2 leading to inhibition of protein synthesis.

NAD^+ + peptide diphthamide $\rightleftharpoons$ Nicotinamide + peptide N-(ADP-D-Ribosyl) diphthamide

We will discuss ADP ribosylation in Chapter 38 (Refer to page no. 588).

POST-TRANSLATIONAL MODIFICATIONS

Post-translational modifications (PTMs) are the structural and chemical modification of a protein after its synthesis. Co-translational modifications take place simultaneously during the synthesis inside the endoplasmic reticulum, e.g. Protein folding.

Types

Phosphorylation	Addition of a PO_4 group to Serine, Threonine and Tyrosine residues – most common PTM
Glycosylation	N-Glycosylation – addition of sugar residues to **amide** nitrogen of asparagine O-Glycosylation - addition of sugar residues to **OH** group of serine, threonine and tyrosine *(Chapter 41) (Refer to page no. 604)*
Hydroxylation	Addition of hydroxyl group to proline and lysyl residues, e.g. Collagen (requires vitamin C)
γ-Carboxylation	γ-Carboxylation of glutamic acid residues (requires Vit K)
Biotinylation	Addition of biotin to ε- amino group of lysine residue of apocarboxylases *(Chapter 33) (Refer to page no. 522)*
Prenylation	Addition of lipids (isoprenylation, farnesylation, geranylation), helps to target the proteins to membranes.
Glypiation	GPI anchor formation *(Chapter 41) (Refer to page no. 607)*
Lipoylation	Addition of lipoic acid, e.g. E_2 component of PDH
Proteolytic cleavage	Zymogen activation *(Chapter 5) (Refer to page no. 68)*
Others:	Acetylation, Methylation, Sulfation, ubiquitination, Glutamylation, Disulphide bond formation

Role of PTMs

- Generates heterogeneity in proteins, i.e. increases proteomic diversity. In other words, it helps in utilising identical proteins for different cellular functions in different cell types.
- Regulates the of activity of protein
- Contributes in protein targeting

PROTEIN FOLDING

In Chapter 2 *(Refer to page no. 17)*, we have studied that primary structure of proteins determines the higher order of structure. In his Nobel Prize winning *in vitro* experiment, Christian B. Anfinsen showed that unfolded ribonuclease enzyme could refold on the removal of the denaturing agents. This does not mean that all the unfolded proteins can refold spontaneously. Folding and refolding of proteins require the help of other proteins *in vivo*.

Chaperones

Chaperones are proteins that **assist the proper folding** of other proteins. They **prevent the aggregation** of newly-synthesized (nascent) polypeptide chains as soon as they start emerging from the ribosomes. They bind to **hydrophobic regions** of the proteins and prevent their hydrophobic interactions. Many previously recognized heat shock proteins (Hsps) are now found to be chaperones. But not all chaperones are Hsps, and not all Hsps are chaperones.

As protein synthesis takes place in mitochondria also, chaperones are utilized in mitochondria as well.

Chaperonins are proteins that assist chaperones.

Examples of Chaperones and Chaperonins

- ❏ Hsp60 is a mitochondrial chaperonin
- ❏ Hsp 70, 90 and 100
- ❏ GroEL is a bacterial chaperonin
- ❏ Calnexin – anchored to the ER membrane
- ❏ Calreticulin – soluble analog of calnexin
- ❏ Protein disulfide isomerase
- ❏ Peptidyl-cis-trans isomerase

Chemical Chaperones versus Molecular Chaperones

- ❏ Glycerol, Trehalose *(Chapter 9)* *(Refer to page no. 103)*, DMSO, etc. are chemicals that enhance protein folding and stability. So, they are known as chemical chaperones. They are used in vitro to maintain the stability of proteins (like antibodies) during storage.
- ❏ Hsp70 and others are called molecular chaperones.

Unfolded Protein Response

Protein folding is not 100% efficient. Protein misfolding does happen. These misfolded proteins are unfolded, and are destroyed by the ubiquitin-proteasomal system (UPS) *(Chapter 11.1)* *(Refer to page no. 175)*.

When there are excessive misfolded proteins in the endoplasmic reticulum (ER stress), a series of cellular responses known as unfolded protein responses is initiated.

- ❏ Protein synthesis is halted
- ❏ Increased production of molecular chaperones
- ❏ Degradation of misfolded proteins by UPS

If the above responses fail to restore the normal function of the cell, apoptotic pathway is initiated.

Diseases Due to Abnormal Protein Folding

Protein misfolding diseases are also known as proteopathies or protein conformational disorders or disorders of proteostasis.

Protein conformational disorders	Major aggregating protein
Alzheimer's disease	Amyloid β peptide (Aβ), Tau protein
Prion diseases	Prion protein
Parkinson's disease and other synucleinopathies	α-Synuclein
Tauopathies	Microtubule-associated protein tau (Tau protein)
Huntington's disease and other poly Q diseases (Chapter 20) *(Refer to page no. 376)*	Proteins with tandem glutamine expansions
Alexander disease	Glial fibrillary acidic protein (GFAP)
Familial amyloidotic neuropathy, Senile systemic amyloidosis	Transthyretin
Pulmonary alveolar proteinosis	Surfactant proteins
Cystic Fibrosis	Cystic fibrosis transmembrane conductance regulator (CFTR)
Sickle cell disease	Hemoglobin

PRION DISEASE

Prion diseases are protein conformational disorders. Unlike the other proteopathies, prion diseases are infectious. Cellular prion protein (PrPc) is a normal, GPI-anchored membrane protein in the body. Its exact function is yet to be identified. PrPc is rich in α helices. PrP^{sc} *(sc stands for scrapie)* is an abnormal form of PrP^c with beta sheets (depicted by arrows in the following diagram).

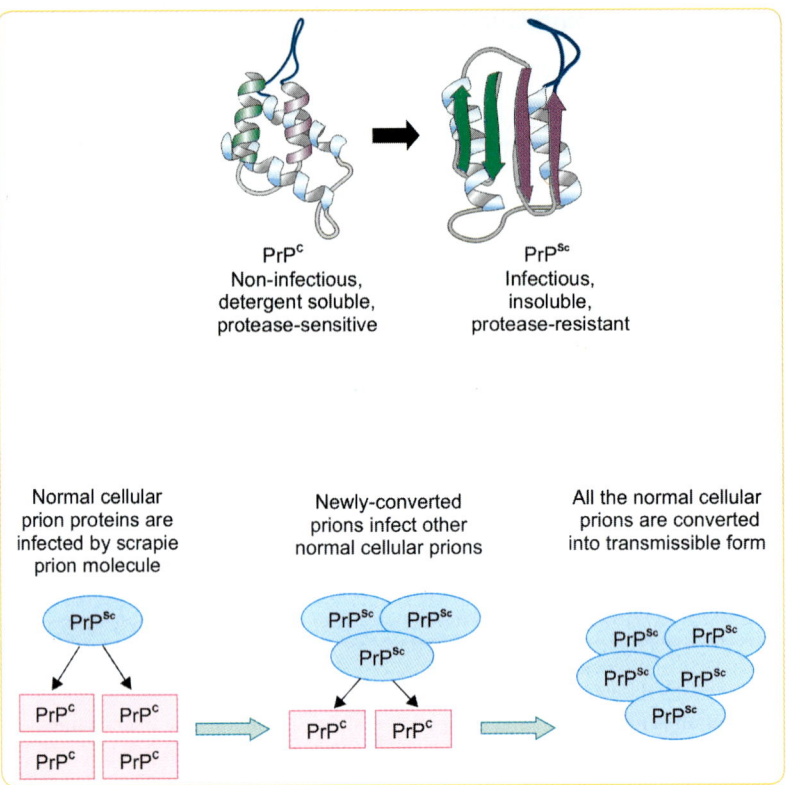

PrP^c
Non-infectious,
detergent soluble,
protease-sensitive

PrP^{Sc}
Infectious,
insoluble,
protease-resistant

Normal cellular
prion proteins are
infected by scrapie
prion molecule

Newly-converted
prions infect other
normal cellular prions

All the normal cellular
prions are converted
into transmissible form

List of Human Prion Diseases

❑ Creutzfeldt-Jakob Disease, Classic
❑ Fatal Familial Insomnia
❑ Variably protease-sensitive prionopathy
❑ Kuru

Animal Prion Diseases

❑ Bovine Spongiform Encephalopathy (cattle)
❑ Scrapie (sheep)
❑ Chronic wasting disease (deer)

ALZHEIMER'S DISEASE

Alzheimer's disease is a protein conformational disease with aggregation of neurofibrillary tangles and plaques. **Tangles contain tau protein whereas the plaques contain β-amyloid**.

Alzheimer's disease is a polygenic disease involving many genes.

Genes Involved

❑ Amyloid precursor protein (APP)
❑ Presenilin-1 and Presenilin-2
❑ ApoE$_4$

APP is encoded by chromosome 21. This explains why a patient with down syndrome invariably develops Alzheimer's disease if they survive beyond the age of 40.

Competency

❑ BI7.2   Describe the processes involved in translation

Key Points

- In humans, there are about 41tRNAs in cytoplasm and 21tRNAs in mitochondria.
- tRNA is made up of 73-93 nucleotides; clover-leave shaped in the secondary structure; inverted, L-shaped in tertiary structure.
- Aminoacyl tRNA synthetases have proof-reading activity.
- The translation is described in three stages—Initiation, elongation, termination. Each stage requires the involvement of specific protein factors.
- All prokaryotic mRNAs contain a unique sequence in the 5′ untranslated region, known as Shine-Dalgarno (SD) sequence which base pairs with 16S rRNA. This helps in the positioning of the start codon (AUG).
- Peptidyl transferase is a ribozyme.
- Formation of one peptide bond requires 4 ATP equivalents.
- Kozak sequence is the eukaryotic equivalent of Shine-Dalgarno sequence.
- Almost all eukaryotic mRNAs are circularized before translation.
- Selenocysteine is the 21st amino acid coded by the stop codon UGA (stop codon reassignment).
- Puromycin is a tyrosinyl tRNA analogue; inhibits translation in both prokaryotes and eukaryotes.
- Diphtheria toxin causes ADP ribosylation of elongation factor-2.
- Chaperones are proteins that assist the proper folding of other proteins. Chaperonins are proteins that assist chaperones.
- The misfolded proteins are unfolded and are destroyed by the ubiquitin-proteasomal system. Excess accumulation of misfolded proteins initiates unfolded protein response in the endoplasmic reticulum.
- Protein misfolding diseases are also known as proteopathies or protein conformational disorders or disorders of proteostasis.
- Non-translating mRNAs can form ribonucleoprotein particles that accumulate in cytoplasmic organelles termed P bodies.

SELF-ASSESSMENT

Short Answer Questions

1. Write two differences between prokaryotic and eukaryotic translation process. Give one example of an antibiotic/compound which inhibits translation,
   a. Only in prokaryotes
   b. Only in eukaryotes
   c. Both in prokaryotes and eukaryotes and specify the mechanism of action.

2. Which signal in prokaryotes ensures the beginning of translation at the initiation codon and not at an internal AUG in mRNA?

3. What is the function of EF-Tu in translation?

4. Enumerate any four post-translational modifications of proteins and state their role in protein function and stability.

5. Draw a labelled diagram of secondary structure of tRNA and indicate the function of each arm. What is tertiary structure of tRNA?

6. How does the charging of a specific tRNA with the correct amino acid take place? How will you account for less than 61 types of tRNAs present in the cell?

7. What is the energy requirement for the formation of a peptide bond? Justify your answer.

8. Differentiate prokaryotic and eukaryotic translation process.

9. What is the advantage of circularization of mRNA before eukaryotic translation?

10. Describe the biosynthesis of selenocysteine.

11. Which amino acid apart from the 20 standard amino acids is incorporated into a protein cotranslationally? Name two amino acids formed by posttranslational modifications in proteins.

12. What are chaperones and chaperonins? In which cellular process are they involved?

13. Name any four diseases that arise due to abnormal protein folding.

14. Name any two human prion diseases.

15. Describe the pathogenesis of prion disease.

Chapter 25 Translation

Multiple Choice Questions

1. **What is present on the 3′ end of all functional, mature tRNAs?**
   a. The cloverleaf loop
   b. The anticodon
   c. The sequence 5′CCA3′
   d. The codon

2. **Components of 50S ribosomal subunit are:**
   a. 5S RNA, 5.8S and 28S RNA
   b. 16S RNA and 23S RNA
   c. 23S RNA and 5.8S RNA
   d. 23S RNA and 5S RNA

3. **Components of 60S ribosomal subunit are:**
   a. 5S RNA, 5.8S and 28S RNA
   b. 16S RNA and 23S RNA
   c. 23S RNA and 5.8S RNA
   d. 23S RNA and 5S RNA

4. **16s rRNA is a component of:**
   a. 30S subunit   b. 40S subunit
   c. 50S subunit   d. 60S subunit

5. **Which of the following RNA is not involved in translation?**
   a. mRNA   b. tRNA
   c. rRNA   d. snRNA

6. **In prokaryotes mRNA binds with the ribosome by interaction of Shine-Dalgarno sequence of mRNA with the:**
   a. 16S rRNA   b. 23S rRNA
   c. 18S rRNA   d. 5S rRNA

7. **Which one of the following is an tyrosine tRNA analogue that causes premature chain termination?**
   a. Cycloheximide
   b. Puromycin
   c. Paromomycin
   d. Erythromycin

8. **Which of the following is an example of post-translational modification?**
   a. Glutamate to Glutamine
   b. Proline to Hydroxyproline
   c. Aspartate to Asparagine
   d. Leucine to Isoleucine

9. **A 20-year-old man with a microcytic anemia is found to have an abnormal form of β-globin (Hemoglobin Constant Spring) that is 172 amino acids long, rather than the 141 found in the normal protein. Which of the following codon change is consistent with this abnormality?**
   a. CGA → UGA   b. GAU → GAC
   c. GCA → GAA   d. UAA → CAA

10. **During translation, initiator tRNA binds to:**
    a. A site   b. P site
    c. E site   d. All of the above

11. **All of the following statements are true, EXCEPT:**
    a. Activation of amino acid by amino acyl tRNA synthetase require 2 high-energy phosphate bonds
    b. Entry of amino acid into A site requires hydrolysis of 1 GTP
    c. Translocation of amino acid from A site to P site require hydrolysis of 1 GTP
    d. Peptide bond formation by peptidyl transferase enzyme require hydrolysis of 1 GTP

12. **All of the following antibiotics are 50S inhibitors, EXCEPT:**
    a. Chloramphenicol b. Erythromycin
    c. Linezolid   d. Streptomycin

13. **Which of the following toxin inhibits the peptidyl transferase?**
    a. Ricin   b. Diphtheria toxin
    c. Pertussis toxin d. Amanitin

14. **Which of the following is the prokaryotic translocase?**
    a. EF-G   b. EF-H
    c. EF-Ts   d. EF-Tu

15. **Which of the following is not a protein misfolding disorder?**
    a. Alzheimer's disease
    b. Creutzfeldt Jakob disease
    c. Cystic fibrosis   d. Tuberculosis

16. **Selenocysteine insertion sequence is a stem loop structure present in**
    a. mRNA   b. rRNA
    c. tRNA   d. lncRNA

ANSWERS

1. c.	**2.** d.	**3.** a.	**4.** a.	**5.** d.	**6.** a.
7. b.	**8.** b.	**9.** d.	**10.** b.	**11.** d.	**12.** d.
13. a.	**14.** a.	**15.** d.	**16.** a.		

Protein Targeting

Proteins are synthesized in the cytoplasm. But, they are required in various locations of the cell, and some are required outside the cells too. For example, $Na^+K^+ATPase$ is required in the plasma membrane, catalase is required in the peroxisomes, and insulin needs to be secreted out. Transporting proteins to the appropriate locations is known as protein targeting/trafficking. Defect in protein targeting leads to various disorders.

It was first recognized by **Blobel** in 1970 that specific signals in the protein itself are involved in the targeting of proteins to various locations. He was awarded the Nobel Prize in 1999 "for the discovery that proteins have intrinsic signals that govern their transport and localization in the cell."

To begin with, synthesis of all proteins takes place in the cytosolic-free ribosome. If certain signals are there, protein synthesis is halted, and ribosomes are translocated to the membrane of endoplasmic reticulum membrane, and protein synthesis resumes.

The first factor that determines the location of the protein is whether the protein is synthesized in the cytosolic-free ribosomes or the endoplasmic membrane-bound ribosomes.

- Proteins synthesized in the free-ribosomes undergo cytosolic pathway of protein sorting.
- Proteins synthesized in the endoplasmic reticulum membrane-bound ribosomes undergo Rough-ER pathway of protein sorting.

	Cytosolic Pathway	Rough ER pathway (Also know as Secretory pathway)
Proteins are synthesized in the	Cytosolic (free) polyribosomes	Endoplasmic reticulum membrane-bound polyribosomes
Proteins are targeted to	Mitochondria, nuclei, peroxisomes, cytosol	Plasma membrane, ER membrane, Golgi membrane, Lysosome or secreted out of cells
Type of translocation	Post-translational	Usually Co-translational

SIGNAL SEQUENCES THAT TARGET THE PROTEIN TO VARIOUS LOCATIONS

Signal sequence	Specific organelle targeted
Mannose-6-phosphate	Lysosome
N-terminal signal peptide	Endoplasmic reticulum
Carboxyl-terminal KDEL (Lys-Asp-Glu-Leu) sequence	Lumen of endoplasmic reticulum
Di-acidic sequences (e.g. Asp-X-Glu) in membrane proteins	Golgi membranes
Nuclear localization signal	Nucleus
Peroxisomal targeting sequence	Peroxisome
None	Absence of signal sequence retains the protein in cytosol

MECHANISM OF TARGETING INSIDE ENDOPLASMIC RETICULUM

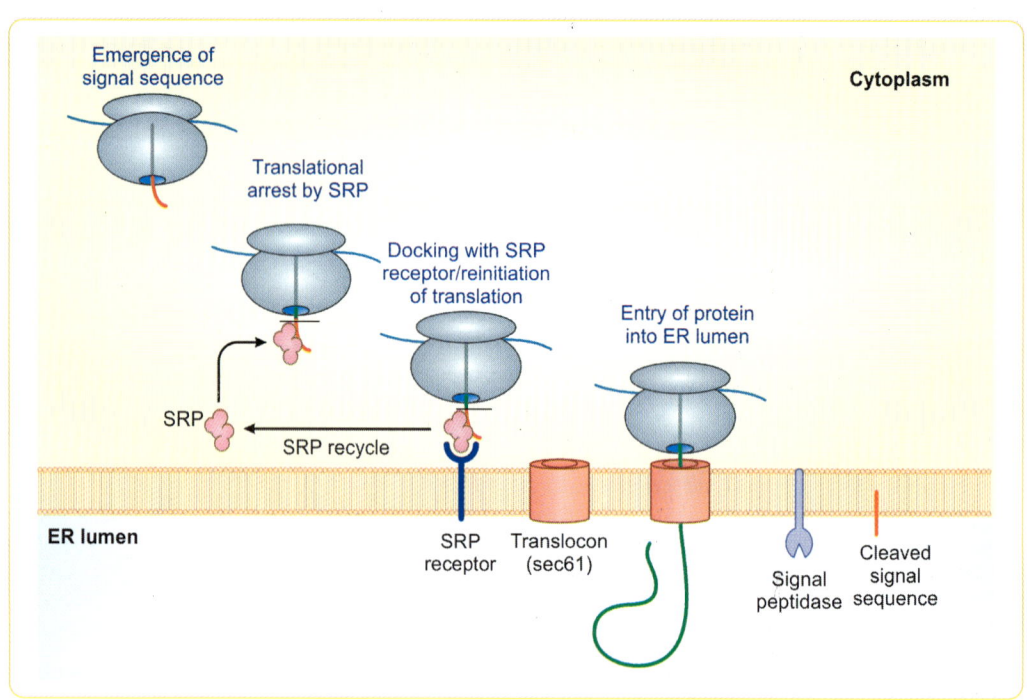

- ❏ Secretory proteins contain a **N-terminal signal peptide** flanked by hydrophobic residues.
- ❏ You know that proteins are always synthesized from N-terminal to C-terminal. So, N-terminal signal peptide comes out first. This signal sequence is recognized by cytosolic signal recognition particle (SRP)
- ❏ SRP is a ribonucleoprotein. It contains six proteins and one RNA (7SRNA).
- ❏ SRP arrests the translation and brings the partially synthesized polypeptide with ribosome to the ER membrane.
- ❏ The SRP-Ribosome complex binds to SRP receptor on the ER membrane. SRP is released. Signal sequence enters into the ER lumen through a channel called translocon (sec61), and the translation continues.
- ❏ The signal peptide is cleaved by signal peptidase and further cleaved by proteases in the ER.
- ❏ The further fate of protein is different depending upon the additional signals in the protein.
  - ○ The protein can be anchored in the ER membrane
  - ○ The protein can permanently stay in the ER lumen
  - ○ The protein can undergo glycosylation in ER and Golgi.

○ The protein can be secreted through secretory vesicles in the following route: Endoplasmic Reticulum → Golgi Apparatus→ Secretory Vesicle → Plasma Membrane → Release into the external environment

Preproalbumin → Proalbumin → Albumin

Albumin is a constitutively secreted protein. In the ER, signal peptidase cleaves Preproalbumin to produce Proalbumin. **Furin** cleaves a hexapeptide from proalbumin to produce albumin.

MANNOSE-6-PHOSPHATE IS A SIGNAL FOR ENTRY INTO LYSOSOMES

Golgi apparatus is the primary site of protein sorting. Mannose-6-phosphate acts as a signal to target enzymes into the lysosome. Lysosomal enzymes are glycoproteins.

The phosphate group to mannose is added by the enzyme **UDP-GlcNAc-1-phosphotransferase**. N-Acetyl glucosamine (GlcNAc) is removed later to reveal mannose-6-phosphate which is recognized by a lectin (mannose-6-phosphate receptor) and targeted into lysosomes (Fig. 1).

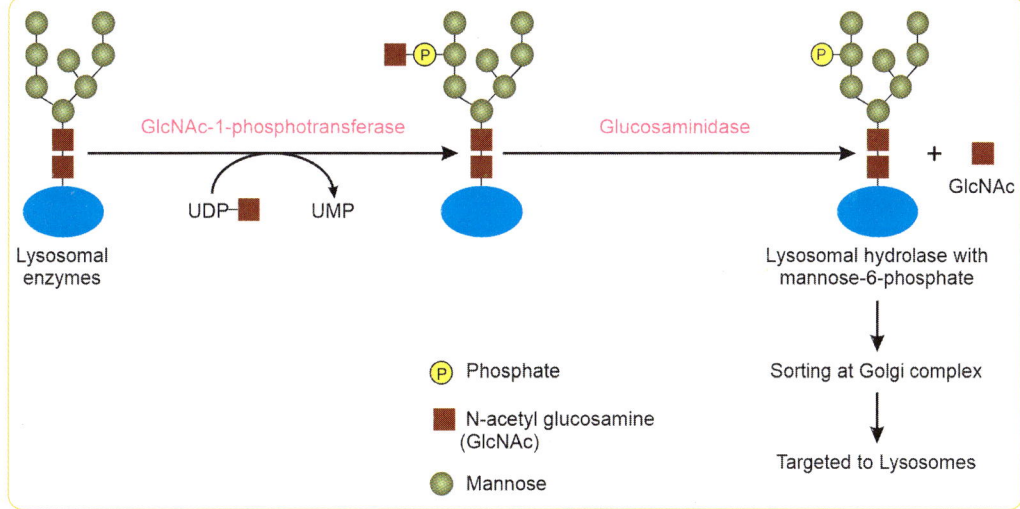

Fig. 1: GlcNAc: N-Acetyl glucosamine, UDP: Uridine diphosphate, UMP: Uridine monophosphate, NAGPA: N-acetylglucosamine-1-phosphodiester alpha-N-acetylglucosaminidase

I-cell Disease

Defective GlcNAc phosphotransferase, resulting in the absence of the Mannose-6-P signal for lysosomal localization of certain hydrolases leads to I-cell disease.

So, instead of being targeted to lysosomes, lysosomal enzymes are secreted out of the cells.

I-cell disease is characterized by severe progressive psychomotor retardation and a variety of physical signs (coarse facies), with death often occurring in the first decade of life.

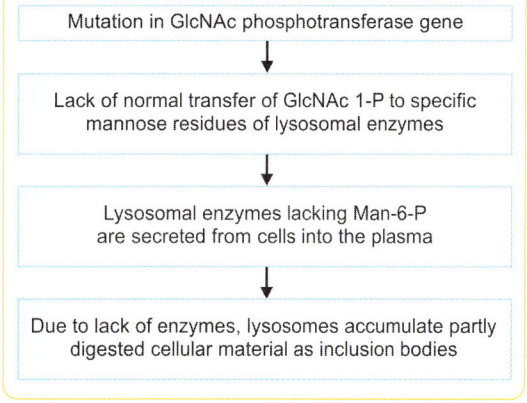

Mutation in GlcNAc phosphotransferase gene

↓

Lack of normal transfer of GlcNAc 1-P to specific mannose residues of lysosomal enzymes

↓

Lysosomal enzymes lacking Man-6-P are secreted from cells into the plasma

↓

Due to lack of enzymes, lysosomes accumulate partly digested cellular material as inclusion bodies

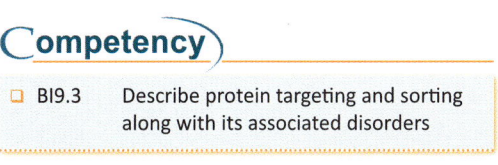

Competency

❏ BI9.3   Describe protein targeting and sorting along with its associated disorders

Key Points

❑ Golgi apparatus is involved in protein sorting.
❑ Proteins synthesized in the cytosolic free ribosomes are targeted to either mitochondria, nuclei and peroxisomes or they remain in the cytosol.
❑ Signal recognition particle is a ribonucleoprotein containing 7S RNA.
❑ The enzyme furin is involved in the conversion of proalbumin to albumin.
❑ Mannose-6-phosphate is the signal for targeting into the lysosome.
❑ I-cell disease is due to defective GlcNAc phosphotransferase.
❑ The mannose-6-phosphate receptor is a lectin.

SELF-ASSESSMENT

Short Answer Questions

1. Explain how newly-synthesized proteins are targeted to lysosomes? Name the enzyme defective in I-cell disease.

2. Write a short note on the signal hypothesis of protein targeting.

Multiple Choice Questions

1. **"Proteins contain information (a signal) that targets them to appropriate intracellular/ extracellular location" – this was first recognized by:**
   a. Blobel   b. Boyd
   c. Berg   d. Bore

2. **All of the following proteins synthesized by free-ribosomes, Except:**
   a. Proteins that remain in cytosol
   b. Lysosomal enzymes
   c. Mitochondrial proteins
   d. Peroxisomal enzymes

3. **Presence of which of the following signals targets a protein permanently in the lumen of endoplasmic reticulum?**
   a. N-terminal signal peptide
   b. Carboxy-terminal KDEL sequence
   c. Mannose-6-phosphate
   d. Di-acidic sequence

4. **A protein is being synthesized on free ribosomes. Presence of which of the following signal sequence will translocate the protein into cytosol?**
   a. Amino-terminal sequence
   b. Di-acidic sequence
   c. KDEL sequence   d. None of the above

5. **Which of the following is the correct sequence of the secretory or exocytotic pathway of protein sorting?**
   a. Endoplasmic Reticulum → Golgi Apparatus→ Secretory Vesicle → Release into the external environment
   b. Golgi Apparatus→ Endoplasmic Reticulum → Plasma Membrane → Release into the external environment
   c. Plasma Membrane → Endoplasmic Reticulum → Golgi Apparatus→ Release into the external environment
   d. Plasma Membrane → Golgi Apparatus→ Endoplasmic Reticulum → Release into the external environment

6. **Signal recognition particle is a:**
   a. Protein   b. Ribonucleoprotein
   c. Lipoprotein   d. RNA

7. **The enzyme defective in I-cell disease is:**
   a. Mannose kinase
   b. Mannose-6-phosphate transferase
   c. GlcNAc transferase
   d. GlcNAc phosphotransferase

ANSWERS

1. a.	2. b.	3. c.	4. d.	5. a.	6. b.
7. d.					

27 CHAPTER

Regulation of Gene Expression

Gene expression is the expression of a functional product from genes. The functional product is not always proteins. Non-protein coding genes are also expressed as functional RNAs (tRNA, snRNA, miRNA, etc.) *We have already studied the definition of gene in Chapter 20 (Refer to page no. 373).*

Genes can be broadly divided into types based on their expression pattern:

- ❑ Constitutively expressed genes, also known as house-keeping genes are **transcribed at a constant level** regardless of the cellular environment, e.g. Glycolytic enzymes, Ribosomal RNA.
- ❑ Regulated (inducible/repressible) genes are transcribed depending upon the environmental conditions. Their expression is highly variable, e.g. UDP-glucuronosyl transferase.

Remember

The basic theme of gene expression is "Trans-acting factors control the expression of cis-regulatory elements."

- ❑ **Cis-regulatory elements** are DNA sequences present on the same molecule of DNA (chromosome) as the gene they regulate.
- ❑ For example, promoter is a cis-regulatory element usually located upstream of the transcribed region. Enhancers are cis-regulatory elements that can be located upstream, downstream, within the introns, or even relatively far away from the gene, they regulate.
- ❑ **Trans-acting factors** regulate the expression of genes distant from the gene from which they were transcribed.
- ❑ For example, transcription factors produced by one chromosome can regulate the gene on a different chromosome.
- ❑ Not all trans-acting factors are proteins. Micro RNAs and long non-coding RNAs can also be considered trans-acting factors.

☐ Whether a gene is transcribed or not is determined by DNA-protein interactions. *In Chapter 23 (Refer to page no. 413), we have learned the general and special transcription factors required for transcription.* Let us first discuss the DNA binding sequences in these proteins.

DNA BINDING MOTIFS

DNA binding motifs mediate the binding of DNA binding proteins to specific regions of DNA.

Helix-turn-helix (HTH) Motif

Helix turn helix motif is made up of two α helices connected by a short stretch of amino acids, i.e. turn *(not loop)*. A fixed angle is maintained between these two helices. *(vid. The difference between turn and loop – Chapter 2) (Refer to page no. 20)*

One of the helices binds to the major groove of DNA and is known as recognition helix. *(most of the DNA binding proteins bind to the major groove)*

Examples of HTH motif-containing DNA binding proteins:
☐ E coli: Lac repressor, CAP (catabolite gene activator protein)
☐ Humans: Homeobox proteins, Oct 2

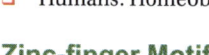

Zinc-finger Motif

In Chapter 2 (Refer to page no. 16), we have studied about domains and motifs. As the name suggests, zinc finger motif contains zinc which stabilizes the protein fold, e.g. Proteins of steroid receptor family, WT1 (Wilm's tumor protein) and sp1 (binds to GC box)

Significance: Mutation in zinc finger motif leads to vitamin D dependent type II rickets.

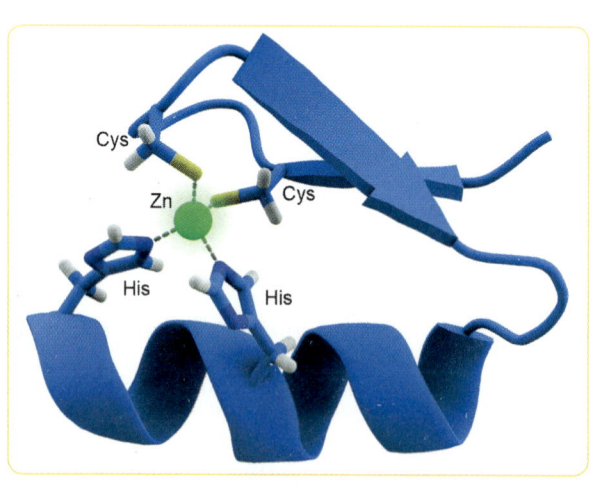

Fig. 1: Zinc finger motif

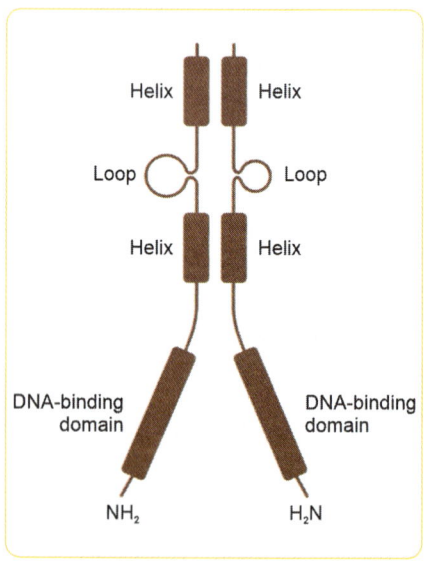

Fig. 2: Helix loop helix

Basic Helix-loop-helix (HLH) Motif

❑ Helix-loop-helix is NOT a DNA-binding domain. It is a dimerization domain.
❑ Proteins with HLH also contains a separate DNA binding domain made up of basic amino acids.

For example, Hypoxia-inducible factor (HIF), MyoD, MITF (Microphthalmia-associated transcription factor).

Basic-region Leucine Zipper

❑ Leucine zipper is an alpha helix in which every 7th amino acid is leucine.
❑ Leucine zipper is a dimerization motif. Basic-region is involved in DNA binding.

For example, c-myc family transcription factors, CRE binding protein (CREB), Retina leucine zipper (NRL)

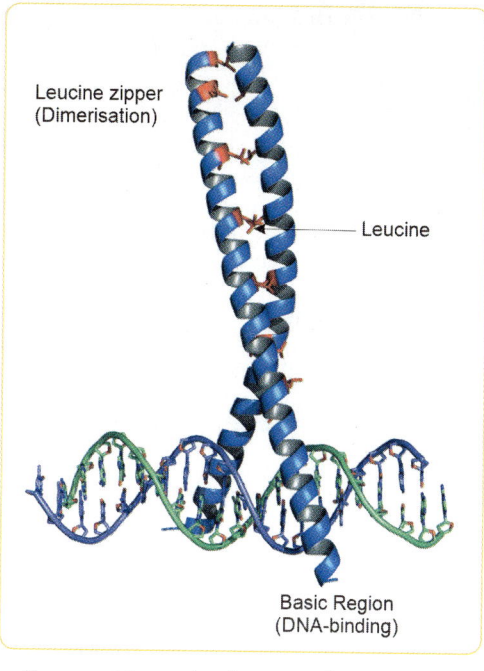

Leucine zipper
(Dimerisation)

Leucine

Basic Region
(DNA-binding)

REGULATION OF GENE EXPRESSION IN PROKARYOTES

Regulation of gene expression in both prokaryotes and eukaryotes takes place mostly at the level of transcription. In prokaryotic regulation, we will limit our discussion to Lac Operon, Attenuation in tryptophan operon and Riboswitches.

LAC OPERON

Jacob and Monod coined the term "operon". Operons are clusters of polycistronic structural genes with related functions regulated by the common promoter. In Chapter 23 *(Refer to page no. 410)*, we have studied that prokaryotic mRNAs are polycistronic, i.e. one mRNA codes for multiple polypeptides. Proteins involved in a particular pathway are usually coded by a single polycistronic mRNA. This makes the regulation of the pathway easier.

Genes that encode for the polycistronic mRNA are known as structural genes. All the structural genes require only one promoter. Genes that regulate the expression of structural genes are known as regulatory genes. Also, there is a regulatory DNA element between the promoter and structural genes known as operator. Operator site should be free for the movement of RNA polymerase. Operator region is unique for prokaryotic operons. Eukaryotes do not contain operator elements in the genome.

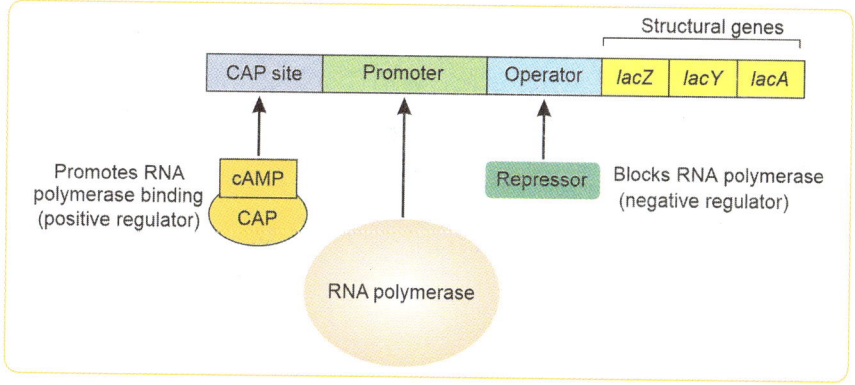

Fig. 3: Repressor is produced by *lacI* gene which is upstream of Lac operon

Chapter 27 Regulation of Gene Expression

449

As illustrated in Fig. 3, Lac operon consists of three structural genes involved in the utilization of lactose and other regulatory elements.

Structural Genes

- *Lac Z* – codes for β-galactosidase that hydrolyzes lactose to galactose and glucose. **It also isomerizes lactose to allolactose.**
- *Lac Y* – codes for permease that transports lactose inside the cells.
- *Lac A* – codes for acetylase that is involved in acetylation and detoxification of molecules entering along with lactose through permease.

Regulatory Gene

- *Lac I* – codes for the repressor protein. This gene is **constitutively** expressed.

Regulatory DNA Elements

- **Promoter:** RNA polymerase binding site
- **Operator:** Repressor binding site; binding of the repressor protein to operator prevents the movement of RNA polymerase.

The interaction between cyclic AMP Receptor Protein (CRP), also known as Catabolite Activator Protein (CAP) and RNA polymerase is essential for the expression of Lac operon. So, **CAP acts as a positive regulator of Lac operon**. CAP exists as a dimer. CAP alone cannot bind to DNA. Only CAP-cAMP complex can. cAMP is produced by adenylyl cyclase enzyme when glucose is not available.

Expression Pattern of Lac operon Under Different Metabolic Conditions

> *Remember*
>
> For the optimal expression of Lac operon, operator should be free and CAP site should be occupied.

When Glucose is Available and Lactose is not Available

- Repressor protein constitutively produced by *lac I* gene binds to the operator site and doesn't allow the RNA polymerase to move.
- In the presence of glucose, cAMP is not produced → No binding of CAP-cAMP to the DNA → RNA polymerase cannot bind to the promoter efficiently.
- Because of both these reasons, structural genes of Lac operon are not expressed.

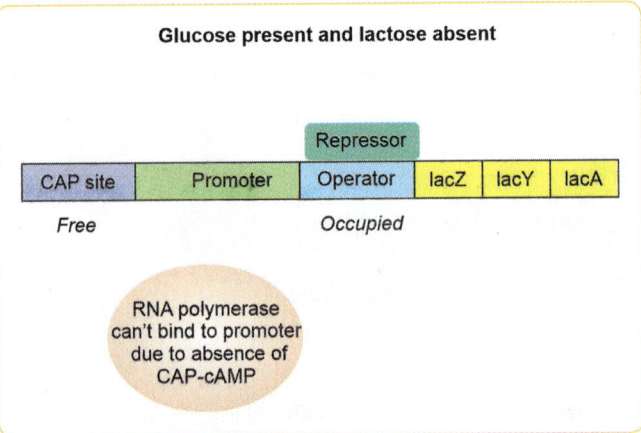

Glucose present and lactose absent

Section III Molecular Biology

When Glucose is not Available and Lactose is Available

○ Lactose enters the cells through permease which was expressed earlier. The lactose entered is converted to allolactose by the minimal amount of β-galactosidase already present inside the cells. Allolactose binds to the repressor protein. Allolactose-bound repressor cannot bind to the operator site. So, the operator site is free and the RNA polymerase can freely move.

Glucose absent and lactose present

cAMP / CAP	RNA polymerase				
CAP site	Promoter	Operator	lacZ	lacY	lacA
Occupied		Free			

Repressor — Allolactose

○ When glucose is not available, cAMP is produced by adenylyl cyclase. cAMP binds to CAP. CAP bound cAMP binds to the Cap site. CAP-RNA polymerase interaction is vital for the efficient activity of RNA polymerase.

○ Thus, there is a maximal expression of structural genes of Lac operon.

When both Glucose and Lactose are Available

In the presence of glucose, cAMP is not produced. There is no binding of CAP-cAMP to the DNA → RNA polymerase cannot bind to the promoter efficiently → Even though the operator site is empty, Lac operon is only minimally expressed *(Note: there is no absolute shutdown)* because of the absence of CAP-RNA polymerase interaction → Lactose is only minimally utilized despite the fact that it is available in considerable amount.

Both glucose and lactose present

CAP site	Promoter	Operator	lacZ	lacY	lacA
Free		Free			

RNA polymerase can't bind to promoter due to absence of CAP-cAMP

ATTENUATION IN TRYPTOPHAN OPERON

❑ Tryptophan operon codes for five structural genes – enzymes involved in tryptophan synthesis.
❑ Tryptophan operon also codes for a peptide sequence known as leader peptide which is translated before these five structural enzymes.
❑ Attenuation is a regulatory mechanism in which translation of a leader peptide affects transcription of a downstream structural gene (e.g. 5 structural genes of tryptophan operon).
❑ Leader sequence contains two codons for tryptophan. When tryptophan concentration is adequate, translation proceeds fast and results in the formation of stem-loop structure in the RNA which terminates transcription. (Recall, in prokaryotes, transcription is coupled with translation).
❑ Tryptophan operon is also regulated by repression. Tryptophan acts as a corepressor and binds to a repressor which in turn binds to operator and inhibits the transcription.

RIBOSWITCHES

Riboswitches are regulatory elements in the messenger RNA that modulate the expression of the gene in response to specific metabolite binding.
For example, when the FMN level is high, it binds to mRNAs of enzymes involved in Riboflavin synthesis and causes premature termination of transcription.

Chapter 27 Regulation of Gene Expression

REGULATION OF GENE EXPRESSION IN EUKARYOTES

Regulation of gene expression is quite complex in eukaryotes. This is partly because of the condensation of DNA by chromatin. Accessibility to the DNA is the first challenge to the gene expression. Unlike prokaryotes, transcription and translation are temporally and spatially separated in eukaryotes. This allows more points of regulation (Fig. 4). We will discuss the regulation of gene expression in eukaryotes under the following headings:

❑ Gene amplification
❑ Gene rearrangement
❑ Differential RNA processing
❑ Transport of mRNA from the nucleus to cytoplasm
❑ Regulation of mRNA stability
❑ Regulation at the translational level
❑ Regulation by miRNA

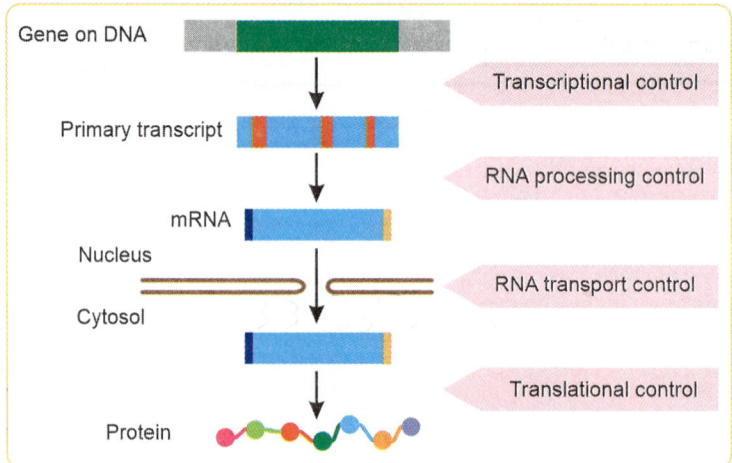

Fig. 4: Points of regulation of gene expression - By ArneLH - Own work, CC BY-SA 3.0, https://commons. wikimedia.org/w/index.php?curid=5815139

GENE AMPLIFICATION

Gene amplification refers to increase in the number of copies of a gene without a proportional increase in other genes. In Drosophila, chorion genes are amplified during eggshell synthesis. In humans, gene amplification is seen in many cancer cells.

Examples:
❑ *MYC* gene amplification in many cancer cells.
❑ **Amplification of dihydrofolate reductase (*DHFR*) gene is responsible for methotrexate resistance.**

GENE REARRANGEMENT

The genome is same in all the somatic cells of the body except in plasma cells and mature T-cells. Generation of antibody diversity involves rearrangement of heavy and light chains of immunoglobulin genes. Rearrangement involves irreversible deletion of certain gene segments. The similar rearrangement takes place in T-cell receptor gene segments.

As you might have guessed, cancer cells also undergo gene rearrangement.

DIFFERENTIAL RNA PROCESSING

With few exceptions, the genome is same in all the somatic cells. But the transcriptome is different. Around 25,000 proteogenic genes code for more than two lakh proteins. This is due to differential RNA processing, i.e. Alternate RNA splicing, RNA editing, **Alternate promoter utilization, Alternate polyadenylation.** *We have already discussed these in Chapter 23* *(Refer to page no. 408).*

TRANSPORT OF mRNA FROM THE NUCLEUS

DNA is transcribed into mRNA in the nucleus. Protein synthesis (translation) takes place in the cytoplasm. So, the mRNA has to exit the nucleus to the cytoplasm through nuclear pore complex.

REGULATION OF mRNA STABILITY

❑ The half-life of mRNA in prokaryotes is just a couple of minutes. But, in eukaryotes, the half-life varies from minutes to hours. Structures present in the 3′ and 5′ untranslated region of mRNA regulate the stability. Length of polyA tail also determines the half-life.

REGULATION AT THE TRANSLATIONAL LEVEL

Reciprocal regulation of synthesis of proteins involved in iron metabolism can be given as an example for this. Ferritin and transferrin receptor 1 *(Chapter 34)* *(Refer to page no. 538)* expression levels are reciprocally regulated at the level of translation initiation in their responses to changes in the intracellular iron levels. When the cellular iron level is low, transferrin receptor 1 (TfR1) is needed on the cell surface and ferritin, the storage protein is not needed inside the cell. mRNAs of ferritin and tfR1 contain a stem-loop structure known as iron response element (IRE).

❑ IRE of ferritin mRNA is located at its 5′ UTR. (**Ferritin → Five**)
❑ IRE of transferrin receptor is located at its 3′UTR. (**TfR → Three**)

Recall aconitase, the enzyme of TCA cycle *(Chapter 7)* *(Refer to page no. 88).* It contains iron-sulphur complex. When the cellular iron levels are low, FeS complex dissociates from aconitase. This creates a binding site for IRE in aconitase. So, aconitase without FeS complex is known as Iron response element binding protein (IRP).

When intracellular iron levels are low, IRP binds to the IRE of both ferritin and TfR1 mRNAs to produce different effects.

❑ IRP-IRE interaction prevents the translation of ferritin mRNA. *(storage protein is not required when the iron levels are low)*
❑ IRP-IRE interaction stabilizes the mRNA of transferrin receptor and leads to more intracellular iron absorption.

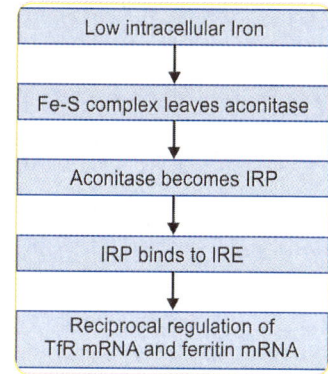

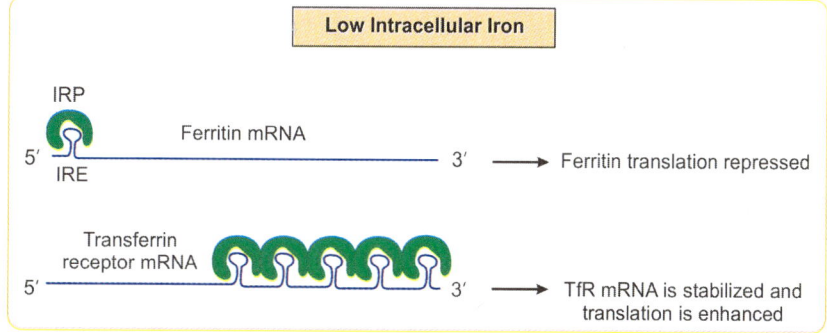

When the intracellular iron levels are high, the opposite of what explained above happens.

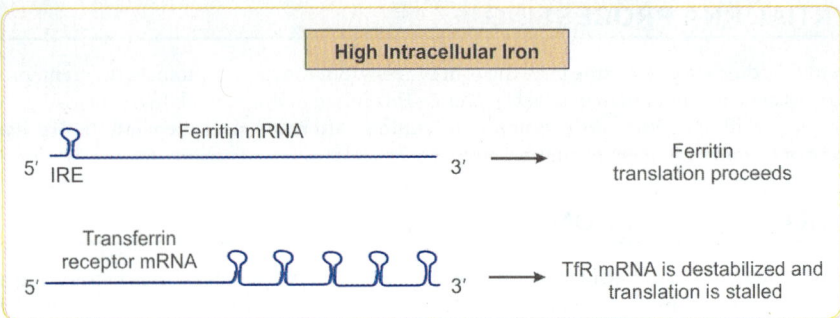

REGULATION BY MICRO RNAs

❏ MicroRNAs are small (22 nucleotides long), endogenous, non-coding RNAs involved in regulation of gene expression by post-transcriptional gene silencing (PTGS)/RNA interference.

Formation

There are two ways in which miRNAs are formed.
1. From intronic DNA of the host gene - introns producing miRNA are known as mirtrons.
2. From miRNA coding genes

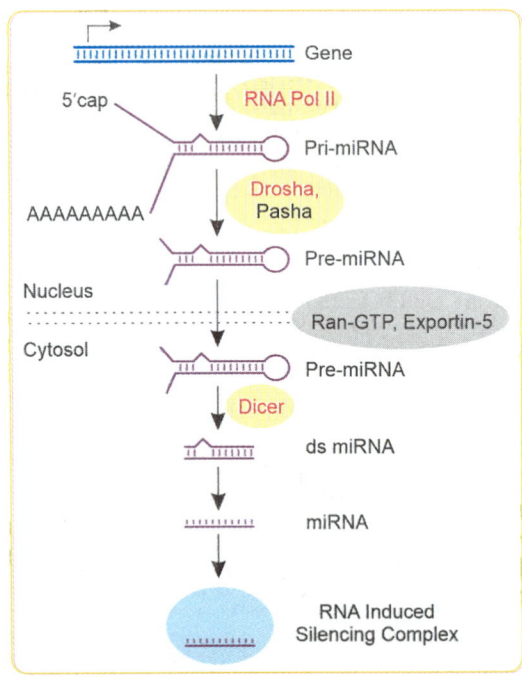

❏ miRNA gene is transcribed by RNA polymerase II to produce primary miRNA (pri-miRNA). Like other RNAs transcribed by RNA pol II, 5' cap and a 3' polyA tail are added to pri-miRNA.
❏ MicroRNA processing complex processes the pri-miRNA to release ~70 nucleotide stem–loop pre-miRNA. This complex consists of Drosha (endonuclease) and its binding partner DGCR8/Pasha.

- Pre-miRNA is translocated to the cytoplasm by exportin-5 with the help of Ran, a monomeric GTPase. *(c.f. Rab, another monomeric GTPase is involved in receptor-mediated endocytosis)*
- Pre-miRNA is processed into ~21-22 nucleotide double-stranded miRNA by dicer endonuclease.
- One of the strands is loaded into RNA-induced silencing complex (RISC) with the help of proteins such as Argonaute.
- RISC complex binds to the target mRNA and causes either repression or degradation.
- miRNA base pairs with the **3′ untranslated region** of mRNA
  - If the base pairing is **imperfect**, it will cause translational **repression**. miRNA-mRNA complexes are stored in **P-bodies** in the cytoplasm.
  - If the base pairing is **perfect**, it will lead to **degradation** of mRNA.

It is critical to distinguish micro RNAs from Small interfering RNA (siRNA).

Feature	miRNA	siRNA
Source	Endogenous	Mostly exogenous
Binding region in mRNA	3′ untranslated region	Anywhere
Complementarity to mRNA	Imperfect or Perfect	Perfect (100% complementarity)
Fate of mRNA	Translational repression, storage in P bodies or destruction	Mostly destruction of mRNA

Micro RNAs are Regulated by Circular RNAs

- Micro RNAs regulate the mRNA stability. Micro RNAs, in turn are also regulated. One way of regulation of miRNA is through circular RNAs (CiRNA).
- CiRNAs are produced by a different form of splicing known as **back splicing**.
- CiRNAs contain binding site for miRNAs. Multiple miRNAs can bind to one circular RNA. Thus ciRNAs act as miRNA sponges.

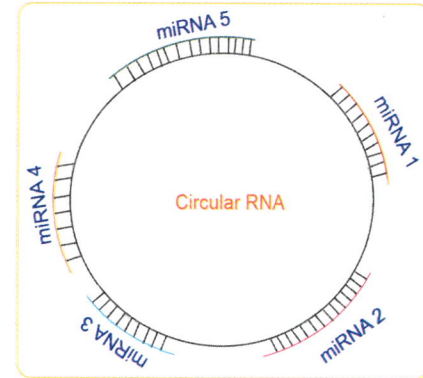

EPIGENETICS

Epigenetics is the study of heritable change in gene expression without change in the nucleotide sequences.

Major epigenetic mechanisms:
- DNA methylation
- Histone modifications
- Alteration in chromatin structure by non-coding RNAs

We have already discussed histone code and alteration in chromatin structure in Chapter 19 *(Refer to page no. 364)*. Histone code is a type of epigenetic code. We have just completed the discussion of microRNAs in this chapter. We will complete this chapter with the discussion on DNA methylation.

In Humans, Most of the CpG Dinucleotides are Methylated

Most of the CpG (C followed by G in the same strand; 'p' denotes phosphodiester bond) dinucleotides in the human genome are methylated. Unmethylated CpG nucleotides of prokaryotic pathogens are recognized by toll-like receptor 9.

CpG islands (CGIs) are regions of the genome that contain a large number of CpG dinucleotide repeats. Most CpG islands are found near promoters. Reversible methylation of these CpG in the promoter region provides a way of turning off/on the gene expression. **Methylation of cytosine represses the gene expression.**

Chapter 27 Regulation of Gene Expression

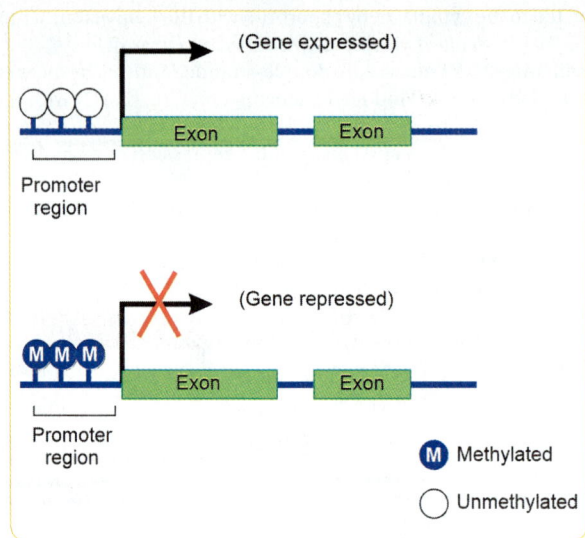

5-methylcytosine is Known as 5ᵗʰ Base

- ❑ Methylation of cytosine is done by DNA methyl transferase—**code writer**
- ❑ Methylcytosine residues are recognized by methyl cytosine binding domain proteins – **code reader**
- ❑ Methylation is removed by TET (ten elven translocase) methylcytosine dioxygenase – **code eraser**

Harrison's Corner

"The defect in Rett syndrome is in the MECP2 gene that codes for methyl cytosine binding protein (MECP). Aberrant methylation results in abnormal gene expression in neurons, which are otherwise normally developed" (Ref. Harrison 19th Ed., p.431)

Role of DNA Methylation

- ❑ Reversible methylation of promoter is a normal way of regulation of gene expression in a time-specific (temporal) and spatial (cell-type specific) manner.
- ❑ DNA methylation is involved in X-chromosome inactivation and genomic imprinting.
- ❑ **Aberrant hypermethylation of promoters of tumor suppressor genes lead to cancer.**

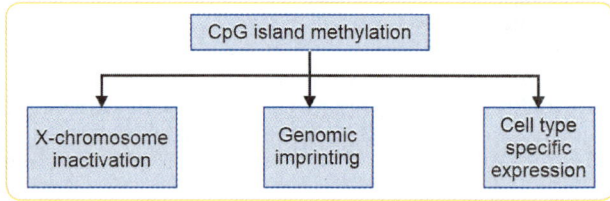

EPITRANSCRIPTOMICS

Epitranscriptomics is the change in gene expression without change in RNA sequence. Methylation of RNA (N^6-methyladenosine (m^{6}A) and N^1-methyladenosine (m^{1}A) are some well known RNA modifications. Epitranscriptomics is an emerging field.

Competency

- ❑ BI7.3   Describe regulation of gene expression

Key Points

- The basic theme of gene expression is "Trans-acting factors control the expression of cis-regulatory elements."
- In both prokaryotes and eukaryotes, gene expression is primarily regulated at the level of transcription.
- Zinc finger motif is present in DNA-binding proteins like steroid hormone receptors.
- Leucine zipper motif is involved in dimerization.
- CAP is a positive regulator of Lac operon.
- Attenuation is seen in tryptophan operon.
- Epigenetics is the alteration in the gene expression patterns without altering the underlying DNA nucleotide sequence.
- DNA methylation at cytosine residue and acetylation of lysine of histone are the two major epigenetic changes.
- Micro RNAs (miRNA) and long non-coding RNA (lncRNA) are involved in the regulation of gene expression.
- miRNA base pairs with the 3' untranslated region of mRNA.
- Hypermethylation of promoter regions of tumor suppressor genes is found in many cancers.
- Unmethylated CpG islands are found near promoters.

SELF-ASSESSMENT

Short Answer Questions

1. Mention the different structural motifs present in DNA binding regulatory transcription factors. A point mutation in which of these motifs and regulatory protein is associated with Vitamin-D resistant rickets?

2. What are miRNAs? With an illustration, explain how are they formed in the cell? Explain their mechanism of action and role in cell function.

3. Describe the role of short single-stranded RNA in the regulation of gene expression.

4. Describe RNA interference or post-transcriptional gene silencing.

5. What is an operon? Explain why Lac operon is not induced by lactose if glucose is present in the medium.

6. What is the role of CAP in Lac operon?

7. Enumerate the salient features of the enhancer elements in DNA. Explain how they stimulate the transcription even from long genomic distances.

8. What do you understand by term 'antisense therapy'?

9. With a schematic illustration explain the molecular basis of reciprocal regulation of translation of ferritin mRNA and transferrin receptor mRNA by the availability of iron inside the cell.

10. Define epigenetics. Name any two epigenetic mechanisms. What do you understand by the following terms: Epigenetic code writers, readers and erasers?

11. What do you understand by the term 'epitranscriptomics'?

12. What is the role of DNA methylation in normal cellular processes and in cancer?

Multiple Choice Questions

1. **At which of the following levels, the predominant regulation of gene expression in both prokaryotes and eukaryotes takes place?**
   a. DNA replication   b. Transcription
   c. Translation   d. Post-translational

2. **The promoter sequence is an example of:**
   a. Cis-acting element   b. Trans acting factor
   c. Coactivator   d. Mediator

3. **Which of the following gene is constitutively expressed in *E. coli*?**
   a. Lac A   b. Lac I
   c. Lac Y   d. Lac Z

4. **Which one of the following cis-acting elements typically resides adjacent to or overlaps with many prokaryotic promoters?**
   a. Regulatory gene   b. Structural gene(s)
   c. Repressor   d. Operator

5. **The ZYA region of the Lac operon will be maximally expressed if:**
   a. Cyclic AMP levels are low
   b. Glucose and lactose are both available
   c. The attenuation stem-loop can form
   d. The CAP site is occupied

6. To which of the following protein family does steroid hormone receptors belong to:
a. Helix-loop-helix proteins
b. Helix-turn-Helix proteins
c. Leucine zipper proteins
d. Zinc finger proteins

7. All of the following are DNA binding domains found in transcription factors, Except:
a. Zinc finger
b. Helix-turn-Helix
c. Helix-loop-helix
d. Basic domains

8. Multiple proteins from a single gene may be generated by:
a. Alternative splicing
b. Mutations in the promoter of the gene
c. RNA interference
d. mRNA capping

9. When E. coli is grown in a medium containing both glucose and lactose, which of the following binds to the components of Lac operon?
a. Only CAP
b. Only Lac repressor
c. Both CAP and the lac repressor
d. Neither CAP nor the lac repressor

10. What is the approximate size of miRNA molecules?
a. 11 bp
b. 22 bp
c. 56 bp
d. 75 bp

11. Which of the following statement is TRUE about iron response element?
a. Stem-loop structure in the messenger RNA
b. Located at the 5′ untranslated region of transferrin mRNA
c. Located at the 3′ untranslated region of ferritin mRNA
d. IRE-binding protein causes degradation of IRE

12. Which of the following is NOT an epigenetic change?
a. Acetylation of histone
b. Methylation of DNA
c. Methylation of histone
d. Point mutation

13. All of the following are epigenetic changes, Except:
a. Genomic imprinting
b. Heterochromatin formation
c. Robertsonian translocation
d. X-chromosome inactivation

14. All of the following are true about histone modifications, Except:
a. Histone acetylation refers to addition of acetyl group to lysine of histone
b. Histone methylation refers to addition of methyl group to cytosine of histone
c. Histone phosphorylation refers to addition of phosphoryl group to serine of histone
d. Histone ubiquitination refers to addition of ubiquitin to lysine of histone

15. Which of the following sequence in our genome is usually found hypermethylated at most of the times?
a. Promoters
b. CpG islands
c. Enhancers
d. Operators

16. A defect in which of the following protein leads to Rett syndrome?
a. Histone acetyl transferase
b. Methyl cytosine binding protein
c. Ten Eleven translocase
d. DNA methylase

17. All of the following are true about microRNAs Except:
a. Type of non-coding RNA involved in regulation of gene expression
b. Works by binding to the 3′ untranslated region of mRNA
c. miRNA precursors contain cap as well as polyA tail
d. Binding of miRNA to mRNA always leads to destruction

18. All of the following are ways of regulation of gene expression in eukaryotes, Except:
a. Attenuation by Operon
b. Gene amplification
c. Gene rearrangement
d. Control of mRNA stability

ANSWERS					
1. b.	2. a.	3. b.	4. d.	5. d.	6. d.
7. c.	8. a.	9. d.	10. b.	11. a.	12. d.
13. c.	14. b.	15. b.	16. b.	17. d.	18. a.

Recombinant DNA Technology and Molecular Techniques

Chapter Outline

In the previous chapters, we have discussed a few techniques like karyotyping and PCR. In this chapter, we will discuss recombinant DNA technology and some more molecular techniques. As there are voluminous books available for detailed knowledge of molecular techniques, we will limit our discussion to a basic and exam-oriented level.

RECOMBINANT DNA TECHNOLOGY

Recombinant DNA technology (RDT) enables joining together of DNA molecules from two different species that are introduced into a host organism to produce new, useful genetic combinations that are of biological importance. Mass production of human proteins like Insulin, Interferons, clotting factor and certain vaccines are made possible by RDT.

The history of RDT began with the isolation of restriction enzymes (molecular scissors) by Smith and Nathans in 1970. This enabled Herb Boyer and Stanley Cohen to create the world's first recombinant DNA organism in 1973. *So, let us discuss restriction enzymes first.*

RESTRICTION ENDONUCLEASES

- ❑ Restriction enzymes (RE) also known as restriction endonucleases are a part of the innate immune system of bacteria.
- ❑ RE cleave the infecting bacteriophage (virus) DNA into pieces and protect the bacteria.
- ❑ Each RE recognizes a short, double-stranded DNA sequence (4 to 6 bp) which is usually palindromic.
- ❑ When the cleavage by the RE is symmetric, **blunt-ended or flush-ended DNA fragments** are produced.
- ❑ When the cleavage by RE is asymmetric, the resulting fragments can circularize. These ends are known as **sticky or cohesive** ends.

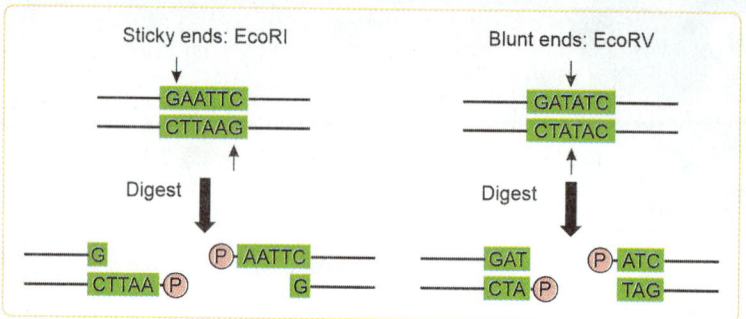

- Sticky ends can be converted to blunt (flush) ends with the use of 3'→5' exonuclease activity of **Klenow fragment** (*E. coli* DNA polymerase without 5'→3' exonuclease activity) or T4 DNA polymerase.
- Blunt ends can be converted to sticky ends by homopolymer tailing with the use of terminal transferase enzyme.

Nomenclature: Restriction enzymes are named after the bacteria from which they are derived.

For example, EcoRI – *Eco* denotes *E. coli*; *R* denotes the strain; Roman numeral I denotes that it is the first restriction enzyme identified in that particular strain of bacteria.

Some other restriction enzymes: BamHI, Hind III, Pac I, etc.

Types of Restriction Enzymes

Type I – Cleaves DNA at random sites far away from the recognition site
Type II – Cleaves DNA close to or within the recognition site; used in molecular biology.
Type III – Cleaves outside the recognition site (intermediate between type I and II)
Type IV – Cleaves modified (e.g. methylated) DNA
Isoschizomers are different restriction enzymes that recognize the same sequence and cleave at the same sites.
Neoschizomers are different restriction enzymes that recognize the same sequence but cleave at different sites.

Methylation of Adenine Protects the Bacterial Restriction Sites

- It is possible that bacteria may also have recognition sequence(s) for its own restriction enzymes.
- To prevent self-destruction, bacteria methylates adenine present in the recognition sites.
- Restriction enzyme-producing bacterium will die if it does not contain methylase activity.

Uses of Restriction Enzymes in Molecular Biology

1. Construction of DNA libraries and DNA cloning
2. DNA fingerprinting
3. Detection of Restriction Fragment Length Polymorphism (RFLP)
4. Generation of restriction map of a particular organism

DNA LIBRARY

Molecular cloning refers to the isolation of a DNA sequence (often a gene) from any species, and its insertion into a vector for propagation, without alteration of the original DNA sequence.

The human diploid genome is composed of 6 billion base pairs and isolation of a specific gene for the purpose of insertion into a vector each time is a laborious process. So, DNA libraries are used. There are two types of gene libraries – genomic library and cDNA library (Fig. 1).

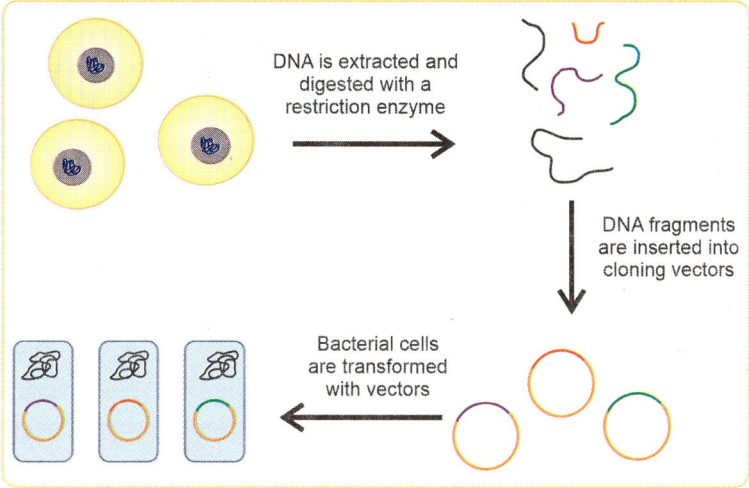

Fig. 1: Schematic overview of the construction of the genomic library by Aluquette - Own work, CC BY-SA 3.0, https://commons.wikimedia.org/w/index.php?curid=25760209

Two Libraries: cDNA Library versus Genomic Library

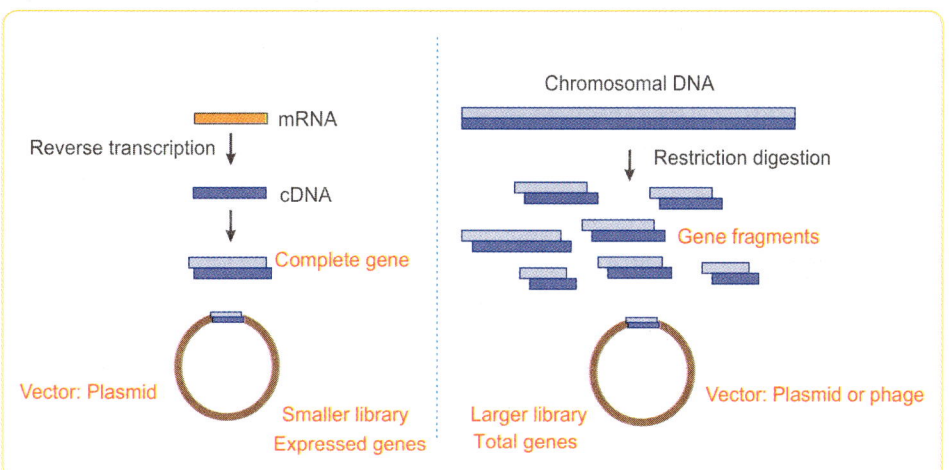

Genomic library	cDNA library
Genomic DNA isolated from an organism is digested by restriction enzymes into fragments and ligated (using DNA ligase) to vector and multiplied.	RNA is reverse transcribed into cDNA and ligated to vector and multiplied.
The total genome will be present in the genomic library.	Only those genes which are expressed will be present.
Non-coding sequences like introns, promoters will be present.	No promoters or introns.
Appropriate for small-sized genome	Appropriate for large genome
Useful for genome analysis, promoter studies, etc.	Useful for analysis of coding regions, gene studies, etc.

The presence of a particular sequence of a DNA in a genomic library is confirmed by Southern blotting technique. *Let us begin a brief discussion on blotting techniques.*

BLOTTING TECHNIQUES

❑ Blotting techniques are used for the detection of the presence of a particular sequence of DNA, RNA or protein in the given sample.
❑ The word blotting refers to the transfer and immobilization of molecules of interest (nucleic acids or proteins) onto a solid support (Nylon, Nitrocellulose or Polyvinylidene difluoride membrane).

Steps Common to all the Blotting Techniques

1. Electrophoretic separation of proteins or nucleic acids
2. Transfer and immobilization of proteins or nucleic acids to the membrane
3. Addition of labelled probe (Antibody for proteins, complementary oligonucleotides for nucleic acids) to the target molecule on the membrane. The ^{32}P radiolabel is used for nucleic acids. ^{125}I radioisotope and chemiluminescent label are added to antibodies.
4. Visualization of the bound probes on the membrane using autoradiography or chemiluminescent detection.

Blotting technique	Detection of
Southern	DNA
Northern	RNA
Western	Protein
Far Western blot	Protein-Protein interaction
Eastern	Post-translational modifications (Glycosylation, Lipidation)
South-Western	DNA-Protein interaction

South – Dosa (DNA); North – Roti (RNA); West – Pizza (Protein)

Southern Blotting

❑ Southern blotting is a technique to detect the presence of a **particular sequence of DNA.**
❑ Invented by Edwin Southern. *Western blot and northern blots are just named like that. These are not invented by western or northern (Fig. 2)!*

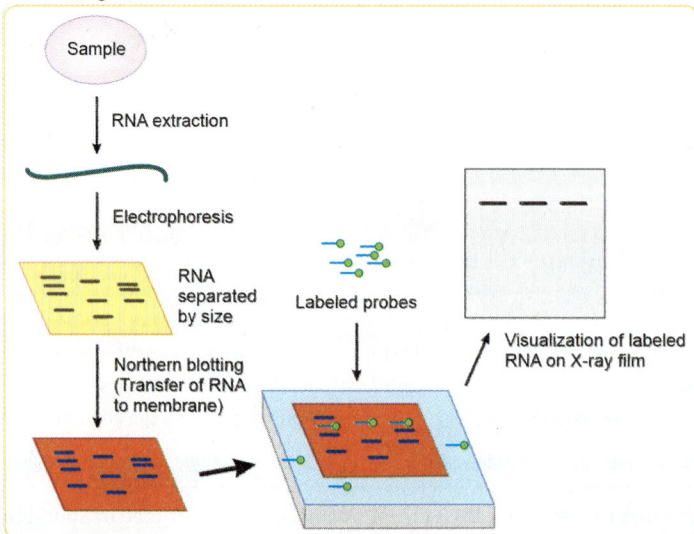

Fig. 2: Schematic illustration of Northern blotting. In Southern blotting, the samples are DNA instead RNA

Section III Molecular Biology

Applications of Southern Blotting

- Identification of a specific gene sequence among thousands of fragments of DNA
- Diagnosis of infectious disease like HIV, HCV infection
- Diagnosis of drug resistance genes e.g. Multi drug-resistant tuberculosis (MDRTB)
- Identification of gene mutations, deletions, duplications, and gene rearrangements involved in diseases

Now, let us go back to the discussion of how recombinant DNA is created.

FORMATION OF CHIMERIC DNA

- Hybrid or Chimeric DNA is the recombinant DNA containing DNA from 2 different organisms – usually human and bacteria.
- DNAs of both organisms are cut with the same restriction endonuclease and sealed with DNA ligase.
- If the fragments are of sticky ends, there is a chance of self-ligation. So, the 5′ phosphate group is removed using the alkaline phosphatase enzyme.
- Later, polynucleotide kinase is added to phosphorylate the 5′ OH group.
- Ligation is achieved using the enzyme DNA ligase which requires ATP.

CLONING VECTORS

A cloning vector is a DNA molecule that is used to carry the desired (foreign) gene into a host cell which can replicate inside the bacterial or yeast cell and produces many copies of itself and the desired gene.

Essential features of the cloning vector:

Feature	Purpose
Should possess the origin of replication (Ori) sequence	Enables autonomous replication inside the host
Should contain multiple cloning sites (restriction sites) also known as polylinker that can be cleaved by multiple restriction enzymes	Enables insertion of foreign DNA
Marker genes like antibiotic resistance genes	Enables in selecting transformed host-cells

Example: **pBR322** is an artificially constructed plasmid vector that contains Ori sequence, 40 restriction sites (multiple cloning sites) and 2 marker genes (tetracycline and ampicillin resistance genes).

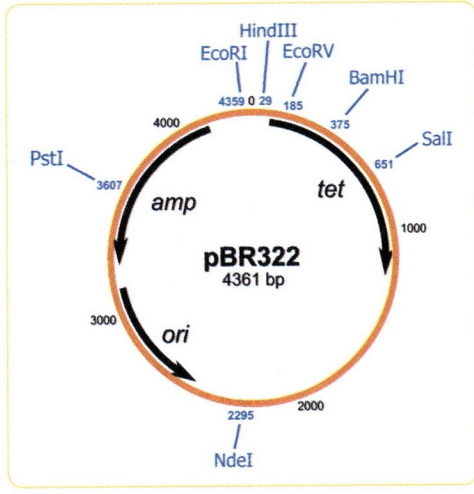

Nomenclature of pBR322
- *p – denotes that it is a plasmid.*

Chapter 28 Recombinant DNA Technology and Molecular Techniques

- BR – stands for Boliver and Rodriguez, the laboratory developed this plasmid.
- 322 – is a number given to distinguish this plasmid from others developed in the same laboratory, e.g., pBR325, pBR327, pBR328, etc. are other plasmids.

Some Commonly used Cloning Vectors

Vector	Description	DNA insert size (kb)
Plasmid	Extrachromosomal circular DNA molecule that autonomously replicates inside the bacterial cell	0.01–10
λ phage	Part of phage DNA can be replaced by foreign DNA without affecting the life cycle of bacteriophage.	10–20
Cosmid	Extrachromosomal circular DNA molecule that combines features of plasmids and phage	35–50
Bacterial artificial chromosomes (BAC)	Artificially constructed DNA based on bacterial fertility (F) plasmid	75–300
Yeast artificial chromosomes (YAC)	The artificial chromosome that contains telomeres, the origin of replication, a yeast centromere, and a selectable marker for identification in yeast cells	100–1000 kb

Expression Vector

- Expression vectors are different from cloning vectors. The protein-coding genes introduced in the expression vectors are transcribed and translated, hence the name expression vector.
- In addition to the sequences found in the cloning vector, expression vectors should contain sequences like a strong promoter (viral promoters), ribosomal entry sites (Kozak's sequence).

INTRODUCTION OF RECOMBINANT PLASMIDS

Once the recombinant plasmid has been made, it is introduced into a bacterium. The process by which foreign DNA is introduced into a bacterial cell is known as transformation.

> **Know the difference:**
> **Transformation**: Cloning in prokaryotes
> **Transfection**: Cloning in Eukaryotic cells (mnemonic: E in transfection, Eukaryotes)

Methods for Introduction of a Transgene

- Incubating the bacterial cells under cold conditions in the presence of a divalent cation (e.g. Ca^{++}) along with the plasmid makes the bacterial cells **competent** for exogenous DNA intake. When a heat shock is given, the divalent cations generate coordination complexes with the negatively charged DNA molecules, facilitating the uptake of exogenous DNA.
- **Electroporation** in which the cells are briefly shocked with an electric pulse. This leads to transient pore formation in the plasma membrane, through which the plasmid can enter the cell; the pore is later repaired by various repair mechanisms.
- **Microinjection** using a glass pipette is usually used in somatic nuclear cell transfer to inject the male pronucleus into the fertilized egg. This is utilized in reproductive cloning or therapeutic cloning.
- **Gene gun** is a biolistic particle delivery system used **mainly in plants** to deliver exogenous DNA.

SELECTION OF RECOMBINANT PLASMIDS

- The transformed bacteria are then grown in the presence of an appropriate marker, so as to select the colonies containing the plasmid with gene of the insert.

- For example, if the recombinant plasmid contains the Ampicillin resistance gene, they are grown in medium containing Ampicillin. Bacteria without the recombinant plasmid will die.
- Another method for selection of appropriate colonies with recombinant plasmids is blue-white screening.

APPLICATIONS OF MOLECULAR CLONING

- **Production of recombinant proteins** like Insulin, erythropoietin, clotting factors, etc. and vaccines like hepatitis B vaccine which contains HBsAg (Hepatitis B surface Antigen).
- **Production of DNA library**: DNA library is a collection of DNA fragments that have been cloned into vectors so that researchers can identify and isolate the DNA fragments that interest them for further study.
- **Site-directed mutagenesis**: The sequence of a protein can be altered at specific amino acid residues by mutating the specific codons in the coding sequence which can be then cloned and expressed. This is helpful in understanding the pathogenesis of genetic diseases, mapping the active site amino acids etc.
- **Gene therapy**: Preparation of lentiviral/adenoviral vectors with the correct insert coding for the protein to be expressed involves molecular cloning.

 High Yield

Tools of RDT

Tool	Role
Type II restriction enzymes	Cleaves DNA at specific sites
DNA ligase	Phosphodiester bond formation between two DNA fragments.
Klenow fragment (DNA polymerase I without 5' to 3' exonuclease activity)	Used in PCR reactions before the advent of Taq polymerase
Polynucleotide kinase	Catalyses the transfer of the γ-phosphate from ATP to the 5′-hydroxyl terminus of double and single-stranded RNA and DNA or oligonucleotides.
Shrimp Alkaline phosphatase	Removes the phosphate group from 5' end. This prevents the self-annealing of sticky ends of DNA produced by restriction enzymes.
Reverse transcriptase	RNA dependent DNA polymerase used for cDNA synthesis.
RNase H	Removes RNA from DNA-RNA hybrid
Exonuclease III	Removes nucleotides from 3' end.
Phage λ exonuclease	Removes nucleotides from 5' end.
Terminal nucleotidyltransferase	Adds multiple nucleotides to the 3' end. Used in homopolymer tailing – a technique to convert blunt ends to sticky ends.

This finishes the discussion of RDT. Now, we will start the discussion of other various molecular techniques.

RESTRICTION FRAGMENT LENGTH POLYMORPHISM

- In Chapter 20 *(Refer to page no. 373)*, we have seen that the human genome contains around 14 million SNPs
- These SNPs can create a new restriction sequence or abolish an existing restriction sequence thus the length of restriction fragments varies due to the SNP.
- Presence of VNTR and STR *(Chapter 20)* *(Refer to page no. 376)* will also change the length of restriction fragments.

- **Restriction fragment length polymorphism refers to an inherited difference in the pattern of restriction enzyme digestion.**

  RFLPs are produced by the following changes in recognition sites:
  - Single base changes – mutations or SNPs
  - Insertion or deletions (INDELs)
  - Variable number of tandem repeats (VNTR)

- In RFLP analysis, the DNA fragment is digested by one or more restriction endonucleases; depending on the presence or absence of the restriction sites for the specific restriction endonucleases, the nucleotide changes in the DNA can be profiled.

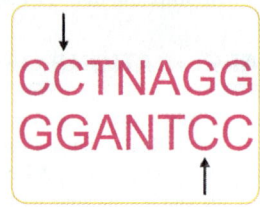

  Let us see an example:

- The genetic defect in sickle cell anemia is a point mutation (transversion) where T is substituted with A in the coding region of the β-haemoglobin gene. This substitution results in a change in the codon from GAG (glutamic acid) to GUG (valine).

  The figure below shows the restriction site of *MstII* corresponding to the 6th position of beta globin. 'N' in the sequence refers to any nucleotide.

  The DNA in and around the β-globin gene has 3 restriction sites for *MstII*.

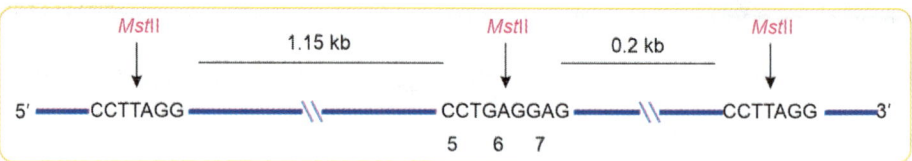

One of these restriction sites is lost in sickle cell anemia (GAG to GTG, i.e. Glutamate to Valine).

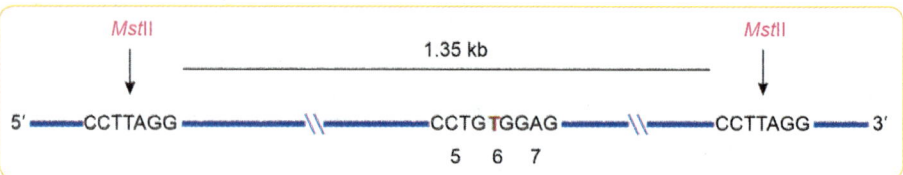

Thus, the restriction pattern will be different in a DNA sample from a normal individual, compared to those with sickle cell trait or sickle cell disease. This can be detected using Southern Blot or PCR (Fig. 3).

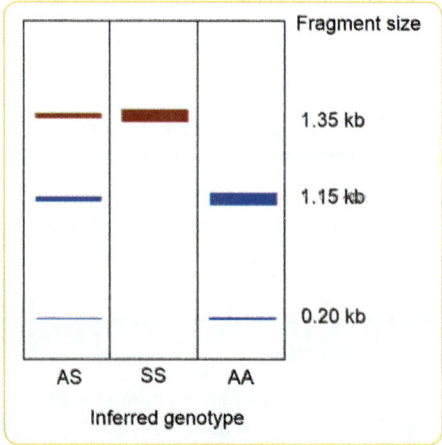

Fig. 3: Mst II restriction pattern of beta-globin gene in sickle cell trait (AS), sickle cell disease (SS) and a normal individual (AA), when detected using Southern blotting

PCR-RFLP

In PCR-RFLP, PCR is performed to amplify a particular region of DNA and the PCR products are subjected to digestion using restriction endonucleases. Since the sequence of interest in the DNA is amplified to significance amounts, the products can be visualized under UV light after an agarose gel electrophoresis, thus avoiding the radioactive hazards that are involved in Southern blotting.

DNA FINGERPRINTING

DNA fingerprinting was first established by Alec Jeffreys. In this technique, genomic DNA is subjected to digestion by a specific set of restriction endonucleases. Humans share a genome sequence similarity of 99.9%. This means the DNA sequence of two individuals differs in 1 in every 1000 nucleotides. Some of these sites that differ may be restriction sites. So, the restriction pattern of DNA from two individuals generally differs. This is another example of RFLP.

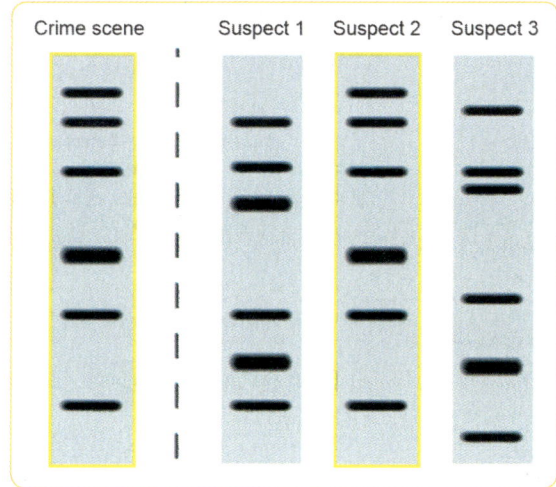

DNA fingerprinting was used to establish a link between biological evidence and a suspect in a criminal investigation. A DNA sample taken from a crime scene is compared with a DNA sample from a suspect. If the two DNA profiles are a match, then the evidence came from that suspect. Conversely, if the two DNA profiles do not match, then the evidence cannot have come from the suspect.

Since RFLPs are inherited, DNA fingerprinting can also be used to establish paternity.

The procedure is almost the same as the RFLP analysis using Southern Blot which we discussed earlier, except for the fact that multiple enzymes are used here, and different samples are compared.

Note: For DNA fingerprinting, RFLP using VNTR has been replaced by the analysis of Microsatellite repeats/ STR (short tandem repeats) (Chapter 20) *(Refer to page no. 376)* **and DNA sequencing in forensic science.**

DNA SEQUENCING

SANGER SEQUENCING

- ❑ Primary structure of DNA, i.e. sequence of nucleotides from 5′ to 3′ end is achieved by DNA sequencing technique.
- ❑ Frederick Sanger got his second Nobel Prize for his work on DNA Sequencing.
- ❑ Sanger's sequencing (Chain termination method) is based on the controlled termination of replication.
- ❑ Deoxynucleotides lack oxygen at 2′ carbon of ribose and Dideoxynucleotides lack both 2′ and 3′ oxygen.

- As you know, DNA polymerase forms a phosphodiester bond between 5' OH group of incoming nucleotide and 3' OH group of existing nucleotide chain.
- If a dideoxynucleotide is incorporated in the growing DNA chain by DNA polymerase, the incoming nucleotide can't be added, i.e. chain termination happens.
- Sanger technique requires DNA primer, DNA polymerase, deoxy NTPs, dideoxy NTPs, Magnesium ions. Sanger sequencing is automated with the use of fluorescently tagged dideoxy NTP and capillary electrophoresis.

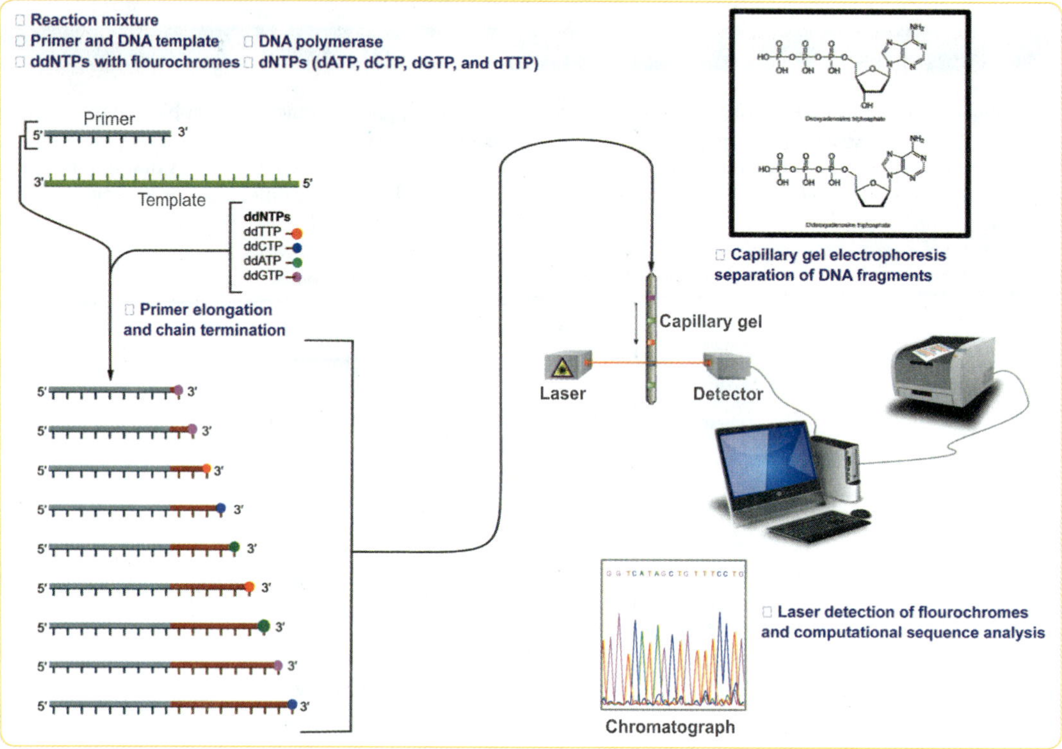

□ Reaction mixture
□ Primer and DNA template   □ DNA polymerase
□ ddNTPs with flourochromes □ dNTPs (dATP, dCTP, dGTP, and dTTP)

Primer
5' 3'

3' 5'
Template

ddNTPs
ddTTP
ddCTP
ddATP
ddGTP

□ Primer elongation and chain termination

5' 3'
5' 3'
5' 3'
5' 3'
5' 3'
5' 3'
5' 3'
5' 3'
5' 3'

Laser

Capillary gel

Detector

□ Capillary gel electrophoresis separation of DNA fragments

□ Laser detection of flourochromes and computational sequence analysis

Chromatograph

NEXT-GENERATION SEQUENCING

Next-Generation Sequencing (NGS) refers to all methods other than Sanger sequencing which uses ultra-high throughput, massively parallel sequencing methods. NGS technology has made sequencing cost lower and has opened the gateway for numerous possibilities. One such possibility is personalized/precision medicine, i.e. the use of genomic data to tailor medical care to individuals based on their genetic makeup. *More details of this topic are beyond the scope of this book.*

GENE EDITING

- **Gene knockout:** Permanent disruption of a gene
- **Gene knockdown:** Temporary inhibition of expression of the gene using siRNA mediated RNA interference (RNAi)
- **Gene knock-in:** Introduction of a new gene or replacement of a gene with other

GENE EDITING TOOLS

Gene editing is required for the creation of transgenic organisms—transgenic bacteria, plants and animals.

Section III Molecular Biology

- ❑ PCR-mediated site-directed mutagenesis is the technique to change the DNA sequence of a particular gene in bacteria and yeast.
- ❑ For eukaryotic cells, recombination-based methods like Zinc finger nucleases, TALENs (transcription activator-like effector nucleases), Cre-Lox and CRISPR/Cas9 are used.

CRISPR/Cas9 System

CRISPR is Clustered Regularly Interspersed Short Palindromic Repeats. **Cas9 is** CRISPR associated endonuclease. This is a part of the bacterial immune system. CRISPR and Cas9 act together to degrade the target DNA, i.e. bacteriophage DNA. This system is modified and **used for targeted genome editing (see p. 396).**

REVIEW

Technique	Purpose
Polymerase chain reaction *(Chapter 22)* *(Refer to page no. 400)*	Amplification of DNA
DNA sequencing	To find out mutations and single nucleotide polymorphism
RNA sequencing	Quantification of gene expression
Southern blotting	To find and document the presence of a particular sequence of DNA in the sample using a probe
Fluorescence in situ Hybridization *(Chapter 19) (Refer to page no. 366)*	To identify the positions of genes on chromosomes and analyze chromosomal abnormalities
RNA microarray	To quantify gene expression
SNP array	To find the common single nucleotide polymorphisms in the genome.
DNA footprinting	To identify the sequence of DNA that bind to proteins (DNA-protein interaction) *in vitro*.
Chromatin Immunoprecipitation-chip (ChIP-chip)	To identify the *in vivo* binding sites of specific proteins in the chromatin. (ChIP followed by DNA microarray)
ChIP-Seq	To identify the exact sequence of DNA binding sites of specific proteins in chromatin. (ChIP followed by high-throughput next-generation sequencing)
CRISPR/Cas	Targeted genome editing
DNA fingerprinting	To establish the identity of a person using specific DNA fragments – RFLP, VNTR, Short Tandem Repeats
Gene knockout	Complete ablation of expression of the gene
Gene knockdown	Reduction in the amount of gene expression
Gene knock-in	Introduction of a gene into a particular location of the genome

Techniques to quantify gene expression through RNA quantification	❑ Real-time PCR ❑ RNA microarrays
Techniques to quantify gene expression through protein quantification	❑ Western blot ❑ Immunohistochemistry ❑ Protein microarray
Techniques to detect DNA-protein interaction	❑ DNAse footprinting ❑ Electrophoretic mobility shift assay (EMSA) also known as Gel shift assay ❑ ChIP-sequencing ❑ Yeast one-hybrid system
Techniques to detect Protein-protein interaction	❑ Co-immunoprecipitation ❑ Surface plasmon resonance ❑ Fluorescence resonance energy transfer ❑ Yeast Two-hybrid system

Competency

☐ BI7.4   Describe applications of molecular technologies like recombinant DNA technology, PCR in the diagnosis and treatment of diseases with genetic basis

Key Points

☐ Blotting techniques are used for the detection of the presence of a particular sequence of DNA, RNA or protein in the given sample.

☐ Western, Southern and Northern blotting techniques are used for the detection of protein, DNA and RNA respectively.

☐ Hybrid or Chimeric DNA is the recombinant DNA containing DNA from two different organisms – usually human and bacteria.

☐ After digestion by restriction endonucleases, DNA strands can be joined again by DNA ligase.

☐ Blue-white screening is used to identify the bacterial clone containing the desired gene insert.

☐ Restriction fragment length polymorphism refers to an inherited difference in the pattern of restriction enzyme digestion.

☐ Comparative genomic hybridization (microarray) can find out the differential expression of genes in normal and cancer cells.

☐ Microarray is used for detection of variation in DNA sequence and Gene expression.

☐ Next-Generation Sequencing (NGS) refers to all methods other than Sanger sequencing which uses ultra-high throughput, massively parallel sequencing methods.

☐ CRISPR/Cas9 system is used for targeted genome editing.

SELF-ASSESSMENT

Short Answer Questions

1. Classify stem cells based on their dividing capacity. Give one example for each.
2. Discuss how induced pluripotent cells can revolutionize the field of human organ transplantation.
3. With a schematic diagram, illustrate the Somatic Cell Nuclear Transfer Technique.
4. Can type 2 Diabetes Mellitus be treated by gene therapy? Justify your answer.
5. What is antisense therapy? Give one example.

Multiple Choice Questions

1. **How does restriction enzymes protect the bacteria against viral infection?**
   a. By cleaving the bacterial DNA
   b. By cleaving the DNA of infecting viruses
   c. By methylating the viral DNA
   d. By methylating the viral DNA

2. **Which of the following is the correct sequence of the procedure followed in southern blot?**
   a. Electrophoretic separation of DNA → Hybridization with probe → Transfer and immobilization
   b. Electrophoretic separation of DNA → Transfer and immobilization → Hybridization with a probe
   c. Hybridization with probe → Transfer and immobilization → Electrophoretic separation
   d. Transfer and immobilization of proteins → Hybridization with probe → Electrophoretic separation of DNA

3. **Which of the following is the reason for which ddNTPs are used in sanger DNA sequencing?**
   a. ddNTPs are less preferred over dNTPs by DNA polymerase
   b. ddNTPs are preferred over dNTPs by DNA polymerase
   c. ddNTPs can be made fluorescent
   d. ddNTPs prevent the incorporation of next nucleotide by DNA polymerase

4. **Blue-white screening is used for:**
   a. Quantification of gene expression
   b. Detection of DNA-protein interaction
   c. Detection the bacterial colonies with recombinant plasmid
   d. Detection of the mutagenic potential of a chemical

5. **Which of the following is the best method among the given for DNA fingerprinting?**
   a. RFLP-PCR for SNPs
   b. RLP-PCR for VNTR
   c. Short tandem repeat (STR) analysis
   d. RFLP followed by Southern blot

6. **Which of the following is the true statement about restriction enzymes?**
   a. Are produced by viruses
   b. Cut only single-stranded DNA
   c. Can cut the DNA anywhere
   d. Restrict the expression of viral DNA

7. **What strategy in transcription factor research allows for the simultaneous identification of all of the genomic sites bound by a given transcription factor under a given set of physiological conditions?**
   a. Fluorescence energy transfer (FRET)
   b. DNAse I sensitivity
   c. Chromatin immunoprecipitation-sequencing (ChIP-seq)
   d. Flourescence insitu Hybridisation (FISH)

8. **HindIII is a restriction endonuclease. Which of the following is most likely to be the recognition sequence for this enzyme?**
   a. AAGAAG   b. AAGAGA
   c. AAGCTT   d. AAGGAA

9. **Klenow fragment lacks:**
   a. 5'→3' polymerase   b. 3'→5' exonuclease
   c. 5'→3' exonuclease d. All of the above

10. **In which among of the following vector, largest DNA segments can be cloned?**
    a. Plasmids
    b. Bacteriophage
    c. Cosmids
    d. Bacterial Artificial Chromosomes

11. **Southern blot can be done using ___ as samples.**
    a. DNA
    b. RNA
    c. Protein
    d. Carbohydrate

12. **CRISPR/Cas9 mediated gene editing involves which of the following DNA repair mechanisms?**
    a. Base excision repair
    b. Nucleotide excision repair
    c. Non-homologous end joining
    d. Mismatch repair

13. **The enzyme used to "flush" the sticky ends of DNA is:**
    a. Klenow fragment
    b. Polynucleotide kinase
    c. Alkaline phosphatase
    d. Primase

14. **All of the following methods can be used to introduce a gene into cells, Except:**
    a. Transfection of competent cells
    b. Electroporation
    c. In situ hybridization
    d. Microinjection

15. **All of the following enzymes are used in recombinant DNA technology, Except:**
    a. Alkaline phosphatase
    b. Alanine transaminase
    c. Reverse transcriptase
    d. DNA ligase

16. **Which of the following is necessarily to be present in expression vector but not in cloning vector?**
    a. Origin of replication
    b. Restriction site
    c. Ribosomal entry site
    d. Selectable marker

ANSWERS

1. b.	2. b.	3. d.	4. c.	5. c.	6. d.
7. c.	8. c.	9. c.	10. d.	11. a.	12. c.
13. a.	14. c.	15. b.	16. c.		

Stem Cells and Gene Therapy

STEM CELLS

Cells present in our body can be classified into three types based on their dividing capacity.

Type	Description	Example
Labile cells	**Continuously dividing** cells	Hematopoietic cells, Epithelial cells of skin and GIT, etc.
Stable cells	**Quiescent cells that** multiply only when needed. They are at the G_0 phase of cell cycle. When stimulated, they can enter the G1 phase.	Hepatocytes, Vascular endothelial cells, Smooth muscle, etc.
Permanent cells	**Non-dividing** cells which have exited the cell cycle forever.	Neurons and cardiac myocytes

STEM CELLS

Stem cells are the undifferentiated cells that can form any cell type (**differentiation**) as well as to replicate into more stem cells (**proliferation**). Telomerase is responsible for the long-term self-renewal of stem cells.

Observe the asymmetric division - A stem cell can be divided into two cells with different property, i.e. a mature cell and a stem cell.

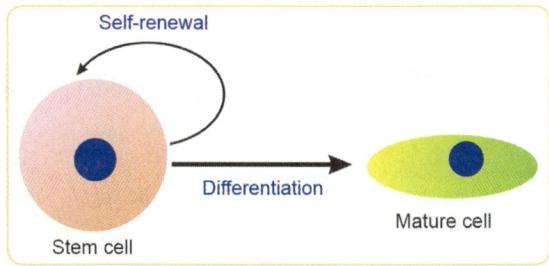

Self-renewal

Differentiation

Stem cell

Mature cell

CLASSIFICATION OF STEM CELLS

Based on the Differentiation Potential

Type	Differentiate into
Totipotent cells	All cell types, including **placenta**
Pluripotent cells	All cell types, except placenta
Multipotent cells	Limited lineages
Unipotent	Single lineage

Based on the Source

Type	Obtained from	Potential
Embryonic stem cells (ESC)	Inner cell mass of blastocyst-derived from unused embryos in vitro fertilization clinics. (Ethical issues are there)	Totipotent
Adult/somatic stem cells	Bone marrow	Multipotent
Induced pluripotent stem cells (iPSC)	Somatic cells treated with transcription factors	Pluripotent

Somatic Cell Nuclear Transfer Technique (SCNT) is used to Obtain Embryonic Stem Cells

Unused embryos from IVF clinics are the most common source of embryonic stem cells but to harvest the embryonic stem cell creates an ethical problem.

In SCNT, the nucleus of the somatic cell is removed and the rest of the cell discarded. At the same time, the nucleus of an egg cell is removed. The nucleus of the somatic cell is then inserted into the enucleated oocyte. After being inserted into the oocyte, the somatic cell nucleus is reprogrammed by the cytoplasmic factors of the host cell.

These cytoplasmic factors erase the molecular memory of the somatic cells. Such cells can then be used to harvest pluripotent stem cells or to transfer a blastocyst into a carrier and develop an organism in vivo.

INDUCED PLURIPOTENT STEMS CELLS

❑ Shinya Yamanaka of Kyoto University pioneered the technique of induced pluripotent stem cells and was awarded the Nobel Prize in the year 2013.
❑ Mature, terminally differentiated adult cells like fibroblasts on the introduction of specific genes (c-myc, Sox2, Klf4, Oct4) through viral vectors developed stem-cells like property.
❑ These pluripotent stem cells can be differentiated into desired cell line using specific cytokine growth factors.
❑ This technique surpasses the ethical constraints as no embryonic cells are involved.
❑ The main limitation is the carcinogenic potential of the Yamanaka factors (c-myc, Sox2, Klf4, Oct4). All are oncogenes.

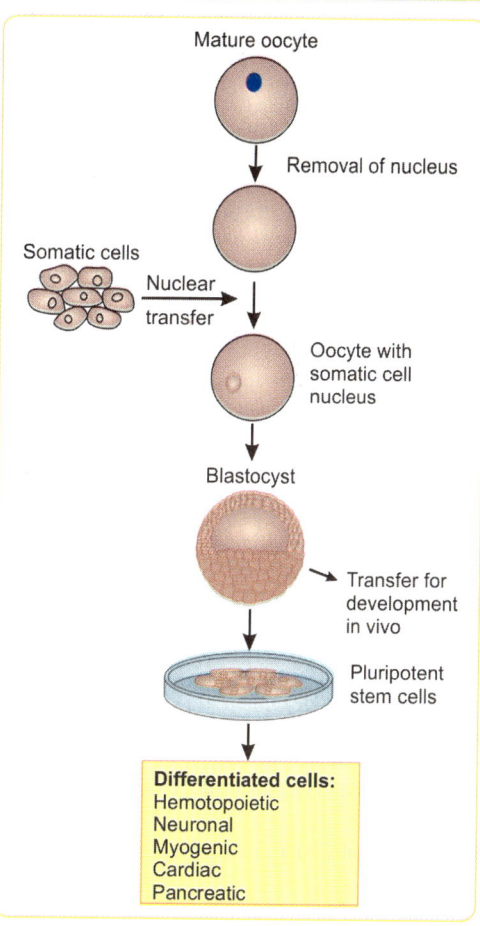

Mature oocyte

Removal of nucleus

Somatic cells

Nuclear transfer

Oocyte with somatic cell nucleus

Blastocyst

Transfer for development in vivo

Pluripotent stem cells

Differentiated cells:
Hemotopoietic
Neuronal
Myogenic
Cardiac
Pancreatic

Chapter 29 Stem Cells and Gene Therapy

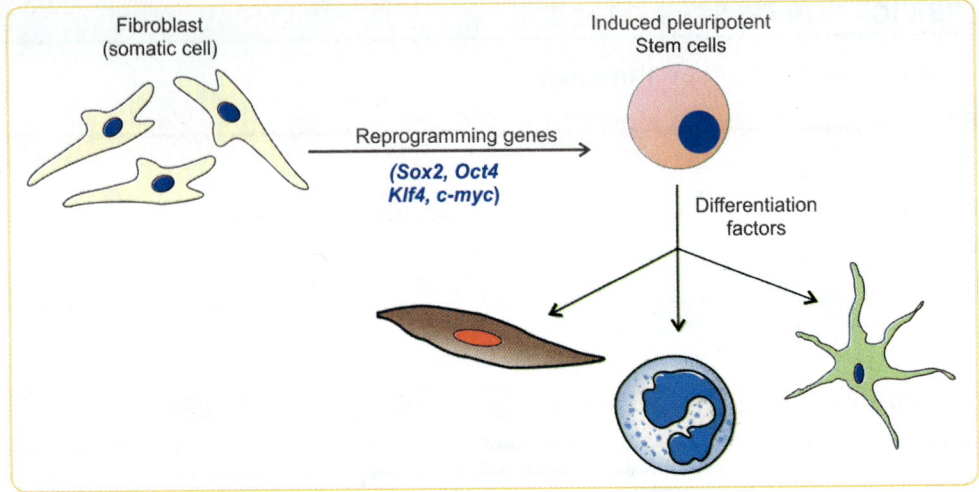

GENE THERAPY

Gene therapy is the treatment of a genetic disease by insertion of a missing gene or mutated gene.

Single gene disorders are best candidates for gene therapy.

Adenosine deaminase deficiency SCID was the first disease treated by gene therapy.

Types of Gene Therapy

Based on the type of cells in which gene is inserted:
- Germline gene therapy–illegal
- Somatic gene therapy

Based on the effect of gene therapy:
- Gene augmentation–simple addition of gene
- Gene replacement also known as corrective gene therapy—based on homologous recombinant repair that will replace the malfunctioning gene causing dominant negative effect
- Specific inhibition of gene expression using antisense mRNA, siRNA and ribozymes
- Targeted cell death—Toxic genes are introduced explicitly into cancer cells to cause death of cancer cells

Steps in Gene Therapy

- Select the gene of interest
- Insert it into vectors—Viral vectors are the most desirable
- Introduction into the target cells

Viruses in Gene Therapy

Viral vectors must be replication-defective to prevent uncontrolled spreading in the body. They should not possess undesirable properties.

Viral vectors can be divided into:
- Retro virus: Malone Murine Leukemia Virus (MMLV)
- Non-retro viruses: Adeno virus, Adeno Associated Virus, Herpès simplex virus

Viral Vectors	
Advantages	Disadvantages
❑ Better uptake of genes. ❑ Safe as the virus is replication deficient. ❑ Suitable for a variety of diseases.	❑ Random insertion of a gene can disrupt the function normal genes; i.e. Insertional mutagenesis.

Herpes virus is the preferred vector for transferring genes to nerve cells.

Delivery Methods for Viral Vectors

Ex Vivo Approach

❑ Cells (fibroblasts, Hematopoietic cells) extracted from the patient are grown in vitro in culture medium.
❑ Viral vector is used on the cultured cells, and genetically modified cells are reintroduced into the body.
❑ Retrovirus like Malone Murine Leukemia Virus (MMLV) is most suitable for this delivery.

In Vivo Approach

❑ The viral vector is directly allowed to infect the affected cells, or non-viral system is used.
❑ **Adenovirus** is most suitable for viral vector this type of delivery.

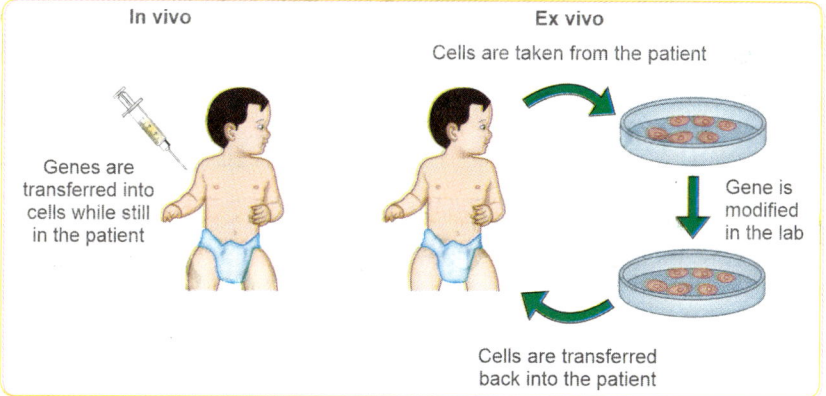

In vivo — Genes are transferred into cells while still in the patient

Ex vivo — Cells are taken from the patient — Gene is modified in the lab — Cells are transferred back into the patient

High Yield

Gene transfer techniques:

❑ **Transformation:** Introduction of genetic materials **into bacteria**
❑ **Transfection:** Introduction of genetic materials **into eukaryotic cells** (e.g. fungi, plant, or animal cells)
❑ **Transduction:** Introduction of genetic materials **using viruses**
❑ **Lipofection:** Introduction of genetic materials **using liposomes**

Retroviruses in Gene Therapy

Retroviruses are RNA viruses. Inside the host cells, viral RNA is converted to DNA by viral reverse transcriptase and incorporated into the host genome with the help of viral integrase.

The maximum size of the insert that can be carried by the retroviral vector is 36 kb.

Retro viral vectors in use:
❑ Malone Murine Leukemia Virus (MMLV)
❑ HIV-1 derived lentivirus

Limitation

❑ Site of integration cannot be controlled → insertional mutagenesis
❑ Cannot be used for non-dividing cells

Adenovirus in Gene Therapy

❑ dsDNA virus, replication-deficient
❑ Commonly used for in vivo approach
❑ Doesn't integrate into the genome; thus, there is no chance of insertional mutagenesis
❑ Remain as episome (extrachromosomal DNA)
❑ The maximum size of the insert that can be carried by an adenoviral vector is 36 kb.

Limitation

❑ The effect is transient
❑ Adenovirus capsid elicits a strong immune response *(An American teen named Jesse Gelsinger succumbed to adenoviral immune response during his clinical trial for OTC deficiency in the year of 1999. This was the first documented death in gene therapy clinical trial and created a huge public outcry regarding the safety of gene therapy in that country.)*

Comparison of Adenovirus and Retrovirus

❑ MMLV use rapidly dividing cells as targets. So, non-dividing cells cannot be used as target cells.
❑ Adenovirus can infect non-dividing cells also. But gene expression declines gradually.

ANTISENSE THERAPY (NON-VIRAL GENE THERAPY)

Antisense oligonucleotides are being developed for many genetic diseases. Some of the FDA approved drugs are:

Drug	Mechanism	Use
Fomivirsen	Inhibits CMV replication through interaction with CMV mRNA	Cytomegalovirus retinitis
Mipomersen	Antisense RNA for apolipoprotein B-100 mRNA	Familial hypercholesterolemia
Eteplirsen	Binds to the mutated portion of dystrophin mRNA and causes exon skipping during translation. This exon is not crucial for the function of the protein.	Duchenne muscular dystrophy

Key Points

❑ Stem cells are the undifferentiated cells that can form any cell type (differentiation) as well as to replicate into more stem cells (proliferation).
❑ Induced pluripotent stem cells are derived by introducing certain genes (*c-myc, Sox2, Klf4, Oct4*) into adult somatic cells.
❑ Adenosine deaminase deficiency SCID was the first disease treated by gene therapy.
❑ Adenovirus and retroviruses are commonly used as viral vectors in gene therapy.
❑ Fomivirsen, Mipomersen and Eteplirsen are examples of antisense therapy (non-viral gene therapy).

SELF-ASSESSMENT

Short Answer Questions

1. Classify stem cells based on their dividing capacity. Give one example for each.

2. Discuss how induced pluripotent cells can revolutionize the field of human organ transplantation?

3. With a schematic diagram, illustrate the Somatic Cell Nuclear Transfer Technique.

4. Can type 2 Diabetes Mellitus be treated by gene therapy? Justify your answer.

5. What is antisense therapy? Give one example.

Multiple Choice Questions

1. **Which of the following is an example of the stable cell?**
   a. Hematopoietic cells
   b. Vascular endothelial cells
   c. Intestinal epithelial cells
   d. Neurons

2. **All are labile cells, except:**
   a. Kupffer cells   b. Intestinal cells
   c. Keratinocytes   d. Hepatocytes

3. **A fertilized egg can develop into all the types of cells in the body and is therefore considered to be:**
   a. Multipotent   b. Pluripotent
   c. Totipotent   d. Unipotent

4. **What of the following is the most common source of embryos for stem cell harvest?**
   a. Cells cultured in Petri dishes
   b. Embryos removed from animals
   c. Unused embryos from IVF clinics
   d. Unused sperm frozen from sperm banks

5. **Which one of the following genes was NOT a part of transcription factors used to generate induced pluripotent stem (iPS) cells from mouse skin fibroblasts by Shinya Yamanaka?**
   a. Oct4   b. Sox2
   c. c-jun   d. Kfl4

6. **Which is the first human disease targeted by gene therapy?**
   a. Cystic fibrosis   b. Hemophilia
   c. Thalassemia
   d. Severe combined immunodeficiency disease

7. **Which of the following is the most common gene delivery system for In vivo gene therapy?**
   a. Micro injection   b. lipofection
   c. Adeno viral vectors d. Electroporation

8. **All of the following are true regarding gene augmentation therapy, except:**
   a. Suitable for single gene disorders
   b. Random insertion of a healthy gene
   c. Suitable for multi-gene disorders
   d. No recombination event is required

9. **The possibility of introducing correct version of the defective gene in all cells of the individual is achieved by:**
   a. Germ-Line Therapy
   b. Somatic Cell Therapy
   c. Gene Augmentation Therapy
   d. Corrective Gene Therapy

10. **A nasal spray containing adenovirus carrying a functional human CFTR gene is used to treat cystic fibrosis. This is an example of ____ gene therapy.**
    a. in situ   b. in vivo
    c. ex vivo   d. vaccine

ANSWERS

1. b.	2. d.	3. c.	4. c.	5. c.	6. d.
7. c.	8. c.	9. a.	10. d.		

Chapter 29 Stem Cells and Gene Therapy

Medical Genetics

Medical genetics involves the application of genetics to medical care, including research on the causes and inheritance of genetic disorders, and their diagnosis and management.

AN ALLELE IS AN ALTERNATIVE FORM OF A GENE

An allele is an alternative form of a gene that is located at a specific position on a specific chromosome.

Usually, we inherit two alleles for any gene, each from father and mother. There are some exceptions to this, e.g. there are two genetic loci for α globin, and thus four genes in diploid cells.

HARDY-WEINBERG EQUILIBRIUM

The frequency of a particular allele of a gene (allelic frequency) in a population remains constant from one generation to the next **in the absence of disturbing factors.**

If p and q are the two alleles for a particular gene:

$$p^2 + q^2 + 2pq = 1$$

pp – homozygous for p allele; qq – homozygous for q allele; pq – heterozygous

Criteria to be fulfilled to apply Hardy-Weinberg equilibrium in humans:
- Population is large
- Mating is random
- Allele frequencies are equal in the sexes
- There is no migration, mutation or selection
- Natural selection is not operating on the population

MENDELIAN INHERITANCE

AUTOSOMAL DOMINANT INHERITANCE

Homozygotes as well as heterozygotes for the affected gene develop the disease; Both sexes are equally affected, and there is no skipping of generation. If one parent is affected, there is 50% chance of offspring also to be affected.

Incomplete penetrance: Normal phenotype even in persons harboring abnormal genotype.

Variable expressivity: Differing clinical features or phenotype among individuals carrying the same gene allele or genotype. The variable expression can be explained by other genetic variations, epigenetic factors and, or environmental factors.

Examples of Autosomal Dominant Disorders

- ❏ All porphyrias except congenital erythrocytic porphyria
- ❏ Familial hypercholesterolemia
- ❏ Most disorders involved with structural proteins, e.g. Marfan's, Achondroplasia, Osteogenesis imperfecta, Neurofibromatosis, Von-Willebrand disease, Hailey-Hailey disease (benign familial pemphigus), Tuberous sclerosis, Von Hippel Lindau, MEN syndrome
- ❏ Hyper IgE–Recurrent Infection Syndrome (Job's Syndrome-STAT3 mutation)
- ❏ Limb-girdle muscular dystrophy type 1

Dominant Negative Effect

- ❏ Mutation in one allele affecting the function of a normal allele.
- ❏ Multimeric proteins composed of different polypeptides are affected by the dominant negative mutation.
- ❏ For example, abnormal collagen chain formed in osteogenesis imperfecta interacts with the normal chain causing procollagen suicide.

Codominance

- ❏ Codominance is a **Mendelian inheritance** in which both of the alleles of a gene pair contribute to the phenotype. *This is the list made after referring to the standard textbooks.*

Examples

- ❏ A and B blood groups (c.f. O blood group is recessive). (*Ref. Harper 30th Ed., p. 697*)
- ❏ Duffy antigen
- ❏ Hemoglobinopathies
- ❏ MHC antigens
- ❏ Familial hypercholesterolemia (*Ref. Harrison 20th Ed., p. 2894*)
- ❏ Z and M alleles of Alpha1 antitrypsin (α1 -AT) gene
- ❏ Apo A-I$_{Milano}$
- ❏ Microsatellite repeats also known as short tandem repeat (STR) sequences
- ❏ BChE gene coding for pseudocholinesterase

AUTOSOMAL RECESSIVE INHERITANCE

- ❏ The disease manifests only in homozygotes
- ❏ Heterozygotes are carriers
- ❏ Skipping of generations is possible
- ❏ Completely penetrant and expression is uniform
- ❏ Consanguinity should be avoided to prevent autosomal recessive disease getting manifested

Examples

Most of the metabolic disorders are inherited in autosomal recessive mode.
- All urea cycle disorders, except Ornithine Transcarbamoylase (OTC) deficiency
- All lysosomal storage disorders, except Fabry and Hunter
- All clotting factor deficiencies, except factor VIII, IX
- Chediak-Higashi syndrome (LYST mutation)
- Limb-girdle muscular dystrophy type 2
- SCID due to Adenosine Deaminase (ADA) deficiency

X-LINKED RECESSIVE INHERITANCE

- Females are usually carriers. Males are affected.
- There is no father to son transmission since the father transmits only his Y chromosome to son.
  - ALA synthase deficiency (X-linked sideroblastic anemia)
  - Fabry disease
  - Hunter syndrome
  - Hemophilia A and B
  - Paroxysmal nocturnal hemoglobinuria
  - Glucose-6-phosphate dehydrogenase deficiency
  - Color blindness
  - Ocular albinism
  - Lesch-Nyhan syndrome
  - Androgen insensitivity syndrome
  - Duchenne and Becker's muscular dystrophy
  - Wiskott-Aldrich syndrome(WASP mutation)
  - OTC deficiency (Hyperammonemia type II)
  - Menkes disease (ATPase 7A defect)
  - Barth syndrome (3-Methylglutaconic aciduria type II)
  - SCID due to IL2RG mutation

Disorders with Autosomal Dominant or X-linked Inheritance

- Alport syndrome

Disorders with Autosomal Recessive or X-linked Inheritance

- Chronic Granulomatous disease (NADPH oxidase deficiency) – 70% X-Linked, 30% Autosomal Recessive

Disorders with Autosomal Dominant, Recessive or X-linked Inheritance

- Retinitis pigmentosa
- Nephrogenic Diabetes insipidus

X-LINKED DOMINANT INHERITANCE

There is no father to son transmission. So, the affected males will have normal sons and affected daughters.
- Vit D resistant rickets
- Rett syndrome
- Incontinentia pigmenti
- Fragile-X-Syndrome
- Fabry disease
- Adrenoleukodystrophy
- Alport syndrome

Note: According to Harrison, Fragile-X-syndrome is X-linked dominant.

LYON'S HYPOTHESIS

Males have only one X-chromosome whereas females have two. To maintain the equivalence of the expression of X-linked genes in males and females, one of the X-chromosome in females is inactivated during the embryonic life. This inactivation is a random process in somatic cells. Either the maternally inherited or paternally inherited X-chromosome undergoes inactivation. This condensed, inactivated X-chromosome is known as 'Barr body'.

Phenotype	Genotype	Number of Barr bodies
Male	46 XY	0
Female	46 XX	1
Female with Swyer syndrome	46 XY	0
Female with Turner syndrome	45 X0	0
Male with Klinefelter syndrome	47 XXY	1
Poly X female	47 XXX	2

Molecular Basis of X-inactivation

XIST gene → long-noncoding (lnc) RNA → binds to the same X-chromosome from which it is produced → leads to DNA methylation and heterochromatin formation → Barr body

Skewed lyonization: X-linked diseases can be manifested in carrier females because of skewed lyonization, i.e. the normal X-chromosome is inactivated more in numbers compared to the mutated one.

arrison's
Corner

Hemophilia carriers can have a wide range of factor VIII values (22–116%) due to random inactivation of the X-chromosomes (lyonization). (*Ref. Harrison 20th Ed., p. 833*)

NON-MENDELIAN INHERITANCE

Disorders with Non-Mendelian Inheritance
- ❑ Disorders of mitochondrial inheritance
- ❑ Disorders due to Genomic imprinting, Uniparental Disomy (UPD)
- ❑ Trinucleotide repeat disorders *(discussed in Chapter 20)* *(Refer to page no. 376)*

MITOCHONDRIAL INHERITANCE

Each cell contains multiple mitochondria, and each mitochondrion contains multiple mitochondrial DNA (polyploidy).

Chapter 30 Medical Genetics

Characteristics of Mitochondrial Inheritance

Maternal inheritance	Heteroplasmy	Random Replicative Seggregation
		Normal Mutant
There is no father to son transmission.	Different types of mitochondria in the same cell accounting for the difference in the severity of the disease.	Random distribution of normal and mutant mtDNA to the offsprings is responsible for the variable expression and reduced penetrance.

COMPLEX GENETIC DISORDERS

Polygenic disorders occur due to defect in many genes.

Disorders with multifactorial inheritance are traits with significant genetic and environmental interplay, e.g. Type II Diabetes mellitus, Coronary heart disease, Cleft-lip, Cancer, Asthma, Gout, Schizophrenia, Alzheimer's disease, Osteoporosis.

Epistasis refers to the gene-gene interaction. The mutation in one gene can mask the effect of a mutation in another gene when both mutations are present in the same organism or cell.

GENOMIC IMPRINTING

❑ Genomic imprinting is an **epigenetic modification** exclusive to mammals in which the expression of certain genes is determined by the parent of origin, i.e. Whether or not they are inherited maternally or paternally.

❑ **Methylation of cytosine of DNA is involved in genomic imprinting.**

❑ Because of this, only one working copy of certain genes are present in our genome. The other copy is inactivated (imprinted) by methylation of cytosine residues.

In other words:
- ○ Both alleles of non-imprinted genes are active, i.e. they are biallelic.
- ○ But in imprinted genes, either of the maternal or paternal alleles is active, i.e. imprinted genes are only expressed in a monoallelic fashion (*Ref. Harrison 20, p. 3352*)

❑ Genomic imprinting occurs during gametogenesis. Parental imprint is erased in the fetal gonads and re-established based on the sex of the fetus. Erasure of imprinting is done by demethylase enzymes.

❑ Imprinting is heritable and can be limited to a subset of tissues.

❑ Genomic imprinting is an example of non-Mendelian inheritance.

❑ Imprinting is the reason that parthenogenesis ('virgin birth') does not occur in mammals. Two complete female genomes cannot produce viable young because of the imprinted genes.

❑ Hydatidiform moles contain 46 chromosomes of paternal origin. Ovarian teratoma contains 46 chromosomes of maternal origin.

❑ **Uniparental disomy (UPD)** is the inheritance of two chromosomes or chromosomal regions from the same parent. Uniparental disomy of imprinted genes or deletion of working copy of imprinted gene leads to diseases.

Section III Molecular Biology

Disorders Associated with Genomic Imprinting

Disorder	Normally expressed copy	Molecular basis of the disease	Clinical features
Angelman syndrome	Maternal	❑ Deletion of maternal copy ❑ Uniparental disomy of paternal copy which is imprinted/inactive (Paternal UPD15)	Mental retardation, seizures, ataxia and hypotonia
Prader Willi syndrome	Paternal	❑ Deletion of paternal copy – 70% of cases ❑ Uniparental disomy of imprinted maternal copy (Maternal UPD 15) – 25% of cases ❑ Imprinting defect – <5% of cases	↓ fetal activity, obesity, Hypotonia, mental retardation, Short stature and hypogonadotropic hypogonadism
Beckwith-Wiedemann syndrome	Paternal	Paternal UPD11	Organomegaly, Exomphalos, hemihypertrophy, risk of Wilms' Tumor and other embryonal malignancies
Transient neonatal diabetes	Paternal	Paternal UPD6	Apparent remission occurs by 3 months but there is a tendency for children to develop diabetes in later life.
Russell-Silver syndrome	Maternal	Maternal UPD7 and UPD11	Failure to thrive, normal head growth, Dysmorphic features like triangular face, clinodactyly
Albright's hereditary osteodystrophy (Pseudohypo-para-thyroidism IA)	Maternal (tissue specific)	Monoallelic loss-of-function mutations in the GNAS1 gene → constitutive activation of Gsα → ↑ cAMP	Pseudo hypoparathyroidism Low calcium and high phosphates. short stature, round facies, mental retardation, short 4th and 5th metacarpals(knuckle knuckle dimple dimple disease).
Pseudopseu-do-hypopara-thyroidism	Paternal		Only physical features of Albrights hereditary osteodystrophy without hormone resistance.

Competency

- ❑ AN73.3 Describe the Lyon's hypothesis
- ❑ AN74.1 Describe the various modes of inheritance with examples
- ❑ AN74.4 Describe the genetic basis & clinical features of Achondroplasia, Cystic Fibrosis, Vitamin D resistant rickets, Haemophilia, Duchene's muscular dystrophy & Sickle cell anaemia
- ❑ AN75.3 Describe the genetic basis & clinical features of Prader Willi syndrome, Edward syndrome & Patau syndrome
- ❑ AN75.4 Describe genetic basis of variation: polymorphism and mutation

Key Points

- ❑ An allele is an alternative form of a gene.
- ❑ The frequency of a particular allele of a gene (allelic frequency) in a population remains constant from one generation to the next in the absence of disturbing factors.
- ❑ Lyon's hypothesis explains the formation of the Barr body.
- ❑ X-linked diseases can be manifested in carrier females because of skewed lyonization.
- ❑ Maternal inheritance, Heteroplasmy and replicative segregation are the features of mitochondrial inheritance.
- ❑ Epistasis refers to gene-gene interaction. The mutation in one gene can mask the effect of a mutation in another gene when both mutations are present in the same organism or cell.
- ❑ Methylation of cytosine of DNA is involved in genomic imprinting.
- ❑ Uniparental disomy (UPD) is the inheritance of two chromosomes or chromosomal regions from the same parent.

Short Answer Questions

1. Give two examples each for
   a. Disorders with autosomal dominant inheritance
   b. Disorders with autosomal recessive inheritance
   c. Disorders with X-linked recessive inheritance
   d. Disorders with X-linked dominant inheritance

2. What is a dominant-negative effect? Give an example.

3. Enumerate three disorders with Non-Mendelian Inheritance.

4. What is codominant inheritance? Give any two examples.

5. Enumerate the characteristics of mitochondrial inheritance.

6. What is genomic imprinting? Name a disorder related to genomic imprinting.

Multiple Choice Questions

1. **All of the following diseases are inherited in autosomal dominant mode, except:**
   a. Congenital erythrocytic porphyria
   b. Familial hypercholesterolemia
   c. Job's syndrome
   d. Marfan's syndrome

2. **Which is NOT an X-linked dominant disease?**
   a. Vitamin D dependant rickets
   b. Rett syndrome
   c. Incontinentia pigmenti
   d. Fragile X syndrome

3. **The phenomenon of procollagen suicide is described in relation to**
   a. Ehler-Danlos syndrome
   b. Scurvy
   c. Osteogenesis imperfecta
   d. Achondroplasia

4. **Differential expression of the same gene depending on parent of origin is referred to as**
   a. Genomic imprinting
   b. Mosaicism
   c. Anticipation
   d. Non-penetrance

5. **Premutation is seen in**
   a. Genomic imprinting
   b. Trinucleotide repeat disorders
   c. Mitochondrial inheritance
   d. Mosaicism

6. **Anticipation is seen in**
   a. Genomic imprinting
   b. Trinucleotide repeat disorders
   c. Mitochondrial inheritance
   d. Mosaicism

7. **Uniparental disomy is not seen in**
   a. Bloom syndrome
   b. Angelman syndrome
   c. Silver Russell

8. **AB blood group is inherited as**
   a. Autosomal recessive
   b. Codominance
   c. Non-Mendelian inheritance
   d. Pseudodominance

9. **Single gene disorder which does not follow Mendelian inheritance:**
   a. Sickle cell anemia
   b. Marfan syndrome
   c. Fragile X-syndrome
   d. Retinoblastoma

10. **Which of the following disorder is inherited in Non-Mendelian mode?**
    a. Hurler syndrome
    b. Hers disease
    c. Hunter syndrome
    d. Huntington's disease

ANSWERS

1. a.	**2.** a.	**3.** c.	**4.** a.	**5.** b.	**6.** b.
7. a.	**8.** b.	**9.** c.	**10.** d.		

Cancer Biology

Cancer is a genetic disorder of somatic cells. Most cancers are not heritable except a few, like Li-Fraumeni syndrome (germline p53 mutation) and some cancer-susceptibility sndromes (DNA repair defects).

HALLMARKS OF CANCER CELLS

- ❑   Self-sufficiency in growth signaling
- ❑   Insensitivity to growth supressing signals
- ❑   Evasion of apoptosis
- ❑   Limitless replicative potential
- ❑   Sustained angiogenesis
- ❑   Local tissue invasion and distant metastasis
- ❑   Deregulated metabolism
- ❑   Evading the immune system

Let us discuss the basic concepts that will help understand these hallmarks.

CELL CYCLE

To understand the process of carcinogenesis, it is important to understand the cell cycle first. Cell cycle is a series of events in which a single cell grows and divides into two daughter cells. Interphase is the stage between two mitoses. DNA synthesis and duplication take places in the S-phase. There are some gaps before and after the S phase in which DNA replication does not happen. These time gaps are denoted as G_1 and G_2 phases. These gaps

are crucial for maintaining the integrity of DNA. The time period of various phases of cell cycle differ according to the cell types.

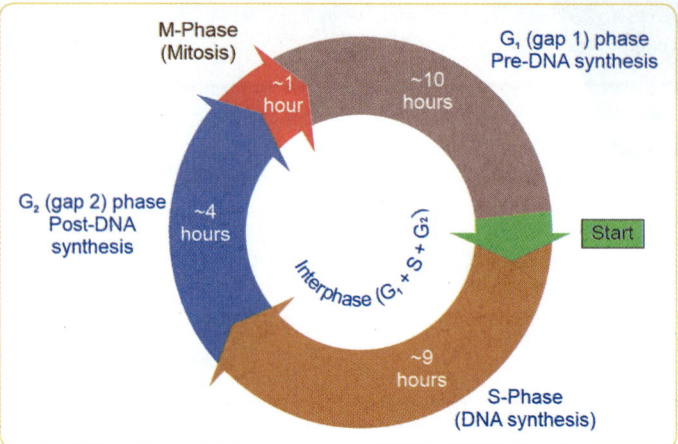

Phases of cell cycle: $G_1 \rightarrow S \rightarrow G_2 \rightarrow M$
Interphase = $G_1 + S + G_2$

Remember

❑ G_2M phase, especially M phase is the most radiosensitive phase of cell cycle.
❑ S phase is the least radiosensitive phase.

Cyclin-Dependent Kinases

❑ Cyclin-dependent kinases (CDKs) are serine/threonine kinases involved in the progression of cell cycle. There are nine CDKs identified in humans.
❑ CDKs are activated by binding of small proteins known as cyclins, e.g. cyclin D activates CDK4 and CDK6.
❑ The level of CDKs is stable throughout the cell cycle. But the level of cyclins fluctuates throughout by ubiquitin-proteasomal degradation thus causing fluctuation in the activity of specific CDK based on the phase of cell-cycle.
❑ There are specific cyclins and CDKs combination involved in specific phase of cell cycle as shown in the following table, e.g. Cyclin D-CDK4/6 complex phosphorylates RB protein and allows the transition of cell cycle from G1 to S phase.
❑ CDKs are regulated by other proteins, e.g. p16 and p21 are CDK inhibitors.

Stage	Events	Cyclins involved	Cdk (cyclin dependent kinase)
G_1	**Transcription and translation happen.** No DNA Replication happens.	D_1, D_2, D_3 E	4,6 2
S (DNA synthesis)	**Doubling of amount of DNA.** Transcription and translation also happen.	A	2
G_2	**No DNA Replication occurs.** Transcription and translation continues to happen.	G2/M transition is mediated by cyclinB-cdk1.	
M phase	Mitosis (nuclear division) and cytokinesis (cell division) produces 2 daughter cells.	B, B_3	1

Cell Cycle Checkpoints

It is important to regulate the transition from one phase of the cell cycle to the other. These points of transitions are known as 'cell cycle checkpoints.' The following are the three checkpoints.

1. G1/S Checkpoint—checks for the integrity of DNA and whether the cellular conditions are favourable for the S-phase
2. G2/M transition—checks the integrity of the replicated DNA
3. Mitotic/spindle checkpoint—checks whether all chromosomes are correctly oriented on the mitotic spindle

RB Protein Guards the Transition from G_1 to S phase

❑ Progression from G_1 to S phase requires a transcription factor called E2F *(don't confuse this with EF-2)*.
❑ Retinoblastoma protein is produced by the RB tumor suppressor gene on chromosome 13. **RB binds to E2F** and prevents the action of E2F.
❑ When there is a stimulus for cell cycle, **cyclin D/cdk4 or cyclin D/cdk6 complex phosphorylates RB**.
❑ **Phosphorylated RB cannot bind to E2F** and E2F is free to activate the genes required for cell cycle progression, i.e. S phase cyclins, DNA polymerase, etc.
❑ So, RB protein is known as **Governor of cell-proliferation** or the 'gatekeeper of G1 to S transition.'
❑ Loss of function mutation of both alleles of RB gene leads to retinoblastoma.

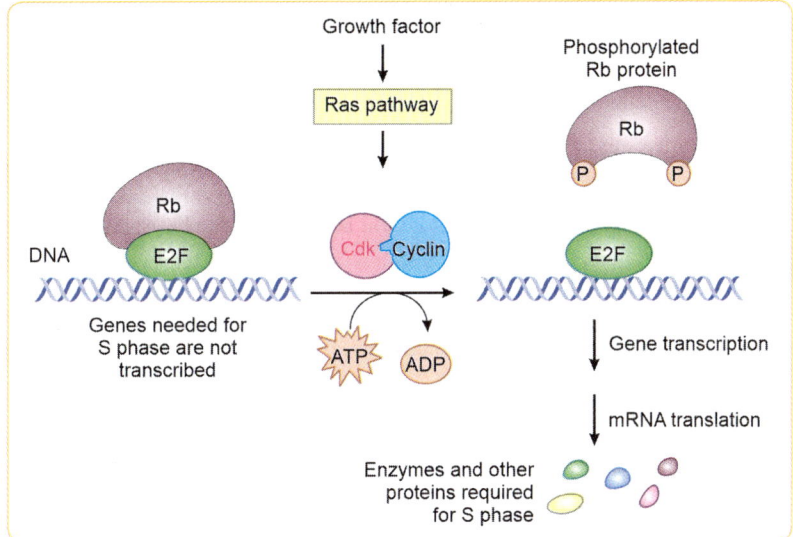

ONCOGENES AND ONCOPROTEINS ARE RESPONSIBLE FOR THE SELF-SUFFICIENCY IN GROWTH SIGNALING OF CANCER CELLS

Proto-oncogenes are genes involved in normal cellular growth and development. Gain of function mutation of proto-oncogenes leads to cancer.

"An oncogene is a mutated/activated proto-oncogene whose protein product is involved in transformation of a normal cell to a cancer cell."

Oncogene	Normal cellular function	Associated neoplasm
RAS	Monomeric GTPase	N-ras – Neuroblastoma K-ras– Leukemia, Lymphoma
C-MYC	Transcriptional regulation	Lymphomas
HER2	Receptor tyrosine kinase	CA Breast
BCL2	Anti-apoptotic (Bcl-2 normally inhibits Bax)	Leukemia, Lymphoma
BCR/ABL	Fusion tyrosine kinase	Chronic Myeloid Leukemia (CML)

Mechanism of Activation of Proto-oncogene to Oncogenes

- ❑ Point mutation, e.g. c-ras activation
- ❑ Gene amplification, e.g. DHFR
- ❑ Promoter insertion by virus, e.g. c-myc activation by Rous Sarcoma Virus
- ❑ Enhancer insertion by virus
- ❑ Chromosomal translocation, e.g. BCR-ABL translocation

Point Mutation

- ❑ Ras is active when it is bound to GTP. GTPase activity of Ras limits its duration of action. Point mutation of Ras leads to loss of GTPase activity. So, Ras becomes persistently active (Gain of function).
- ❑ HER2 is the epidermal growth factor receptor with tyrosine kinase activity. Point mutation leads to activation of receptor even in the absence of ligand.

Gene Amplification

In Chapter 27 *(Refer to page no. 452)*, we have seen that gene amplification is a physiological process during the development of Drosophila melanogaster (fruit fly).

Gene amplification is found in many types of cancer cells. The most notable is the amplification of the gene for dihydrofolate reductase after methotrexate therapy. This provides resistance to methotrexate.

Viral Promoter Insertion

Viral promoters are strong. Insertion of viral promoter upstream of an oncogene, like c-myc activates the oncogene. This is known as insertional mutagenesis. Rous sarcoma virus acts by promoter insertion.

Chromosomal Translocation

ABL1 (Abelson tyrosine kinase) gene on chromosome 9 codes for a tyrosine kinase which is involved in cell division. *BCR* (Breakpoint cluster region protein) gene on chromosome 22 codes for a serine/threonine kinase.

Reciprocal translocation between chromosome 9 and chromosome 22, i.e. t(9:22) leads to the production of a chimeric protein with tyrosine kinase overactivity. t(9:22) translocation is known as Philadelphia chromosome and is diagnostic of chronic myeloid leukemia.

Imatinib, a small molecule acts as inhibitor of fusion tyrosine kinase and is used in the treatment of chronic myeloid leukemia.

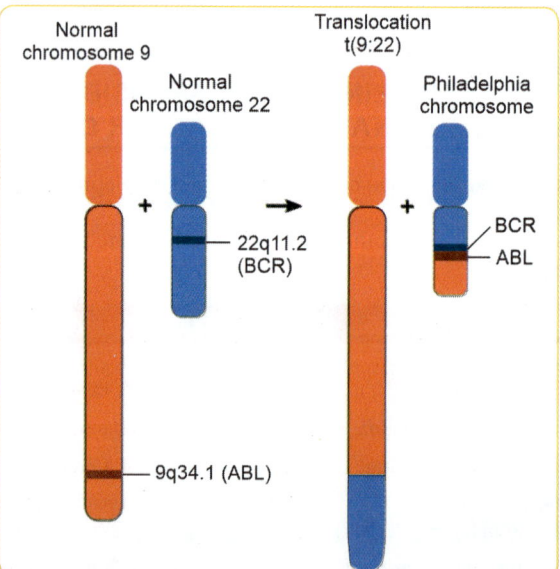

https://upload.wikimedia.org/wikipedia/commons/c/c6/Schematic_of_the_Philadelphia_Chromosome.svg

MUTATION IN THE TUMOR SUPPRESSOR GENES IS RESPONSIBLE FOR INSENSITIVITY TO GROWTH SUPRESSING SIGNALS

Tumor suppressor genes are involved in the regulation of cell functions like cell cycle control, DNA repair, etc. So, these genes prevent cancer development if they function normally. Even if there is a mutation in one allele of a tumor suppressor gene, function of other allele prevents the development of cancer. For a cancer to develop, both alleles should become inactive.

Gene	Chromosome	Normal cellular function	Associated tumor
RB	13	G1/S checkpoint control (governor of cell Proliferation)	Retinoblastoma
P53	17	Guardian of genome	Many tumors
VHL	3	Ubiquitin ligase for hypoxia-inducible factor	Renal cell carcinoma

TP53 is the Guardian of Genome

- ❏ TP53 (Tumor protein 53) gene is a tetrameric protein. It is a transcription factor for genes involved in DNA repair, cell cycle arrest and apoptosis.
- ❏ 53 in P53 refers to the molecular weight of the protein, i.e. 53 kilo Dalton.
- ❏ In normal cells, the half-life of p53 is low. This is because p53 is bound by Mdm2, a ubiquitin ligase (E_3) and degraded in proteasomes.
- ❏ When there is DNA damage, DNA damage sensor proteins like **ATM** cause phosphorylation of p53.
- ❏ MDM-2 cannot bind to phosphorylated p53. So, half-life of p53 is prolonged.
- ❏ p53 increases the transcription of numerous genes. We will look at a few.

Gene product	Role	Effect
GADD45 (Growth Arrest and DNA Damage)	Binds to Proliferating cellular nuclear antigen, i.e. PCNA (DNA sliding clamp that acts as a processivity factor for DNA polymerase δ)	Processivity is lowered; DNA replication is slowed. This gives the cell time to repair the damage
p21	Inhibits cyclin dependent kinases	**Cell-cycle arrest** and gives time to repair damage
Bax (Bcl-2-associated X protein)	Positive regulator of apoptosis (Proapoptotic)	If the damage is not repaired, the cell undergoes apoptosis
Apaf1 (Apoptotic protease activating factor 1)	Scaffold protein involved in apoptosis	

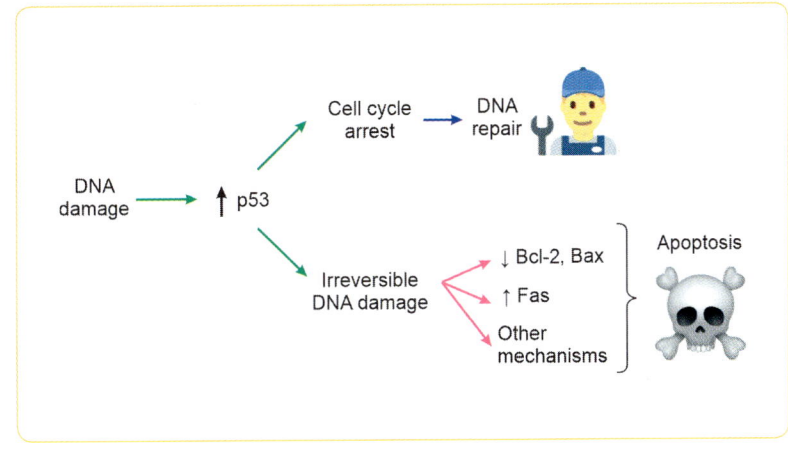

Proapoptotic genes	Anti-apoptotic genes
❑ p53	BCL-2 family
❑ Fas	❑ Bcl-2
❑ BCL-2 family:	❑ Bcl-XL
○ Bax	❑ Bcl-W
○ Bad	
○ Bid	
○ Bak	
○ Bik	

p53 mutation is the most common mutation in cancer cells. Mutation of p53 leads to evasion of apoptosis. Limitless replicative potential of cancer cell is due to increased telomerase activity. *Angiogenesis and metastasis of tumor are beyond the scope of this book. Warburg effect (Chapter 10.2)* *(Refer to page no. 128)* *illustrates the altered metabolism in tumor cells.*

Protooncogene Versus Tumor Suppressor Genes

	Proto-oncogene	Anti-oncogene (Tumor suppressor genes)
Function	Normal growth and development	Regulation and restraining of normal cell growth
Number of alleles that needs to be mutated for development of cancer	One	Both (recessive manner)
Effect of mutation on the gene function	Activation (Gain of function)	Inactivation (Loss of function)
Type of cells affected and inheritance	Somatic cells; hence non-heritable	Somatic and Germ cells; hence heritable sometimes
Tissue preference	Little	Strong (e.g. RB mutation affects retina only)
Example	*MYC, RAS, HER-2/neu, Cyclin -E*	*TP53, RB, VHL*

All Mutagens are NOT Carcinogens

A mutagen is any substance or agent that can cause a mutation or change in the sequence or structure of DNA. Many mutagens are carcinogens. Ames test is used test to screen for mutagenic potential of a drug or any chemical.

Ames Test

The Ames test is a frequently used method which utilizes bacteria to test whether a particular chemical can cause mutations in the DNA of the test organism. Compounds like benzo[a]pyrene, are not mutagenic themselves but their metabolic products are mutagens. That is why the rat liver extract containing the CYP450 system is added in the Ames test.

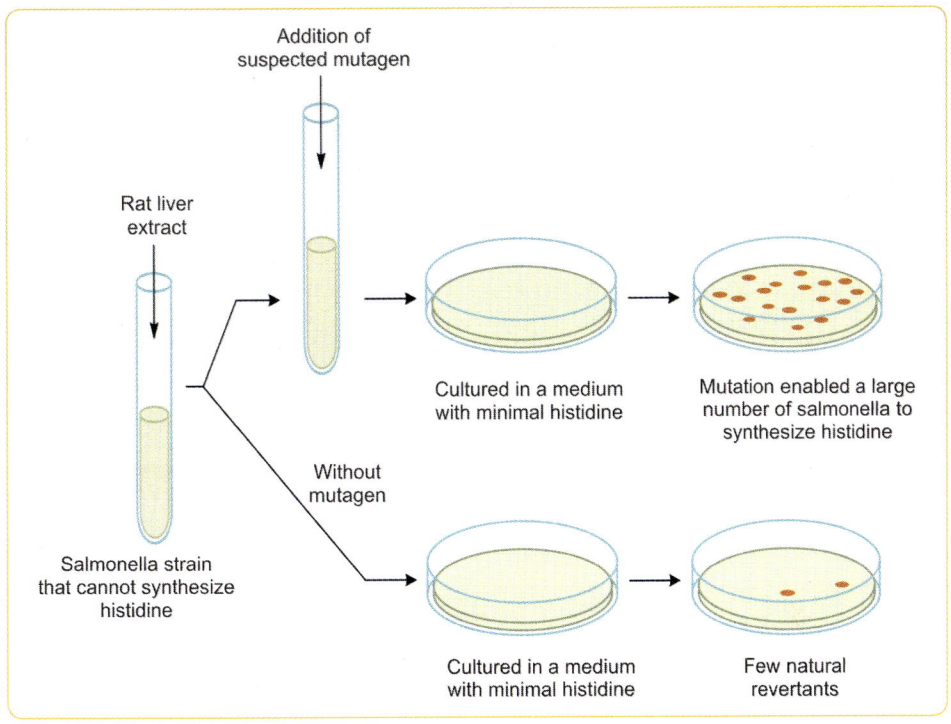

Addition of suspected mutagen

Rat liver extract

Cultured in a medium with minimal histidine

Mutation enabled a large number of salmonella to synthesize histidine

Without mutagen

Salmonella strain that cannot synthesize histidine

Cultured in a medium with minimal histidine

Few natural revertants

Biological Carcinogens

Pathogen	Mechanism	Associated neoplasms
Bacteria		
Helicobacter pylori	Chronic inflammation	Carcinoma and lymphoma of stomach
Fungus		
Aflatoxin B produced by Aspergillus flavus	DNA damage	Hepatocellular carcinoma
Parasite		
Schistosoma hematobium	Chronic inflammation	Urinary bladder cancer
Fasciola hepatica (liver fluke worm)	Chronic inflammation	Cholangiocarcinoma
DNA virus		
Epstein Barr Virus - EBV	Polyclonal B-cell proliferation	Burkitt lymphoma
Hepatitis B Virus - HBV	Inactivation of p53	Hepato cellular carcinoma
Human Papilloma virus - HPV	❏ Inhibition of p53 by HPV E6 protein ❏ Inhibition of RB by HPV E7 protein	Squamous cell carcinoma
RNA Virus		
Hepatitis C virus - HCV	Chronic inflammation	Hepato cellular carcinoma
Human T-lymphotropic virus 1 – HTLV 1	Polyclonal T-cell proliferation	T-cell leukemia and lymphoma

TUMOR MARKERS

A tumor marker is a substance produced by a tumor or by the non-tumor host cells in response to a tumor, which is used to differentiate a tumor from normal tissue, or to detect the presence of a tumor based on measurements in the blood or secretions.

Categories of Tumor Markers

- ❑ Enzymes
- ❑ Hormones
- ❑ Oncofetal antigens
- ❑ Carbohydrate markers
- ❑ Blood group antigens
- ❑ Proteins
- ❑ Receptors
- ❑ Genes

Category	Tumor marker	Associated cancer
Carbohydrate markers	CA-125	CA ovary, CA breast
Blood group antigen (Sialylated Lewis X)	CA-19-9	CA pancreas
Hormones	hCG (human chorionic gonadotrophin)	Germ cell tumors Choriocarcinoma
	Calcitonin	Medullary carcinoma of thyroid
Enzymes	PSA (Prostate specific antigen)	CA Prostate
	Neuron specific enolase	Neuroblastoma Small cell carcinoma lung
	CEA (Carcinoembryonic antigen)	Pancreas, lung, stomach
Oncofetal Antigens	AFP (Alpha Feto protein)	Hepatocellular carcinoma Germ cell tumors
Hormone receptors	HER2 (human epidermal growth factor receptor 2)	CA Breast
Other Proteins	Monoclonal Ig	Multiple myeloma
	S100 beta	Melanoma
Characteristic gene mutation	BCR/ABL (Philadelphia chromosome)	Chronic myeloid leukemia

Applications of Tumor Markers

- ❑ Screening of cancer: Serum prostate specific antigen levels are used in screening of prostate cancer. Diagnosis by tumor markers alone is not advisable since many tumor markers lack specificity.
- ❑ Prediction of therapeutic response:
  - ○ HERr-2/*neu* amplification in breast carcinoma responds to Herceptin therapy
  - ○ Estrogen receptor positive breast carcinoma responds to tamoxifen therapy
- ❑ Monitoring the effectiveness of cancer therapy
- ❑ Detecting tumor recurrence or remission

ANTICANCER DRUGS

Class	Examples	Role/Mechanism of action
Alkylating agents	Cyclophosphamide Cisplatin	Transfer alkyl groups to nucleophilic sites on DNA bases resulting in abnormal cross-linking and DNA strand breakage
Antimetabolites	Methotrexate	Dihydrofolate reductase inhibitor
	5-Flourouracil	Suicidal inhibitor of thymidylate synthase
Microtubule inhibitors	Vincristine Vinblastine	Inhibits the microtubules during cell cycle
Topoisomerase inhibitors	Campothecin Irinotecan	Topoisomerase I inhibitor
	Etoposide Teniposide	Topoisomerase II inhibitor
Antitumor antibiotics	Actinomycin D/Dactinomycin	DNA intercalating agent; prevents DNA replication
	Bleomycin	Binds to DNA and cause free radical damage
Monoclonal antibodies	Trastuzumab (Herceptin)	Anti-HER2
	Rituximab	Anti-CD20
	Bevacizumab	Anti-VEGF (Vascular Endothelial Growth Factor)
Enzymes	Asparaginase	Depletes asparagine required for protein synthesis

Anticancer Drugs and the Effect on Cell Cycle

Anticancer drugs can be classified into cell-cycle nonspecific and cell cycle specific agents. S-phase inhibitors inhibit the DNA synthesis. Mitotic tubule inhibitors inhibit M phase. All the antitumor antibiotics act in a cell cycle nonspecific manner except bleomycin.

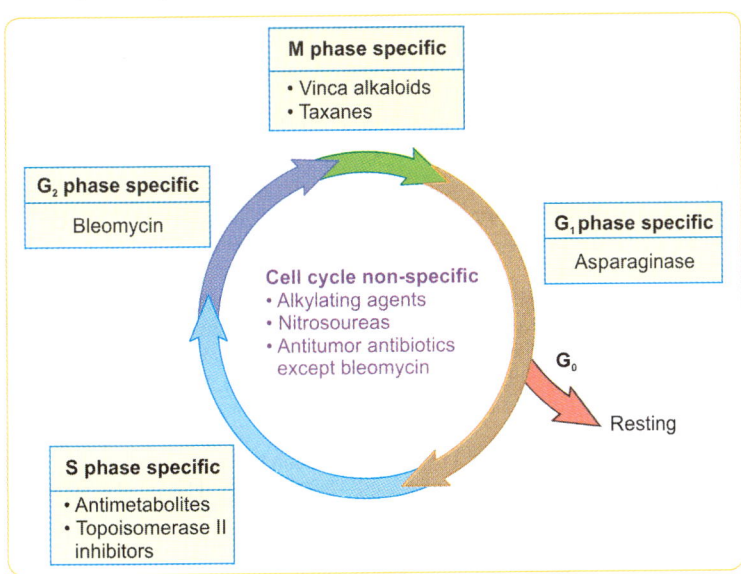

Methotrexate is an Antimetabolite

Antimetabolites are drugs which inhibit nucleotide synthesis by interfering with a specific metabolite. **Methotrexate inhibits dihydro folate reductase** (DHFR) which regenerates active, reduced folate from the oxidized folate

(Chapter 11.5) (Refer to page no. 203). Tetrahydrofolate is important for dUMP to dTMP conversion *(Chapter 15) (Refer to page no. 313)*. So, methotrexate causes **thymineless death of cells** at S-phase.

Cancer cells develop resistance to methotrexate due to the following mechanism:
- ❑ Decreased uptake of methotrexate
- ❑ ↑ **levels of Dihydro folate reductase (DHFR) due to DHFR gene amplification**
- ❑ Synthesis of altered form of DHFR that has reduced affinity for methotrexate
- ❑ ↑ **efflux** of methotrexate from the cell through **p170 glycoprotein,** a transmembrane ATP binding casette (ABC) protein encoded by the MDR1 gene.

Role of Folinic acid in methotrexate treatment has been already discussed in Chapter 33 (Refer to page no. 525).

CANCER AND IMMUNE SYSTEM

People with immunodeficiencies are more prone to cancer. This suggests that surveillance by immune system prevents the malignant transformation. In contrast, chronic inflammation leads to cancer. This is due to persistent reactive oxygen and nitrogen species, cytokines, chemokines, reactive aldehydes, and growth factor in the local environment at the site of chronic inflammation. As we have already discussed, Helicobacter pylori induced inflammation leads to stomach cancer. Obesity leads to a low-grade chronic inflammatory state in the body thus increasing the risk of cancer. This should motivate you to stay fit and active.

Cancer Vaccines

Cancer vaccines prevent the development of cancer or help in the regression of tumor.
- ❑ HPV vaccine protects against the human papilloma virus mediated cervical cancer in females.
- ❑ Hepatitis B and C vaccines prevent the HCV-mediated liver cancer.
- ❑ Dendritic cell-based cancer vaccines are being prepared for many cancers.
- ❑ Immunotoxins are hybrid antibodies carrying toxic molecules like diphtheria toxin *(Diphtheria toxin causes ADP-ribosylation of EF-2 and shuts down protein synthesis)*. Immunotoxins are used to selectively kill the cancer cells.

IMPORTANT CONCEPTS IN CANCER BIOLOGY

Cancer Stem Cells

- ❑ Cancer stem cells are immortal cells found within a tumor that can both self-renew by dividing and give rise to many cell types that constitute the tumor and can therefore form tumors.
- ❑ Cancer stem cells can be formed by malignant transformation of a stem cell or cancer cell acquiring stem cell property.
- ❑ Cancer stem cells are resistant to routine chemotherapy that kills the other malignant cells.
- ❑ Cancer stem cells are believed to be responsible for the recurrence of tumor after chemotherapy.
- ❑ So, targeted killing of cancer stem cells is an exciting ongoing research in many laboratories across the world.

Tumor Microenvironment

- ❑ A tumor tissue not only contains cancer cells but also non-cancer cells.
- ❑ Extracellular matrix of the tumor consists of blood vessels, cells of immune system, fibroblasts, bone marrow-derived inflammatory cells, lymphocytes and signaling molecules. This constitutes the tumor microenvironment.
- ❑ Cancer-associated fibroblasts also known as tumor-associated fibroblasts secrete cytokines and promote the tumor growth.
- ❑ M_2 macrophages promote tumor whereas M_1 macrophages have anti-tumor property.

Epithelial to Mesenchymal Transition (EMT)

- ❑ EMT is a process by which epithelial cells lose their cell polarity and cell-cell adhesion, and gain migratory and invasive properties to become mesenchymal stem cells.

- EMT is crucial for embryonic development and is also implicated in malignant transformation.
- Inhibitors of EMT are potential anti-cancer drugs.

MicroRNAs – Emerging Cancer Biomarkers

- The term **oncomiR** is used for the cancer-associated micro RNAs.
- miRNA levels are altered in cancer cells. So, they are found to be potential biomarkers in cancer detection.
- Both oncogenic and tumor suppressor miRNAs have been described.
- "miR-15 and miR-16 target the BCL2 oncogene and are found to be deleted or downregulated in the many tumors." *(Ref. Harrison 19th Ed., p. 670)*

Cancer and Epigenetics

We have already discussed epigenetics in Chapter 27 (Refer to page no. 455).
- Aberrant hypermethylation of promoter region of tumor suppressor genes (e.g. BRCA1) leads to cancer.
- Azacytidine and its deoxy derivative decitabine are DNA methylase inhibitors used in acute myeloid leukemia and myelodysplastic syndrome.
- Histone deacetylase inhibitors are used in the treatment of cancers, like multiple myeloma.

Competency

❏	BI7.1	Outline the cell cycle
❏	BI10.1	Describe the cancer initiation, promotion oncogenes & oncogene activation. Also focus on p53 & apoptosis
❏	BI10.2	Describe various biochemical tumor markers and the biochemical basis of cancer therapy
❏	PA7.2	Describe the molecular basis of cancer

Key Points

- DNA replication takes place at the S-phase of the cell cycle.
- The most radiosensitive phase of the cell cycle is the G2M phase and the least radiosensitive is S phase.
- RB gene is the governor of cell proliferation. It regulates the G1 to S cell-cycle transition.
- RAS, C-MYC, HER2, BCL2 are some protooncogenes.
- Imatinib is an inhibitor of fusion tyrosine kinase inhibitor used in chronic myeloid leukemia.
- TP53 gene (the guardian of the genome) is a transcription factor for genes involved in DNA repair, cell cycle arrest and apoptosis.
- Li-Fraumeni syndrome is due to germline p53 mutation.
- Bcl-2 is antiapoptotic. Bax is proapoptotic.
- Multi-drug resistance protein/p170 glycoprotein (ABC transporter) is responsible for methotrexate resistant by causing efflux of the drug.
- Calcitonin is a tumor marker for medullary carcinoma of the thyroid.

Short Answer Questions

1. Which gene has been designated as guardian of the genome? Why?

2. What are tumor suppressor genes? Mention two examples and explain the molecular mechanism of action of each.

3. Tabulate the difference between proto-oncogenes and tumor suppressor genes.

4. Define protooncogenes and oncogenes and briefly write how protooncogenes are converted into oncogenes.

5. Name one enzyme and one hormone which are used as tumor markers and indicate the malignancy in which they are clinically used.

6. Some cancer patients during treatment with the drug methotrexate develop resistance to it. Explain its molecular basis.

7. Explain why the rat liver extract is added during the Ames test.

8. Name two cellular processes that p53 can activate or repress in response to DNA damage.

Clinical Case-based Questions

1. A 40-year-old females presented to the clinic with a lump in the right breast. Her mammography revealed a breast mass with numerous microcalcifications suggestive of breast cancer. Her sister had died of metastatic breast cancer at the age of 35. she also recalled that one of her aunts had died long ago with great abdominal distension (probably ovarian cancer).

   a. Which gene is most probably associated with the above clinical presentation?

   b. Explain the mechanism of action (MOA) of any other gene belonging to the same category.

2. A cancer patient was on the treatment of methotrexate for several months. On review, PET scan metastasis was observed to increase even after treatment.

   a. What is the reason behind the progression of cancer even after chemotherapy?

   b. What is the biochemical mechanism behind this?

Multiple Choice Questions

1. In which of the following phase of cell cycle, DNA replication takes place?
   - a. G_1 phase
   - b. S phase
   - c. G_2 phase
   - d. M phase

2. The progression of cell cycle is halted if the cell's DNA is damaged. At which cell-cycle checkpoint this happens?
   - a. G_1-S
   - b. S-G_2
   - c. M-G1
   - d. G_0-G_1

3. What is the role of Bcl-2 and Bax in apopotosis?
   - a. Bax inhibits and Bcl-2 stimulates
   - b. Bax stimulates and Bcl-2 inhibits.
   - c. Both Bax and Bcl-2 inhibit
   - d. Both Bax and Bcl-2 stimulate

4. In which of the following process, p170 glycoprotein is involved?
   - a. Initiation of Apoptosis
   - b. Angiogenesis and metastasis
   - c. Insensitivity to growth inhibitory signal
   - d. Efflux of anticancer drug

5. All of the following are tumor suppressor genes, Except:
   - a. RB
   - b. P53
   - c. VHL
   - d. HER-2

6. Which of the following is marker of medullary carcinoma thyroid?
   - a. Thyroxine
   - b. Calcitonin
   - c. Paratharmone
   - d. Melatonin

7. To which of the following transcription factor RB protein binds to?
   - a. EF2
   - b. E2F
   - c. Cyclin D
   - d. Cyclin E

8. TP53 protein is a:
   - a. Transcription factor
   - b. Product of protooncogene
   - c. Growth factor
   - d. Cytokine

9. 53 in P53 refers to:
   - a. 53% of P100
   - b. 53 kDa
   - c. UV absorption at 530 nm
   - d. Order of discovery

ANSWERS

1. b.	2. a.	3. b.	4. d.	5. d.	6. b.
7. b.	8. a.	9. b.			

IV
SECTION

Nutrition

Section Outline

General Principles of Nutrition and Macronutrients

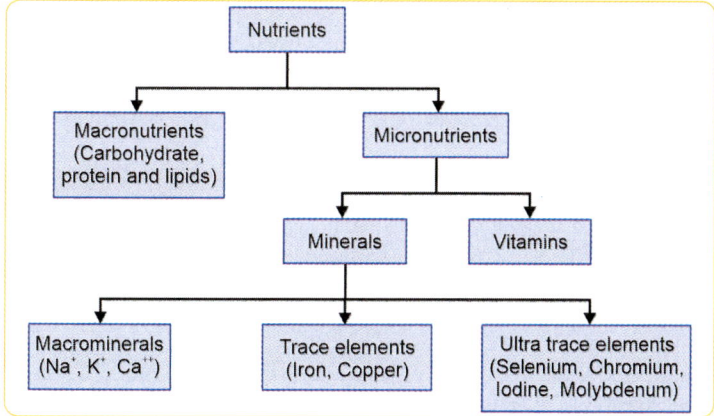

Corner

Ultratrace elements are defined as those needed in amounts <1 mg/day. (*Ref. Harrison 20th Ed., p. 2319*)

CALORIFIC VALUES OF MACRONUTRIENTS

Calorific value is the amount of heat produced by burning 1 g of food stuff completely in the presence of O_2.
Measured by: Bomb calorimeter
Unit: Cal/g; kilocalories are written as Cal.

Food stuff	Calorific value (Cal/g)
Carbohydrates	4.0
Proteins	4.2
Fat	9.0
Alcohol	7.0

CONCEPT CORNER

Movement of electrons in the ETC ultimately produces ATP. Fats are electron-rich. So, they can produce more NADH and $FADH_2$ compared to carbohydrates. That is why calorific value of fat is higher than that of carbohydrates.

Calories Provided by Alcohol is an Example of Empty Calories

❑ Alcohol is a bad food providing empty calories.
❑ Empty calories are calories not associated with nutrients like vitamins and minerals.

BASAL METABOLIC RATE (BMR)

The BMR is the rate of energy expenditure by the body during complete physical, mental and digestive rest (*basal conditions*). During measurement, the person lies in supine position, kept awake throughout and temperature is maintained at 20–25°C.
Measured by: Calorimetry
Unit: kcal/m² body surface area/hour.

Factors Affecting BMR

Lean body mass (Lean body mass is the weight of a person's body that is not fat) determines the BMR. So muscular people have more BMR and obese people will have less BMR. Because of relatively more fat, *females tend to have less BMR* compared to men. Aging is associated with fatty infiltration of muscle tissue → ↓BMR.

Conditions Affecting BMR

↑BMR	↓ BMR
Fever (pyrexia)	Protein energy malnutrition
Hyperthyroidism	Hypothyroidism
Cold environment	Addison's disease
Cushing's syndrome	Starvation
Cachexia	
Exercise	

Note: Both cachexia and PEM are associated with loss of body weight. But the effect on BMR is opposite.

SPECIFIC DYNAMIC ACTION (SDA)/DIETARY THERMOGENESIS

Specific dynamic action/diet induced thermogenesis/thermic effect of food is the extra energy needed by the body for digestion and absorption of the food.

If we eat 25 g of protein, we expect release of 100 (25 × 4) kcal energy. But in reality, only 70 kcal energy is released. This 30% energy is used for digestion, absorption, etc. (This explains the basis of weight loss by low calorie, high protein diet)

Significance of knowing SDA: When we calculate diet plan, additional calories equivalent to SDA has to be included.

Macromolecule	SDA value (%)
Carbohydrate	7
Lipid	12
Protein	30
Mixed diet	10

RESPIRATORY QUOTIENT (RQ)

The RQ is the molar ratio of CO_2 produced to oxygen consumed. It varies for different type of macromolecules in food.

$$RQ = \frac{\text{Volume of } CO_2 \text{ produced}}{\text{Volume of } O_2 \text{ consumed}}$$

Macromolecule	RQ value	Explanation
Glucose $(C_6H_{12}O_6)$	1	$\dfrac{6CO_2}{6O_2}$
Proteins	0.8	Not completely oxidized. Value is empirical.
Palmitic acid $(C_{16}H_{32}O_2)$	0.7	$\dfrac{16CO_2}{23O_2}$
Mixed diet	0.85	

Compared to carbohydrates, fats are highly reduced molecules. So, they require more oxygen for getting oxidized. *RQ value can go more than 1 during heavy exercise!*

 Clinical Correlation

Respiratory quotient in patients of COPD

Chronic Obstructive Pulmonary Disease (COPD) is an inflammatory disease that causes chronic obstruction of airflow. In 1992, a study found that carbohydrate-rich food increases production of CO_2, leading to an increase in respiratory rate and eventually respiratory failure in patients of COPD. In contrast, fat-rich diet decreases the production of CO_2, leading to improvement in respiration.

LIMITING AMINO ACIDS

Complete proteins contain all the essential amino acids, e.g. egg, milk and soya. Incomplete proteins lack one or more essential amino acids. Amino acids that are absent or inadequately present in the incomplete protein are known as limiting amino acids. If incomplete proteins are taken as a staple diet, there will be limitation of growth.

Food item	Limiting amino acid
Cereals	Lysine
Pulses	Methionine ± cysteine
Wheat	Lysine and threonine
Maize	Lysine and tryptophan

Cereals lack lysine. Pulses lack methionine. This is the rationale behind making foods with combination of cereals and pulses, e.g. Idly, Dal with Roti.

RECOMMENDED DIETARY ALLOWANCE

Recommended dietary allowance (RDA) is the average daily dietary nutrient intake level sufficient to meet the nutrient requirement of nearly all (97–98%) healthy individuals in a particular life stage and gender group.

RDA = EAR + 2 SD; (EAR is estimated average intake; SD is standard deviation)

ESTIMATED AVERAGE INTAKE

Estimated average intake (EAR) is the daily intake value of a nutrient that is estimated to meet the nutrient requirement of half the healthy individuals in a life stage and gender group.

METHODS FOR EVALUATION OF NUTRITIONAL VALUE OF PROTEINS

- ❑ Biological value
- ❑ Net protein utilization
- ❑ Protein efficiency ratio
- ❑ Relative nitrogen utilization
- ❑ Net protein ratio
- ❑ Nitrogen growth index

$$\text{Biological value} = \frac{\text{Nitrogen retained}}{\text{Nitrogen absorbed}} \times 100$$

Biological value (BV) of animal proteins is higher than plant proteins.

Among animal proteins, **egg protein has the highest biological value** (100%).

As BV calculation does not include the amount of nitrogen lost during digestion, it is not a good marker for nutritional value of proteins.

$$\text{Net protein utilization} = \frac{\text{Nitrogen retained}}{\text{Nitrogen ingested}} \times 100$$

$$\text{Protein efficiency ratio} = \frac{\text{Weight gained}}{\text{Weight of protein consumed}}$$

Nitrogen Content of Protein

The protein content of foods can be determined based on total nitrogen content. Kjeldahl's method was used to detect nitrogen content. It has been estimated that animal proteins contain 16% of nitrogen. In other words, 100 g of protein will contain 16 g of nitrogen. Based on this, protein content of the food can be calculated by the following formula:

Protein content = 6.25 × Nitrogen content

NITROGEN BALANCE

- ❑ Healthy adults will be at a state of nitrogen equilibrium, i.e. Nitrogen intake = Nitrogen excreted.
- ❑ A high protein intake does not lead to positive nitrogen balance since high-protein intake also leads to protein degradation.
- ❑ Large intake of proteins is not advised since protein synthesis and protein degradation are energy-expensive processes.

Positive nitrogen balance	Negative nitrogen balance
Intake is more than loss (Nitrogen intake > nitrogen excreted)	Loss is more than the intake (Nitrogen excreted > Nitrogen intake)
❑ Childhood growth	❑ Infection
❑ Pregnancy	❑ Surgery
❑ Lactation	❑ Burns
❑ Recovery after illness (Convalescence)	❑ Trauma
❑ Anabolic steroid intake	❑ Malnutrition
	❑ Hyperthyroidism, Cushing syndrome

DIETARY CARBOHYDRATES

Starch is the most abundant form of carbohydrate we take in food. Components of our staple food (rice, wheat and potatoes) are rich in starch.

Starch is digested by α-amylase, which is found in saliva and pancreatic secretions. α-amylase cleaves the α (1→4) glucosidic linkages of the starch, producing smaller molecules like maltose. Disaccharidases present in the brush borders of the intestine cleave the disaccharides into monomeric units.

GLYCEMIC INDEX

Glycemic index (GI) is the extent to which a food substance raises the blood glucose level compared to an equivalent amount of reference carbohydrate. Glycemic index ranges from zero to one. Glycemic index of pure glucose is 100.
Foods with high glycemic index (70–99): Corn flakes, white bread
Foods with low glycemic index: Most fruits and vegetables (except potatoes and water melon)

Significance

❑ Foods with high glycemic index cause rapid increase in blood glucose and are detrimental to diabetics.
❑ Foods with low glycemic index release glucose more slowly and steadily and are beneficial for diabetics.

GLYCEMIC LOAD

Glycemic index does not consider the amount of carbohydrate in a particular food. This is why the concept of glycemic load is developed.

$$\text{Glycemic load} = \frac{\text{Glycemic index} \times \text{Carbohydrate (g) content per portion}}{100}$$

Similar to the glycemic index, the glycemic load of a food can be classified as low, medium, or high.
Low: 10 or less; Medium: 11–19; High: 20 or more.
It is advisable to keep the daily glycemic load under 100.

DIETARY FIBERS

Fibers are organic compounds, mostly polysaccharides which cannot be digested by human digestive enzymes because of the presence of beta glycosidic linkages. Intestinal bacteria ferment a part of these fibers into lactate, hydrogen and short chain fatty acids (e.g. butyrate).

Butyrate has a role as an anti-inflammatory agent and thus decreases the chance of colon cancer. Thus, butyrate is considered as a *tumor suppressor metabolite*.

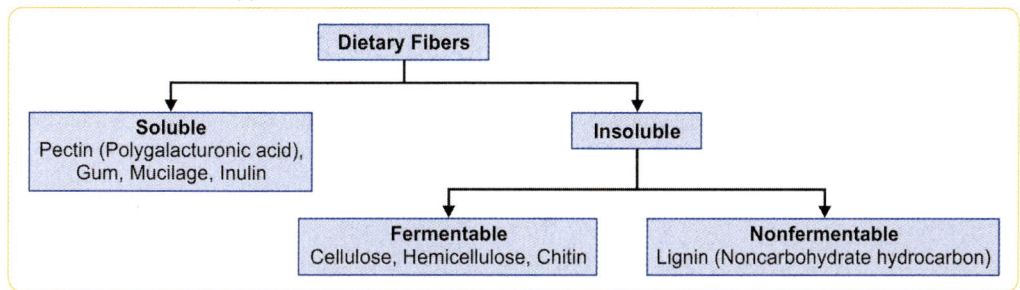

- Lignin is not a carbohydrate at all. Thus, it is neither digested nor fermented and it is a superior dietary fiber.
- Inulin is a fructosan.
- Lignin is a phenolic compound.

Beneficial Effects

- Soluble fibers retain more water in colon and produce soft stools → *Prevents constipation*
- As the transit time of stool is reduced, there is reduced stasis of toxins. Butyric acid produced by microbial fermentation is antiproliferative → *prevents colon cancer.*
- Certain dietary fibers are called *prebiotics* as they are needed for the growth of beneficial intestinal flora.
- Dietary fibers bind to bile salts and prevent their enterohepatic circulation → Excretion of cholesterol in the form of bile salts → ↓*plasma cholesterol*
- **Improve glucose tolerance**
  - ○ Soluble fibers—beneficial with regard to cholesterol reduction and glucose tolerance
  - ○ Insoluble fibers—beneficial with regard to colonic function.

Disadvantages

- Bind to essential minerals and interferes in their absorption.
- Excess ingestion causes bloating and flatulence.
- Certain patients of inflammatory bowel disease should avoid fibers.

Know the Difference

Probiotics are live microorganisms that when administered in adequate amounts confer a health benefit on the host, e.g. *Lactobacillus, Bifidobacterium, Saccharomyces boulardii.*
Prebiotics are nondigestible food ingredients that stimulate the growth and/or activity of commensal bacteria in the digestive system, e.g. Fructooligosaccharides like Inulin (present in garlic, onion)

RESISTANT STARCH

Resistant starch escapes the digestion in the small intestine. Starch enclosed within intact cell walls, starch granules and starch with high amylose content are resistant to small intestinal digestion to a variable degree. In healthy individuals, resistant starch is passed to the large intestine. Similar to dietary fibers, microbial fermentation of resistant starch in the large intestine produces volatile short-chain fatty acids like acetate and butyrate.

Metabolic effects of resistant starch:
- ↑ Satiety
- ↓ Postprandial glycemic and insulinemic responses
- ↓ Plasma cholesterol and triglyceride concentrations
- ↓ Fat storage

Foods rich in resistant starch:
- Wheat
- Raw Banana
- Potatoes
- Corn flakes

LACTOSE INTOLERANCE

Intolerance to lactose containing dairy products occurs due to decreased small intestinal brush border enzyme lactase (β-galactosidase). Lactase enzyme is highly expressed in intestine during the period of breastfeeding. After weaning, level of lactase gradually declines.

1° lactose intolerance	2° lactose intolerance
❑ Most of the adults have decreased amount of lactase. (Hypolactasia after weaning) ❑ Autosomal recessive congenital lactase deficiency	"Destruction of intestinal brush borders" ❑ Bacterial overgrowth ❑ Enteritis ❑ Celiac and Crohn's disease

↓ Lactase →↑Lactose goes to colon → fermented to short chain fatty acids and hydrogen gas by bacteria → flatulence

C/f: Flatulence, abdominal cramps, explosive diarrhea

R$_x$: Avoidance of lactose containing dairy products and substituting dairy products with lactose-free formulations like soy-based milk.

TREHALOSE INTOLERANCE

❑ Trehalose is a nonreducing disaccharide containing 2 glucose molecules linked by α (1→1) glucosidic linkage.
❑ Trehalose is present in young mushrooms; digested by the intestinal brush-border trehalase.
❑ People deficient in trehalase are intolerant to mushrooms.

STEATORRHEA

Definition: Excretion of >7 g of fat in the stool during a 24-hour period while the person is on a diet containing ≤100 g of fat per day. There will be a history of greasy and foul-smeling stools that leave an oily residue in the toilet bowl, increased flatulence and weight loss.

Steatorrhea Occurs due to Maldigestion and Malabsorption of Lipids

Causes

❑ **Liver diseases:** Impaired synthesis of bile salts that are essential for emulsification of fats in the intestine during digestion.
❑ **Pancreatic insufficiency:** Pancreatitis, cystic fibrosis → ↓ lipase
❑ **Hypersecretion of gastrin and HCl:** Zollinger-Ellison syndrome—↓ pH of intestinal lumen → Inactivation of pancreatic lipase
❑ **Short bowel syndrome:** Ileal resection leads to ↓Enterohepatic circulation of bile salts
❑ **Bacterial overgrowth:** ↓Bile acids throughout the intestine
❑ **Abetalipoproteinemia (Bassen-Kornzweig syndrome):** ↓ Chylomicron formation → Impaired absorption of fat

Diagnosis: Quantitative estimation of stool fat.

BALANCED DIET

A balanced diet is one which *provides all the nutrients in required amounts and proper proportions.* It can easily be achieved through a blend of the following four basic food groups:
1. Cereals, millets and pulses
2. Vegetables and fruits
3. Milk and milk products, egg, meat and fish
4. Oils, fats, nuts and oilseeds

The quantity of food needed to meet the nutrient requirements varies with age, gender, physiological status and physical activity.

Proportion of Calorie Source for Balance Diet

❑ 50–60% of total calories from carbohydrates, preferably from complex carbohydrates
❑ 10–15% from proteins
❑ 20–30% from both visible and invisible fat

Chapter 32 General Principles of Nutrition and Macronutrients

In addition, a balanced diet should provide dietary fiber, antioxidants and phytochemicals. Antioxidants such as vitamins C and E, β-carotene, riboflavin and selenium protect the human body from free radical damage. Other phytochemicals such as polyphenols, flavones, etc., also afford protection against oxidant damage. Spices like turmeric, ginger, garlic, cumin and cloves are rich in antioxidants.

Considerations to be Kept while Selecting Balanced Diet

- ❑ Food should be locally available
- ❑ It should be cost-effective
- ❑ Should fit into the local food habits of the people
- ❑ Should be palatable and digestible
- ❑ The formulation of balanced diet should also take into consideration physiological conditions like pregnancy, lactation or diseases. Special diet is needed for people with celiac disease.
- ❑ Additional allowances should be made when required.

PEOPLE WITH CELIAC DISEASE ARE GLUTEN INTOLERANT

- ❑ Gliadin is a component of gluten protein present in wheat, barley, and rye.
- ❑ Gliadin elicits antigenic response in these patients.
- ❑ People with celiac disease are invariably lactose intolerant (secondary hypolactasia) too.
- ❑ In addition to life-long gluten restriction, dietary lactose and fat restriction are also needed at the initial stages of treatment.

TOTAL PARENTERAL NUTRITION

Total parenteral nutrition (TPN) is the process of providing processed nutrients to the patient through the venous blood completely bypassing the GI tract.

Indications: Diseases in which intake of food is not possible through GI tract, like inflammatory bowel disease and short bowel syndrome.

Components: Many commercial preparations are available. Regimens can be modified according to the need of the patient. *Dietary fibers are not included in TPN.*

Complications

- ❑ Catheter site thrombosis
- ❑ Air embolism
- ❑ Biochemical abnormalities:
  - ○ Hypokalemia
  - ○ Hypomagnesemia
  - ○ Hypophosphatemia
  - ○ Hypercalcemia
  - ○ Hyperglycemia
  - ○ Hyperlipidemia
  - ○ Acid-base abnormalities

Monitoring of Patients on TPN

- ❑ Proper input/output chart for fluid balance
- ❑ 6-hourly capillary blood glucose checking
- ❑ Daily checking of urea and electrolyte values
- ❑ Folate and B_{12} levels every 2 weeks if the patient is on long-term TPN.

PROTEIN ENERGY MALNUTRITION

Protein-energy malnutrition is a nutritional deficiency resulting from either inadequate energy (caloric) or protein intake and manifesting in either marasmus or kwashiorkor. The difference between two is given in the following table:

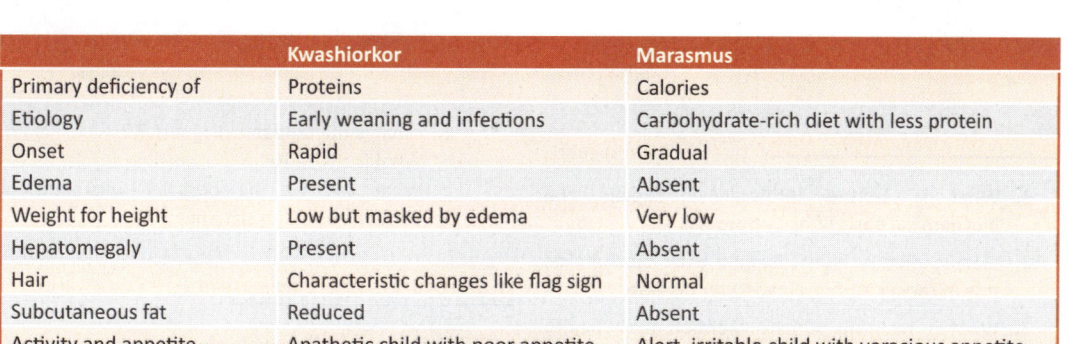

	Kwashiorkor	Marasmus
Primary deficiency of	Proteins	Calories
Etiology	Early weaning and infections	Carbohydrate-rich diet with less protein
Onset	Rapid	Gradual
Edema	Present	Absent
Weight for height	Low but masked by edema	Very low
Hepatomegaly	Present	Absent
Hair	Characteristic changes like flag sign	Normal
Subcutaneous fat	Reduced	Absent
Activity and appetite	Apathetic child with poor appetite	Alert, irritable child with voracious appetite
Cortisol levels	↓	↑

TOXINS ASSOCIATED WITH FOOD ITEMS

In addition to nutrients, certain plants also contain toxic components. Recall the toxin present in litchi fruit that can cause severe hypoglycemia (Chapter 13.1) *(Refer to page no. 248)*. The following table shows a list of well-known toxins containing plants, the toxin and the disease caused. This knowledge will help you to educate the common public.

Disease	Toxin	Source
Epidemic dropsy	Sanguinarine	*Argemone mexicana* (adulteration in mustard oil)
Endemic ascites	Pyrrolizidine alkaloid	Panicum miliare millet contaminated with seeds of crotalaria
Lathyrism	Beta oxalyl Amino Alanine (BOAA)	*Lathyrus sativus* (kesari dhal)

Competency

☐ BI8.1   Discuss the importance of various dietary components and explain importance of dietary fiber
☐ BI8.2   Describe the types and causes of protein energy malnutrition and its effects

Key Points

- ☐ Ultratrace elements are defined as those needed in amounts <1 mg/day.
- ☐ The calorific values of Carbohydrates, Proteins, Fat and Alcohol are 4.0, 4.2, 9.0 and 7.0 kcal/g respectively.
- ☐ Calories provided by alcohol are empty calories due to the absence of associated micronutrients.
- ☐ The BMR is the rate of energy expenditure by the body during complete physical, mental and digestive rest.
- ☐ Specific dynamic action/diet-induced thermogenesis/thermic effect of food is the extra energy needed by the body for digestion and absorption of the food.
- ☐ The RQ is the molar ratio of CO_2 produced to oxygen consumed. RQ values of glucose, proteins and palmitic acid are 1, 0.8 and 0.7 respectively.
- ☐ Complete proteins contain all the essential amino acids, e.g. egg, milk and soya.
- ☐ Amino acids that are absent or inadequately present in the incomplete protein are known as limiting amino acids, e.g. Cereals lack lysine and pulses lack methionine.
- ☐ Among animal proteins, egg protein has the highest biological value (100%).
- ☐ The dietary fiber lignin is not a polysaccharide.
- ☐ Resistant starch is the starch that resists the digestion in the small intestine.
- ☐ Steatorrhea is defined as the excretion of >7 g of fat in the stool during a 24-hour period while the person is on a diet containing ≤100 g of fat per day.
- ☐ Fibers are not included in the total parenteral nutrition.

Short Answer Questions

1. What is lactose intolerance? Explain the biochemical basis of its symptoms.

2. Classify dietary fibers. Explain their biochemical role in cancer and cardiovascular disease.

3. What is resistant starch? Mention their health benefits.

4. Explain why combining different food items often improves the Biological Value (BV) of dietary proteins.

5. Explain why the requirement of protein depends on its quality. How is the quality of protein determined?

6. Define basal metabolic rate. What is the major factor determining the BMR of a person? Name any two conditions with increased and decreased BMR.

7. If we eat 25 g of protein, we expect the release of 100 (25 × 4) kcal energy. But, only 70 kcal energy is released. How can you explain the difference?

8. What is respiratory quotient (RQ)? What is the significance of RQ in a patient of chronic obstructive disease?

9. Explain the concept of limiting amino acids with suitable example.

10. Differentiate
    a. Glycemic load and Glycemic index
    b. Kwashiorkor and marasmus

11. What is the glycemic index (GI)? What is the significance of GI in a patient of diabetes?

12. Mention any four causes of steatorrhea.

13. What are the possible complications one should expect in a patient undergoing total parenteral nutrition?

Clinical Case-Based Questions

1. A 25-year-old man is presented to the medicine OPD with complaints of progressive weight loss. He gave a history of sticky foul-smeling stools. He also gave a history of prolonged hospitalization due to gall stone induced chronic pancreatitis.

a. What is the probable diagnosis?
b. What is the cause of foul-smeling stools?
c. What can be the probable micronutrient deficiencies in this patient?

Multiple Choice Questions

1. **Basal metabolic rate is closely dependent on:**
   a. Body surface area b. Lean body mass
   c. BMI   d. Height

2. **Energy expenditure in resting state depends upon:**
   a. Lean body mass   b. Adipose tissue
   c. Heart rate   d. None of the above

3. **BMR is low in:**
   a. Hyperthyroidism   b. Obesity
   c. Cold exposure   d. Exercise

4. **Thermogenic effect or specific dynamic action is highest for:**
   a. Protein   b. Fat
   c. Carbohydrates   d. Mixed diet

5. **Calorific value of alcohol is:**
   a. 4 kcal/g   b. 5 kcal/g
   c. 6 kcal/g   d. 7 kcal/g

6. **Recommended dietary allowance is calculated by:**
   a. Estimated average intake + 2 standard deviation
   b. Estimated average intake – 2 standard deviation
   c. Estimated average intake ± 2 standard deviation
   d. 2 × estimated average intake

7. **Which of the following fruit juice helps in preventing urinary tract infection?**
   a. Blueberry   b. Cranberry
   c. Mulberry   d. Raspberry

8. **Which of the following phytochemical is responsible for the cardioprotective effects of red wine?**
   a. Sorbitol
   b. Resveratrol
   c. Resorcinol
   d. Beta carotene

9. A person on total parenteral nutrition was given 62.5 g of protein. Calculate the amount of nitrogen present in this amount of protein.
   a. 32 g
   b. 6.25 g
   c. 10 g
   d. 100 g

10. All of the following are components of total parenteral nutrition, except:
    a. Amino acids
    b. Fats
    c. Vitamins
    d. Fiber

11. Inulin like fructosans are used as prebiotics as they are nondigestible because of the:
    a. Presence of glucosidic linkages
    b. Beta configuration of anomeric carbon
    c. Low pH of the stomach
    d. Presence of alpha-glycosidic linkages

12. Among the following food items, which one has the highest "Glycemic Index"?
    a. Corn flakes
    b. Brown rice
    c. Ice-cream
    d. Whole wheat bread

13. Which of the following source of protein has the highest biological value?
    a. Egg
    b. Cow milk
    c. Gelatin
    d. Soy milk

14. Which of the following plant components is not fermented by gastrointestinal microorganisms?
    a. Lignin
    b. Cellulose
    c. Hemi-cellulose
    d. Pectin

15. Which of the following amino acid is a limiting amino acid in pulses?
    a. Leucine
    b. Lysine
    c. Methionine
    d. Glutamine

16. Respiratory quotient of a person on a predominant carbohydrate diet will be:
    a. 1
    b. 0.7
    c. 0.8
    d. 1.2

17. What is the normal quantity of fecal fat in a healthy individual?
    a. <7 g/day
    b. <10 g/day
    c. <12 g/day
    d. <15 g/day

18. Lactase is a:
    a. α glucosidase
    b. β glucosidase
    c. α galactosidase
    d. β galactosidase

19. Which of the following in an example of a prebiotic?
    a. Yogurt
    b. Inulin
    c. Creatinine
    d. Fish oil

20. All of the following are conditions associated with negative nitrogen balance, except:
    a. Burns
    b. Convalescence
    c. Infection
    d. Malnutrition

21. Which of the following food item does not contain gluten and is acceptable for patients with celiac disease to consume?
    a. Barley
    b. Rice
    c. Rye
    d. Wheat

22. Steatorrhea occurs due to maldigestion of:
    a. Alcohol
    b. Carbohydrates
    c. Lipids
    d. Protein

ANSWERS

1. b.	**2.** a.	**3.** b.	**4.** a.	**5.** d.	**6.** a.
7. b.	**8.** b.	**9.** c.	**10.** d.	**11.** b.	**12.** a.
13. a.	**14.** a.	**15.** c.	**16.** a.	**17.** a.	**18.** d.
19. b.	**20.** b.	**21.** b.	**22.** c.		

Micronutrients: Vitamins

Vitamins are a heterogenous group of organic micronutrients required in very small amounts for the normal bodily functions. The human body cannot synthesize most of them and therefore vitamins must be taken in the diet. Based on their chemical nature, vitamins can be broadly classified into fat-soluble and water-soluble.

The initially discovered vitamins were amines, hence the old name "vitamines" (vital amines). Later, the "e" from "vitamine" was removed when it was realized that not all vitamins are amines.

	Fat-soluble vitamins	Water-soluble vitamins
Members	A, D, E and K	B complex and Vitamin C
Stored in the body	Yes	No, except Vitamin B_{12}
Requirement of bile salts for absorption	Yes	No
Coenzyme role	No, except Vitamin K	Yes, each member has coenzyme activity
Toxicity on excessive intake	Yes, usually	No, usually

SOME VITAMINS CAN BE SYNTHESIZED BY THE BODY; SOME VITAMINS ARE SYNTHESIZED IN THE BODY

Niacin is not a true vitamin since it can be produced from tryptophan by metabolic process. 60 mg of tryptophan is equivalent to 1 mg of Niacin.

Vitamin D is not strictly a vitamin since it can also be synthesized by the body.

Certain vitamins like **Vitamin K and Biotin** are synthesized in the body but not by the body. They are synthesized by the intestinal flora.

VITAMERS ARE DIFFERENT FORMS OF A VITAMIN

Most vitamins exist in multiple forms with different degree of activity and nutritional availability. These chemical analogues of a vitamin are known as vitamers. Except Biotin, all the vitamins have more than one vitamer.

Vitamin	Vitamers
A	Retinal, Retinol, Retinoic acid
K	Phylloquinone, Menaquinone, Menadione
E	α, β, γ and δ tocopherol

IMPORTANT RING STRUCTURES AND COMPONENTS OF VITAMINS

Vitamin	Ring Structure
A	β-ionone
D	Cyclo Pentano Perhydro phenanthrene (CPP) ring of steroids
E	Chromane ring in Tocopherol
K	Naphthoquinone
Vitamin A, D, E and K are made up of isoprene units	
B_1	Thiazole and substituted pyrimidine
B_2	Isoalloxazine ring
B_6	Pyridine ring
Folate	Pteroylglutamate
B_{12}	Corrin ring
Biotin	Imidazole and thiophene

COENZYME FORMS OF B-COMPLEX VITAMINS

Vitamin	Name	Coenzyme forms
B_1	Thiamine	Thiamine pyrophosphate/diphosphate
B_2	Riboflavin (Warburg yellow enzyme)	Flavin Adenine Dinucleotide (FAD) Flavin Mono Nucleotide (FMN)

Contd...

Vitamin	Name	Coenzyme forms
B_3	Niacin	NAD NADP
B_5	Pantothenic acid	Coenzyme-A (CoA)
B_6	Pyridoxine	Pyridoxal phosphate (PLP)
B_7	Biotin	Biocytin
B_9	Folic acid	Tetrahydrofolate (THFA)
B_{12}	Cobalamine	Adenosyl cobalamine Methyl cobalamine

VITAMIN A

There are three vitamers and one precursor of Vitamin A.

Retinol – contains alcohol (OH) group	Storage form of vitamin A Retinol is stored as retinyl esters (esterified to long-chain fatty acid) in liver.
Retinal – contains aldehyde (CHO) group	Retinol dehydrogenase mediates the reversible conversion of retinol to retinal. Retinal is the prosthetic group of visual pigments.
Retinoic acid – contains acidic (COOH) group	Aldehyde dehydrogenase mediates the irreversible oxidation of retinal to retinoic acid. Involved in growth and cellular differentiation—important for maintaining the integrity of epithelial cells.
β-carotene (provitamin A): plant source of A	Plant source of vitamin A. Theoretically, 1 molecule of β-carotene can be cleaved to 2-molecules of retinal. But, this conversion is inefficient in humans. β-carotene acts a lipid-phase antioxidant in the membranes. 12 µg of β-carotene = 1 µg Retinol activity equivalent (RAE)

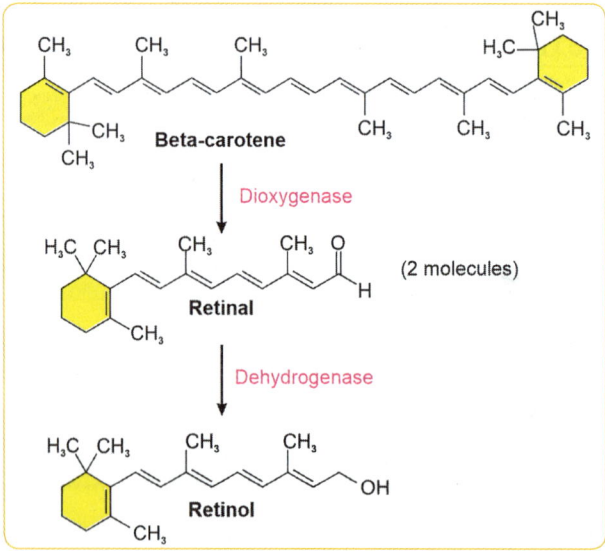

Fig. 1: Dioxygenase is involved in the cleavage of β-carotene to retinal

Figure 1 shows how Dioxygenase is involved in the cleavage of β-carotene to retinal.

Role of Vitamin A in Vision

Retinal is the prosthetic group of the visual pigment.
Rhodopsin = Opsin (protein) + 11-cis-retinal

Rhodopsin is a 7-transmembrane, serpentine, G-protein coupled receptor. The coupled G-protein is called as transducin.

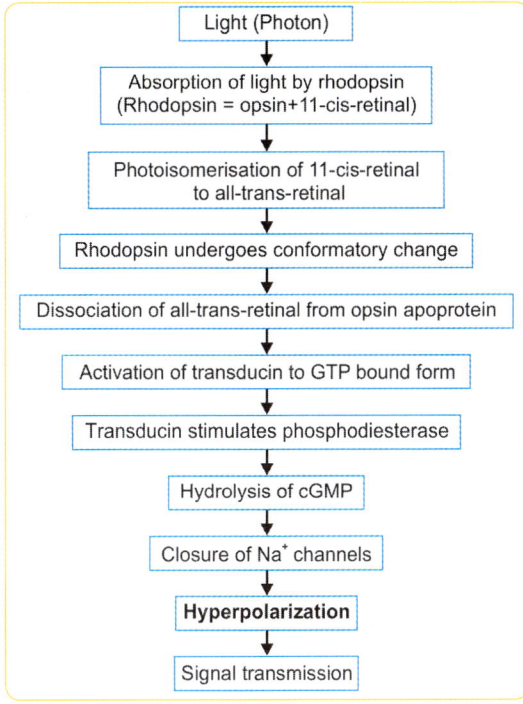

Wald's visual cycle explains the mechanism of regeneration of active 11-*cis* retinal from inactive all-*trans* retinal. Regeneration can happen through either of the two mechanisms:

i. Isomerization in the retinal pigment epithelium itself.

ii. Trans-retinal enters the blood and is converted into active form in liver and goes back to retina.

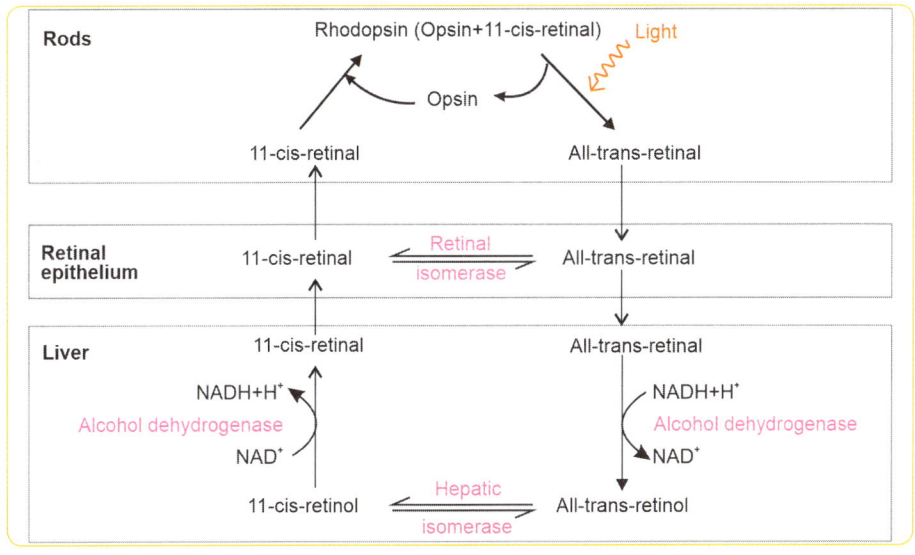

Note: In liver, vitamin A is stored in the hepatic stellate (Ito) cells located in the space of Disse

Deficiency Manifestations of Vitamin A

Ocular Changes

❑ Night blindness
❑ Xerophthalmia (dryness of conjunctiva)
❑ Bitot's spots (white patches of keratinized epithelium appearing on the sclera)
❑ Keratomalacia (softening of the cornea)
❑ Corneal scarring.

Changes to Loss of Integrity of Mucosal Epithelium

❑ Diarrhea, Dysentery and Respiratory Disease
❑ Children with Vitamin A deficiency are more susceptible to Measles.

Excess of Vitamin A is Toxic

❑ Acute toxicity manifestations: increased intracranial pressure (pseudotumor cerebri), vertigo, diplopia, seizures, and exfoliative dermatitis and death.
❑ Chronic toxicity manifestations: dry skin, cheilosis, glossitis, vomiting, alopecia, osteoporosis, hypercalcemia, lymph node enlargement, hyperlipidemia, amenorrhea, and features of pseudotumor cerebri.
❑ In experimental animals, it is found that both deficiency and hypervitaminosis of vitamin A reduce the stability of hepatic **lysosomes** and increase the release of acid hydrolases from lysosomes.

Tretinoin is All-trans Retinoic Acid (ATRA)

Tretinoin (all-trans retinoic acid (ATRA) is used for the treatment of acne and acute promyelocytic leukemia. We have already seen that retinoic acid is important for differentiation of epithelial cells. Tretinoin forces the differentiation of cancer cells.

Isotretinoin is 13-cis-retinoic Acid

❑ Isotretinoin is primarily used for the treatment of severe acne.
❑ Because of its teratogenic effect, it should be avoided in the periconceptional period.

VITAMIN E

Vitamin E family contains tocopherols and tocotrienols. There are 4 vitamers α, β, γ and δ in each group. α-tocopherol is the most active form (Fig. 2).
Dietary sources: Wheat germ oil, Sunflower oil, Nuts, etc.

Fig. 2: Structure of α-tocopherol–Can you appreciate the isoprene side chain?

Absorption and Transport

❑ Like all fat-soluble vitamins, bile salts are necessary for the absorption into lymphatic system.
❑ Vitamin E is incorporated into chylomicrons along with triglycerides.
❑ Stored in liver parenchymal cells, adipose tissue and adrenal glands.
❑ Exported out of the liver in association with VLDL.

$$R^{\cdot} + \text{Vitamin E}_{(reduced)} \longrightarrow R + \text{Vitamin E}_{(oxidized)}$$

Function

❑ Lipid phase, chain breaking anti-oxidant in the cell membrane and lipoproteins. First line of defense against lipid peroxidation.

- Maintains the fluidity of the membrane by preventing the oxidation of PUFA.
- In animals, vitamin E is important in reproductive function.

Vitamin E deficiency is seen in

- Chronic malabsorption syndromes, like celiac disease, cystic fibrosis and biliary atresia
- Total parenteral nutrition
- Abetalipoproteinemia

Premature infants are more prone to deficiency due to inadequate body stores and impaired absorption. Hemolytic anemia is the usual clinical presentation.

Excess intake of vitamin E, i.e. **Hypervitaminosis E leads to risk of bleeding**, especially in people taking aspirin and anticoagulants.

Both vitamin E and selenium have antioxidant function. So, vitamin E has sparing action on selenium.

Harrison's Corner

"Vitamin E appears to inhibit protein kinase C–mediated platelet aggregation and nitric oxide production."
(*Ref. Harrison 20th Ed., p.404*)

VITAMIN K

Vitamers of Vitamin K

- Phylloquinone (K_1)—found in plants
- Menaquinones (K_2)—produced by intestinal bacteria
- Menadione (K_3)—synthetic compound

Vit K Helps in Posttranslational γ Carboxylation of Glutamate Residues

Vitamin K is the cofactor for γ-**carboxylation of glutamate** residues of the following proteins to produce γ-carboxy glutamate (Gla):

- Clotting factors – II, VII, IX, X
- Protein C, Protein S, Protein Z
- Growth arrest-specific 6 (Gas6) protein
- Osteocalcin also known as bone Gla protein
- Periostin
- Matrix Gla Protein
- Transthyretin (Prealbumin)

Cranial nerves carrying parasympathetic supply – III, VII, IX, X
Vit K dependent clotting factors – II, VII, IX, X

Clinical Correlation

Menaquinone is synthesized by the intestinal flora. Intestine of newborn is sterile and the placental transport of vitamin K is poor. So, newborns, especially preterm ones are prone to bleeding. This is known as **hemorrhagic disease of new born.** This is prevented by injection of vit K1 at the time of birth.

Warfarin Acts as an Anticoagulant by Inhibiting the Epoxide Reductase

❏ As you can see from the image, vitamin K epoxide reductase (VKOR) is necessary for the regeneration of active vitamin K.
❏ Warfarin inhibits epoxide reductase.
❏ Vitamin K is the antidote in warfarin poisoning or toxicity.

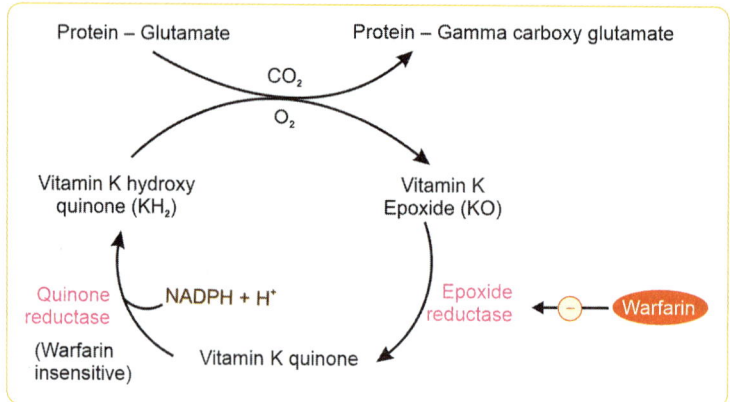

 Clinical Correlation

Vitamin K is needed for the γ-carboxylation of osteocalcin. **Long-term warfarin therapy is associated with reduced bone mass** (*Ref. Harrison 20th Ed., p. 2497*).
Vitamin K is the antidote for warfarin poisoning (c.f. Protamine sulfate is the antidote for heparin poisoning).

▶ youtu.be/ncgUmh5U4fk or scan QR code

**arrison's**
          Corner

In the absence of Vitamin K, γ-carboxylation of prothrombin is decreased. Des-γ-carboxy prothrombin also known as protein induced by vitamin K absence (PIVKA-II) is elevated in vitamin K deficiency or warfarin therapy. Elevated levels of PIVKA-II are also seen patients of hepatocellular carcinoma. *(Ref. Harrison 19th ed. p.547)*

VITAMIN D

1,25-dihydroxy vitamin D_3 also known as calcitriol is the biologically active form of vitamin D.
Calcitriol is formed in our body through series of reactions in skin, liver and kidney (Fig. 3).

❏ In the skin, on exposure to Ultraviolet-B rays of sun light or other sources, 7-dehydrocholesterol, an intermediate in the cholesterol synthesis is converted to cholecalciferol (contains single OH group).
❏ In the liver mitochondria, cholecalciferol is hydroxylated at the 25th carbon to produce 25-(OH) cholecalciferol (contains two OH groups)—the **major circulating form of vitamin D.** Half-life of 25-(OH) cholecalciferol is 3-weeks.
❏ 25-(OH) D_3 is hydroxylated at 1st position by 1 α hydroxylase enzyme to produce the most biologically active form of vitamin D—calcitriol (1,25-dihydroxy vitamin D). This conversion takes place in PCT of kidney, macrophages and placenta.
❏ Parathyroid hormone promotes the 1α hydroxylation.

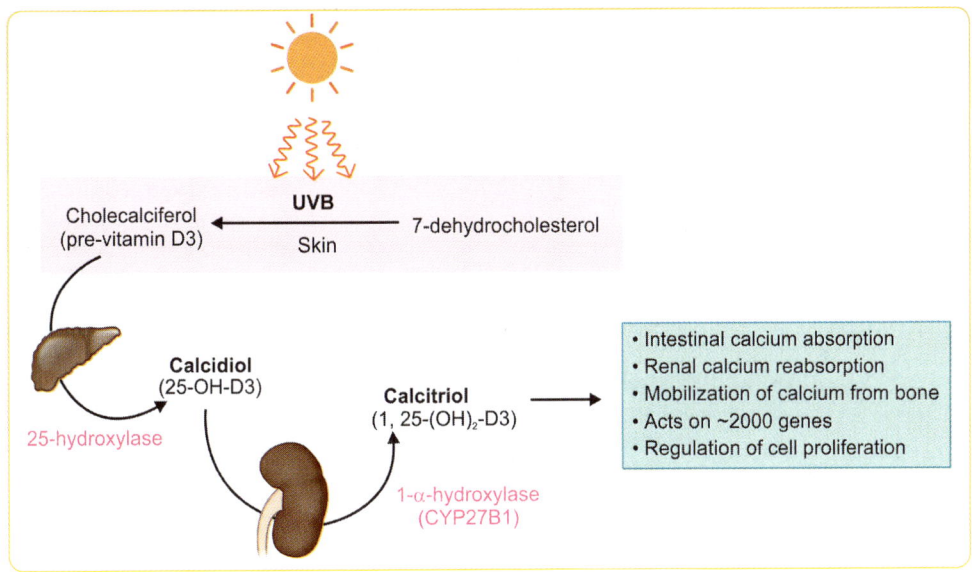

Fig. 3: Synthesis of Calcitriol

Vitamin D is more Like a Hormone

❑ It is produced in the body (skin, liver, kidney) and acts at distant organs (bone, intestine, kidney and many).
❑ It acts through (nuclear) receptors similar to hormones.
❑ Synthesis of vitamin-D is autoregulated according to the requirement.

Role of Vitamin D in Calcium Homeostasis

Intestine: ↑transcription of TRPV5, TRPV6, and Calbindin protein → ↑Ca and PO_4 absorption in small intestine
Bone: Mineralization of bone when calcium level is normal.
Kidney: ↑transcription of TRPV5 and TRPV6 → ↑Ca and PO_4 reabsorption

Other Roles of Vitamin D

❑ Vitamin D has various pleiotropic effects like regulation of the renin-angiotensin-aldosterone system, adiposity, energy expenditure, and pancreatic cell activity.
❑ Vitamin D is found to influence the activity of more than 2000 genes.

Measurement of Vitamin D in Clinical Laboratory

❑ 25-(OH) cholecalciferol is measured to assess the vitamin D status.
❑ Even though the sun is showing enough mercy on our nation, vitamin D deficiency has been a cause for concern in our country. Skin complexion, poor sun exposure, vegetarian food habits and lower intake of vitamin D-fortified foods could be attributed to the high prevalence of vitamin D deficiency.
Vitamin D deficiency: Vitamin D deficiency leads to Rickets in children and Osteomalacia in adults.

WATER-SOLUBLE VITAMINS

I have made a comprehensive table for all the water-soluble vitamins. Additional points about each vitamin is also given following the table. Make use of the table. I want you to know everything given in this table.

Vitamin	Coenzyme form	Reactions catalyzed	Deficiency
B_1	Thiamine pyrophosphate (TPP)	**Only 4 reactions:** 3 Oxidative Decarboxylation reactions: ❏ Pyruvate Dehydrogenase ❏ α ketoglutarate dehydrogenase ❏ Branched chain **ketoacid dehydrogenase** and ❏ Transketolase reaction Phosphorylation of Cl channel in central nervous system	Beriberi – 2 types ❏ Wet—High-output cardiac failure ❏ Dry—Symmetric peripheral nerve Wernicke encephalopathy and Korsakoff psychosis in alcoholics with chronic thiamine deficiency.
B_2	FAD	❏ Above-mentioned 3 Oxidative Decarboxylation reactions ❏ L-Amino acid oxidases ❏ **Fatty acyl-CoA dehydrogenase** ❏ **Succinate dehydrogenase** ❏ **Mitochondrial** glycerol-3-phosphate dehydrogenase ❏ Xanthine oxidase ❏ Glutathione reductase	Ulcers in angle of mouth (angular stomatitis), lips (**cheilosis**) and tongue (glossitis), seborrheic dermatitis, vascularization of cornea **Oculo-Orogenital Syndrome** is due to deficiency of vitamin B_2 and B_6.
	FMN	❏ D-Amino acid oxidases ❏ NADH-CoQ reductase (complex I) ❏ FMN reductase	
B_3	NAD⁺	(Pyruvate, Malate, Lactate, Isocitrate, Glutamate, Glyceraldehyde-3 phosphate, alpha keto glutarate, Branched ketoacid) dehydrogenase **Cytoplasmic** glycerol-3-phosphate dehydrogenases	
	NADP⁺	❏ G6PD and 6 phosphogluconate dehydrogenase of HMP shunt ❏ Malic enzyme ❏ Cytosolic Isocitrate dehydrogenase ❏ Glutamate dehydrogenase	Pellagra 3 Ds – Dermatitis, Diarrhea and Dementia, 4th D: Death if untreated.
	NADPH	**Reductive biosynthesis and xenobiotic metabolism** ❏ Enoyl-CoA reductase, Ketoacyl-CoA reductase (Fatty acid synthase) ❏ HMG-CoA reductase ❏ Folate reductase ❏ Ribonucleotide reductase via thioredoxin reductase ❏ Glutathione reductase ❏ Cytochrome P450 (mixed function oxidase) ❏ NADPH Met Hb reductase ❏ NO synthase ❏ NADPH oxidase ❏ Aldose reductase	**Casal collar or Casal necklace dermatitis** Causes of Pellagra: ❏ Nutritional ❏ Carcinoid syndrome ❏ Hartnup disease ❏ Drugs—Isoniazid

Adenine was considered as vitamin B_4. After the realization that adenine can be synthesized in the body, it was removed from the vitamin list leaving a gap in the numerical order of vitamins.

Contd...

Vitamin	Coenzyme form	Reactions catalyzed	Deficiency
B_5	Pantothenate	Component of coenzyme A and phosphopantetheine of acyl carrier protein.	Burning foot syndrome also known as. Grierson-Gopalan syndrome presents with severe burning of the feet, hyperesthesia, and vasomotor changes like excessive sweating. Ocular changes like scotoma and amblyopia can also occur.
B_6	PLP	**Carbohydrate metabolism:** ❏ Glycogen Phosphorylase **Amino acid metabolism:** ❏ All Transaminases – ALT, AST ❏ Cystathionine β synthase ❏ Cystathioninase ❏ Various amino acid decarboxylases ❏ Kynureninase **Heme synthesis** ❏ ALA synthase **Lipid metabolism:** ❏ Synthesis of arachidonic acid ❏ Synthesis of Sphingosine (Serine palmitoyltransferase and sphingosine-1-phosphate lyase) **Steroid hormone action:** ❏ Removes hormone receptor complex from DNA and terminates the action of hormone. **Formation of niacin from tryptophan**	❏ Sideroblastic Anemia – Heme synthesis is affected ❏ Peripheral neuropathy ❏ Early stroke due to Homocysteinemia. ❏ Generalized weakness ❏ Seizures (B6 is required for amino acid decarboxylation reactions, e.g. GABA synthesis)
B_7	Biotin	**Carboxylation reactions:** ❏ Pyruvate carboxylase ❏ Propionyl--CoA carboxylase ❏ Acetyl -CoA carboxylase ❏ 3-methylcrotonyl-CoA carboxylase ❏ Biotinylation reactions: ❏ Histone biotinylation is involved in regulation of gene expression	**Avidin** present in raw white egg has high affinity to biotin. This property is utilized in ELISA and other techniques. Multiple carboxylase deficiency is treated with biotin.

Inositol was the vitamin B_8.

B_9	Tetrahydro folate (THFA)	One-carbon carrier *See one-carbon metabolism*	Megaloblastic anemia; Deficiency in pregnant women leads to neural tube defects like spina bifida.

Para-aminobenzoic acid (PABA) was the vitamin B_{10}.

B_{12}	2 coenzyme forms known as **'cobamide coenzymes'**		❏ Subacute combined degeneration of spinal cord ❏ Megaloblastic anemia (due to folate trap) ❏ Methylmalonic aciduria
	(5'-deoxy) Adenosyl cobalamine	Mitochondrial **M**ethyl **m**alonyl -CoA mutase	
	Methyl cobalamine	Cytoplasmic **Meth**ionine synthase	

Contd...

Vitamin	Coenzyme form	Reactions catalyzed	Deficiency
B_{12} dependent Leucine amino mutase enzyme is not present in humans.			
Vit-C	As a reducing substance	❏ Helps in iron absorption by reducing Fe^{3+} iron to Fe^{2+} form. ❏ Helps in folate utilization by reducing folate to THFA.	
	Anti-oxidant	❏ Aqueous phase anti-oxidant	
	Co-enzyme	**Copper containing hydroxylases:** ❏ Dopamine β hydroxylase - synthesis of catecholamines **α-ketoglutarate linked iron containing hydroxylases:** ❏ Proline and Lysine hydroxylase – collagen maturation ❏ Proline hydroxylase – formation of Osteocalcin and C1q component of complement ❏ Aspartate β-hydroxylase – protein C modification ❏ Trimethyl lysine hydroxylases - carnitine synthesis	**Scurvy:** ❏ Defective post-translational modification of collagen leads to loss of tensile strength → Bleeding gums, Ecchymoses, Easy bruisability and poor wound healing ❏ Defective osteoid formation in bone

THIAMINE

Thiamine Pyrophosphate (TPP) is the cofactor needed for only 4 biochemical reactions in the body (the above). TPP is involved in energy metabolism. Deficiency of TPP will affect the link reaction and TCA cycle. This leads to reduced ATP production.

Beriberi (Sinhalese Word Meaning "I cannot") is Due Deficiency of Thiamine

Causes

❏ Alcoholism is the most common cause of beriberi because of poor intake of B_1 and inhibition of thiamine absorption by alcohol.
❏ Intake of polished rice (Polishing leads to loss of the vitamin-rich aleurone layer).

Types of Beriberi

❏ **Wet (cardiovascular)** beriberi – Impaired myocardial energy metabolism and peripheral vasodilation → ↑cardiac output → stimulation of Renin-Angiotensin system as vasodilation is perceived as relative loss of volume → salt and water retention → edema
  ○ **Soshin** beriberi – acute type of wet beriberi resulting in fulminant cardiac failure
❏ **Dry** beriberi – symmetric polyneuropathy, loss of deep tendon reflexes and ultimately complete paralysis
  ○ Wernicke encephalopathy – acute condition with ophthalmoplegia (gaze palsy), ataxia and cognitive dysfunction
  ○ Korsakoff psychosis – chronic condition with anterograde amnesia, i.e. failure to convert short-term memory to long-term memory. So, the patient fills the gap in memory by making up events (confabulation).
❏ **Infantile** beriberi – seen in infants of thiamine-deficient mothers as their milk is devoid of thiamine.
Wernicke-Korsakoff syndrome shows features of both Wernicke encephalopathy and Korsakoff psychosis.
Diagnosis: ↑ RBC Transketolase activity after adding TDP to erythrocyte lysate in vitro.

Treatment

❑ Injection of thiamine in acute stages is followed by oral supplementation.

 Clinical Correlation

Q. Assume: You are an intern. On one Sunday afternoon, an alcoholic person is brought to you in a state of disorientation. His blood glucose level is low. You decide to infuse 25% Dextrose. When you are about to start the drip, your consultant stops you and directs you to give something first. What could that be?

A. Intravenous infusion of glucose in thiamine-deficient patients should always be preceded by injection of thiamine otherwise glucose injection in thiamine-deficient patient will precipitate lactic acidosis.

PYRIDOXINE

Coenzyme Role of Pyridoxine

Pyridoxal 5'-phosphate (PLP) is the coenzyme form of pyridoxine. It is usually tightly attached (prosthetic group) to the apoenzyme via Schiff base formed between the aldehyde group of PLP and ε-amino group of lysine of the apoenzyme.

Pyridoxal 5'-phosphate (PLP) serves in a wide variety of chemical reactions involving amino acid (major role), carbohydrate (only glycogen phosphorylase) and Lipid metabolism. *See the above table for the list of reactions.*

In most of the PLP catalyzed reactions, aldehyde group of PLP forming a Schiff base is the basis of catalysis. But, in glycogen phosphorylase reaction, it is the phosphate group that plays the role.

Both deficiency and excess of pyridoxine will lead to peripheral neuropathy.

Therapeutic Roles of Pyridoxine

❑ Treatment of pellagra – (Niacin synthesis requires Pyridoxine)
❑ Prevention and treatment of INH (isoniazid) toxicity
❑ Copper-chelating agent penicillamine increases the risk of pyridoxine deficiency. So, it is supplemented in patients with Wilson's disease treated with penicillamine
❑ High doses of B_6 may help to activate cystathionine β synthase enzyme in patients of homocystinuria
❑ Certain cases of X-linked sideroblastic anemia respond to B_6.
❑ Pyridoxine-dependent seizures in infants won't respond to any anticonvulsant therapy except pyridoxine.

NIACIN (B_3)

Niacin is Not a Vitamin

Essential amino acid tryptophan can be converted into Niacin (B_3) in our body (60 mg tryptophan = 1 mg Niacin). As, it can be synthesized in the body by metabolism, strictly speaking, Niacin is not a Vitamin.

Deficiency of one vitamin can lead to deficiency of the other

Kynureninase, one of the enzymes involved in the conversion of tryptophan to niacin is Pyridoxal phosphate (B_6) dependant. Thus, deficiency of B_6 can lead to deficiency of B_3.

See the table for the coenzyme role of niacin

Causes of Pellagra

❑ Decreased intake of B_6
❑ Maize (corn) is used as staple food – Niacin is in bound form, and hence unavailable.
❑ Use of sorghum as staple food → **High amount of leucine** inhibits the tryptophan to Niacin conversion
❑ Drugs causing B_6 deficiency: Isoniazid (anti-tubercular drug)
❑ Carcinoid syndrome → diversion of tryptophan into serotonin production
❑ Hartnup disease → Defect in the absorption of tryptophan

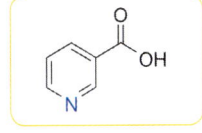

Fig. 4: Nicotinic acid

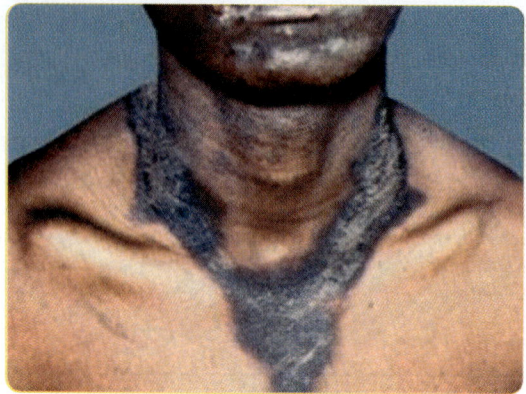

Fig. 5: The Casal collar or Casal necklace is an erythematous pigmented skin rash in the distribution of a broad collar (dermatomes C3 and C4) seen in patients with pellagra.

Carcinoid Syndrome Leads to Pellagra-like Symptoms

A carcinoid tumor is a neuroendocrine tumor of the gut and respiratory tract. Tumor cells secrete lots of serotonin which is responsible for the symptoms of flushing, diarrhea, wheezing and abdominal cramps. All the tryptophan is diverted into serotonin synthesis in carcinoid. This leads to decreased conversion of tryptophan to niacin and thus results in pellagra (Fig. 5).

arrison's
Corner

Mechanism of Isoniazid-induced Pellagra

"Isoniazid inhibits pyridoxal phosphokinase". The biologically active form of pyridoxine is pyridoxal phosphate. Thus, Isoniazid inhibits the formation of the active form of B_6.
We have already learned that B_6 is needed for the formation of Niacin (B_3).
(Ref. Harrison 20th Ed., p. 2687)

BIOTIN

Biotin (vitamin H) is the covalently attached cofactor (prosthetic group) for certain carboxylases. Biotin is linked to the carboxylase enzymes by an amide bond between the terminal carboxyl group of the biotin side chain and the ε amino group of a lysine residue of the enzyme.

The combined biotin and lysine side chains act as a long flexible arm that allows the biotin ring to access the active site of the enzyme efficiently (Fig. 6).

Fig. 6: Apocarboxylase + Biotin = Holocarboxylase

During the catalysis, biotin itself first undergoes carboxylation. This requires ATP. CO_2 is in the form of bicarbonate.

Biotin-Dependent Carboxylases

Enzyme	Role
Pyruvate carboxylase	The enzyme catalyzing the major anaplerotic reaction; key enzyme of Gluconeogenesis.
Propionyl-CoA carboxylase	Metabolism of branched-chain amino acids and odd chain fatty acids
	The key enzyme of fatty acid synthesis
Methylcrotonyl-CoA carboxylase	Metabolism of leucine

Biotin-Independent Carboxylases

Not all carboxylases are biotin-dependant. These are the carboxylase enzymes that do not require Biotin.
- ❑ AIR carboxylase → Purine synthesis
- ❑ CPS I → Urea synthesis
- ❑ CPS II → Pyrimidine synthesis
- ❑ Vit-K dependent gamma-carboxylase

> **Recent Update**
>
> **Role of Biotin in the cell cycle:**
> Biotinylation of histone proteins regulates gene function through epigenetic mechanisms.

Biotin-Avidin Interaction is one of the Strongest Interactions

Avidin is a biotin-binding protein. It is present in egg white. Another form of avidin produced by Streptomyces bacteria is known as streptavidin. The dissociation constant of avidin-biotin non-covalent interaction is 10^{-13} M. Thus this is one of the strongest known non-covalent interactions in the universe!
This principle is utilized in many biochemical techniques, some of them are:
- ❑ Enzyme-linked immunosorbent assay (ELISA)
- ❑ Chemiluminescence assay
- ❑ Affinity chromatography
- ❑ Cell-surface labelling
- ❑ Fluorescence-activated cell sorting (FACS)

Biotin is used in the Treatment of Multiple Carboxylase Deficiency

Multiple carboxylase/biotinidase deficiency is an autosomal recessive disorder due to a defect in holocarboxylase synthase or biotinidase deficiency.

FOLIC ACID

Folic acid is pteroyl-L-glutamate (Pteridine+PABA+Glutamate). It is the carrier of one-carbon groups.

Reactions Requiring Folate

- ❑ Conversion of dUMP to dTMP
- ❑ Synthesis of Serine from Glycine by the reversible conversion
- ❑ Regeneration of methionine by the methionine synthase reaction (homocysteine to methionine)
- ❑ C_2 and C_8 of purine ring

Methyl Folate Trap

Trapping of folate in the form of less utilizable N^5-CH_3-THFA due to deficiency of methylcobalamin is known as folate trap.

CH$_3$-THFA is the most reduced form of THFA, and it is not interconvertible to other forms like methylene and formyl THFA. Moreover, production of CH$_3$-THFA is an irreversible reaction. The only way to regenerate active forms of THFA is the following B$_{12}$ catalyzed reaction

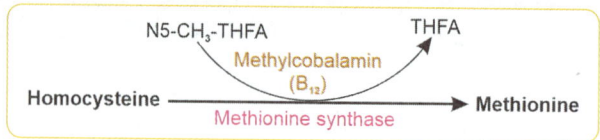

Therefore, Vit B$_{12}$ deficiency leads to functional folate deficiency.

FIGLU Excretion Test (Histidine Load Test) for Folate Deficiency

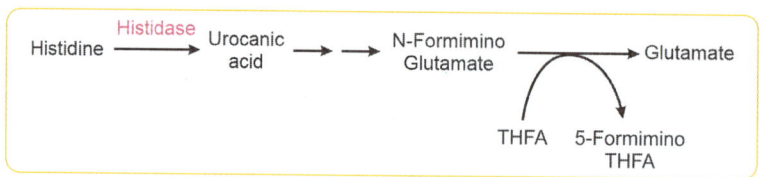

N-formimino glutamate (FIGLU) is a molecule formed during the catabolism of Histidine, one of the donors of one-carbon groups. Tetrahydro folate (THFA) is the acceptor of formimino group from histidine. If folic acid is deficient, this one carbon transfer won't happen and FIGLU will be excreted through the urine.

↑ FIGLU excretion following a load of histidine indicates folate deficiency.

These days, exact measurement of serum folate levels is done by methods like electrochemiluminescence method.

Folate Analogues are used as Anti-cancer, Anti-bacterial and Anti-malarial Drugs

The enzyme dihydrofolate reductase plays a crucial role in the formation of deoxythymine monophosphate from deoxyuridine monophosphate.

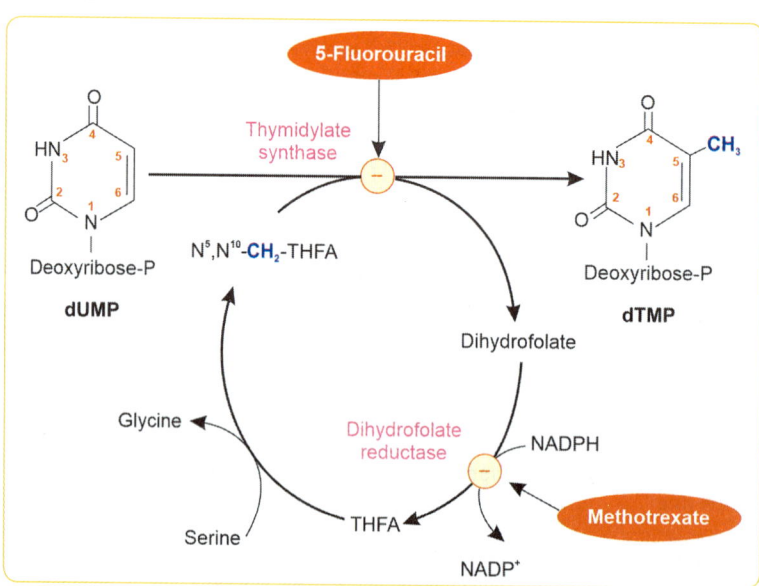

Drug	Class	Inhibition of
Methotrexate	Anti-cancer	**Human** Dihydrofolate reductase (DHFR)
Proguanil, Pyrimethamine	Anti-malarial	**Malarial** DHFR
Trimethoprim	Anti-bacterial	**Bacterial** DHFR

Bacteria can Synthesize Folate

Bacteria can synthesize all the 20 amino acids and many vitamins including folate. Sulphonamides resemble para aminobenzoic acid (PABA) (Fig. 7). That is why they inhibit the synthesis of folate in bacteria and have no harmful effect on humans as we get the vitamin through diet.

Sulfanilamide   PABA   Dihydrofolate

Fig. 7: Sulphonamides resemble PABA which is a component of folate. By Fdardel - Own work, CC BY-SA 3.0, https://commons.wikimedia.org/w/index.php?curid=15621310

Folinic Acid is used to Rescue Normal Cells After Methotrexate Therapy

❑ Folinic acid is 5-formyltetrahydrofolate. It is also known as **leucovorin** and citrovorum factor.
❑ Folinic acid can be directly converted to N^5, N^{10}-methylene tetrahydrofolate.
❑ When methotrexate is used for chemotherapy, normal cells are also affected.
❑ Folinic acid is administered to rescue the normal cells from the toxic effect of methotrexate.

VITAMIN B$_{12}$

Peculiarities of Vitamin B$_{12}$

❑ The only water-soluble vitamin that is stored in the body
❑ Synthesized exclusively by microbes
❑ The dietary source is exclusively of animal origin
❑ Rare compound with direct metal-carbon bond

Structure

B$_{12}$ possesses a cobalt-containing **corrin** ring *(as shown in adjacent figure.).* Similar to the iron of heme, cobalt makes 6 valances. 4 of the 6 valances are occupied by the nitrogens of the pyrrole rings of corrin ring. 5th valance is bound to the side chain dimethyl benzimidazole (DMB) moiety. 6th valance is free to be occupied. The group occupying this valance determines the name of the cobamide coenzyme.

R = 5'-deoxyadenosyl, Me, OH, CN

6th valence	Name	Importance
Cyanide	Cyano cobalamine	**Synthetic** form
5'-deoxy adenosyl	5'-deoxy adenosyl cobalamine	Only two biologically active cobamide coenzymes
Methyl	Methyl cobalamine	

Dietary Source

B_{12} is only of animal origin, and thus vegans are at high risk of Vit B_{12} deficiency. Clams and liver are the richest dietary sources. Milk and curd contain a low amount of Vit B_{12}.

Absorption

B_{12} in the food is in protein-bound form. In the stomach, pepsin degrades the bound proteins and releases B_{12}, which then binds to **salivary** B_{12}—binding proteins called Haptocorrins/cobalophillins/**R**-binders. Haptocorrins protect the B_{12} from acidic medium of stomach.

R-B_{12} complexes are transported to the duodenum and processed by pancreatic proteases; this releases B_{12}, which attaches to intrinsic factor (IF) secreted from the parietal cells of the gastric fundic mucosa.

IF-B_{12} complex passes to the distal ileum and binds to the cubam receptor, which leads to endocytosis of the complex. This receptor cubam is made up of cubulin and amnionless.

Note: Free B_{12} cannot bind to the receptor.

Mnemonic for Sites of absorption
Divya Is Just Feeling Ill, Bro
Duodenum – Iron, Jejunum -Folate; Ileum-B$_{12}$

Transport

The absorbed B_{12} is bound to transport proteins called **transcobalamins**, which then deliver it to the liver and other cells of the body. B_{12} is the only water-soluble vitamin to be stored in the body. Stored B_{12} in liver can last for years.

Coenzyme Role in Metabolism

Coenzyme form	Location	Reaction
5'-deoxy adenosyl cobalamine	Mitochondrial	Methyl malonyl CoA mutase and Leucine amino mutase (not in humans)
Methyl cobalamine	Cytoplasm	**Methi**onine synthase

Importance of Reactions

Only two reactions in the human body need Vit B_{12}. These reactions are important.

❏ **Methionine synthase reaction**:

Methylation of homocysteine to methionine is the only way of regeneration of active THFA from methyl THFA. So, deficiency of B_{12} leads to trapping of THFA in less utilized CH_3-THFA – a phenomenon known as folate trap. Homocysteinemia is treated by high dose folate.

❏ **Methyl malonyl-CoA mutase reaction**:

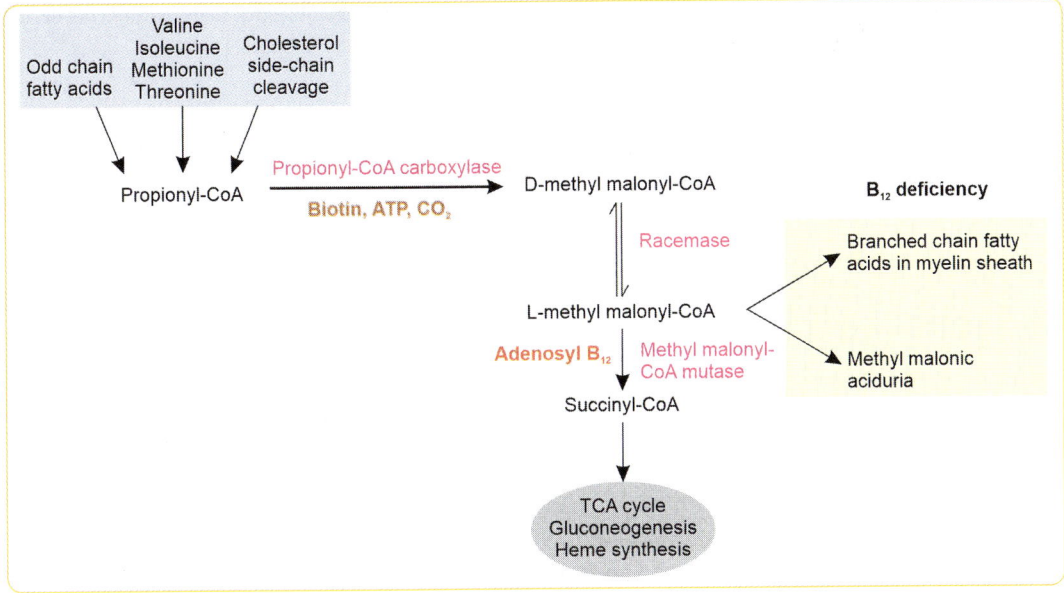

❏ Leucine amino mutase *(This enzyme is not present in humans)*

$$(2S)\text{-}\alpha\text{-leucine} \leftrightarrow (3R)\text{-}\beta\text{-leucine}$$

B₁₂ Deficiency Leads to Functional Folate Deficiency and Megaloblastic Anemia

❏ Deficiency of vitamin B_{12} leads to trapping of folate in the form of N^5-CH_3THFA, a less utilizable form.
❏ Folate is the one carbon donor involved in many steps of nucleotide synthesis. One such step is as follows:

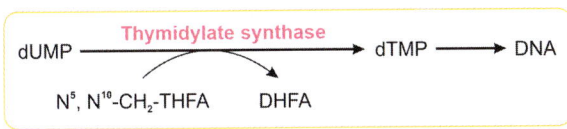

❏ Deficiency of B_{12} → Functional deficiency of Folate → ↓ Thymine triphosphate synthesis → Impaired nuclear maturation while cytoplasmic maturation is unaffected during cell division (nuclear-cytoplasmic asynchrony) resulting in megaloblastic cells → impaired RBC production and destruction of defective RBCs in the marrow (ineffective erythropoiesis) → Macrocytic anemia

B₁₂ Deficiency Leads to Neurological Symptoms—Subacute Combined Degeneration of Spinal Cord

The exact mechanism of neuropathy in B12 deficiency is not clear. The possible reasons are as follows:
❏ In B12 deficiency, methyl malonyl CoA cannot be converted to succinyl-CoA. Accumulated methyl malonyl-CoA is used for fatty acid synthesis to produce abnormal branched-chain fatty acids which are incorporated into myelin sheath.
❏ B_{12} deficiency reduces the level of S-Adenosyl methionine required for transmethylation reactions of neurotransmitter synthesis. So, B_{12} deficiency reduces the level of neurotransmitters.

Folate Supplementation May mask the Vitamin B12 Deficiency

❏ High intakes of folic fortified food and dietary folate supplements may mask the macrocytic anemia of B_{12} deficiency.
❏ Folic acid may produce a hematological response in vitamin B_{12} deficiency but may aggravate the neuropathy and also precipitate subacute combined degeneration of the spinal cord.

Chapter 33 Micronutrients: Vitamins

527

❏ Large doses of folic acid alone should therefore not be used to treat megaloblastic anemia unless the serum vitamin B_{12} level is documented to be normal.

Harrison's
Corner

Imerslund-Grasbeck syndrome is an autosomal recessive disease due to a defect in either cubulin or amnionless; presents with megaloblastic anemia and proteinuria. (*Ref. Harrison 20th Ed., p.645*)

 Video lecture on Vitamin B_{12} - youtu.be/Ku588t1yf_Q or scan QR code

ASSESSMENT OF DEFICIENCY OF VITAMINS

Vitamin	Detection
Thiamine-B_1	↑ RBC Transketolase activity after adding TDP to erythrocyte lysate in vitro
Riboflavin- B_2	↑ RBC glutathione reductase activity after adding FAD in vitro
Pyridoxine-B_6	↑ RBC transaminase activity after adding PLP in vitro Xanthurenic aciduria (Tryptophan loading test) due to impairment of B6 dependent kynureninase
Cobalamine-B_{12}	Measurement of B12 levels in the blood Methyl malonyl aciduria
Folate-B_9	Measurement of folic acid levels in the blood FIGLU excretion test (Histidine loading lest)

Competency

❏ BI6.5 Describe the biochemical role of vitamins in the body and explain the manifestations of their deficiency
❏ PA15.1 Describe the metabolism of vitamin B12 and the etiology and pathogenesis of B12 deficiency
❏ IM23.3 Discuss and describe the aetiology, causes, clinical manifestations, complications, diagnosis and management of common vitamin deficiencies
❏ PE12.2 Describe the causes, clinical features, diagnosis and management of deficiency/excess of vitamin A
❏ PE12.3 Identify the clinical features of dietary deficiency/excess of vitamin A
❏ PE12.6 Discuss the dietary sources of vitamin D and their role in Health and disease
❏ PE12.11 Discuss the RDA, dietary sources of vitamin E and their role in Health and disease
❏ PE12.14 Describe the causes, clinical features, diagnosis, management and prevention of Deficiency of Vitamin K
❏ PE12.17 Identify the clinical features of vitamin B complex deficiency

Key Points

- ❏ Vitamin K is the only fat-soluble vitamin with coenzyme function.
- ❏ Vit K is involved in γ-carboxylation of glutamate of clotting factors II, VII, IX, X, Protein C, S, Z, and Osteocalcin.
- ❏ Vitamin B_{12} is present only in foods of animal origin.
- ❏ Vitamin B_{12} is the only water-soluble vitamin stored in the body.
- ❏ Casal-necklace pattern (skin lesions in the neck) is seen in pellagra.
- ❏ Riboflavin deficiency leads to circumcorneal vascularization.
- ❏ The similarity between vitamin C and vitamin K is that both are involved in post-translational modification of proteins.
- ❏ As patients with Hartnup disease (neutral amino acid transporter defect) cannot absorb tryptophan, they develop pellagra.
- ❏ 60 mg of tryptophan = 1 mg Niacin; So, Niacin is NOT a true vitamin.
- ❏ Vitamin B6 is required for the decarboxylation and transamination of amino acids.
- ❏ Deficiency of selenium can be treated with Vitamin E, and the vice versa is also true. Selenium has a sparing effect on vitamin E.
- ❏ Major source of vitamin K is bacterial synthesis in the large intestine. As the gut of newborn is sterile, they are prone to hemorrhagic disease of newborn unless vitamin K is injected.
- ❏ The active form of vitamin D is 1,25 (OH)2 cholecalciferol. 25-α-hydroxylation takes place in liver, and then 1-α-hydroxylation takes place in the kidney.
- ❏ 24,25-dihydroxy cholecalciferol is an inactive metabolite.
- ❏ Tretinoin is all-trans-retinoic acid. Isotretinoin is 13-cis-retinoic acid.
- ❏ Imerslund-Grasbeck Syndrome occurs due to defect in cubulin or amnionless; presents with megaloblastic anemia.

SELF-ASSESSMENT

Short Answer Questions

1. Name a provitamin form of vitamin A. Add a note on sources and deficiency symptoms of vitamin A.

2. Mention the dietary sources of vitamin A. Explain its role in normal vision.

3. Write the function of vitamin D in intestine and bones. Discuss the mechanism of its action in the intestine.

4. Why is vitamin D considered as a hormone? Discuss its role in the regulation of calcium levels.

5. How is vitamin E transported from intestine to other tissues? Draw vitamin K cycle, mention where the cycle is blocked by warfarin, an anticoagulant drug.

6. Which vitamin deficiency is associated with hemorrhagic disease of the newborn? Explain the biological function of this vitamin.

7. Name the vitamin deficient in beriberi? Which intermediate will accumulate? Explain why? Name a test to diagnose this vitamin deficiency.

8. A chronic alcoholic patient admitted to the medicine ward is found to have hypoglycemia. For this patient, an intramuscular injection of thiamine was administered first before the intravenous infusion of 20% dextrose. What is the biochemical basis of administration of thiamine?

9. How does vitamin B1 deficiency lead to lactic acidosis?

10. Explain why patients with carcinoid syndrome are often presented with pellagra-like symptoms. Name one urinary metabolite that is measured for the diagnosis of this condition.

11. Name the physiologically active forms of cobalamin and write the reactions in which they participate.

12. A patient with a blood picture of megaloblastic anemia, when treated with folic acid alone, suffered from aggravated neuronal symptoms. Explain.

13. Draw the vitamin K cycle. Explain why vitamin K is injected in the cases of warfarin poisoning.

14. Explain with the help of a schematic diagram how photoexcitation of rhodopsin leads to the generation of an impulse in the optic nerve.

15. Explain the biochemical basis of bleeding gums and petechiae in scurvy.

16. Why is vitamin D called a prohormone?

17. How does vitamin E differ from vitamin C as antioxidants?

18. Which two vitamin deficiency states, are presented with a blood picture of macrocytic anemia? Explain the pathogenesis.

19. Give two biochemical reactions where vitamin B6 acts as a cofactor. Explain why the appearance of Xanthurenic acid in the urine (after tryptophan loading) is considered as an indicator of vitamin B_6 deficiency.

20. Name the vitamin deficiency state is associated with Xanthurenic aciduria. Why is pellagra a frequent accompaniment in this condition?

21. Give the biochemical functions of niacin with examples and the manifestation of its deficiency.

22. Describe the sources, digestion, absorption and blood transportation of Vit. B_{12} (cobalamin) in humans. Write two reactions occurring in the body which involve Vit B_{12} as a coenzyme.

23. Explain why there is a high risk of developing vitamin K deficiency in premature infants.

24. Name two enzymes and reaction catalyzed, which are present in RBCs that can be estimated as a marker of deficiency of two water-soluble vitamins.

Clinical Case-Based Questions

1. A 45-year old vegan male was presented to the OPD with non-specific complaints of tingling and numbness in his extremities. The attending physician observed mild pallor in the eyes and ordered some routine investigations which revealed increased MCV and hypersegmented neutrophils on a peripheral blood smear.
   a. What is the possible diagnosis?
   b. What is the biochemical basis of the neurological features in this patient?
   c. What is the pathogenesis behind the hematological manifestations?

2. A 32-year-old bodybuilder has decided to go on a diet consisting of raw egg whites with the misconception to ensure high proteins for his muscle growth. After a few weeks, he experienced easy fatigability after moderate exercise. He visited the nearby hospital and is found to be hypoglycemic. After correcting his blood glucose levels, the physician prescribed a vitamin supplement and sent the patient for nutritionist counselling.
   a. Which is the deficient vitamin in this patient and what is the cause?
   b. Name 3 enzymes that require this vitamin
   c. In a person with decreased activity, which enzyme could have been the cause of hypoglycemia?

3. A 44-year-old vegan consulted to his physician with the complaint of fatigue, muscle weakness, and depression. The initial laboratory profile included a hematological analysis that showed that he had macrocytic anemia. A bone marrow aspirate showed hyperproliferative marrow with megaloblasts. Further analyses revealed that his serum folic acid level was 1.2 ng/mL (normal 2.5 to 20), his serum B_{12} level report was awaited, but his serum iron level was normal.
   a. What is the probable diagnosis?
   b. Despite folate is present in the green leafy vegetable diet of this vegan, explain (with biochemical reactions) the basis of how this person is suffering from folate deficiency.

4. A 32-year-old male reported to the emergency with active bleeding from the nose. History revealed that he had been on Orlistat (pancreatic and hepatic lipase inhibitor) for weight reduction for the past two years. He had started Orlistat without the advice of any practitioner and had lost 20 kg of body weight. There was no history of hypertension, bleeding disorder or any other medical illness. No such bleeding episode occurred in the past. There was no abnormality detected upon local examination of the nose.
   a. What is the relationship between Orlistat and epistaxis in this person?
   b. The person is taking orlistat for the past 2-years. Why bleeding didn't happen earlier? Why now?
   c. Can biliary obstruction also produce similar kind of symptoms? Justify your answer.

Multiple Choice Questions

1. Which of the following vitamin is excreted in urine?
   a. Vitamin C
   b. Vitamin A
   c. Vitamin D
   d. Vitamin K

2. **Vitamin A intoxication causes injury to:**

   a. Lysosomes   b. Mitochondria
   c. Endoplasmic reticulum
   d. Microtubules

3. **Isotretinoin is:**

   a. All trans retinoic acid
   b. All cis retinoic acid
   c. 13 trans retinoic acid
   d. 13 cis retinoic acid

4. **Hypervitaminosis E leads to:**

   a. Hypercalcemia
   b. Raised intracranial tension
   c. Increased bleeding time
   d. Psychosis

5. **Which of the following is the major circulating form of Vitamin D?**

   a. $25\text{-}(OH)\,D_3$ b. $1,25\text{-}(OH)_2\,D_3$
   c. $24,25\text{-}(OH)_2\,D_3$ d. $1,24\text{-}(OH)_2\,D_3$

6. **Which of the following is the most biologically active form of Vitamin D?**

   a. $25\text{-}(OH)\,D_3$ b. $1,25\text{-}(OH)_2\,D_3$
   c. $24,25\text{-}(OH)_2\,D_3$ d. $1,24\text{-}(OH)_2\,D_3$

7. **Which of the following vitamin is metabolically produced by the body?**

   a. Biotin   b. Vitamin K
   c. Niacin   d. Pyridoxine

8. **Which of the following amino acids is a precursor of Vitamin B3?**

   a. Tyrosine   b. Threonine
   c. Tryptophan   d. Phenylalanine

9. **Which of the following is the synthetic vitamer of vitamin K?**

   a. Phylloquinone   b. Menaquinone
   c. Menadione   d. All of the above

10. **Which of the following tocopherol is the most active form of vitamin E?**

    a. Alpha   b. Beta
    c. Gamma   d. Delta

11. **Which one of the following amino acid residue is carboxylated by Vitamin K?**

    a. Aspartate   b. Glutamate
    c. Tryptophan   d. Tyrosine

12. **Vitamin K dependent clotting factors are:**

    a. Factor III and XI   b. Factor IX and X
    c. Factor V and VIII   d. Factor XI and XIII

13. **All of the following proteins contain gamma-carboxyglutamate residue, EXCEPT:**

    a. Osteocalcin   b. Osteopontin
    c. Prothrombin   d. Periostin

14. **Which of the following vitamin can perform its coenzyme function without phosphorylation?**

    a. B_1 b. B_2
    c. B_5 d. B_6

15. **Which of the following is a regulator of growth and differentiation?**

    a. Retinol
    b. Retinal
    c. Retinoic acid
    d. Retinol-binding protein

16. **Why does Vit B6 deficiency lead to anemia?**

    a. ALA synthase is affected
    b. Iron absorption is affected
    c. RBC life span is reduced
    d. B6 deficiency leads to bleeding

17. **The coenzyme not derived from vitamins is:**

    a. Coenzyme A   b. Coenzyme Q
    c. TPP   d. PLP

18. **PLP is not required for:**

    a. Cystathionine β synthase
    b. Glutamate decarboxylase
    c. Aspartate transaminase
    d. Glucose 6 phosphate dehydrogenase

19. **Which of the following is a non-vitamin coenzyme?**

    a. Coenzyme A   b. Tetrahydrobiopterin
    c. NADP   d. FAD

20. **All of the following vitamins are antioxidants, EXCEPT:**

    a. Vitamin A   b. Vitamin C
    c. Vitamin D   d. Vitamin E

21. **Inhibition of which of the following vitamin is caused by Isoniazid?**

    a. Niacin   b. Pyridoxine
    c. Thiamine   d. Folate

22. **Select the TRUE statement about vitamin B12:**

    a. Does not act as cofactor
    b. Involved in the transfer of amino groups
    c. Requires a specific glycoprotein for its absorption
    d. Dietary requirement is met by intake of plant products

23. **Thiamine deficiency causes decreased energy production because....**

    a. It is required for the process of transamination
    b. It is a cofactor in carboxylation reactions
    c. It is a co-enzyme for transketolase in pentose phosphate pathway
    d. It is a coenzyme for pyruvate dehydrogenase and α ketoglutarate dehydrogenase

24. **Biotin is used in the treatment of:**
   a. Multiple carboxylase deficiency
   b. Keto acid dehydrogenase deficiency
   c. Transketolase deficiency
   d. Multiple dehydrogenase deficiency

25. **Deficiency of which of the following vitamins leads to lactic acidosis?**
   a. Riboflavin
   b. Thiamine
   c. Niacin
   d. Pantothenic acid

26. **Oculo-Orogenital Syndrome is due to deficiency of:**
   a. Riboflavin
   b. Niacin
   c. Vit. A
   d. Vit. C

27. **Folinic acid is:**
   a. 5-formyl tetrahydrofolate
   b. 5-formyl dihydrofolate
   c. N5, N10-methylene tetrahydrofolate
   d. N5, N10-methylene dihydrofolate

28. **Excess of which of the following vitamin can cause hemolytic anemia?**
   a. Vit A
   b. Vit D
   c. Vit E
   d. Vit K

29. **Which of the following vitamins is used to increase the HDL levels?**
   a. Folic acid
   b. Nicotinic acid
   c. Thiamine
   d. Pyridoxine

30. **Deficiency of which of the following vitamins is associated with progressive keratinization of the cornea?**
   a. Vitamin A
   b. Vitamin C
   c. Niacin
   d. Pyridoxine

31. **Which one of the following is involved in the transfer of methyl, methylene and formyl groups in metabolic reactions?**
   a. Folic acid
   b. Nicotinic acid
   c. Thiamine
   d. Pyridoxine

32. **Wernicke encephalopathy is due to chronic dietary deficiency of:**
   a. Folic acid
   b. Nicotinic acid
   c. Thiamine
   d. Pyridoxine

33. **Vitamin required for the activity of transaminases is:**
   a. Folic acid
   b. Nicotinic acid
   c. Thiamine
   d. Pyridoxine

34. **Vitamin B12 is absorbed in the:**
   a. Stomach
   b. Duodenum
   c. Jejunum
   d. Ileum

35. **Which of the following vitamins are needed for the conversion of propionyl-CoA to succinyl-CoA?**
   a. Folate and Biotin
   b. B12 and Biotin
   c. Pyridoxine and B_{12}
   d. Folate and B_{12}

36. **Which of the following reaction is affected in patients of scurvy?**
   a. Carboxylation
   b. Acetylation
   c. Hydroxylation
   d. Transamination

37. **All of the following vitamins have known coenzyme function EXCEPT:**
   a. Thiamine
   b. Niacin
   c. Folic acid
   d. Tocopherol

38. **A patient with biliary atresia is more prone to the deficiency of:**
   a. Vit B12
   b. Vit E
   c. Vit C
   d. Niacin

39. **Most common presentation of Imerslund-Graesbeck syndrome is:**
   a. Megaloblastic anemia
   b. Microcytic hypochromic anemia
   c. Hereditary spherocytosis
   d. Hereditary elliptocytosis

40. **Excess dietary intake of carbohydrates increases the RDA level of:**
   a. Thiamine
   b. Vit E
   c. Vit C
   d. Pantothenate

ANSWERS

1.	a.	2.	a.	3.	d.	4.	c.	5.	a.	6.	b.
7.	c.	8.	c.	9.	c.	10.	a.	11.	b.	12.	b.
13.	b.	14.	b.	15.	c.	16.	a.	17.	b.	18.	d.
19.	b.	20.	c.	21.	b.	22.	c.	23.	d.	24.	a.
25.	b.	26.	a.	27.	a.	28.	d.	29.	b.	30.	a.
31.	a.	32.	c.	33.	d.	34.	d.	35.	b.	36.	c.
37.	d.	38.	b.	39.	a.	40.	a.				

34
CHAPTER

Micronutrients: Minerals

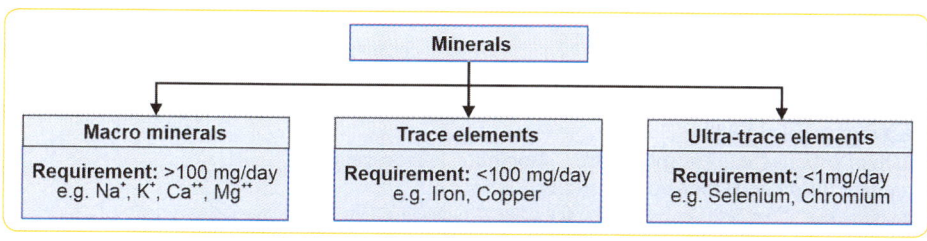

Minerals		
Macro minerals	**Trace elements**	**Ultra-trace elements**
Requirement: >100 mg/day e.g. Na^+, K^+, Ca^{++}, Mg^{++}	**Requirement:** <100 mg/day e.g. Iron, Copper	**Requirement:** <1mg/day e.g. Selenium, Chromium

CALCIUM

Human body contains about one kilogram of calcium. The normal level of calcium in serum is 8.5 to 10.5 mg/dl. This calcium includes both ionized and non-ionized calcium.

Ionized calcium is the biologically active form. Normal level of ionized calcium is 4.5 to 5.6 mg/dl.

Albumin Bound Calcium and the Influence of Blood pH on Albumin Binding to Calcium

Much of the non-ionized calcium, i.e. 40 to 45% of total calcium (4 mg/dL) is bound to albumin.

Changes in the albumin concentration lead to changes in the measured total calcium concentration. But the ionized calcium can be normal. That is why hypoalbuminemia won't produce tetany even if the total calcium is low (Fig. 1).

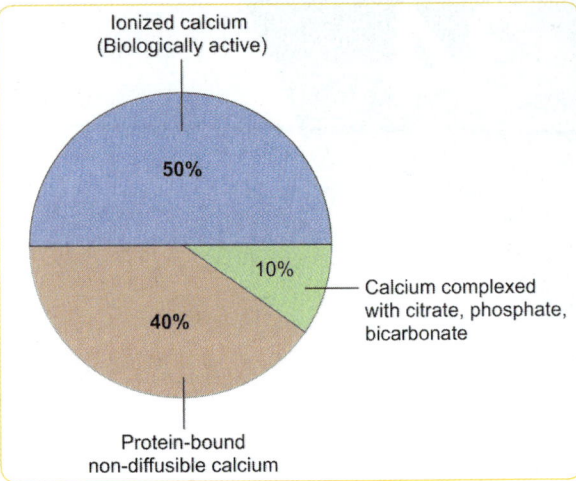

Fig. 1: Different forms of circulating calcium

Calcium binding to albumin is pH dependent. Acute changes in the pH lead to acute change in the calcium binding and cause changes in ionized calcium. But, total calcium will be normal. This is why measurement of ionized calcium is more important (Fig. 2).

❑ Acidosis → More positive charge in albumin → Decreases the calcium binding → Increases the ionized calcium
❑ Alkalosis → Less positive charge in albumin → Increases the calcium binding → Decreases the ionized calcium

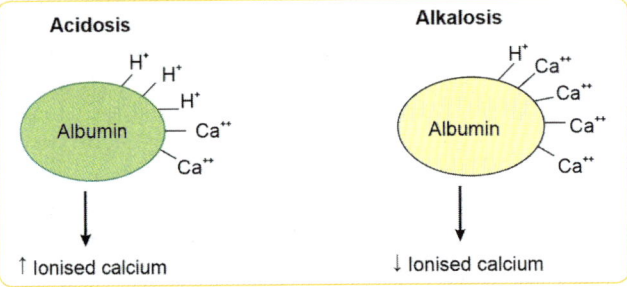

Fig. 2: Variation in albumin level may affect the total calcium. Ionized calcium is unaffected

Factors promoting calcium absorption	Factors inhibiting calcium absorption
Vitamin D	Phytates and Oxalates
Paratharmone (PTH)	High dietary phosphate
Gastric acidity	Alkaline pH of stomach
	Excess dietary fiber
	Fatty acid malabsorption

Biological Role (Functions) of Calcium

❑ Calcium hydroxyapatite [$Ca_{10}(PO_4)_6(OH)_2$] is the mineral component of bone and teeth
❑ Coagulation factor IV is nothing but calcium
❑ Neuromuscular coordination—transmission of nerve impulse and muscle contraction
❑ 2nd messenger in cell signaling
❑ Activation of enzymes, e.g. Phospholipase C, Isocitrate dehydrogenase
❑ Important for fusion and release of transport vesicles.

Calcium Binding Proteins and their Roles

- ❑ Calbindin—Intestinal absorption of calcium
- ❑ Calnexin and Calreticulin—Molecular chaperone involved in protein folding
- ❑ Calmodulin—small calcium protein/domain that modulate the activity of many enzymes
- ❑ Calcineurin—calcium and calmodulin dependent serine/threonine protein phosphatase involved in immune signaling.
- ❑ Calsequestrin—calcium storage protein in sarcoplasmic reticulum of muscle

Some Enzymes Regulated by Calcium/Calmodulin

- ❑ Adenylyl cyclase
- ❑ Guanylyl cyclase
- ❑ Nitric oxide synthase
- ❑ Glutamate decarboxylase
- ❑ Phosphatidyl Inositol 3 kinase
- ❑ Myosin light chain kinase
- ❑ cAMP phosphodiesterase
- ❑ Pyruvate dehydrogenase
- ❑ Phospholipase A_2
- ❑ Glycogen phosphorylase kinase

Causes of hypercalcemia	Causes of hypocalcemia
❑ Hyperparathyroidism	❑ Hypoparathyroidism
❑ Multiple myeloma and other osteolytic diseases	❑ Chronic Renal failure
❑ Metastatic carcinoma of bone	❑ Malabsorption
❑ Thiazide diuretics	❑ Acute pancreatitis
❑ Vitamin D intoxication	

- ❑ Thiazide diuretics are given in patients with hypercalciuria to prevent renal stones.
- ❑ Calcium level is normal in patients of Paget disease. There will be only elevated alkaline phosphatase activity.
- ❑ Acute pancreatitis leads to saponification (Calcium soap formation) and hence hypocalcemia.

Factors Regulating the Level of Calcium/Calcium Homeostasis

Hormone	Intestine	Bone	Kidney	Net effect
Vitamin D	↑transcription of TRPV5 and TRPV6 and Calbindin protein → ↑Ca and PO_4 absorption in small intestine	Bone mineralisation	↑transcription of TRPV5 and TRPV6 → ↑Ca and PO_4 reabsorption	↑ serum $[Ca^{2+}]$ and $[PO_4^{3-}]$
PTH	PTH is needed for 1-α-hydroxylase, enzyme that produces active Vit D_3	↑ bone resorption	Absorption of calcium and excretion of phosphate	↑ serum $[Ca^{2+}]$ ↓ serum $[PO_4^{3-}]$
Calcitonin	None (exact role is still debatable)	↓blood Ca and PO_4 Inhibits osteoclastic bone resorption	↑excretion of calcium and phosphate in urine	↓ serum $[Ca^{2+}]$

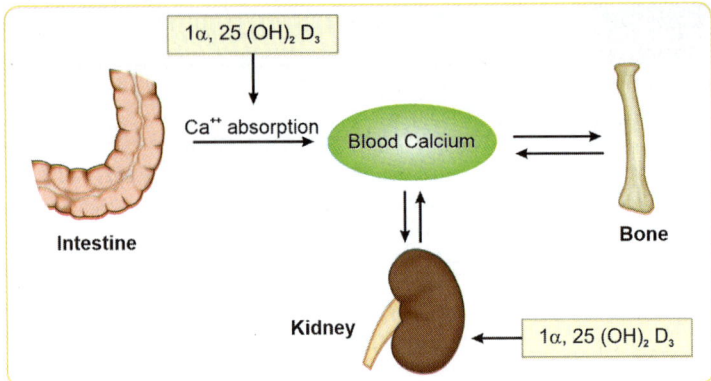

The concentration of calcium in the cytosol is very low, i.e. around 100 nanomole/L. Raised intracellular calcium leads to cell death by activation of endonucleases, proteases and phospholipase A_2.

Laboratory Estimation of Calcium

- ❑ Serum total calcium is measured by dye-binding method (O-cresolphthalein or Arsenazo-III dye)
- ❑ Serum ionized calcium is measured by ion-selective electrodes
- ❑ Intracellular calcium is measured using dyes like Fura-2

PHOSPHORUS

Role of Phosphorous

- ❑ 85% of body phosphorous is in combination with calcium to form mineral components of bones and teeth
- ❑ Component of nucleic acids
- ❑ Role in energy utilization and transfer via ATP and high energy phosphate compounds
- ❑ Phospholipids
- ❑ Enzyme regulation by covalent phosphorylation/dephosphorylation
- ❑ Regulation of oxygen delivery through 2,3-BPG
- ❑ Phosphate buffer system is an intracellular buffer and most predominant urine buffer; its urinary excretion increases with acidosis.

Causes of Altered Levels of Phosphate

Hyperphosphatemia	Hypophosphatemia
❑ Renal failure	❑ Hyperparathyroidism
❑ Hypoparathyroidism	❑ Renal disorders of phosphate reabsorption.
❑ Hypervitaminosis D	❑ Rickets (Vitamin D deficiency)
❑ Tissue trauma	❑ Ingestion of large amount of antacids
❑ Tumor lysis syndrome	

Note: Tumor lysis syndrome is associated with hyperphosphatemia, **hypocalcemia**, hyperuricemia, hyperkalemia. Laboratory estimation of inorganic phosphorus is done by **Fiske and Subbarao** method.

IRON

Iron and copper are transition elements, i.e. they can accept as well as donate electrons depending on their oxidation state. Ferrous iron is an electron donor whereas ferric iron is electron acceptor.

Biochemical Functions of Iron

- Transport of gases—Ferrous iron of heme transports oxygen. *Heme iron is oxygenated, not oxidized (Chapter 3)* *(Refer to page no. 27)*.
- Cellular respiration—Cytochromes *(Chapter 6)* *(Refer to page no. 79)* form the important component of electron transport chain. Heme iron undergoes reversible oxidation.
- Iron containing enzymes and iron activated enzymes *(Chapter 11.7)* *(Refer to page no. 221)* play an important role in **redox reactions**.

Human Body has no Mechanism for Excreting Iron

- Iron is so precious that nature has preserved it in the body as much as possible. No active way of iron excretion exists in the body. Absorption of iron at the level of intestinal mucosal cells is the only way of controlling body iron level.
- Loss of iron from body occurs during bleeding, shedding of epithelial cells of skin and mucosa. Menstrual bleeding and blood loss due to pregnancy predisposes females to iron deficiency.
- Hook worm (*Ancylostoma*) infestation leads to iron loss.

Dietary Sources of Iron

- Non-vegetarian sources: Liver, kidney, heart, red meat, egg yolk. **Heme iron is better absorbed than non-heme iron.**
- Vegetarian: Beans, peas, whole wheat, spinach, Moringa, Traditional sweeteners made in iron vessels (jaggery and palm sugar), foods cooked in iron vessels

Absorption of Iron and Mucosal Block Theory

- Iron is absorbed in the proximal small intestine, i.e. duodenum
- Heme iron is absorbed easily by a separate heme transporter (Heme carrier protein) and is oxidized by Heme oxygenase *(Chapter 11.7)* *(Refer to page no. 221)* to release iron
- Phytates and phosphates in the vegetarian diet chelate non-heme iron
- Non-heme iron enters the enterocyte through divalent metal transporter 1 (DMT1) also known as natural resistance macrophage-associated protein type 2 (Nramp-2)
- Non-heme iron is present as ferric form in the diet. DMT1 can absorb only ferrous (divalent) iron. So, ferric iron in the diet needs to be converted to ferrous form
- Ferric to ferrous conversion is done by:
  - ○ Vitamin C and gastric hydrochloric acid reduces the ferric iron to ferrous
  - ○ Membrane associated duodenal cytochrome b5 reductase (DCytb) also helps in reduction of ferric iron (Fig. 3)
- It is important to note that DMT1 is not specific for the transport of iron. It also transports other divalent metal ions like copper and zinc.

Chapter 34 Micronutrients: Minerals

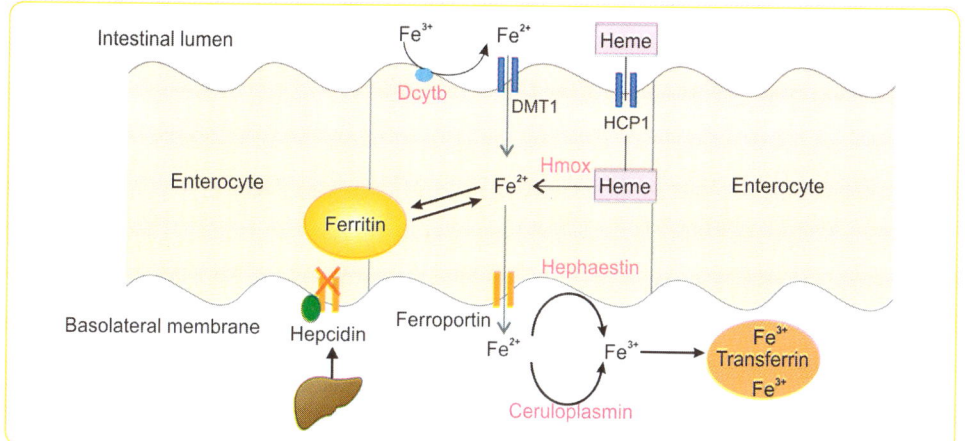

Fig. 3: HCP1 – Heme carrier protein, Hox – Heme oxygenase, DCytb – Duodenal cytochrome b; Appreciate that Hepcidin (blue circle) blocks ferroportin by causing internalization of hepcidin-ferroportin complex

Fate of Ferrous Iron

- Ferrous iron entered inside the duodenal cell is converted back to **ferric form for storage** in ferritin (around 4000 atoms are stored in one ferritin).
- When the storage is sufficient, **ferrous iron leaves the cell via basolateral membrane through ferroportin**.
- Transferrin is the transporter of iron. For transport, ferrous iron needs to be oxidized to ferric iron. This is done by hephaestin, a membrane associated ferroxidase protein.
- **In the blood, ceruloplasmin, a copper containing ferroxidase** oxidizes any ferrous iron to ferric iron
- Transferrin is the transporter of iron in blood. It can carry 2 molecules of ferric ion. (apo transferrin + 2 Fe^{3+} = Holo-transferrin)

Transferrin Receptor is Expressed in all the Cells

- Transferrin receptor is a transmembrane glycoprotein receptor involved in the receptor-mediated endocytosis of holotransferrin.
- **The acidic pH in the endosomes is responsible for the release of iron from transferrin inside**.
- STEAP3, a metallo reductase reduces the ferric iron to ferrous. DMT in the endosomal membrane exports the ferrous iron into cytosol.
- The iron can be incorporated into various iron containing proteins or stored in ferritin.

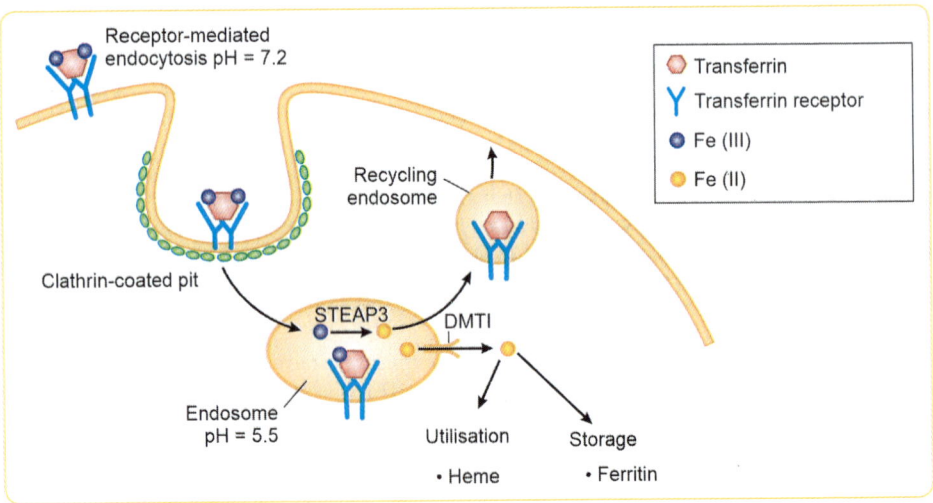

Two Types of Transferrin Receptors Play Different Roles

There are two types of transferrin receptors – TfR1 and TfR2. TfR1 is expressed in all the cells and have more affinity to holotransferrin. The expression of TfR1 and ferritin are reciprocally regulated at the level of gene expression by iron response protein which binds to the stem loop structures (**iron-response element**) present in the mRNA of these proteins *(Chapter 27) (Refer to page no. 453)*.

TfR2 has low affinity to holotransferrin. So, holotransferrin binds to TfR2 when TfR1 is saturated, i.e. when the iron is adequate.

Mucosal Block Theory

We have already learnt that no active way of excretion of iron exists in the body. The level of iron in the body is regulated at the level of absorption. Mucosal block theory explains how iron absorption from the duodenal mucosa to blood is inhibited when there is enough iron. **Hepcidin, a protein produced by liver** causes degradation of ferroportin. Thus, the iron that has entered the enterocytes is not absorbed into the blood. When the enterocyte is shed, iron is also lost.

Regulation of iron absorption is done by **iron sensing complex of liver**. This is made up TfR1, TfR2 and a transmembrane protein, HFE (Human hemochromatosis protein).

TfR1 is usually bound by HFE. Binding of holotransferrin to TfR1 displaces the HFE which binds to TfR2. As we have already learnt, holotransferrin binds to TfR2 only when the TfR1 is saturated. When TfR2 binds to holotransferrin and HFE, hepcidin production is initiated by Extracellular Signal-regulated Kinase-1 (ERK1)/Mitogen-Activated Protein Kinase (MAPK) pathway.

Defect in Hepcidin Production Leads to Hereditary Hemochromatosis

Any defect in hepcidin production leads to uncontrolled duodenal absorption of iron. As there is no way of active excretion, iron accumulates resulting in a condition known as hemochromatosis. **Mutation in the HFE gene is the most common.** Other genes mutated are hepcidin, TfR2, HJV(hemojuvelin), or ferroportin.

Iron Overload is Harmful to the Body

Free ferrous iron transforms the hydrogen peroxide molecule into more reactive hydroxyl radical through Fenton reaction *(Chapter 39) (Refer to page no. 591).*

$$Fe^{2+} + H_2O_2 \rightarrow Fe^{3+} + OH^\cdot + OH^-$$

This ferric iron is reduced back to ferrous form by other hydrogen peroxide molecules. So, this sets up a toxic cascade. That is why iron is always maintained in ferric state during transport and storage.

❑ Hemochromatosis is a condition in which free iron level is increased.
❑ Ceruloplasmin converts ferrous iron to ferric iron by its ferroxidase activity. Aceruloplasminemia leads to **hemochromatosis**.

Hemosiderin is Derived from Ferritin

Hemosiderin is not a separate chemical entity. It is actually derived from ferritin. It is a more stable, less available storage form of iron as compared to ferritin.

Hemosiderosis versus Hemochromatosis

	Hemosiderosis	Hemochromatosis
Accumulation of	Hemosiderin in reticuloendothelial system	Hemosiderin all over the body
Associated with tissue injury	–	+

Biochemical Basis of Anemia of Chronic Disease

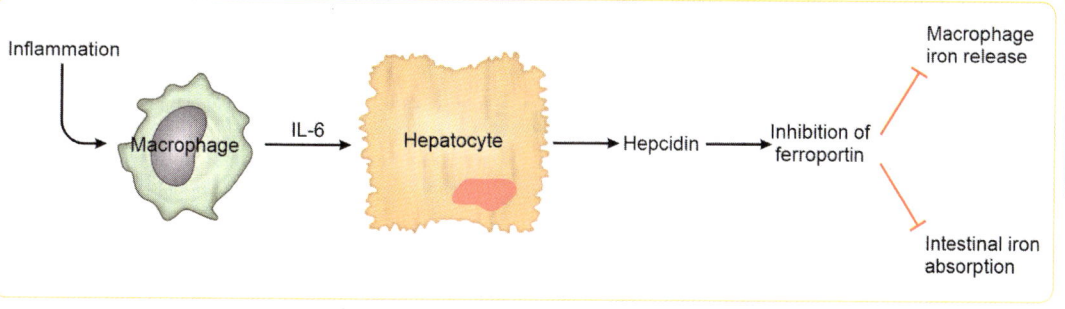

❑ Binding of Inflammatory cytokine IL-6 to IL-6 receptor in liver induces the production of hepcidin and other acute phase reactants like C-reactive protein.
❑ Hepcidin causes degradation of ferroportin in the macrophages also. This leads to accumulation of iron-in macrophage and anemia.

COPPER

In our diet, copper is present in shellfish, liver, nuts, legumes, bran, and organ meats. Copper is primarily absorbed in the duodenum.

Copper-Containing Enzymes

- Lysyl oxidase—crosslinking of collagen and elastin
- Tyrosinase—melanin synthesis
- Superoxide dismutase—anti-oxidant (scavenging of superoxide radicals)
- Cytochrome c oxidase—cellular respiration
- Cerulplasmin (Ferroxidase) and Ferroportin—Iron metabolism

Diseases of Copper Metabolism

	Menkes disease	Wilson's disease
Defective gene and its product	*ATP7A* – ATPase 7 (X chromosome)	*ATP7B* (chromosome 13)
Inheritance	X-linked recessive	Autosomal Recessive
Pathogenic defect	Intestinal absorption of copper	Defective biliary excretion of copper → accumulation of copper in liver → spillage into circulation due to liver damage
Blood and urinary copper	↓	↑
Clinical Features	Kinky hair and those of copper deficiency	Kayser–Fleischer ring in cornea, Liver cirrhosis, Neurological manifestations
Ceruloplasmin levels	Low since there is very less copper to incorporate into apoceruloplasmin.	Low since ATP7B mediates the transport of copper for the incorporation of six copper atoms into apoceruloplasmin in Golgi apparatus.

Wilson's Disease

Autosomal recessive disorder due to mutation in the gene for biliary copper transporter **ATPase7B** which is coded by chromosome 13.

Pathogenesis

Defective biliary excretion of copper → accumulation of copper in liver → spillage into circulation → ↑ Blood and urinary copper and deposition of copper in tissues.
Clinical features:
- **Hepatolenticular degeneration** – accumulated copper damages liver and lentiform nucleus of basal ganglia →Cirrhosis of liver and neuropsychiatric manifestations
- Kayser–Fleischer ring (**KF ring**) is a golden-colored ring in cornea due to deposition of copper in **Descemet's membrane**

Diagnosis

- ↑ urinary copper excretion
- ↓ serum Ceruloplasmin levels (accumulated copper in liver prevents conjugation of copper to Ceruloplasmin)
- Liver biopsy with copper quantitation is the "gold standard" method.

Treatment

- Penicillamine and Trientine – chelates copper by forming co-ordination complex
Zinc – reduces the intestinal absorption of copper and increases the level of copper-sequestering metallothionein

- ↓ in Ceruloplasmin is NOT the cause of Wilson's disease.
- Ceruloplasmin is NOT the transporter of copper. It is actually albumin.
- **Ceruloplasmin contains ferroxidase activity**. It converts ferrous iron to ferric iron helping in iron transport. That is why aceruloplasminemia leads to hemochromatosis.

ZINC

Zinc-containing Enzymes

Around 300 enzymes require Zinc. Some of them are:
- Carbonic anhydrase
- Alcohol dehydrogenase
- ALA dehydratase
- Alkaline phosphatase
- Glutamate dehydrogenase
- Lactate dehydrogenase
- Matrix Metalloproteinases

Physiological Role

- Storage and release of insulin. Storage granules of insulin are made up of 6 insulin molecules bound with two zinc ions.
- Zinc finger motifs of DNA binding proteins
- Important for growth and wound healing
- Absolutely needed for spermatogenesis
- Component of salivary polypeptide gustin (carbonic anhydrase VI) that is needed for development of taste buds.

Features of Zinc Deficiency

- Characteristic lesions in mouth, anal canal is seen (periorificial dermatitis)
- Mild deficiency: Hypogeusia (decreased taste sensation), Impaired immune function
- Chronic severe deficiency: Hypogonadism, Dwarfism, hypopigmented hair
- Acrodermatitis enteropathica is an autosomal recessive disorder with defective Zn absorption; presents with diarrhea, alopecia, muscle wasting, depression, irritability, and a rash involving the extremities, face, and perineum

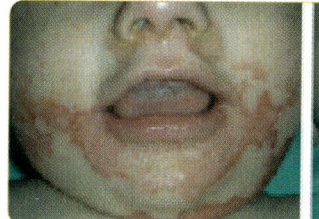
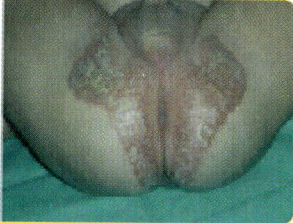

Fig. 4: Periorificial lesions in a patient of Acrodermatitis Enteropathica

Therapeutic Uses

- Administration of Zinc along with oral rehydration solution reduces the severity and duration of diarrhea in children.
- Zinc reduces the intestinal absorption of copper by increasing the level of metallothionein therefore used in Wilson's disease.

MAGNESIUM

Approximately 55% of the body's total magnesium is in bones along with calcium and phosphorous, 45% is intracellular and the rest is extracellular. One-third of the magnesium in ECF is albumin bound.

Role of Magnesium

❑ Component of bone: 55% of the body's total magnesium is in bones along with calcium and phosphorous.
❑ **Cofactor role:**
   Mg stabilizes the negative charges of β and gamma phosphates of ATP. In all reactions, ATP takes part as ATP-Mg complex or ATP-Mn complex.
   ○ All kinases, Enolase, Glucose-6-phosphatase, Pyruvate carboxylase, Alkaline phosphatase, DNA polymerases and RNA polymerases require Mg.
❑ Chlorophyll of plants contains magnesium which is comparable to iron of heme.

SELENIUM

Selenium is a component of Selenocysteine - 21st amino acid. The following are the selenium containing enzymes involved in redox reactions.
❑ Glutathione peroxidase
❑ **Thioredoxin reductase**
❑ **5′ deiodinase**
   Selenium is also a component of selenoprotein P – secretory glycoprotein involved in selenium homeostasis.
   As selenium is having **anti-oxidant** activity, it has **sparing effect on Vitamin E**, i.e. Vitamin E deficiency can be corrected with selenium and vice versa.

Deficiency

❑ **Keshan disease** → Dilated cardiomyopathy
❑ **Kashin–Beck disease** → Endemic osteochondropathy

Toxicity

❑ Selenosis – **Garlic breath** odour due to the pulmonary excretion of volatile Se metabolites

CHROMIUM

❑ **Chromium enhances the insulin sensitivity** probably by increasing insulin receptor mediated signaling.
❑ Dietary source: yeast, meat, and grains
❑ Chromium deficiency leads to impaired glucose tolerance; toxicity (most commonly due to occupational exposure) leads renal failure, dermatitis, pulmonary cancer.

Remember

Role of Metals on Insulin
❑ Zinc is needed for storage of insulin.
❑ Calcium is needed for release of insulin.
❑ Chromium is needed for inulin signaling and action.

MOLYBDENUM

Molybdenum is a prosthetic group of:
- Sulfite oxidase
- Xanthine oxidase

Molybdenum deficiency may result in neurological abnormalities; toxicity leads to reproductive and fetal abnormalities.

FLUORIDE

Normal fluoride content of water should be between **0.5 to 0.8 mg/dl**. Fluoride is used in toothpaste to prevent dental caries.

Fluorosis

Fluorosis results from presence of excess fluoride (3 to 5 mg/L) in drinking water. Fluoride gets deposited in skeletal tissue as *flurohydroxyapatiite* crystals which are resistant to osteoclastic resorption.
- >1–2 mg/L → Dental fluorosis
- >3–6 mg/L→ Skeletal fluorosis
- >10 mg/L → crippling fluorosis

Causes

- Fluorosis is endemic in certain parts of our nation due to high fluoride content of drinking water and plants. Sorghum concentrates fluoride.
- Occupational exposure

Types

- Dental fluorosis
- Skeletal fluorosis

Clinical Features

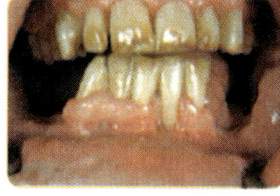

Fig. 5: Dental fluorosis

Dental fluorosis: mottled enamels of permanent teeth (Fig. 5)
Skeletal fluorosis:
- Usually males are affected.
- Upper vertebrae are usually affected.
- Stiffness of major joints due to calcification of ligaments
- Compression of vertebral space causes radiculomyelopathy. Tingling sensation (paraesthesia), muscle paralysis are seen.

Treatment

- There is no specific treatment. Decompression surgery is done in patients with spinal involvement to relieve the symptoms.

Prevention

- Alternative water supply with low fluoride.
- Removal of existing fluoride from water: Nalgonda technique is a method developed by National Environmental Engineering Research Institute (NEERI) for defluoridation.
- Avoiding fluoride-containing toothpaste for children.

Competency

- BI6.9   Describe the functions of various minerals in the body, their metabolism and homeostasis
- PA14.1   Describe iron metabolism
- PE13.1   Discuss the dietary sources of Iron and their role in health and disease
- BI6.10   Enumerate and describe the disorders associated with mineral metabolism
- DR17.4   Enumerate and describe the various changes in Zinc deficiency
- DR17.2   Enumerate and describe the various skin changes in Vitamin B complex deficiency

Key Points

- Alkalosis leads to decrease in ionized calcium; predisposes to tetany.
- Hepcidin causes degradation of ferroportin.
- DMT-1 is not specific for iron. It also transports selenium, zinc, Calcium, etc.
- Ceruloplasmin is a copper-containing ferroxidase. It oxidizes iron from ferrous to ferric state.
- The human body has no mechanism for excreting iron.
- Wilson's disease and Menke disease are due to defect in ATP7B and ATP7A respectively.
- Glutathione peroxidase, Thioredoxin reductase and 5' deiodinase contain selenium.
- Zinc is needed for storage of insulin. Calcium is needed for the release of insulin. Chromium is needed for insulin signaling and action.
- Nalgonda technique is a household method of defluoridation.
- Lead is not an essential mineral.

SELF-ASSESSMENT

Short Answer Questions

1. Define Ultratrace elements. Give one example.
2. Hypoalbuminemia leads to hypocalcemia but not tetany, Why?
3. What is the biochemical basis of development of tetany in a patient with alkalosis?
4. Name any four functions of calcium in the body.
5. Name any three calcium-binding proteins and their role.
6. What is the role of ceruloplasmin in iron metabolism?
7. Describe in detail the sources, absorption, functions and factors regulating blood calcium level. Discuss one clinical condition with abnormal blood calcium level.
8. Enumerate the proteins involved in iron absorption. Briefly discuss the role of these proteins in regulating iron absorption using diagrams. Add a note on mucosal block theory.
9. Name the proteins involved in the storage and transport of iron respectively. What changes are observed in the patients of iron deficiency anemia?
10. What is the recommended level of fluoride in drinking water? Justify the statement, "Fluoride may act as a double-edged sword in our body".

Multiple Choice Questions

1. All of the following enzymes are regulated by calcium/calmodulin, except:
   a. Hexokinase
   b. Nitric oxide synthase
   c. Pyruvate dehydrogenase
   d. Phosphatidyl Inositol 3 kinase

2. Which of the following is a small calcium binding protein that modifies the activity of many enzymes and other proteins in response to changes of Ca^{2+} concentration?
   a. Cyclin
   b. Calmodulin
   c. Collagen
   d. Kinesin

3. **All of the following are causes of hypocalcemia, EXCEPT:**
   a. Acute pancreatitis
   b. Chronic renal failure
   c. Hypoparathyroidism
   d. Vitamin D intoxication

4. **All of the following are seen in tumor lysis syndrome, except:**
   a. Hypercalcemia
   b. Hyperkalemia
   c. Hyperphosphatemia
   d. Hyperuricemia

5. **Iron is absorbed in:**
   a. Stomach
   b. Duodenum
   c. Jejunum and ileum
   d. Colon

6. **All of the following proteins are involved in the metabolism of iron, except:**
   a. Transthyretin
   b. Ceruloplasmin
   c. Ferritin
   d. Hepcidin

7. **Which of the following is a non-essential mineral?**
   a. Iron
   b. Lead
   c. Manganese
   d. Sodium

8. **Which of the following is a non-specific regulator of iron metabolism?**
   a. Ferroportin
   b. Ferritin
   c. DMT-1
   d. Transferrin

9. **Selenium is required for all of the following enzymes, EXCEPT:**
   a. Glutathione reductase
   b. Glutathione peroxidase
   c. 5' deiodinase
   d. Thioredoxin reductase

10. **Which of the following molecules delivers iron to tissues by binding to specific cell surface receptors:**
    a. Ferritin
    b. Ferredoxin
    c. Hemosiderin
    d. Transferrin

11. **Which of the following enzyme contains molybdenum prosthetic group?**
    a. Adenylyl cyclase
    b. Xanthine oxidase
    c. Uricase
    d. Adenine deaminase

12. **Keshan disease is due to deficiency of:**
    a. Chromium
    b. Molybdenum
    c. Selenium
    d. Nickel

13. **Deficiency of which of the following metal leads to hypogeusia?**
    a. Copper
    b. Zinc
    c. Calcium
    d. Phosphorous

14. **Which of the following mineral can induce the synthesis of metallothionein?**
    a. Zinc
    b. Calcium
    c. Iron
    d. Phosphorus

15. **Which of the following gene is defective in Wilson's disease?**
    a. ATP7A
    b. ATP7B
    c. ATP7C
    d. ATP7D

16. **Which of the following trace elements is important for conversion of procollagen to collagen?**
    a. Selenium
    b. Copper
    c. Zinc
    d. Magnesium

17. **Which of the following major minerals is a component of biological membranes?**
    a. Potassium
    b. Sodium
    c. Calcium
    d. Phosphorus

18. **Which of the following mineral has similar action like vitamin E?**
    a. Calcium
    b. Iron
    c. Selenium
    d. Magnesium

Chapter 34 Micronutrients: Minerals

ANSWERS					
1. a.	**2.** b.	**3.** d.	**4.** a.	**5.** b.	**6.** b.
7. b.	**8.** c.	**9.** a.	**10.** d.	**11.** b.	**12.** c.
13. b.	**14.** a.	**15.** b.	**16.** b.	**17.** d.	**18.** c.

V

SECTION

Special Topics

Section Outline

35 CHAPTER

Plasma Proteins

In humans, except albumin, all the plasma proteins are glycoproteins. Predominant protein in plasma is albumin.
- Total protein—6–8 g/dL
- Albumin—3.5–5 g/dL
- Globulin—2.5–3.5 g/dL
- A:G ratio—1.5 to 2.5 : 1

PLASMA VERSUS SERUM

	Plasma	Serum
Description	Fluid portion of blood after the cells are removed by centrifugation	Fluid component of blood after the blood is allowed to clot. i.e. Serum = Plasma – fibrinogen
Anticoagulants needed	Heparin EDTA Citrate solutions	None
Analytes measured	Glucose, coagulation profile	Most of the analytes

Many of the plasma proteins are produced by liver, except:
- Immunoglobulins (produced by plasma cells)
- Clotting factor III also known as tissue factor (produced by endothelium), factor VIII and Von-Willebrand factor (produced by platelets and endothelium)

MEASUREMENT OF PROTEINS IN CLINICAL BIOCHEMISTRY LABORATORY

- ❑ Total protein is usually measured by Biuret method.
- ❑ Albumin is usually measured by bromocresol green (BCG) or bromocresol purple (BCP) dye binding method.
- ❑ The amount of globulin is calculated by the difference between total protein and albumin.

METHODS OF SEPARATION OF PLASMA PROTEINS

- ❑ Differential centrifugation
- ❑ **Salt fractionation**
- ❑ **Electrophoresis**
- ❑ Chromatography

Salt Fractionation

This is based on the difference in solubility of proteins with different salt concentrations. Proteins are charged poly-ions—that is why they are surrounded by a coat of water.

If the concentration of salt is high, the ions of salt will compete for the water molecules bound to the proteins in solution. Without water molecules to associate with, the proteins will seek complementary charges on other proteins, thus aggregating together and precipitating out of solution. This is known as *salting out.*

If you see the following figure, you can notice that when the salt concentration is optimum the protein solubility is maximum. This is known as salting in. This is because the charged atoms of the salt associate with the complementary charges of the protein and thus preventing the protein-protein interaction and precipitation.

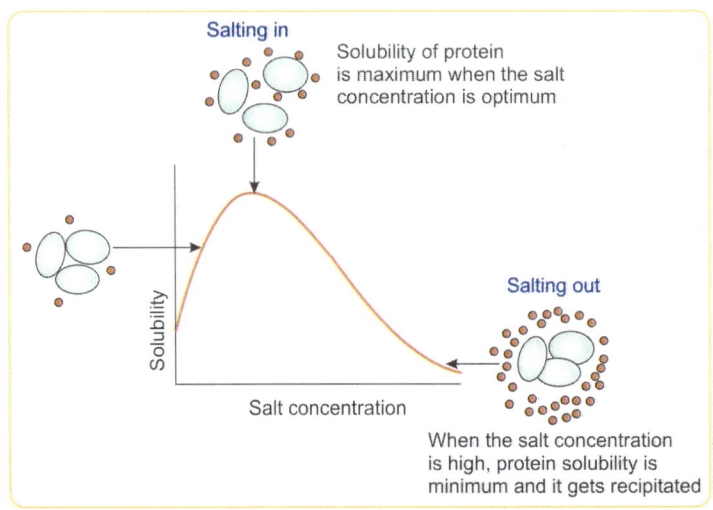

Salts used:
- ❑ Ammonium sulfate—most common.

Half saturation and full saturation:
- ❑ On 1/5th saturation with ammonium sulfate, fibrinogen is best precipitated.
- ❑ On 1/3rd saturation, high-molecular weight globulins get precipitated. It is called euglobulin.
- ❑ On ½ saturation, total globulins get precipitated out. It is called pseudoglobulin.
- ❑ Albumin is precipitated out by full saturation (Memory aid: All – Full).

Separation of Serum Proteins is Done by Electrophoresis at Alkaline pH

Electrophoretic separation of plasma proteins is done at an alkaline *pH of 8.6* using *Tris-barbital buffer*. As this pH is higher than the pI of most amino acids, proteins carry net negative charge at this pH and *move towards*

the anode. Separated proteins are stained with stains like *Coomassie brilliant blue and amido black. Five major bands* are observed when agarose is used as the support medium. More number of bands are obtained using other support mediums for electrophoresis.

Proteins like histones contain a lot of positively charged amino acids and they would not move towards anode.

In clinical chemistry laboratory, the main indication for doing serum protein electrophoresis is for the detection of multiple myeloma. So, we will discuss the diagnosis of multiple myeloma in detail in the later part of this chapter.

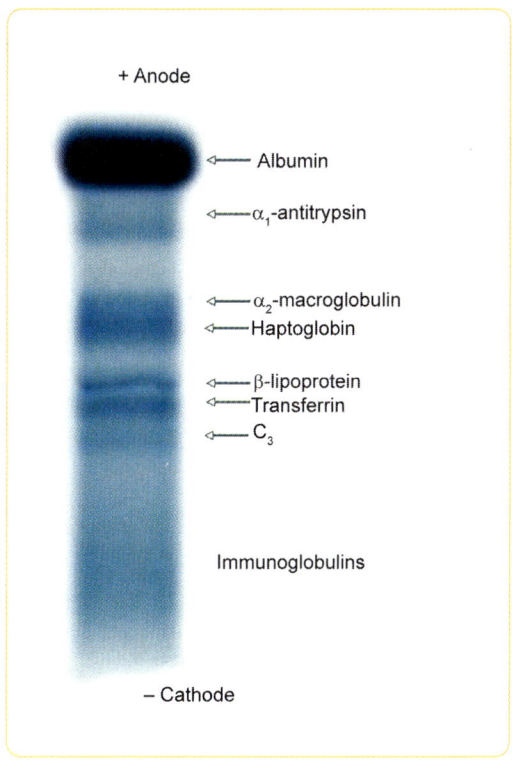

+ Anode

←—— Albumin

←——α_1-antitrypsin

←——α_2-macroglobulin
←——Haptoglobin

←——β-lipoprotein
←——Transferrin
←—— C_3

Immunoglobulins

– Cathode

Fig. 1: Electrophoretogram showing the protein components of serum using polyacrylamide support medium

Region	Components	Function and clinical significance
Before albumin	Retinol binding protein (RBP)	❑ Vitamin A transport ❑ Marker of visceral protein mass (Ref. Harrison 19th Ed., p. 462)
	Prealbumin (Transthyretin)	❑ Transport of thyroid hormones and RBP ❑ Deposited in familial amyloid polyneuropathy ❑ Marker of short-term nutrition as half-life is 12 hours
Albumin	Albumin	Transporter, oncotic pressure maintenance, marker of long-term nutrition, etc.
α_1	High density lipoprotein	Reverse cholesterol transport
	α_1 antitrypsin	❑ Elastase inhibitor; defect results in emphysema and cirrhosis ❑ Smoking causes oxidation of methionine 358 (critical residue) of α_1-antitrypsin
	α_1 acid-glycoprotein	❑ Also known as orosomucoid ❑ Transporter of progesterone and many drugs

Contd...

Region	Components	Function and clinical significance
α₂	Haptoglobin	❏ Binds to free hemoglobin (hemopexin binds to free heme) ❏ Decreased serum level signifies intravascular hemolysis
	α₂ –macroglobulin	❏ Very high-molecular weight component ❏ Protease inhibitor ❏ Raised levels are seen in nephrotic syndrome
	Ceruloplasmin	❏ Oxidizes ferrous iron to ferric (Ferroxidase) ❏ Low levels are seen in Wilson's disease ❏ Albumin is more important in copper transport
β₁	Transferrin	❏ Transports 2 Fe³⁺ ❏ 1/3rd saturated with iron normally ❏ Negative acute-phase reactant
	Low density lipoprotein	Supplies cholesterol to peripheral tissues
	C₄	Complement protein
β₂	β₂ microglobulin	❏ Component of MHC-I ❏ Prognostic marker in multiple myeloma ❏ Component of long-standing renal dialysis related amyloid (Aβ2M)
	C₃	Complement protein
	IgA	Secretory antibody that protect GIT and airways
β-γ junction	Fibrinogen*	❏ Arrests the bleeding by forming fibrin clots ❏ Positive acute phase protein responsible for raised erythrocyte sedimentation rate
γ globulins	IgG	Major antibody of body fluids
	IgM	Antibody of primary immune response
	C-reactive protein	❏ 'C' refers to capsular polysaccharide of *Pneumococcus* ❏ Major acute phase protein ❏ Initiates classical complement pathway

Fibrinogen is not present in serum; only seen in plasma protein electrophoresis.

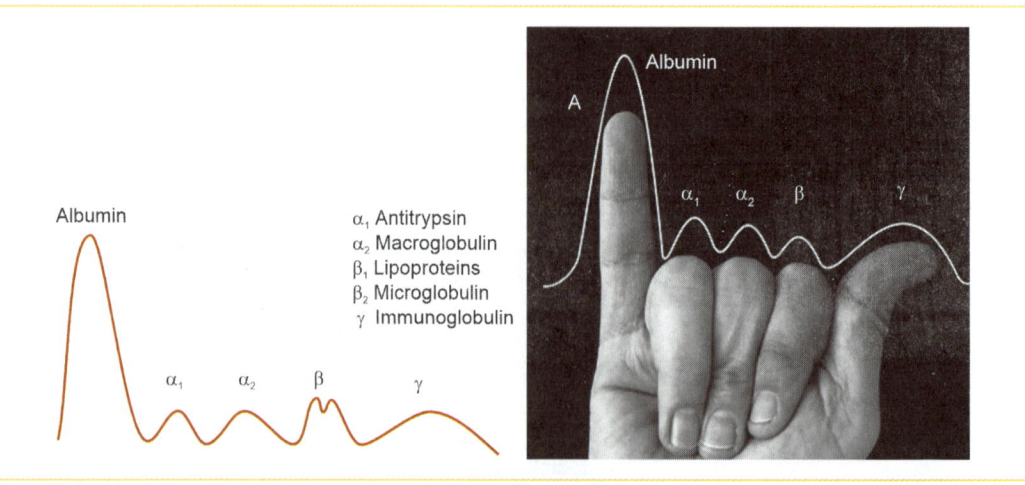

Albumin

Albumin

A

α₁ Antitrypsin
α₂ Macroglobulin
β₁ Lipoproteins
β₂ Microglobulin
γ Immunoglobulin

α₁ α₂ β γ

Fig. 2: An easy way to remember the components and their proportions

Section V Special Topics

Densitometry is used for Quantitation of Electrophoresed Proteins

- What you are seeing in Fig. 2 is the densitometric result of electrophoresed serum proteins.
- Area under the curve is used to calculate the concentration of the component.
- If you notice, albumin gives a peak while globulin region is broader.
- Albumin gives a peak in densitometry because it is a single protein.
- Globulin region consists of many proteins, that is why it is broader.

γ Globulins

- Not all immunoglobulins are γ globulins since IgA is in β_2 position.
- Not all γ globulins are immunoglobulins since C-reactive protein (CRP) is in γ band.

Reversal of A : G Ratio

- A : G ratio is reversed in chronic liver diseases due to decrease in the rate of albumin synthesis.

ELECTROPHORETIC PATTERN OF SERUM PROTEINS IN VARIOUS DISORDERS

Electrophoretic Pattern in Nephrotic Syndrome (Fig. 3)

- **Reduction in albumin** because of massive albuminuria. This stimulates proteins synthesis in liver. But, low molecular weight albumin is excreted.
- **Increase in α_2 macroglobulin** since it is of high-molecular weight and retained in the plasma.
- Minimal reduction in gamma globulins.

Electrophoretic Pattern in Liver Disease (Fig. 4)

- Severe reduction in albumin
- Reduction in α_1 and α_2 bands
- ↑ Polyclonal immunoglobulins leads to β-γ bridging

Multiple Myeloma

Multiple myeloma is a malignancy of B-cells. The tumor arise from the single clone of B-cells. So, it is also known as monoclonal gammapathy.

Investigations for the Diagnosis of Multiple Myeloma

- Urinary Bence Jones proteins
- Serum Free light-chain (sFLCs) assay
- Serum electrophoresis for detection and M band
- Immunofixation to identify the type of immunoglobulin
- Bone marrow biopsy

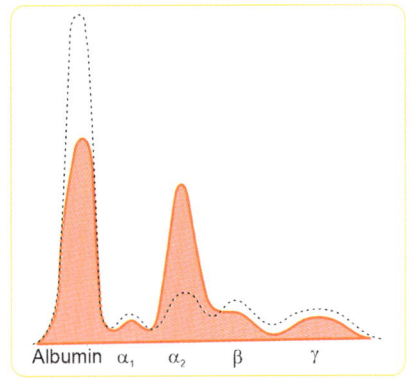

Fig. 3: Electrophoretic pattern in nephrotic syndrome

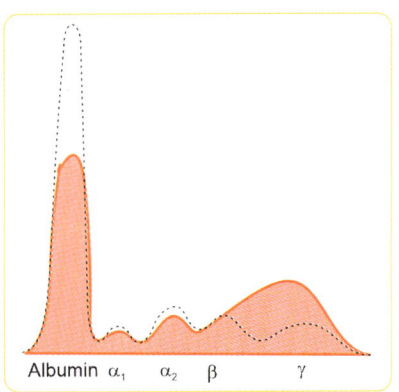

Fig. 4: Electrophoretic pattern in liver disease

Chapter 35 Plasma Proteins

Diagnostic Criteria for Multiple Myeloma

❏ Bone marrow plasma cells >10%
❏ M protein—positive
❏ Features of end-organ damage—hyper **C**alcemia, **R**enal failure, **A**nemia and **B**one lesions (mnemonic: CRAB)

Bence Jones Proteins are Abnormal Monoclonal Light Chains Excreted in Urine

❏ Bence Jones proteins are nothing but immunoglobulin light chains excreted in the urine in patients of multiple myeloma.
❏ On heating the urine sample above 60°C, these proteins get precipitated and turbidity appears; disappears on further heating to 80°C; reappears on cooling.

Electrophoretic Pattern in Multiple Myeloma (Fig. 5)

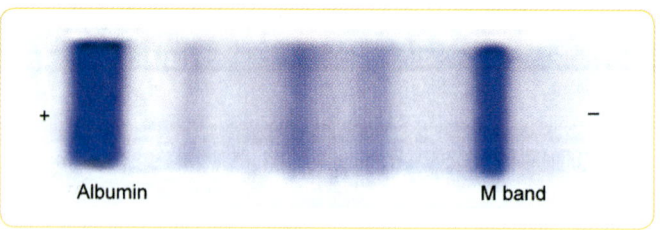

Fig. 5: Serum protein electrophoresis showing M band

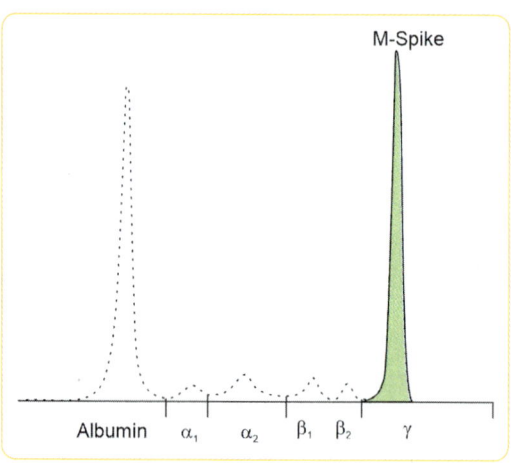

Fig. 6: Densitometry showing sharp, single peak—M band

❏ Clear, sharp band in the γ region known as M band (Fig. 6)
❏ Sharp band is because of the monoclonal nature of the antibody produced by a single clone of malignant plasma cells.

Immunofixation is Done to Detect the Type of Antibody

The purpose of immunofixation is to detect the type of abnormally produced immunoglobulin.

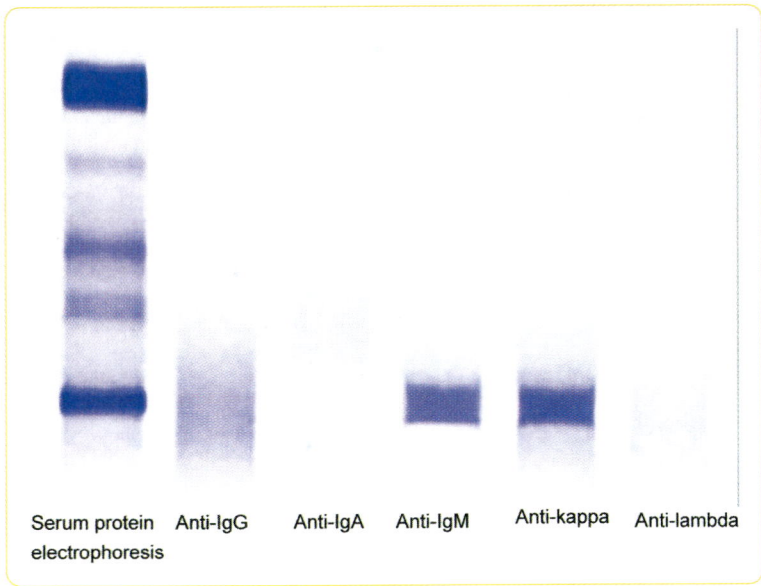

| Serum protein electrophoresis | Anti-IgG | Anti-IgA | Anti-IgM | Anti-kappa | Anti-lambda |

Let me explain the image:
- ❏ There are 6 lanes shown in the image. First lane is the ordinary serum protein electrophoresis. You can notice the M band.
- ❏ The other lanes are showing immunofixation—electrophoresis is done first and antibodies against antibodies are added later followed by staining.
- ❏ You can notice that anti-IgM antibody and anti-kappa antibodies react with the serum to produce a band.
- ❏ This means, the abnormally produced antibody contains μ heavy chains and κ light chains.

ALBUMIN

Functions of Albumin

- ❏ Major contributor (80%) to the plasma oncotic pressure. Hypoalbuminemia leads to edema.
- ❏ Transporter of endogenous and exogenous substances
  - ○ **Endogenous:** Nonesterified free fatty acids, bilirubin, metals (calcium, copper and magnesium), hormones (thyroxine, steroid), nitric oxide
  - ○ **Exogenous:** Drugs (warfarin, benzodiazepines and NSAIDs)
- ❏ **Acid-base balance:** Albumin acts as buffer.
- ❏ **Nutritional function:** Albumin can be utilized by cells as a source of amino acids.

Therapeutic Uses of Albumin

- ❏ Albumin infusion is done in burns patients to restore the protein
- ❏ Albumin infusion has been tried in neonatal jaundice to prevent the unconjugated bilirubin from entering brain.

Analbuminemia is Compatible with Life

We have seen the important roles albumin has been playing so far. But it is surprising to know that analbuminemia is a benign condition with mild edema.

ACUTE PHASE PROTEINS

Acute-phase proteins are a group of plasma proteins whose plasma concentrations increase or decrease in response to inflammation. They are of two types—positive and negative.

- ❑ The level of positive acute phase proteins increases during acute inflammation.
- ❑ The level of negative acute phase proteins decreases during acute inflammation.

Positive acute phase proteins	Negative acute phase proteins
C-reactive protein	Albumin
Serum amyloid A protein	Prealbumin
Fibrinogen	Transferrin
Haptoglobin	Thyroxine binding globulin
α_1 antitrypsin	
α_2 macroglobulin	

Transferrin and ferritin are discussed in Chapter 34 (Refer to page no. 538)

❑ PY2.2 Discuss the origin, forms, variations and functions of plasma proteins

Key Points

- ❑ In humans, except albumin, all the plasma proteins are glycoproteins.
- ❑ Plasma – Fibrinogen = Serum
- ❑ Albumin is precipitated out by full saturation whereas globulin is precipitated by half-saturation.
- ❑ During electrophoresis at alkaline pH, most proteins carry a net negative charge and move toward the anode.
- ❑ Smoking causes oxidation of methionine residues of α1 antitrypsin.
- ❑ Reversal of A : G ratio seen in chronic liver diseases.
- ❑ Bence Jones proteins are immunoglobulin light chains excreted in the urine in patients of multiple myeloma.

SELF-ASSESSMENT

Short Answer Questions

1. Differentiate plasma and serum.
2. Name any two-negative acute-phase proteins.
3. With a schematic diagram, explain the concept of salting out of proteins.
4. Which method is used for the estimation of total protein and albumin in the clinical laboratory?
5. Explain why hypoalbuminemia leads to edema.
6. Outline the principle of separation of serum proteins by electrophoresis. Draw the typical pattern of electrophoretic bands seen in nephrotic syndrome and cirrhosis of the liver.

Multiple Choice Questions

1. In humans, all of the following are glycoproteins, except:
   a. Albumin
   b. Ceruloplasmin
   c. IgA
   d. Transferrin
2. Which of the following is used to separate serum from blood?
   a. Heparin
   b. EDTA
   c. Citrate
   d. None

3. **Prealbumin is named so, because:**
   a. It has the electrophoretic mobility anodal to albumin
   b. It is a precursor of albumin
   c. It is previously considered as albumin
   d. It is the pre-pro albumin

4. **Maximum of how many iron molecules can be transported by one transferrin?**
   a. $2\ Fe^{3+}$ b. $3\ Fe^{2+}$
   c. $2\ Fe^{2+}$ d. $3\ Fe^{3+}$

5. **Which of the following protein binds to free heme?**
   a. Ceruloplasmin
   b. Haptoglobin
   c. Hemopexin
   d. Hemosiderin

6. **Which of the following protein has ferroxidase activity?**
   a. Albumin   b. Ceruloplasmin
   c. Haptoglobin   d. Transferrin

7. **All of the following plasma proteins are produced by liver, except:**
   a. Albumin   b. Fibrinogen
   c. Haptoglobin   d. Immunoglobulin

8. **All of the following clotting factors are produced by liver, except:**
   a. II   b. III
   c. VII   d. IX

9. **Which of the following amino acid residue of α_1 antitrypsin is oxidized by smoking?**
   a. Cysteine   b. Methionine
   c. Serine   d. Phenylalanine

10. **C in CRP stands for:**
    a. Choline of corynebacterium
    b. Concanavalin A
    c. Capsular polysaccharide of pneumococci
    d. Capsular polysaccharide of *corynebacterium*

11. **Edema in Nephrotic syndrome is due to:**
    a. Na and water restriction
    b. Increased venous pressure
    c. Hypoalbuminemia
    d. Hyperlipidemia

12. **Which of the following acute phase protein is responsible for raised erythrocyte sedimentation rate during inflammation?**
    a. Albumin
    b. Fibrinogen
    c. Prealbumin
    d. Ceruloplasmin

13. **Which of the following is a negative acute phase protein?**
    a. Haptoglobin
    b. Transferrin
    c. Ceruloplasmin
    d. Serum amyloid A protein

Cell, Cell Membrane and Cytoskeleton

CELL AND SUBCELLULAR ORGANELLES

Cell is the structural and functional unit of all living organisms. Human body is made up of around 37 trillion (3.7×10^{13}) cells. Cellular organization is different in prokaryotes and eukaryotes.

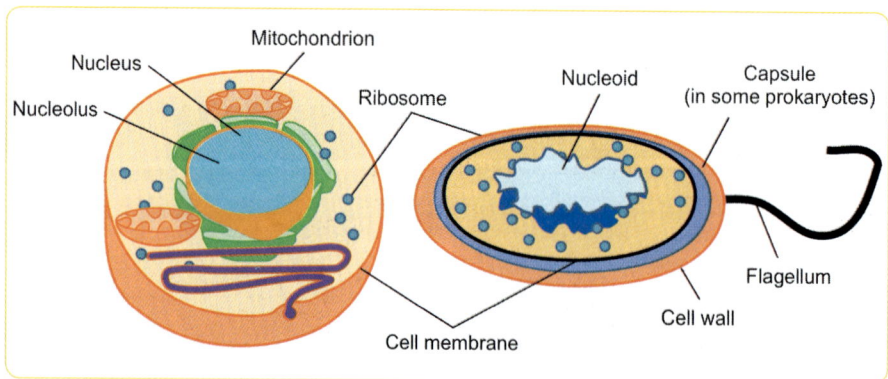

	Prokaryotic	Eukaryotic cell
Cell size	Usually smaller	Variable
Nuclei	No nuclear membrane or nucleoli	True nucleus is present
Membrane bound organelles	Absent	Present (e.g. Mitochondria)
Sterols in plasma membrane	Absent	Present
Type of ribosomes	70S (50S and 30S)	80S (60S and 40S)
Location of ETC enzymes	Embedded in plasma membrane	Inner mitochondrial membrane
Cell division	Simple binary fission	Complex cell cycle with mitosis
Extra-chromosomal DNA	Plasmid DNA	Mitochondrial DNA and Chloroplast DNA

MARKER ENZYMES ARE USED IN THE IDENTIFICATION OF AN ORGANELLE DURING SUBCELLULAR FRACTIONATION

A cell contains many organelles. These organelles are isolated by subcellular fractionation using **differential centrifugation**. Repeated centrifugation of homogenized cells at progressively higher speeds result in separation of organelles on the basis of their size and density. The larger and denser compounds like nuclei and mitochondria move down rapidly during centrifugation. Organelles with lesser density remain in the supernatant and are subjected to successive centrifugation steps in higher speed.

Marker enzymes are used to identify the isolated organelles in this process.

Organelle		Marker enzyme
Plasma Membrane		Na+K+ATPase 5' Nucleotidase Adenylyl cyclase
Mitochondria	Outer membrane	Monoamine oxidase
	Inner membrane	Succinate dehydrogenase Cytochrome C oxidase
	Intermembrane space	Creatine kinase Adenylate kinase
	Matrix	Glutamate dehydrogenase
Endoplasmic Reticulum (Liver)		Glucose-6-phosphatase
Nucleus		DNA, Histone
Peroxisomes		Catalase
Golgi apparatus		Galactosyl transferase α Mannosidase II
Lysosomes		Cathepsin C, Acid phosphatase

Urate oxidase/uricase is a marker of peroxisomes of non-primates. Uricase is absent in humans.

Know the Difference:

Cytoplasm	Cytosol
Contents inside the cell membrane other than the nucleus. i.e. Cytoplasm = Protoplasm –Nucleoplasm	Fluid part of cytoplasm other than the organelles. i.e. Cytosol = Cytoplasm - membrane bound organelles

Glucose-6-phosphatase is a cytoplasmic enzyme, not a cytosolic enzyme. Clear?

ENDOPLASMIC RETICULUM

Microsomes

❑ Microsomes are actually artefacts.
❑ They are nothing but reformed from pieces of the endoplasmic reticulum (ER) formed during exposure to excess physical force like sonication and ultracentrifugation.

Endoplasmic Reticulum forms a Network Link Between Nucleus and Cell Membrane

❑ Endoplasmic reticulum is in continuation with nuclear membrane.
❑ Ribosome studded ER is known as rough ER.
❑ Rough endoplasmic reticulum is involved in post-translational modification of proteins.
❑ Smooth endoplasmic reticulum is involved in storing calcium in muscles, elongation of fatty acids, triglyceride synthesis and steroidogenesis.

Glucose-6-phosphatase is the Marker Enzyme of ER

❑ Golgi apparatus is the sorting center of the cell (Fig. 1).
❑ Composed of stacks of membrane-bound structures.
❑ Cis-Golgi is convex and near the nucleus and trans-Golgi is concave and away from the nucleus.
❑ O-linked glycoprotein synthesis takes place exclusively in the Golgi.
❑ N-linked glycoproteins synthesis begins in ER and ends in Golgi.
❑ Post-translationally modified proteins synthesized in the ER are packaged into vesicles, which then fuse with the Golgi apparatus. These cargo proteins are modified and destined for secretion via exocytosis. Thus, Golgi apparatus is the packaging and sorting center of the cell.

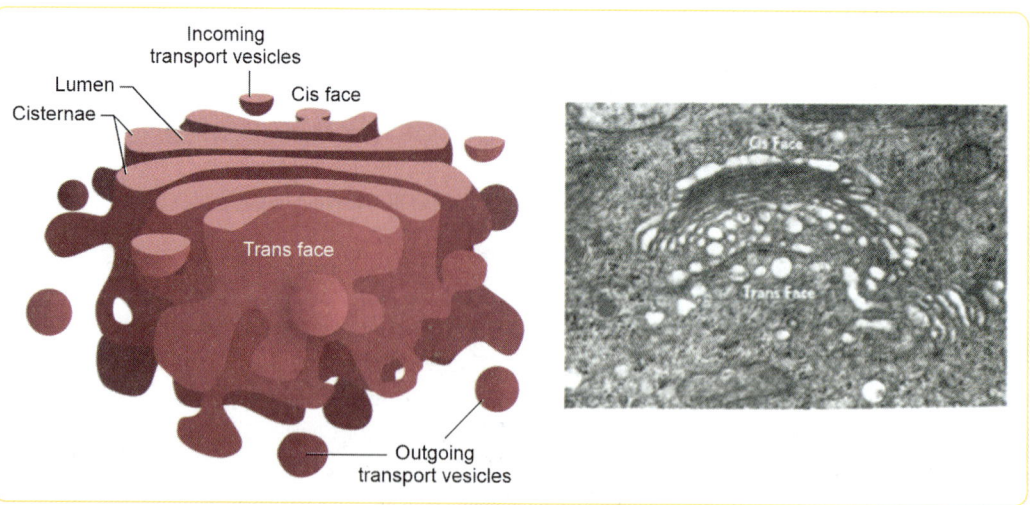

Fig. 1: Observe the convex-forming cis-Golgi and concave-forming trans-Golgi

Section V Special Topics

LYSOSOMES

- ❏ **'Single membrane'** bound organelles containing various hydrolytic enzymes.
- ❏ Produced from budding of Golgi apparatus.
- ❏ While Golgi apparatus is found in both plant and animal cells, lysosome is found only in animal cells.
- ❏ Lysosomal membrane contains a **proton pump** that maintains acidic pH inside the organelle.
- ❏ Both intracellular and extracellular substances are degraded by lysosomes.
- ❏ As the lysosomal enzymes catalyze hydrolysis in **acidic medium**, they are also known as acid hydrolases

Lysosomal enzymes	
Nucleases	DNase, RNase
Proteases	Cathepsins, Collagenase
Lipases	Sphingomyelinases, Esterase
Amylases	β-glucosidase, Hexosaminidase A, α-mannosidase, α-fucosidase, Sialidase, acid maltase, etc.

Types of Lysosomes

- ❏ 1° lysosome—sac-like structures containing enzymes synthesized by the rough endoplasmic reticulum.
- ❏ 2° lysosome—fusion of 1° lysosome with phagosome; engulfed materials are progressively digested by the enzymes.

Functions

Heterophagy	Digestion of extracellular materials by the process of endocytosis (Phagocytosis and Pinocytosis)
Autophagy	Physiological destruction of the cell
Apoptosis	Partial and selective lysosomal membrane permeabilization (LMP) induces apoptosis
Fertilization	Acrosome (giant lysosome) of sperm head ruptures and releases enzymes on the surface of the egg

MITOCHONDRIA

- ❏ Mitochondria are the double-membrane bound organelles found in eukaryotic cells. They are prokaryotic in origin. They contain their own DNA and protein synthesizing machinery (tRNA, rRNA and ribosome).
- ❏ Each cell contains multiple mitochondria. The number of mitochondria is higher in tissues with high rate of aerobic metabolism.
- ❏ The outer membrane of mitochondria is permeable to most of the substances, but inner membrane is selectively permeable.
- ❏ The lipid to protein ratio of inner mitochondrial membrane is 1:1 as it is rich in protein.
- ❏ The inner membrane is folded into finger-like projections known as Cristae in which proteins of electron transport chain are embedded. Cristae increase the total surface area.

Labels: ATP synthase particles, Intermembrane space, Matrix, Cristae, Ribosome, Granules, Inner membrane, Outer membrane, Deoxyribonucleic acid (DNA)

Functions of Mitochondria

- Internal respiration: Production of ATP by electron transport chain
- Fatty acid oxidation, TCA cycle, part of heme synthesis and part of gluconeogenesis
- Elongation of fatty acids takes place predominantly in ER and to some extent in mitochondria also
- Mediator of intrinsic pathway of apoptosis via leakage of cytochrome c

PEROXISOMES

Peroxisomes arise from the budding of smooth endoplasmic reticulum or division of pre-existing peroxisomes.

Functions of Peroxisome

- Plasmalogen synthesis
- α-oxidation of branched chain fatty acids (phytanic acid) and β-oxidation of very long chain fatty acids
- Bile salt synthesis
- Detoxification of alcohol through catalase pathway in liver

THE CYTOSKELETON

Cytoskeleton is a cellular structure that helps cells maintain their shape and internal organization. It also provides mechanical support that enables cells to carry out essential functions like cell-division, anchorage and movement.

FUNCTIONS OF CYTOSKELETON

- Giving shape and strength to the cell
- Intracellular transport
- Beating of cilia
- Amoeboid movement of phagocytic cells

THREE CLASSES OF CYTOSKELETAL ELEMENTS

Type	Size (nm)	Examples
Microfilaments	5–7	Actins
Intermediate filaments	8–10	Lamins, Keratins, Vimentin, Desmin, Neurofilaments
Microtubules	23–25	Tubulins

Microfilaments are dynamic structures which undergo rapid assembly and disassembly. ATP binding promotes the addition of actin monomers. Polymerization of G-actin (globular actin) produces F-actin (filamentous actin).

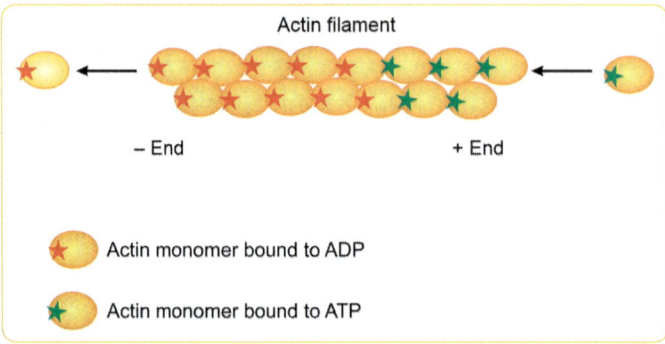

MOLECULAR MOTORS

Molecular motors are the vehicles carrying the cellular cargo. They run upon the cytoskeletal framework.

For example, Myosin is the motor, actin filaments are the tracks on which myosin moves, and ATP is the fuel that powers movement. So, myosin is an ATPase.

Molecular motor	Cytoskeletal railroad	Fuel used	Function
Myosin	Actin	ATP	Muscle contraction
Dynamin	Actin	**GTP**	Receptor mediated endocytosis
Kinesin	Tubulin	ATP	**Anterograde** transport (e.g. neuronal transport)
Dynein	Tubulin	ATP	**Retrograde** transport (e.g. cilia)

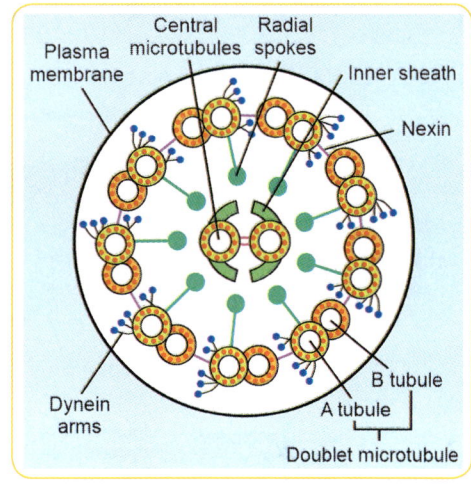

Fig. 2: Eukaryotic flagellum showing the 9+2 arrangement of microtubules. Dynein arms are shown in blue.

❏ Sperm is the only flagellated cell in human body.

Dynein Arm Defect Leads to Primary Ciliary Dyskinesia/Kartagener Syndrome

Primary ciliary dyskinesia (PCD), also called immotile ciliary syndrome or Kartagener syndrome is an autosomal recessive disorder due to dynein arm defect.

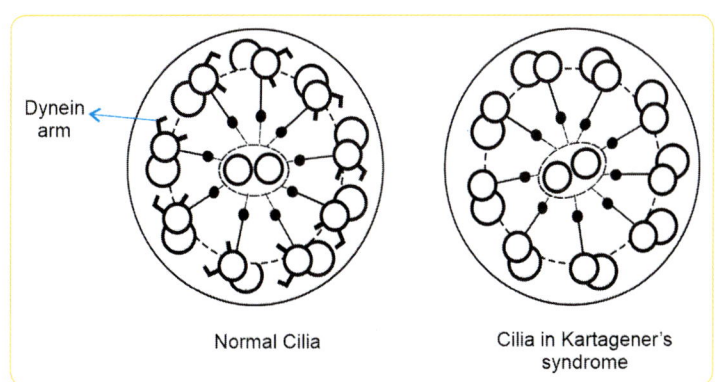

Normal Cilia   Cilia in Kartagener's syndrome

CELL MEMBRANE

FLUID MOSAIC MODEL

Fluid mosaic model proposed by Singer and Nicolson in 1972 is an accepted model for cell membrane

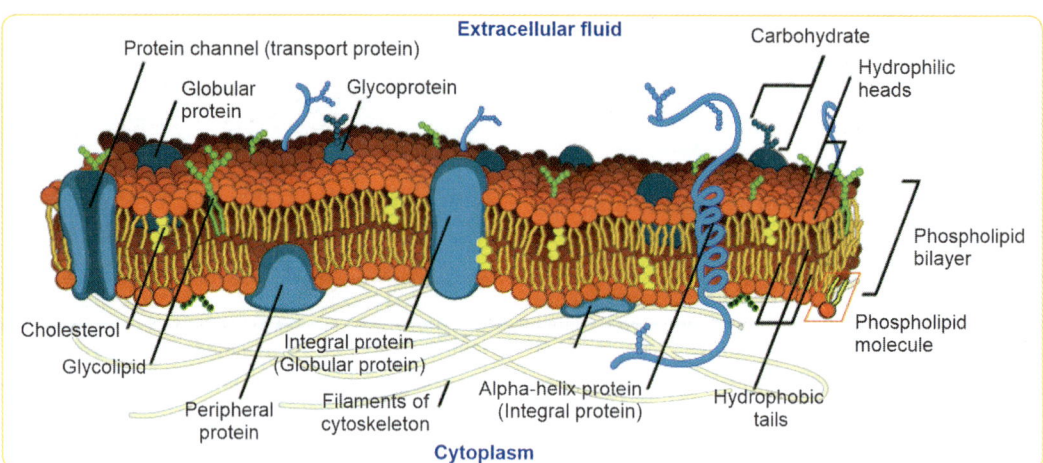

- ❏ The polar portion of the phospholipid is illustrated as globular head and the non-polar fatty acid portion is illustrated as the tail.
- ❏ Can you observe the kink in the fatty acid tail? This is because of the cis double bonds. So, fatty acids with cis double bonds increase the membrane fluidity whereas trans fatty acids and saturated fatty acids decrease it.
- ❏ Membrane fluidity is important for processes like cell division, endocytosis and communication across cells.
- ❏ Fluid model visualizes integral membrane proteins as iceberg floating in the sea of phospholipid molecules based on the observation with fluorescent microscopic studies.
  Salient points of this model are:

Cell Membrane is a Lipid Bilayer

- ❏ Hydrophilic head group of phospholipids is in contact with the fluid compartments, the ECF and ICF.
- ❏ Hydrophobic fatty acid tail groups are in contact with each other.
- ❏ Cis double bonds in the fatty acids produce a kink and contributes to fluidity.
- ❏ Triglycerides are not a component of cell membrane.

Two Types of Membrane Proteins

- ❏ Integral proteins interact with membrane to significant amount so that removal of this proteins requires detergent. Some integral proteins are transmembrane proteins that traverses the entire length of the membrane once or many times.
- ❏ Peripheral proteins interact less with the membrane and their removal requires only concentrated salt solutions. Peripheral proteins can be found on either side of the membrane.

Membrane is Asymmetric

- ❏ Outer and inner leaflets of the membrane differ in their both lipid and protein composition.
- ❏ Outer leaflet: made up of Lecithin (phosphatidyl choline) and sphingomyelin (*memory aid: LS ratio of respiratory distress syndrome*). Carbohydrate moieties of glycoconjugates protrude only outward towards ECF.
- ❏ Inner leaflet: made up of phosphatidyl serine and phosphatidyl ethanolamine (cephalin).

Melting Temperature of Membrane

Transition temperature is the temperature at which there is a transition from ordered structure of fatty acid tails to unordered.

❑ Chain length of fatty acids $\alpha\, T_m$
❑ Degree of unsaturation $\alpha\, 1/T_m$
❑ Cis double bonds contribute to fluidity.

Role of Cholesterol

❑ Rigid steroid ring system interferes with the lateral movement of fatty acid chains.
❑ At lower temperatures: ↑ fluidity
❑ At higher temperatures: ↓fluidity

LIPID RAFT AND CAVEOLAE

Lipid rafts are areas in the plasma membrane that are rich in cholesterol, sphingolipids, and GPI-anchored transmembrane proteins (Chapter 41) *(Refer to page no. 606)*. The lipid raft is more tightly packed than the surrounding plasma membrane. Lipid raft can freely diffuse in the plane of the plasma membrane and are dynamic in nature (assemble and disassemble). Thus, lipid raft forms a microdomain within the plasma membrane. Lipid raft in involved in cellular signaling.

Caveolae are distinct from lipid rafts. As the name suggests, caveolae are cup-shaped invaginations in the plasma membrane. Caveolae are associated with endocytosis, cell signaling, and the entry of pathogens into the cell. Like lipid rafts, caveolae also contain high-cholesterol and sphingolipid. Unlike the lipid rafts, caveolae contain the protein caveolin-1.

TRANSMEMBRANE REGIONS OF PROTEINS ARE MADE UP OF α HELICES OR β BARRELS

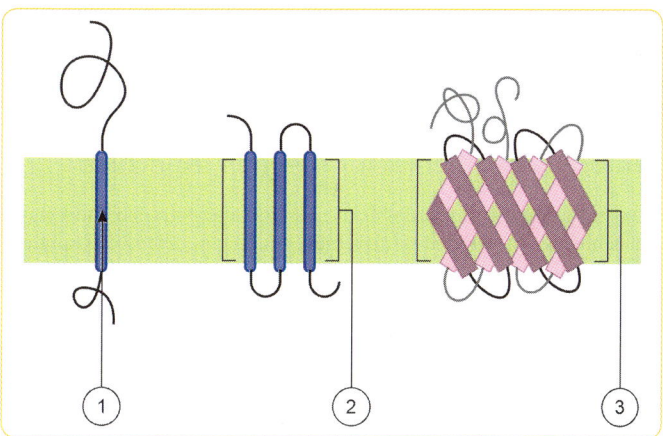

Fig. 3: Schematic representation of transmembrane proteins: 1. A single transmembrane α-helix 2. A polytopic transmembrane α-helical protein 3. A polytopic transmembrane β-sheet protein

Transmembrane Proteins May Span the Membrane Just Once or Many Times

❑ Glycophorin is a transmembrane protein in the RBC that span the membrane just once and known as single pass transmembrane protein (Fig. 3).

- G-protein coupled receptors span the membrane 7 times. That is why they are known as seven transmembrane (7TM), multi pass, serpentine receptors.
- Glucose transporters (GLUT) span the membrane 12 times.

Membrane Spanning Regions of a Transmembrane can be Predicted

- An alpha helix that span the membrane consist of approximately 20 amino acid residues. Most of these amino acids are hydrophobic amino acids.
- Hydropathy index is a number that shows how hydropathic an amino acid is.
- Hydropathy plot values are helpful in recognizing the membrane spanning regions of a proteins. Hydropathy scale value more than 20 kcal/mol suggests that hydrophobic sequence is a transmembrane segment.

Aquaporin Water Channels are Transmembrane Proteins

- Water can freely diffuse across the cell membrane. But the rate is low. Aquaporins increases the rate of water flow.
- ADH/AVP hormone acts on V2 receptor and increases the insertion of aquaporin into the late DCT and collecting ducts.

Reduced or absent aquaporin-2 insertion into the membrane of renal principal cells leads to diabetes insipidus. *(Ref. Harrison, 20th ed. p.296)*
Aquaporin-4 antibodies are diagnostic of Neuromyelitis Optica. *(Ref. Harrison, 20th ed. p.3178)*

Anchoring of Proteins to Membrane is Facilitated by Addition of Lipids to the Proteins

Addition of isoprenyl group, cholesterol and GPI anchor helps in the covalent attachment of certain proteins to one of the leaflets of the membrane.

MEMBRANE TRANSPORT

Permeability of Membrane is Different for Different Molecules

Lipid bilayer is more permeable to small molecules and gases and less permeable to ions (Fig. 4). Permeability of membrane to various molecules is expressed as permeability coefficient expressed as nanomole per second.

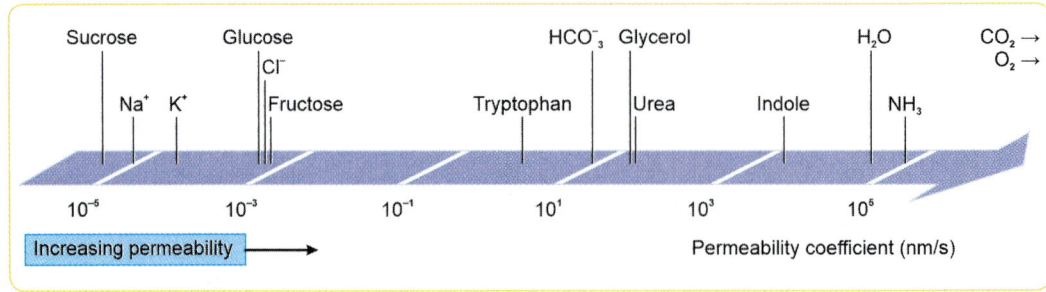

Fig. 4: Lipid bilayer is most impermeable to sodium

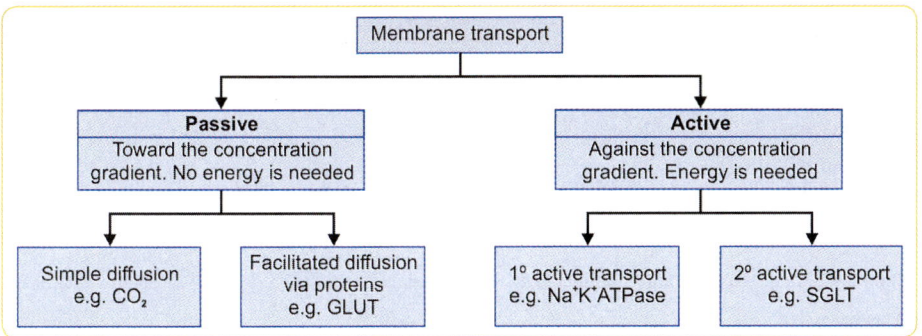

ACTIVE TRANSPORT

❑ Active transport is ATP-requiring uphill transport of molecules against their concentration gradient
❑ Transfer of molecules against their electrochemical (uphill movement) gradient by a membrane carrier protein using the energy provided by ATP hydrolysis.

Based on the Direction of Movement

❑ Symport: transport of 2 different molecules in the same direction, e.g. SGLT1 and 2
❑ Antiport: transport of 2 different molecules in 2 different direction, e.g. Na⁺K⁺ATPase

Based on the Way Energy is Used

Two types: 1° active transport and 2° active transport

1° Active Transport

❑ Transport of solute is **directly coupled with ATP hydrolysis** by the protein which contains ATPase activity, e.g. Na⁺K⁺ATPase
❑ Inhibitors of Na⁺K⁺ATPase: Digoxin (drug) and Ouabain (poison)

2° Active Transport

❑ Sodium Glucose transporter (SGLT) is an example (Fig. 5). Na⁺K⁺ATPase pumps sodium out of the cell and creates a concentration gradient for sodium. This gradient is used to transport glucose and sodium.
❑ Inhibitors of SGLT-2 in renal tubules (canagliflozin, dapagliflozin, and empagliflozin) are anti-diabetic drugs as they prevent glucose reabsorption.

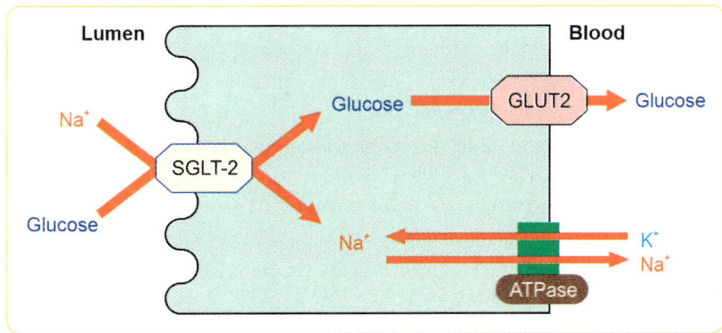

Fig. 5: Sodium-Glucose transporter is an example of secondary active transport

Chapter 36 Cell, Cell Membrane and Cytoskeleton

Features	Facilitated diffusion	Active transport
Movement of molecules	Down the electrochemical gradient (Downhill movement)	Against the electrochemical gradient (Uphill movement)
Need of ATP	No	Yes
Another name of the carrier protein/transporter	Transporters or channels	Pumps
Speed	Fast	Fastest mode
Example	GLUT	Na$^+$K$^+$ATPase

Competency

- ☐ BI1.1   Describe the molecular and functional organization of a cell and its subcellular components
- ☐ PY1.1   Describe the structure and functions of a mammalian cell
- ☐ PY1.5   Describe and discuss transport mechanisms across cell membranes

Key Points

Cell Membrane

- ☐ Triglycerides are NOT a component of the eukaryotic plasma membrane.
- ☐ Lipid rafts are specialized areas of the exoplasmic (outer) leaflet of the lipid bilayer enriched in cholesterol, sphingolipids, and certain GPI anchored proteins.
- ☐ Aquaporin 2 mutation is seen in some patients of nephrogenic diabetes insipidus.
- ☐ CFTR chloride channel is a member of the ABC transporter family.
- ☐ Phosphatidylserine is flipped to the outer leaflet of membrane during apoptosis.
- ☐ Gases like oxygen and CO_2 cross the plasma membrane by passive diffusion.
- ☐ Na$^+$K$^+$ ATPase is an example of primary active transport. SGLT2 is an example of secondary active transport.
- ☐ Na$^+$K$^+$ ATPase is stimulated by thyroxine; inhibited by ouabain and digoxin.
- ☐ Mitochondrial cytochrome c is a mobile electron carrier. Cytosolic cytochrome c mediates apoptosis.
- ☐ Facilitated diffusion requires carrier proteins but no energy.
- ☐ The plasma membrane is freely permeable to Urea, CO_2, O_2, H_2O, and alcohol.

Cytoskeleton

- ☐ The cytoskeleton is a cellular structure that helps cells maintain their shape and internal organization.
- ☐ Cytoskeleton consists of three major class of protein elements, namely microfilaments, intermediate filaments and microtubules.
- ☐ Microfilament is formed by actin family of proteins
- ☐ Microtubules are the largest cytoskeletal element made up of tubulin proteins.
- ☐ Microtubule inhibitors are used in the therapy of cancer.
- ☐ Cytoskeletal Molecular motors are the vehicles carrying the cargo. They run upon the cytoskeletal framework.
- ☐ Myosin is the motor protein that moves along actin filaments using the energy of ATP hydrolysis while Kinesin is the motor protein that moves along tubulin using the energy of ATP hydrolysis
- ☐ Dynamin requires free energy of hydrolysis of GTP.
- ☐ Primary ciliary dyskinesia (PCD), also called immotile ciliary syndrome or Kartagener syndrome is an autosomal recessive disorder due to dynein arm defect.

Short Answer Questions

1. What is the usefulness of membrane fluidity? What is the role of cholesterol on membrane fluidity?

2. Describe the salient features of the fluid mosaic model.

3. Draw a labelled diagram to depict secondary active transport of glucose across the membrane of an enterocyte.

4. What is the molecular defect in Kartagener syndrome?

Multiple Choice Questions

CELL AND CYTOSKELETON

1. Which of the following is seen in both prokaryotic and eukaryotic cell?
   a. Nucleoli
   b. Mitochondria
   c. Lysosome
   d. Ribosomes

2. During subcellular fractionation, Liver tissue is homogenized and differential centrifugation was done to isolate organelles. Presence of which of the following enzymatic activity will confirm the organelle as endoplasmic reticulum?
   a. Adenylate kinase   b. Catalase
   c. Galactosyl transferase
   d. Glucose-6-phosphatase

3. All of the following are marker enzymes of plasma membrane, except:
   a. 5' Nucleotidase   b. Adenylate kinase
   c. Adenylyl cyclase   d. Na⁺K⁺ATPase

4. Defect in which of the following leads to Kartagener's syndrome?
   a. Actin   b. Dynamin
   c. Dynein   d. Myosin

5. Which of the following is a microfilament?
   a. Actin   b. Vimentin
   c. Keratin   d. Desmin

6. All of the following are cytoskeletal motors, except:
   a. Dynein
   b. Actin
   c. Myosin
   d. Dynamin

7. Which of the following organelles is involved in sorting of proteins in the cell?
   a. Endoplasmic reticulum
   b. Golgi apparatus
   c. Peroxisome
   d. Lysosome

8. All of the following processes occur in mitochondria, except:
   a. Elongation of fatty acids
   b. Oxidation of fatty acids
   c. Synthesis of fatty acids
   d. Synthesis of protein

9. The lipid protein ratio of inner mitochondrial membrane is:
   a. 1:2   b. 2:1
   c. 1:1   d. 1:3

10. By which of the following processes peroxisomes are produced?
    a. Vesicular budding from golgi apparatus
    b. Budding from endoplasmic reticulum
    c. Budding from mitochondria
    d. Modification of lysosomes

11. Free energy of hydrolysis of GTP is required for the function of which of the following cytoskeletal element?
    a. Actin
    b. Myosin
    c. Kinesin
    d. Dynamin

12. Fluidity of the membrane is maintained by all of the following, except:
    a. Cholesterol
    b. Unsaturated cis fatty acids
    c. Integral proteins
    d. Length of the hydrocarbon chain

CELL MEMBRANE AND TRANSPORT

13. All of the following are components of eukaryotic cell membrane, except:
    a. Triglycerides   b. Carbohydrates
    c. Lecithin   d. Cholesterol

14. Which of the following has the lowest membrane permeability coefficient?
    a. Na⁺   b. K⁺
    c. Glucose   d. Urea

15. A lipid bilayer is permeable to:
a. Urea
b. Potassium
c. Glucose
d. Sodium

16. How many amino acids are present in a single transmembrane alpha helical region of a transmembrane protein?
a. 20
b. 30
c. 40
d. 50

17. Hydropathy plot values are helpful in recognizing the membrane spanning regions of a proteins. Which of the following value in hydropathy plot can suggest that hydrophobic sequence is a transmembrane segment?
a. >20 kcal/mol
b. < 20 kcal/mol
c. >10 kcal/mol
d. <10 kcal/mol

18. Glucose transporters (GLUTs) are integral membrane proteins that traverse the lipid bilayer several times. How many membranes spanning alpha helices are present in GLUT proteins?
a. 1
b. 7
c. 10
d. 12

19. All of the following are true about Na⁺K⁺ ATPase of the plasma membrane, except:
a. Pumps three Na⁺ out and pumps two K⁺ into cells
b. Inhibited by ouabain and digitalis
c. Consumes significant amount of cellular ATP
d. Belongs to ABC transporter family

20. Na⁺-glucose cotransporter is an example for:
a. Facilitated diffusion
b. Passive diffusion
c. Primary active transport
d. Secondary active transport

21. All of the following are true statements regarding membrane asymmetry, except:
a. Phosphatidylcholine is located mainly in the outer leaflet
b. Phosphatidylethanolamine is located mainly in the inner leaflet
c. Sphingomyelin is located mainly in the outer leaflet
d. Sugar moieties of the glycosphingolipids and glycoproteins protrude towards the inner surface

22. Which of the following is the only way for a substance to get accumulated against its electrochemical gradient?
a. Facilitated diffusion
b. Passage through ion channels
c. Diffusion through a uniport
d. Active transport

23. Gases like O_2 and CO_2 cross plasma membrane by:
a. Specific gas transporter
b. Facilitated diffusion
c. Passive diffusion
d. Primary active transport

24. Sodium-Potassium pump is an example of:
a. Simple diffusion
b. Facilitated diffusion
c. Secondary active transport
d. Primary active transport

ANSWERS

1. d.	2. d.	3. b.	4. c.	5. a.	6. b.
7. b.	8. c.	9. c.	10. b.	11. d.	12. c.
13. a.	14. a.	15. a.	16. a.	17. a.	18. d.
19. d.	20. d.	21. d.	22. d.	23. c.	24. d.

Physical Chemistry and Acid-Base

pH

- pH is the negative logarithm (base-10, not natural logarithm) of the hydrogen ion concentration in **mol/L**.
- Change of pH by 0.3 unit indicates doubling or halving of hydrogen ion concentration.
- Normal arterial blood pH of the body: 7.35–7.45
- As the biochemical reactions in the body take place in an aqueous environment and enzymes catalyze most of these reactions, it is important to maintain the narrow range of pH since any change in pH will alter the enzymatic activities.

True acidity	Titratable (Total) acidity
Measure of free H⁺ ions in the solution	Total of free and dissociable H⁺ ions in solution
Depends on the degree of ionization. Hence, weak acids have low true acidity and strong acids have high true acidity	Does not depend on the degree of ionization
Measured by pH meter	Measured by titration against alkali

Isoelectric pH (pI)

Isoelectric pH is the pH at which there is no net electrical charge on the molecule, i.e. the molecule exists as Zwitter ion. So, the molecule won't have electrophoretic mobility and biomolecules like proteins get precipitated at this pH. Isoelectric focussing *(Chapter 43)* *(Refer to page no. 628)* is based on pI.
- If the pH is above the isoelectric pH → molecule will carry net negative charge
- If the pH is below the isoelectric pH → molecule will carry net positive charge

 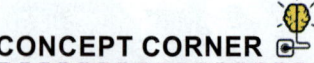
Biochemical basis of curdling of milk

Milk is an emulsion type of colloid with pH of about 6.5-6.7. Casein is the milk protein with an isoelectric pH (pI) of 4.6. So, it exists as negatively charged micelle and is soluble at the pH of 6.5. Addition of acidic substances like lemon, vinegar or curd reduces the pH of milk to 4.6 which matches the isoelectric pH of casein → casein gets precipitated.

Calculation of Isoelectric pH of Amino Acids

To find out the isoelectric pH, we have to calculate the **average of pKa values on either side of the Zwitter ion form**.

$$pI = \frac{pK_1 + pK_2}{2}$$

But amino acids like lysine and aspartate have extra ionizable group in side-chain. So, they have 3 pKa values. How to calculate pKa for lysine and aspartate? Let us see (Fig. 2).

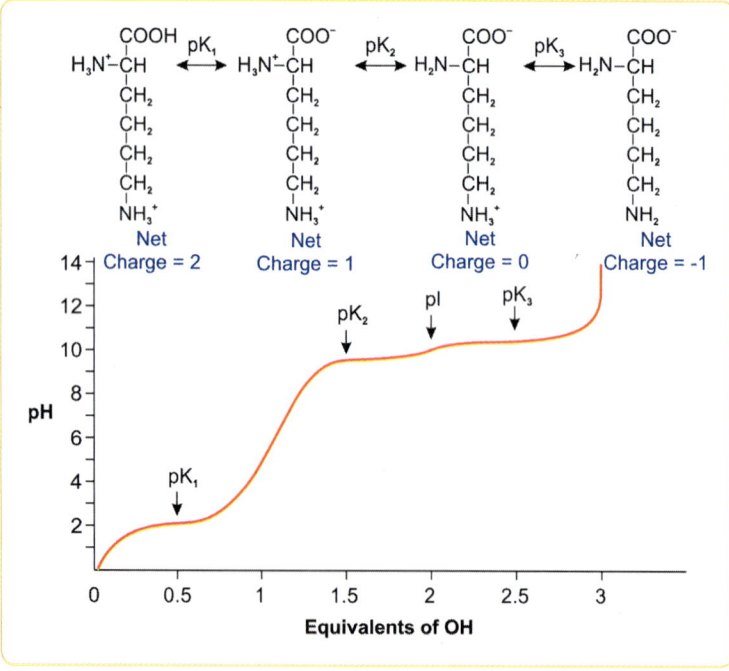

Fig. 1: Titration curve of lysine is shown. Zwitter ion formation takes place between pK_2 and pK_3. Thus, pI is the average of pK2 and pK_3

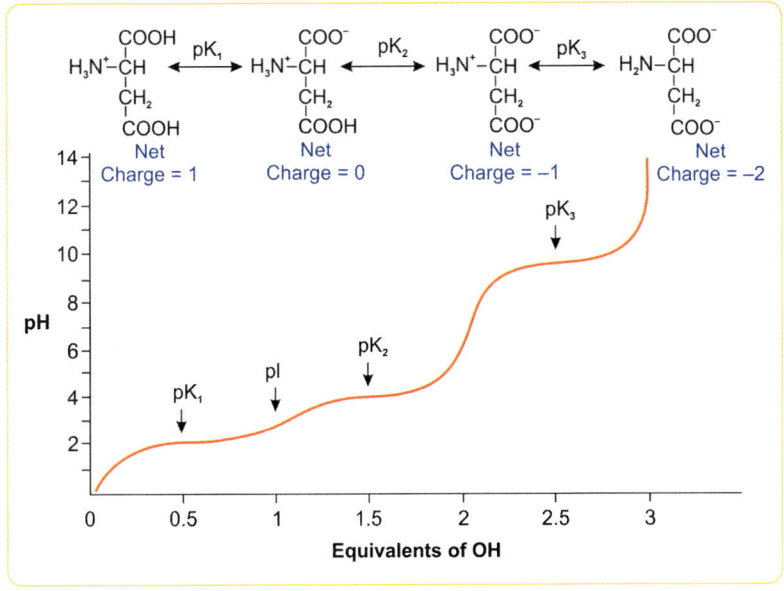

Fig. 2: Titration curve of aspartate is shown. Zwitter ion form happens between pK_1 and pK_2. Thus, pI is the average of pK_1 and pK_2

Buffers Resist the Change in pH

Blood pH is maintained in a narrow range of 7.35–7.45 for normal cellular activities and maintenance of ionic concentrations in the ECF. The pH can change by the addition of excess alkali (HCO_3^-) or acids (lactate, pyruvate, ketones).

Buffers are solutions which can resist the change in pH when a small amount of acid or alkali is added. They are mixtures of weak acids and the salt formed with a strong base or weak base and salt formed with a strong acid, e.g. Bicarbonate buffer is made up of H_2CO_3 and $NaHCO_3$.

H_2CO_3 is the weak acid and $NaHCO_3$ is the salt formed with NaOH, strong base.

We call $NaHCO_3$ as conjugate base. *(In exam, if they ask to identify the conjugate base, remove H^+ from the acid, you'll get the conjugate base)*

BUFFER SYSTEMS OF BLOOD

Buffer system	Constituents (conjugate base/conjugate acid)
Bicarbonate buffer	$HCO_3^-/H_2CO_3(CO_2)$
Plasma protein buffer	Protein/Protein-H^+
Hemoglobin buffer	Hb/Hb-H^+
Phosphate buffer	$HPO_4^{2-}/H_2PO_4^-$

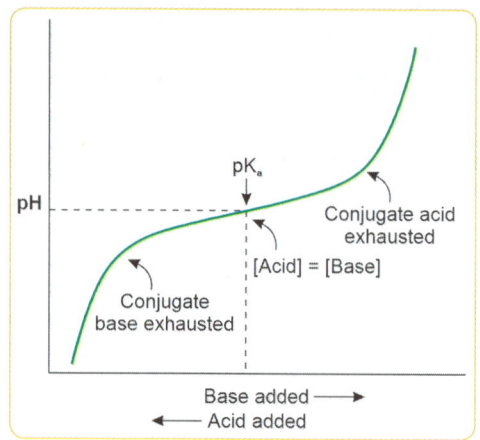

Henderson-Hasselbalch Equation Provides the Relationship of pH to pK

$$pH = pKa + \log \frac{[\text{salt}]}{[\text{acid}]}$$

- When the components of the buffer are fully ionized, i.e. [salt] = [acid] or [deprotonated form] = [protonated form], pH = pK (log 1 = 0). Thus, **a buffer is most effective when the pH is equal to the pK$_a$.**
- Effective pH range of a buffer = pK$_a$ ± 1

Let us apply the formula to bicarbonate buffer.

pK$_a$ of HCO$_3$ buffer is 6.1.

Concentration of salt, i.e. $[HCO_3^-]$ = 24 mEq/L

Concentration of acid, i.e. $[H_2CO_3]$ = solubility coefficient of CO$_2$ × pCO$_2$

= 0.03 × 40 = 1.2 mEq/L

$$pH = 6.1 + \log \frac{24}{1.2}$$

$$= 6.1 + 1.3$$

$$= 7.4$$

Thus, even though the pK$_a$ of bicarbonate buffer is far away from the physiological pH, bicarbonate buffer is an effective buffer at physiological pH because of the ratio of bicarbonate to carbonic acid.

Bicarbonate Buffer is Known as the Open-end Buffer

- Bicarbonate buffer is said to be open at both the ends since the level of both the constituents (HCO$_3$ and pCO$_2$) of this buffer can be altered by the body.

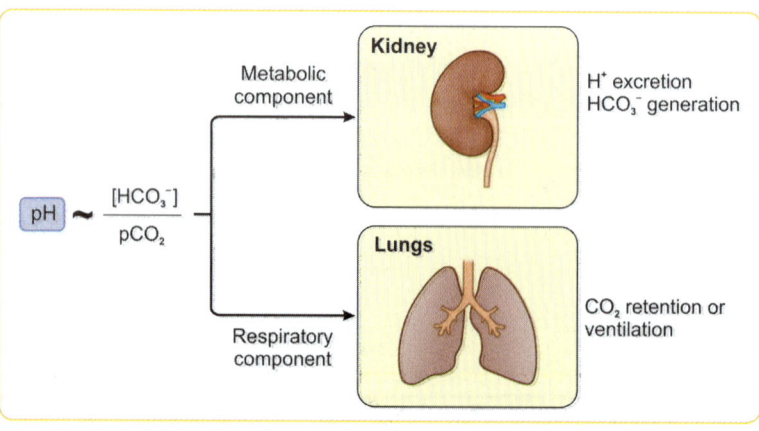

Hemoglobin as a Buffering Agent

Histidine has the pKa value of 6.1 which is near the physiological pH of 7.4. Buffering action of hemoglobin is due to histidine residues. Deoxygenated hemoglobin is a better proton acceptor than the OxyHb.
- ❑ At the level of tissues, where CO_2 (HCO_3) is more, hemoglobin accepts H^+.
- ❑ At the level of the lungs, where O_2 is more, hemoglobin releases H^+ and combines with O_2.

ANION GAP

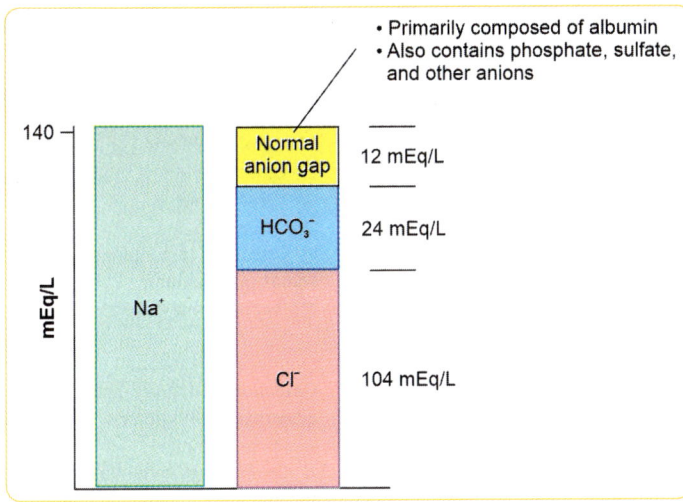

- ❑ Plasma anion gap represents the **unmeasured anions** in the plasma. It is calculated by the difference in the measured cations (positively charged ions) and the measured anions.

$$\text{Anion Gap} = [Na^+] - \{[Cl^-] + [HCO_3^-]\}$$

- ❑ Plasma proteins, lactate, acetoacetate and sulphate constitutes the unmeasured anions.
- ❑ The normal anion gap (AG) is 8–10 mEq/L. (*Ref. Harrison 19th ed., p.64e-1*)
- ❑ **Use:** Anion gap measurement helps in differentiating the causes of acid-base disorders.

Types of Acidosis based on the anion gap

Normal Anion Gap Acidosis (Anion gap is normal because of increased level of Chloride)
- ❑ Renal Tubular acidosis (RTA)
- ❑ Gastrointestinal loss of HCO_3^-
  - • Diarrhea, Fistula, urinary diversion
- ❑ Iatrogenic (large volume saline infusion)

High Anion Gap Acidosis

- ❑ Ketoacidosis (starvation, diabetic, metabolic)
- ❑ Uremic syndrome (accumulation of phosphate)
- ❑ Lactic acidosis (accumulation of lactate)
- ❑ Toxins (Salicylate, Ethylene Glycol, Methanol) KULT is the mnemonic

Decreased Anion Gap is seen in

- ❑ Hypoalbuminemia
- ❑ Multiple Myeloma

Chapter 37 **Physical Chemistry and Acid-Base**

ARTERIAL BLOOD GAS ANALYSIS

(This section has been contributed by Dr. Ashikh Seethy)

Arterial blood gas (ABG) refers to a battery of tests that includes estimation of pH, oxygen tension (P_aO_2), carbon-dioxide tension (P_aCO_2), bicarbonate concentration (HCO_3), and oxygen saturation of hemoglobin (S_aO_2) in arterial blood. Specialized instruments are also available for measurement of methemoglobin and carboxyhemoglobin levels.

INDICATIONS FOR ARTERIAL BLOOD GAS (ABG) ANALYSIS

ABG analysis is performed to evaluate the acid-base status and the oxygen-carbon dioxide gas exchange in an individual. Some indications for ABG analysis are as follows:

Gas exchange abnormalities	Acid-base disturbances
Acute and chronic pulmonary disease	**Metabolic acidosis**
❑ Chronic obstructive pulmonary disease (COPD)	❑ Lactic acidosis
❑ Bronchial asthma	❑ Renal failure
❑ Pneumonia	❑ Ketoacidosis
❑ Pulmonary edema	❑ Poisoning (e.g. methanol, ethylene glycol)
❑ Pulmonary embolism	**Metabolic alkalosis**
Acute respiratory failure	❑ Hypokalemia
❑ In chronic obstructive pulmonary disease and bronchial	❑ Vomiting
asthma	❑ Sodium bicarbonate administration
❑ Acute respiratory distress syndrome (ARDS)	**Detection and quantification of the levels of**
❑ Trauma	**abnormal hemoglobins**
Cardiovascular diseases	❑ CO poisoning
❑ Congestive heart failure	❑ Methemoglobinemia
❑ Shock	
Monitoring acute and chronic oxygen therapy	
Sleep disorder studies	

SAMPLE COLLECTION FOR ABG

The arterial blood is obtained usually by percutaneous needle puncture. This procedure which employs a needle and a syringe is preferred in patients who require only a limited number of ABG samples (Intra-arterial catheters/arterial cannulas are employed when recurrent sampling is required). The preferred sites of needle puncture are radial artery, brachial artery, femoral artery, axillary artery, and dorsalis pedis artery, in that order. In infants, the sample can also be collected from the temporal artery.

The blood sample is collected aseptically, observing the universal precautions. It is utmost pertinent to collect the sample anaerobically. This means that the sample should not come in contact with atmospheric air at no point during and after the collection.

Heparin is the anticoagulant of choice for ABG sample. Heparin can be in the form of lyophilized heparin or heparin solution (1000 units/mL). When the latter is used, the syringe is flushed with heparin solution, not allowing any liquid to remain behind. Lyophilized heparin is preferred because any inadvertent persistence of liquid heparin (which has PO_2 and PCO_2 values of atmospheric air) will dilute the sample, leading to erraneous results.

Precautions

❑ The samples should be collected under strict aseptic conditions.
❑ Contact of the sample with air should be prevented.
❑ Heparin must be completely removed from the syringe (when heparin solution is used).
❑ The sample should be transported to the analyzer without any delay. This is very important especially in patients with leucocytosis and thrombocytosis, as WBCs and platelets use up oxygen and generate

carbon dioxide, as the analysis is delayed. Delay also will lead to diffusion of gases through the plastic syringes. To avoid the delay in transportation, ABG analysis is nowadays done as POCT (Point of Care Testing), placing the analyzers at the point of sample collection (like ICUs and wards).

❏ Samples should NOT be transported on ice when plastic syringes are used, since low temperatures will lead to contraction of polymer molecules, thus widening the pores in the material (in the atomic-scales). This will enhance the diffusion of gases, leading to spurious results.

❏ Mixing of arterial and venous blood should be avoided since the reference intervals for parameters like P_aO_2 are different for arterial and venous samples.

Contraindications

❏ Abnormal modified Allen's test (for sampling from the radial artery).
❏ Local infection at the site of arterial puncture.
❏ Peripheral vascular disease involving the artery chosen for sampling.
❏ Severe thrombocytopenia.

Modified Allen's Test

A modified Allen's test measures competency of collateral circulation. The patient is instructed to clench his or her fist. Occlusive pressure is applied to both the ulnar and the radial arteries using the examiner's fingers, to obstruct blood flow to the hand. While occlusive pressure is applied, the patient is made to relax his or her hand. It is ensured that the palm and fingers have blanched.

The occlusive pressure is released on the ulnar artery alone, to determine whether the modified Allen test is positive or negative.

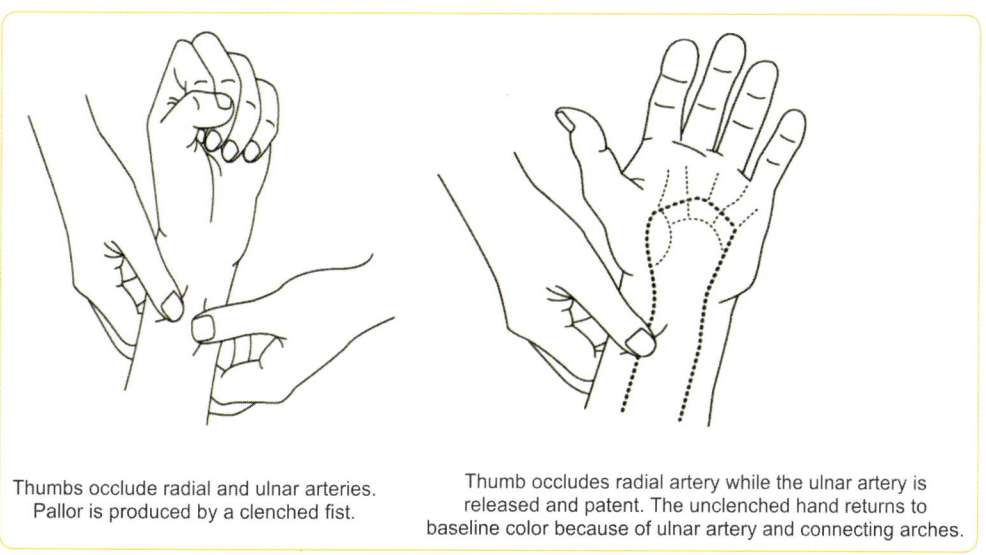

Thumbs occlude radial and ulnar arteries. Pallor is produced by a clenched fist.

Thumb occludes radial artery while the ulnar artery is released and patent. The unclenched hand returns to baseline color because of ulnar artery and connecting arches.

Positive modified Allen's test: If the hand is flushed within 5-15 seconds, it indicates that the ulnar artery has good blood flow.

Negative modified Allen's test: If the hand is not flushed within 5-15 seconds, it indicates that ulnar circulation is inadequate or nonexistent; in this situation, the radial artery supplying arterial blood to that hand should not be punctured.

Complications

Complications of ABG sampling using needle puncture are usually mild and include local pain, hematoma formation, bleeding, vasovagal response, vessel embolism (rare) and infection (rare)

Chapter 37 Physical Chemistry and Acid-Base

REFERENCE INTERVALS

- pH: 7.35 to 7.45
- P_aCO_2: 35 to 45 mm Hg
- HCO_3^-: 22 to 30 mEq/L
- P_aO_2: 67-104 mm Hg
- S_aO_2: 91-100%
- Carboxyhemoglobin and methemoglobin: ≤1%

*The oxygen saturation and P_aO_2 should always be interpreted in the context of F_iO_2 (fraction of inspired oxygen). Natural air, which has 21% oxygen, has an F_iO_2 of 0.21. The value of F_iO_2 differs when patients are on oxygen supplementation. Generally, the expected P_aO_2 (in mm of Hg) is five times the F_iO_2. Further, these values should be interpreted considering the previous reports (if available) – e.g. a resting S_aO_2 value of 95% is abnormal in a patient with previous resting S_aO_2 value of 99%.

INSTRUMENTATION

ABG analyses, these days, are performed by automated analyzers. The analytes that are estimated include pH, P_aO_2 and P_aCO_2. The estimation of these analytes are done with the help of electrochemical sensors, and employs ion selective electrodes and the principles of potentiometry (for pH and P_aCO_2) and amperometry (for P_aO_2).

Microprocessors of blood gas instruments estimate S_aO_2 from measured pH, PO_2, and Hb using empirical equations. S_aO_2 can be measured spectrophotometrically also, as the wavelength of light absorbed by oxyhemoglobin and deoxyhemoglobin is different. The ratio of oxyhemoglobin to total hemoglobin (oxyhemoglobin + deoxyhemoglobin) gives the S_aO_2.

The bicarbonate concentration is calculated once the pH and P_aCO_2 values are obtained, using the Henderson-Hasselbalch equation.

$$pH = pK_a\,(H_2CO_3) + Log_{10}\,\frac{[HCO_3^-]}{[H_2CO_3]}$$

$pK_a\,(H_2CO_3) = 6.1$

$[H_2CO_3] = K_{HCO_2} \times P_aCO_2;$ where K_H is the Henry's constant for the solubility of CO_2 in blood, which is 0.03 mmol/L/mm Hg.

Hence: $pH = 6.1 + Log_{10}\,\dfrac{[HCO_3^-]}{0.03 \times P_aCO_2}$

On rearranging, $[HCO_3^-]$ can be obtained. Thus in most of the ABG analysis reports, bicarbonate is denoted as $cHCO_3^-$, where 'c' denotes that it is 'calculated'.

INTERPRETATION

The primary change and the compensation in various acid-base disorders are summarized below:

Disorder	pH	Primary change	Compensation
Respiratory acidosis	↓	P_aCO_2 ↑	HCO_3^- ↑
Metabolic acidosis	↓	HCO_3^- ↓	P_aCO_2 ↓
Respiratory alkalosis	↑	P_aCO_2 ↓	HCO_3^- ↓
Metabolic alkalosis	↑	HCO_3^- ↑	P_aCO_2 ↑

Interpretation of an ABG report involves addressing the following questions:
- Is the ABG report valid?
- Is there acidosis or alkalosis present?
- Is the disturbance metabolic or respiratory?
- Is there appropriate compensation?

- ❑ What is the anion gap?
- ❑ Does the report correlate clinically?

Is the ABG REPORT VALID?

The consistency of the ABG report is assessed with the help of the Henderson Hasselbalch equation.

$[H^+]$ in nanomoles/L $= 24(P_aCO_2)/[HCO_3^-]$

If the values do not corroborate, the ABG report is not valid.

Is there ACIDOSIS or Alkalosis Present?

Look at the pH.

pH <7.35 → Acidosis

pH > 7.45 → Alkalosis

It should be noted that acidosis or alkalosis may be present even if the pH is within the reference interval. (This happens in cases where body mechanisms compensate for acidosis or alkalosis).

Is the Disturbance Metabolic or Respiratory?

Primary respiratory disorders, the pH and P_aCO_2 change in opposite directions.
Primary metabolic disorders the pH and P_aCO_2 change in the same direction.

Acidosis	Respiratory	pH ↓	P_aCO_2 ↑
Acidosis	Metabolic	pH ↓	P_aCO_2 ↓
Alkalosis	Respiratory	pH ↑	P_aCO_2 ↓
Alkalosis	Metabolic	pH ↑	P_aCO_2 ↑

Is There Appropriate Compensation?

When acidosis or alkalosis set in, body mechanisms start acting to compensate, so as to restore the blood pH to normal. Respiratory disorders are compensated by metabolic responses and vice versa. Compensatory mechanisms return the pH toward, but not to, the normal value. In chronic respiratory alkalosis, pH may even reach the normal value due to compensation.

Disorder		Expected Compensation
Metabolic Acidosis		P_aCO_2 will ↓ 1.25 mm Hg per mmol/L ↓ in $[HCO_3^-]$
Metabolic Alkalosis		P_aCO_2 will ↑ 6 mm Hg per 10 mmol/L ↑ in $[HCO_3^-]$
Respiratory Acidosis	Acute	$[HCO_3^-]$ will ↑ 0.1 mmol/L per mm Hg ↑ in P_aCO_2
	Chronic	$[HCO_3^-]$ will ↑ 0.4 mmol/L per mm Hg ↑ in P_aCO_2
Respiratory Alkalosis	Acute	$[HCO_3^-]$ will ↓ 0.2 mmol/L per mm Hg ↓ in P_aCO_2
	Chronic	$[HCO_3^-]$ will ↓ 0.4 mmol/L per mm Hg ↓ in P_aCO_2

It should be noted that no disorder of acid-base balance is over-compensated. Inappropriate compensatory responses (inadequate or excessive) should lead to a suspicion of a mixed disorder (A mixed disorder is characterized by the presence of two or more separate primary acid-base abnormalities occurring in the same patient).

What is the Anion Gap?

We have already discussed this.

Does the Report Correlate Clinically?

Finally, ABG reports should always be interpreted in the context of the clinical history and examination findings, and should correlate with the underlying cause of acid-base imbalance. If there is no clinical correlation, a repeat sampling and report should be obtained; other underlying causes should be considered.

OTHER CONCEPTS

GIBBS DONNAN EQUILIBRIUM/DONNAN MEMBRANE EQUILIBRIUM

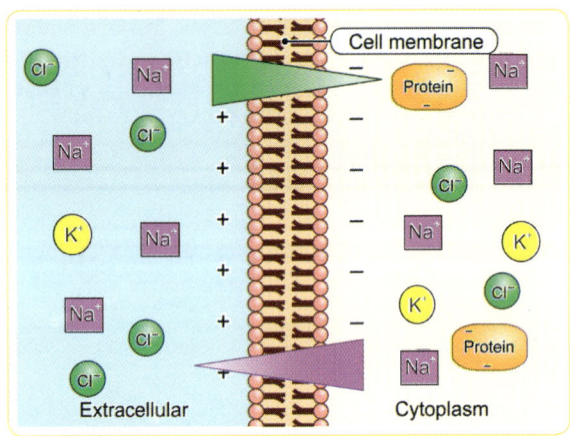

Unequal distribution of diffusible ions across a selectively permeable membrane because of presence of non-diffusible ion (proteins) in one of the compartments.

CSF Chloride Concentration is Higher than Plasma Chloride

- ❑ Protein concentration of CSF is lesser than that of plasma. To maintain the Donnan membrane equilibrium [Cl⁻] is increased.
- ❑ Increased CSF protein seen in TB meningitis is associated with decreased CSF chloride levels.

Plasma Oncotic Pressure

Oncotic pressure or **colloid osmotic pressure** is the osmotic pressure exerted by plasma proteins. Normal value: 25 to 30 mmHg

Clinical Significance

Albumin is the major determinant (80%) of oncotic pressure because of its **low molecular weight (69,000 Da) and the high concentration in plasma**. **Albumin retains water** and electrolytes in vascular compartment. **Hence Hypoalbuminemia leads to edema**.

Isotonic solution	Isosmotic solution
Isotonic solution is one that has the **same non-diffusible solute concentration** as compared to another solution	Isosmotic solution is one that has the **same osmotic pressure** as compared to another solution
0.9% NaCl solution is said to be isotonic to plasma	An isosmotic solution is not necessarily isotonic. 0.3 mol/L urea solution is isosmotic with plasma but not isotonic (since urea in non-diffusible)

STOOL OSMOTIC GAP

Stool osmotic gap is used to differentiate secretory diarrhea (e.g. gastrinoma) from osmotic diarrhea (e.g. Lactose intolerance).

Stool osmotic gap = Stool osmolality – {2 × (stool Na$^+$ + stool K$^+$) mosmol/L

Stool osmolal gap > 50 → osmotic diarrhea

Stool osmolal gap <50 → secretory type diarrhea *(Ref. Harrison 20th ed. p.2244)*

Competency

- ☐ BI6.8   Discuss and interpret results of Arterial Blood Gas (ABG) analysis in various disorders
- ☐ PY1.7   Describe the concept of pH and Buffer systems in the body
- ☐ BI11.16   Observe use of commonly used equipments/techniques in biochemistry laboratory including:
  - • ABG analyser

Key Points

- ☐ pH is the negative logarithm to the base ten of the hydrogen ion concentration in mol/L.
- ☐ Isoelectric pH is the pH at which there is no net electrical charge on the molecule, i.e. the molecule exists as Zwitter ion.
- ☐ Blood pH is maintained in a narrow range of 7.35 to 7.45.
- ☐ A buffer is most effective when the pH is pKa. Effective pH range of a buffer = pKa ± 1.
- ☐ Bicarbonate buffer is the most efficient body buffer since both the components of the buffer can be easily regulated.
- ☐ Normal anion gap is seen in renal tubular acidosis.
- ☐ Heparin is the anticoagulant of choice for the collection of ABG sample.
- ☐ Abnormal modified Allen's test is a contraindication for taking ABG sample from wrist.
- ☐ Acidosis is the blood pH <7.35; Alkalosis is the blood pH > 7.45.

SELF-ASSESSMENT

Short Answer Questions

1. Why is CSF chloride concentration higher than that of plasma? Why CSF chloride level decrease in tubercular meningitis?

2. What is Gibbs-Donnan equilibrium? Mention its physiological importance.

3. Which is the most effective buffer present in plasma? Explain.

4. Bicarbonate: Carbonic acid is an efficient buffer in blood although the ratio is 20:1 – Justify.

5. What is the anion gap? How does it help in the differential diagnosis of a patient with abnormal arterial blood gas readings?

6. Explain the role of hemoglobin as a buffering agent.

7. What are buffers? Discuss any two buffer systems of the body.

8. What is respiratory acidosis? Give two causes.

9. Enumerate the precautions to be followed during sample collection for arterial blood gas analysis.

10. What do you understand by the term 'compensated respiratory acidosis'?

Multiple Choice Questions

1. **Select the TRUE statement regarding Arterial blood gas analysis:**
   a. Specimen should be covered in ice prior to analysis
   b. Interpretation can be done without knowing body temperature
   c. Heparinized blood obtained by venipuncture is used
   d. Indicates acidemia if there is an increase in pH

2. **Calculate the pH of a solution containing Hydrogen ions in the concentration of 5 millimoles/L.**

   a. 1.9   b. 3.3

   c. 2.3   d. 5

3. **A buffer is most effective at its:**

   a. pKa   b. pka+1

   c. pKa-1   d. Any of the above

4. **Which of the following buffer system is known as open end buffer?**

   a. Bicarbonate   b. Phosphate

   c. Hemoglobin   d. Protein

5. **Calculate the pH of a mixture of 0.1 M acetic acid and 0.1 M sodium acetate (pKa of acetic acid is 4.76).**

   a. 4.76   b. 5.76

   c. 5.06   d. 4.06

6. **High anion gap is found in all of the following conditions, except:**

   a. Uremia

   b. Lactic acidosis

   c. Renal tubular acidosis

   d. Methanol poisoning

7. **In which of the following condition, reduced anion gap is found?**

   a. Hyperalbuminemia

   b. Multiple myeloma

   c. Starvation ketosis

   d. Salicylate poisoning

8. **Which one of the following property of albumin is responsible for its maximum contribution to oncotic pressure?**

   a. High molecular weight, low concentration

   b. Low molecular weight, low concentration

   c. High molecular weight, high concentration

   d. Low molecular weight, high concentration

9. **What is the isoelectric point for phenylalanine given the pKa for the COOH group is 1.83 and the NH^{3+} group is 9.13?**

   a. 2.43   b. 4.83

   c. 5.48   d. 9.13

10. **What do you infer from the decrease of pH value from 7.35 to 6.35?**

    a. 10-fold increase in $[H^+]$

    b. 10-fold decrease in $[H^+]$

    c. 2-fold increase in $[H^+]$

    d. 2-fold decrease in $[H^+]$

ANSWERS

1. a.	2. c.	3. a.	4. a.	5. a.	6. c.
7. b.	8. d.	9. c.	10. b.		

Cell-Signaling

Cell signaling is a collection of events that leads to adaptive changes in organisms in response to changes in their external environment. The ability of cells to perceive and correctly respond to their microenvironment is the basis of development, tissue repair, and immunity, as well as normal tissue homeostasis. Errors in signaling interactions and cellular information processing are responsible for diseases such as cancer, autoimmunity, and diabetes.

MODES OF CELL-SIGNALING

1. Signaling by cytoplasmic bridges and gap junctions
2. Signaling by secreted molecules
3. Signaling by plasma membrane: Bound molecules that are not secreted, also known as Juxtracrine signaling – e.g. Integrin-ICAM (Intercellular Adhesion Molecule 1) interaction

We will discuss the first two modes in this chapter.

GAP JUNCTIONS

- Also known as macula communicans or nexus
- Directly connects the cytoplasm of two cells, allowing the passage of various molecules and electrical impulses to directly pass **through a regulated gate** between cells.
- Gap junctions are seen in all tissues except adult skeletal muscle and mobile tissues like erythrocytes and sperms (Fig. 1).

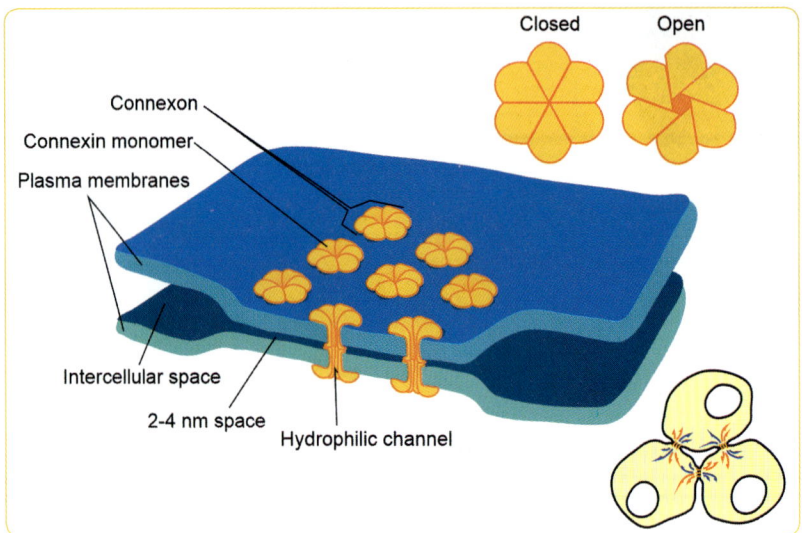

Fig. 1: Appreciate the following in the image. The space between two cells in the gap junction is 2 to 4 nm. One gap junction channel is composed of two connexons (or hemichannels). One connexon is made up of 6 connexins proteins. One connexin protein has four transmembrane domains

SIGNALING BY SECRETED MOLECULES

Signaling by secreted molecules can be further divided into the following three types:

	Description	Example
1. Endocrine signaling	Signaling molecules, called hormones act on target cells distant from their site of synthesis.	Insulin produced by beta cells of pancreas acts on skeletal muscle, adipose tissue, liver, etc.
2. Paracrine signaling	Signaling molecules released by a cell only affect target cells in close proximity to it.	Signaling by fibroblast growth factor and transforming growth factor-β (TGF-β)
3. Autocrine signaling	Cells respond to substances released by themselves.	Cytokines produced by monocytes

Steps Involved in Endocrine Signaling

❏ Synthesis of the signaling molecule (e.g. hormone) by the cell
❏ Release of the hormone by the cell
❏ Transport of the hormone to the target cell
❏ Binding of the hormone to a specific receptor protein and conformational change in the receptor (e.g. dimerization)
❏ Initiation of intracellular signal-transduction cascade by the activated receptor
❏ A change in cellular metabolism, function, structure, or development triggered by the receptor-hormone complex
❏ Deactivation of the receptor
❏ Removal of the signal, which usually terminates the cellular response

Hormones are Grouped into Two Classes

Hormones can be divided into two classes based on their solubility and location of receptors.

Group I hormones	Group II hormones
❑ Lipophilic in nature ❑ Their receptors are usually present intracellularly ❑ Their half-life is long (hours to days), e.g. Steroids, thyroid hormones, calcitriol, retinoic acid.	❑ Hydrophilic in nature ❑ Their receptors are present in the plasma membrane ❑ Their half-life is short (minutes), e.g. Peptide hormones (Insulin) and some amino acid derivatives (catecholamines)

Hormone Receptors

Hormone receptors are broadly classified into the following four types:
1. G-protein coupled receptors (GPCR)
2. Receptor enzymes
   - Receptor tyrosine kinases—Insulin and Epidermal growth factor
   - Receptor Guanylyl cyclases—Atrial natriuretic peptide
3. Gated ion channels *(refer to physiology books)*
4. Intracellular receptors
   - Cytoplasmic receptors (translocate to nucleus upon binding of hormone)
     ➢ Steroid hormone receptors, e.g. Estrogen receptor
     ➢ Vitamin D receptor (VDR)
   - Nuclear Receptors
     ➢ Thyroid hormone receptor
     ➢ Retinoid acid receptor
     ➢ Retinoid X receptor
     ➢ Peroxisome proliferator-activated receptor
     ➢ Liver X receptor and farnesoid X receptor

G-protein Coupled Receptors (GPCRS)

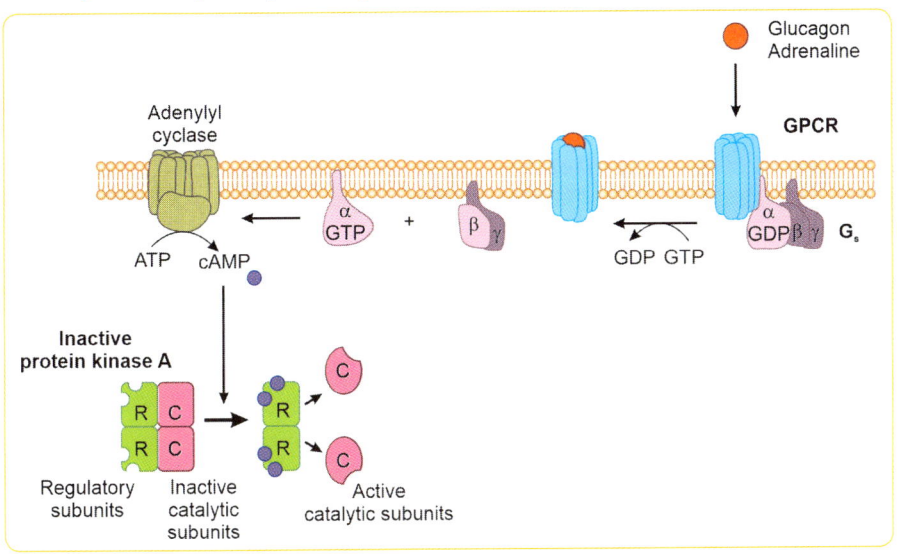

GPCRs are serpentine, seven transmembrane receptors associated with heterotrimeric G-proteins (alpha, beta and gamma). Many of the group II hormones act through GPCRs. Binding of the hormone (e.g. Glucagon/epinephrine) activates the receptor and exchanges the GDP of alpha subunit with GTP.

GTP bound α subunit dissociates from βγ subunits and activates membrane-bound adenylate cyclase to produce the second messenger cyclic AMP. *Further action of cyclic AMP is described under the heading cAMP.*

α subunit cannot activate adenylate cyclase persistently. Its action needs to be terminated. This is achieved by the inherent GTPase activity of α subunit which hydrolyzes the GTP to GDP. Thus, α subunit is inactivated.

Second Messengers

Second messengers are small, diffusible molecules and are generated inside the cell upon binding of hormones (1st messengers) to the membrane receptor. They help in signal transduction from the receptor to the target.

Duration of action of second messenger is controlled by action of various regulating enzymes. For example, phosphodiesterase enzyme inactivates cyclic AMP by converting it to 5'AMP.

	Cyclic AMP system	Phosphoinositol system	cGMP system	Tyrosine kinase system
1st Messengers	❏ Gonadotrophins (FSH & LH) ❏ Corticotroph (ACTH) & CRH ❏ Thyrotroph (TSH) ❏ Parathormone MSH, Glucagon, hCG, Adrenaline (α_2, β_1, β_2) Acetylcholine (M_2)	❏ Some Releasing hormones: TRH, GnRH ❏ Post. Pituitary hormones: Oxytocin & Vasopressin ❏ Gastrin, Angiotensin, Adrenaline (α_1), Acetylcholine (M_1, M_3),	❏ Natriuretic peptides (ANP & BNP) ❏ Nitric oxide	❏ Insulin ❏ Insulin like growth factor ❏ Epidermal growth factor
Receptor	G Protein Coupled receptor–G_s (increases cAMP) or G_i (decreases cAMP)	G Protein Coupled receptor–Gq	Natriuretic peptides act through membrane receptors whereas nitric oxide passes through the membrane freely.	Receptor with intrinsic tyrosine kinase activity
2nd messengers	Cyclic AMP	IP_3, DAG, Ca^{2+}	Cyclic GMP	Ras (small monomeric G-protein)
2nd messenger is generated by	Adenylyl cyclase	Phospholipase C	Guanylate cyclase	Ras-Guanine nucleotide exchange factor
2nd messenger activates	Protein kinase A	Diacyl Glycerol activates protein kinase C Inositol triphosphate releases Ca^{2+}.	Protein kinase G	Mitogen Activated Protein kinase family

Significance of Second Messengers in Hormonal Action

❏ Signal amplification: Second messengers **amplify the signal** produced by the first messengers
❏ Signal integration: Second messengers can integrate information from multiple independent upstream signals and thus there can be a cross-talk between the second messenger pathways.

Cyclic AMP

❏ 3', 5' cAMP is one of the intracellular second messengers.
❏ Produced from ATP by membrane-associated adenylyl cyclase enzyme on stimulation by α_s subunit of G protein.
❏ cAMP activates Protein kinase A. Proteins kinase A is a heterotetramer with 2 regulatory and 2 catalytic subunits (R_2C_2). Active sites of catalytic subunits are blocked by pseudosubstrate sequence provided by regulatory subunits. Binding of cAMP to regulatory subunits causes dissociation of R_2 from C_2. Catalytic subunits (C & C) are now free to phosphorylate serine/threonine residues of proteins.

- cAMP is inactivated by phosphodiesterase by cleavage of 3'-5 phosphodiester bond. cAMP is converted to 5'AMP.
- Adenylyl cyclase is inhibited by α_i subunit of G protein.

Cyclic GMP

- Only Nitric oxide and Natriuretic peptides [Atrial natriuretic peptide and B-type (Brain) natriuretic peptide] utilize cyclic GMP as the second messenger.

Phosphodiesterase terminates the action of second messengers – cAMP and cGMP

- Phosphodiesterases are the enzymes that degrade the 3', 5' cyclic AMP or 3', 5' cyclic GMP to 5' AMP and 5' GMP respectively.
- Specificity of the phosphodiesterase enzyme varies. Certain types act on either cAMP or cGMP only while some can act on both.
- Inhibitors of various types of phosphodiesterase are used in medicine.

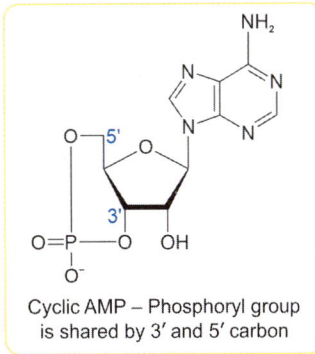

Cyclic AMP – Phosphoryl group is shared by 3' and 5' carbon

Fig. 2: Cyclic AMP – Phosphoryl group is shared by 3' and 5' carbon

Inhibitor	Phosphodiesterase type inhibited	Use
Caffeine, Theophylline, Theobromine and Aminophylline	All types	Stimulant, Asthma
Sildenafil (Viagra)	PDE 5 (cyclic GMP specific)	Erectile dysfunction
Apremilast	PDE 4	Psoriatic arthritis
Inamrinone	PDE 3	Congestive cardiac failure

Receptor Enzymes

Receptor Tyrosine Kinases

As the name suggests, receptor tyrosine kinases have intrinsic tyrosine kinase activity. Binding of the growth factor/ligand induces dimerization of the receptor and each monomer phosphorylates the other. This is known as cross-phosphorylation. Phosphorylation of tyrosine creates a new docking site for proteins that contain the phosphotyrosine binding domain (SH_2 domain).

Examples of receptor tyrosine kinases:

- Insulin receptor *(Chapter 10.7)* *(Refer to page no. 168)*
- Epidermal growth factor receptor
- Fibroblast growth factor receptor (FGFR)
- Vascular endothelial growth factor receptor (VEGFR)

Receptor guanylyl cyclases

Atrial Natriuretic factor (ANF) is synthesized, secreted, and released by cardiac myocytes in response to volume overload. This hormone acts through binding to receptor guanylyl cyclase which on binding produces cyclic GMP.

Guanylin, a 15-amino acid polypeptide that is secreted by goblet cells of colon also acts through receptor guanylyl cyclase. It regulates electrolyte and water transport in intestinal and renal epithelia.

Non-receptor Tyrosine Kinases (nRTKs)

It is important to differentiate non-receptor kinases from receptor tyrosine kinases. Non-receptor tyrosine kinases are cytosolic enzymes. More than thirty nRTKs have been identified in human cells. Most of the nRTKs are involved in the signal transduction in activated T- and B-cells of the immune system.

Examples of nRTKs
- Btk (Bruton's tyrosine kinase)
- ABL1 (Abelson murine leukemia viral oncogene homolog 1)
- JAK (Janus kinases)

Chapter 38 Cell-Signaling

Note: Binding of growth hormone, prolactin and cytokines to their receptors leads to dimerization of the receptor and further signaling through JAK-STAT pathway.

TOXINS AND CELL SIGNALING

Cholera Toxin Causes ADP-ribosylation of α_s Subunit of G-protein

Mechanism of cholera toxin involves overproduction of cyclic AMP. Cholera toxin is made up of A and B subunits. 'A' subunit causes ADP-ribosylation of Gs subunit. This prevents the GTPase activity of Gs and thus adenylate cyclase is persistently activated.

Pertussis Toxin Causes ADP-ribosylation of Alpha Subunit of Gi_2 Type of G-protein

Pertussis toxin catalyses the transfer of ADP ribosyl moiety from NAD$^+$ to the cysteine residue of α_{i2} subunit of Gi (inhibitory) type of G protein. This covalent modification of αi results in loss of GTP-GDP exchange.

α_i **normally decreases the activity of adenylyl cyclase**. As ADP ribosylation inhibits the action of α_i adenylyl cyclase activity is not inhibited. Moreover, α_s activity also goes unopposed → Persistent activation of adenylate cyclase and overproduction of cAMP

Mechanism of action of diphtheria toxin has been explained in Chapter 25 (*Refer to page no. 437*).

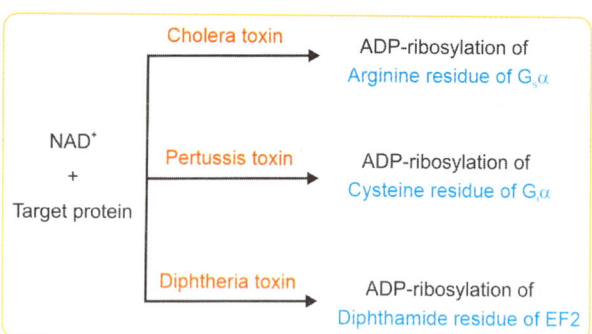

PARACRINE SIGNALING

In paracrine signaling, signaling molecules released by a cell affect only target cells in close proximity to it. There are four major families which acts via paracrine signaling.
1. Fibroblast growth factor (FGF) family
2. Hedgehog family
3. Wnt family
4. TGF-β superfamily

AUTOCRINE LOOPS

Autocrine mediator produced by the cell can act on the same cell to enhance or inhibit the production of the same autocrine mediator. This creates an autocrine loop.

Example of Autocrine Loops

Physiological
- ❑ Regeneration of liver
- ❑ Proliferation of antigen stimulated lymphocytes

Pathological
- ❑ **Autocrine loops in cancer:** Growth factors and their receptors are overproduced by tumor cells and thus stimulating the proliferation of cancer cells.

Competency

❑ PY1.3   Describe intercellular communication

Key Points

❑ A gap junction is composed of two connexons (hemichannels). One connexon is made of six connexins. One connexin is made up of four transmembrane domains.

❑ Group I hormones are lipophilic in nature whereas group II hormones are hydrophilic in nature.

❑ G-protein coupled receptor traverses the membrane seven times.

❑ cGMP is the second messenger for nitric oxide, Atrial natriuretic peptide and B-type natriuretic peptide.

❑ Sildenafil (Viagra) inhibits phosphodiesterase type 5 and prolongs the half-life of cGMP.

❑ The insulin receptor is an example of a receptor with tyrosine kinase activity.

❑ Janus kinases, Bruton's tyrosine kinase are some examples of non-receptor tyrosine kinases.

❑ Cholera toxin causes ADP-ribosylation of α subunit of stimulatory type G-protein.

❑ Pertussis toxin causes ADP-ribosylation of alpha subunit inhibitory type G-protein.

SELF-ASSESSMENT

Short Answer Questions

1. Name two hormones which use cGMP as the 2nd messenger? Explain the role of G-proteins in their mechanism of action. What is the effect of cholera toxin on G-proteins?

2. Name a hormone whose receptors are located on the plasma membrane and explain its mechanism of signal transduction.

3. Describe the mechanism of action of
   a. Pertussis toxin   b. Cholera toxin

4. What is the biochemical basis of massive fluid loss in cholera?

5. Mention the similarities and differences between the receptors and mode of signal transduction by thyroid and steroid group of hormones.

6. What is the action of phosphodiesterase enzyme? With suitable example, mention the importance of phosphodiesterase inhibitors.

7. Differentiate receptor tyrosine kinase and non-receptor tyrosine kinase.

Clinical-Based Question

1. A 3-year-old girl child was brought by her mother emergency room complaining of the difficulty of breathing, difficulty in swallowing. On examination, dense gray-colored pseudomembranous was found on the throat. It was found that the child was not fully immunized with vaccines.
   a. What is your probable diagnosis?
   b. Describe the molecular basis of the disease.

Long Answer Question

1. What is the structure of the insulin receptor? Which intrinsic enzymatic activity is associated with this receptor? What are the downstream effects of its activation? Explain with an illustration. Name two other growth factor receptors with similar associated enzymatic activity.

Multiple Choice Questions

1. **All of the following are second messengers, except:**
   a. cAMP   b. Diacyl glycerol
   c. Guanylyl cyclase   d. IP_3

2. **Which of the following is the action of α subunit of G protein?**
   a. Binding to ligand
   b. Breakdown of GTP to GDP
   c. Conversion of GDP to GTP+
   d. Exchange of GDP for GTP

3. **All of the following hormones act mainly through membrane receptor, except:**
   a. Insulin   b. Glucagon
   c. Ghrelin   d. Thyroxin

4. **All of the following are examples of receptor tyrosine kinases, except:**
   a. Insulin receptor
   b. Glucagon receptor
   c. Epidermal growth factor receptor
   d. Fibroblast growth factor receptor (FGFR)

5. **The second messenger for Atrial natriuretic peptide is:**
   a. Cyclic AMP
   b. Cyclic GMP
   c. Phosphatidyl Inositol triphosphate
   d. Inositol triphosphate

6. **All of the following are examples of non-receptor tyrosine kinases, except:**
   a. Janus kinase
   b. Casein kinase
   c. Zeta-chain-associated protein kinase 70
   d. Bruton's tyrosine kinase

7. **Which of the following hormone acts through Phosphoinositol system?**
   a. Follicular stimulating hormone
   b. Parathyroid hormone
   c. Adrenocorticotropic hormone
   d. Oxytocin

8. **Which of the following enzyme produces cyclic AMP?**
   a. Adenylate kinase
   b. Adenylyl cyclase
   c. Phosphodiesterase
   d. 5′ Nucleotidase

9. **Which of the following is the mechanism of action of cholera toxin?**
   a. ADP-ribosylation of elongation factor-2
   b. ADP-ribosylation of stimulatory type α subunit of G-protein
   c. ADP-ribosylation of inhibitory type α subunit of G-protein
   d. ADP-ribosylation of q type α subunit of G-protein

10. **Ras is a:**
    a. Receptor tyrosine kinase
    b. Non-receptor tyrosine kinase
    c. Heterotrimeric G-protein coupled to receptor
    d. Monomeric GTPase

11. **Which of the following activity is found in insulin receptors?**
    a. Tyrosine kinase
    b. GTPase
    c. ATPase
    d. Serine/threonine kinase

ANSWERS

| **1.** c. | **2.** b. | **3.** d. | **4.** b. | **5.** b. | **6.** b. |
| **7.** d. | **8.** b. | **9.** b. | **10.** d. | **11.** a. | |

39
CHAPTER

Free Radicals and Antioxidants

Chapter Outline

- Free Radicals
  - Definition
  - Comparison with Reactive Oxygen Species
  - Sources
  - Products Formed by Reduction of Oxygen
  - Properties
  - Harmful Effects
  - Physiological Role
  - Markers of Oxidative Stress
- Antioxidants
  - Classification
  - Dietary Antioxidants
  - Reactions Catalyzed
  - Toxicity

FREE RADICALS

Definition of Free Radical

A free radical is a molecular species that contains an *unpaired electron* in its *outermost* atomic orbital.

As there is an unpaired electron, free radicals extract electrons (usually as hydrogen atoms) from other compounds to complete their orbital. This initiates a chain reaction.

The word "free" in free radical refers to their ability of independent existence. Most free radicals are short-lived. The half-life of hydroxyl radical is 10^{-9} seconds.

Definition of Reactive Oxygen Species (ROS)

Reactive oxygen species are oxygen-containing compounds that are highly reactive free radicals or compounds that are readily converted to oxygen-free radicals in the cell.

Free Radical versus Reactive Oxygen Species

H_2O_2 is not a free radical. It is classified as a ROS because it can generate the hydroxyl radical ($OH^{\bullet}$) through the following reactions:

- **Fenton reaction**

$$Fe^{2+} + H_2O_2 \longrightarrow Fe^{3+} + OH^{\bullet} + OH^{-}$$

- **Haber-Weiss Reaction**

$$O_2 + H_2O_2 \longrightarrow O_2 + OH^{\bullet} + OH^{-}$$

Sources of Free Radicals

Endogenous	Exogenous
❑ Electron leakage from mitochondrial electron transport chain	❑ Cigarette smoke
❑ Xanthine oxidase	❑ Environmental pollutants
❑ Peroxisomal oxidation of fatty acids	❑ Radiation
❑ Respiratory burst catalyzed by NADPH oxidase	❑ Certain drugs, pesticides
❑ Arachidonate pathways	❑ Ozone
❑ Ischemia-Reperfusion Injury	
❑ Fenton's reaction	

Products Formed by Reduction of Oxygen with Electrons

Ö=Ö	Ö=Ö	H—Ö=Ö—H	Ö—H	H—Ö—H
Molecular oxygen (O_2)	Superoxide anion radical ($\cdot O_2^-$)	Hydrogen peroxide (H_2O_2)	Hydroxyl radical ($^\cdot OH$)	Water (H_2O)

$$O_2 \xrightarrow{e^-} \cdot O_2^- \xrightarrow[2H^+]{e^-} H_2O_2 \xrightarrow{e^-} {}^\cdot OH + OH^- \xrightarrow[2H^+]{e^-} 2H_2O$$

Look at the number of electrons in each molecule depicted in the illustration. One unpaired electron is present in superoxide and hydroxyl radicals. There is no unpaired electron in hydrogen peroxide. So, it is a reactive oxygen species, not a free radical.

Superoxide	Reduction of oxygen by a *single* electron
Hydrogen peroxide	Reduction of oxygen with *two* electrons
Hydroxyl radical	Reduction of oxygen with *three* electrons
Water	Reduction of oxygen with *four* electrons

Properties of Free Radicals

❑ Highly reactive
❑ Very short half-life
❑ Generate new radicals by chain reaction
❑ Cause damage to biomolecules, cells and tissues.

Harmful Effects of Free Radicals on Biomolecules

❑ **Proteins:** Oxidation of sulfhydryl groups and modification of amino acids
❑ **Lipids:** Peroxidation of polyunsaturated fatty acids
❑ **Nucleic acids:** Base modifications → mutations, double stranded DNA breaks.

Free Radicals and Diseases

❑ **Cardiovascular diseases:** Oxidized low-density lipoprotein (LDL) is produced by free radical mediated damage of LDL. It is taken up by macrophage scavenger receptors (SRB1) in an unregulated manner and leads to atherosclerotic diseases.
❑ **Inflammatory diseases:** Free radicals and inflammation go hand in hand. Inflammation produces free radicals and excessive free radicals produce inflammatory changes.
❑ **Neoplasia:** DNA damage caused by free radicals leads to neoplasia.
❑ Many other diseases like *diabetes mellitus, cataract* are associated with free radicals.
❑ **Aging process is mainly free-radical mediated.**

Physiological Role of Free Radicals

Free radicals are not always harmful. They are needed for the following physiological processes:
- Microbial killing by activated phagocytes through *respiratory burst* mediated by NADPH oxidase
- Iodination of tyrosine in thyroxin synthesis
- Nitric oxide (NO) radical is involved in cellular signaling
- Blastocyst implantation
- Apoptosis

Markers of Oxidative Stress

Researchers measure the following markers to quantify the total oxidative stress.

Damage	Marker
Oxidative DNA damage	8-oxo/8-hydroxyguanine
Lipid peroxidation	**Malondialdehyde**, isoprostanes, 4-hydroxynonenol
Oxidative protein damage	Nitrotyrosine

ANTIOXIDANTS

Antioxidants scavenge free radicals. They:
- Prevent the transfer of electron from O_2 to other organic molecules
- Stabilize free radicals
- Terminate free radical reactions

Classification of Antioxidants

Types based on the Chemical Nature

- **Antioxidant enzymes:** Superoxide dismutase, catalase, glutathione peroxidase
- **Scavenging substances:**
  - Lipid phase acting
    - Vitamin E (α-tocopherol)
    - β-carotene
    - Ubiquinone/Coenzyme Q (Q_{10})
    - Lycopene
  - Aqueous phase acting:
    - Vitamin C (Ascorbic acid)
    - Uric acid
    - Bilirubin

Types based on the Mode of Action

- **Preventive antioxidants** (reduce the rate of chain initiation):
  - Catalase
  - Chelators of metal ions—Ceruloplasmin, transferrin
- **Chain breaking antioxidants** (interfere with chain propagation):
  - Vitamin E
  - Superoxide dismutase
  - Uric acid
  - Polyphenols

CONCEPT CORNER

The enzyme uricase in non-primate mammals converts uric acid to water-soluble allantoin. In primates, uricase enzyme is absent and thus insoluble uric acid builds up in the body to some extent. Uric acid acts as an antioxidant in the blood. This may explain the increased longevity of the primate mammals compared to nonprimates.

Ascorbic acid cannot be synthesized in primates due to lack of **L-gulonolactone oxidase**. Loss of one antioxidant, i.e. ascorbic acid might have been compensated by the production of another antioxidant, i.e. uric acid during primate evolution.

Chapter 39 Free Radicals and Antioxidants

Dietary Antioxidants

For a normal healthy person, it is always better to take dietary antioxidants instead of pharmaceutical supplements.

Dietary antioxidants	Rich dietary sources
Vitamin C	Citrus fruits
Isoflavonoids	Soy, beans, peanuts
Vitamin E	Fish, meat, leafy vegetables
Epigallocatechin gallate (EGCG)	Green tea
Selenium	Nuts, meat, fish, spinach
Carotenoids	Leafy vegetables, plums, tomatoes, watermelon, carrots
Resveratrol	Grape skin, red wine

Reactions Catalyzed by Antioxidant Enzymes

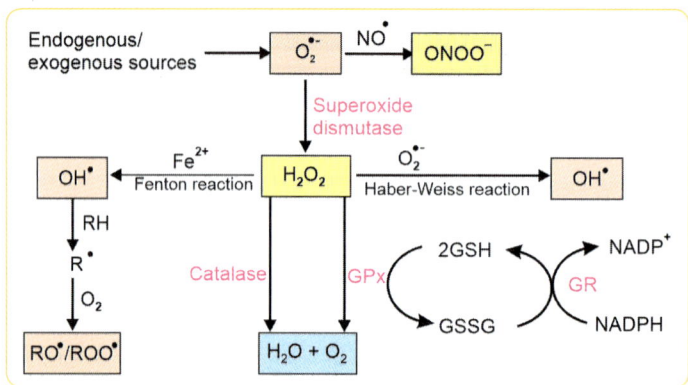

Fig. 1: Sources of free radicals and the action of antioxidant enzymes.
Abbreviations: GPx, glutathione peroxidase; GR, glutathione reductase

Comparison of Catalase and Glutathione Peroxidase

Both catalase and glutathione peroxidase (GPx) detoxify H_2O_2. But the chemical reaction is different. Moreover, these enzymes are located in different cellular compartments. Catalase is in peroxisomes, GPx is in cytosol. Peroxisomes are present in all nucleated cells and platelets. RBCs do not contain peroxisomes. So, they are dependent on GPx, and G6PD deficiency leads to free radical damage of RBCs on exposure to oxidizing agents.

All the Antioxidant Enzymes are Metalloenzymes

Transition metals can pass through various oxidative states because of their ability to gain or lose electrons easily. Thus, enzymes involved in redox reactions utilize transition elements as prosthetic group.

Antioxidant enzyme	Prosthetic group
Catalase	Heme
Superoxide dismutase	Copper, zinc, manganese (*see* the image below)
Glutathione peroxidase	Selenium

Antioxidants can Become Prooxidants in High Doses and in the Presence of Excess Iron/Copper

Antioxidants are **double-edged sword**. At physiological concentration, they are needed for the defence against free radicals. But, at high concentrations and in the presence of excess iron and copper, antioxidants can behave as prooxidants.

Corner

Vitamin C should not be supplemented in iron excess states because it generates free radicals. (*Ref. Harrison 20th Ed., p.698*)

Competency

- ❑ BI7.6   Describe the anti-oxidant defence systems in the body
- ❑ PY11.7   Describe and discuss physiology of aging; free radicals and antioxidants

Key Points

- ❑ Hydroxyl radical is the most potent reactive oxygen-free radical.
- ❑ Ferrous iron converts the H_2O_2 to the most toxic hydroxyl radical via Fenton reaction.
- ❑ Vitamin E is the most potent chain-breaking antioxidant.
- ❑ Vitamin A (b carotene), vitamin C and vitamin E are the antioxidant vitamins.
- ❑ Malondialdehyde is a marker of lipid peroxidation.
- ❑ Irrational use of antioxidants should be avoided as many antioxidants can also act as prooxidants.
- ❑ Uric acid and bilirubin are endogenous antioxidants.

SELF-ASSESSMENT

Short Answer Questions

1. Define free radicals. Write three physiological functions of free radicals in humans.

2. Mention the reaction products when oxygen molecule in the cell is reduced by 1, 2, 3 and 4 electrons.

3. Mention any two physiological roles of free radicals.

4. What is superoxide anion? How is it formed and destroyed in the body?

5. Differentiate preventive and chain-breaking antioxidants. Give an example for each.

6. Name any four dietary antioxidants and their dietary source.

7. Name any two antioxidant enzymes and write the reaction catalyzed by them.

8. What advise will you give to a person who is taking an excessive amount of antioxidant medications to rejuvenate his body?

Multiple Choice Questions

1. **Which of the following is produced when oxygen is reduced with 3 electrons?**
   a. Superoxide radical
   b. Hydrogen peroxide
   c. Hydroxyl radical
   d. Water

2. **Which one of the following is a chain-breaking antioxidant?**
   a. Glutathione peroxidase
   b. Selenium
   c. Superoxide dismutase
   d. Catalase

Chapter 39   Free Radicals and Antioxidants

595

3. Which of the following is the most potent radical?

   a. Superoxide radical b. Hydrogen peroxide
   c. Hydroxyl radical   d. Superoxide anion

4. Which of the following prevents the peroxidation of lipids in the plasma membrane?

   a. Glutathione   b. Tocopherol
   c. Ascorbate   d. Urate

5. All of the following are antioxidant vitamins, except:

   a. Vitamin A   b. Vitamin C
   c. Vitamin D   d. Vitamin E

6. During phagocytosis, the metabolic process called respiratory burst involves the activation of:

   a. Oxidase
   b. Hydrolase
   c. Peroxidase
   d. Dehydrogenase

7. Enzyme responsible for respiratory burst reaction is:

   a. Dehydrogenase   b. Peroxidase
   c. Hydroxylase   d. NADPH oxidase

8. Which of the following element is known to influence the body's ability to handle oxidative stress?

   a. Calcium   b. Sodium
   c. Potassium   d. Selenium

9. Glutathione peroxidase contains:

   a. Chromium   b. Manganese
   c. Zinc   d. Selenium

10. All of the following are true statements about glutathione, except:

   a. It converts hemoglobin to methemoglobin
   b. It scavenges free radicals and superoxide ions
   c. It is a tripeptide
   d. It conjugates xenobiotics

ANSWERS

| 1. c. | 2. c. | 3. c. | 4. b. | 5. c. | 6. a. |
| 7. d. | 8. d. | 9. d. | 10. a. | | |

40
CHAPTER

Water and Electrolytes

Water is the elixir of life. In humans, total body water is about 60% of body weight in adults and about 75% in children. 2/3rd of the total body water is intracellular and 1/3rd is extracellular. Females have lesser body water than males because of relatively high fat. Total body water decreases in the elderly irrespective of the gender because of fatty infiltration of muscles.

In this chapter, we will very briefly discuss the abnormalities of sodium and potassium. *Calcium and phosphate have already been discussed in Chapter 34* *(Refer to page no. 534 and 536)*.

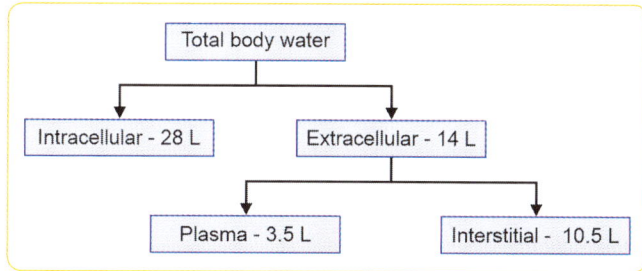

METABOLIC WATER

Water produced during the oxidation of macromolecules is known as metabolic water. Around 200 mL of metabolic water is produced in body everyday. Lipids produce the maximum amount of metabolic water since they are more reduced.

Kangaroo rat is an animal that never drinks water. It meets its water needs through metabolic water.

Kangaroo rat

HORMONES INVOLVED IN WATER HOMEOSTASIS

Hormone	Source	Action
Arginine-Vasopressin/Anti-diuretic hormone	Secreted by supraoptic hypothalamic nuclei; stored and released from neurohypophysis	Insertion of aquaporin-2 in the luminal surface of collecting tubules
Aldosterone	Zona glomerulosa of adrenal cortex	Na^+ and water retention; K^+ excretion
Renin-Angiotensin system	Renin - Juxtaglomerular (JG) cells Angiotensinogen – Liver Angiotensinogen converting enzyme – Lungs	Aldosterone secretion Vasoconstriction Thirst stimulation

HYPOKALEMIA

Hypokalemia is plasma Potassium level <3.5 mmol/L or mEq/L. *(1 mmol of Na^+ and K^+ is equal 1 mEq, while 1 mmol of Ca^{++} is equal 2 mEq.)*

Causes

Intracellular shift of K^+	Loss of potassium:
❑ Insulin administration ❑ Alkalosis (memory aid: K Loss)	❑ Renal loss—Renal tubular acidosis, Cushing syndrome, Hyperaldosteronism, Hypomagnesemia, Loop and thiazide diuretics, Amphotericin B ❑ Non-renal loss—Diarrhea, Laxative abuse, vomiting, excessive sweating

Clinical features: Irritability, fatigue, confusion, muscle cramps, abdominal distention due to reduced intestinal motility (ileus)

ECG findings: Depressed ST segment, flattening of the T wave, appearance of a prominent U wave, Prolongation of PR interval

Management

- ❑ Treatment of underlying cause
- ❑ Replacement of K^+ either by oral or IV.
- ❑ Continuous ECG monitoring for arrythmia.
- ❑ IV infusion is reserved for those patients with K^+ <2.5 mEq/L or those with arrythmia or those with severe symptoms.

HYPERKALEMIA

Fig. 1: Potassium-rich food items

Hyperkalemia is plasma [K⁺] >5.5 mmol/L.

Causes

- Renal causes:
  - Renal failure
  - Hypoaldosteronism
  - **Drugs:** Potassium sparing diuretics, Cyclosporine, Digoxin

(**Note:** Digoxin causes hyperkalemia but hypokalemia potentiates digoxin toxicity)

- Intracellular to extracellular shift
  - Acidosis–e.g. Diabetic ketoacidosis
  - Tissue Trauma
    - Muscle injury (3/4 of the total potassium is inside skeletal muscles)
    - Burns

Clinical Features

- Muscle weakness
- Flaccid paralysis
- Cardiac arrhythmia
- **Cardiac arrest in diastole** if uncorrected (c.f. Hypercalcemia causes cardiac arrest in systole)

ECF findings: Tall T waves, widening of QRS complex, loss of P waves

Management of Hyperkalemia

First aim is to **protect the heart** if ECG changes are present. Calcium gluconate is given to reduce the membrane excitability. *(Calcium gluconate does not reduce potassium)*

To Decrease the K⁺ Levels

- **Insulin with glucose**: Insulin causes **intracellular shift of K⁺**. Glucose is given along to prevent hypoglycemia.
- β_2-adrenergic agonists like salbutamol (albuterol) as nebulization → promotes intracellular uptake of K⁺.
- Potassium binding resins like kayexalate (sodium polystyrene sulfonate) is administered via rectal route when the patient is stable.

- In severe cases → Hemodialysis is done
- NaHCO$_3$ has no role in management of hyperkalemia.

CAUSES OF HYPONATREMIA (<135 MMOL/L)

1. **Hypovolemic hyponatremia** (associated with ECF volume depletion - Total body water ↓ and Total body sodium ↓↓)
   i. Gastrointestinal losses, e.g. diarrhoea (when corrected with hypotonic fluids)
   ii. Skin losses, e.g. sweating (when corrected with hypotonic fluids)
   iii. Diuretic therapy
   iv. Mineralocorticoid deficiency
   v. Osmotic diuresis, e.g. Diabetic ketoacidosis
2. **Hypervolemic hyponatremia** (associated with ECF volume excess and edema-Total body water ↑↑ and total body sodium ↑.
   i. Congestive heart failure
   ii. Cirrhosis
   iii. Nephrotic syndrome
3. **Euvolemic hyponatremia** (associated with **normal ECF volume** -Total body water ↑ and Total body sodium: unchanged.
   i. Syndrome of inappropriate anti-diuresis (SIADH)
4. Pseudo hyponatremia – Not seen with modern laboratory techniques
   i. Hyperproteinemia
   ii. Hyperlipidemia

CAUSES OF HYPERNATREMIA (>145 MMOL/L)

1. Associated with an increase in ECF volume–Total body water ↑ and Total body sodium ↑↑)
   ○ Administration of hypertonic saline or sodium bicarbonate
2. Associated with no increase in ECF volume - Total body water ↓↓ and Total body sodium ↓)
   i. Insensible water loss
   ii. Gastro-intestinal water losses
   iii. Diabetes insipidus (water loss is due to decrease in ADH secretion or lack of renal response to ADH due to V$_2$ receptor mutation or Aquaporin-2 mutation)
   iv. Disorders of thirst mechanism.

LABORATORY MEASUREMENT OF ELECTROLYTES

Sample collection: Plasma is the preferred sample. For the separation of plasma, anticoagulants like EDTA and citrate can't be used since they come in the form of potassium EDTA and sodium citrate. Moreover, EDTA may chelate calcium. Lithium Heparin can be used for plasma separation.

Analytical methods:
- Flame photometry–Outdated method
- Atomic absorption spectroscopy–reference method
- Ion selective electrodes–Most commonly used method in clinical laboratories

Ion Selective Electrode (ISE) utilize the principle of potentiometry. ISE membrane allows the entry of only a particular type of ion. For example,
- Special glass membrane allows only Na$^+$
- Membrane with **Valinomycin allows only K$^+$**

Entry of the electrolyte generates a potential proportional to the concentration of the electrolyte based on the Nernst equation.

$$E = \frac{RT}{zF} \frac{\ln C_0}{\ln C_i}$$

E = Membrane potential
R = Universal gas constant
T = Temperature (Kelvin)
z = Charge on the ion
F = Farday's constant
ln = Natural log
C_0 = Concentration of ions outside the cell
C_i = Concentration of ions inside the cell

Since R (gas constant), Temperature K (kelvin), n (valency), F (Faraday's constant), E° (standard potential) are constant, the potential difference (E) is determined by the concentration of ions inside and outside of the semipermeable membrane.

Recommended Dietary Intake of Sodium and Potassium

❑ Sodium should be taken less than 2 g or less than 5 g of salt (NaCl) per day.
❑ Potassium should be taken more than 3.5 g/day.

Competency

❑ BI6.7   Describe the processes involved in maintenance of normal pH, water & electrolyte balance of body fluids and the derangements associated with these
❑ PY1.6   Describe the fluid compartments of the body

Key Points

❑ Water produced during the oxidation of macromolecules is known as metabolic water.
❑ ADH (Vasopressin) triggers the insertion of aquaporin 2 water channels into the membrane of renal collecting ducts.
❑ Hypokalemia is plasma Potassium level <3.5 mmol/L or mEq/L.
❑ Hypokalemia is plasma Potassium level >5.5 mmol/L or mEq/L.
❑ In hyperkalemia, the first aim is to protect the heart with calcium gluconate if ECG changes are present.
❑ Insulin causes an intracellular shift of K^+.
❑ SIADH causes euvolvemic (normal ECF volume) hyponatremia.

Chapter 40 **Water and Electrolytes**

Short Answer Questions

1. Why hyperkalemia is a medical emergency? Mention two causes of hyperkalemia. What is the treatment approach of hyperkalemia?

2. What is the basis of the use of Insulin in the treatment of hyperkalemia?

3. What is hyponatremia? What are its causes?

Clinical Case-Based Questions

1. A 55-year-old man one-carbon for 7 hours in a railway accident. He sustained severe injuries including crush injuries to both the thighs, fracture of the pelvis and scalp lacerations. His pulse was 130/min and blood pressure was 60/40 mm Hg. Lab findings are:

   Na^+ = 142 mmol/L   Serum urea = 38 mg/dL

   K^+ = 7.6 mmol/L   Serum creatinine = 1.2 mg/dL

   a. What is the probable diagnosis?
   b. What are the cardiac manifestations of the above abnormality?

2. A patient with 40% burns is admitted to the burns ward. She is being administered hypertonic fluids. After one day she exhibits features of CNS dysfunction, confusion, irritability, etc. She was immediately infused with saline and the symptoms improved over time.

   a. What was the probable abnormality?
   b. How will you confirm your diagnosis?

Multiple Choice Questions

1. In which of the following person, percentage of total body water is highest?
   a. One-year-old female
   b. Seventy-year-old female
   c. Twenty-year-old male
   d. Twenty-year-old female

2. During Digoxin treatment, level of which of the following electrolyte must to be monitored?
   a. Na^+ b. Ca^{2+}
   c. K^+ d. Mg^{2+}

3. Which of the following is the cause of euvolemic hyponatremia?
   a. Congestive cardiac failure
   b. Diarrhea
   c. SIAD   d. Diabetes insipidus

4. All of the following are associated with hyperkalemia, except:
   a. Diabetic ketoacidosis
   b. Renal tubular acidosis
   c. Hypoaldosteronism d. Renal failure

5. What is the optimum intake of sodium per day?
   a. <5 gram   b. <3.5 gram
   c. <2.5 gram   d. <2 gram

6. 66% of the total body water is present in which of the following body compartment?
   a. Intracellular fluid   b. Extracellular fluid
   c. Interstitial fluid   d. Intravascular fluid

ANSWERS

| 1. a. | 2. c. | 3. c. | 4. b. | 5. d. | 6. a. |

41
CHAPTER

Glycoproteins and Proteoglycans

Glycoconjugates are carbohydrates covalently linked with other chemical species such as proteins, peptides and lipids. Glycoproteins, proteoglycans and glycolipids are the glycoconjugates. We have discussed glycolipids in Chapter 12 *(Refer to page no. 242)*. In this chapter, we will discuss glycoproteins and proteoglycans.

GLYCOPROTEINS

All the plasma proteins except human albumin are glycoproteins. Glycoproteins are proteins to which branched oligosaccharide chain is covalently attached. The carbohydrate content of glycoprotein is highly variable depending upon the type of glycoproteins.

Glycoproteins are involved in diverse functions.

Function	Example
Structural molecule	Collagens
Lubricant	Mucins
Transport molecule	Transferrin, Ceruloplasmin
Immune system	Immunoglobulins, Histocompatibility antigens, Blood group determinants
Hormone	Human Chorionic Gonadotrophin, Thyroid stimulating hormone

Contd...

Function	Example
Enzymes	Alkaline phosphatase
Blood clotting	Fibrinogen
Cell surface recognition	Lectins

GLYCOPROTEINS VERSUS PROTEOGLYCANS

Glycoprotein	Proteoglycans
Oligosaccharide chains are short and highly branched	Oligosaccharide chains are long and unbranched
Amino sugars are present. Uronic acid is absent	Both are present
No repeating disaccharide unit	Disaccharide repeat unit (-Amino sugar-Hexuronic acid-)$_n$ is present with some variations
Most cell surface proteins and secretary proteins are glycoproteins	Most connective tissue proteins are proteoglycans

THREE TYPES OF GLYCOPROTEINS

Based on the way the oligosaccharide chains are attached, Glycoproteins are of three types:
1. O-linked glycoproteins
2. N-linked glycoproteins
3. GPI-anchored proteins

Know the Difference:
- ❑ Glycation - non-enzymatic addition of sugars to the proteins
- ❑ Glycosylation - enzymatic addition of sugar residues to the proteins
- ❑ Glypiation - enzymatic addition of GPI anchor to the proteins

N-linked and O-linked Glycoproteins

O-glycosidic linkage involves OH group of amino acids whereas N-glycosidic linkage involves amide group of asparagine. O-glycosidic linkage usually involves serine and threonine. The only O-glycoprotein in which O-glycosylation involves tyrosine is glycogenin, primer for glycogen synthesis. Collagen is a glycoprotein in which hydroxylysine is involved in glycosylation.

	N-linked glycoproteins	O-linked glycoproteins
Amino acid and the functional group to which glycosidic linkage is made	**Amide group** of asparagine	**OH** group of serine, threonine, tyrosine, hydroxyl lysine
Location of synthesis	Rough endoplasmic reticulum and Golgi apparatus	Only in Golgi apparatus
The way oligosaccharides are added	En bloc transfer of dolichol-linked oligosaccharide chain to the asparagine residue in the lumen of the ER and further trimming in the Golgi	Sugars are added sequentially to the amino acids involved
Requirement of Dolichol phosphate	Yes	No
Inhibitor	Swainsonine (inhibits α-mannosidase II) Tunicamycin	–
The pattern of oligosaccharide attachment	N-glycosidic linkages are separated in between by a stretch of amino acids	Sugar residues are clustered nearby
Examples	Plasma proteins	Glycophorin, Mucin

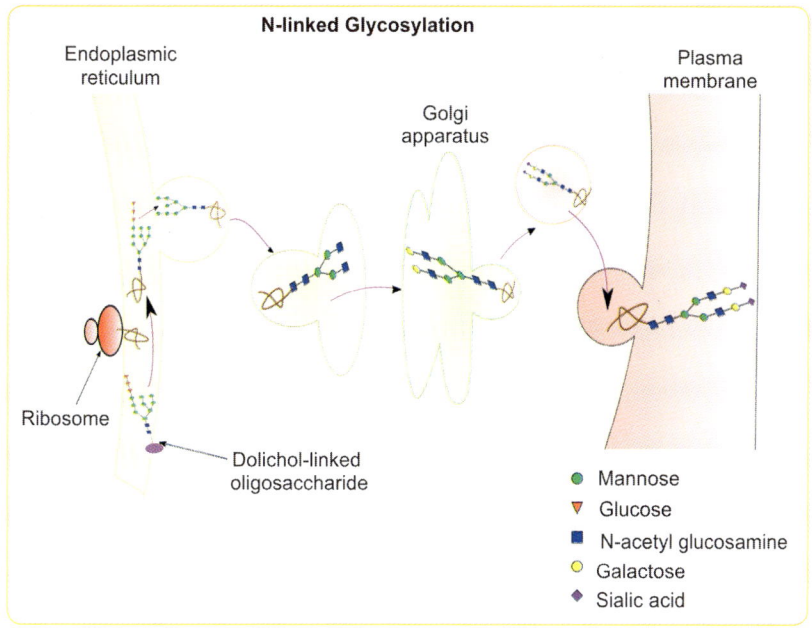

Fig. 1: I want you to observe this: The role of dolichol; N-linked glycosylation taking place both in ER and Golgi apparatus
Source: Dna 621 - Own work, CC BY-SA 3.0, https://commons.wikimedia.org/w/index.php?curid=31824610

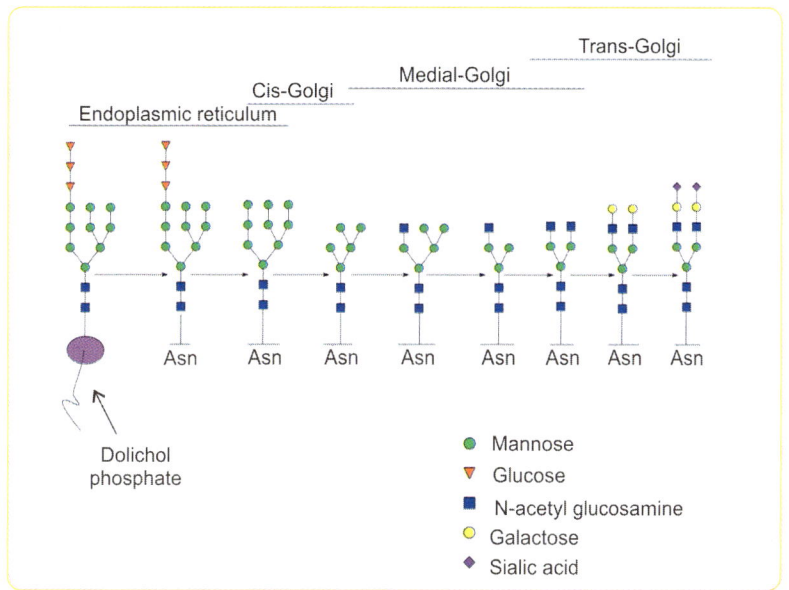

Fig. 2: Trimming and processing of N-linked glycoproteins in the Golgi apparatus obeserve that most N-linked glycoproteins contain sialic acid as a terminal residue. Loss of this sialic acid leads to degradation of the protein by hepatic
Source: DNA 621 - Own work, CC BY-SA 3.0, https://commons.wikimedia.org/w/index.php?curid=31825198

GPI Anchored Proteins are Attached to the Outer Leaflet of Plasma Membrane

❑ GPI stands for Glycosyl phosphatidylinositol.
❑ Phosphatidylinositol portion anchors the protein to the outer leaflet of the membrane.
❑ Glycosyl/oligosaccharide chain portion provides the flexibility to the attached protein.
❑ Phosphoethanolamine links the oligosaccharides and the protein.
❑ GPI anchored proteins are abundant in lipid rafts, which are involved in receptor-mediated signal transduction pathways and membrane trafficking.

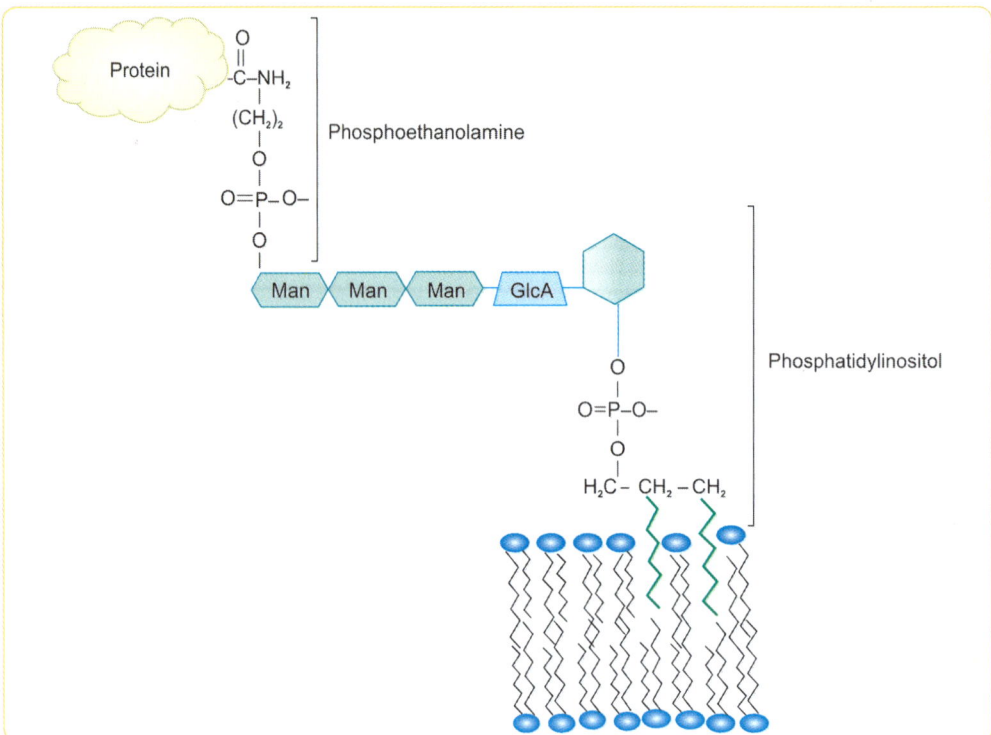

GPI Anchored Proteins-examples

❑ 5′-Nucleotidase
❑ Acetylcholinesterase (RBC membrane)
❑ Alkaline phosphatase (intestinal, placental)
❑ Angiotensin-converting enzyme
❑ CD 55 and CD 59 (RBC membrane)
❑ PrP^c – cellular prion protein

Ectoenzymes

Ectoenzymes are catalytic membrane proteins with their active sites outside the cell. Many of the ectoenzymes are GPI anchored proteins as mentioned above.

DEGRADATION OF GLYCOPROTEINS

Asialoglycoprotein Receptor is Involved in the Degradation of Plasma Glycoproteins

Glycoproteins contain sialic acid (Neuraminic acid) as their terminal residue. Due to wear and tear, this sialic acid is lost in circulation after some time. Hepatic asialoglycoprotein receptors bind and interanlise these asialoglycoproteins. They are degraded by the lysosomal proteases.

Defective Degradation of Glycoproteins Leads to Glycoproteinoses

❑ Lysosomal hydrolases like α-neuraminidase, β-galactosidase, β-hexosaminidase are involved in the degradation of oligosaccharide chains during glycoprotein turnover.
❑ Enzyme defects lead to abnormal degradation of glycoproteins and accumulation of partially degraded glycoproteins in tissues, known as Glycoproteinoses.
❑ Glycoproteinoses are classified into lysosomal storage disorders.
❑ All the glycoproteinoses are associated with mental retardation and inherited as an autosomal recessive trait.

arrison's Corner

Antibodies against asialoglycoprotein receptors (hepatic lectin) are found in autoimmune hepatitis.
(*Ref. Harrison 20th Ed., p.2396*)

PAROXYSMAL NOCTURNAL HEMOGLOBINURIA (PNH) IS DUE TO DEFECTIVE GPI ANCHORAGE

❑ PNH is an acquired stem cell disease due to a defect in phosphatidylinositol glycan class A (*PIGA*) gene.
❑ *PIGA* is involved in the GPI anchor formation. Defect in *PIGA* leads to the absence of CD55 and CD59 on the RBC Membrane. (*Note: Synthesis of CD55 and 59 is normal, only the membrane anchorage is affected.*)
❑ CD55 and CD59 are important regulators of the complement pathway.
❑ CD 55 (DAF) is a.k.a. decay accelerating factor since it accelerates the decay of C3 convertase of the complement pathway.
❑ CD59 (MIRL or Protectin) is a.k.a. membrane inhibitor of

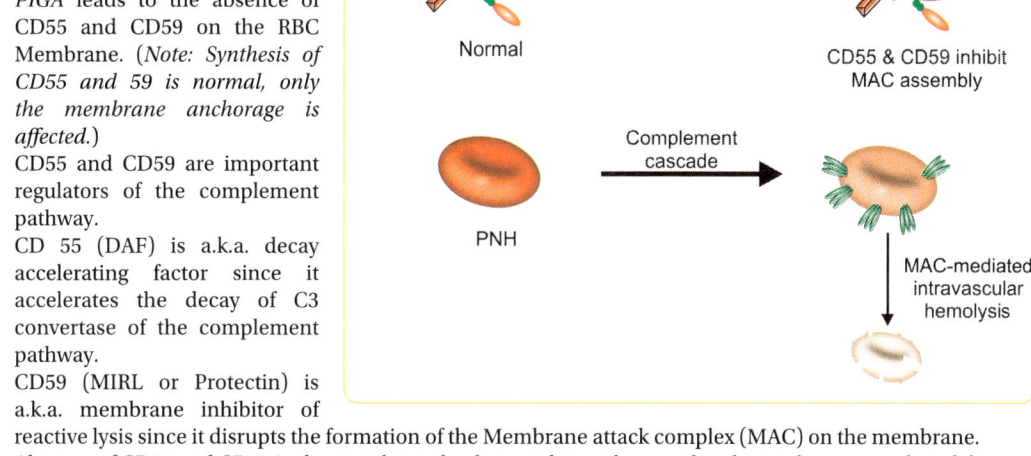

reactive lysis since it disrupts the formation of the Membrane attack complex (MAC) on the membrane.
❑ Absence of CD55 and CD59 in the membrane leads to undue and unregulated complement-mediated damage of RBCs → Intravascular hemolysis and hematuria.
❑ **Eculizumab**, a monoclonal antibody against C5 is used to prevent the MAC formation.

GLYCOME IS BIGGER THAN PROTEOME

"Glycome" refers to the complete set of glycans and glycoconjugates that cells produce under specified conditions (time, space and environment). "Glycomics" is the study of glycome.

We have already learnt that the reason behind proteome being bigger than genome is differential RNA processing.

The reason behind glycome being bigger than proteome is the possibility of multiple combination of glycosidic linkages between sugars.

Example

❑ In our body, Glycine can be linked with another glycine only via peptide bond.
❑ Glucose can be linked with another glucose via various bonds: alpha (1,1), alpha (1,4), alpha (1,6), beta (1,4) and so on.

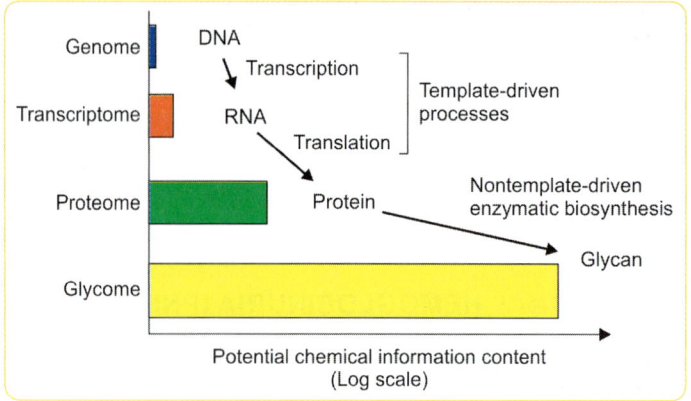

LECTINS

Lectins are proteins that bind to carbohydrates. For example, calnexin, calreticulin, concanavalin A, mannose-binding lectin, asialoglycoprotein receptor, mannose-6-phosphate receptors.

Importance

❑ In affinity chromatography, lectins are used in the stationary phase to purify glycoproteins (Chapter 43) *(Refer to page no. 630)*
❑ Calreticulin, a lectin, binds to monoglucosyl residues of nascent protein during their folding (Chapter 25) *(Refer to page no. 438)*
❑ Absence of mannose-6-phosphate in the lysosomal enzymes leads to I-cell disease (Chapter 26) *(Refer to page no. 445)*

PROTEOGLYCANS

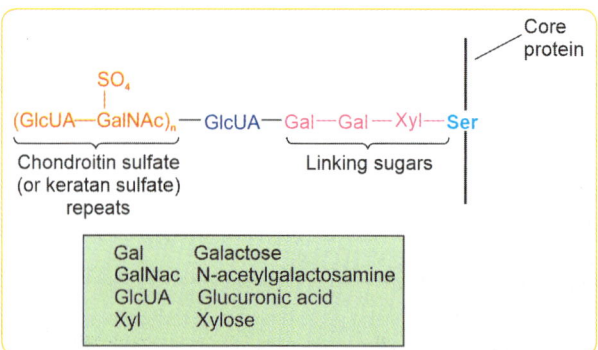

Fig. 3: Schematic structure of a proteoglycan. Glycosaminoglycans are attached through trisaccharide linkage to the core protein. Look at the disaccharide repeat and the site of Sulfation.

Proteoglycans are glycosaminoglycans covalently attached to a core protein. Proteoglycans were also called as mucopolysaccharides.

Glycosaminoglycans are unbranched copolymers of a specific disaccharide unit. This disaccharide unit typically consists of a hexosamine and a hexuronic acid *(with some exceptions)*.

❑ Hexosamine refers to the 6-carbon amino sugars. These are usually acetylated and sulphated. Therefore, they are negatively charged and can hold large amount of water.

❑ Hexuronic acid refers to 6-carbon uronic acid e.g. Glucuronic acid or Iduronic acid

❑ Iduronic acid is the epimer of glucuronic acid.

Hyaluronic acid is a polymer of N-acetylglucosamine and Glucuronic acid, repeated up to 10,000 times.

CLASSES OF GLYCOSAMINOGLYCANS

There are six classes of glycosaminoglycans.

Glycosaminoglycan	Repeating disaccharide unit	Tissue distribution
1. Hyaluronic acid	N-acetylglucosamine and Glucuronic acid	Skin, synovial fluid, bone, cartilage, vitreous Humor, embryonic tissues
2. Chondroitin sulfate	N-acetylgalactosamine and Glucuronic acid	Cartilage, bone, central nervous system
3. Keratan sulfate I and II	N-acetylglucosamine and Galactose	Cornea, cartilage, loose connective tissue
4. Heparin	Glucosamine and Iduronic acid	Mast cells, skin, liver, lung
5. Heparan sulfate	Glucosamine and Glucuronic acid	Skin, renal basement membrane
6. Dermatan sulfate	N-acetylgalactosamine and Iduronic acid/Glucuronic acid	Skin and many tissues

Hyaluronic Acid is Unique in Many Ways

Hyaluronic acid is neither sulphated nor covalently attached to the core protein. It is the only glycosaminoglycan (GAG) found in bacteria. Other GAGs are found only in animals.

EXAMPLES FOR PROTEOGLYCANS

❑ Syndecan
❑ Glypican
❑ Aggrecan
❑ Lumican
❑ Perlecan

ROLE OF PROTEOGLYCANS

❑ Hydrated glycosaminoglycans provide the support for connective tissue.
❑ They can hold large amount of water and provide cushioning and lubrication.
❑ Selectivity of glomerular filtration barrier – albumin is repelled by negative charges of sulphated glycosaminoglycans (explained below).

THERAPEUTIC USES OF GLYCOSAMINOGLYCANS

▶ Video on heparin: youtu.be/6vHm7vX1GUk or scan QR code

❑ Heparin acts an anticoagulant by activating antithrombin III and thus inactivating thrombin and factor Xa.
❑ Heparin displaces lipoprotein lipase enzyme from the endothelial wall.
❑ Danaparoid is an anticoagulant. It is a mixture of heparan sulfate, dermatan sulfate, and chondroitin sulfate.
❑ In osteoarthritis, glucosamine supplements are given to replenish the glycosaminoglycans. But there is no strong evidence for this treatment. *(Ref. Harrison 20th ed., p.2631)*

NEPHROTIC SYNDROME IS DUE TO LOSS OF NEGATIVE CHARGE ON THE GLOMERULAR BASEMENT MEMBRANE

❑ Heparan sulphate and other proteoglycans are responsible for the negative charge of the glomerular basement membrane
❑ In addition, epithelial and endothelial cells are also negatively charged because of sialoglycoproteins in the glycocalyx.
❑ Loss of this negative charge leads to the excretion of albumin in urine.

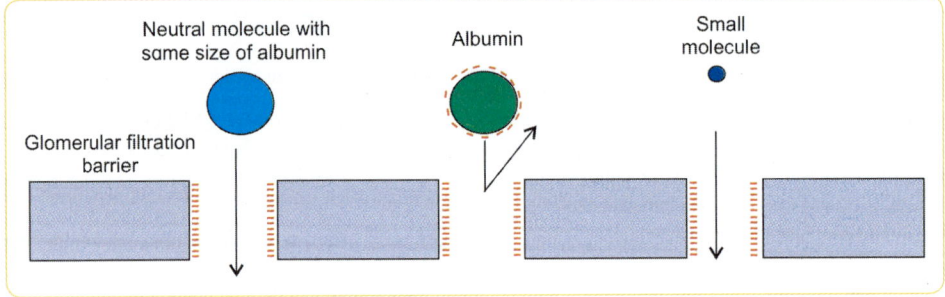

Fig. 4: The negative charges of basement membrane repel albumin

MUCOPOLYSACCHARIDOSES

Mucopolysaccharidoses are rare inborn errors in the degradation of glycosaminoglycans; Characterized by mental retardation and/or skeletal abnormalities. They are classified under lysosomal storage disorders.

MPS type	Defective enzyme	Syndrome	GAG accumulating	Features
I	α-L-Iduronidase	Hurler	Dermatan sulfate, heparan sulfate	Corneal clouding, mental retardation, coarse facial features, hepatosplenomegaly
II	Iduronate sulfatase	Hunter	Dermatan sulfate, heparan sulfate	Mental retardation
IIIA	Heparan sulfate-N-sulfatase	Sanfilippo A		
IIIB	α-N-Acetylglucosaminidase	Sanfilippo B		
IIIC	α-Glucosaminide Nacetyltransferase	Sanfilippo C	Heparan sulfate	Developmental delay and motor dysfunction
IIID	N-Acetylglucosamine 6-sulfatase	Sanfilippo D		
IV A	Galactosamine 6-sulfatase	Morquio A	Keratan sulfate, Chondroitin-6- sulfate	Short stature and skeletal abnormalities
IV B	β-Galactosidase	Morquio B	Keratan sulfate	
VI	N-Acetylgalactosamine 4-sulfatase	Maroteaux-Lamy	Dermatan sulfate	Curvature of the spine, short stature, skeletal dysplasia, cardiac defects

Contd...

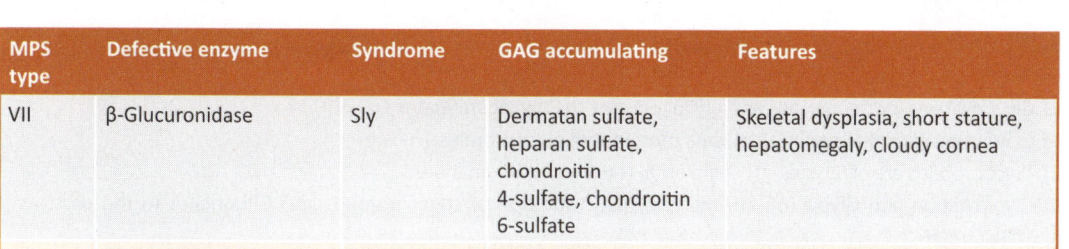

MPS type	Defective enzyme	Syndrome	GAG accumulating	Features
VII	β-Glucuronidase	Sly	Dermatan sulfate, heparan sulfate, chondroitin 4-sulfate, chondroitin 6-sulfate	Skeletal dysplasia, short stature, hepatomegaly, cloudy cornea
IX	Hyaluronidase	Natowicz	Hyaluronic acid	Short stature, arthralgia

Note: MPS-V and MPS-VIII are no longer in use as designations for any disease.

Inheritance

Hunter syndrome is inherited as X-linked recessive trait. All other MPS are inherited as autosomal recessive traits. *(For the list of X-linked disorders, refer to Chapter 30)* *(Refer to page no. 480)*

Treatment

❏ Hematopoietic stem cell transplantation before 2.5 years of age prevents cognitive dysfunction in MPS I.
❏ Intravenous enzyme replacement therapy is available for few MPS types
  ○ Laronidase (polymorphic variant of the human enzyme alpha-L-iduronidase produced by recombinant DNA technology) for MPS type I
  ○ Idursulfase alpha (purified form of human iduronate-2-sulfatase) for MPS type II

Key Points

❏ Glycoproteins, proteoglycans and glycolipids are the glycoconjugates
❏ In humans, except albumin, all plasma proteins are glycoproteins.
❏ Glycoproteins are of 3 types - 1) O-linked 2) N-linked 3) GPI-anchored.
❏ Glycosylation and Glypiation are the enzymatic addition of sugar residues and GPI anchor to the proteins respectively.
❏ O-glycosidic linkage involves OH group of amino acids whereas N-glycosidic linkage involves amide group of asparagine.
❏ GPI anchored proteins are abundant in lipid rafts, which are involved in receptor-mediated signal transduction pathways and membrane trafficking.
❏ Ectoenzymes are catalytic membrane proteins with their active sites outside the cell.
❏ Asialoglycoprotein Receptor is involved in the Degradation of Plasma Glycoproteins.
❏ Paroxysmal Nocturnal Hemoglobinuria (PNH) is due to defect in the PIGA gene; CD55 and CD59 are not GPI anchored to RBC membrane.
❏ Lectins are proteins that bind to carbohydrates.
❏ Proteoglycans are glycosaminoglycans covalently attached to the core protein.
❏ Glycosaminoglycans are unbranched copolymers of a specific disaccharide unit.
❏ Hyaluronic acid is neither sulphated nor covalently attached to the core protein.
❏ Glucosamine supplements are given with the aim of replenishing the glycosaminoglycans in patients of osteoarthritis.
❏ Nephrotic syndrome is due to loss of negative charge on the glomerular basement membrane.
❏ Except hunter syndrome, which is inherited in X-linked recessive trait, all the other mucopolysaccharidoses are inherited as autosomal recessive traits.

SELF-ASSESSMENT

Short Answer Questions

1. Differentiate
   a. Glycation and Glycosylation
   b. Glycoproteins and proteoglycans
   c. N-linked glycoproteins and O-linked glycoproteins
2. What do you understand by the term 'GPI anchored proteins'? Give two examples of GPI anchored proteins.
3. What is the molecular defect in paroxysmal nocturnal hemoglobinuria? Describe the pathogenesis. Name the monoclonal antibody used in the treatment of this condition.
4. What are lectins? Give any two examples. Mention their importance.
5. Mention any two therapeutic roles of glycosaminoglycans.

Multiple Choice Questions

1. **All of the following are glycoproteins, EXCEPT:**
   a. Transferrin
   b. Human chorionic gonadotropic hormone
   c. LDL-receptor
   d. Human albumin
2. **All of the following are features of glycoproteins, Except:**
   a. Highly-branched oligosaccharide
   b. Presence of amino sugar
   c. Absence of glucuronic acid
   d. Presence of disaccharide repeat unit
3. **In N-linked glycoproteins, to which of the following amino acids, oligosaccharides are covalently attached?**
   a. Glutamine   b. Asparagine
   c. Acetyl lysine   d. Serine

4. **Dolichol is required for the synthesis of:**
   a. N-linked glycoproteins
   b. O-linked glycoproteins
   c. Glycosaminoglycans
   d. GPI anchored proteins

5. **Synthesis of O-linked glycoproteins takes place in:**
   a. Golgi apparatus
   b. Endoplasmic reticulum
   c. Endoplasmic reticulum and Golgi apparatus
   d. Lysosomes

6. **Trimming and further modification of N-linked glycoproteins takes place at:**
   a. Golgi apparatus   b. Endoplasmic reticulum
   c. Peroxisomes   d. Lysosomes

7. **All of the following are GPI anchored proteins, EXCEPT:**
   a. 5'-Nucleotidase
   b. Erythrocyte membrane acetylcholinesterase
   c. Intestinal alkaline phosphatase
   d. LDL receptor

8. **Eculizumab is a monoclonal antibody used in patients of Paroxysmal nocturnal hemoglobinuria. The target of this drug is:**
   a. CD55   b. CD59
   c. C5   d. CD65

9. **Which of the following is a non-enzymatic process?**
   a. Glycation   b. Glycosylation
   c. Glypiation   d. Prenylation

10. **Lectins are:**
    a. Proteins that bind to carbohydrates
    b. Carbohydrates that bind to proteins
    c. Proteins that bind to lipids
    d. Lipids that bind to proteins

11. **All of the following are Glycoproteinoses, Except:**
    a. Fucosidosis   b. Sanfilippo A syndrome
    c. Sialidosis   d. α-mannosidosis

12. **Proteoglycans are:**
    a. Branched oligosaccharides covalently attached to core protein
    b. Sugar residues covalently attached to core protein
    c. Glycosaminoglycans noncovalently attached to core protein
    d. Glycosaminoglycans covalently attached to core protein

13. **Hyaluronic acid is a:**
    a. Glycosaminoglycan
    b. Protein
    c. Proteoglycan   d. Glycoprotein

14. **Which of the following glycosaminoglycan contains iduronic acid?**
    a. Hyaluronic acid   b. Keratan sulfate I
    c. Keratan sulfate II   d. Heparin

15. **Which of the following is true about hyaluronic acid?**
    a. Sulphated
    b. Covalently attached to the core protein
    c. Found in capsule of bacteria
    d. Contains iduronic acid

16. **All of the following are true regarding proteoglycans, EXCEPT:**
    a. Chondroitin sulfate is a glycosaminoglycan
    b. They hold less amount of water
    c. They are made up of sugar and amino acids
    d. They carry charge

17. **Heparin acts as an anticoagulant by:**
    a. Activating antithrombin III
    b. Inactivating antithrombin III
    c. Activating thrombin
    d. Activating factor Xa

18. **High concentration of hyaluronic acid is seen in:**
    a. Vitreous humor   b. Cartilage
    c. Cornea   d. Aqueous humor

19. **Which of the following enzymes is coded by X-chromosome?**
    a. α-L-Iduronidase   b. Iduronate sulfatase
    c. β-Galactosidase   d. Hyaluronidase

20. **Hurler syndrome is due to deficiency of:**
    a. α-L-Iduronidase   b. Iduronate sulfatase
    c. β-Galactosidase   d. Hyaluronidase

Chapter 41 **Glycoproteins and Proteoglycans**

ANSWERS

1. d.	**2.** d.	**3.** b.	**4.** a.	**5.** a.	**6.** a.
7. d.	**8.** c.	**9.** a.	**10.** a.	**11.** b.	**12.** d.
13. a.	**14.** d.	**15.** c.	**16.** b.	**17.** a.	**18.** a.
19. b.	**20.** a.				

Extracellular Matrix

Extracellular matrix (ECM) is the collection of molecules found around and in between cells. Fibrous proteins, proteoglycans *(Chapter 41)* *(Refer to page no. 604)* and water form the bulk of extracellular matrix. The major proteins of the extracellular matrix are collagen, elastin, fibrillin, fibronectin and laminin.

The components of ECM are secreted by fibroblast in most connective tissues. In specialized connective tissues like cartilage and bone, they are secreted by chondroblasts and osteoblasts respectively.

COLLAGEN

Collagens are the most abundant groups of proteins in mammals. There are various types of collagens. These are synthesized and secreted by fibroblasts and undergo various post-translational modifications.

FOUR LEVELS OF COLLAGEN STRUCTURE

Level	Description
1°	(-Glycine–X–Y-)n
2°	**Left-handed helix** (NOT α-helix), Each turn of helix contains 3.3 amino acyl residues.
3°	Right-handed **triple helix**
4°	**Quarter staggered** arrangement

EVERY 3rd AMINO ACID IN COLLAGEN IS GLYCINE

Primary structure of collagen is Gly-X-Y, repeated multiple times. This means every 3rd amino acid of collagen is glycine. The side chain of glycine is just a hydrogen atom, which is small enough to fit inside the narrow part of the triple helix where the alpha chains meet. Any other side chain would be too bulky to fit in this space and may cause steric hindrance that would destabilise the triple helical structure.

X and Y can be any other amino acids. Proline and Hydroxy proline are frequently seen in X and Y position respectively. Lysine/hydroxylysine can also be found in Y position. Proline and hydroxy proline provide rigidity to the collagen.

HELIX IN COLLAGEN IS NOT ALPHA HELIX

Proline is an **imino** acid. It cannot be accommodated in α-helix (and the helix in collagen is NOT α-helix). The abundance of glycine in collagen favors the formation of the left-handed helix with 3.3 amino acids in one turn (c.f. α-helix is right-handed and contains 3.6 amino acids in one turn).

NOT ALL COLLAGENS FORM FIBRIL STRUCTURE

Most collagen proteins are fibrillar in nature and some are non-fibrillar. Basement membrane contains non-fibrillar collagen with mesh-like structure. e.g. Type IV collagen of the basement membrane.

NON-COLLAGEN COLLAGENS

The Gly-X-Y structure is not unique to collagen. It is also found in certain other proteins like C1q complement component, pulmonary surfactant proteins (SP-A & SP-D). These are known as non-collagen collagens.

THERE ARE 28 TYPES OF COLLAGEN MADE UP OF 30 POLYPEPTIDE CHAINS

As we have already learned, collagen refers to a group of proteins. There are 28 types of collagens recognised so far. These 28 types are made up of same or different collagen polypeptides. There are 30 such polypeptide chains (α chains) coded by separate genes.

The name of collagen contains 4 parts. First part is always 'COL' indicating collagen. Second part is a number denoting the type of collagen. The third part is always 'A' denoting alpha chain. Fourth part is a number denoting the type of α-chain.

Alpha chain name contains 3 part. First part is always α, indicating the alpha chain. Second part is a number denoting the type of alpha chain. The third part is a Roman numeral written within bracket, which denotes the type of collagen.

For example,

❑ Type I collagen is a heterotrimer made up of two α1(I) chains and one α2 (I) chain.
  ○ α1(I) chain is coded by the *COL1A1* gene
  ○ α2(I) chain is coded by the *COL1A2* gene
❑ Type II collagen is a homotrimer made up of three α1(II) chain coded by the *COL2A1* gene.
❑ Type III collagen is also homotrimer made up of three α1(III) chain coded by the *COL3A1* gene.
❑ Type IV collagen family of proteins comprises six isomeric chains, designated α1(IV) to α6(IV).

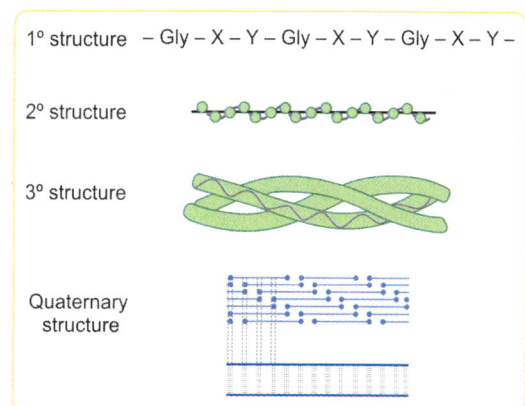

1° structure – Gly – X – Y – Gly – X – Y – Gly – X – Y –

2° structure

3° structure

Quaternary structure

Major types of collagen	
Type	**Tissue distribution**
Fibril-forming	
I	Skin, bone, tendon, blood vessels, cornea, **Fibrocartilage**
II	Cartilage, Intervertebral disk, Vitreous body
III	Blood vessels, fetal skin, granulation tissue, **Hypertrophic scar and keloid**
Network-like	
IV	Basement membrane
X	Hypertrophic cartilage
Anchoring fibrils	
VII	Dermal-epidermal junction
Fibril-associated collagens with interrupted triple helices (FACITs)	
IX	Cartilage
Non-collagen proteins	C1q, Lung surfactant protein, acetylcholine esterase

Remember

- ☐ Type I collagen is the abundant collagen type in bone and fibrocartilage.
- ☐ Type II collagen is the abundant collagen type in hyaline cartilage.
- ☐ Type III collagen is present in granulation tissue and keloid.
- ☐ Type IV collagen is present in basement membrane.

FORMATION OF COLLAGEN

In our body collagen is produced by fibroblasts and some other cells. After the ribosomal synthesis, collagen undergoes extensive post-translational modifications. For easy understanding, the events of collagen formation can be divided into intracellular and extracellular.

INTRACELLULAR EVENTS

- ☐ Like other secretory proteins, synthesis of collagen begins in the free ribosome as a 'prepro' form *(Chapter 26) (Refer to page no. 445).*
- ☐ Signal sequences are recognized by signal recognition particle. Protein synthesis is temporarily halted and the nascent polypeptide with ribosome is brought to the endoplasmic reticulum for further synthesis and processing.
- ☐ Removal of the signal peptide from preprocollagen by signal peptidase produces procollagen **in the endoplasmic reticulum.**

$$\text{Preprocollagen} \xrightarrow{\text{Signal peptidase}} \text{Procollagen + Signal peptide}$$

- ☐ In the endoplasmic reticulum, hydroxylation of lysine and proline of collagen is mediated by lysyl hydroxylase and prolyl hydroxylase respectively. Both these enzymes require vitamin C (ascorbate) and α-ketoglutarate as cofactors.
- ☐ Some of the hydroxylysine residues of procollagen are glycosylated.
- ☐ Amino and carboxy-terminal of procollagen contain regions rich in cysteine residues. These cysteine residues are involved in disulphide bond formation between three procollagen chains. This is known as "registration". This assists in the formation of the triple helix.
- ☐ Procollagen with triple helix is secreted out of the cells.

EXTRACELLULAR EVENTS

❏ **Aminoproteinase** and **carboxyl proteinase** remove the extension peptides at the amino and carboxyl terminal end of the procollagen to produce tropocollagen.

❏ Tropocollagen contains around 1000 amino acids per chain.

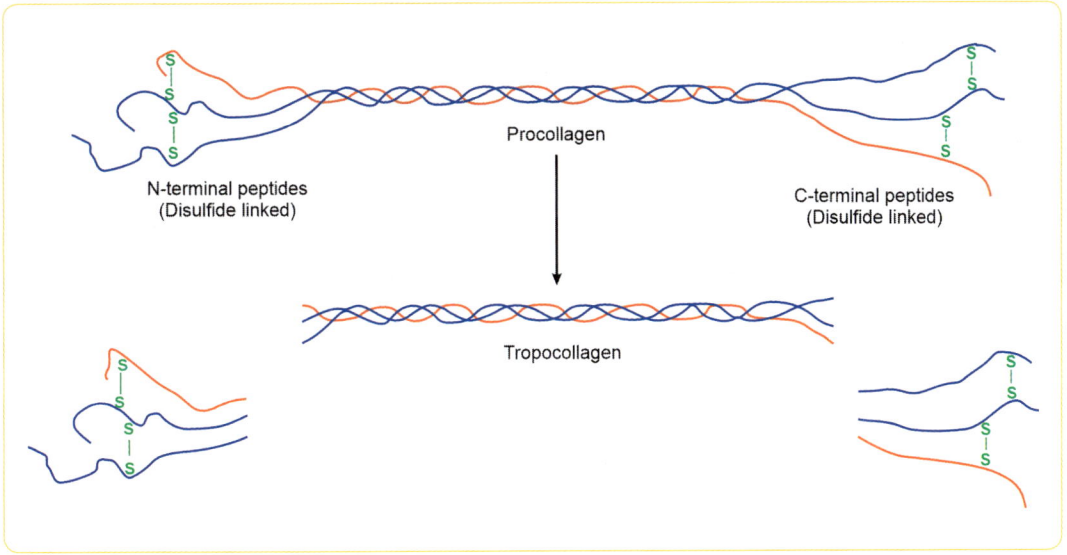

❏ Copper-containing lysyl oxidase oxidatively deaminates the ε-amino groups of some but not all lysine and hydroxylysine residues to produce allysine with a reactive aldehyde group. The allysine is involved in the formation of collagen crosslinks in two ways:

1. Two of the allysine molecules undergo aldol condensation to produce aldol cross-link (shown in the figure in the next page).

2. One allysine aldehyde reacts with ε-amino groups of lysine/hydroxylysine residues to form Schiff base. The Schiff bases undergo further chemical rearrangements to produce the stable covalent cross-links.

❏ Covalent cross-linking produces the quarter staggered alignment of collagen. There are regularly spaced gaps between the triple helical molecules in the array. This produces the banded appearance of collagen.

Quarter-staggered array of collagen molecules

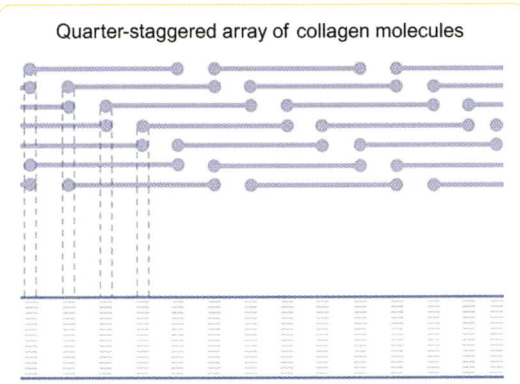

Banded appearance of fibrillar collagen

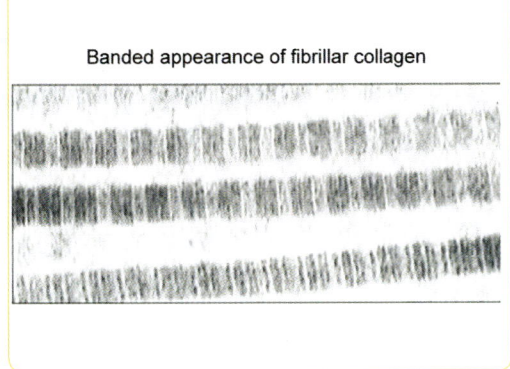

Chapter 42 Extracellular Matrix

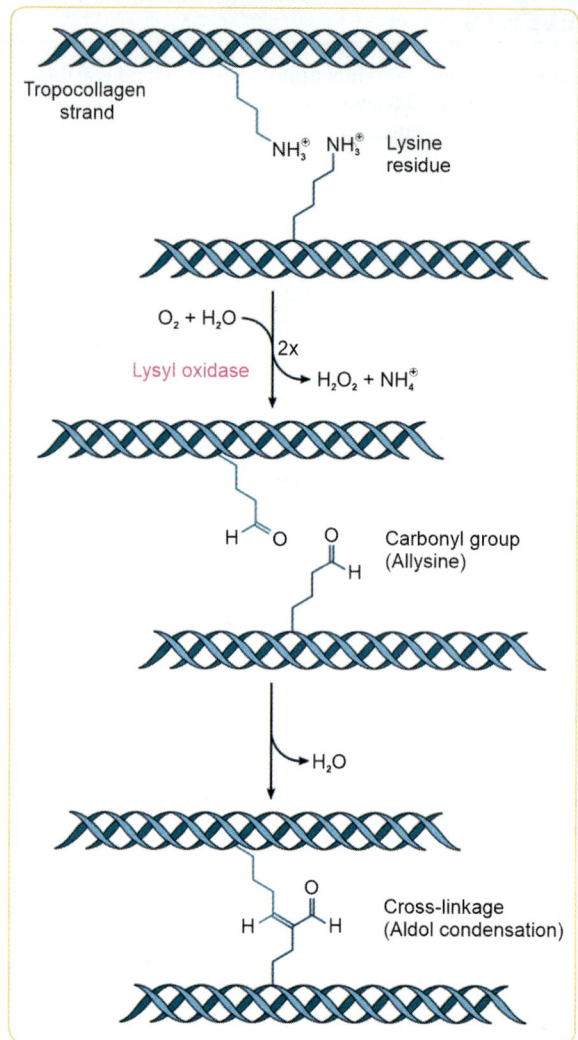

Tropocollagen strand

$NH_3^\oplus$ $NH_3^\oplus$ Lysine residue

$O_2 + H_2O$

Lysyl oxidase 2x

$H_2O_2 + NH_4^\oplus$

Carbonyl group (Allysine)

H_2O

Cross-linkage (Aldol condensation)

DISORDERS DUE TO DEFECTIVE COLLAGEN

Collagen gene	Disease due to defect
COL1A1, COL1A2	**Osteogenesis imperfecta** type 1 (Brittle bone disease) Osteoporosis Ehlers-Danlos syndrome - arthrochalasia subtype
COL2A1	Severe chondrodysplasia Osteoarthritis
COL3A1	Ehlers-Danlos syndrome - vascular subtype
COL4A3-COL4A6	**Alport syndrome** (autosomal and X-linked)
COL5A1, COL5A2, COL1A1	**Ehlers-Danlos syndrome** – classical subtype
COL7A1	Epidermolysis bullosa – dystrophic type
COL3A1, tenascin XB (TNXB)	Ehlers-Danlos syndrome - hypermobility subtype

Type I collagen is the most abundant collagen in bone. So, Serum propeptide of type I procollagen is used as a marker of bone formation. Urinary and serum cross-linked N-terminal and C-terminal telopeptides are used as markers of bone resorption. *(Ref. Harrison 20th Ed., p.2950)*

ELASTIN

As the name suggests, elastin is the protein giving elasticity to certain tissues. A large amount of elastin is present in lung, large arteries like aorta, and elastic ligaments. A smaller amount of elastin is found in skin, ear cartilage, and several other tissues. Elastin is synthesised as a soluble monomer called tropoelastin. Elastin gene is located in the long arm of chromosome 7.

Like collagen, elastin also undergoes post-translational modifications like collagen. Lysyl oxidase oxidatively deaminates the ε-amino group of lysine to produce allysine. However, the cross-linking is different in elastin. Three allysyl side chains and one lysyl side chain from the same or neighbouring polypeptides condense to form **desmosine** tetra-functional cross-link.

Feature	Collagen	Elastin
Genetic types	28 types coded by 30 genes	Only one genetic type
Presence of triple helix	Yes	No
Presence of (Gly-X-Y)$_n$ repeating structure	Yes	No
Presence of hydroxylysine	Yes	No
Presence of carbohydrates	Yes	No
Nature of intramolecular cross-links	Aldol and Schiff base	Desmosine
Presence of extension peptides during biosynthesis	Yes	No

Chapter 42 Extracellular Matrix

DISORDERS DUE TO DEFECTIVE ELASTIN

❑ **William syndrome**—Supravalvular aortic stenosis, nervous system involvement
❑ **Pulmonary emphysema**
❑ **Cutis laxa**—Loose, redundant skin folds, decreased elasticity of the skin, and general connective tissue involvement
❑ **Aging of skin**

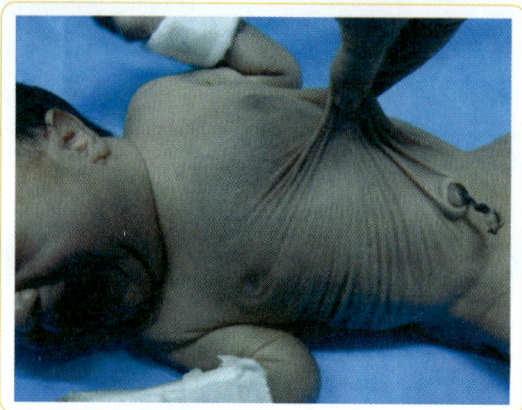

Fig. 1: A child with cutis laxa

FIBRILLIN

There are three types of Fibrillins – 1, 2 and 3. Fibrillin 1 is the predominant type. Fibrillins are the structural components of microfibrils *(Don't confuse with microfilaments)*. Microfibrils provide a scaffold for disposition of elastin.

MARFAN SYNDROME IS DUE TO A DEFECT IN FIBRILLIN 1

❑   Marfan syndrome is caused by *FBN1* mutations on chromosome 15.
❑   Fibrillin normally binds to transforming growth factor β (TGF-β). Mutated fibrillin cannot bind to TGF-β. This leads to elevated TGF- β.
❑   TGF-β controls proliferation, differentiation and other functions in many cell types.
❑   Elevated levels of TGF-β contribute to the clinical features of disproportionately long arms and legs, ectopia lentis (dislocation of the ocular lens), arachnodactyly (abnormally long and slender fingers), and aortic dilatation.

Section V Special Topics

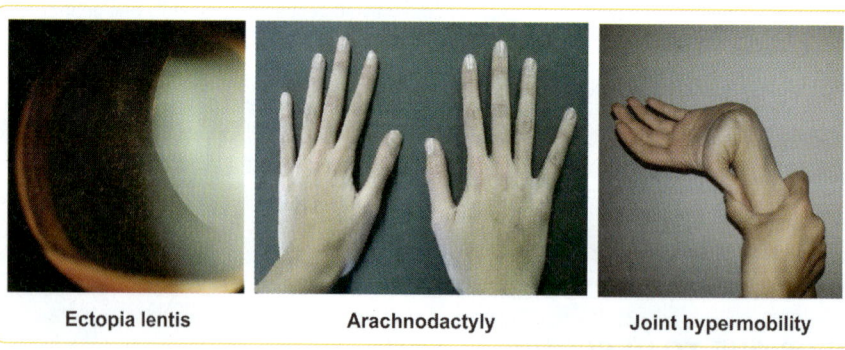

| Ectopia lentis | Arachnodactyly | Joint hypermobility |

FIBRONECTIN

Fibronectin is a high-molecular-weight, disulphide-linked homodimeric glycoprotein. It is coded by a single gene but there are many isoforms that exist due to alternative splicing (Chapter 23) *(Refer to page no. 416)*. Fibronectin

binds to integrins located in the cell membrane. It also binds to components of ECM like collagen and heparan sulfate. In addition to ECM, fibronectin is also found in plasma since it is produced not only by fibroblasts but also by hepatocytes, endothelium, monocytes.

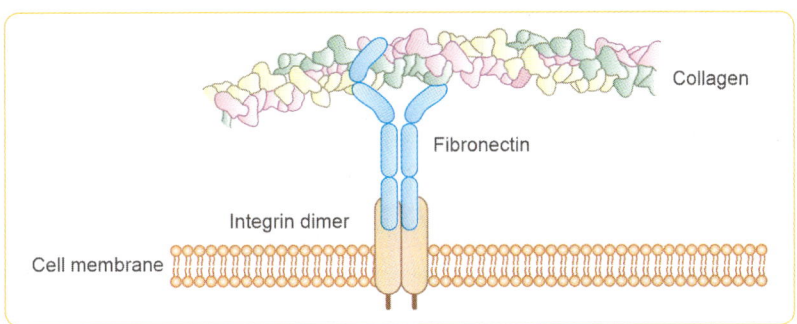

LAMININ

Laminins are high-molecular-weight, heterotrimeric ($\alpha\beta\gamma$) glycoproteins. Each monomer is coded by multiple genes (5 genes for α-chains, 4 for β, and 3 for γ). The gene products combine in multiple ways to produce 15 different types of laminins.

Role of Laminin:
❑ The most abundant glycoproteins of the basal lamina of the basement membrane
❑ Serves as an attachment of the cell to other components of ECM
❑ Essential for cell proliferation, differentiation, and migration

Note: Lamin (Chapter 36) *(Refer to page no. 562)* is different from laminin

ENZYMES DEGRADING ECM

MATRIX METALLOPROTEINASES

Matrix metalloproteinases (MMPs) are zinc-containing endopeptidases that degrade ECM proteins. They are involved in the remodelling of ECM. So far, 23 MMP enzymes have been discovered. These are written with the numerical suffix after the word MMP. MMP-1, MMP-8 and MMP-13 degrade collagen and are known as collagenases.

Note: All collagenases are MMPs

MMPs are involved in the physiological processes like angiogenesis, wound healing and bone remodelling as well as in pathological processes like cancer metastasis and chronic inflammatory diseases.

ELASTASE

Elastases are a group of serine protease that can degrade elastin. In humans, there are eight different elastases. One of them is MMP-12 (macrophage metalloelastases). All elastases are not MMPs.

Neutrophil elastase, one among the elastases needs to be discussed because of the clinical significance. The activity of elastase is kept in control by Serpins (Serine Protease Inhibitors). α1-antitrypsin (α1-AT) is one of the Serpins. Deficiency of α1-AT leads to the unopposed activity of elastase and damage leading to various pathologies like pulmonary emphysema, pancreatitis, and gallstones.

Competency

☐ BI9.1   List the functions and components of the extracellular matrix (ECM)
☐ BI9.2   Discuss the involvement of ECM components in health and disease

Key Points

☐ (Gly-X-Y)n is the primary structure of type I collagen.
☐ C1q complement component and pulmonary surfactant proteins are non-collagen collagens.
☐ Vitamin C is required for lysyl and prolyl hydroxylases.
☐ Lysyl oxidase requires copper.
☐ Type IV collagen is the most abundant collagen in the basement membrane.
☐ Alport syndrome is due to a mutation in the alpha 3, 4 or 5 chain of type IV collagen.
☐ Desmosine cross-links are found in elastin.
☐ Fetal fibronectin is used in the diagnosis of preterm labor.
☐ Marfan syndrome is due to a mutation in the fibrillin gene.
☐ Matrix metalloproteinases (MMPs) are zinc-containing endopeptidases that degrade ECM proteins.

SELF-ASSESSMENT

Short Answer Questions

1. Explain the role of ascorbic acid (vitamin C) in the synthesis of mature collagen. Enumerate the various deficiency manifestations of vitamin C.

2. Describe the extracellular events involved in the maturation of collagen.

3. Describe the primary structure of collagen.

4. Differentiate the structure of collagen and elastin.

5. What is the molecular defect in Marfan syndrome?

6. Explain why copper and vitamin C deficiency may result in abnormal connective tissue formation?

Clinical Case-based Questions

1. A male neonate delivered to a 33-year-old female had excessive fragile bones. After a few months baby suffered from multiple fractures and skeletal deformities.

    a. What is a provisional diagnosis?
    b. Mention any other two collagen abnormalities along with a mutational defect in them.

Multiple Choice Questions

1. All of the following are components of extracellular matrix, except:
    a. Collagen   b. Actin
    c. Elastin   d. Fibrillin

2. Which of the following is non-collagen collagen?
    a. Fibrinogen   b. Protein C
    c. Protein S   d. C1q complement

3. All are true about collagen, except:
    a. Vitamin C is required for the maturation of the collagen chain
    b. Alpha helix of collagen is right-handed
    c. Every 3rd amino acid of collagen is glycine
    d. Proline provides rigidity to collagen

4. Desmosine tetra-functional cross-link is present in:
    a. Collagen
    b. Elastin
    c. Fibrillin
    d. Fibronectin

5. Which of the following enzyme is NOT involved in the post-translational modification of tropoelastin?
   a. Prolyl hydroxylase
   b. Lysyl hydroxylase
   c. Lysyl oxidase
   d. All of the above

6. Mutation of which of the following gene leads to Marfan syndrome?
   a. Collagen   b. Elastin
   c. Fibrillin   d. Fibronectin

7. Mutation of which of the following gene leads to Epidermolysis bullosa, dystrophic type?
   a. COL1A1   b. COL3A1
   c. COL7A1   d. COL10A1

8. Which of the following amino acids of procollagen is glycosylated?
   a. Serine   b. Lysine
   c. Hydroxylysine d. Hydroxyproline

9. How many amino acid residues are present in each turn of collagen helix?
   a. 2.3   b. 3.3
   c. 2.6   d. 3.6

10. The defect in X-linked Alport syndrome is:
    a. α5 chain of type IV collagen
    b. α4 chain of type V collagen
    c. α5 chain of type I collagen
    d. α1 chain of type V collagen

11. Which of the followings is absent in collagen?
    a. Glycine   b. Lysine
    c. Desmosine   d. Proline

12. All of the following are features of collagen, Except
    a. It is a fibrous protein
    b. The triple helical structure is essential for its function
    c. Phenylalanine is more abundant than glycine
    d. Prolyl and Lysyl hydroxylases are involved in post-translational modification

ANSWERS

1. b.	**2.** d.	**3.** b.	**4.** b.	**5.** b.	**6** c.
7 c.	**8** c.	**9** b.	**10** a.	**11** c.	**12.** c.

Tools of Proteomics

The proteome is defined as the complete set of proteins expressed by a cell at a given point of time. It varies with cell type, developmental stage and in the disease conditions. Proteomics is the study of the structure and function of these proteins. Early detection of diseases using peptide biomarkers and identification of drug targets are some of the applications of clinical proteomics.

Around 25,000 protein-coding genes produce one-lakh proteins and these proteins further undergo various post-translational modifications. Thus, the **proteome is larger than the genome**.

The proteomic techniques we will be discussing in this chapter can be divided into 3 major categories as follows:

Techniques for purification/isolation of proteins	Techniques for quantification/ identification of proteins	Techniques for identification of protein structure
❏ Cell lysate preparation (to break the membrane lipids) ❏ Salt fractionation ❏ Ultracentrifugation ❏ Chromatography	❏ Spectrophotometric estimation ❏ Electrophoresis ❏ Chromatography ❏ Western blotting ❏ ELISA	❏ X-ray crystallography ❏ NMR spectroscopy ❏ Mass spectrometry

Let us first learn how to isolate and identify the proteins. Physicochemical properties of proteins vary based on the number and type of amino acids. Therefore, they can be separated based on their molecular size, solubility, electrical charge, isoelectric pH (pI), adsorption properties and specific affinity towards ligands.

TECHNIQUES FOR PURIFICATION/ISOLATION

CELL LYSATE PREPARATION

If the protein of interest is in the blood, it is easy to separate it. Whereas, if it is inside the cell, we need to lyse the cell first. Membranes cover the cells and internal organelles. Therefore, the first step is to break open the membrane lipids and separate them (Separation of membrane proteins can also be done) using physical (sonication), chemical (detergents) or enzymatic (lysozymes) methods.

Proteases are released during cell lysis. So, it is essential to add protease inhibitors and keep the sample on the ice (low temperature) to prevent the proteolysis. During cell lysis, there is a chance for SH group of proteins getting oxidised. To prevent this, SH group-containing reducing agents like dithiothreitol (DTT) is added. Optimal pH is maintained using appropriate buffers to prevent the premature precipitation of proteins.

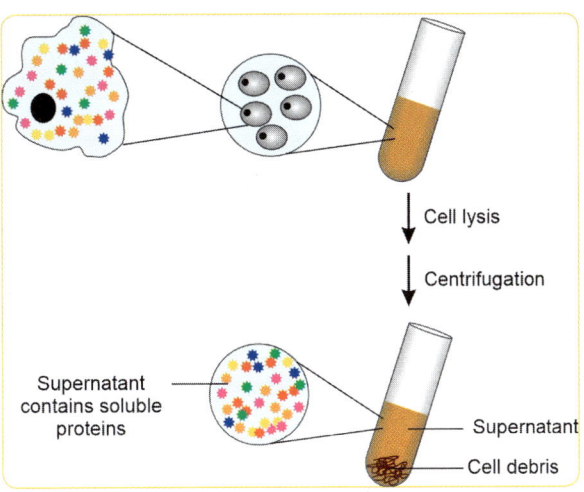

DIFFERENTIAL CENTRIFUGATION IS USED FOR THE SEPARATION OF PROTEINS OF DIFFERENT ORGANELLES

The cell lysate contains disrupted cell membrane, cellular organelles and many cytosolic (soluble) proteins. If the protein of interest is inside a specific organelle, we first need to seperate the organelle. The organelles are separated from the soluble proteins by differential centrifugation.

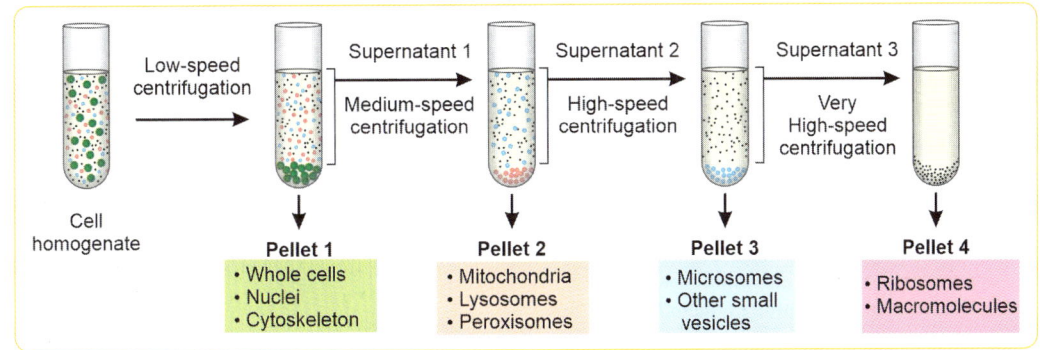

Differential centrifugation involves repeated centrifugation at progressively increasing speed and time. For example, if the protein of interest is in the mitochondrial fraction, the cell lysate is first centrifuged at 500 revolutions per minute (rpm) for 10 minutes removing nuclei, debris etc. The supernatant contains other organelles including the mitochondria, which can be pelleted at 10,000 rpm for 20 minutes.

SALT FRACTIONATION IS THE PRELIMINARY STEP TO REMOVE IMPURITIES

This is a crude way of separation of proteins based on the solubility of proteins under different concentrations of salt like ammonium sulphate. *The technique has already been explained with the illustration in Chapter 35* *(Refer to page no. 550)*. Proteins separated by salt-fractionation are further purified by dialysis to remove the excess salts.

TECHNIQUES FOR ISOLATION, IDENTIFICATION AND QUANTIFICATION

ELECTROPHORESIS

Electrophoresis is the movement of a charged particle under the influence of an electric field. Electrophoresis can be used for proteins as well as nucleic acids.

The velocity (v) of migration of the molecule is,

$$v = \frac{Eq}{f}$$

E = Strength of electric field

q = Net electric charge on the molecule

f = Frictional coefficient, which depends upon the size, shape of the molecule and the viscosity of the solution

From this equation, we can infer the following:

❑ Increasing the voltage will accelerate the movement of molecules. (But, very-high voltage is not preferred as high heat may denature the protein).

❑ Larger molecules move slowly (Albumin will move faster compared to globulin)

❑ Even if two molecules have the same mass, the globular shaped molecule will move fast due to less friction.

Types of Electrophoresis

Type	Example	
Zone Electrophoresis *(done with a support medium)*	❑ **Paper** Electrophoresis	
	❑ **Cellulose acetate** electrophoresis	
	❑ **Gel** Electrophoresis	❑ Agarose gel electrophoresis (AGE) ❑ Polyacrylamide gel electrophoresis (PAGE) ❑ Sodium dodecyl sulphate – PAGE (SDS-PAGE) ❑ Isoelectric Focusing *(using pH gradient)* ❑ Immuno Electrophoresis *(Ag-Ab reaction plus electrophoresis)*
Moving Boundary Electrophoresis	❑ *Electrophoresis in U-shaped tube* ❑ Capillary Electrophoresis *(done without support medium in free moving solution)*	

Clinical Applications

❑ Quantitative analysis of specific serum proteins

❑ Identification and quantification of Hemoglobin and its subclasses

❑ Identification of monoclonal antibodies (M band) in serum and urine

❑ Separation and quantification of major lipoproteins

❑ Isoenzyme analysis: LDH, CK, ALP

❑ Western Blot (identification and quantification of proteins)

❑ Southern Blot (identification of a particular sequence DNA)

Gel Electrophoresis is Based on Molecular Sieving

Gel electrophoresis utilises polymerised gels (agarose, polyacrylamide) as a support medium. The polymerized gel is porous. The pore size is inversely proportional to the concentration of the gel.

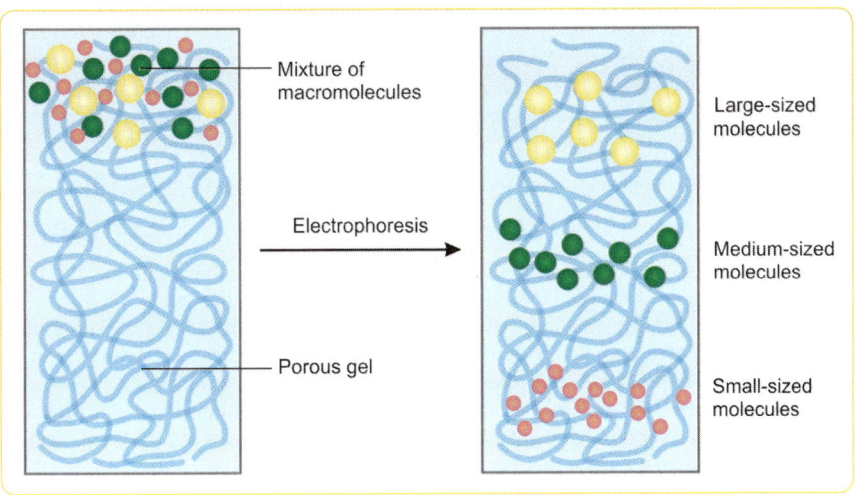

Fig. 1: Illustration of molecular sieving

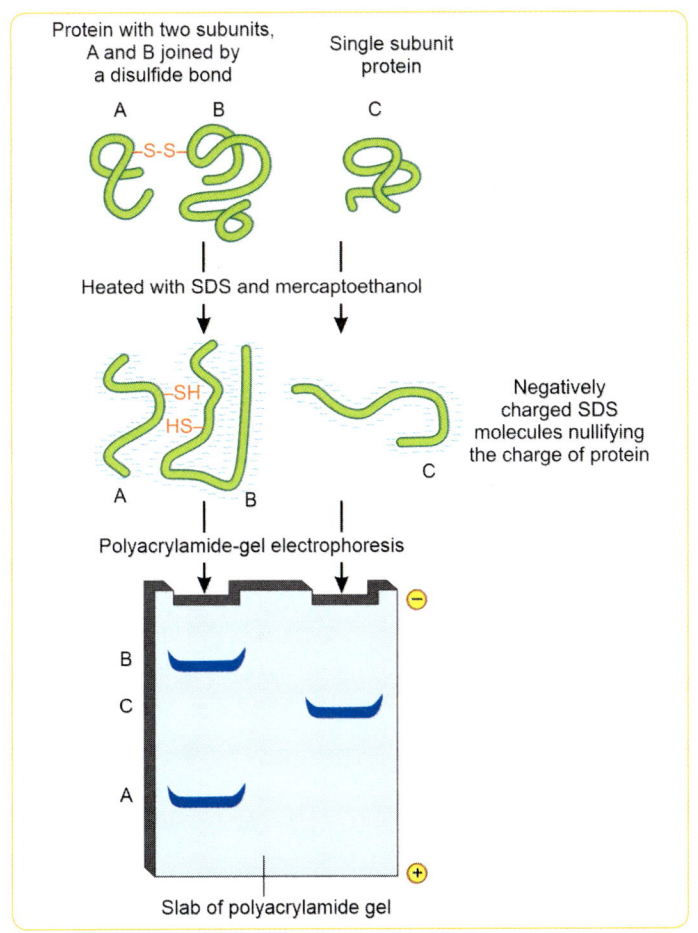

Fig. 1A: Fig 1a SDS- PAGE

SDS-PAGE Separates Protein Purely Based on their Mass (Fig. 1A)

Component	Role
Acrylamide	Monomer for polymerisation (gel formation)
Bisacrylamide	Cross-linking agent for polymerisation
β-mercaptoethanol	Breaks the disulphide bonds and **converts the multisubunit proteins into monomers**
SDS	One molecule of SDS binds to two amino acids and thus gives a uniform charge to the polypeptide

SDS is sodium dodecyl sulphate. It is an anionic detergent. SDS is mixed with acrylamide and bisacrylamide to make the gel. β-mercaptoethanol is added to the sample before loading into the gel.

As SDS provides uniform negative charge over the protein molecule and masks the inherent charge of the polypeptide, the movement of protein in SDS-PAGE is purely dependent on the mass of the protein. So, SDS-PAGE is used for the determination of relative molecular weight (Mr) of a protein.

Native PAGE is a term used for polyacrylamide gel electrophoresis without using denaturing agents like SDS. If a protein shows a band corresponding to 100 kDa in native-PAGE but 25 kDa in SDS-PAGE, it suggests that the protein is an oligomeric protein with four identical subunits, i.e. homotetramer. *(Note: It can be 4 different subunits with 25kDa molecular weight. The nature of subunit can be confirmed by protein sequencing).*

Isoelectric Focusing (IEF)

Isoelectric pH is the pH at which a molecule exists as Zwitterion, i.e. the net charge on the molecule is zero. At pI, there is no electrophoretic movement of the molecule. Isoelectric focusing is the separation of proteins based on their pI. Different protein molecules can have different pI.

In IEF, a continuous pH gradient is created using polyampholytes and the sample is applied in the presence of an electrical field. While moving in the gel, when the protein encounters its pI, it stops moving. IEF is useful for the separation of isoenzymes and isoforms (Chapter 44) *(Refer to page no. 638)*.

Two-dimensional Electrophoresis = IEF + SDS-PAGE

In 2-D electrophoresis, proteins are separated in two directions. First, based on their pI by isoelectric focusing and next based on their mass by SDS-PAGE. Different proteins can have the same pI. So, a single band of IEF (1st dimension) can produce multiple bands when subjected to SDS-PAGE (2nd dimension).

2-D electrophoresis is a powerful tool of proteomics. *Western blotting and other blotting techniques were described in Chapter 28 (Refer to page no. 462).*

CHROMATOGRAPHY

Chromatography is a separation technique that separates molecules based on their **differential distribution** between a stationary and mobile phase. Russian botanist Tswett developed this technique. He separated the plant pigments first hence the name chromatography. *Chromatography can separate colorless substances also!*

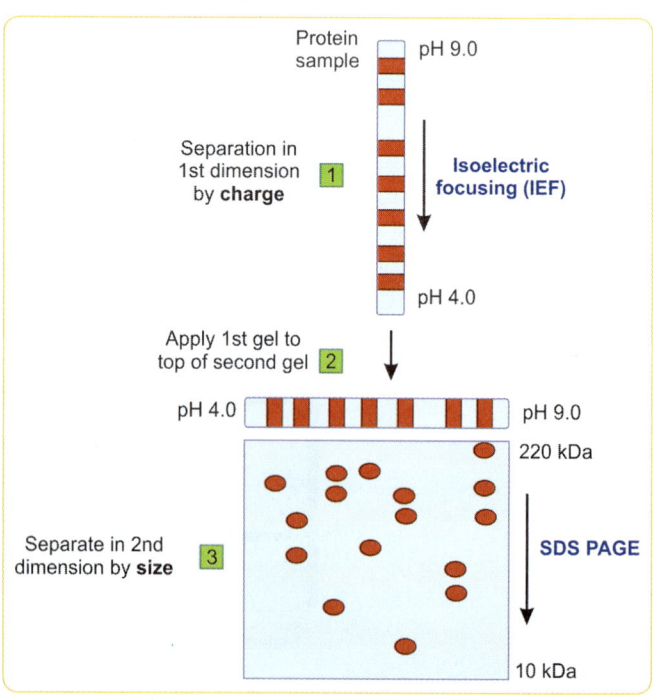

Types of Chromatography

❑ Based on the nature of stationary and mobile phases:
  ○ Planar – Stationary phase is coated on a planar structure
    ➢ **Paper** chromatography
    ➢ **Thin layer** chromatography
  ○ Column - Stationary phase is packed into a column. Depending upon the physical nature of the mobile phase, there are two subtypes:
    ➢ **Gas** chromatography: Mobile phase is gas.
    ➢ **Liquid** chromatography: Mobile phase is liquid. In High-Performance/pressure Liquid Chromatography **(HPLC)**, stationary is made up of finely divided silica particles.

Types of Chromatography Based on the Mechanism of Separation

Technique	Property of molecule used for separation
Partition chromatography (e.g. paper chromatography)	Relative solubility
Ion – Exchange Chromatography	Charge
Affinity chromatography	Specific interaction/Biological function
Gel filtration/size-exclusion chromatography	Molecular size

Size-exclusion (Gel-filtration) Chromatography

Size-exclusion Chromatography separates proteins based on their size. Proteins are applied over the beads containing pores. The beads are made of agarose/dextrin. Beads with different pore sizes are available commercially.

The larger proteins cannot enter the pores of the beads. They move through the available space in between beads and comes out (elutes) faster. Smaller proteins enter through the pore of each bead. They spend more time in passing through the beads and come (elute) later.

Ion Exchange Chromatography

❑ The stationary phase is made up of charged beads in a column.
❑ The beads can be either positively or negatively charged.
❑ If positively charged beads are used, the technique is known as anion exchange chromatography since the beads retain the anionic proteins which can be eluted (separated by washing) at a later stage.
❑ If negatively charged beads are used, the technique is known as cation exchange chromatography since the beads retain the cationic proteins which can be eluted at a later stage.

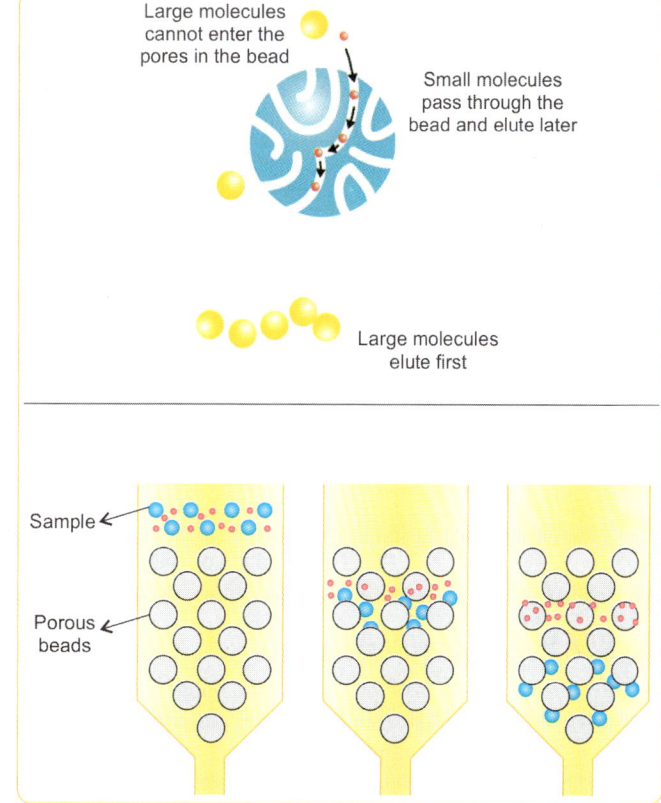

Large molecules cannot enter the pores in the bead

Small molecules pass through the bead and elute later

Large molecules elute first

Sample

Porous beads

- The proteins bound to beads can be eluted by increasing the salt (NaCl) concentration in the mobile phase. This weakens the charge-charge interactions. Proteins with less charge density elute first.

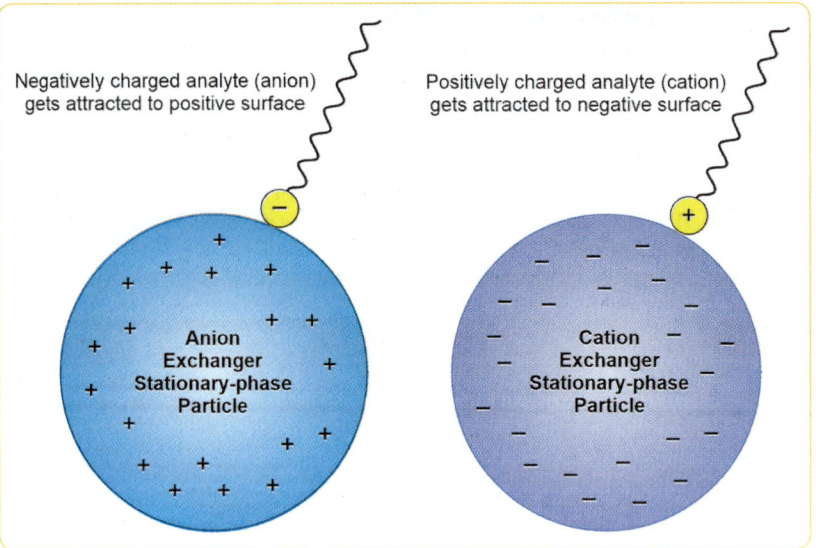

Negatively charged analyte (anion) gets attracted to positive surface

Positively charged analyte (cation) gets attracted to negative surface

Anion Exchanger Stationary-phase Particle

Cation Exchanger Stationary-phase Particle

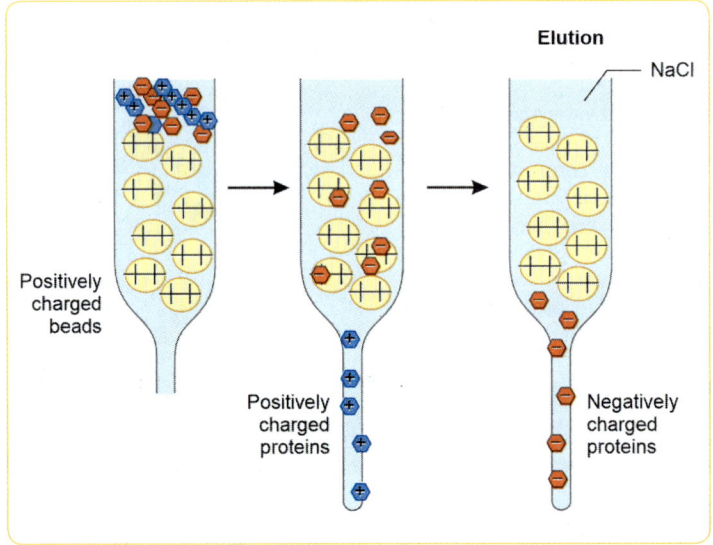

Elution

NaCl

Positively charged beads

Positively charged proteins

Negatively charged proteins

Fig. 2: Illustration of anion-exchange chromatography

Affinity Chromatography

Affinity chromatography is a purification technique based on the highly specific interactions such as antigen-antibody, enzyme-substrate, receptor-ligand interactions etc. Reversible nature of this interaction is helpful in elution of the desired component from the mixture.

The substance to be purified is passed in the mobile phase. Stationary phase contains the affinity ligand for the substance to be purified.

Mobile phase	Stationary phase
Lectins	Carbohydrates
Receptor	Ligand
Enzymes	Substrates, substrate analogues, cofactors, inhibitors
Histidine tagged recombinant protein	Nickel ion (Ni^{2+})
GST tagged recombinant protein	Glutathione-Sepharose
DNA binding protein	Specific DNA
Protein-A tagged recombinant protein	Immunoglobulin G-Sepharose
Streptavidin	Biotin conjugated protein
Oligo dT	Poly A tail containing RNAs

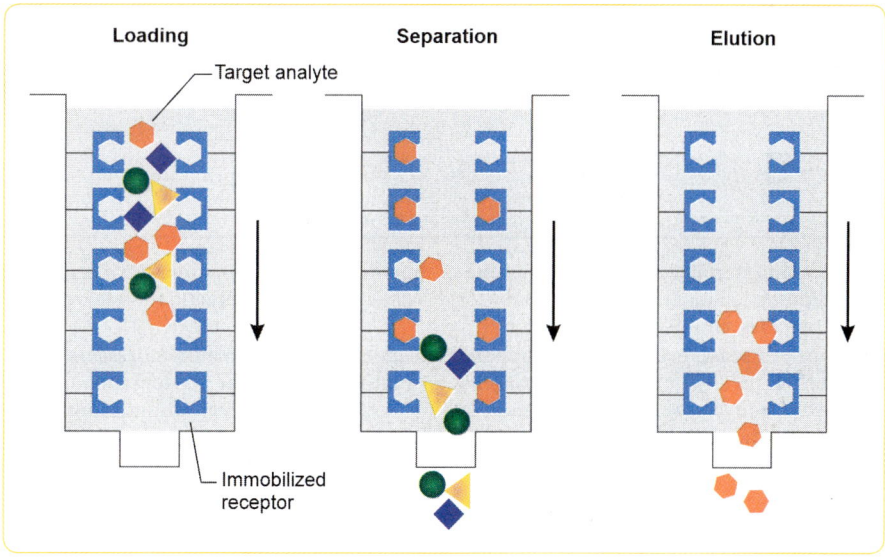

Fig. 3: Affinity chromatography based on receptor-ligand interaction

GST Tag Affinity Chromatography

❑ Assume that human proteins can be synthesised in the bacterial system using recombinant DNA technology (Chapter 28) *(Refer to page no. 459)*.

❑ Affinity chromatography is mainly used to purify recombinant proteins.

❑ Recombinant proteins can be tagged with desirable amino acid sequences like GST tag (The sequence of GST is inserted into the 5′ or 3′ end of the expression vector)

❑ GST refers to the glutathione S-transferase (GST) enzyme. Glutathione is the substrate of GST.

❑ GST tagged recombinant proteins in the mobile phase when passed through the glutathione linked agarose beads (stationary phase), there will be a specific binding. Other unwanted proteins do not bind to the stationary phase.

❑ By adding excess glutathione, recombinant proteins can be separated from the stationary phase.

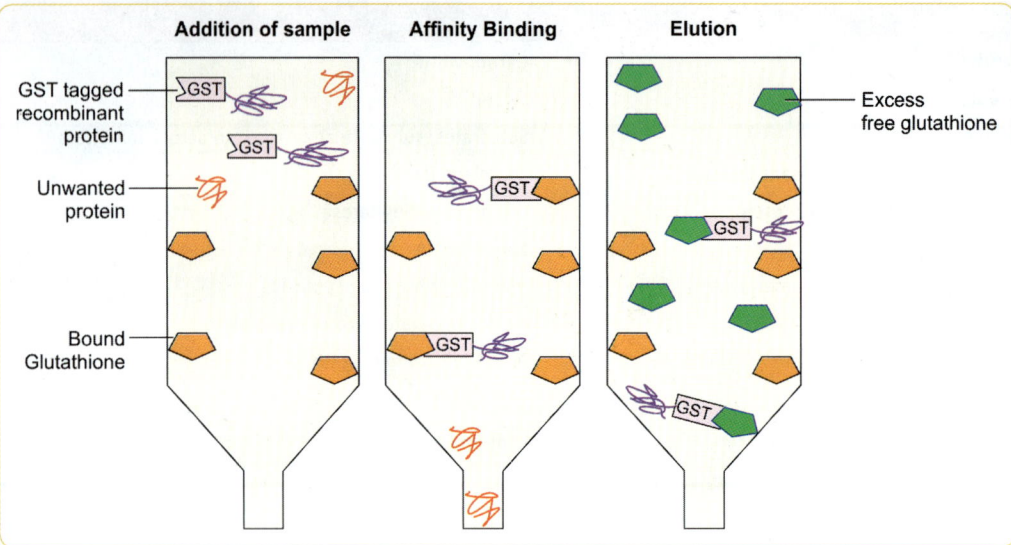

Addition of sample	Affinity Binding	Elution

GST tagged recombinant protein

Unwanted protein

Bound Glutathione

Excess free glutathione

ELISA

Enzyme-Linked Immunosorbent Assay (ELISA) is one of the most common immunoassays performed in both research and diagnostic laboratories. ELISA is used for the detection and quantitation of biological substances such as proteins, peptides, antibodies, and hormones.

Before starting the discussion on ELISA, it is important to understand what an immunoassay is? Immunoassay is an analytical technique used for the quantification of an analyte based on the antigen-antibody reaction. We know that antigen-antibody interaction is highly specific. This specificity and the high affinity of antibodies for their antigens, coupled with the ability of antibodies to cross-link antigens, allow the identification and quantitation of specific substances by a variety of immunoassay techniques.

Labels in Immunoassay

The advantage of immunoassay is that either the antigen or the antibody can be labelled. This label can be an enzyme, radioactive compound, chemiluminescent or fluorescent. The labelled component of an immunoassay is sometimes called the tracer.

Types of Immunoassays

❑ Radioimmunoassay (RIA)
❑ Enzyme-linked immunosorbent assays (ELISA)
❑ Fluoroimmunoassay (FIA)
❑ Chemiluminescence immunoassay (CLIA)

Constituents of Labelled Immunoassay

❑ Specific antibody/antibodies (labelled)
❑ A method to separate the unbound components from bound immune complex
❑ A method for detection of the label
❑ Standards or Calibrators

Types of ELISA

1. Direct ELISA
2. Indirect ELISA
3. Sandwich ELISA
4. Competitive ELISA

Comparison of the Four Types of ELISA

Type of ELISA	Well coated with	Detects in serum	Example
Direct	Antigen	Antigen	Beta hCG. Usually only qualitative
Indirect	Antigen	Antibody	Anti HBV antibody
Sandwich	Antibody	Antigen	HIV: p24 antigen, Thyroid stimulating Hormone
Competitive	Antigen	Antigen	Hapten detection, T3 and T4 hormones

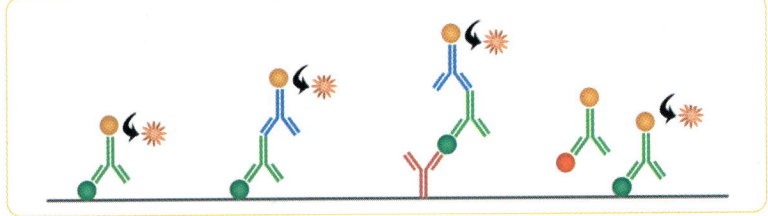

For a detailed explanation:

▶ Radio Immunoassay - https://youtu.be/9vN9N1dt0VQ or scan QR code

ELISA - https://youtu.be/W-cSTxUorFc or scan QR code

MEASUREMENT OF PROTEIN CONCENTRATION

Protein concentration can be measured by Biuret, Bradford, Lowry, and BCA (bicinchoninic acid) assays. All methods are based on the development of color upon adding reagents. Bradford method uses a dye called Coomassie brilliant blue. Other three methods use reagents containing copper. All four methods denature the protein sample. So, the samples cannot be reused for any other processes.

Another simple method for estimation of the amount of protein is based on the UV absorbance. Conjugated double bonds in aromatic amino acids (phenylalanine, tyrosine, and tryptophan) absorb UV light at 280 nm. Tryptophan is the amino acid that absorbs the maximum amount of light compared to other amino acids at 280 nm. The advantage of this method is that the protein is not denatured and can be used for some other purpose.

TECHNIQUES FOR IDENTIFICATION OF PROTEIN STRUCTURE

X-RAY CRYSTALLOGRAPHY

X-ray crystallography and nuclear magnetic resonance (NMR) spectroscopy are the two methods used **for the study of three-dimensional** protein structure at the molecular level.

X-ray crystallography is used to determine the protein structure in atomic detail with high-resolution. The resolution depends on the wavelength. Let us compare the X-ray crystallography with other visualization methods.

Incident rays	Wavelength (λ)	Visualization
Visible light	400–700 nm	Cells and subcellular organelles
Electron	2.5 pm	Finer details of subcellular organelles, Macromolecules surrounding cells extracellular matrix (ECM)
X-ray	1 Å	Atomic detail of macromolecules and their order of arrangement

As the name suggests, X-ray crystallography involves crystallization of proteins. Crystallization is done to fix the order of arrangement of atoms.

Steps in X-ray Crystallography

1. Protein purification
2. Protein crystallization
3. Exposing X-rays and collection of the diffraction pattern
4. Structure solution (phasing)
5. Structure determination (model building and refinement)

Limitation of X-ray crystallography is that only crystallisable proteins can be studied. Nuclear magnetic resonance (NMR) spectroscopy is used to overcome this limitation.

Pioneer women in crystallography:
- **Dorothy Crowfoot Hodgkin** was awarded Nobel prize for her crystallographic studies.
- Watson and Crick proposed their model based on the crystallographic studies of **Rosalind Franklin**. As she passed away prematurely at the age of 37 years, she could not be honoured with Nobel prize.

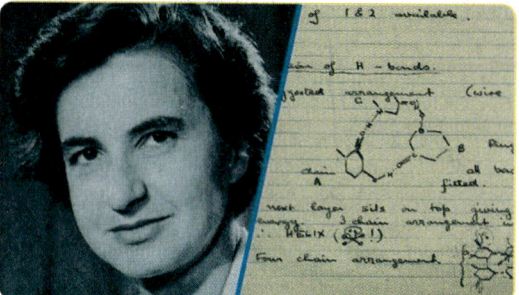

Fig. 4: Rosalind Franklin and page of her notes indicating a double helix structure of DNA. Franklin is one of the unsung hero of science like Yellapragada Subbarow

NUCLEAR MAGNETIC RESONANCE (NMR) SPECTROSCOPY

Detailing of NMR spectroscopy is beyond the scope of this book. So, I am giving only the simplified basics of NMR here. This is just for your understanding purpose.

If an atom has unpaired particles in its nucleus (e.g. ^{1}H, ^{13}C), it will have a spin. These nuclei are intrinsically magnetic because of the spin. When an external magnetic field is applied to these intrinsically magnetic nuclei, protons align either parallel or opposite to the magnetic field. Parallel orientation is known as α-spin and the opposing orientation is known as β-spin. α orientation has the lowest energy as it is parallel to the applied field. ΔE is the energy difference between these two spins.

If ΔE is provided in the form of an external radiofrequency wave, nuclei in the α-state get the energy and obtain the β-state. This is known as nuclear magnetic resonance (NMR). This resonance is determined by the electrons surrounding the nuclei since, they oppose the applied electrical field. Thus, different atoms will have different NMR. Moreover, nuclear spin is affected by the neighbouring atoms also.

Like X-ray crystallography, NMR produces data about the relative position of atoms of a molecule from which structure of the whole molecule is deduced.

MASS SPECTROMETRY

Mass spectrometer (MS) identifies and quantifies analytes based on their molecular mass and charge. For analysis by a mass spectrometer, the analyte of interest must be converted into ions. These ions are separated by an

electromagnetic field and detected by a quantitative method and processed into mass spectra based on their mass to charge ratio (m/z).

It is important to note that MS can be coupled with chromatography and NMR spectroscopy for superior results.

Components of MS Instrument

1. Sample inlet
2. Ion source
3. Mass Analyser
4. Detector
5. Data analysis system

Chemical Ionization (CI), Electron ionization (EI), Electro-Spray Ionisation (ESI), Photoionization and Matrix assisted laser desorption ionization (MALDI) are some of the ionisation techniques used in MS.

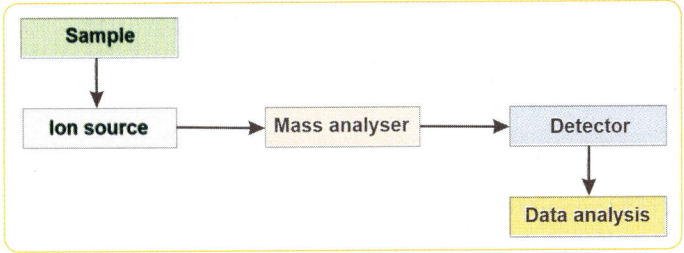

Fig. 5: Schematic of mass spectrometry Instrumentation

Tandem Mass Spectrometry (MS/MS)

In a tandem mass spectrometer, the ionised sample enters the first mass spectrometer (MS1) and only selected ions are allowed to enter the second mass spectrometer (MS2). The detection takes place at the MS2.

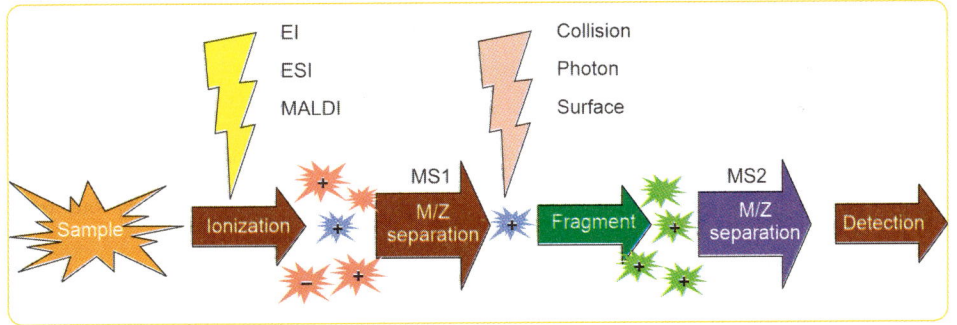

□ BI11.16 Observe use of commonly used equipments/techniques in biochemistry laboratory including:
  • Protein electrophoresis
  • ELISA

Protein separation technique	Property used
Gel electrophoresis	Molecular weight, Charge
SDS-PAGE and Size exclusion chromatography, Ultracentrifugation	Molecular weight
Ion exchange chromatography	Charge
Isoelectric focusing	Isoelectric pH
Affinity chromatography	Specific bioaffinity
Hydrophobic interaction chromatography	Polarity
Salt fractionation	Solubility
NMR spectroscopy	Nuclear Spin
X-ray crystallography	Scattering or diffraction of X-rays by electron clouds of atoms

Key Points

- ❑ SDS-PAGE separates proteins purely based on their mass.
- ❑ Polyampholytes are used in isoelectric focusing to create a continuous pH gradient.
- ❑ Two-dimensional electrophoresis = IEF + SDS-PAGE
- ❑ Ion-exchange chromatography is used to separate charged molecules.
- ❑ Affinity chromatography is used to separate recombinant proteins with tags like histidine tag.
- ❑ Competitive ELISA is used for the detection of haptens like thyroxin.
- ❑ Gas chromatography is used for the separation of volatile compounds like lipids.
- ❑ Mass spectrometry can be used for the determination of the primary structure of proteins including the detection of post-translational modifications.
- ❑ X-ray crystallography and NMR spectroscopy are used for the study of three-dimensional protein structure at the molecular level.

SELF-ASSESSMENT

Short Answer Questions

1. What do you understand by proteomics? Give two applications of proteomics in medicine.

2. What is the principle behind the separation of proteins by isoelectric focusing?

3. Briefly describe the technique to separate proteins based on their isoelectric pH.

4. Briefly describe the electrophoretic technique to separate proteins purely based on their molecular weight.

5. Name two techniques for determination of the molecular weight of proteins. Give the principle of any one.

6. What do you understand by the term '2-D electrophoresis'?

7. Enumerate the types of chromatographic techniques based on the mechanism of action of separation. Describe any one of them in detail with a schematic diagram.

8. Which chromatographic technique is best suited for the purification of recombinant proteins? Explain the technique with an appropriate illustration.

9. Write briefly the principle of 'gel-filtration chromatography'.

10. With a schematic diagram, describe the various types of ELISA.

Multiple Choice Questions

1. **All of the following techniques can separate proteins, except:**
   a. Electrophoresis
   b. Ultracentrifugation
   c. Gas chromatography
   d. Salt fractionation

2. **All of the following are true about electrophoresis, except:**
   a. Increasing the voltage will accelerate the movement of molecules up to some extent
   b. Larger molecules move slowly compared to smaller molecules
   c. The globular-shaped molecule will move faster compared to its irregular shaped counterpart with the same mass
   d. Frictional coefficient does not depend upon the size and shape of the molecule

3. **A protein shows a band corresponding to 100 kD in native-PAGE but two bands of 20 kD and 30 kD in SDS-PAGE. What is your inference about the nature of protein?**
   a. Homodimer
   b. Homotetramer
   c. Heterodimer
   d. Heterotetramer

4. **Which of the following is used in the second direction of two-dimensional electrophoresis?**
   a. Polyampholytes
   b. Sodium dodecyl sulphate
   c. Carboxymethylcellulose sodium
   d. Polystyrene beads

5. **Recombinant proteins can be best purified by:**
   a. Salting out
   b. SDS-PAGE
   c. Affinity chromatography
   d. Isoelectric focusing

6. **Which of the following is the LEAST preferred method to separate Isoenzymes?**
   a. Agarose gel electrophoresis
   b. Gel filtration chromatography
   c. Ion-exchange chromatography
   d. Affinity chromatography using the substrate as a ligand

7. **Which of the following separation technique is based on the ability of amino acid/proteins to behave like zwitterions?**
   a. Gel permeation chromatography
   b. Mass spectrometry
   c. Isoelectric focusing
   d. Ion exchange chromatography

8. **Protein concentration can be determined by the following group/bond of protein absorbing light in the UV range:**
   a. Acidic groups   b. Disulphide bonds
   c. Prosthetic groups d. Aromatic groups

9. **Detection of proteins by which of the following methods does NOT affect its function or after which of the following techniques the protein sample can be reused for further applications?**
   a. SDS-PAGE
   b. 2-D- electrophoresis
   c. UV absorbance at 280 nm
   d. Heat coagulation

10. **Which of the following separation techniques is based on the molecular size of the protein?**
    a. Chromatography on a carboxymethyl (CM) cellulose column
    b. Affinity chromatography using nickel column
    c. Gel filtration chromatography using porous beads
    d. Chromatograph on a diethylaminoethyl (DEAE) cellulose column

11. **Which of the following type of ELISA is suitable for the estimation of thyroxine hormone?**
    a. Direct   b. Indirect
    c. Sandwich   d. Competitive

Chapter 43 Tools of Proteomics

ANSWERS

1. c.	**2.** d.	**3.** d.	**4.** b.	**5.** c.	**6.** d.
7. c.	**8.** d.	**9.** c	**10.** c.	**11.** d.	

44

CHAPTER

Clinical Enzymology

ISOENZYMES

- ❏ Isoenzymes (Isozymes) are the **different enzymes** catalyzing the **same chemical reaction.**
- ❏ As they catalyze the same chemical reaction, isoenzymes have same enzyme commission number. E.g. EC 2.7.1.1 denotes all types of hexokinase including glucokinase (hexokinase type IV).
- ❏ Some isoenzymes are a result of gene duplication (e.g. hexokinase and glucokinase), and some are due to differential combination of subunits (e.g. Lactate dehydrogenase and Creatine kinase).
- ❏ They have **different primary structure** → different electrophoretic mobility → recognised by different antibodies. (can be separated by electrophoresis and isoelectric focusing).
- ❏ They have different K_m and V_{max} values.
- ❏ Their regulatory mechanisms are also different.
- ❏ They are produced by **different genes** in same chromosome or different chromosomes.

Allozymes

Allozymes are different from isoenzymes. They are allelic variants of an enzyme produced from same gene locus.

Isoforms

Isoforms are the different forms of the **same enzyme** that arise due to post translational modifications, e.g., Phosphorylated and non-phosphorylated enzyme.

Both isoenzymes and isoforms can be understood using alkaline phosphatase as an example.
- ❏ Intestinal and Placental alkaline phosphatase are isoenzymes. They are coded by different genetic loci.
- ❏ Liver, Bone and renal alkaline phosphatases are isoforms. They are coded by same genetic loci but undergo differential glycosylation.

ENZYMES IN THE DIAGNOSIS OF DISEASES

Enzymes found in the plasma can be differentiated into two types – functional and non-functional plasma enzymes.

Functional plasma enzymes	Non-functional plasma enzymes
Enzymes that have specific function in blood plasma. Enzymes and their substrates are ordinarily present in plasma of normal individuals.	Enzymes that do not have any function in plasma. Released as a result of normal wear and tear of cells.
E.g. Lipoprotein lipase, Pseudocholinesterase, Proenzymes of coagulation and complement pathway	E.g. LDH, Creatine kinase, AST (SGOT), ALT (SGPT)
The absence of these enzymes indicates pathology. e.g., Lipoprotein lipase is deficient in patients of Familial hyperchylomicronemia. Pseudocholinesterase deficiency results in succinylcholine apnea.	In pathological conditions, levels of these enzymes increase.

Let us discuss the various isoenzymes and their clinical utility.

ISOENZYMES OF LACTATE DEHYDROGENASE

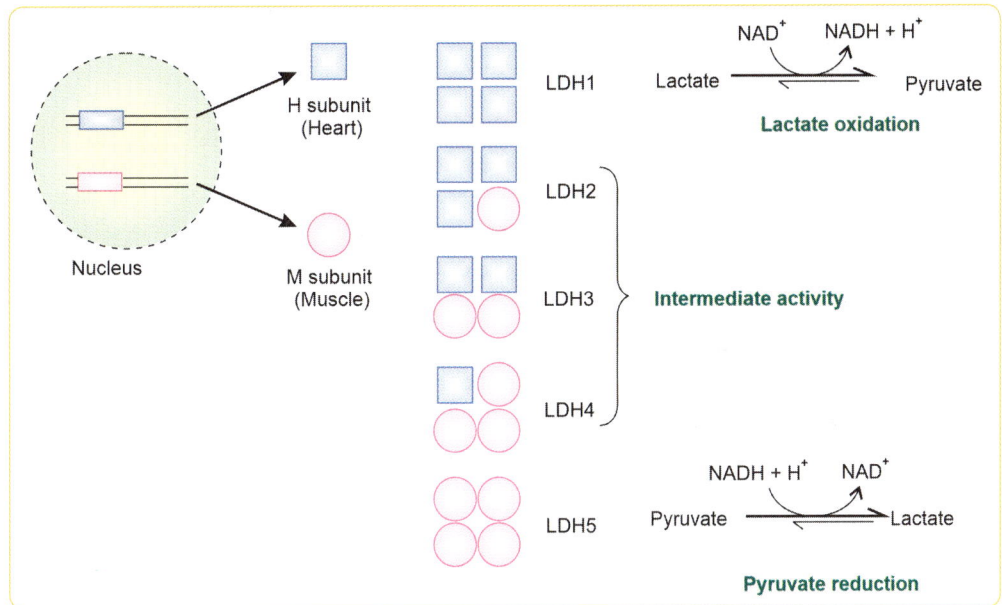

Fig. 1: H and M subunits originate from different genetic loci. LDH2, 3 and 4 are known as hybrid isozymes since they are made of different subunits

H and M subunits originate from different genetic loci. LDH2, 3 and 4 are hybrid isozymes since they are made of different subunits.

Lactate dehydrogenase (LDH) is a tetramer of different combinations of H and M subunits. Based on the combination of H and M subunits, five isoenzymes are present. Two of them are made of H or M subunits purely. Three of them are hybrid enzymes made up of both H and M subunits. The numbering of isoenzymes is based on their electrophoretic mobility. H subunit contains more negatively charged amino acids, so it moves fastest towards the anode hence, LDH1 moves fastest, and LDH5 is the slowest.

Isozyme	Quaternary structure	Predominant* tissue in which present	% in normal serum
LDH$_1$	HHHH (H$_4$)	Heart	33
LDH$_2$	HHHM (H$_3$M$_1$)	RBC	45
LDH$_3$	HHMM (H$_2$M$_2$)	Brain	18
LDH$_4$	HMMM (H$_1$M$_3$)	Liver	3
LDH$_5$	MMMM (M$_4$)	Skeletal muscle	1

*Each isoenzyme is found in at least three organs. To make things simple, I have given the predominant tissue only.

Flipped LDH Pattern in Myocardial Infarction

As shown in the table above, LDH$_2$ is the predominant LDH isoenzyme in the serum of a healthy person. In myocardial infarction, damage to myocyte releases more LDH$_1$ into the blood. Therefore, LDH$_1$ becomes the predominant LDH isoenzyme in the serum of a patient with myocardial infarction. This is known as flipped pattern of LDH.

LDH is Raised in Hemolytic Anemia

LDH-2 is the fraction that causes increased LDH levels in hemolytic anemia.

CREATINE KINASE

Creatine phosphate is the phosphagen in vertebrates. Creatine kinase (CK, previously CPK) catalyzes the Lohman reaction (Chapter 4) *(Refer to page no. 47)*. Creatine kinase is a dimer. Based on the subunit (B and M) combination, there are three isoenzymes of CK.

Isozyme	Quaternary structure	Tissue distribution	Marker of
CK1	BB	Brain and smooth muscle	Brain injury
CK2	MB	Predominantly cardiac muscles	Myocardial infarction
CK3	MM	Predominantly skeletal muscles	Muscular dystrophy

CK-MB is a Valuable Marker of Reinfarction

CK-MB appears in serum within 6 hours after MI and is cleared in 2 days. A persistence of CK-MB level in serum indicates an extension of the infarction to other areas or reinfarction.

Historically, AST was the first enzyme to be used as a marker of myocardial infarction. Due to lack of specificity, it is seldom used for this purpose currently.

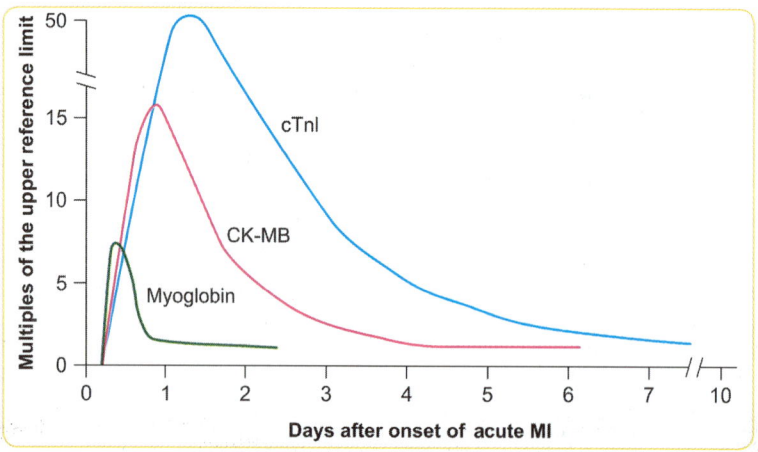

Fig. 2: Myoglobin is the first marker (not enzyme) to get elevated in MI. Compare the pattern of CK-MB and cardiac Troponin I (cTnI). Troponins are not useful to diagnose re-infarction. Aspartate transaminase and Lactate dehydrogenase are seldom used these days to diagnose Myocardial Infarction. Therefore, their pattern is not shown

Cardiac Troponins (cTn)

Troponins are non-enzyme proteins associated with actin filaments of skeletal and cardiac muscle. Troponins are absent in smooth muscles. Troponins modulate the interactions between actin and myosin. There are three troponins.

Troponin T	Tropomyosin binding
Troponin C	Calcium binding
Troponin I	Inhibitory to actin

Cardiac troponin I (cTnI) and Troponin T are more useful in the diagnosis of myocardial infarction. Troponin I raises within 6 hours, peaks at 72 hours and remains elevated for a week.

Detection: Point of care testing devices based on lateral flow immunoassay is available. (*Point of care testing is testing outside the laboratory premises especially at the bedside of patient, emergency room etc.*)

Note: Recent studies have found that cTnI to be useful in the diagnosis of reinfarction also.

ENZYMES IN DISORDERS OF THE LIVER

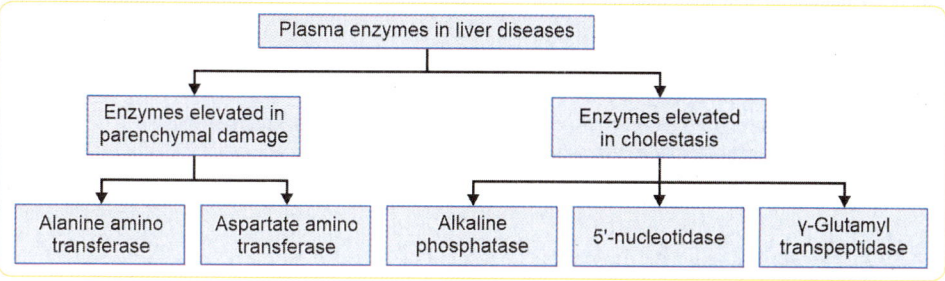

Locations of Liver Enzymes

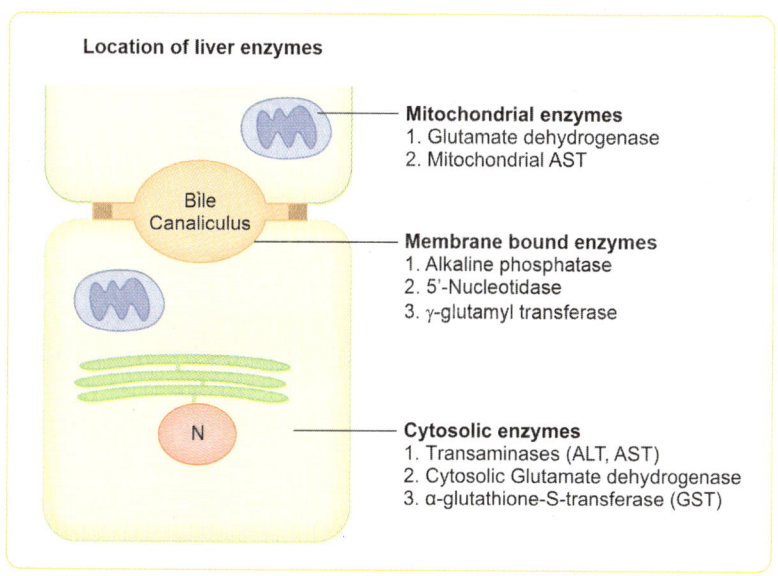

Enzymes Elevated in Cholestasis

Alkaline Phosphatase

❏ Routinely measured in clinical laboratories for liver function than GGT and 5'nucleotidase.
❏ Alkaline phosphatase is a GPI-anchored enzyme found in biliary canaliculi. Bile is a surfactant. During biliary obstruction, regurgitated bile causes the release of ALP from the membrane of biliary canalicular membrane and there is increased production also.
❏ In addition to Level of ALP is increased in children (due to bone growth) and pregnancy (due to placenta)

❏ ALP is a crude marker of zinc level in the body.
❏ Regan isoenzyme is a placental-type alkaline phosphatase that is expressed in some human tumors, particularly in gonadal and urologic cancers. Presence of this isozyme indicates a reversal to or re-emergence of fetal pattern in the tumors. This is due to derepression of placental ALP in tumors.

5' Nucleotidase

❏ Membrane-bound, GPI anchored enzyme.
❏ Though widely distributed in the body, rise in serum levels primarily reflects liver disease.
❏ Highly sensitive marker for intra and extra-hepatic cholestasis.
❏ Highly specific for the hepatobiliary disease. NOT increased in children or pregnancy (unlike ALP).

Gamma Glutamyl Transferase (GGT)

❏ Situated on the surface of hepatocyte canalicular membrane (major fraction) and microsomes (minor fraction).
❏ Highly sensitive marker for intra and extra-hepatic biliary obstruction.
❏ Elevated levels of GGT are also seen in alcoholic liver disease (microsomal fraction is induced by alcohol).

Markers of Parenchymal Liver Disease

Aspartate aminotransferase (AST) a.k.a. Serum glutamate oxaloacetate transaminase (SGOT) and **alanine amino transferase** (ALT) a.k.a. Serum glutamate pyruvate transaminase (SGPT), are the markers of parenchymal liver disease. AST activity is not entirely cytosolic. **10% of AST activity in serum is due to the mitochondrial fraction of the enzyme.**

De Ritis Ratio (AST/ALT Ratio)

❏ AST/ALT is normally < 1.0
❏ When AST/ALT > 2.0, it is more likely to be associated with alcoholic hepatitis, Cirrhosis or hepatocellular carcinoma

Parameter	Normal	Prehepatic	Hepatic	Post-hepatic
Direct (conjugated) bilirubin (mg/dL)	0.1 - 0.3	N	↑	↑↑
Indirect (unconjuaged) bilirubin (mg/dL)	0.2 - 0.7	↑↑	↑	↑
Urine bilirubin	---	---	+	++
Urine urobilinogen (mg/24 hr)	0 - 4	↑↑	Normal or Decreased	Absent
ALT and AST (IU/L)	7-41 and 12-38	N	↑↑	↑

Contd...

Section V Special Topics

Parameter	Normal	Prehepatic	Hepatic	Post-hepatic
ALP (IU/ L)	33-96	Normal	↑	↑↑
5'NT (IU/L)	0-11	N	↑	↑↑
GGT (IU/L)	9-58	N	↑	↑↑

We have already discussed bilirubin and urobilinogen in Chapter. 11.7 (Refer to page no. 226).

SERUM AMYLASE

Amylase breaks the α(1,4) glycosidic linkages present in polysaccharides. Animal amylases are known as α-amylases since they act on α-glycosidic bonds. Amylases require calcium ions and are activated by chloride and bromide.

There are two isoenzymes of serum amylase:
1. P-type (pancreatic)
2. S-type (Salivary)

Sources of amylase: Normal sources of serum amylase are:
❑ Major sources: Pancreas, salivary gland
❑ Minor sources: Liver, small intestine, kidney, fallopian tube
❑ Amylase is normally found in urine

Causes of Raised Amylase Activity

Pancreatic diseases	Non-Pancreatic disorders
Acute pancreatitis	Salivary gland disorders – Mumps and salivary calculi
Chronic ductal obstruction	Renal insufficiency
Pancreatic trauma	Tumor hyperamylasemia
Pancreatic carcinoma (25 to 50% cases)	Carcinoma of the lung
	Carcinoma of the esophagus
	Breast carcinoma
	Ovarian carcinoma
	Peritonitis and Perforation

Serum lipase is more specific than amylase since amylase is raised in many disorders other than pancreatitis.

Remember

❑ **Acute pancreatitis:** Acutely raised amylase
❑ **Pancreatic pseudocyst:** Persistently raised amylase
❑ **Chronic pancreatitis:** Low amylase

ACID PHOSPHATASE (ACP)

Acid Phosphatases include all phosphatases with optimal activity below a pH of 7.0.

ACP is present predominantly in lysosomes. Extra-lysosomal acid phosphatases are also found in many cell types. The highest activity of ACP is seen in prostate, bone (osteoclasts), spleen, platelets and erythrocytes.

Based on the tartrate sensitivity, ACP is of two types
❑ Tartrate-sensitive (Inhibited by L-(+)-Tartaric acid): Present in lysosomes and prostate.
❑ Tartrate-resistant Acid phosphatase (TRAP): Present in bone, erythrocytes and leucocytes.

Chapter 44 Clinical Enzymology

Causes of elevated acid phosphatase
❑ Carcinoma prostate
❑ Osteoclastic phase of Paget's disease
❑ Hyperparathyroidism
❑ Osteoclastoma
❑ Lysosomal storage disorders – e.g. Gaucher's disease

Tartrate Resistant Acid Phosphatase (TRACP/TRACP)

❑ TRACP exists in two isoforms - 5a and 5b.
❑ TRACP 5a is secreted by macrophages and dendritic cells and is increased in patients with rheumatoid arthritis.
❑ TRACP 5b is produced from osteoclast; a **marker of bone resorption**.
❑ Expression of TRACP is increased in certain pathological conditions such as Leukemic Reticuloendotheliosis (Hairy Cell Leukemia), Gaucher's Disease, HIV-induced Encephalopathy, Osteoclastoma and in osteoporosis and metabolic bone diseases.
❑ Anti-TRACP antibody labels the cells of **Hairy Cell Leukemia** (HCL) with a high degree of sensitivity and specificity.

REVIEW

Disorder	Enzyme elevated
Pancreatitis	Amylase Lipase (more specific)
Cholestasis	Alkaline phosphatase Gamma glutamyl transferase 5' Nucleotidase
Myocardial infarction	Creatine kinase – MB
Hemolytic anemia	Lactate dehydrogenase
Viral hepatitis	Aminotransferases (ALT and AST)
Muscular dystrophies	Creatine kinase – MM
Paget's disease, Rickets	Alkaline phosphatase
Alcoholic liver disease	Gamma glutamyl transferase 5' Nucleotidase
Carcinoma Prostate	Acid phosphatase

ENZYMES AS THE TARGET OF DRUGS

Drug	Target enzyme	Use
Allopurinol	Xanthine oxidase	Gout
Statins	HMG CoA reductase	Hypercholesterolemia
Theophylline	Phosphodiesterase	Asthma
Propylthiouracil	Thyroid peroxidase, 5' deiodinase	Hyperthyroidism
Disulfiram	Aldehyde dehydrogenase	De-addiction of alcohol
Acetazolamide	Carbonic anhydrase	Glaucoma
Captopril, Enalapril	Angiotensin Converting Enzyme	Hypertension
Finasteride	5 α-reductase	Benign Prostatic Hyperplasia

THERAPEUTIC ENZYMES (ENZYMES AS DRUGS)

Enzyme	Reaction	Use
Streptokinase and Urokinase	Plasminogen → plasmin	Thrombolysis in myocardial infarction
Pepsin	Protease	Digestive supplement
Asparaginase	L-Asparagine + H_2O → L-aspartate + NH_3	Leukaemia
Hyaluronidase	Hyaluronate hydrolysis	Adjuvant with local anaesthetics
Serratiopeptidase	Protease	Anti-inflammatory
Rasburicase	Uric acid → allantoin	Tumor lysis syndrome
Collagenase	Collagen degradation	Wound debridement

IMMOBILIZED ENZYMES

Immobilized enzymes are enzymes physically confined to a defined region of space with retention of their catalytic activities, and which can be used repeatedly and continuously.

Immobilized enzymes are used in various diagnostic purposes, e.g. **Urine dipstick test** uses glucose oxidase and peroxidase enzymes adsorbed (immobilised) on a cellulose support. The chemical reactions that take place on the dipstick pad are:

$$\text{Glucose} + O_2 \xrightarrow{\text{Glucose oxidase}} \text{D-glucono-δ-lactone} + H_2O_2$$

$$H_2O_2 + \text{Chromogen} \xrightarrow{\text{Peroxidase}} \text{Oxidised chromogen (coloured)} + H_2O$$

Similar immobilised enzymes are used for the testing of galactose, urea etc.

Competency

- ❑ BI2.5   Describe and discuss the clinical utility of various serum enzymes as markers of pathological conditions
- ❑ BI2.6   Discuss use of enzymes in laboratory investigations (Enzyme-based assays)
- ❑ BI2.7   Interpret laboratory results of enzyme activities & describe the clinical utility of various enzymes as markers of pathological conditions

Key Points

- ❑ Isoenzymes are different enzymes catalyzing the same chemical reaction.
- ❑ Isoforms are the different forms of the same enzyme that arise due to posttranslational modifications.
- ❑ Lipoprotein lipase, Pseudocholinesterase, Proenzymes of coagulation, enzymes of the complement pathway are some functional plasma enzymes.
- ❑ Flipped LDH pattern is seen in myocardial infarction.
- ❑ CK-MB is a marker of reinfarction.
- ❑ Regan isoenzyme is a placental-type alkaline phosphatase expressed in some human tumors.
- ❑ Streptokinase, Urokinase and Rasburicase are a few examples of therapeutic enzymes.

Chapter 44 Clinical Enzymology

Short Answer Questions

1. What are isoenzymes? How do they differ from isoforms?

2. What is the normal level of serum transaminases? Mention their clinical significance.

3. What are non-functional plasma enzymes? What is the significance of identifying specific isoenzymes in clinical practice? Explain with appropriate examples

4. Name the enzymes elevated in obstructive liver disease.

5. What is De Ritis Ratio? What is the significance of this?

6. Name any four non-pancreatic disorders that lead to elevation of serum amylase levels.

7. What is the significance of the estimation of Tartrate Resistant Acid Phosphatase (TRACP/TRACP)?

8. Name any four therapeutic enzymes and their use.

9. What do you understand by immobilised enzymes? Mention their diagnostic importance.

Multiple Choice Questions

1. **All of the following are true about isoenzymes, except:**
   a. They have a different primary structure
   b. They have same enzyme commission number
   c. They catalyze the same chemical reaction
   d. They are always coded by the same gene

2. **Which of the following is a functional plasma enzyme?**
   a. Lipoprotein lipase
   b. Creatine kinase
   c. Lactate dehydrogenase
   d. Alanine transaminase

3. **The presence of which of the following isoenzymes in serum indicates a derepression of genes?**
   a. LDH-5
   b. Regan's isoenzyme
   c. Glucokinase
   d. Pancreatic amylase

4. **Th most abundant LDH isoenzyme in a healthy adult is:**
   a. LDH-1
   b. LDH-2
   c. LDH-3
   d. LDH-4

5. **In which of the following conditions, flipped LDH isoenzyme pattern is seen?**
   a. Choriocarcinoma
   b. Hemolytic anaemia
   c. Muscular dystrophy
   d. Myocardial infarction

6. **All of the following enzymes are used to assess the liver function, except:**
   a. Aspartate transaminase
   b. Alanine transaminase
   c. Alkaline phosphatase
   d. Alpha Amylase

7. **Aspartate Transaminase (AST) is not a specific biomarker for myocardial infarction because:**
   a. AST is not heat-stable
   b. AST levels are elevated in liver damage
   c. It does not increase in sufficient amount for detection
   d. It is difficult to estimate AST levels in serum

8. **Which of the following is useful to assess reinfarction in a patient of MI?**
   a. Lactate dehydrogenase
   b. Creatine kinase – MB
   c. Troponin T
   d. Troponin C

9. **All of the following enzymes are predominately elevated in cholestasis, except:**
   a. γ-Glutamyl transpeptidase
   b. 5'-nucleotidase
   c. Alkaline phosphatase
   d. Alanine aminotransferase

ANSWERS

1. d.	2. a.	3. b.	4. b.	5. d.	6. d.
7. b.	8. b.	9. d.			

45

CHAPTER

Organ Function Tests

Chapter Outline

'Organ function tests' is a collective term for a variety of individual tests/procedures used to evaluate an organ function

PANCREATIC FUNCTION TEST

Direct tests

- Secretin/CCK stimulation test
- Lundh standardized meal test

Indirect tests

- Fecal enzyme assays
  - Fecal Elastase-1 assay
  - Chymotrypsin
- Serum enzyme assays
- Fat absorption test
  - C^{13}-MTG test
  - Fluorescein dilauryl test
- NBT-PABA test
- Pancreatic Schilling test

Direct tests involve the stimulation of the pancreas with the administration of a meal or hormonal secretagogues after which duodenal fluid is collected and analyzed to quantify normal pancreatic secretory content (i.e. enzymes, HCO_3^-). **Indirect tests** measure the consequences of pancreatic insufficiency.

SECRETIN-CCK TEST

- ❏ Invasive test that requires passage of a triple lumen tube under radiological guidance.
- ❏ The patient is kept on fasting for 6 hours with the tube.
- ❏ Pancreatic secretion is collected and analyzed after an IV injection of secretin.
- ❏ Considered as the **gold-standard** test for pancreatic insufficiency.

FECAL ELASTASE ASSAY

Elastases are elastin degrading enzymes (Chapter 42) *(Refer to page no. 621)*. There are eight different elastases in humans. Elastase-1 is a pancreas-specific enzyme. It is not degraded in the gut and excreted as such in feces. The concentration of elastase in feces is 5 to 6 times that of concentration in the pancreatic juice. It can be quantified by ELISA. Thus, it serves as a useful indicator of exocrine pancreatic sufficiency. As this is non-invasive, this test is suitable for pediatric patients.

SCHILLING TEST

Schilling test helps to determine/differentiate the etiology of vitamin B_{12} malabsorption. A radioactive form of B12 was administered orally and its appearance in the urine was used as a sign of absorption. Radioactive B12 is no longer available, and Schilling tests are no longer performed. *NBT-PABA test and schilling tests are obsolete these days.*

INTESTINAL FUNCTION TEST

D-XYLOSE ABSORPTION TEST IS A TEST FOR PROXIMAL SMALL INTESTINAL MUCOSAL FUNCTION

D-Xylose is a 5-carbon sugar of plant origin, which is absorbed predominantly in proximal small intestine by a passive carrier-mediated system. Normally two-thirds (60%) of the orally taken xylose is absorbed. Out of this two-third, half is metabolized in liver, and the rest is excreted in urine. This means normally 30% of the orally taken xylose is excreted in urine. If 25-gram oral D-Xylose is given, urinary excretion will be at least 4.5 g in 5 hours. Decreased urinary excretion indicates malabsorption of monosaccharides.

LIVER FUNCTION TESTS (LFT)

The liver plays a key role in a variety of metabolic, synthetic and excretory functions. The following are the commonly performed tests for the assessment of liver function.

Routine LFTs	Special LFTs

❏ Serum bilirubin (Total and Conjugated), urinary bilirubin, urinary urobilinogen
❏ **Serum enzymes:**
  ○ Alanine aminotransferase (ALT/SGPT)
  ○ Aspartate aminotransferase (AST/SGOT)
  ○ Alkaline phosphatase
  ○ Gamma glutamyl transferase/transpeptidase
  ○ 5' Nucleotidase
❏ Serum total protein and Albumin
❏ Prothrombin time, clotting time
	❏ Viral Markers, e.g. Hepatitis B surface antigen assay
	❏ Immunological Markers, e.g. Smooth muscle antibodies for autoimmune hepatitis
	❏ Assay for metabolic defects – Serum and urinary copper level in Wilson's disease

  We have already discussed most of the components of LFT. Bilirubin in 11.7, Serum enzymes in the previous chapter, Serum protein and albumin in Chapter 35 *(Refer to page no. 549)*. Let's learn prothrombin time now.

PROTHROMBIN TIME

Prothrombin time is a coagulation test that measures the activity of extrinsic clotting pathway which is widely used to monitor oral anticoagulant therapy.

International Normalized Ratio (INR) is used to maintain the uniformity of PT results.

INR = (PT patient/PT normal)[ISI;]

ISI - International Sensitivity Index

Normal value of PT: 11-13 seconds

Table: PT value can be used to differentiate hepatic and post-hepatic jaundice

	Hepatocellular jaundice	Post-hepatic jaundice
Prothrombin time	↑	↑↑
Reason	Inability to form clotting factors	Malabsorption of vitamin K
Measurement of PT after vitamin K injection	Same as earlier	Returns to normal

RENAL FUNCTION TEST

The most common tests used in the clinical biochemistry laboratory for the assessment of renal function are:

❏ Serum Urea
❏ Serum Creatinine
❏ Serum Sodium
❏ Serum Potassium

We have already discussed sodium and potassium in Chapter 40 (Refer to page no. 598). The methods used for the estimation of urea and creatinine will be discussed in the next chapter. Now, let us discuss renal clearance.

RENAL CLEARANCE

Clearance is the volume of plasma that is the kidneys completely clear of a particular substance per unit of time. Each substance/drug will have specific clearance value.

Clearance = {Urine concentration (mg/dL) × Urine volume (mL/min)}/*Plasma concentration (mg/mL)*

Clearance = UV/P

Methods to Measure Renal Clearance

❏ Urea clearance
❏ Creatine clearance
❏ Inulin clearance
❏ Uric acid clearance

ESTIMATION OF CLEARANCE

Urea clearance is the amount of plasma that is cleared off **urea** excreted by kidneys in one minute.

Measures glomerular filtration rate.

Technique

❏ 24-hour collection of urine is made.
❏ Blood sample is taken for the estimation of urea/creatinine.
❏ Urea is estimated by urease method or Diacetyl monoxime method and creatinine by Modified Jaffe's (alkaline picrate) method.

Urea Clearance

$$\text{Urea clearance} = \frac{\text{Urinary urea (mg/dL)} \times \text{Urine volume (mL/min)}}{\text{Serum urea (mg/mL)}}$$

Unit: mL/minute

As urea is passively absorbed in renal tubules, urea clearance depends upon the urinary flow rate.

- ❑ If the urinary flow rate is ≥2 **mL**/minute, the clearance calculated is called **maximal** urea clearance. Normal value: 75 mL/min
- ❑ If the urinary flow rate is ≤2 **mL**/minute, the formula is modified and the clearance calculated is called **standard** urea clearance. Normal value: 54 mL/min

$$\text{Standard urea clearance} = \frac{U\sqrt{V}}{P}$$

From the value obtained, the percentage is derived using the normal value. These percentages can give an idea of the extent of renal damage. ≥70% clearance is considered to be normal.

Disadvantages of Urea Clearance

- ❑ Affected by dietary protein intake
- ❑ Depends upon urinary flow rate

Creatinine Clearance

$$\text{Creatinine clearance} = \frac{\text{Urinary creatinine (mg/dL)} \times \text{Urine volume (mL/min)}}{\text{Serum creatinine (mg/dL)}}$$

Advantages of Creatinine Clearance Over Urea Clearance

- ❑ Not affected by dietary protein intake. Only dependent upon muscle mass
- ❑ Excretion of creatinine is constant in health in an individual and varies little from day to day
- ❑ Independent of urine flow
- ❑ Not absorbed by kidneys
- ❑ Renal tubular secretion is minimal

Estimated GFR Calculation

To overcome the inconvenience of 24 hours urine collection, few formulas have been developed to estimate the GFR using serum creatinine and some other easily measurable parameters.

Cockcroft-Gault Formula

$$\text{eGFR} = \frac{(140\text{-age in years}) \times \text{Weight (kg)} \times \text{K}}{72 \times \text{Serum creatinine (mg/dL)}}$$

K is a constant. The value of K is 0.85 for women and 1 for men.

MDRD (Modification of Diet in Renal Disease Study) Formula

eGFR = 186 × Serum Cr$^{-1.154}$ × age$^{-0.203}$ × 1.212 (if the patient is black) × 0.742 (if female)

Urea clearance	Creatinine clearance
Underestimates GFR	Overestimates GFR
Inferior to creatinine clearance.	Better than urea clearance.
Depends upon the flow rate. That is why two formulas - standard and maximal urea clearance	Independent of flow rate
Depends upon dietary protein intake	Independent of dietary proteins.

Comparison of clearance of various substances

Fate of filtered substance in the renal tubules	Relationship of clearance and GFR	Example
Neither reabsorbed nor secreted	Clearance = GFR	Inulin
Reabsorbed and secreted	Clearance ≈ GFR	Creatinine (somewhat overestimates the GFR)
Partially reabsorbed	Clearance < GFR	Urea
secreted and not reabsorbed	Clearance > GFR	PAH

Note: PAH clearance is used to measure renal plasma flow.

INULIN CLEARANCE IS THE BEST METHOD OF ESTIMATING GFR

❑ Inulin is a homopolysaccharide of fructose.
  ○ Present in garlic, onion etc.
  ○ Neither absorbed nor secreted
❑ Inulin clearance is a near accurate estimate of GFR
❑ Not used widely as it has to be exogenously given to the patient

CYSTATIN C IS A MARKER OF GFR

❑ Cystatin C is a new endogenous marker of GFR superior to serum creatinine.
❑ Cystatin C is a cysteine protease inhibitor, which is produced by all the nucleated cells of the body in a constant amount irrespective of age, sex and muscle mass.
❑ Cystatin C is eliminated predominantly by kidney. Thus, raised level of serum cystatin C indicates impaired GFR.
❑ Therefore, serum Cystatin C is a better indicator of GFR than serum creatinine. It is more sensitive for the detection of mild to moderate impairment in renal function.
❑ One should keep in mind that cystatin C levels are elevated in inflammation and malignant conditions also.

BLOOD UREA NITROGEN (BUN)

Blood urea levels are sometimes represented as BUN. BUN indicates the amount of nitrogen that is present in the form of urea (NH_2CONH_2). Molecular Weight of urea is 60 (14+2+12+16+14+2). Out of this 48% (28/60) is in the form of Nitrogen.

Urea [mg/dL] = BUN [mg/dL] * 2.14
BUN [mmol/L] = Urea [mmol/L]

FRACTIONAL EXCRETION OF SODIUM

The Fractional Excretion of Sodium (FENa) determines if renal failure is due to pre-renal or renal pathology. FENa is the percentage of the sodium filtered by the kidney, which is excreted in the urine.

$$FENa = \frac{(Urine_{Na}/Serum_{Na})}{(Urine_{Cr}/Serum_{Cr})} \times 100$$

❑ FENa <1% indicates prerenal pathology
❑ FENa >2% indicates acute renal pathology
Proteinuria and microalbuminuria will be discussed in the next chapter.

BIOCHEMICAL BASIS OF ANEMIA AND HYPOCALCAEMIA IN PATIENTS WITH CHRONIC RENAL FAILURE

Anemia

Main Reason

- **Erythropoietin** is a hormone produced by interstitial fibroblasts in the kidney. This is a potent stimulator of erythropoiesis in bone marrow.
- In CRF, there is ↓ erythropoietin → Normocytic normochromic anemia
- This is why erythropoietin is administered in patients with CRF to correct anemia.

Other Reasons

- Uremia induced inhibitors of erythropoiesis
- Shortened red blood cell lifespan
- ↑ blood loss due to platelet dysfunction, repeated phlebotomies

Hypocalcemia

- In CRF, there is **impairment of 1 α hydroxylation of 25-OH** cholecalciferol to calcitriol. This leads to reduced active vitamin D_3 → ↓ intestinal absorption of calcium
- In CRF, there is **reduced excretion of phosphates** by the kidney and the increased serum phosphate binds to calcium, depositing it to bone, resulting in hypocalcemia.

THYROID FUNCTION TESTS (TFTS)

Thyroid stimulating hormone (TSH), Free T3 and Free T4 are the three most commonly performed TFTs. Autoantibodies like thyroperoxidase autoantibodies (anti-TPO) and thyroglobulin autoantibodies (anti-Tg) are measured when autoimmune etiology is suspected.

TFTs are done by immunoassays like ELISA and chemiluminescence immunoassay. The latter is superior to the former.

A simple algorithm used in the interpretation of TFTs is given below:

Competency

- BI6.13   Describe the functions of the kidney, liver, thyroid glands
- BI6.14   Describe the tests that are commonly done in clinical practice to assess the functions of these organs (kidney, liver, thyroid glands)
- BI6.15   Describe the abnormalities of kidney, liver, thyroid glands
- PY8.4   Describe function tests: Thyroid gland and pancreas
- PY7.8   Describe and discuss Renal Function Tests
- PY4.8   Describe and discuss pancreatic exocrine function tests and liver function tests
- IM10.17 Describe and calculate indices of renal function based on available laboratories including FENa (Fractional Excretion of Sodium) and CrCl (Creatinine Clearance)

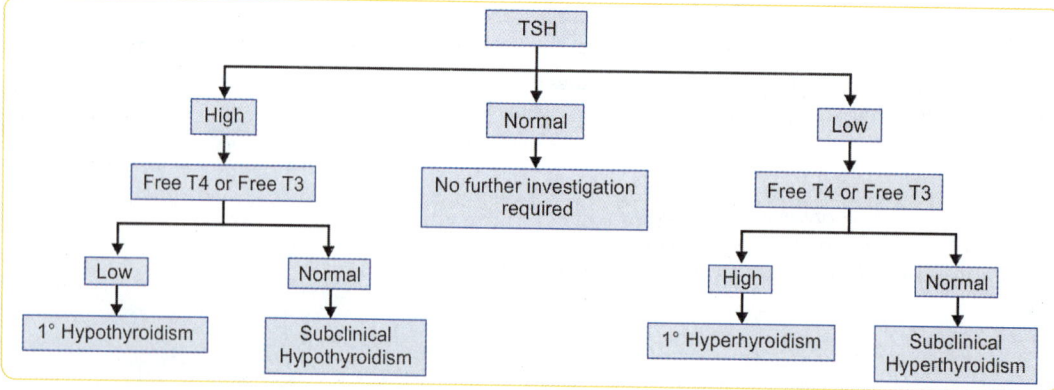

Key Points

- ❏ D-Xylose absorption test is a test for small intestinal function.
- ❏ NT-PABA test and fecal elastase assay are tests for pancreatic function.
- ❏ International Normalised Ratio is used to maintain the uniformity of prothrombin time results.
- ❏ Clearance is the volume of plasma that the kidneys completely clear of a particular substance per unit of time.
- ❏ Creatinine clearance overestimates the GFR.
- ❏ Cystatin C is an endogenous marker of GFR superior to serum creatinine.
- ❏ BUN [mmol/L] = Urea [mmol/L]
- ❏ Thyroid function tests are done by immunoassays like ELISA and chemiluminescence immunoassay.

SELF-ASSESSMENT

Short Answer Questions

1. How can you differentiate hepatic jaundice and post-hepatic jaundice with the help of the result of prothrombin time?

2. What is the advantage of creatinine clearance over urea clearance?

3. To assess the function of which organ cystatin C is used? Explain the biochemical basis.

4. If the blood urea nitrogen is 40 mmol/L in a patient, what is the value of urea?

5. Explain the biochemical basis of anemia and hypocalcaemia in patients with chronic renal failure.

6. Name one hormone whose assay in the blood will help in differentiating primary and secondary hypothyroidism. Explain.

7. Write a simple algorithm for interpretation of thyroid function tests.

Clinical Case-Based Questions

1. A 15-year-old boy was admitted with edema, loss of appetite, nausea, vomiting, breathlessness and fatigue. He had a history of similar symptoms 5 years back. O/E puffiness of face was observed and his urine analysis revealed heavy proteinuria, specific gravity- 1.020, turbidity ++ and Albumin 3+

   a. What is the provisional diagnosis?
   b. Mention kidney function tests and other parameters that will be deranged in this clinical picture.

Multiple Choice Questions

1. **Which of the following is FALSE about urea clearance?**
   a. Influenced by dietary protein intake
   b. Influenced by urinary flow rate
   c. Inferior to creatinine clearance
   d. Overestimates the GFR

2. **In the assessment of which of the following organ's function, cystatin C is used?**
   a. Kidney   b. Liver
   c. Heart   d. Intestine

3. **To diagnose which of the following deficiency Schilling test is used?**
   a. Vitamin B12
   b. Folate
   c. Iron
   d. Vitamin C

4. **Which of the following is a direct test of pancreatic function?**
   a. NBT-PABA assay
   b. Fecal Elastase-1 assay
   c. Schilling test
   d. Secretin stimulation test

ANSWERS

| **1.** d. | **2.** a. | **3.** a. | **4.** d. |

Practical Biochemistry

BLOOD COLLECTION TUBES

In this final chapter of this book, I want you to know about the color coding of various blood collection tubes used in clinical practice. This will help you to choose the appropriate tube for appropriate laboratory investigation.

Blood collection tube	Color of the cap	Application
EDTA tube	Lavender	Hemogram, HbA1C estimation, Haemoglobin electrophoresis DNA and RNA isolation
Citrate tube	Blue	Prothrombin time, ESR estimation by the Westergren method
Heparin tube	Green	Karyotyping
Plain vial (no anticoagulant)	Red	Most biochemical investigations – Liver function test (Bilirubin, Total protein, Albumin, AST, ALT, ALP), Renal function test (Urea, Creatinine, Sodium, Potassium), Enzymes (LDH, amylase), Uric acid etc.
Serum separator gel tube	Yellow	
Fluoride vial/glucose vial	Grey	Glucose estimation

▶ youtu.be/-OksYhAt5c4 or scan QR code

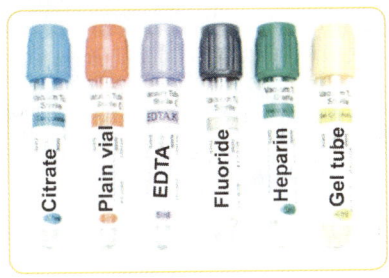

Citrate | Plain vial | EDTA | Fluoride | Heparin | Gel tube

COLORIMETRY

Photometry is the measurement of intensity of light. Colorimetry and spectrophotometry are two major photometric techniques. Colorimetry is the measurement of the intensity of color. In Biochemistry practical classes, we use colorimetry to measure the amount of substances (analytes) like glucose, urea and creatine, which are colorless. Therefore, the first step is to make these substances into a colored complex using chemical reactions. The intensity of color will be directly proportional to the amount of substrate. Using known concentrations of the substrate, a standard curve can be prepared. The concentration of the sample can be found using the standard curve.

We want to measure only the color produced by the substance. Therefore, we subtract the color of all the reagents by blanking.

❑ **Blank** solution contains all the reagents except the analyte of interest.
❑ **Standard** solution is the analyte with known concentration.
❑ **Test** solution is the analyte with unknown concentration, which needs to be measured.

BEER-LAMBERT'S LAW

Beer's Law

The amount of light absorbed (A) by a colored solution is directly proportional to the **concentration (C)** of the light absorbing substance.

$$A \alpha C$$

Lambert's Law

Amount of light absorbed by a colored solution is directly proportional to the **thickness or path length (*Lambert's law → length*)** of the absorbing material.

$$A \alpha L$$

Beer-Lambert's law: When combining both laws,

$$A \alpha CL$$
$$A = \varepsilon CL;$$

ε = molar extinction coefficient; constant for each analyte.

Unit of Concentration is $mol\ m^{-3}$.

Unit of Length is m.

Hence, the unit of extinction coefficient is $m^2\ mol^{-1}$.

Relationship between absorption and Transmittance: $A = 2 - log_{10} \% T$

This is why when you take reading in colorimetry, it is always between zero to two.

Limitations of Beer-Lambert's Law

❑ Cannot be used when the concentration of the analyte is too high or too low.
❑ Only colored substances can be measured.
❑ Only clear solutions can be used. There should not be any light scattering.

Components of a Colorimeter

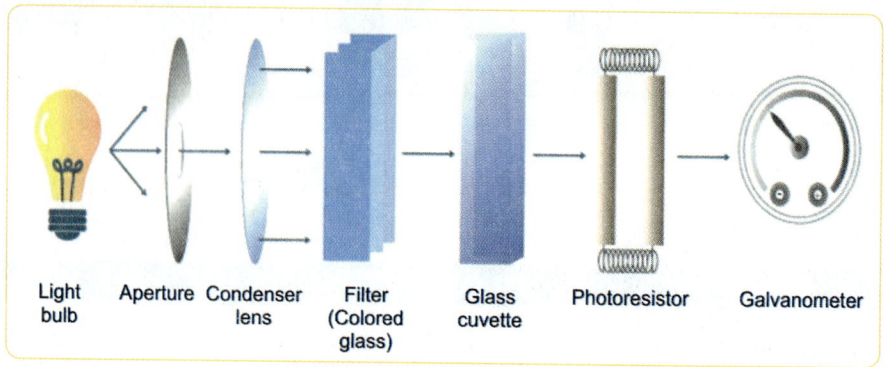

| Light bulb | Aperture | Condenser lens | Filter (Colored glass) | Glass cuvette | Photoresistor | Galvanometer |

Component	Description and function
Light source	Tungsten lamp (emits polychromatic light at visible spectrum).
Slit	Narrows the path of light.
Condensing lens	Light rays are made parallel
Filter	Converts the polychromatic light to monochromatic. Colored glasses are used to absorb all the colors except one. Color of the filter to be used is usually complementary to the color of the solution.
Cuvette	Used to keep the colored solution. Made up of glass, square-bottomed cuvettes are better than rounded ones. The colored solution absorbs the light (absorption), and the rest is emitted (transmittance).
Photocell	The photoelectric effect happens based on the intensity of the transmitted light
Galvanometer	Measures the electric current and detect the percentage transmittance.
Motherboard (in new, digital colorimeters)	❑ Converts the percentage transmittance to absorbance. ❑ Storage of reading is also possible.

Color of solution	Filter (complementary color)
Red-orange	Blue-blue green
Blue	Red
Green	Red
Yellow	Violet

SPECTROPHOTOMETRY

In clinical laboratories, spectrophotometry is used instead of colorimetry because it can measure the absorbance over a wide range.

	Colorimetry	Spectrophotometry
Light source	Tungsten lamp	Deuterium lamp
Monochromator	Colored glass filters	Prism/grating
Wavelength of light	Visible range only	Wide range — Visible, UV and IR
Nature of the sample	Only colored sample can be used	Even colorless samples can be used
Cost	Comparatively inexpensive	Expensive
Operative skill	Requires less skilled personnel	Requires skilled personnel

REVIEW OF QUALITATIVE TESTS

Test	Purpose
Molisch test	General test for carbohydrates
Iodine test	Test for polysaccharides
Benedict's	For reducing sugars
Barfoed's	To differentiate monosaccharides from disaccharides
Seliwanoff's	Specific test for keto sugars like fructose
Biuret	All peptides with at least two peptide bonds
Ninhydrin	All compounds with free amino group
Xanthoproteic	All aromatic amino acids with benzene ring
Millions test	Tyrosine (phenol group)
Hopkin-Cole-Adam's	Tryptophan (indole group)
Sakaguchi	Arginine (Guanidino)
Pauly's	Histidine (Imidazole)
Liebermann Burchard	Cholesterol and other sterols

OSAZONES AND THEIR SHAPE

Osazone	Shape
Glucose and Fructose	Needle
Lactose	Powderpuff/woolly cotton ball
Maltose	Sunflower

QUANTITATIVE METHODS USED IN THE LABORATORY

Analyte	Method used in practical class (Colorimetric)	Method used in the clinical laboratory (Spectrophotometric)
Glucose	Alkaline copper reduction method. (Asatoor and King method)	Hexokinase method (most specific) GOD-POD method
Urea	Diacetyl monoxime method	Urease method
Creatine	Jaffe's method (end point)	Modified Jaffe's method (rate-kinetic) Creatinase method
Total proteins	Biuret method	Biuret method
Albumin	Bromo cresol green (BCG) dye binding method	Bromo cresol green (BCG) dye binding method
Uric acid	Caraway's method	Uricase method
Calcium	o-cresolphthalein method	o-cresolphthalein method
Phosphate	Fiske-Subbarao's method	Fiske-Subbarao's method
Bilirubin	Malloy and Evelyn method (based on Vandenberg reaction)	Modified Jendrassik and Grof method (based on Vandenberg reaction)
Cholesterol	Zak's method Zlatkis method	Cholesterol oxidase-peroxidase method

Yellapragada Subbarow, the unsung hero, discovered the function of ATP as an energy source in the cell, developed methotrexate and discovered tetracyclin. Although he conducted his research in the USA, his contributions helped the entire world including his motherland.

BIURET TEST

Biuret test is a general test for proteins. It is used for both qualitative detection and quantitative measurement. Biuret molecule is a product of condensation of 2 (Bi) urea molecules. Molecules with at least two peptide bonds will react with alkaline copper sulphate to produce a purple colored co-ordination complex. Free amino acids will not give the +ve test.

Biuret reagent does not contain biuret! It is so named because biuret gives a +ve test with this reagent.

Components of Biuret Reagent

❑ NaOH – provides alkaline medium
❑ $CuSO_4$ – provides Cu^{2+} for coordination compound
❑ KI – oxidising agent
❑ Sodium potassium tartrate – preservative

In the quantitative method, the color developed is measured at 540 nm.

PROTEIN PRECIPITATION METHODS

In your practical sessions, you might have performed any one of these protein precipitation methods. Let's discuss the principle of each method.

Test and procedure	Principle
Precipitation by Heavy Metals	
Mercuric sulphate or lead acetate solution is added drop by drop to 3 mL of protein solution.	At pH above their pI, proteins exist as anions. The heavy metal cations neutralise the negative charge on the surface and cause precipitation of proteins.
	Clinical Correlation
	In lead and mercury poisoning, egg albumin is given orally to remove heavy metals from the intestine.
Precipitation by alkaloidal reagents	
❑ Precipitation by sulphosalicylic acid ❑ Precipitation by trichloroacetic acid ❑ Precipitation by Esbach's reagent (Esbach's reagent is the equal volume of Picric acid and citric acid)	Alkaloidal reagents are acids that can combine with alkaloids (Alkaloids are organic bases from plants). Alkaloidal reagents combine with protein to form insoluble protein salts, e.g. Protein
Precipitation by denaturation	
Precipitation by mineral acids – Heller's test 3 mL of concentrated HNO_3 is added protein solution carefully along the sides of the test tube. Precipitation is observed at the junction.	Mineral acids disrupt the ionic bonds and hydrogen bonds.

Contd...

Test and procedure	Principle
Heat coagulation test Only the top portion of 5 mL of protein solution taken in a test tube is heated. 1 or 2 drops of 1% acetic acid is added while heating.	Loss of higher order of structures other than primary structure.
Precipitation by dehydration	
Salt fractionation with ammonium sulphate Half saturation: To 2 mL of protein solution equal volume of $(NH_4)_2SO_4$ solution is added. Full saturation: If no precipitation develops in half saturation, solid $(NH_4)_2SO_4$ is added to saturate fully.	Addition of salts and electrolytes removes the water of hydration around the protein colloid. Albumin is smaller in size and has a large surface area and large water of hydration → requires a higher salt concentration for precipitation → full saturation (memory aid: All – full) Laboratory use: Salting procedure is used to purify proteins

URINE ANALYSIS

Positive Test	Inference	Further notes
Benedict's (**5 mL Benedict's** + 8 drops urine)	Presence of reducing substance, suggestive of glycosuria, possibly diabetes mellitus.	Don't directly conclude that patient is suffering from DM. Know about other reducing sugars and non-carbohydrate reducing substances like Vit. C, Glutathione.
Rothera's Test	Ketonuria – starvation Ketoacidosis	Beta hydroxybutyrate is not a ketone and cannot be detected by this test.
Benedict's + Rothera's	Diabetic ketoacidosis (DKA)	
Bile salts + Bile Pigments	Obstructive Jaundice	Do the Fouchet's test first to avoid last minute tension.
Heller's or Heat coagulation test	Proteinuria	Denaturation is sometimes reversible. Coagulation is NOT.
Benedicts + Hellers/Heat coagulation	Diabetic nephropathy	

BENEDICT'S TEST

Benedict's test is used to detect the presence of reducing sugars. Any sugar with free aldehyde or keto group will give a +ve Benedict's test. Reducing sugars under alkaline conditions from enediols, which are powerful reducing agents.

$$Na_2CO_3 + 2H_2O \rightarrow 2NaOH + H_2CO_3$$
$$2NaOH + CuSO_4 \rightarrow Cu(OH)_2 + Na_2SO_4$$
$$Cu(OH)_2 \rightarrow CuO + H_2O$$
$$\text{D-Glucose} + 2CuO \rightarrow \text{D-Gluconic acid} + Cu_2O\downarrow$$

Components of Benedict's Reagent and their Role

- ❑ $Na_2CO_3 \rightarrow$ provides mild alkaline condition
- ❑ $CuSO_4 \rightarrow$ provides Cu^{2+} for reduction
- ❑ Sodium citrate → stabilising agent – prevents precipitation of Cu^{2+}

Semi-quantitative test: By fixing the amount of Benedict's reagent and the sample, Benedict's test is made semi-quantitative. 5 mL of Benedict's solution is heated first to check for auto reduction. Then eight drops (2 mL) of urine sample is added and heated for 2 minutes.

Green color	+ (up to 0.5 g%)
Green precipitates	++ (0.5 to 1 g%)
Green to yellow precipitates	+++ (1 to 1.5 g%)
Yellowish orange precipitates	++++ (1.5 to 2 g%)
Brick red precipitates	>2 g%

Benedict's test is non-specific: Reducing substances like glutathione, uric acid, ascorbic acid will also give positive Benedict's test

ROTHERA'S NITROPRUSSIDE TEST

❑ Rothera's test is a test for detection of ketone bodies.
❑ Principle: In alkaline medium, freshly prepared Sodium nitroprusside forms a purple colored compound with ketone bodies.
❑ Procedure: 5 mL of urine is saturated with solid ammonium sulphate, and 0.2 mL of the freshly prepared sodium nitroprusside solution is added. The solution is mixed well. To this, 0.5 mL of NH_3 solution is added slowly along the sides of the test tube. A purple ring at the junction of the two liquids indicates the presence of ketone bodies.

Note: β-**Hydroxybutyrate does not give a positive Rothera's test** as it does NOT contain a keto group. It contains a **secondary alcohol** group.

FOUCHET'S TEST FOR BILIRUBIN

When urine is mixed with a solution of BaSO4, bilirubin gets adsorbed onto it. It is then filtered over a filter paper and dried. To this, a drop of Fouchet's reagent ($FeCl_3$ and CCl_3COOH) is added. $FeCl_3$ oxidises bilirubin to green colored biliverdin. CCl_3COOH stabilises $FeCl_3$.

HAY'S SULPHUR TEST FOR BILE SALTS

Take two test tubes one containing distilled water and the other containing urine having bile salts (test). Sprinkle a few particles of sulfur in both tubes. Sulfur particles will float in the 'control' tube, while they will settle down in the 'test' tube because of the surfactant action of bile salts.

PROTEINURIA

Proteinuria is classified into the following four types. The first three are due to renal pathology and the last one is due to post-renal pathology.

Type	Defect	Example	Characteristics
Glomerular	Loss of Glomerular filtration barrier	Nephrotic syndrome	Massive albuminuria
Tubular	Renal tubular damage	Interstitial nephritis due to heavy metal poisoning	Low molecular weight proteinuria
Overload	Level of certain proteins exceeding the tubular reabsorption capacity.	Hemolysis	Hemoglobinuria
		Rhabdomyolysis e.g. Crush Injuries	Myoglobinuria
		Multiple Myeloma	Bence-Jones Proteinuria
Post-renal	Due to lower urinary tract pathology	Urinary tract infections, tumours.	Pus cells and RBC cast

Bence-Jones Proteins are Light Chains Excreted in Urine of Patients with Multiple Myeloma

Multiple myeloma is a neoplastic condition of bone marrow with the clonal proliferation of **plasma cells**. Immortalised B-cells produce **monoclonal** antibodies. Excess light chains are excreted in urine and are known as Bence-jones proteins (BJP). They may of either κ or λ type, never both as the proliferation is monoclonal.

Heat Precipitation Test for BJP

Urine sample is collected from the patients suspected of multiple myeloma. When heated gradually in boiling water bath, Bence-Jones protein precipitates at temperatures between 40° and 60°C but dissolve at and above 100°C. Upon cooling, a precipitate will reappear around 60°C only to dissolve again below 40°C.

Confirmatory Test

❑ **Serum protein electrophoresis for detection of M band** is the confirmatory test for multiple myeloma.
❑ **Immunofixation** is done to find out the antibody type (exact detail of heavy and light chain).

Microalbuminuria

In a day (24 hr), only less than 30 mg of albumin gets excreted in the urine of a healthy individual. Anything more than that is abnormal. But the dye binding methods used for albumin estimation can detect only when the amount is >300 mg. **Albumin excretion in the range of 30 to 300 mg/day is known as microalbuminuria**. This is an early indicator of diabetic nephropathy. Early detection and treatment can prevent progression to persistent albuminuria and renal failure.

Microalbuminuria is assessed by highly sensitive immunological techniques like Immunonephelometry, Immunoturbidimetry and Radioimmunoassay.

Parameter	Normal level	Microalbuminuria	Macroalbuminuria
Urine albumin excretion rate (mg/24 hr)	<30	30–300	>300
Urine albumin excretion rate (µg/minute)	<20	20–200	>200
Spot urine albumin to creatine ratio (ACR) (mg/gm)	<30	30–300	>300

As the 24-hour collection of urine is difficult, spot collection of urine is practised. If we just measure the spot albumin, we will get falsely high value when the urinary sample is concentrated. Measurement of urinary creatinine is done to nullify the osmolarity difference. Result is expressed as milligram of albumin excreted per gram of creatinine in the urine.

Microalbuminuria is also seen in urinary tract infection, dehydration, fever, and after vigorous exercise.
Note: *Microalbumin refers to small amount of albumin, not small size of albumin!*

BIOMEDICAL WASTE MANAGEMENT

Biomedical waste is any waste generated during the diagnosis, treatment or immunization of human beings or in research activity. Biomedical waste carries a higher potential for infection and injury than any other type of waste. Hence, proper disposal of biomedical waste is of utmost importance. It is important to segregate the waste in the laboratory itself before it is gets piled up. Biomedical Waste (Management and Handling) Rules, 2016 recommends the use of appropriate color-coded containers in which the various types of biomedical waste must be disposed.

Chapter 46 Practical Biochemistry

General non-infectious waste

Contaminated Plastics (Recyclable)

Glassware:
Broken or discarded and contaminated glass including medicine vials & ampoules

Waste sharps including Metals

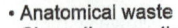

- Anatomical waste
- Chemotherapeutic drugs
- Soiled waste
- Expired Medicines
- Soiled linen

Bag/Box/Container	Type of waste
Black plastic bag	Municipal waste such as office waste (e.g. paper waste), kitchen waste, food waste and other non-infectious waste.
Yellow Plastic bags	Human anatomical waste, Animal waste, Discarded Medicines/Cytotoxic drugs, Soiled waste (items contaminated with blood and body fluids including cotton, dressings, soiled plaster casts, linen beddings, other material contaminated with blood)
Red Plastic bags	All contaminated recyclable plastics should go in this bag, e.g. blood stained plastic syringe.
Puncture-proof, tamper-proof and leak-proof containers	All sharps including needles, syringes with fixed needles
Cardboard box with blue markings	Contaminated Glassware (vials, ampoules) and metallic body implants should go in this.

Competency

- ☐ BI11.6 Describe the principles of colorimetry
- ☐ BI11.7 Estimation of serum creatinine and creatinine clearance
- ☐ BI11.20 Identify abnormal constituents in urine, interpret the findings and correlate these with pathological states
- ☐ PE21.11 Perform and interpret the common analytes in a Urine examination

Key Points

- ☐ Flouride vial (Grey top vial) is used for the estimation of blood glucose.
- ☐ Biuret reagent does not contain biuret.
- ☐ Rothera's test is positive in starvation and diabetic ketosis.
- ☐ Urea [mg/dL] = BUN [mg/dL] * 2.14
- ☐ Benedict's test is a semiquantitative test.
- ☐ Bence-Jones Proteins are light chains excreted in the urine in patients with multiple myeloma.
- ☐ Albumin excretion in the range of 30 to 300 mg/day is known as microalbuminuria.

SELF-ASSESSMENT

Short Answer Questions

1. Enumerate the color coding of any four blood collection tubes and their intended use.
2. Name the components and principle of Colorimeter
3. Compare colorimetry with spectrophotometry.
4. Explain why oral egg albumin is administered in heavy metal poisoning.
5. Explain why Benedict's test is a semi-quantitative test.
6. Suggest a simple bedside test to detect the presence of light chains in the urine of a patient suspected to have multiple myeloma? What are the further confirmatory tests you would like to proceed with?
7. Define microalbuminuria. What is the clinical significance of this condition? Name one technique used for the detection of microalbuminuria.
8. What are all the waste items to be disposed in yellow and red bags respectively?

Multiple Choice Questions

1. **What is the unit of molecular extinction coefficient used in beer-lambert law?**
   a. $m \, mol^{-1}$
   b. $m^2 \, mol^{-1}$
   c. $m^3 \, mol^{-1}$
   d. $mol \, m^{-2}$

2. **All of the following are components of colorimeter, except:**
   a. Colored glass filters
   b. Cuvette
   c. Grating
   d. Tungsten bulb

3. **The amino acid detected by Sakaguchi reaction is:**
   a. Basic amino acid
   b. Branched chain amino acid
   c. Imino acid
   d. Sulphur containing amino acid

4. **All of the following are components of Biuret reagent, except:**
   a. Biuret   b. Copper sulphate
   c. Potassium Iodide   d. Sodium hydroxide

5. **What is your interpretation when heat coagulation test and Benedict's test are positive on urine analysis?**
   a. Diabetic ketoacidosis
   b. Diabetic nephropathy
   c. Starvation ketosis
   d. Diabetes mellitus

6. **Which of the following anticoagulant is used in the blood collection tube used in the red-capped vial?**
   a. Pottasium EDTA   b. Sodium Citrate
   c. Sodium Fluoride   d. None of the above

Chapter 46 Practical Biochemistry

ANSWERS

1. b.	2. c.	3. a.	4. a.	5. b.	6. d.

Review

Section Outline

Appendix (Review)

NAMED REACTIONS, CYCLES AND PATHWAYS, MOLECULES

NAMED REACTIONS

Lohman reaction	Creatine phosphate + ADP $\rightarrow$ ATP + Creatine
Fenton reaction	$Fe^{2+} + H_2O_2 \longrightarrow Fe^{3+} + OH^{\bullet} + OH^-$
Haber-Weiss reaction	${}^{\bullet}O_2^- + H_2O_2 \rightarrow {}^{\bullet}OH + OH^- + O_2$
Maillard reaction	Reaction between amino acids and reducing sugars – leads to advanced glycation end products

NAMED CYCLES AND PATHWAYS

Cahill cycle	Glucose-Alanine cycle
Cori's cycle	Reutilisation of lactate (produced by RBC and exercising muscle) by liver in gluconeogenesis
Embden Meyerhof Pathway	Glycolysis
Krebs-Henseleit cycle	Urea cycle
Leloir Pathway	Galactose breakdown
Meister cycle	Glutathione mediated absorption of neutral amino acids
Rapaport-luebering shunt	2,3 BPG shunt

NAMED MOLECULES

Warburg yellow enzyme	Riboflavin
Edman's reagent	1-fluoro-2,4-dinitrobenzene (FDNB)
Sanger's reagent	Phenyl isothiocyanate (PITC)
Leventhal's paradox	A thought experiment related to protein folding
Klenow fragment	E. coli DNA polymerase without 5' $\rightarrow$ 3' exonuclease activity
Rossman fold	NAD(P)H binding domain of certain dehydrogenases
Cori's ester	Glucose 1-phosphate

PIONEERS OF BIOCHEMISTRY

Pioneer	Discovery
Alec Jeffreys	DNA fingerprinting
Andrew fire and Craig Mello	siRNA
Arber, Smith and Nathans	Restriction enzymes
Arthur Kornberg	DNA polymerase
Avery, Macleod and McCarty	DNA is the information molecule/genetic material
Barbara Mcclintock	Transposons
Blobel	Signal sequence hypothesis
Dorothy Hodgkin	Protein crystallography
Frederick Sanger	Sequencing of 1° structure of bovine insulin and sequencing of nucleotides (He got Nobel prize twice!)
Goldstein and Brown	LDL receptor and its relation to familial hypercholesterolemia
Griffith	Transformation experiment
Jacob and Monad	Operon model
James Lind	Scurvy and citrus fruit trial in HMS Salisbury
Kary B Mullis	Polymerase Chain Reaction
Kohler and Milstein	Monoclonal antibodies by Hybridoma technique
Linus Pauling and Robert Corey	2° structure of protein
Lohmann	Discovered ATP in biochemical reactions
Nirenberg, Khorana and Holley	Genetic Code
Paul berg	Recombinant DNA technology
Robert W. Holley	Sequencing of tRNA
Rosalind Franklin	X-ray crystallographer whose work helped in the determination of the structure of DNA
Ruska	Electron microscope
S.N. Dei	Cholera toxin discovery
Shinya Yamanaka	Induced pluripotent stem cells (iPSC)
Susumu Tonegawa	Gene rearrangements in immunoglobulins
Thomas Cech and Sidney Altman	Ribozyme
Tiselius	Electrophoresis
Tswett	Chromatography
Venkatraman Ramakrishnan	Ribosome structure
Yalow and Berson	Radioimmunoassay (RIA)
Yellapragada Subbarao	Discovered that ATP is the energy currency of the cell

Amino acid in plasma	Glutamine
Anterior pituitary hormone	Growth hormone
Biological forms in the world	Polysaccharides
Nucleoprotein	Histone

Contd...

Amino acid in plasma	Glutamine
Cell type in the human body	Erythrocyte (RBC)
Type of collagen in basement membrane	Type IV
Type of collagen in the body	Type I
Type of collagen in the cartilage (except white fibrocartilage)	Type II
Type of collagen in the bone and white fibrocartilage	Type I
Constituent of body	Water
Glycoprotein in basement membrane	Laminin
Glycosaminoglycan	Chondroitin sulphate
Heteropolysaccharides in the body	Glycosaminoglycan
Immunoglobulin	IgG
Lipid in chylomicron	Triacylglycerol
Membrane proteins of RBC	Glycophorin and Band 3 anionic transporter
Osmotically active component of the plasma	Sodium
Peripheral membrane protein of RBC	Spectrin
Platelet receptors	GPIIb-GPIIIa complex
Prokaryotic DNA polymerase	DNA polymerase I
Protein in HDL	Apo A-I (70% of weight) followed by Apo AII
Protein in the human body	Collagens
Saturated fatty acid in circulation	Palmitic acid
Sigma factor in E. coli	Sigma 70
Stop signal for transcription termination	RNA hairpin
Tocopherol in extrahepatic tissues	α-tocopherol

THE MOST COMMON GENETIC DISORDERS

Gene disorder worldwide	Thalassemia
Enzyme deficiency (enzymopathy)	G6PD (mostly asymptomatic)
Qualitative Hemoglobinopathy	Sickle cell anemia
Mutation in cystic fibrosis	ΔF508 (Deletion of phenylalanine at 508th position)
Viable chromosomal disorder	Down syndrome (21 trisomy)
2nd most common autosomal trisomy resulting in live birth	Edward syndrome (18 trisomy)
Mutation in galactosemia in Caucasian population	Q188R (replacement of glutamine by arginine)
Gene mutated in congenital adrenal hyperplasia	CYP21A2 (21-α hydroxylase)
Mutation leading to permanent neonatal diabetes	KCNJ11 (ATP sensitive K^+ channel)
Inherited urea-cycle defect	OTC deficiency
SCID	X-linked SCID
Saturated fatty acids present in the cell	Palmitic acid (C16) and stearic acid (C18)
Type of plasma membrane receptor	GPCR (G-Protein Coupled receptor)

Appendix (Review)

669

Contd...

Gene disorder worldwide	Thalassemia
Type of prosthetic groups, cofactors for enzymes	Metal ions
Covalent modification regulating enzyme activity	Phosphorylation dephosphorylation.
Fatty acid in natural fats	Oleic acid
Chronic liver disorder worldwide	Non-alcoholic fatty liver disease (NAFLD)
Cause of Conjugated Hyperbilirubinemia	Obstruction in the Biliary Tree
Primary immunodeficiency	IgA deficiency
Sterol in the membranes of animal cells	Cholesterol
Lysosomal storage disease	Gaucher's disease
Cause of proteinuria	Loss of integrity of the glomerular basement membrane (glomerular proteinuria)
Clotting factor deficiency	Factor VIII
Hereditary bleeding disorder	❑ Von Willebrand disease ❑ Bernard-Soulier syndrome
Gene mutated in cancers	P53
Coagulopathy/inherited thrombophilia	Factor V Leiden
2° structure of proteins	α-helix
Sphingolipid found in mammals	Sphingomyelin
DNA binding motif	Helix turn helix
Insoluble fiber in diet	Cellulose
Form of the DNA double helix	B-DNA
Type of point mutation	Transition
Separation method used in proteomic study	Two-dimensional gel electrophoresis
Oncogene involved in the development of human cancers	RAS
Cause of preventable blindness in children	Vit A deficiency
❑ Toxin producing dilated cardiomyopathy ❑ Environmental teratogen ❑ Cause of cirrhosis in the Western world	Alcohol
Mode of inheritance of congenital malformations	Multifactorial inheritance
Inborn error of fatty acid oxidation	Medium-chain fatty acyl CoA dehydrogenase deficiency
Genetic error of amino acid transport	Cystinuria
Porphyria	Porphyria cutanea tarda
Hepatic porphyria	Acute intermittent porphyria
Porphyria in children	Erythropoietic protoporphyria
Defective enzyme in homocystinuria	Cystathionine β-synthase
Acceptor in transaminase reactions	α - ketoglutarate (2-oxoglutarate)
Trinucleotide repeat	CAG
Level of gene regulation in eukaryotes	Transcription initiation
Cause of non-ketotichyperglycinemia	P protein mutation
Cause of insulin resistance	Obesity

Contd...

Gene disorder worldwide	Thalassemia
Inherited cause of intellectual disability	Fragile X syndrome
Inherited platelet dysfunction	Glanzmann thrombasthenia
Biochemical abnormality in congenital hypertrophic pyloric stenosis	Hyponatremic hypokalemic metabolic alkalosis with paradoxical aciduria.
Inherited nonspherocytic hemolytic anemia	Pyruvate kinase deficiency
Hereditary hemolytic anemia	Hereditary spherocytosis (G6PD deficiency is usually asymptomatic)
Type of chromosomal translocation	Robertsonian
Vitamin deficiency in the United States	Folate (B$_9$)
Inborn error in bile acid synthesis	3β-hydroxy Δ5 C27-steroid oxidoreductase (HSD3B7)
Amino acids found in beta turns	Glycine and Proline
Gene mutation in hemochromatosis	HFE C282Y
Feedback/homeostatic systems in the body	Negative feedback
Renal stones in children	Cysteine
Cause of cobalamin deficiency	Pernicious anemia

REGULATORY/RATE LIMITING STEPS OF METABOLIC PATHWAYS

Pathway	Enzyme catalysing the rate-limiting step
Glycolysis	PFK-1
Glycogen synthesis	Glycogen synthase
Glycogenolysis	Glycogen Phosphorylase
β-oxidation of fatty acids	CPT-I
Fatty acid synthesis	Acetyl-CoA Carboxylase
Ketone body synthesis	HMG-CoA synthase
Cholesterol synthesis	HMG-CoA reductase
De novo Purine synthesis	PRPP synthetase is rate limiting; PRPP-glutamyl amidotransferase catalyzes the committed step
De novo Pyrimidine synthesis	CPS II
Heme synthesis	ALA synthase I
HMP shunt	G6PD
TCA cycle	Isocitrate dehydrogenase
Urea cycle	CPS I
Bile acid synthesis	7-α-hydroxylase (CYP7A1)
Polyamine synthesis	Ornithine decarboxylase
Catecholamine synthesis	Tyrosine hydroxylase
Triacylglycerol synthesis	Diacylglycerol acyltransferase
Testicular steroidogenesis	STAR protein mediated uptake of cholesterol

THE FIRST IN HISTORY

Genome to be sequenced	Bacteriophage φX174
Genome that belongs to a free-living organism to be sequenced	*H. influenzae*
Sequence of human chromosome released	Ch. 22
Protein to be sequenced (by sanger)	'Bovine' Insulin
Metabolic pathway discovered	Glycolysis
Metabolic cycle to be discovered	Urea Cycle
Disease treated by gene therapy	ADA deficient SCID
Ribozyme discovered (by Cech)	26S rRNA
Molecular machine recognised	Ribosome

BIOCHEMICAL TESTS

Test	Analyte detected
Alcian blue spot test	Urinary glycosaminoglycans
Barfoed's test	Monosaccharide
BCG dye binding method	Albumin
Benedict's test	All reducing substances (Reducing sugar, Uric acid, Ascorbate etc.)
Bial's test	Pentose sugars
Biuret reaction	Amino acids and proteins (Both Qualitative and quantitative)
Ehrlich aldehyde test	Urobilinogen
Ferric chloride test	Phenylketonuria, Tyrosinemia, Alkaptonuria, MSUD
Fouchet's test	Bilirubin in urine (Qualitative test)
Gerhardt's test, Rothera's test	Acetoacetate (ketone body)
Hay's sulphur test	Bile salts in urine
Molisch test	All carbohydrates
Ninhydrin reaction	Proteins with minimum of 2 peptide bonds (qualitative).
Sakaguchi test	Arginine
Seliwanoff's test	Fructose
Shake test (foam stability test) of amniotic fluid	Fetal lung maturity assessment
Sulkowitch test	Urinary Calcium
Vandenberg test	Differentiates conjugated and unconjugated bilirubin in serum (Quantitative test)